FISHES AND FISHERIES OF NEVADA

Ira La Rivers

Foreword by
Gary Vinyard and James E. Deacon

Current Nomenclature and Status of Nevada Fishes by
Craig Stockwell

University of Nevada Press
Reno   Las Vegas   London

Fishes and Fisheries of Nevada by Ira La Rivers was originally published by the Nevada State Fish and Game Commission, in 1962. The 1994 University of Nevada Press edition reproduces the original except for the following changes: the front matter has been modified to reflect the new publisher; a foreword to the new edition by Gary Vinyard and James E. Deacon has been added; and Craig Stockwell has provided the current nomenclature and status of Nevada fishes.

The paper used in this book meets the requirements of American National Standard for Information Sciences—Permanence of Paper for Printed Library Materials, ANSI Z39.48–1984. Binding materials were selected for strength and durability.

Library of Congress Cataloging-in-Publication Data

La Rivers, Ira, 1915–1977.
    Fishes and fisheries of Nevada / Ira La Rivers ; foreword by Gary Vinyard and James E. Deacon ; current nomenclature and status of Nevada fishes by Craig Stockwell.
        p.   cm.
Originally published: Carson City : Nevada State Fish and Game Commission, 1962.
    Includes bibliographical references and index.
    ISBN 0-87417-256-X
    1. Freshwater fishes—Nevada. 2. Fisheries—Nevada. I. Title.
SH222.N3L3   1994
597.092'9793—dc20   94-32098
                                                                                               CIP

University of Nevada Press, Reno, Nevada 89557 USA
Copyright © 1994 by the University of Nevada Press
All rights reserved
Printed in the United States of America

9 8 7 6 5 4 3 2 1

TABLE OF CONTENTS

Foreword to New Edition 1
Current Nomenclature and Status of Nevada Fishes 5
Introduction 7
Acknowledgements 9
I Chronology 11
II Palenotology 35
III Physiography 65
IV Rivers 91
V Lakes and Major Reservoirs 119
VI The Fishes 193
    1. Checklist 193
    2. Illustrative Key 203
    3. Accounts of Species 239
VII Appendix 593
VIII Bibliography 713
IX Pluvial Lakes and Streams of Nevada 774
X Index 775

Ira La Rivers

Enjoy thy stream, O harmless fish;
And when an angler for his dish,
Through glutton's vile sinne,
Attempts, the wretche, to pull thee out,
God give thee strength, O gentle trout,
To pull the raskall in!

> From *The Scottish Field,* January, 1945,
> as quoted by Carl and Clemens

FOREWORD TO NEW EDITION

A primary reason for the lasting interest in *Fishes and Fisheries of Nevada* is the subject matter. Unique faunas around the world attract special attention, and Nevada's fishes are among the most notable. Thirteen of the state's 47 native fish species occur nowhere else in the world. Two of its endemic species are extinct, and one or more subspecies of 26 additional native species are considered endangered, threatened, or of special concern by the Endangered Species Committee of the American Fisheries Society. Eight of the ten species or subspecies noted as extinct in the Current Nomenclature and Status of Nevada Fishes to this volume were already gone when *Fishes and Fisheries of Nevada* was first published in 1962. Many had only recently disappeared, and La Rivers did not record their demise. Although the Colorado squawfish is the only species to have disappeared from the state's fauna in the past three decades, many more have suffered severe reductions in distribution and abundance.

Publication of *Fishes and Fisheries of Nevada* occurred after a decade that saw seven of Nevada's ten known fish extinctions. The appropriation of surface and groundwater for agriculture and the introduction of nonnative species for recreation were primary causes for these early losses. Publication of this book also came just prior to the most explosive population growth our state has experienced. Nevada's population stood at 317,000 in 1962. Over the subsequent three decades, concentrated urban growth in Las Vegas and Reno brought the population of the state to nearly 1.4 million.

In his book La Rivers documented the introductions of many species of nonnative fishes into Nevada's waters. This is probably the second most serious cause of decline of our native fish fauna. While "unofficial" introductions have created numerous problems for native fishes, the introductions sponsored by state and federal wildlife agencies have been the most devastating. From the 1882 "Call for Carp" from Nevada noted by La Rivers, to the more recent stream poisonings to develop sport fisheries based on nonnative species, Nevada's native fishes continue to be assaulted by official introductions of exotic fishes.

Fishes and Fisheries of Nevada could only have been produced by a consummate naturalist. Ira La Rivers's catholic interests and vivid character were shaped and nurtured by his early life. He came to Nevada from San Francisco at age three with his mother, who moved among the mining camps and ranches of Nevada as a camp cook. La Rivers received a bachelor's degree from the University of Nevada in 1937 and a Ph.D. from the University of California at Berkeley in 1948. He joined the faculty of the University of Nevada Biology Department in 1948 and remained until his death in 1977.

At the University of Nevada he taught herpetology, limnology, ichthyology, entomology, general zoology, and fish hatchery management. He made extensive collections of geological specimens, algae, insects, fish, amphibians, and reptiles, published in each of these fields, and established his own journal, *The Occasional Papers of the Biological Society of Nevada*. Many who knew La Rivers still recount stories of his colorful habits in the field and laboratory and of his intimate knowledge of the state.

Fishes and Fisheries of Nevada has had a long and useful life. It bridges the gap between the earlier naturalist period and the modern environmentalist period. It contains a huge store of information about conditions at the time of its publication and provides clues to understanding the changes occurring in our native fish fauna. The breadth of material covered by La Rivers is exceptional. The book is much larger, both in terms of coverage and total content, than similar books documenting fish fauna in states having many more species. He addressed the geological, historical, and ecological settings, as well as the taxonomic treatment more common to works of that time. In this effort he foreshadowed many of the more recent attempts to produce ecosystem-wide or community-wide analyses of communities in the Great Basin.

In the early 1960s a transition was occurring in ecological thought and environmental awareness. Qualitative, observational approaches to understanding the natural world were being replaced by quantitative methods. Today, ecosystem modeling, ecological energetics, evolutionary ecology, and molecular systematics offer opportunities for both reductionist and synthetic understanding of our fish fauna that were unavailable to La Rivers. The new subdiscipline of conservation biology integrates several areas within ecology, population biology, resource management, and economics.

A growing environmental awareness, stimulated in no small measure by Rachel Carson's book *Silent Spring*, which also appeared in 1962, led to the enactment of a historic and unprecedented body of environmental legislation, including the National Environmental Policy Act; the Endangered Species Act; the Federal Water Pollution Control Act; the Clean Air Act; the Wilderness Act; the Federal Insecticide, Fungicide, and Rodenticide Act; the Resource Conservation and Recovery Act; the Federal Land Policy and Management Act; and the National Forest Management Act.

Despite this legislation, most of the 85 percent of Nevada's land managed by the federal government is considered degraded to some extent. Unsustainable agricultural and recreational water use, coupled with greatly increased urban demands for water, continue to degrade and eliminate aquatic habitats throughout the state. Use and abuse fall most heavily on riparian and aquatic habitats, areas essential for protection and preservation of fishes.

Nearly all native fish have disappeared from the lower Virgin River. The bonytail is all but gone from the Colorado River. The last native fish habitat in Pahrump Valley went dry in the mid-1970s. Extinctions of Raycraft Ranch and Pahrump Ranch poolfish and of the Las Vegas dace are directly attributable to unsustainable groundwater pumping. A tiny population of the endemic White River spinedace hangs on in a single spring on the Wayne Kirch Wildlife Management Area. With 58 percent of our extant fish fauna recognized as endangered, threatened, or of special concern (see Current Nomenclature and Status of Nevada Fishes, Table 2), we are poised on the threshold of a potentially greater period of extinction than that which occurred in the 1950s.

There have been considerable efforts to sustain or improve the status and survival prospects for some native fishes in recent times. In 1976 the U.S. Supreme Court handed down a precedent-setting decision that for the first time brought groundwater under the doctrine of prior rights and recognized a dependent relationship between groundwater and surface water. In the resolution of that case, Devil's Hole, sole natural habitat for the Devils Hole pupfish, was protected from excessive groundwater pumping, which would have eliminated both the habitat and the fish. That case and the body of environmental legislation enacted since, provide tools for protection of native fishes and their habitats undreamed of thirty-two years ago. Today, instead of ignoring our native fishes, proponents of massive water projects to support explosive urban growth, mining development, recreation, and agriculture, must accommodate the fishes' requirements for survival.

Although it has not yet achieved a restoration to historic abundance levels, the cui-ui, with considerable assistance from the U.S. Fish and Wildlife Service, is again spawning in the lower Truckee River. Populations of the Lahontan cutthroat trout are still much reduced over most of their former range, however, a new recovery plan for this fish is nearing completion, and their reestablishment in some streams of their native range has been successful. The primary stimulus for the creation of the Moapa and Ash Meadows National Wildlife Refuges was the protection of endemic fishes within their native habitats. The Nature Conservancy has played a very important role with its efforts to preserve native fishes in Ash Meadows, Condor Canyon, Ruby Valley, and Soldier Meadows. The Desert Fishes Council, formed during the period of controversy over Devil's Hole, continues to focus its efforts on conservation and research of native fishes. The Nevada Department of Wildlife has taken a sympathetic stance toward protection and management needs for Nevada's native fishes. That attitude follows a tradition established in the 1950s by Chief of Fisheries Thomas Trelease, a student of Ira La Rivers. All these recent preservation efforts would have been enormously difficult in La Rivers's day, and their occurrence demonstrates a growing awareness of the problems associated with conserving native fishes.

La Rivers's book is being reissued thirty-two years after its initial publication at a time of growing concern with issues of biodiversity in the Great Basin. The unique fish fauna of Nevada is being increasingly threatened by unsustainable water management practices, accelerating habitat loss, and the continuing invasion by nonnative species. Reversing these processes will be among the major concerns of biologists in the coming decades. Such efforts must be grounded in an understanding of the basic natural history information presented by La Rivers, which provides a foundation for applying more modern analytical techniques and making use of the insights developed during the past three decades.

We applaud the University of Nevada Press for bringing back *Fishes and Fisheries of Nevada* and hope that its greater circulation will increase interest in and understanding of the rich aquatic resources of our state.

> Gary Vinyard
> University of Nevada, Reno
> and
> James E. Deacon
> University of Nevada, Las Vegas
>
> November, 1994

CURRENT NOMENCLATURE AND STATUS OF NEVADA FISHES

During the last three decades the fish fauna of Nevada has changed considerably (Deacon and Williams 1984, Sigler and Sigler 1987). Numerous species of nonnative fish have become established in Nevada (Table 1), and new species of native Nevada fish can be added to the Nevada list. These native fish include species that were discovered in Nevada subsequent to 1962 as well as a few species that have been recently described.

Changes in the taxonomic status of Nevada fishes have occurred during the last three decades (Table 1). The names furnished here follow those provided in Robins et al. (1991) and Mayden et al. (1992). Most changes involve adjustments in alignment of genera; however one major realignment has occurred between the Cyprinodontidae and the Goodeidae. I follow Parenti (1981) who determined springfish (*Crenichthys* sp.) and poolfish (*Empetrichthys* sp.) to be members of the family Goodeidae. Smith and Stearley (1989) showed that the rainbow, cutthroat, and golden trout were more closely aligned with the salmon (*Oncorhynchus*). Debate has also arisen regarding the taxonomic status of the tui chub. Bailey and Uyeno (1964) resolved this debate by reducing *Siphatales* to a subgeneric level within the genus *Gila*. However, debate remains on whether *Gila* is monophyletic (Coburn and Cavender 1992). Both *Gila* and *Rhinichthys* have many undescribed subspecies and the resolution of the relationships within these groups remains a large task.

Table 1 includes all species referred to in chapter 6. I provide the current names for fish species that have been renamed. Subspecies are not listed unless the scientific name has changed. This table includes 30 new species (20 nonnative and 10 native species) that were not listed by La Rivers.

Unfortunately, numerous native fishes have undergone serious declines (see Table 2) in the last three decades (Williams et al. 1989). This general decline among native fish can be attributed primarily to the degradation of habitat and the introduction of nonnative fish (Williams et al. 1989). The American Fisheries Society (Table 2) recognizes many Nevada fishes as endangered, threatened, or of special concern (Williams et al. 1989). Furthermore, 22 of these fish are currently listed (Table 2) as threatened or endangered species under the Endangered Species Act (50 CFR 17.11 and 17.12).

Table 3 lists 11 species, not listed by La Rivers, that were introduced to Nevada but have not become established (Deacon and Williams 1984).

Craig A. Stockwell
Biodiversity Research Center
University of Nevada, Reno
November, 1994

TABLE 1. REVISED LIST OF NEVADA FISHES

Former Common Name (New Common Name)	Scientific Name in La Rivers	Taxonomic Revision or New to List
Aguillidae: Freshwater Eels		
eel[+]		*Anguilla rostrata* (Lesueur)
Clupeidae: Herrings and Shads		
American shad*	*Alosa sapidissima* (Wilson)	
Mississippi threadfin shad (threadfin shad)	*Dorosoma petenense* (Günther)	
Salmonidae: Trout, Salmon, and Whitefish		
lake whitefish[!]*	*Coregonus clupeaformis* (Mitchill)	
chum salmon*	*Oncorhynchus keta* (Walbaum)	
silver salmon (coho salmon)[a]*	*Oncorhynchus kisutch* (Walbaum)	
Kokanee red salmon[a(*)]	*Oncorhynchus nerka* (Walbaum)	
king salmon (chinook salmon)[N]	*Oncorhynchus tshawytscha* (Walbaum)	
mountain whitefish[N!]	*Prosopium williamsoni* (Girard)	
golden trout*	*Salmo aguabonita* Jordan	*Oncorhynchus aguabonita* (Jordan)
cutthroat trout[N†]*	*Salmo clarki* Richardson	*Oncorhynchus clarki* (Richardson)

Continued on next page

Note: [a] See Deacon and Williams 1984 for current status.

[N] Nevada native; [†] various subspecies occur in Nevada; [+] new to list; [!] family realignment; *These species appear in the Supplementary Annotated List of Unsuccessfully Introduced Species on page 198.

Table 1—*Continued*

Former Common Name (New Common Name)	Scientific Name in La Rivers	Taxonomic Revision or New to List
rainbow trout[N†*]	*Salmo gairdneri* Richardson	*Oncorhynchus mykiss* (Walbaum)
Atlantic salmon[a*]	*Salmo salar* Linnaeus	
brown trout	*Salmo trutta* Linnaeus	
bull trout[N+]		*Salvelinus confluentus* (Suckley)
brook trout	*Salvelinus fontinalis* (Mitchill)	
Dolly Varden trout (Dolly Varden)[N]	*Salvelinus malma* (Walbaum)	
lake trout	*Salvelinus namaycush* (Walbaum)	
arctic grayling[!*]	*Thymallus signifer* (Pallas)	*Thymallus arcticus* (Pallas)
Esocidae: Pikes		
northern pike[+]		*Esox lucius* Linnaeus
Cyprinidae: Carps and Minnows		
chiselmouth[N]	*Acrocheilus alutaceus* Agassiz and Pickering	
goldfish	*Carassius auratus* Linnaeus	
red shiner[+]		*Cyprinella lutrensis* (Baird and Girard)
Asiatic carp (common carp)	*Cyprinus carpio* Linnaeus	
Soldier Meadows dace (desert dace)[N]	*Eremichthys acros* Hubbs and Miller	
Alvord chub[N+]		*Gila alvordensis* Hubbs and Miller

Continued on next page

Table 1—*Continued*

Former Common Name (New Common Name)	Scientific Name in La Rivers	Taxonomic Revision or New to List
Utah gila[N]	*Gila atraria* (Girard)	
leatherside chub[N+]		*Gila copei* (Jordan and Gilbert)
humpback chub[N+]		*Gila cypha* Miller
Colorado gila (roundtail chub)[N†]	*Gila robusta* Baird and Girard	
swiftwater Colorado gila (bonytail)[N]	*Gila robusta elegans* Baird and Girard	*Gila elegans* Baird and Girard
Virgin River roundtail chub (Virgin River chub)[N+]		*Gila seminuda* Cope and Yarrow
White River spinedace[N]	*Lepidomeda albivallis* Miller and Hubbs	
Pahranagat spinedace[N]	*Lepidomeda altivelis* Miller and Hubbs	
Colorado River spinedace (Virgin River spinedace)[N†]	*Lepidomeda mollispinis* Miller and Hubbs	
Moapa dace[N]	*Moapa coriacea* Hubbs and Miller	
golden shiner	*Notemigonus crysoleucas* (Mitchill)	
Sacramento blackfish	*Orthodon microlepidotus* (Ayres)	

Continued on next page

[N] Nevada native; [†] various subspecies occur in Nevada; [+] new to list; [!] family realignment; [*] These species appear on the Supplementary Annotated List of Unsuccessfully Introduced Species on page 198.

Table 1—*Continued*

Former Common Name (New Common Name)	Scientific Name in La Rivers	Taxonomic Revision or New to List
fathead minnow[+]		*Pimephales promelas* Rafinesque
woundfin[N]	*Plagopterus argentissimus* Cope	
Colorado squawfish[N]	*Ptychocheilus lucius* Girard	
northern squawfish[N]	*Ptychocheilus oregonensis* (Richardson)	
relict dace[N+]		*Relictus solitarius* Hubbs and Miller
longnose dace[N+]		*Rhinichthys cataractae* (Valenciennes)
Las Vegas dace[N+]		*Rhinichthys deaconi* Miller
speckle dace (speckled dace)[N†]	*Rhinichthys osculus* (Girard)	
Columbia redshiner (redside shiner)[N†]	*Richardsonius balteatus* (Richardson)	
Lahontan redshiner (Lahontan redside shiner)[N]	*Richardsonius egregius* (Girard)	
tui chub[N†]	*Siphatales bicolor* (Girard)	*Gila bicolor* (Girard)
tench[*]	*Tinca tinca* (Linnaeus)	
Catostomidae: Suckers		
Utah sucker[N]	*Catostomus ardens* Jordan and Gilbert	
bridgelip sucker[N]	*Catostomus columbianus* (Eigenmann and Eigenmann)	

Continued on next page

Table 1—*Continued*

Former Common Name (New Common Name)	Scientific Name in La Rivers	Taxonomic Revision or New to List
flannelmouth sucker[N]	*Catostomus latipinnis* Baird and Girard	
biglip sucker (largescale sucker)[N]	*Catostomus macrocheilus* Girard	
Wall Canyon sucker[N+]		*Catostomus* sp.
Tahoe sucker[N]	*Catostomus tahoensis* Gill and Jordan	
Warner sucker[N+]		*Catostomus warnerensis* Snyder
cui-ui lakesucker (cui-ui)[N]	*Chasmistes cujus* Cope	
White River mountainsucker (desert sucker)[N]	*Pantosteus intermedius* (Abbott)	*Catostomus clarki intermedius* Baird and Girard
Lahontan moutainsucker (mountainsucker)[N]	*Pantosteus lahontan* (Cope)	*Catostomus platyrhinchus* (Cope)
razorback sucker[N]	*Xyrauchen texanus* (Abbott)	
Ictaluridae: Bullhead Catfish		
white catfish	*Ictalurus catus* (Linnaeus)	*Ameiurus catus* (Linnaeus)
black bullhead[†]	*Ictalurus melas* (Rafinesque)	*Ameiurus melas* (Rafinesque)
yellow bullhead[+]		*Ameiurus natalis* (Lesueur)
brown bullhead	*Ictalurus nebulosus* (Lesueur)	*Ameiurus nebulosus* (Lesueur)

Continued on next page

[N] Nevada native; [†] various subspecies occur in Nevada; [+] new to list; [!] family realignment; [*] These species appear on the Supplementary Annotated List of Unsuccessfully Introduced Species on page 198.

Table 1—*Continued*

Former Common Name (New Common Name)	Scientific Name in La Rivers	Taxonomic Revision or New to List
channel catfish		*Ictalurus punctatus* (Rafinesque)

Loricariidae: Armored Catfish

suckermouth catfish+		*Hypostomus plecostomus* (Linnaeus)

Fundulidae: Topminnows

rainwater killifish+		*Lucania parva* (Baird and Girard)
plains killifish+		*Plancteros zebrinus* (Jordan and Gilbert)

Goodeidae: Goodeids[1]

butterfly goodeid+		*Ameca splendens* Miller and Fitzsimons
White River springfish[N!†]	*Crenichthys baileyi* (Gilbert)	
Railroad Valley springfish[N!]	*Crenichthys nevadae* Hubbs	
Pahrump poolfish[N†!]	*Empetrichthys latos* Miller	
Ash Meadows poolfish[N!]	*Empetrichthys merriami* Gilbert	

Cyprinodontidae: Pupfishes

Devil pupfish (Devils Hole pupfish)[N]	*Cyprinodon diabolis* Wales	
Amargosa pupfish[N†]	*Cyprinodon nevadensis* Eigenmann and Eigenmann	

Continued on next page

Table 1—*Continued*

Former Common Name (New Common Name)	Scientific Name in La Rivers	Taxonomic Revision or New to List
Poeciliidae: Livebearers		
mosquitofish	*Gambusia affinis* (Baird and Girard)	
black molly (sailfin molly) shortfin molly+	*Mollienesia latipinna* (Lesueur)	*Poecilia latipinna* (Lesueur) *Poecilia mexicana* Steindachner
guppy+		*Poecilia reticulata* Peters
swordtail (green swordtail)	*Xiphophorus helleri* Heckel	
moonfish (Platy) (southern platyfish)	*Xiphophorus maculatus* (Günther)	
Percichthyidae: Temperate Bass+		
white bass+		*Morone chrysops* (Rafinesque)
striped bass+		*Morone saxatilis* (Walbaum)
Centrarchidae: Sunfish		
Sacramento perch	*Archoplites interruptus* (Girard)	
green sunfish	*Lepomis cyanellus* Rafinesque	
pumpkinseed+		*Lepomis gibbosus* (Linnaeus)
bluegill sunfish (bluegill)	*Lepomis macrochirus* Rafinesque	

Continued on next page

N Nevada native; †various subspecies occur in Nevada; +new to list; !family realignment; *These species appear on the Supplementary Annotated List of Unsuccessfully Introduced Species on page 198.

Table 1—*Continued*

Former Common Name (New Common Name)	Scientific Name in La Rivers	Taxonomic Revision or New to List
redear sunfish[+]		*Lepomis microlophus* (Günther)
smallmouth bass	*Micropterus dolomieu* Lacepède	
spotted bass[+]		*Micropterus punctulatus* (Rafinesque)
largemouth bass[†]	*Micropterus salmoides* (Lacepède)	
white crappie	*Pomoxis annularis* Rafinesque	
black crappie	*Pomoxis nigromaculatus* (Lesueur)	

Percidae: Perch

yellow perch	*Perca flavescens* (Mitchill)	
walleye[+]		*Stizostedion vitreum* (Mitchill)

Cichlidae: Cichlids[+]

convict cichlid[+]		*Cichlasoma nigrofasciatum* (Günther)
zebra mbuna[+]		*Pseudotropheus zebra* (Boulenger)
spotted tilapia[+]		*Tilapia mariae* (Boulenger)

Cottidae: Sculpins

Baird sculpin (mottled sculpin)[N]	*Cottus bairdi* Girard	
Belding sculpin (Paiute sculpin)[N]	*Cottus beldingi* Eigenmann and Eigenmann	

TABLE 2. NEVADA FISH SPECIES THAT ARE EXTINCT, ENDANGERED, THREATENED, OR OF SPECIAL CONCERN

Common Name	Species Name	ESA Listing	General Status	Reason for Decline
Salmonidae: Trout, Salmon and Whitefish				
Lahontan cutthroat trout	*Oncorhynchus clarki henshawi* (Gill and Jordan)	TH	TH	1, 4
Bonneville cutthroat trout	*O. c. utah* (Suckley)		EN	1, 4
Alvord cutthroat trout	*Oncorhynchus clarki* ssp.		EX	4
Humboldt cutthroat trout	*O. c.* ssp.		SC	1
Warner Valley redband trout	*Oncorhynchus mykiss* ssp.		SC	1, 4
Interior redband trout	*O. m. gibbsi* (Suckley)		SC	1, 2, 4
king salmon (chinook salmon)	*Oncorhynchus tshawytscha* (Walbaum)		EXN	
bull trout	*Salvelinus confluentus* (Suckley)		SC	1, 4
Cyprinidae: Carps and Minnows				
desert dace	*Eremichthys acros* Hubbs and Miller	TH	TH	1, 5

Continued on next page

ESA (Endangered Species Act) codes: EN–endangered, TH–threatened.
Status categories: EX–extinct, EXN–extinct in Nevada, EN–endangered, TH–threatened, SC–special concern.
Reason for the Decline: 1–habitat modification; 2–overuse because of commercial, recreation, scientific, or educational purposes; 3–disease; 4–human-induced problems because of introduction of nonnatives, predation, and competition; 5–restricted range.
Source: Deacon and Williams 1984, Miller et al. 1989, and Williams et al. 1989.

Table 2—*Continued*

Common Name	Species Name	ESA Listing	General Status	Reason for Decline
Alvord chub	*G. alvordensis* Hubbs and Miller		SC	1, 4
Fish Creek Springs tui chub	*Gila bicolor euchila* Hubbs and Miller		TH	1, 4, 5
Sheldon tui chub	*G. b. eurysoma* Williams and Bond		SC	1, 5
Independence Valley tui chub[a]	*Gila bicolor isolata* Hubbs and Miller		EX	4
Newark Valley tui chub	*G. b. newarkensis* Hubbs and Miller		SC	1, 5
Lahontan Creek tui chub	*G. b. obesa* (Girard)		SC	1
humpback chub	*G. cypha* Miller		EXN	1
bonytail chub	*G. elegans* Baird and Girard	EN	EN	1, 4
Pahranagat roundtail chub	*G. robusta jordani* Tanner	EN	EN	1, 4
Virgin River chub	*G. seminuda* Cope and Yarrow	EN	EN	1
White River spinedace	*Lepidomeda albivallis* Miller and Hubbs	EN	EN	1, 4
Pahranagat spinedace	*Lepidomeda altivelis* Miller and Hubbs		EX	4, 5
Virgin River spinedace	*Lepidomeda mollispinis mollispinis* Miller and Hubbs		TH	1, 4

Continued on next page

Note: [a]Recent survey work suggests that the Independence Valley tui chub may not be extinct. During 1992 tui chubs were observed in the outflows of numerous springs within Independence Valley (Heinrich 1993).

Table 2—*Continued*

Common Name	Species Name	ESA Listing	General Status	Reason for Decline
Big Spring spinedace	*L. m. pratensis* Miller and Hubbs	TH	EN	1, 4, 5
Moapa dace	*Moapa coriacea* Hubbs and Miller	EN	EN	1, 3, 4, 5
woundfin	*Plagopterus argentissimus* Cope	EN	EN	1, 3, 4
Colorado squawfish	*Ptychocheilus lucius* Girard		EXN	1, 3, 4
relict dace	*Relictus solitarius* Hubbs and Miller		SC	1
Las Vegas dace	*Rhinichthys deaconi* Miller		EX	1, 5
Independence Valley speckled dace	*Rhinichthys osculus lethoporus* Hubbs and Miller	EN	EN	1, 4, 5
Moapa speckled dace	*R. o. moapae* Williams		TH	1, 3, 4
Ash Meadows speckled dace	*R. o. nevadensis* Gilbert	EN	EN	1, 4
Clover Valley speckled dace	*R. o. oligoporus* Hubbs and Miller	EN	EN	1, 4, 5
Grass Valley speckled dace	*R. osculus reliquus* Hubbs and Miller		EX	1, 4, 5

Continued on next page

ESA (Endangered Species Act) codes: EN–endangered, TH–threatened.
Status categories: EX-extinct, EXN–extinct in Nevada, EN–endangered, TH–threatened, SC–special concern.
Reason for the Decline: 1–habitat modification; 2–overuse because of commercial, recreation, scientific, or educational purposes; 3–disease; 4–human-induced problems because of introduction of nonnatives, predation, and competition; 5–restricted range.

Table 2—*Continued*

Common Name	Species Name	ESA Listing	General Status	Reason for Decline
Amargosa River speckled dace	*R. o.* ssp.		SC	1, 5
Preston speckled dace	*R. o.* ssp.		SC	1, 4, 5
Catostomidae: Suckers				
White River sucker	*Catostomus clarki intermedius* (Tanner)		EN	1
Wall Canyon sucker	*Catostomus* sp.		SC	1, 5
Warner sucker	*Catostomus warnerensis* Snyder	TH	EN	1, 4
cui-ui	*Chasmistes cujus* Cope	EN	EN	1
razorback sucker	*Xyrauchen texanus* (Abbott)	EN	EN	1, 2, 4
Goodeidae: Goodeids				
Preston springfish	*Crenichthys baileyi albivallis* Williams and Wilde		EN	4, 5
White River springfish	*C. b. baileyi* (Gilbert)	EN	EN	1, 3, 4
Hiko springfish	*C. b. grandis* Williams and Wilde	EN	EN	1, 4
Moapa springfish	*C. b. moapae* Williams and Wilde		TH	1, 4
Moorman springfish	*C. b. thermophilus* Williams and Wilde		TH	1, 4, 5
Railroad Valley springfish	*C. nevadae* Hubbs	TH	TH	1, 4

Continued on next page

Table 2—*Continued*

Common Name	Species Name	ESA Listing	General Status	Reason for Decline
Raycraft Ranch poolfish	*Empetrichthys latos concavus* Miller		EX	4, 5
Manse Ranch poolfish	*E.l. latos* Miller	EN	EN	1, 4, 5
Pahrump Ranch poolfish	*E. l. pahrump* Miller		EX	1, 4, 5
Ash Meadows poolfish	*E. merriami* Gilbert		EX	1, 4, 5
Cyprinodontidae: Pupfish				
Devils Hole pupfish	*Cyprinodon diabolis* Wales	EN	TH	1, 5
Ash Meadows pupfish	*C. nevadensis mionectes* Miller	EN	TH	1, 4
Warm springs pupfish	*C. n. pectoralis* Miller	EN	EN	1, 4, 5

TABLE 3. UNSUCCESSFUL FISH INTRODUCTIONS

Common Name	Species Name
Osteoglossidae: Osteoglossids	
Arawana	*Osteoglossum bicirrhosum* (Vandelli)
Cyprinidae: Carps and Minnows	
grass carp	*Ctenopharyngodon idella* (Valenciennes)
blacktail shiner	*Cyprinella venusta* (Girard)
Mohave tui chub	*Gila bicolor mohavensis* (Snyder)
Clariidae: Labyrinth Catfish	
walking catfish	*Clarias batrachus* (Linnaeus)
Cyprinodontidae: Pupfish	
Amargosa pupfish	*Cyprinodon nevadensis amargosae* Miller
Cichlidae: Cichlids	
Rio Grande cichlid	*Cichlasoma cyanoguttatum* (Baird and Girard)
banded cichlid	*Heros severus* (Heckel)
golden mbuna	*Melanochromis auratus* (Boulenger)
blue mbuna	*M. johanni* (Eccles)
redbelly tilapia	*Tilapia zilli* (Gervais)

REFERENCES

BAILEY, R. M., AND T. UYENO. 1964. Nomenclature of the blue chub and the tui chub, Cyprinid fishes from the western United States. Copeia 1964:238-239.

COBURN, M. M., AND T. M. CAVENDER. Interrelationships of North American Cyprinid fishes. In Mayden, R. L., ed., Systematics, historical ecology and North American freshwater fishes. Stanford: Stanford Univ. Press, 1992:328-373.

DEACON, J. E., AND J. E. WILLIAMS. 1984. Annotated list of the fishes of Nevada. Proc. Biol. Soc. Wash. 97:103-118.

HUBBS, C. L., R. R. MILLER, AND L. C. HUBBS. 1974. Hydrogeographic history and relict fishes of the north-central Great Basin. Mem. Calif. Acad. Sci. 7:1-259.

HEINRICH, J. 1993. Native nongame fish program progress report: January 1, 1992 through December 31, 1992. Nevada Department of Wildlife, Carson City.

MAYDEN, R. L., B. M. BURR, L. M. PAGE, AND R. R. MILLER. The native freshwater fishes of North America. In Mayden, R. L., ed., Systematics, historical ecology and North American freshwater fishes. Stanford: Stanford University Press, 1992:827-867.

MILLER, R. R., J. D. WILLIAMS, AND J. E. WILLIAMS. 1989. Extinctions of North American fishes during the past century. Fisheries 14(6):22-38.

PARENTI, L. R. 1981. A phylogenetic and biogeographic analysis of cyprinodontiform fishes (Teleostei, Atheninomorpha). Bull. Amer. Mus. Nat. Hist. 168:335-557.

ROBINS, C. R., R. M. BAILEY, C. E. BOND, J. R. BROOKER, E. A. LACHNER, R. N. LEA AND W. B. SCOTT. 1991. Common and scientific names of fishes from the United States and Canada. American Fisheries Society Special Publication 20.

SIGLER, W. F., AND J. W. SIGLER. Fishes of the Great Basin: a natural history. Reno: University Nevada Press, 1987.

SMITH, G. R., AND R. F. STEARLEY. 1989. The classification and scientific names of rainbow and cutthroat trouts. Fisheries 14:4-10.

WILLIAMS, J. E., J .E. JOHNSON, D. A. HENDRICKSON, S. CONTRERAS-BALDERAS, J. D. WILLIAMS, M. NAVARRO-MENDOZA, D. E. MCALLISTER, AND J. E. DEACON. 1989. Fishes of North America Endangered, Threatened, or of Special Concern. 1989. Fisheries 14:2-20.

INTRODUCTION

To one unfamiliar with the region, the arid lands of Nevada with their meager and sometimes uncertain water supplies, would seem a fleeting habitat for fish. In the minds of most people there is something incongruous in the association of deserts and fishes—they do not seem to go well together. Deserts are water impoverished; the water they do contain often seems too sparse, or alkaline, or too ephemeral and isolated from other waters, not only to support fish but to allow of any reasonable explanation how the fish could have arrived there.

But facts, which are prone to speak for themselves, have proven otherwise to the biologist who studies fish. During an historical collecting period of over 100 years, a sizeable number of species of fish have been collected and described from water ranging from large lakes and streams to small pools and springs. Since water, disregarding its Great Basin potential for increased salt and alkali is water, whether it occurs in the desert or in an area of considerable rainfall, finding relatively rich faunas of fishes in the larger lakes and streams of a desert area is not remarkable, except insofar as one speculates upon the ways in which fishes came to these isolated drainage systems.

For the tiny and often unique fishes living in small, isolated warm springs systems which are scattered widely over Nevada, a solution to the problem of arrival and survival is more obscure, and involves a competent knowledge of geologic changes which have taken place in the region in times past and the consequent shifting and interaction between the various drainages of Nevada and those of adjacent areas.

Hand-in-hand with the mountain building and erosional changes which have produced the topography characteristic of the Great Basin, past and present, and as such are responsible for the establishment and shifting of the drainage systems, go the climatic effects upon such systems. Climate, again, is locally monitored by topography, which is, within its positional limitations (latitude, longitude and altitude), the single most definitive force affecting drainage systems.

Many of the numerous habitats in which Nevada fishes have become established are found only in desiccating closed basins such as the type of which the Great Basin is largely constituted. The limnology, or study of the physical, chemical and biological characteristics, of these variable waters is not only of academic interest in providing comparative information on water as an environment for living plants and animals, but lends itself to the more immediately applicable economic field of fishery biology in which it is important to know, among other things, the unit fish yield from different types of waters, in order to properly manage and harvest, for pleasure or for food, these ever dwindling resources.

The marked isolation of Nevada's major drainage systems from each other and from the larger numbers of small spring-fed pools and streams which do not flow beyond the confines of their own separate valleys establishes a situation in which small, disconnected populations of fishes rapidly evolve away from their parent stock to the point where the differences between the isolated stock and the ancestral stock become of generic importance. Specific and subspecific differentiations are, of course, even proportionately more common to such situations.

The beginnings of the study of Great Basin ichthyology are intimately tied to the early history of the region. The principal Federal expeditions of the 1850s, 1860s and 1870s all included one to several naturalists whose job was collecting the fauna and flora of interest. Often such collecting was subsidiary to the individual's primary assignment as surgeon to the expedition. Earlier explorers, such as John C. Frémont (whom historians have conflictingly characterized as the "Pathfinder of the West" and the "Follower of other men's trails") collected as circumstances permitted, and made notes on many plants and animals they could not collect. The first definitely known whites in the area, such mountain men as Smith and Walker and the partisan Ogden, certainly fished at every opportunity, but left few records of their achievements.

The economic beginnings were almost as early. The creation of the office of Fish Commissioner for Nevada followed closely the establishing of the United States Fish Commission in 1871, and the introduction of non-native game fishes began immediately and has continued to the present day.

The following pages will attempt to show as much of the highlights and supporting details of Nevada ichthyology as can reasonably be contained in a single volume.

ACKNOWLEDGMENTS

In the eleven years since this effort began (on a Forest Service lookout in Sierra Valley), so many people have contributed—directly and indirectly, many even unknowingly—that a standard acknowledgment format is a difficult achievement. Worthy contributants will doubtless be overlooked, but only in small numbers. And it may perhaps be that some will wander with seeming aimlessness through these pages. However, all are pertinent in one way or another.

In another way, the aid I have gotten from others will be more-or-less evident, both as to "whom" and "how much," in the treatment given individuals in the text. I have tried to be complete in this respect.

Needless to say, my appreciation of the efforts of others is boundless in some cases since the following pages represent very little originality but rather should be regarded as a compendium of what is already known. I have ever been mindful of the contributions of the craftsmen of Ichthyology. While I have had the brashness to disagree with the masters on minor points (learning, meanwhile, that some masters regard nothing as minor), on the whole the solid substance of these pages is to be regarded to their credit, not mine.

Undoubtedly the treatment of Nevada Fishes is incomplete; my attack on the problem of producing a limited encyclopedia of our fish fauna is certainly colored not only by my own biases but by my limited resources. I have been guided as much by what phases of the subject I am personally interested in, and what I can use later in my projected activities in the field, as I have by what is considered standard in faunal treatments.

Now for specific details—

The following have read and commented on some or all of the typescript: Robert C. Allan, Ray Corlett, Theodore Frantz, V. Kay Johnson, Dr. E. Richard Larson, Dr. Robert R. Miller, Robert Sumner and Thomas Trelease.

I have borrowed illustrations from:

California Department of Fish and Game, and Leo Shapovalov; Mrs. Nettie Sutcliffe Cooper; Cranbrook Institute of Science; Theodore Frantz; George Hardman; Dr. Carl L. Hubbs; Iowa State Conservation Commission, and James R. Harlan; Ernest Mack; Michigan Department of Conservation, and John Gray; Nevada Fish and Game Commission; Nevada Highway Department, and Adrian Atwater; Nevada State Planning Board; Ohio State University Press, Milton B. Trautman and Weldon A. Kefauver; Oregon State Game Commission, and Harold P. Smith; Pennsylvania Fish Commission, Johnny Nicklas and Keen Buss; Dr. Osgood Smith; Smithsonian Institution, Geology Department, Dr. David H. Dunkle; Robert Sumner; Thomas J. Trelease; United States National Archives; University of California Press, and Erhard Rostlund; University of Chicago Press (Goode Series of Base Maps of North America); University of Michigan Press, and Dr. Robert R. Miller; University of Utah Press (fold-out map); University of Washington Press, and W. M. Read; Dr. George F. Weisel; Wisconsin Conservation Department, Jens von Sivers and N. H. Hoveland.

Specific source materials have come from:

Boston Public Library; California State Library; Theodore Frantz; Kay Johnson; Burton Ladd; Library of Congress; George W. McCammon; Richard B. Millin; Nevada Fish and Game Commission; Shirl Coleman; Nils Nilsson; E. P. Pister; Stanford Libraries; Robert Sumner; Thomas J. Trelease; United States Fish and Wildlife Service at Portland, Oregon; and Leo Laythe.

Of the illustrations made for this work, those by Silvio Santina merit special mention. Additional drawings were made by John B. Harris and James E. Neider.

Specimens, living and fossil, have been contributed by many people to the University of Nevada Biology Department museum collections. Prominent among these have been former Nevada biology students Theodore Frantz, Kay Johnson, Robert Sumner and Thomas Trelease. Fossil fishes have been given or loaned by R. E. Brady, Manuel Chabagno, Dr. E. Richard Larson and Jack Manhire.

Most of the algae cited in this report were identified by Dr. Francis Drouet.

Finally, three separate financial grants by the Max C. Fleischmann Foundation of Nevada to the Nevada State Fish and Game Commission have made possible the publication of this bulletin; corollary grants-in-aid by the University of Nevada Research Committee have helped with typing, proofreading and illustrations.

Knowing that *nemo solus satis sapit,* it is appropriate in closing this summation, to point out our indebtedness to the sound and painstaking contributions of Dr. Carl L. Hubbs and Dr. Robert R. Miller. Over nearly the past 30 years they have, singly and together, brought knowledge of the Nevada fish fauna to an organizational stage where it is readily and clearly available to all.

To my wife Marian I am grateful for many unrewarding hours of galley-proofing. To Drs. Fred A. Ryser, Jr., and Hugh N. Mozingo, I must acknowledge page-proof held in the finest Homeric and Linnaean traditions.

As for myself, if, after using the work, one is inclined to say *"redolet lucerna,"* I can only reply that 11 years of this have left me *au bout de son latin.* About halfway through this effort, when certain people would inquire when I was going to finish it, I could only shrug my shoulders and say *"auribus teneo lupum. Quien sabe?"*

Auf wiedersehen,

May 1, 1962
Verdi, Nevada.

CHAPTER I

CHRONOLOGY
(The Highlights)

A—Prehistoric

PERMIAN TIMES

200,000,000 B. C. Large sharks of the genus *Helicoprion* swam in marine waters which extended as shallow seaways over what is now Nevada.

TRIASSIC TIMES

170,000,000 B. C. Other Sharks and their relatives, the Placoderms, were plentiful in the continental marine seas which continued to flood Nevada.

CRETACEOUS TIMES

120,000,000 B. C. Distant relatives of modern ten-pounders have left remains of this age in rocks of east-central Nevada.

MIOCENE TIMES

20,000,000 B. C. From what must have been an extensive fauna of freshwater fishes, we have the remains of a modern sucker, a minnow and a pirate perch.

PLIOCENE TIMES

10,000,000 B. C. In its comparative nearness to the modern scene, this period shows greater evidences of its rich heritage of fishes; shiners, suckers, sticklebacks and killifishes have left their imprints on the rocks.

PLEISTOCENE TIMES

50,000 B. C. These recent deposits have yielded fragments of the same fishes now living in the Lahontan watershed. These have included the Cutthroat Trout, Tui Chub, Lakesucker and Belding Sculpin as Pleistocene fossils, species identical to those alive today.

10,000 B. C. Archeological excavations have shown that the Indians of Lovelock Cave and the vicinity of Humboldt Sink were living for protracted periods on the same fish we catch now.

B—Historic

1776. The Franciscan friar, Francisco Garcés, seems to have been the first known European to explore any part of what is now Nevada, apparently entering the extreme southern tip. Fishing was everywhere good in those days, and although fish very often supplied a considerable part of their food, these early wanderers seldom made specific mention of them.

The better known priest, Silvestre Velez de Escalante, came very close to the southeastern border of the State in the same year, but is not definitely known to have entered Nevada.

1792. Johann Julius Walbaum described the King Salmon (*Oncorhynchus tshawytscha*) and the Dolly Varden Trout (*Salvelinus malma*) —both native to Nevada waters—from Siberian specimens of the Kamchatka Peninsula.

1825. In mid-summer of this year, Peter Skene Ogden, a Hudson's Bay Company partisan (trapping brigade leader), trapped along the upper Humboldt River. His route in lay along the Owyhee River. As the story goes, one of the party took an Indian wife in the area and from her name came "Mary's River" by which the stream was first known.[1] Ogden, in his journal of this trip, never referred to the stream as anything but the "unknown river," although various historians credit him with calling it "Mary's River."

Jim Bridger and others were supposed to have trapped westward into Nevada and met the Ogden party this same year. From this time on, the traffic of "Mountain Men" was heavy enough to rapidly trap out Humboldt beaver in the next few years.

1826. Jedediah S. Smith and company, Mountain Men, went from Great Salt Lake to California along the Virgin River in the extreme southeastern part of Nevada. He returned the following year across central Nevada from San Francisco's vicinity back to Salt Lake.

1828. Ogden returned to the upper Humboldt.

1829. This year saw the third and most extensive of Ogden's trips to the Humboldt River. He followed the river to its point of disappearance in the lake below the site of Lovelock. Here Indians told him of two streams to the southwest, which would later be called the Truckee and the Carson, but he never saw them. Then he turned east and crossed central Nevada to the great bend of the Sevier River in Utah and worked his way down the Virgin River and Colorado River much as Smith had done three years before, and into California.

In his journal of the trip, Ogden added to his store of names for the Humboldt: "Unknown River is known as Swampy River or Paul's River," the last in allusion to his trapper Joseph Paul who died along its banks.

1830. William Wolfskill came down the Virgin River and along the Nevada side of the Colorado River on his way from central Utah to the pueblo of Los Angeles.

1831. One historian's account (Morgan 1943: 45–46) that John Work, another Hudson's Bay Company partisan, reached "Ogden's River" in this year and could not trap because of flood waters, cannot

---

[1]One of the tributaries of the Humboldt is still called Mary's River.

be substantiated. It was Work's original plan to do this, but illness in his brigade forced him to take a more westerly and direct route to California, passing just outside the northwestern tip of Nevada, many miles from the Valley of the Humboldt (Cleland 1950: 330).

1832. George Nidever and party, fresh from the battle of Pierre's Hole in eastern Idaho, trapped on the Humboldt River (which they called both "Ogden's" and "Mary's" River), and returned eastward to winter in the rendezvous at Green River Valley, Wyoming.

Milton Sublette, one of the best known Mountain Men, ventured by the headwaters of the Humboldt, swinging north to the Snake.

1833. Joseph Reddeford Walker, a member of Captain Benjamin Louis Eulalie de Bonneville's trapping force, set out from the vicinity of Great Salt Lake with some 35 men, one being Nidever. Their journey to California was chronicled by both Nidever and Zenas Leonard.

The route was down the Humboldt to its sink—there is some uncertainty about the direction of travel from here, but the best consensus routes the party down to a lake and up its river to the Sierran crest (both lake and river would later be named for Walker).

During their attempts to trap the Humboldt, which they called "Barren River," they had several skirmishes with Indians, or massacres, depending upon whose viewpoint you take, the trappers' or the Indians'. They returned across central Nevada the next year; to posterity, in addition to the lake and river, Walker left a mountain pass in his name, all honors later given him by Frémont.

Christopher "Kit" Carson is reputed to have trapped down the Humboldt River this year with a Hudson's Bay Company group headed by Thomas McKay (Bancroft 1890: 45). This may be true, but there is evidence that Carson did not get to the Sink of the Humboldt for when the waters of Pyramid Lake broke before the eyes of Frémont and his guide Carson eleven years later, both at first thought they were looking at Mary's Lake (i.e., Humboldt Lake or Sink), although Frémont immediately doubted this because of the discrepancy between the rugged basin of Pyramid and the flat, open, marshy description he had of the Sink.

(An additional irregularity becomes patent in Frémont's mention that Carson's favorite criterion for whether a new stream might flow into the Pacific Ocean was the presence of beaver cuttings. According to Carson, the lack of such cuttings indicated interior drainage with no outlet to the sea; apparently, while trapping the Humboldt, Carson and others must have believed it to eventually flow to the sea and so could not have made any extensive explorations of the Sink below Lovelock where it disappeared into the desert sands and clays. We can only surmise that Carson was somewhat colorful in relating his qualifications as a Great Basin guide to Frémont when signing on with the latter—a not uncommon characteristic of guides, then and now.)

1834. Walker returned eastward across central Nevada.

1836. Sir John Richardson, a famous English naturalist, published his "Fauna Boreali-Americana," the third volume of which dealt with fishes. Among other species, he described for the first time several which range into Nevada:

Cutthroat Trout (*Salmo clarki*.[2])
Northern Squawfish (*Ptychocheilus oregonense*.)
Columbia Redshiner (*Richardsonius balteatus*.)

These were collected for Richardson by an enterprising young doctor, Meredith Gairdner, attached to the Hudson's Bay Company post on the Columbia River. For him, Richardson named the Rainbow Trout (*Salmo gairdneri*).

1838. Kit Carson again came down the Humboldt with a Hudson's Bay Company brigade from Fort Hall in Idaho (Morgan 1943: 46).

1841. A typical early emigrant group of this year was that of the Bidwell-Bartleson party from Missouri, who found the lower Humboldt River dry—to them it was still "Mary's River." They worked southward and up the Walker River (their "Balm River") and over the Sierra Nevadas.

The Bartleson segment, separating from the main party, was sustained by fish and pine nuts obtained from Indians before coming up with Bidwell again just prior to the passage of the mountains, which they made after abandoning their wagons.

1843. Joe Walker brought a wagon train to California down the Humboldt and up the Walker River. The wagons were buried at Mono Lake.

John Charles Frémont, the "Great Pathfinder" or the "Follower of other men's trails," whichever view of history you prefer, began the expedition which would take his party through Nevada.[3]

1844. Frémont and Carson discovered Pyramid Lake on January 10th; five days later, the party was at the mouth of the inflowing stream, eating fish and proclaiming this the "Salmon Trout River." From there they followed south to the Big Bend of the river at modern Wadsworth, thence to the Carson and Walker Rivers, ascending the latter and crossing the Sierras into California. Frémont returned eastward this same year from southern California into Nevada, meeting Walker at Las Vegas and entering Utah up the Virgin River.

"Old Greenwood" (Caleb Greenwood), a remarkable and ancient Mountain Man of 81 years, later in the year guided a party of "emigrators" down the Humboldt, up Frémont's Salmon Trout River and across the cliffs of North Pass which would become a graveyard for the ill-starred Donners a few years later.

It was the Greenwood party who christened the stream the "Truckee" River after an Indian who directed them to it—and Frémont's name was relegated to the limbo of his reports.

1845. Frémont brought a large party into Nevada from the east;

---

[2]Named for William Clark of the Lewis and Clark 1804–1806 expedition to the Columbia River.
[3]By the time Frémont started out, the broad reaches of the West were so cluttered with other people's trails that he had difficulty finding a fresh one for himself.

near the headwaters of Mary's River, as people still called the Humboldt, the group divided. Frémont and Carson, with a few men, made southwest for Walker Lake. Joe Walker, with the bulk of the party under Theodore Talbot, followed down the Humboldt River—noting occasional fishing Indians—and rendezvoused with Frémont at Walker Lake. Here the expedition split again, the Walker-Talbot segment quartering south for a crossing of the mountains, while Frémont went north with a small group, arriving at the Truckee and going over the Sierra Nevada Mountains on what was now a fairly well established emigrant trail.

Greenwood and Solomon Sublette both piloted companies to California over North Pass.

Lansford Warren Hastings, of cutoff ill-fame, came down the Humboldt and made the passage of the Sierras in the last days of November.

1846. Hastings, and the Mountain Man Jim Clyman and the now 83-year-old Caleb Greenwood, crossed Nevada as a single party from west-to-east. Hastings was seeing and studying for the first time the cutoff to California he had recommended in his 1845 book to the oncoming wagon trains. They went down the Truckee, up the Humboldt and across to the Great Salt Lake.

The Applegates, Jesse and Lindsey, broke trail from Oregon southeasterly to the Humboldt, thence explored into northeastern Nevada and out of the State.

Joe Walker went east along the Humboldt route with stolen California horses for the eastern market.

The Harlan-Young group made the Humboldt-Truckee trip.

This was the year of the luckless Donner Party, whose unenviable fame transformed "Truckee Lake" into Donner Lake and "North Pass" into Donner Pass.

1847. The mutinous Frémont was finally removed from the confused California scene and brought east for court-martial; his first glimpse of the Humboldt River, which he proceeded to name, was while traveling its dusty banks under arrest.

Many others were passing back and forth this year, much of the traffic being Mormon.

1849. The Golden Army came down an alkali Humboldt Valley from which the grass had been grazed by stock and burned by the sun, to the Forty-mile Desert below Lovelock—a trail marked by fired wagons, stock decaying in the water-course and crude graves by the wayside.

1850. The big migration to the goldfields began to taper off by the end of the year.

1853. J. Soulé Bowman collected specimens of the Lahontan Tui Chub (*Siphateles bicolor obesus*) from the Humboldt River, which Charles Girard described in 1856. Some of Bowman's Snake River drainage redshiners (*Richardsonius balteatus*) were mixed in error with Humboldt collections and described in 1856 by Girard as "*Tigoma*

FIG. 1. Dr. Charles Girard, who described many of the fishes native to Nevada in the middle of the last century. Born in Mülhausen, France, March 9, 1822, Girard was educated in Switzerland and came to the United States in 1847 mit das schweitzer Professor Louis Agassiz. Later, and for many years, Girard was an assistant to, and associate of, Professor Spencer Fullerton Baird at the Smithsonian Institution. In 1856 Girard obtained his medical degree from Georgetown College. Returning to France in 1865, he entered the practice of medicine, and died in Paris January 29, 1895.

humboldti." Bowman, a promising collector related to S. F. Baird, died of typhoid fever in San Francisco later in the same year.

Spencer F. Baird, Assistant Secretary of the Smithsonian Institution, and Charles Girard, also of that institution, described two fishes —also native to Nevada—from specimens collected farther south by United States Army expeditions:

Colorado Gila[4] (*Gila robusta*)
Flannelmouth Sucker (*Catostomus latipinnis*)

1854. Girard described the Sacramento Perch (*Archoplites interruptus*) from California which would later be transplanted into Nevada waters with much success.

Lieutenant E. G. Beckwith crossed Nevada in the summer—leaving Salt Lake City April 4th, and arriving in the Sacramento Valley September 12th—on a survey for a Pacific railroad route, collecting biological specimens enroute.

1855. Louis Agassiz, the famous Swiss naturalist transplanted to the United States (where he bloomed profusely), characterized the Chiselmouth (*Acrocheilus*) and the Squawfish (*Ptychocheilus*) as technical genera.

W. P. Gibbons described our commonest type of Rainbow Trout (*Salmo gairdneri irideus*) from the vicinity of San Francisco.

1856. Girard described several fishes important to the Nevada scene:
Biglip Sucker (*Catostomus macrocheilus*)
Lahontan Tui Chub (*Siphateles bicolor obesus*)
Speckle Dace (*Rhinichythys osculus*)
Utah Gila (*Gila atraria*)
Colorado Squawfish (*Ptychocheilus lucius*)
Mountain Whitefish (*Prosopium williamsoni*)

He also characterized the genus *Richardsonius* (in honor of Sir John) to which the redshiners belong.

1859. Captain J. H. Simpson of the United States Army Engineers came across central Nevada to Carson City in search of a satisfactory wagon route—the report of this expedition was not issued until seventeen years later.

Girard published his description of the Lahontan Redshiner (*Richardsonius egregius*) from Humboldt River specimens collected by Creutzfeldt[5] (Kreuzfeld) of the 1854 Beckwith expedition.

1860. Charles C. Abbott described, among other things, the Razorback Sucker (*Xyrauchen texanus*) which is part of the Colorado River fauna.

---

[4]Pronounced "Hee-la."
[5]Although Girard consistently credits Creutzfeldt as the collector of Nevada material for this expedition, the botanist was actually killed by Indians in Utah before the group entered Nevada. It is not known whether Beckwith or the surgeon-geologist for this segment of the railroad survey, Dr. J. Schiel, collected the Humboldt River material.

1868. J. G. Cooper contributed the zoölogy chapter to T. F. Cronise's book on the natural wealth of California, and reviewed the important known fishes.

1872. Edward Drinker Cope, one of the leading zoölogists of his day, described the fossil Nevada Amyzon Sucker (*Amyzon mentalis*) from Elko County.

1874. Cope made known two unique little genera of fishes which are members of the Nevada fauna:
Woundfin (*Plagopterus argentissimus*)
Spinedace (*Lepidomeda vittata*)

1875. Cope and H. C. Yarrow reported on a collection of fishes from several Western States, including Nevada, taken on the Wheeler Surveys of the 100th Meridian; among other species, the Mountainsucker genus (*Pantosteus*) was characterized for the first time.

1876. The Simpson report finally appeared on the results of his 1859 trip across Nevada.

1878. David Starr Jordan (1851–1931), a student of Agassiz, began his series of published studies which were to extend over the ensuing 50 years and touch at many points on the fish fauna of Nevada. Between the presidencies of Indiana and Stanford Universities, Jordan did much to lay the groundwork for American ichthyology and eventually became generally recognized as the country's premier ichthyologist.

In this year, with the older Theodore Gill, Jordan described the Tahoe Sucker (*Catostomus tahoensis*) and the Lahontan Cutthroat Trout (*Salmo clarki henshawi*) from Lake Tahoe. With H. W. Henshaw, Jordan also reported on several Nevada specimens of various species collected in 1875, 1876 and 1877 by the Wheeler Surveys West of the 100th Meridian.

H. G. Parker, first Fish Commissioner for the State of Nevada and earlier Superintendent of Indian Affairs for the State, noted the experimental introduction of Quinnat Salmon into Nevada waters.

1879. Parker, in the first biennial report of the State Fish Commissioner, related his 1877 planting of Sacramento Perch (*Archoplites interruptus*) and Brown Bullhead (*Ictalurus nebulosus*) from the Sacramento River into Washoe Lake in southern Washoe County. Bullheads also went into the Truckee, Carson and Humboldt Rivers.

Cope briefly drew attention again to the Nevada *Amyzon* fish fossil beds in Elko County.

1881. Parker's classic statement originated in this year's biennial report of the Fish Commissioner: "Carp, as food fish, have no superior; when our streams are stocked with them the people of the State will possess as grand a luxury as found in the waters of those States celebrated for the abundance and variety of their fish; besides, carp should be as plentiful to our people as chickens to the table of the prudent farmer; the fact of raising them in small ponds, say from twenty-five feet and upwards square, or in springs on the many farms in this State, renders this fish most popular and desirable.

"One of my great aims has been to stock our waters with the best species of carp, and with the very sufficient appropriation at my command I regret the causes that have occasioned the delay I now report and trust the coming season will be more propitious to the ends I have in view."

He concluded with a resumé of a United States Fish Commission report detailing the methods and advantages of carp culture.

Israel C. Russell began his important geological work on Pleistocene Lake Lahontan this year.

1882. I. D. Pasco issued a call for carp from the Toquima Range of northern Nye County. Wrote Pasco: "The big thing is to get a good start (to get the fish), get them to breeding and we will supply and stock the country . . . Last season I persuaded the man above me on my stream not to go to Reese River after trout, because I hoped sooner or later to get carp, and I did not want trout in the stream to eat the young."

1883. Cope published some extensive observations on, among other places, the fishes of Pyramid Lake, describing several new species. Only one of these, the Cui-ui Lakesucker (*Chasmistes cujus*), has stood up to time, the others falling as synonyms of species previously named.

Parker put into print some of the first notes on the natural history and utilization of the Cui-ui:

"Among the most peculiar and least known fish of Nevada is the 'Couewe,' native of and found only in Pyramid and Winnemucca lakes. This fish belongs to the sucker persuasion, averages about four pounds, and when first caught and properly cooked is not unlike our lake suckers, but after remaining out of water four or five hours is worthless for food. During the month of March, the only season of the year when these fish are seen, large numbers are taken and cured by the Indians. It is at this time of the year that one of the natives' holidays is most observed."

Jordan and Charles H. Gilbert issued their "Synopsis of the fishes of the United States," which was important as the first comprehensive manual of keys, descriptions and bibliographies of the fishes of the country. It was the bible of ichthyologists until superceded thirteen years later.

Russell terminated his geological fieldwork in the Lahontan basin during this year.

1884. Richard Ellsworth Call, a prominent malacologist, published a detailed bulletin concerned with the snail fauna of the Pyramid Lake area, a fauna which is extinct at the present time in the lake, due to increased salinity.

1885. Parker ended his role as Fish Commissioner of Nevada with this biennial report—he departed with dignity, planting fish to the last.

Russell's classic monograph on the results of his Lahontan fieldwork appeared this year and still remains the most accurate and comprehensive work on Pyramid Lake and its vicinity.

1887. W. M. Cary, succeeding Parker as Fish Commissioner, began to express vague doubts about the value of carp:

"Professor Bird" [of 'Baird' and Girard] "also very kindly presented me with 295 young carp from the United States fish car, all of which have been distributed to the best advantage possible. While fully appreciating the generous courtesy which prompted the gift, and gladly acknowledging our obligation to the donor, I feel it my duty to say that my observations have not impressed me favorably with the qualities of these fish. They multiply rapidly and attain a considerable size, but in my judgement they are not what can be fairly called a good table fish."

He also pointed out that sawdust mills on the upper Truckee River were causing much damage to fish in the river, and that California was working up resentment over dams installed in the lower Truckee.

1889. Fish Commissioner Cary was tireless in the years 1887–1888 covered by this report. His freshwater eels from Rhode Island got as far as Kansas City before dying. Nothing daunted he turned to other things:

"Believing the soil and waters of this State were adaptable to the growth of the eatable terrapin, I so construed the law defining the duties and powers of Fish Commissioner as to warrant me in the small expenditure of funds for this purpose. I therefore purchased one hundred and eighty and distributed them throughout the State. They have been looked after by gentlemen taking interest in them and I am happy to report every probability of terrapin culture a success."

(Either these were the Pacific Terrapin (*Clemmys marmorata*), which were undoubtedly introduced to the State at one time or another (La Rivers 1942: 66–69), or something else which did not survive—Cary fails to say where he got them.)

The Eigenmanns, Carl H. and Rosa Smith, described the Amargosa Pupfish (*Cyprinodon nevadensis*) from nearby Death Valley.

1891. The Eigenmanns described the Belding Sculpin (*Cottus beldingi*) from Lake Tahoe.

The State Fish Commissionership had passed to George T. Mills. His biennial report of this year tells us of the results of many of the earlier fish plantings. Unlike his predecessor, Mills had no fault to find with the now ubiquitous carp, and noted that many of the State's residents were busy growing them. He was particularly partial to the Brown Bullhead as a food fish.

The Death Valley expedition of the United States Bureau of Biological Survey, composed of such premier naturalists as Vernon Bailey and C. Hart Merriam, was in the field, collecting new species of Nevada fishes in Ash Meadows (Nye County) and Pahranagat Valley (Lincoln County).

1893. The Eigenmanns named the Bridgelip Sucker (*Catostomus columbianus*) from the area north of Nevada.

Gilbert reported on the fishes taken by the Death Valley group, naming two of them as new in honor of the collectors:
White River Springfish (*Crenichthys baileyi*)
Ash Meadows Poolfish (*Empetrichthys merriami*)
Fish Commissioner Mills wrote that the California Fish Commission had succeeded in eliminating the dumping of sawdust into the upper Truckee River by lumber mills, at the same time hinting broadly that they would appreciate it if Nevada would do something about removing the numerous dams and obstructions to fish spawning runs in the lower Truckee.

1895. Mills makes some technical mention of the most important trout in some of our streams.

1896. This year saw the publication of the first volume of Jordan and Barton Warren Evermann's vast "Fishes of North and Middle America" which has remained to date the most complete taxonomic documentation of the fish fauna of this great area.

1897. Mills speaks of the Indians peddling far and wide the Sacramento Perch (*Archoplites interruptus*) of Pyramid and Walker lakes, much as they did the giant Cutthroat Trout:

"Indians carry them through the country by packhorse and wagon, selling them at such low figures as to enable our inhabitants, remote from railroad stations and fish markets, to supply their tables with a good quality of fresh water fish."

But he has changed his mind about carp:

"Several years ago, during the carp furore, the General Government, while not entirely to blame, was *particeps criminis* in foistering upon this State and in polluting our waters with that undesirable fish, the carp.

"True, applications for same were made by many of our citizens, ignorant of the qualities and habits of the fish and unsuspecting as to the ruin their introduction would bring. Time has now established their worthlessness, and our waters are suffering from their presence."

(This was the last biennial report until 1909.)

1898. Volumes Two and Three of the Jordan-Evermann fish catalogue appeared this year.

1900. Jordan and Evermann completed Volume Four, the last, of the catalogue.

Frederic A. Lucas of the United States National Museum described the fossil Esmeralda Shiner (*"Leuciscus" turneri*, now *Richardsonius turneri*) from beds near Fish Lake Valley, Esmeralda County.

1903. Cloudsley Rutter, a student of Jordan's, named the new Lahontan Speckle Dace (*Rhinichthys osculus robustus*) from a tributary of the upper Truckee River, California.

1906. Miss Edna W. Wemple described some fossil sharks from Triassic strata of the Humboldt Range near Lovelock, Pershing County, from material collected by two expeditions of the University of California (1902 and 1905).

FIG. 2. Professor John Otterbein Snyder, *custos rotulorum* of Lahontan fishes, was born in Butler, Indiana August 14, 1867. He was introduced to his life's work, ichthyology, as a student under Professor David Starr Jordan, first at Indiana, then at Stanford University. In 1899 he was appointed to the faculty at Stanford and served until his retirement in 1932. Between 1897 and 1930 he did a great deal of ichthyological field work for the U. S. Fish Commission, one of his projects being the study of Lahontan Basin fishes. Following this, in 1914, he spent a year at the U. S. National Museum, and from 1919 to 1931, Professor Snyder was consultant to the California Fish and Game Department. In 1925, he was director of the U. S. Bureau of Fisheries Marine Laboratory at Woods Hole, Massachusetts. For five years—1931 to 1936—he served as chief of the California Department's Bureau of Fish Conservation. He died August 19, 1943.

1907. Jordan scarcely edged out Oliver P. Hay of the American Museum of Natural History in his describing of the Pliocene Nevada stickleback (*Gasterosteus doryssus*) from a cut of the Derby Canal near Hazen, Churchill County.

1909. The Fish Commissioner of yesteryear—one man planting fish—was gone. In his place stood a triumvirate; the singletons Parker, Cary and Mills gave way to Mills-Yerington-Coryell—three men planting fish.

They credit themselves with the introduction of Steelhead Trout (*Salmo gairdneri gairdneri*), give "instructions for classifying trout" and extol the crawfish:

"With the object of providing food for the introduced varieties of fish as well as a table delicacy for our citizens, the Commission secured, through Hon. H. G. Van Dusen, formerly of the Oregon Fish Commission, thirty dozen crawfish, which arrived in splendid condition, and of which fifteen dozen were planted in the sloughs and quiet waters of the Truckee River near the Truckee Meadows. Eight dozen were planted in the Carson River, and three dozen in Washoe Lake.

"As this shipment was received with so little loss, an additional twenty dozen was secured and placed in Washoe Lake, and as these 'fish' multiply rapidly, we hope in a year or so to have a supply from these waters for other sections of the State. These are probably the Pacific Coast variety of the *Astacus nigescens* (Goode).

"From a former plant made by the Fish Commission of nineteen males and thirty-one females, received through the kindness of Mr. E. E. Hunt, from the Klamath River in May, 1895, and all placed in the Truckee River at Reno, adult specimens were found at Verdi last summer. At the same time and from the Klamath River were deposited five dozen of the salmon fly (*Corydalis*) near Reno, and for some years it has been plentiful, swarming in the month of June, and is used extensively by the anglers on that stream as a successful lure. It is known locally as the 'Junebug.'"

1911. John Otterbein Snyder (1867–1943), another of Jordan's students, and major chronicler of Lahontan fishes, was doing fieldwork in western Nevada with the help of the herpetologist C. H. Richardson. Snyder was associated with Stanford University for 50 years, retiring as professor of zoölogy in 1932.

Fish Commissioners Mills, Yerington and Clarke had an interesting experiment to report:

"This commission has been experimenting for some five years with hybrid trout, a cross between the Lake trout (*Salmo mykiss henshawi*) and the Rainbow trout (*Salmo iridius*), and in March of this year Mr. A. A. Oldenburg, the fish culturist of the commission took specimens of this trout in the Truckee River weighing 2½ pounds and 3½ pounds, respectively. This fish is Nevada's own production, an even stronger fish than the Rainbow trout and possessing extraordinary gameness.

"The ultimate object in the production of this fish is to produce a trout suitable for both streams and lakes, and which we hope will prove an excellent food fish."

1912. Snyder named the Royal Silver Trout (*Salmo regalis*) from Lake Tahoe (= Tahoe Rainbow Trout).

1913. Henry W. Fowler, Philadelphia Academy of Natural Sciences ichthyologist, suggested the new generic name *Pithecomyzon* for the Cui-ui (*Chasmistes cujus*).

Fish Commissioners Mills, Yerington and Clarke continued to praise the virtues of their hybrid trout; planted Washoe Lake catfish into White Pine County; and reported good progress with the rearing, release and subsequent capture by local fishermen of the recently introduced "Royal Chinook salmon"—they contemplated using the Silver Salmon as the next experiment. They were friendly with bass this year—

"As an example as to the value of these fish [Largemouth Blackbass], not only for food but for the destruction of that pest, the carp, we would mention the condition at Fish Lake in Esmeralda County, where formerly the carp were so numerous that they were a nuisance. By the introduction of black bass some years ago, it is now reported that the carp have almost entirely disappeared and that only the large or parent fish are left, and the bass have rapidly multiplied."

1914. Snyder produced a preliminary report on his Lahontan fishes survey, and University of Nevada geologist J. Claude Jones published his first paper containing observations on the geologic history of Lake Lahontan, thirty years after Russell's work.

1915. Fish Commissioners Mills, Yerington and Nixon reported the near completion—after seven years of negotiation between the States of Nevada and California and the Federal Government—of a satisfactory fishladder at Derby Dam. They also concurred with the recently stated policy of the United States Bureau of Fisheries in discontinuing the distribution of spiny-rayed fishes to Western United States applicants for fear of polluting trout waters. Wrote Chairman Mills:

"So far, we have only supplied the bass for large and isolated and land-locked springs, and waters that were not connected with waters containing trout."

1917. Snyder's important paper on the fishes of the Lahontan system came out this year. In it he described all the species known to occur in the waters of this system, illustrated them and named some new ones: Sandbar Sucker (*Catostomus arenarius*), a synonym of the Tahoe Sucker (Pyramid Lake); Tahoe Shiner (*Richardsonius microdon*), which turned out to be a hybrid between the Lahontan Tui Chub and the Lahontan Redshiner (Lake Tahoe); Lake Minnow (*Leucidius pectinifer*), a synonym of the Lahontan Tui Chub (Pyramid Lake); Eagle Lake Trout (*Salmo aquilarum*), either an early introduction of Coast Rainbow stock, or a hybrid between Lahontan Cutthroat Trout and Rainbow Trout (Eagle Lake); and Emerald Trout (*Salmo smaragdus*), either an early implanted rainbow, or an aberrant cutthroat (Pyramid Lake, = Pyramid Rainbow Trout).

Charles R. Eastman, re-working fossil fishes in the United States National Museum collection, discovered and described a new species, the Nevada Killifish (*Fundulus nevadensis*), from the same beds near Hazen, Churchill County containing the Nevada Stickleback.

The Fish Commissioners—Miller, O'Brien and Mills—set the pattern for the austere brevity which was to become so typical of most State reports in the future. The emphasis was on statistical presentation and an almost complete lack of informative writing.

In this year, the State legislature created the office of State Fish and Game Warden, which was filled by C. W. Grover.

1919. C. W. Grover wrote his first biennial report on the activities of his office. He travelled extensively around the State, noted some game violations and remarked that the Derby Dam fish ladder was in good condition but no fish were running in the river [6th and 18th of April].

Pirie Davidson described a new species of Triassic shark from a spine found in the Humboldt Range near Lovelock by a University of California expedition.

O'Brien, Macmillan and Tennant were the Commissioners of Fish issuing the biennial report this year. They contributed some interesting life-history material on the Lahontan Cutthroat Trout; "We operated at the new dam in the Pyramid Indian Reservation at two different times, in an effort to take spawn from the run of Blackspotted, known as 'Tommies.' The run was very poor, and the fish very green at the time they were taken—only about one in fifteen being ready for spawning. An effort was made to keep these fish alive in boxes until ripe enough to spawn, but owing to the number of leeches in the river at this station, all of the fish died within a few days after being placed in the boxes."

1921. The Fish Commission—O'Brien, Johnson and Tennant—was not liking bass this year: "As the well known character of Black Bass and other Spiny Rayed fishes is such that they soon destroy the trout in all waters that they inhabit, it is the policy of this Commission to discourage the planting of this variety of fish in any waters of the State where there is a possibility of them getting into waters stocked with trout."

1923. The State experienced a considerable turnover of Fish Commissioners, now Meyers, Johnson and Dangberg, with apparently little of any moment occurring during this biennium as far as they were concerned. They reported that the floor of the Verdi Fish Hatchery—which had rotted and fallen in—had to be replaced with a concrete floor before hatchery operations could begin again.

Limnologists Kemmerer, Bovard and Boorman, doing chemical, physical and biological fieldwork for the United States Bureau of Fisheries on Northwestern United States lakes, wrote their report and included important data on Lake Tahoe.

1924. Jordan recorded the Belding Sculpin (*Cottus beldingi*) as a Pleistocene fossil from cave beds in the Fallon area.

1925. J. Claude Jones published his major paper on the geologic history of Lake Lahontan—his conclusions were at striking variance with those of all other investigators, prior or to come. He felt the entire story of Lake Lahontan could be told in 3,000 years, rather than the more accepted (and now substantiated) ten times that amount.

Meyers, Johnson and Robbins—Fish Commission—prescribed regulations for Indians handling fish off the reservations; listed State game, refuge and recreational areas; and included a survey report of the Mt. Wheeler area by Hatchery Superintendent J. H. Vogt.

1927. The Fish Commissioners were unchanged from the last biennium, and had many interesting things to report concerning the Black-spotted Trout. Six hundred thousand eggs had been taken at the Numana (Indian Dam above-river from Nixon) field station in 1925, which was only about half of what they had been getting. Low water was blamed, and the following year the sand bars were so extensive at the river mouth that no trout could get out of the lake. Their attempt to seine spawners from the lake resulted in the United States Commissioner of Indian Affairs revoking their permit to operate on the reservation.

1928. Charles T. Brues, Harvard entomologist, sent out the first of two reports on his studies of Western United States hot springs and their faunas. His fieldwork, begun the preceding year, included many Nevada thermal springs.

1929. The Nevada State Fish Commission was reconstituted and added to, now becoming Douglass, Steninger, Edler, Inwood and Phillips with J. P. Morrill as Executive Officer and Superintendent of Hatcheries. Among other things, they mention great difficulty at the Verdi Hatchery because of insufficient uncontaminated water, a complaint which had been recurring since 1919.

About this time, the professional and versatile collector Percy Train,[6] who had worked the Nevada scene for years for fossils and rock specimens, shipped 100 specimens of fossil cyprinid fish from the well-known Jersey Valley beds to an eastern biological supply company.

1930. Joseph H. Wales, California State Fisheries biologist, described the odd little Devil Pupfish (*Cyprinodon diabolis*) from Devil's Hole, a warm limestone pothole of Ash Meadows, Nye County.

Jordan—Evermann—Howard W. Clark published the very useful checklist of fishes . . . of North and Middle America, whose pages have borne the brunt of the author's thumb on innumerable occasions.

1931. The Fish Commission (Douglass, Bradshaw, Edler, Phillips, Getchell) was fond of bass this year: "The commission recently received a full carload of pond fishes from the United States Bureau of Fisheries, consisting of large mouth black bass, yellow perch, crappie and blue-gills (bream). We feel that these varieties of fish, which are held in high esteem as food and game fishes throughout the east and southern

---

[6]Born Helena, Montana, 1876. Died Reno, Nevada, 3 Feb. 1942. Train was a well-known and talented collector who began as a mining engineer and assayer, but spent his later life collecting fossils, mineral specimens and plants for many institutions throughout the country.

States, will find congenial waters and take hold in our State, thereby adding greatly to the fishes we already have and making our waters more productive."

1932. Carl L. Hubbs—then University of Michigan ichthyologist and now at the University of California's Scripps Institution—published a description of the new Railroad Valley Springfish (*Crenichthys nevadae*) from two specimens collected by Charles T. Brues in his 1930 swing around the hot springs of the State. This was the first of a series of reports to be issued by Hubbs during his forthcoming years, based upon field work done largely with his student Robert R. Miller, who co-authored the later items.

The Nevada Fish and Game Commission now consisted of Douglass, Bradshaw, Phillips, Getchell and Doyle, who served, with little overall change, through several biennia. They had hatchery troubles, insufficient means of fish transportation and recommended to the legislature that the Fish and Game Commission be empowered to appoint "a State Game Warden or wardens, said wardens to be appointed at the discretion of the Board."

1935. The same Fish and Game Commission reported on an agreement with the California Fish and Game Commission whereby California was granted the right to take spawn at Marlette Lake (owned by the Virginia and Gold Hill Water Company, who had allowed Nevada to do this for years) in return for which California agreed to assume the Nevada obligation of restocking the lake and its streams. They also brought up officially, for the first time, the problem of dual control of fish and game matters:

"We firmly believe that the present dual control between this Commission and the various counties is obsolete and should be abolished."

United States Bureau of Fisheries biologists M. J. Madsen (now Chief of Fisheries for the Utah Department of Fish and Game) and William F. Carbine examined Pyramid Lake at the request of the Indian Service and noted the major causes of the decline of the once extensive Cutthroat Trout fishery there.

1936. With Barr replacing Bradshaw, the Commission recommended not only that they be given authority to "set fishing and hunting seasons throughout the State," but also "to set bag limits from year to year, rather than having them arbitrarily set by the Legislature." In the matter of fisheries, they report the handling of 2,100,000 Rainbow Trout eggs and 300,000 Brook Trout eggs for the 1934–1935 period.

1937. G. Evelyn Hutchinson published the results of his brief but informative survey of Lahontan waters. His data were the result of six-weeks field work in the summer of 1933, out from Yale University.

1938. The commission was unchanged this year. Both the Verdi and Smith Creek Hatcheries were snowed in during the winter of 1936–1937, and the Washoe County Fish Hatchery in Idlewild Park, Reno, undertook the initial handling of eggs ordinarily consigned to

Verdi. In 1937, some 25,000 "catfish" were planted in Washoe Lake, which was then closed to fishing. As for coming events and their shadows—

"Some thought was given again this year to taking spawn from Pyramid Lake trout for the purpose of rearing some of these fish in other waters. However not enough fish made the run upstream this year. Some research is being carried on by the United States Bureau of Fisheries with regard to Pyramid Lake and its fish."

The first conference on Lake Mead problems mutual to Nevada and Arizona was held at Boulder City.

FIG. 3. The massive "salmon-trout" of Frémont, in a later era (circa 1920) just before their final decline. Courtesy Tom Trelease.

1939. Harry E. Wheeler, University of Nevada geologist, described the quite peculiar Nevada Edestid Shark (*Helicoprion nevadensis*) from the Lower Permian of the Humboldt Range, near Lovelock, Pershing County—it had long passed for a fossil fern around the Department of Geology at the University.

1940. A brief report on the dying Pyramid Lake Cutthroat Trout industry was published by Federal Fishery Biologist F. H. Sumner.

The ichthyologist Francis B. Sumner, of Scripps Institution of Oceanography, issued the results of fieldwork done in eastern Nevada on the physiology of native warm-water fishes. He was assisted by M. C. Sargent.

1941. Messrs. Getchell, Barr, Phillips, Doyle and Powell constituted the commission during the years 1938 to 1940. Little was reported except routine hatchery activity, but they made the following policy observation:

"Each year an increased number of fish are planted in an effort to keep up with the growing numbers of fishermen, but increasing demands for fish indicate that restocking of streams alone is not enough. It is the belief of the Commission that proper screening of ditches, restoration of food to the waters, installation of fish ladders, and control of pollution will be necessary before ideal fishing conditions can be created."

Lore David, of the California Institute of Technology, described the fossil fish *Leptolepis nevadensis*, of apparent Cretaceous age, from distinctive sediments of Newark Cañon, east of Eureka, Nevada.

Harold Peer was appointed State Game Warden in 1939 to work with James C. Savage, United States Game Management Agent, during the migratory bird season.

1942. Francis B. Sumner continued his eastern Nevada studies with Urless N. Lanham, later an entomologist.

Vasco M. Tanner of Brigham Young University, one of the deans of Utah biology, described a new mountainsucker (*Pantosteus intermedius*) from eastern Nevada.

James W. Moffett, a Federal fishery biologist, wrote "A fishery survey of the Colorado River" below Boulder Dam in which some of the gross biological features of the river in that area were considered.

Phillips, Barr, Powell, Baker and Casady reported the planting of 4,000,000 Rainbow, Brook and Cutthroat Trout in the State during 1940–1942 and expansion of the Verdi hatchery. Vernon Mills, State Game Warden replacing Harold Peer of the previous biennium, indicated that pelicans and cormorants were a fishery problem in the new Indian Lakes project near Fallon. Here in very exposed, open and shallow situations, blackbass were apparently being effectively preyed upon by these fish-eating birds. Mills concluded:

"It is significant that a very little shooting will usually move the fish-eating birds on to some place where they can feed on rough fish without doing any particular harm to game species, but if permitted to feed unmolested on concentrations of game fish they can easily undo in one week what it has taken sportsmen and conservationists years to build up."

1943. Moffett wrote "A preliminary report on the fishery of Lake Mead."

1944. The only change in the State Fish and Game Commission was the replacement of Casady by Olin for the duration of the war. The Smith Creek Hatchery, established about 1927 in the Desatoya Range west of Austin, was abandoned and dismantled by the Commission as an uneconomical operation. As a substitute, the individual counties involved—Lander, Nye and Eureka—agreed to operate rearing ponds with the help of the State. Charles Comstock replaced Vernon Mills as State Game Warden. A considerable addition to the Verdi Hatchery water supply was potentially made but pipelining was not completed.

1946. Robert R. Miller and J. Ray Alcorn, the latter a prominent Nevada biologist, published a very useful paper discussing the introduced fishes of the State.

Return of Casady reconstituted the Fish and Game Commission as it was for the 1940–1942 biennium. For the first time, and because of increased water supplies, the Verdi Hatchery instituted a double operation routine, placing the second batch of eggs into their troughs in June after planting fish from the first hatching. The following categories received some planning or actual work:

(1) Greater hatchery production, as mentioned above;
(2) Rearing pond installations—of a county nature;
(3) Fisheries survey. The need for this long overdue analytical treatment of the State's water resources was being sorely felt and some preliminary runs were made.
(4) Introduction of warm-water fishes. Active initial stocking and restocking of suitable waters was being done;
(5) Special research on depleted waters. "The fisheries resource at Pyramid Lake has disappeared due to conditions traceable to man's activity, and the same condition threatens destruction of the resources of Walker Lake." Restocking of Walker Lake was begun, as well as planning for more intensive investigation of all lake waters.
(6) Removal of rough fish. This involved the utilization of carp and other rough fish through the agency of commercial fishermen.
(7) Pollution control. Lessening of the amount of pollution in the long-suffering lower Truckee River was reported.

Under the heading of "Educational Program," it was noted that additions to the course offerings at the University of Nevada was making it possible for that institution to offer a more suitable program for wildlife management. "The State Fish and Game Commission is co-operating with the University in this matter, offering to the University the facilities at the game farm and hatchery for the training of students. In return, the Biology Department of the University has made its research facilities available to the State Commission and is furnishing various students to work with the Commission on research problems and wildlife surveys."

Considerable expansion and re-organization in personnel occurred during this 1944–1946 biennium. A larger and more effective group was headed up by Sessions S. ("Buck") Wheeler, one of the State's pioneer—and, in the opinion of this writer, premier—conservationists. It was Wheeler, a graduate of the University of Nevada in Biology, who saw the need for liaison between the Fish and Game Commission and the University and did something about it. The value of having a professional biologist operate a professional biology program became self-evident. H. Shirl Coleman, another trained biologist, became the first field man for the Commission.

1948. Hubbs and Miller published two papers important to Nevada ichthyology—(1) descriptions of two very interesting relict genera of cyprinid fishes from widely separated locales in the State: the Soldier Meadows Desertfish (*Eremichthys acros*) from a western Humboldt County area and the Moapa Dace (*Moapa coriacea*) from Warm Springs on northern Clark County; and (2) an extensive contribution to the biological aspects of Late Pleistocene activities in the Great Basin, in which the fish fauna, largely of Nevada, was used to illustrate their points. In most respects, this work summed up Hubb's extensive work with Nevada fishes over a period of better than half a generation, much of the work done with his student, Robert R. Miller.

Miller described the Pahrump Poolfish (*Empetrichthys latos*) from Pahrump Valley in Nye County and also showed that the Amargosa Pupfish (*Cyprinodon nevadensis*) population in the Amargosa drainage consisted of several subspecies.

Miller and his father, Ralph G. Miller, added several new records to the State's fish fauna from the Snake River drainage streams of extreme northern Nevada.

The writer began collecting the present representative series of Nevada fishes for the Museum of the Biology Department of the University of Nevada, a collection which now numbers more than 16,000 teaching and research specimens, with some comparative material from other States.

The preceding five-man commission had, in a manner of speaking, "sickened and died." For the first time a 17-man commission came into existence, with one representative from each county. From these commissioners was selected a five-man executive board. Commissioners were: William O. Bay, Dan Evans, Jr., Leroy Casady, Cal Liles, Warren Monroe, Fay Baker, A. C. Barr, Walter Bowler, William Curran, Jack Fogliani, Charles Gilbert, H. C. Helwig, Jack Heward, W. T. Holcomb, William Hollan, Dave Norman and George Starks. The first-named five made up the executive board (as will be the case in subsequent listings below).

Department personnel were considerably increased, a long-overdue improvement. Thomas J. Trelease, a recent fish and game graduate of the University of Nevada, was appointed the first fishery biologist and the present extensive and very productive fishery program for the State began to take shape.

During this biennium, the Verdi Upland Game Bird Farm was built southwest of the hatchery and game law enforcement became the statewide responsibility of the Commission with a force of five State Game Wardens. Another improvement in operating efficiency was a new law giving all revenue derived from the "sale of fishing, hunting, and trapping licenses, and other sources," to the State Commission.

As an indication that fishery planning was sound, the first attempt was made to preserve a "pure strain of Lahontan basin Blackspotted trout" from Walker Lake stock in brood ponds at Verdi. A state rearing pond unit was established in Smith Valley (Lyon County) and the first actually planned investigation began of the fishery resources of the State. Among other items: "Complete experiments are now in progress on Pyramid Lake and if they prove the waters satisfactory for trout life, a restocking program will start immediately."

1949. John Kopec published life-history information on the White River Springfish (*Crenichthys baileyi*) from the Warm Springs area of northern Clark County.

1950. J. B. Kimsey, California fishery biologist, reported on the introduction—presumably by bait fishermen—of four native Lahontan drainage system fishes into headwaters of the Sacramento River drainage on the west side of the Sierra Nevada Mountains.

Nevada State Fish and Game Commissioners were: Casady, Evans, Bay, Liles, Monroe, Tot Ambrose, Earl Branson, James Clark, Ed Culberson, Curran, Robert Dickey, Alex Glock, Hugh Henrichs, Heward, Homer Macucci, Tom McCulloch, and Verne Stever. Buck Wheeler resigned as director and his place was taken by Frank W. Groves, a professionally trained wildlife management graduate of Oregon State College. Fishery was recognized as a separate section for the first time with a total personnel of nine, headed by Tom Trelease. The total warden force had effectively increased to seven wardens and two assistant Game Wardens.

Ira N. Gabrielson, past director of the United States Fish and Wildlife Service and later president of the Wildlife Management Institute in Washington, D. C., was invited to Nevada to make an operational study of the Fish and Game Commission with an aim to further improvement.

Activities and improvements in the fishery field (as well as other areas) now became so numerous that it is not possible here to say much about them. The Smith Valley Rearing Station was near completion; the Spring Creek Rearing Station in eastern White Pine County was initiated; co-operative Kokanee Salmon stocking of Lake Tahoe with the California Fish and Game Commission was begun; the first arrangements were successfully made with the Indians of Summit Lake (Humboldt County) to obtain what presumably were the only Lahontan Cutthroat Trout in existence to provide spawn for future rearings and plantings; a program of stocking warm-water fishes into suitable areas in the State was in effect; and so forth and so on, all to the good.

Tanner, from Utah, described a new species of Gila (Bonytail) Chub (*Gila jordani*) from the White River system of Eastern Nevada.

1951. Fraser and Pollitt, California fishery biologists, reported on recent introductions of the Kokanee Salmon into Lake Tahoe, while Kimsey had some notes on "Kokanee spawning in Donner Lake, etc." Wallis, biologist for the Lake Mead National Recreational Area, had a paper on the status of the fish fauna of his locale. Hubbs and Miller synonymized Snyder's controversial "Sandbar Sucker" (*Catostomus arenarius*) of Pyramid Lake with the common Tahoe Sucker (*C. tahoensis*) of the Lahontan System.

1952. Miller and Morton wrote of the first record of the Dolly Varden Trout (*Salvelinus malma*) from extreme northern Nevada and the first key (La Rivers) and annotated checklist (La Rivers and Trelease) for Nevada fishes appeared. Miller also issued a very useful "Bait fishes of the lower Colorado River, etc." which keyed and illustrated many otherwise obscure native fishes.

1953. Nevada Commissioners were: Monroe, Ambrose, Branson, Wayne Kirch, Don Quilici, Herb Carter, Art Champagne, Myron Clark, D. E. Culberson, Dickey, John Etchart, Evans, Glock, Hobart Leonard, McCulloch, Jewel Turner and Owen Walker. Further additions were made in Fishery—its eight members being augmented by three fishery technicians in the Wildlife Restoration Division. The report for this biennium (1950–1952) was authored visibly by the several heads of departments for the first time. Tom Trelease, Chief of Fisheries, noted that the year 1951 marked the actual organizational establishment of an effective fishery division in that a "co-ordinated operation" was achieved for the first time. Federal aid in the form of the Dingell-Johnson program became available with the voting of matching funds by the Legislature—this is a Federal fund comparable to the Pittman-Robertson aid for game.

The State was divided into five fishery districts for better management; full-fledged investigations were begun on Lakes Mead and Mohave in Clark County; and two fishery biologists were assigned to survey the streams of the State to assess their present conditions and future potentials. Kokanee Salmon planting continued in Lake Tahoe and began in Pyramid and Walker Lakes. Installations were expanded and improved.

1954. Kimsey published a comprehensive report on the "Life history of the Tui Chub, *Siphateles bicolor*, etc." from nearby Eagle Lake in California, which threw much light on the biology of the same species in Nevada lakes.

Nevada Commissioners were: Kirch, Branson, Etchart, Monroe, Jewel Parsons (Turner), Lloyd Boone, Art Biale, Carter, Champagne, Glock, Ned Kendrick, Bill Kottke, Leonard, McCulloch, Tom Papez, Quilici and Walker. Fishery had expanded to ten, with four additional fishery biologists listed under Dingell-Johnson projects. Divisional reports in the main biennial report were now so extensive we will not attempt to further analyze them here. For the interested reader, suffice it to say that the fishery program can now be considered modern and exceptionally forward-looking not only with respect to the immediate needs of the State, but by comparison with what was being accomplished in other States, many with far superior monetary resources than are available in Nevada.

While the only complete set of Nevada Fish and Game biennial reports in existence that the writer is aware of is that in the files of the Reno office of the Commission, these later reports—from 1950 on—can be consulted freely at some additional places, such as the University of Nevada in Reno and the State Library in Carson City.

CHAPTER II—PALEONTOLOGY

CONTENTS

	PAGE
Paleozoic era	37
Permian period	37
The Nevada Edestid Shark	37
Mesozoic era	41
Triassic period	41
The Humboldt Hybodus Shark	41
The Alexander Acrodus Shark	42
The Humboldt Acrodus Shark	42
The Humboldt Placoderm	43
Cretaceous period	44
The Nevada Leptolepis	44
Cenozoic era	47
Tertiary period	47
The Nevada Amyzon Sucker	48
The Nevada Pirate Perch	51
The Esmeralda Redshiner	54
The Nevada Stickleback	56
The Nevada Killifish	61
Quaternary period	63
Modern, living species	63

CHAPTER II
PALEONTOLOGY

From the available paleontologic evidence, the western Great Basin is relatively barren of fish fossils, although, where these do occur, they may be comparatively abundant. However, even with this paucity of material, it is possible to find species representative of the three great life eras within Nevada borders: Paleozoic, Mesozoic and Cenozoic covering, in all, an estimated span of some 200 million years of "Nevada" prehistory.

PALEOZOIC ERA
185,000,000 to
500,000,000
years ago

PERMIAN PERIOD
185,000,000 to
210,000,000
years ago

Our only known fishes for this period are the fragmentary shark remains listed below.

Class *CHONDRICHTHYES* (Elasmobranchii), the Cartilaginous fishes: Sharks, Rays, Skates

Subclass SELACHII, Sharks and Rays

Superorder *Selachoidei*, Sharks

Order HETERODONTIFORMES

Family *EDESTIDAE*, Edestid Sharks

Genus HELICOPRION Karpinsky 1899
(Genotype *H. bessonowi* Karpinsky 1899)

THE NEVADA EDESTID SHARK
(*Helicoprion nevadensis* Wheeler 1939)

Helicoprion nevadensis Wheeler 1939, Journal Paleontology 13(1): 109–112.

The oldest fish fossil the writer has knowledge of in Nevada is that of the peculiar Permian edestid shark (*Helicoprion*). The only portion of these cartilaginous animals to be preserved in the fossil record was their odd tooth whorls, and these have been the subject of considerable conjecture as to position and function. One prominent paleontologist of many years ago visualized this spirally coiled series of teeth as protruding from the back region and being used by the shark as a cutting tool in obtaining food and defending itself.

There is little doubt, however, from what is known of the developmental pattern of sharks' teeth in general, that these are true teeth growing in the jaws and that the larger, older teeth were either shed and replaced by younger teeth growing behind them or represent adult structures while the smaller teeth deeper in the whorl were those which functioned for its immature years.

Helicoprion remains are relatively rare as fossils, and are known from only a few parts of the world. The University of Nevada specimens masqueraded for many years as the impressions of fern leaves in rock, until they were finally recognized as shark teeth.

In 1939, Harry E. Wheeler, then of the University of Nevada Geology Department, described *Helicoprion nevadensis,* new species,

FIG. 4. *Helicoprion nevadensis* Wheeler 1939, type specimen in the collection of the Mackay School of Mines, University of Nevada, Reno. Courtesy Dr. E. Richard Larson.

from a tooth whorl collected in Permian rocks of Humboldt Range near the old mining camp of Rochester, in Pershing County. Another species, *H. sierrensis*, new species, was described at the same time from the adjacent Sierra Nevada Mountains in the Frazier Creek-Feather River Middle Fork area of Plumas County, California.

One of the seemingly important results of the discovery of *Helicoprion* in this region was that of strongly suggesting, at least, that rocks which had hitherto been regarded as Mesozoic in age were actually older than that (i.e., Paleozoic).

However, E. Richard Larson, University of Nevada geologist, pointed out later (1955) in a paper with James B. Scott that the known localities for *Helicoprion* in western North America represent rather widely spaced horizons and that consequently the genus is not the specific indicator it once seemed. Larson and Scott also reported on a new occurrence of a *Helicoprion* tooth spiral from the vicinity of Contact in northern Elko County, a fragment which was specifically unidentifiable but seemed to most closely resemble *Helicoprion ferrieri* (Hay 1907) from the Idaho Phosphoria horizon.

Helicoprion, as a marine type has, of course, no relatives among present-day fresh-water fishes; these sharks seemingly lived in a warm sea which stretched diagonally across western North America from British Columbia and California to Texas and eastern Mexico. At its northwestern end, the seaway was connected with the north Pacific Ocean, and flowed into what is now the Gulf of Mexico at its other extremity.

There is some disagreement among authorities whether edestid sharks belong in this class, or in the class Holocephali, which is represented in our modern fauna by the peculiar chimaeras, or ratfishes.

ORIGINAL DESCRIPTION

"HELICOPRION NEVADENSIS Wheeler, n. sp.

"*Holotype*—University of Nevada, Mackay Museum Paleontological Coll., no. 1001, collected by Elbert A. Stewart in 1929. Plastotypes in the British Museum (Natural History), University of California, and Stanford University.

"*Description*—The single specimen upon which this species is based consists of an external mold showing the major part of very little more than 2½ volutions. The greatest diameter measures 120 mm. None of the coils are in contact, the relative amount of separation being unmeasurable because of a slight distortion of the specimen. The teeth, as best they can be estimated, number about 37 for the first volution, 39 for the second, and 19 for the first half of the third.

"For the purpose of description, and in accordance with Karpinsky's description of the genotype, the teeth are regarded as consisting of three parts (fig. 2); first, the cutting blade, which lies between the apex and the points of contact with adjacent teeth; second, a middle part; and, third, a narrowed base inclined toward the older (smaller) teeth of the series. In the smaller teeth the blade is relatively the longest; while in the largest teeth each of the 3 parts consists of about one-third of the total length. At the end of the first volution, the blade forms about three-eighths of the total length; and the line separating the other 2 divisions is not sharply defined at this stage. In the teeth of the first volution, the narrowed base extends forward to a point directly below the middle of the adjacent tooth; those of the largest teeth, however, extend to the middle of the second tooth in advance. As in most described specimens of *Helicoprion*, all teeth are apposed to one another symmetrically; i.e., the ends of the bases are not in lateral scalelike apposition to those of adjacent segments.

"The teeth at the end of the first volution are 9.4 mm. long, and their greatest width (at base of blade) is 2.7 mm. Those at the end of the first volution are 25.5 mm. long and 6.5 mm. wide. The largest complete tooth is 39 mm. long and 9 mm. wide.

"The apical angle (measured between apex of cutting blade and points of contact with adjacent teeth) is 42° at the end of the first volution, 37° at the 1½-volution stage, 43° at the end of the second, and 45° at 2½ whorls.

"The edge of the cutting blade is finely serrated, the larger ones showing about 15 small denticles on each side of the apex.

"The interdental spaces which separate the middle and narrowed portions of adjacent teeth are relatively wider in the earlier than in the latter growth stages.

"The width of the shaft at the base of the teeth is equal to about one-tenth of the volution height.

"*Comparisons and affinities*—*Helicoprion nevadensis* more closely resembles the genotype, *H. bessonowi* Karpinsky, than any other described species. The principle difference between these two is the relatively greater width of the shaft in *H. nevadensis*. In this respect, the Nevada species takes an intermediate position between *H. bessonowi* and *H. ferrieri* (Hay) from the Phosphoria formation of Idaho and Wyoming. In the genotype the shaft width is equal to about one-fifteenth of the volution height; in *H. nevadensis*, to one-tenth; and in *H. ferrieri*, to one-fifth.

"One other noteworthy difference exists between *H. bessonowi* and *H. nevadensis*. The middle part of the teeth is defined at an earlier stage, and the relative length of this middle part increases more rapidly and to a greater extent in the Artinsk form than in the Rochester specimen. In the former, the middle portion greatly exceeds the length of the blade at the 2½-volution stage; while in the latter, the 2 divisions are about equal. In this feature, also, this new species must be regarded as constituting an intermediate stage between the genotype and *H. ferrieri*.

"*Preservation and matrix*—The fossil, an external mold, is impressed in a fine-grained tuffaceous shale. A petrographic examination by Gianella (1937) reveals that

> "it consists of about 50 per cent quartz, with occasional feldspar fragments; and the remainder is mostly fine fragmental glass. There are also occasional muscovite fragments and minute grains of pyrite. Some fine-grained material appears to be carbonaceous. The largest quartz grain measures 0.18 mm; while the average lies between 0.01 and 0.05 mm. The quartz is very angular, none of it showing any appreciable rounding.

"*Type locality*—U. N. loc. 28, Rochester district, Lovelock quadrangle, Pershing County, Nevada. Map: U. S. Geol. Survey, Bull. 762, pl. 1. Lens of tuff exposed on the north slope of Sunflower Hill, about 100 feet west of the drain running northward from the saddle at B. M. 6671; also, about 100 feet west of a line between B. M. 6671 and B. M. 6237, at a point about 700 feet north of the former. Elevation: 6400 feet. SE¼, SW¼, Sec. 16, T. 28 N., R. 34 E.

"*Formation*—About 4000 feet below the top of the Rochester trachyte, as mapped by Knopf (1924). Anthracolithic (Uralian or Artinskian)" (Wheeler 1939: 109–112).

MESOZOIC ERA
70,000,000 to
185,000,000
years ago

TRIASSIC PERIOD
155,000,000 to
185,000,000
years ago

Again, as in the preceding and older periods, we have only isolated material to represent what must have been the very abundant and extensive fish life of this period, and these are even more fragmentary than the Paleozoic remains.

Family *HYBODONTIDAE*, Hybodontid sharks

The following species of hybodontid sharks were described by Miss Wemple from, in each case, a single tooth. All of the teeth came from what were considered by the describer as Middle Triassic beds in the area between Rochester and Unionville on the east side of the Humboldt Range, Pershing County, Nevada—*Acrodus alexandrae* from Fisher Canyon, and *Acrodus oreodontus-Hybodus nevadensis* from Cottonwood Canyon. The material was collected originally by two expeditions of the University of California (Berkeley), one visiting the region in 1902, and a second returning in 1905. As would be expected from the meagerness of the evidence, there is some doubt as to the validity of the genus *Acrodus*, at least, and its taxonomic position.

Genus HYBODUS Agassiz 1832

THE HUMBOLDT HYBODUS SHARK
(*Hybodus nevadensis* Wemple 1906)

Hybodus nevadensis Wemple 1906, University California Publications Geology 5(4): 72.

ORIGINAL DESCRIPTION

"HYBODUS NEVADENSIS, new sp., Pl. 7, fig. 3

"Type specimen, one detached tooth, No. 10254, Univ. Calif. Col. Vert. Palae. From the upper part of Middle Triassic, Cottonwood Cañon, West Humboldt Range, Nevada.

"Tooth cuspidate, crown relatively low and vertically striated, with 1 principal elevation situated a little away from the middle of the tooth. The lateral prominences are 2 in number. These prominences are well defined and sharply conical. No lateral denticles are present on the opposite side. The root is separated from the crown by a deep groove. The type specimen probably belonged to the symphyseal portion of the jaw, as it shows a high, robust, principal cone" (Wemple 1906: 72).

Genus *ACRODUS* Agassiz 1934

THE ALEXANDER ACRODUS SHARK
(*Acrodus alexandrae* Wemple 1906)

Acrodus alexandrae Wemple 1906, University California Publications Geology 5(4): 71–72.

ORIGINAL DESCRIPTION

"Type specimen, 1 detached tooth, No. 9874, Univ. Calif. Col. Vert. Palae. From the upper part of the Middle Triassic, Fisher Cañon, West Humboldt Range, Nevada.

"The tooth is large, elongated, and with a faint median keel. The crown is low and wider than the root. The overhanging margins are very deeply and sharply serrated. This ornamentation is coarser than in any other species of *Acrodus* known to the writer. Only one of the serrations is connected with the ridges higher up on the crown. The middle of the crown is much wider than the ends, which narrow gradually. The longitudinal crest on the crown is a very narrow but well defined ridge which continues unbroken along the entire surface of the crown. Immediately above the serrated lateral margin the crown is smooth. Half way up the side of the tooth a series of delicate ridges arises and runs into the longitudinal crest" (Wemple 1906: 71–72).

THE HUMBOLDT ACRODUS SHARK
(*Acrodus oreodontus* Wemple 1906)

Acrodus oreodontus Wemple 1906, University California Publications Geology 5(4): 72.

ORIGINAL DESCRIPTION

"Type specimen No. 10251, Univ. Calif. Col. Vert. Palae. From the upper part of the Middle Triassic, Cottonwood Cañon, West Humboldt Range, Nevada.

"The teeth are elongated, depressed, with a median prominence, and a well defined median ridge. No lateral prominences are present. The coronal contour is strongly rounded. Coarse wrinkles converge toward the apex and the longitudinal crest. This convergence is more noticeable on one surface of the tooth. On this surface the distance from the base to the lower edge of the crown is about half the same distance on the other surface. This surface is more coarsely ornamented than the opposite side. The straight longitudinal crest is situated medially on the crown" (Wemple 1906: 72).

Class *PLACODERMICHTHYES*, the Plate-skinned fishes: Placoderms
Order ACANTHODIFORMES
Family *GYRACANTHIDAE*, Gyracanthid Placoderms
Genus COSMACANTHUS Agassiz 1845

THE HUMBOLDT PLACODERM
(*Cosmacanthus humboldtensis* Davidson 1919)

Cosmacanthus humboldtensis Davidson 1919, University California Publications Geology 11(4): 433–435.

This new species of very primitive fish is based on even more controversial material than are the hybodontids—a single, incomplete spine —and was assigned only provisionally to Agassiz's genus *Cosmacanthus* by the describer. There is apparently good reason to suspect that *Cosmacanthus* is a synonym of *Gyracanthus*, which latter is based on better and more complete material.

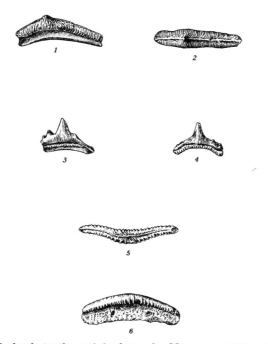

FIG. 5. Fossil shark teeth, mainly from the Mesozoic of Nevada. (1) and (2) *Acrodus oreodontus*, Middle Triassic, Cottonwood Cañon, Humboldt Range. (3) *Hybodus nevadensis*, idem. (4) *Hybodus shastensis*, Upper Triassic, Bear Cove, Shasta County, California. (5) and (6) *Acrodus alexandrae*, Middle Triassic, Fisher Cañon, Humboldt Range (From Wemple 1906).

ORIGINAL DESCRIPTION

"Type specimen, no. 9162, Univ. Calif, Coll. Vert. Palae. From the upper part of the Middle Triassic, Straight Cañon, West Humboldt Range, Nevada.

"The spine (figs. 1 and 2) is of medium size, tapering quite abruptly with a slight curve backward. The greatest length is 59.5 mm. and its greatest width 12 mm. It is bilaterally symmetrical and is made up of a long plain inserted portion which has a deep furrow posteriorly, and a short exserted portion which is partly covered by small closely set sculptured tubercles. Most of the projecting part is closed posteriorly and the furrow extends into it as the medullary cavity. The line separating the smooth base from the ornamental distal portion runs obliquely upward from anterior to posterior at an angle of about 45°. The lateral faces are slightly rounded. The inserted portion tapers to a quite sharp point in which the groove is wide open, shallow and has sharp edges. There is but slight indication of the arrangement of the tubercles in rows" (Davidson 1919: 433–434).

CRETACEOUS PERIOD
70,000,000 to
135,000,000
years ago

A single species is our only Nevada representative for this 65 million year period, but its remains are surprisingly complete in contrast to the fragments of older times.

Class *OSTEICHTHYES* (Teleostomi), the Bony Fishes

Subclass ACTINOPTERYGII, Ray-finned fishes

Superorder *Teleostei*

Order CLUPEIFORMES

Superfamily Leptolepidoidea

Family *LEPTOLEPIDAE*

Genus LEPTOLEPIS Agassiz 1832

THE NEVADA LEPTOLEPIS
(*Leptolepis nevadensis* David 1941)

Leptolepis nevadensis David 1941, Journal Paleontology 15(3): 318–321.

The genus *Leptolepis* occurs in various parts of the world—America, Europe and Africa—in Jurassic and Cretaceous rocks. Nevada specimens described by David may be immature, and the describer could find no apparent close relatives among other known species of *Leptolepis*. The geologic importance of this find in Nevada can be judged from the fact that it seems to definitely date the beds in which it occurs as at least Lower Cretaceous, possibly later Cretaceous. The type conglomerates were originally considered to be Pennsylvanian in age.

Although called "Weber" conglomerates by Hague in 1892 because of a supposed correlation with the Utah Weber, later work by Nolan (1956) redefined this stratum as the Newark Canyon formation. Its Cretaceous age has been further confirmed by gastropod mollusks as well as the fish.

ORIGINAL DESCRIPTION

"Family LEPTOLEPIDAE
"LEPTOLEPIS NEVADENSIS David, n. sp.

"*Holotype*—A specimen $41 + 9 = 50$ mm. in length, no. 10138, California Institute of Technology Coll. Vertebrate Paleontology.

"*Paratypes*—Ten specimens $33 + 7 = 40$ to $48 + 10 = 58$ mm. in length and a number of more or less fragmentary specimens. All from Cornell University locality 38C.

"*Diagnosis*—Body 5.75–6.5 mm. in standard length, head 3.6 mm., orbit a little more than 3 in head; 49 to 50 vertebrae; D. = 14 to 15; A. = 9; V. = 8, below posterior part of dorsal fin base; P. = 14; C. = 38. Nine hypurals, five below, four above median line; two epurals.

"*Description*—Body elongate, depth $5\frac{3}{4}$—($6\frac{1}{2}$) in standard length, caudal peduncle two-thirds greatest depth. Head $3\frac{3}{5}$ in standard length, elongate $1\frac{3}{5}$ as long as deep. Orbit large (possibly ovoid in shape), slightly more than 3 in head.

"The structure of the headbones, in general, does not differ apparently from that described in detail by Rayner (1938). The parasphenoid is always distinct, rising upward toward the front, and situated rather low in most specimens, cutting through the basal part of the orbit. Maxillary with two well-developed supramaxillaries; lower jaw typical for genus, with highly arched dentary. No teeth can be seen. The ceratohyale distinct in several specimens with 10 branchiostegals. Opercular arch as in *Leptolepis*, opercular $1\frac{1}{4}$ as deep as long, with a pronounced diagonal lower border, lower anterior corner sharply pointed.

"Vertebrae 34 to $35 + 15 = 49$ to 50. Vertebrae pierced by notochord, which is covered by a layer of dark ossified material. The vertebrae constricted in middle, where neural and haemal arches originate, the anterior and posterior extremities (zygapophyses?) of the vertebrae projecting into pointed edges. Abdominal neural processes feeble and short, very delicate intermuscular bones lying across; numerous arched and strong ribs extend to ventral border of body.

"Structure of tail end of column clearly shown on a small tail fragment, no. 10131 California Inst. Technology Coll. Vertebrate Paleontology (fig. 1). Structure approaches homocerque tail of modern clupeids; only last six vertebrae taking part in structure of tail have prolonged and strong haemal and neural processes. Urostyle slender, pointed; nine hypurals, two more haemal arches prolonged into base of caudal rays; five hypurals below, three above median line; two small epurals, a horizontal rod preceding first caudal fulcra dorsally and ventrally. The haemal processes of the last five vertebrae extend ventrally from the centra for a distance of one-fifth their length and then only are directed posteriorly at a sharp angle, forming enlarged, dagger-shaped hypurals; the end of the ventral extension forms a sharp projection pointed forward.

"Dorsal rays 14–15; the first two or three short and simple, following rays branched, third or fourth ray longest, two-thirds to three-fourths

FIG. 6. *Leptolepis nevadensis* David 1941. (a) Holotype, No. 10138, California Institute of Technology Coll. Vert. Paleon.
(b) No. 10137, X 2. *Fr.*, frontal; *Dent.*, dentary; *Mx.*, maxillary; *Op.*, opercular; *Par.*, parietal; *pop.*, preopercular; *psph.*, parasphenoid; *smx.*, supramaxillary; *so.*, supraorbital; *sop.*, subopercular (from David 1941).

of head. Dorsal inserted in middle of body or slightly nearer tip of snout. Ventral rays 8; the first strong and simple, fin three-fifths to two-thirds of head, originating below posterior part of dorsal base. Pelvic girdle short, its anterior point not reaching below origin of dorsal. Anal rays 9; longest not longer than ventral, anal fin near to caudal base, distance of anal origin to caudal base 1½ in distance from origin of ventrals to origin of anal. Pectoral with 14 rays, longest ray not longer than ventral, fin situated near ventral border of body. Caudal rays 26 + 6 pair of fulcra, 38; two middle rays spaced from neighboring rays, prolonged above hypurals. Fin three-fourths of length of head, distinctly furcated, middle rays one-half of longest outer rays.

"Body stained dark with an irregular scale-like pattern. No scales are evident" (David 1941: 318–320).

Type locality—"Weber conglomerates" (Newark Cañon formation) east of Eureka, Eureka County, Nevada. Newark Cañon conglomerates are exposed on the north side of the road leading up the cañon, and fish-bearing strata occur at the head of the cañon some five airline miles (about 15 by road) east-northeast of Eureka on the west side of Diamond Mountains.

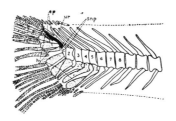

FIG. 7. *Leptolepis nevadensis*. Tail end of vertebral column, X 4. *c*, last vertebral centrum; *ep*, epurals; *hy*, hypurals; *snp*, specialized neural processes; *ur*, urostyle (From David 1941).

CENOZOIC ERA
Present time to
70,000,000
years ago

        TERTIARY PERIOD
        1,000,000 to
        70,000,000
        years ago

                Miocene Epoch
                12,000,000 to
                28,000,000
                years ago

Coming closer to modern times, we find, as we would expect, that fossil fish remains become much more abundant and complete, so that we are able to reconstruct correspondingly clearer pictures of what the fish fauna was like. The evidence we have for this epoch indicates that, while the fishes swimming in Nevada Miocene fresh waters were modern types, they represented species no longer found in our area.

Order CYPRINIFORMES (Ostariophysoidea in part)
Superfamily Cyprinoidea
Family *CATOSTOMIDAE*, Suckers
Genus AMYZON Cope 1872

THE NEVADA AMYZON SUCKER
(*Amyzon mentalis* Cope 1872)

Amyzon mentale Cope 1872A, American Philosophical Society Proceedings 12: 480–481 / 1873, U. S. Geological Survey Territories (Hayden) Sixth Annual Report 1872; 643 / 1874B, U. S. Geological Geographical Survey Territories (Hayden) Annual Report 1873: 461.

Amyzon, Cope, 1879A, American Naturalist 13: 332.

Amyzon, Eastman, 1917, U. S. National Museum Proceedings 52(2177): 288, 292.

Amyzon mentale, Macfarlane, 1923, Fishes the source of petroleum. Macmillan Co.

Amyzon mentalis, Hubbs & Miller, 1948B, University Utah Bulletin, Biological Series, 10(7): 25–26.

Amyzon, Miller, 1958, American Association Advancement Science Publication 51: 193.

Cope and others subsequently described several more extinct species in this genus in various parts of the west. These fossil suckers seem to be very closely allied to the living Buffalo suckers (*Ictiobus*) of the eastern United States, species characterized by an elongated, carp-like back or dorsal fin.

The modern species of Buffalo suckers are also carp-like in their feeding habits, as was probably *Amyzon mentalis*.

The "Osino coal beds" in which remains of the Nevada Amyzon sucker were found, are highly carbonized, fossiliferous beds containing a great many snails and some insects, in addition to the fishes and large amounts of plant materials. In his original description, Cope hazarded only the opinion "Tertiary" for the age of the strata, but in 1879, his comment was: "I have named this epoch that of the Amyzon beds, from the characteristic genus which it includes, and refer it to the later Eocene or early Miocene eras" (p. 332).

Clarence King, director of the famous Fortieth Parallel Survey, in 1878 expressed the opinion that these northeastern Nevada strata were westward extensions of the Green River Eocene formations of eastern Utah, an idea which Cope contradicted (1879). Sharp, many years later (1939A) established the term "Miocene Humboldt Formation" for these beds.

ORIGINAL DESCRIPTION

"AMYZON. Cope. Genus novum Catostomidarum

"Allied to *Bubalichthys*. Dorsal fin elongate, with a few fulcral spines in front, and the anterior jointed rays osseous for a considerable part of the length. A few short osseous rays at front of anal fin. Scales cycloid. Caudal fin emarginate. Mouth rather large, terminal.

"The characters of this genus appear to be those of the *Catostomidae*.

There are three broad branchiostegals. The vertebrae are short, and the haemal spines of the caudal fin are distinct and rather narrow. In one specimen a pharyngeal bone is completely preserved. Not having it before me at the moment, I merely observe that it is slender, and with elongate inferior limb. The teeth are arranged comb-like, are truncate, and number about thirty or forty. This and other portions of the structure will be more fully described when the whole series of specimens is investigated. The bones bordering the mouth above are a little displaced, and the lower jaw projects beyond them, and is directed obliquely upwards. The dentary bone is slender and toothless, and the angular is distinct. The premaxillary appears to extend beneath the

FIG. 8. *Amyzon mentalis* Cope 1872. Type specimen.
Courtesy Smithsonian Institution.

whole length of the maxillary. Should this feature be substantiated, it will indicate a resemblance to Cyprinidae. The maxillary has a high expansion of its superior margin, and then contracts towards its extremity. Above it two bones descend steeply from above, which may be out of position. The preoperculum is not serrate. The superior ribs are well developed.

"This form approaches, in its anterior mouth, the true Cyprinidae through *Bubalichthys*. It is the first extinct form of Catostomidae found in this country.

"AMYZON MENTALE. Cope. Species nova.

"This fish occurs in considerable numbers in the Osino shales, and numerous specimens have been procured. Two only of these are before

me at present. They are of nearly similar length, viz., M. O. .12 and
.105. The most elevated portion of the dorsal outline is immediately in
front of the dorsal fin. From this point the body contracts regularly
to the caudal fin. The dorsal fin is long, and is elevated in front and
concave in outline, the last rays being quite short. They terminate one-
half the length of the fin in front of the caudal fin. The interneural
spines are stout in front and weak behind. Radii, III. 26, and (?) II.
23. There are about twenty-three vertebrae between the first interneural
spine and the end of the series in the former specimen, in which, also,
there are no distinct remains of scales. In the second, scales are preserved, but no traces of lateral line; there are six or seven longitudinal

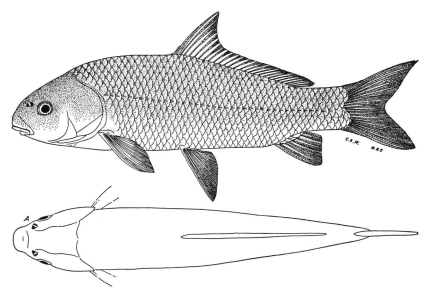

FIG. 9. Black Buffalofish, *Ictiobus niger* (Rafinesque) 1820. This is a type related to *Amyzon*. Reproduced from *The Fishes of Ohio*, by Milton B. Trautman (Columbus, Ohio: Ohio State University Press, 1957), p. 226. Used by permission of the publisher.

rows above the vertebral column. The anal fin is preserved somewhat
damaged; the rays are not very long, and number II. 7. The anterior
interhaemal is expanded into a keel anteriorly. Ventral fins injured.

"The ribs and supplementaries are well developed. The inferior
quadrate is a broad bone, with deep emargination for the symplectic.
Depth No. 2 in front of dorsal fin, M. .025. Length basis of dorsal,
.026" (Cope 1872A: 480–481).

Type locality—"Northeastern portion of Nevada, 25 miles northeast
of Elko, on the Central Pacific Railroad. The outcrop is on the south
side of the low mountain range, bounding Humboldt Valley on the
north. The beds are exposed in a drift and adjacent cutting, and a
shaft 200 feet in depth" (Cope 1872A: 478).

Family *CYPRINIDAE*, Carp and Minnows

Hubbs and Miller (1948B: 26) mention an undescribed cyprinid taken by Percy Train from the highly lucrative Miocene mammal beds of Virgin Valley in extreme northwestern Nevada.[1] They regard the specimen as showing close relationship to modern species in the Great Basin.

Order PERCOPSIFORMES (Salmopercae)

Superfamily Aphredoderoidea

Family *APHREDODERIDAE*, Pirate Perches

Genus TRICHOPHANES Cope 1872

The solitary living Pirate perch of the eastern United States (*Aphredoderus sayanus* (Gilliams) 1824) is a small fish about six inches long, living in slow waters and feeding mainly on insects. Cope's *Trichophanes hians* specimen was that of an even smaller individual, one less than 2½ inches long.

THE NEVADA PIRATE PERCH
(*Trichophanes hians* Cope 1872)

Trichophanes hians Cope 1872A, American Philosophical Society Proceedings 12: 479–480 / 1873, U. S. Geological Survey Territories (Hayden) Sixth Annual Report 1872: 642 / 1878, U. S. Geological Geographical Survey Territories (Hayden) Bulletin 4: 73.

Trichophanes, Cope, 1879A, American Naturalist 13: 332.

Trichophanes hians, Macfarlane, 1923, Fishes the Source of Petroleum. Macmillan Co.

Trichophanes hians, Hubbs & Miller, 1948B, University of Utah Bulletin, Biological Series, 7: 26.

Trichophanes, Miller, 1958, American Association Advancement Science Publication 51: 193.

ORIGINAL DESCRIPTION

"TRICHOPHANES. Cope. Gen. nov.

"Allied to *Erismatopterus*, Cope, and to the family of *Cyprinodontidae*. Dorsal and anal fins short, each with a long and short spinous ray on the anterior margin. Ventrals beneath the dorsal. Operculum, with a longitudinal keel above. Mouth with wide gape, extending beyond orbit. Scales wanting, represented by rigid fringes or hair-like bodies.

"Several important characters of this genus are not very distinctly displayed by the specimen described. This is especially the case with the maxillary region. The premaxillary bone evidently forms a large

---

[1] Hubbs and Miller erroneously use the term "Virgin River" (which is in extreme southern Nevada) for this locality.

part of the arcade of the mouth, but whether the whole, is not certain. The presence of teeth, and number of branchiostegal radii, cannot be stated.

"Other points, more definitely exhibited, are a preoperculum without serrations, directed a little obliquely backwards; a coracoid of little width; an inferior postclavicle with a superior (proximal) conchoidal expansion and long, slender shaft, extending to the anterior extremity of the femora. The latter are quite slender and acuminate anteriorly, and grooved to the apex, but apparently not furcate. They do not present any marked posterior union. Vertebrae not elongate.

"Caudal fin furcate. Interneural spines wanting in front of dorsal fin; those of the anterior rays very strong. Interhaemals of the anterior anal rays similarly strong. Caudal fin embracing 1 vertebra, and supported by separated haemal spines.

"The characters which separate *Trichophanes* from *Erismatopterus*, are seen in the large mouth and short muzzle, and in the peculiar covering of the body. In the former character it resembles some of the *Scopeli*, while the latter is not seen in any genus. The bristle-like bodies are scattered over the whole extent of the fish, excepting the head and the fins, and are arranged in little aggregations, which are irregularly disposed. The processes themselves lie irregularly together, as though free from each other, and are evidently not the impressions of keels of the scales. Traces of other scales are not visible, and the bodies described would suggest the existence of an ossified ctenoid fringe on a less fully calcified scale, or possibly without such basis.

FIG. 10. *Trichophanes hians* Cope 1872. Type specimen.
Courtesy Smithsonian Institution.

"TRICHOPHANES HIANS. Cope. Sp. nov.

"Vertebrae, D. 9; C. 15; six between interneural spine of dorsal, and interhaemal of anal fin. Radii, D. II (?) 6 (soft rays somewhat injured); A. II. 7; V. and P. not all preserved; caudal rays numerous, forming a deeply bifurcate fin. The ventrals reach a little over half way to the anal, and the latter about half way from its basis to that of the caudal fin. The dorsal fin, laid backwards, reaches the line of the base of the first anal ray. The first dorsal ray is a little nearer the end of the muzzle than the origin of the caudal fin. The muzzle is very obtuse, and if the specimen be not distorted, not longer than the diameter of the orbit. The gape extends at least to the posterior line of the orbit. The suborbital region deep posteriorly.

"In its present somewhat distorted condition the specimen measures in

"Total length	0.059
Head	.016
Vertebrae	.029
Caudal fin	.0142
Length dorsal spine	.008
Length anal	.008
Length of hair-like bodies	.0005"

(Cope 1872A: 479–480).

Type locality—Same as for *Amyzon mentalis* above.

Pliocene Epoch
1,000,000 to
12,000,000
years ago

With the arrival of this epoch, the known fish life, although still not representative of species now living in Nevada, is composed largely of types which today live nearby. The extensive lacustrine beds which dot western Nevada so conspicuously, such as the white diatomites, indicate that vast fresh-water lakes existed over much of the area; the west-central part of the state, including the Reno and Fallon regions, was under the waters of Lake Truckee which some authorities believe may have extended as far north as Oregon.

To the south, Lake Esmeralda inundated the Fish Lake-Tonopah section and all these and lesser bodies of water undoubtedly teemed with fish life.

Order CYPRINIFORMES (Ostariophysoidea in part)
Superfamily Cyprinoidea
Family *CYPRINIDAE*, Carps and Minnows
Genus RICHARDSONIUS Girard 1856

THE ESMERALDA REDSHINER
(*Richardsonius turneri* (Lucas) 1900)

Leuciscus turneri Lucas 1900, U. S. Geological Survey Annual Report 21 : 223–224 / 1901, U. S. National Museum Proceedings 23(1212) : 333–334.

Leuciscus turneri, Eastman, 1917, U. S. National Museum Proceedings 52(2177) : 292–293.

Leuciscus turneri, Macfarlane, 1923, Fishes the Source of Petroleum. Macmillan Co.

Leuciscus turneri, Hubbs & Miller, 1948B, University Utah Bulletin, Biological Series 10(7) : 26, 46.

Leuciscus turneri, Dorr, 1954, November meeting abstracts, Geological Society America, p. 41.

This fish was described from fresh-water lake beds originally believed to be Miocene in age, but now, with equal uncertainty, placed by authority as probably belonging to Lower Pliocene times. The type locality lies in beds exposed at the northeast end of the Silver Peak

FIG. 11. *Richardsonius turneri* (Lucas) 1900. Type specimen. Courtesy Smithsonian Institution.

Range in the extreme southwest end of Big Smoky Valley, Esmeralda County, Nevada. These light-colored deposits are usually referred to as the "Esmeralda Formation," and are composed of fine sands, muds and, in many places, the microscopic siliceous shells of minute floating algae or single-celled plants called "diatoms."

A United States geologist named H. W. Turner collected the several specimens which Frederic A. Lucas of the United States National

Museum described as *Leuciscus turneri;* the species was seemingly a fish much like the shiners we find today in the Lahontan system of lakes and streams farther north. With its well-developed tail fin, this rather small fish (about five to six inches in length) seems to have been a strong swimming, open water type.

The generic name "Leuciscus," which was then applied to a number of living minnows in both the New and Old Worlds, is no longer in use for North American species, and there is considerable doubt that it can be retained for Lucas' Esmeralda species. However, the term "Leuciscus turneri" has had rather wide usage as a stratigraphic index, where it is variously considered as pointing to an Upper Miocene or Lower Pliocene age.

Examination of comparative living material from western Great Basin drainages shows the extant genus of shiners, *Richardsonius* (Redshiners), to be structurally closest to *"Leuciscus" turneri.* The latter certainly was not a small Tui Chub (*Siphateles*), as can readily be seen from fin placements; and size, among other things, militates against its being a Speckle Dace (*Rhinichthys*). We know, from recent frog specimens (the Lower Pliocene *Rana johnsoni* La Rivers 1953), that some animals still living in the area, generically go back several million years into the fossil record, and there is no reason at the present time to indicate otherwise for our current fish population. As a consequence, I feel that the genus "Richardsonius" is a more suitable and realistic depository for Lucas' species than anything else we have.

The *"Leuciscus lineatus"* which Lucas used to compare his *L. turneri* with is a name based on unidentifiable material from an unknown locality; it is considered synonymous with the modern *Gila atraria,* the Utah Gila, a member of the Bonneville system of Utah and extreme eastern Nevada.

ORIGINAL DESCRIPTION

"The name *Leuciscus turneri* is proposed for a small fish obtained by Mr. H. W. Turner, of the United States Geological Survey, from the tertiary of the west side of the Big Smoky Valley in the Silver Peak quadrangle, Esmeralda County, Nevada. The type specimen, shown on Pl. XXXI, B, is No. 4302a, Catalogue of Fossil Vertebrates, United States National Museum.

"In its general aspect the fish bears a strong resemblance to such small cyprinoids as *Semotilus* and *Leuciscus,* being of much the same general proportions as *Leuciscus lineatus.* The head, as in that species, is a trifle over $3\frac{1}{2}$ in the total length;[2] depth of head, two-thirds of length. There are 19–20 precaudal vertebrae and 17–18 caudal, while *Leuciscus lineatus* and *Semotilus atromaculatus* have, respectively, 20–17 and 21–18. The tail is slightly forked; the lobes are slightly rounded.

"The anterior end of dorsal is in line with the anterior end of ventrals, and the posterior end of dorsal is in line with the anterior end of anal. In *Leuciscus* the dorsal is directly over the ventrals and in *Semotilus* the dorsal is behind the ventrals. In both *Leuciscus* and

---

"[2]According to Jordan and Evermann the head is $4\frac{1}{4}$ in the total length, but this does not accord with the specimen here used for comparison."

Semotilus the anterior end of the anal is a little back of posterior edge of dorsal. The fin rays are as follows: Dorsal, 9; anal, 10; pectoral, 11–12; ventral, 9; caudal, 23. These may be compared with *Leuciscus lineatus* and *Semotilus atromaculatus* as follows:

	D.	A.	P.	V.	C.
"Leuciscus turneri	9	10	11	9	23
Leuciscus lineatus	9	8	17	9	23
Semotilus atromaculatus	7	8	14	8	21

"The greater number of resemblances are thus seen to be to *Leuciscus lineatus*.

"It is quite probable that the very fine rays of the pectorals have failed to make an impression, which would account for the lesser number of rays in *turneri* as compared with others.

"Epineurals, epihaemals, and epicentrals are present, but there are no apparent traces of epipleurals, nor should there be if the affinities of this fish are as they have been assumed.

"The extreme length of the type specimen, which is of the average size, from tip of nose to center of caudal is $5\frac{1}{8}$ inches; from tip of nose to process of last vertebra, $4\frac{1}{4}$ inches.

"With the exception of a few small fragments, it is the impressions of bones that are preserved and not the bones themselves, and this fish is placed with the Cyprinidae on account of its strong general resemblance to that group of fishes, since the pharyngeal teeth have not in any case been found. For the same reason it is kept in the genus *Leuciscus*, as no sufficiently good characters can be assigned to these specimens to warrant the establishment of a new genus" (Lucas 1900: 223–226).

Family *CATOSTOMIDAE*, Suckers

Hubbs and Miller (1948B: 26, 46) make passing mention of a type of sucker they have examined from the same Esmeralda Formation as *Richardsonius turneri*, but the specimen is so far undescribed.

Order GASTEROSTEIFORMES
Family *GASTEROSTEIDAE*, Sticklebacks
Genus GASTEROSTEUS Linnaeus 1758

THE NEVADA STICKLEBACK
(*Gasterosteus doryssus* (Jordan) 1907)

Merriamella doryssa Jordan 1907A, University California Publications Geology 5(7) : 131–133 / 1908, Smithsonian Miscellaneous Collections 52: 117.

Gasterosteus williamsoni leptosomus Hay 1907, U. S. National Museum Proceedings 32(1528) : 271–273.

Gasterosteus williamsoni leptosomus, Jordan, 1908, Smithsonian Miscellaneous Collections 52: 117.

Gasterosteus doryssus, Jordan 1908, Smithsonian Miscellaneous Collections 52: 117.

FIG. 12. (Upper) Fossil cyprinid fish from Jersey Valley. Courtesy Manuel Chabagno, Golconda Ranch.

FIG. 12. (Lower) View of the Jersey Valley fish quarry looking northerly. Recent work by Deffeyes (1959) gives a probable age of Middle-to-Upper Pliocene, according to Dr. Richard Olsen, for this section. The small cyprinids from these beds have been extensively collected in the past, some by Stanford University. About 1929, Percy Train, a professional collector, shipped 100 fish specimens to a biological supply house. The fish occur near the top of a fissile bed of the zeolite erionite, below which is a conspicuous, supporting bed of gray rhyolite tuff overlying more erionite. Photo courtesy of Dr. Richard Olson.

Gasterosteus doryssus, Eastman, 1917, U. S. National Museum Proceedings 52(2177) : 291.

Gasterosteus doryssus, Hubbs & Miller, 1948B, University Utah Bulletin, Biological Series, 10(7) : 26, 46.

Gasterosteus doryssus, La Rivers, 1953A, Journal Paleontology, 27(1) : 80.

Gasterosteus doryssus, Miller, 1958, American Association Advancement Science Publication 51 : 193.

Not long after the turn of the present century, David Starr Jordan, the country's leading ichthyologist and then chancellor of Stanford University, obtained and described specimens of this stickleback from Coal Valley (then Truckee) diatomaceous fresh-water lake beds exposed along a section of the Truckee-Carson Irrigation District's new Derby

FIG. 13. Hay's type of his *Gasterosteus williamsoni leptosomus* = *Gasterosteus doryssus* of Jordan. Courtesy Smithsonian Institution.

Canal carrying water from the Truckee River eastward into the basin of Lahontan Valley.

In 1907, he applied the name *Merriamella doryssa*, new genus and species, to the fossil in honor of John C. Merriam, the eminent University of California paleontologist who was to do so much work in later years with the fossil mammal fauna of Nevada.

Jordan placed the species in the Silversides family Atherinidae. That was in April. The following month, not knowing of Jordan's work, Oliver P. Hay of the American Museum of Natural History published a description of *Gasterosteus williamsoni deptosomus* based on United States National Museum material from the same locality, collected by Thomas H. Means. Hay's was the more correct and searching analysis, and he felt that the fossil represented a fish not specifically distinct from a stickleback now living along the California coast, the common Three-spined Stickleback (*Gasterosteus williamsoni*).

The following year, Jordan conceded the point that he had badly misinterpreted the taxonomic position of the fish (although not in so many words) and removed it to the family Gasterosteidae. He was careful to indicate that, since his description preceded Hay's by a few weeks, the name should stand as *Gasterosteus doryssus* (Jordan).

Along with this confusion in names went an equal confusion in the stratigraphic picture. There was much carelessness in referring to the supposed age of the "Truckee" (Coal Valley) Formation beds from which these fossils had come, both on the part of Jordan and subsequent writers.

In his original 1907 paper (which was entitled "Fossil Fishes of California") Jordan observed that he was examining, among other specimens, some belonging to the University of California from "a fresh-water deposit of marl rock of Miocene age in a cut in the canal near Truckee, California," which is some 80 miles west of the actual type locality near Hazen, Churchill County, Nevada, from which the specimens came. He was confusing the town of Truckee, California with the geological "Truckee" Formation.

However, in alluding to their possible age, he more correctly called them Miocene, apparently on the advice of Merriam. Hay's quickly following paper used the term "Lahontan beds," and the error was perpetuated without comment in Jordan's 1908 note. Lahontan beds are much younger in time than those of the Coal Valley, being Pleistocene, and this has been the generally quoted age of these fish remains until the author's note of 1953.

In attempting to unravel the paleohydrography of the western Great Basin, Hubbs and Miller (1948B: 26) were led into untenable assumptions by believing that the Hazen Coal Valley beds were Lahontan in age rather than Early Pliocene. This would mean that the western Great Basin, at least, and the Sierra Nevada Mountains, would have to have been elevated more recently than the geologic evidence otherwise seems to indicate. The fossil fishes, representing as they do, coastal lowland forms, were eliminated from the Nevada scene by these upthrusts and forced back to the Pacific Coast. Geologically, there is good evidence that such earth crustal changes took much longer than the relatively short period ascribed to the Pleistocene Epoch.

ORIGINAL DESCRIPTION

"Family ATHERINIDAE

"40. *Merriamella doryssa* Jordan, new genus and species.

"Head, 3½ in length; depth, 6; D. I, I. 8, A. I. 7, C. 12 to 14, P. 11 or 12, vertebrae 15 + 18 = 33.

"Body moderately elongate, formed as in *Atherina;* the head larger and more pointed; mouth rather oblique, the lower jaw prominent, the maxillary apparently extending beyond front of eye; no signs of teeth; eye large, the orbit about three in head; opercles apparently unarmed, the opercle convex and striated; orbital region elevated, the profile depressed over the snout; branchiostegals slender, about six in number; pectoral fin inserted high, fan-shaped, the form apparently symmetrical, its length 2⅗ in head; no distinct traces of ventral fins on any of the four specimens.

"First dorsal composed of a single moderate curved spine, sharply defined in all examples, inserted behind middle of length of pectoral at a distance from gill opening equal to two-thirds of head; length of dorsal spine about five in head. Soft dorsal entirely similar to anal, inserted a little in front of the latter, the anterior rays in both elevated, the height of the longest ray three in head, the base of the fin, two, three in head. Caudal moderately forked; bones slender, mostly very fine. There is no trace of scales in any specimen.

"This species is known to us from four specimens, one and one-half to four inches long, found in the white marl in a cutting of the Truckee River Canal in Nevada" (note Jordan's introductory statement in the

FIG. 14. Three-Spined Stickleback (*Gasterosteus aculeatus* Linnaeus 1758), a modern 3-inch species living in fresh and marine waters of the northern half of the New World Northern Hemisphere. A variable species. Courtesy British Columbia Provincial Museum and Dr. G. Clifford Carl.

text above). "The rock, according to Professor Merriam, is probably of Miocene Age. This is a fresh-water deposit.

"Two of the specimens, one of them the type described above, are much more slender in apparent form than the others, a difference which may indicate difference of species; the depth must have been at least six and one-half times in the length to base of caudal. In two others of equal length the body seems much deeper. In one the depth is about five times in length; in the other about four and one-half, but this last shows evidence of distortion. The technical characters so far as they can be made out seem to be the same in all, and we treat them provisionally as one species, which is probably the case.

"In all the specimens the curved hook-like dorsal spine is very distinct, but in one of them it seems to be preceded by three other spines much more slender and shadowy. These possibly do not really belong to the same specimen, as in the other three the spine is very distinct and stands alone. If these are really additional spines, the generic diagnosis must be adjusted accordingly.

"We propose for this fish the name *Merriamella doryssa*. If the genus is placed among the *Atherinidae*, it will differ by its differentiated dorsal spine, which either stands alone or is preceded by three slenderer ones. It suggests also the genus *Hypoptychus*, a Siberian type of *Ammodytidae*. *Hypoptychus* has no trace of the first dorsal. The genus *Merriamella* has also much in common with the extinct family of *Cobitopsidae*, of the European Oligocene. But *Cobitopsis*, like *Hypoptychus*, has no first dorsal fin, and its ventral fins are present and abdominal" (Jordan 1907A: 131-133).

Order CYPRINODONTIFORMES (Microcyprini, Cyprinodontes)

Suborder Cyprinodontoidei (Poecilioidei)

Superfamily Cyprinodontoidea

Family *CYPRINODONTIDAE*, Pupfish, Springfish, Killifish

Genus FUNDULUS Lacépède 1803

THE NEVADA KILLIFISH
(*Fundulus nevadensis* (Eastman) 1917)

Parafundulus nevadensis Eastman 1917, United States National Museum Proceedings 52(2177) : 291.

Parafundulus, Jordan, 1925B, Stanford University Biological Sciences Publication 4(1) : 43.

Parafundulus nevadensis, Miller, 1945C, Washington Academy Sciences Journal 35(10) : 315.

Fundulus nevadensis, Miller, ibid: 315, 318, 320 / 1958, American Association Advancement Science Publication 51: 193.

Fundulus nevadensis, Hubbs & Miller, 1948B, University Utah Bulletin, Biological Series, 10(7) : 26.

Fundulus nevadensis, La Rivers, 1953A, Paleontology Journal 27(1) : 80.

Fundulus (Parafundulus) nevadensis, Miller, 1955, University Michigan Museum Zoology Occasional Paper 568: 12.

While bringing the fossil fish collection of the United States National Museum up to date, Charles R. Eastman discovered and described this additional species from the Coal Valley Formation exposures near Hazen (also the type locality for *Gasterosteus doryssus*), and offered both the genus and species as new. The specimens upon which he based his names had been collected some years earlier (1905) by the well-known geologist N. H. Darton.

Eastman continued the idea of a Lahontan age for the locality. Jordan (1925: 43) was the first to throw doubt on the validity of Eastman's new genus "Parafundulus," and subsequent writers (Miller, Hubbs and Miller) have rejected it as synonymous with *Fundulus*. Miller subsequently (1945) described and figured four additional cyprinodont fossil fishes from the nearby Death Valley region—*Fundu*-

lus curryi, F. eulepis, F. davidae and *Cyprinodon breviradius*—all of which seem to be older than Pleistocene in age.

These last finds have given us our only evidence to date as to the possible port of entry of the killifishes into Nevada, through the southeastern California area where the land was less elevated than in the north, and where living cyprinodonts are still found.

There is only one *Fundulus* now living on the California coast, *F. parvipinnis* Girard 1854, but several species of *Cyprinodon* occur in California, Nevada and the adjacent southwest.

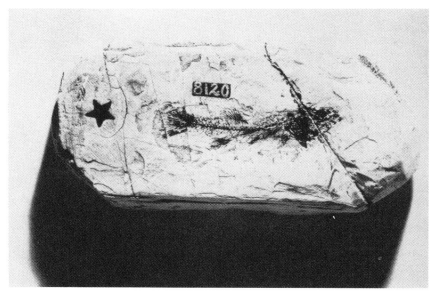

FIG. 15. *Fundulus nevadensis* (Eastman) 1917. Type specimen. Courtesy Smithsonian Institution.

ORIGINAL DESCRIPTION

"PARAFUNDULUS, new genus

"A new genus closely related to existing killifishes, and also to the extinct *Gephyrura*, but distinguished from the latter chiefly by its smaller and less conspicuously marked scales, larger number of dorsal fin rays, and presence of a typural bone. Caudal fin gethyrocercal.

"*Type of the genus—Parafundulus nevadensis*, new species.

"PARAFUNDULUS NEVADENSIS, new species
"Plate 16, fig. 2; plate 17; plate 18, fig. 3

"A small form attaining a total length of about 5.5 cm., in which the length of the head and opercular apparatus is contained three and one-half times. Dorsal comprising 11 rays, supported by an equal number of interspinous bones, and inserted opposite a point midway between the pelvics and anal. Scales small and thin, with fine concentric markings, crossed by a few inconspicuous radiating proximal striae.

"Fin formula: D. 11; C. 23; R. 10; V. 9; P. 11–12.

"The specimen (Cat. No. 8120) selected as type of this species is photographed of the natural size in plate 16, figure 2, and a drawing of it is reproduced in plate 18, figure 3. It is the most perfect of several that were obtained in 1905 by Mr. N. H. Darton, in strata of very white clay near Hazen, Nevada, which have received the name of Lahontan beds. From the same locality a single species of stickleback, known as *Gasterosteus doryssus* Jordan, was described almost simultaneously in 1907 by Drs. D. S. Jordan and O. P. Hay. Besides the type several other examples of this species, shown on plate 17, were collected by Mr. Darton at the same locality, and are now preserved in the collection of the United States National Museum. The writer is

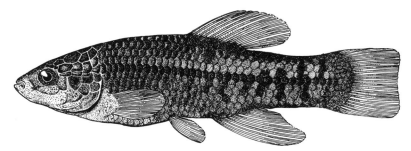

FIG. 16. Southern California Killifish, *Fundulus parvipinnis parvipinnis* Girard 1854 (drawn by William L. Brudon for Miller 1952). Compare with *Fundulus nevadensis*. *F. parvipinnis* occurs from Morro Bay, California to Baja California, in fresh, brackish and salt water. One known attempt to use this species for bait on the Colorado River was recorded by Miller (1952: 36). Courtesy California Department of Fish and Game and Leo Shapovalov.

indebted to his colleague, Mr. John Treadwell Nichols, of the American Museum of Natural History, for helpful suggestions in regard to comparing this form and its scale characters with the existing *Fundulus*.

"*Formation and locality*—Lahontan beds; near Hazen, Nevada" (Eastman 1917: 291).

QUATERNARY PERIOD
Present time to
1,000,000
years ago

Pleistocene Epoch
Present to
1,000,000
years ago

We would expect to have a more complete picture of the rich fish fauna of this most recent of geologic epochs than for the previous time periods, but such is not the case, unless we include the modern relics found in Indian caves. The Pleistocene is often restricted to the period of the four great glaciations occupying most of the million years, plus or minus, while the times we live in are egotistically separated as "Recent," or representative of the "Holocene Epoch."

If we are concerned with the truly fossil fishes of the Ice Ages, then we find the picture to be very sparse indeed. Fragments increase as we approach modern times until we find much dried material in aboriginal remains left in cave deposits.

However, of the material interesting to us, nothing is certainly known to represent species of fishes not now living, in Nevada; i.e., what we find as Pleistocene fossils can all be identified, without much doubt, as remains of fishes still occupying our area.

CUTTHROAT TROUT (*Salmo clarki* Richardson 1836)

Fish bones referrable to this species have been found in Lahontan beds in Lahontan Valley north of Fallon, Churchill County, Nevada, in some profusion.

TUI CHUB (*Siphateles bicolor* (Girard) 1856)

This fish, the common chub of the Lahontan system, has been found associated with the above mentioned trout remains.

LAKESUCKER (*Chasmistes* sp.)

Fossils of this peculiar Great Basin genus of lacustrine sucker have not been found in Nevada, but some interesting crania, presumably belonging here, have been described from Quaternary (= Pleistocene) lakebeds of adjacent Oregon by Cope (1883) and Starks (1907). Our species, the familiar Cui-ui Lakesucker (*Chasmistes cujus* Cope 1883), is one of the unique elements of the Lahontan fauna.

BELDING SCULPIN (*Cottus beldingi* Eigenmann and Eigenmann 1891)

In 1926, Jordan described fossil remains of this living species from a good series sent him by F. B. Headley, then Federal superintendent of irrigation in the Fallon area. The specimens occurred in diatomite slabs and were found in a cave some five miles south of Stillwater, Churchill County, on the east side of Lahontan Valley. In 1883, Cope described several new species of fossil sculpins of this genus from Pliocene beds in Idaho, but his material is too fragmentary to compare it with the Fallon specimens, and there is some doubt that the age of the Nevada fossil *Cottus beldingi* has been correctly determined.

CHAPTER III—PHYSIOGRAPHY

CONTENTS

	PAGE
A—The Modern Backdrop	67
(1) Present Physiography	67
(2) Dynamics	71
B—Past Times	73
(1) Physiographic History of Nevada	73
(2) Pleistocene Lakes of Nevada	78
Lake Lahontan	78
(a) History	79
(1.) Russell	79
(2.) Antevs	79
(3.) Jones	81
(4.) Broecker and Orr	83
(b) Associations	83
Lake Bonneville	85
Other Nevada Lakes	86
(a) Eastern Nevada	86
(1) Steptoe Lake	86
(2) Spring Lake	86
(3) Clover Lake	86
(4) Lake Franklin	86
(b) Southern Nevada	87
(1) Railroad Lake	87
(2) White River	88
(3) Carpenter River	88
(4) Lake Manly	88
(5) Amargosa River	88
(6) Lake Pahrump	89
(7) Fish Lake	89
(c) Central Nevada	89
Lake Toiyabe	89
(d) Northwestern Nevada	89
(1) Lake Meinzer	90
(2) Summit Lake	90
(3) Lake Alvord	90

CHAPTER III
PHYSIOGRAPHY
A—The Modern Backdrop
(1) PRESENT PHYSIOGRAPHY

The term "Great Basin" in the past and present has had several meanings. To some people it was all the reaches between the Rocky Mountains and the Sierra Nevada-Cascade chain from British Columbia north and Mexico south. To others it was a much smaller area.

In early days of the last century, wandering Mountain Men found it an inhospitable land, and ventured on it only occasionally and incidentally, always intent on finding their way elsewhere. It was a desert to all of them.

The "writin' man" Zenas Leonard, scribe of the 1833 Walker expedition, gave us the first account of Mountain Men in Nevada's "desert." He wrote with a combination of imagination and lack of it. His story is High Fancy in its erasure of the typical mountain-valley aspect and unimaginative in its lack of factual detail. He remembered only a great plain reaching out before them.

To those of later times, some of whom became connoisseurs of aridity, it was also largely a barren wasteland.

As for the term "Great Basin:" If it was to be a region of interior drainage—a land of disappointment from the fable of Buenaventura which Stansbury and Frémont sought—it would have to withdraw from the north and south and fall back to tortuous boundaries only a few hundred miles apart. This is the most restrictive meaning of the Great Basin, and under its terms, Nevada does not lie entirely within the Basin, but protrudes at various points—the northeast and southeast parts of the State are not then technically in the Basin since these areas drain into, respectively, the Columbia and Colorado Rivers and thence to the sea.

While the concept of interior drainage is the most logical definition in keeping with the "basin" idea, it does not show the true biological picture in most respects. There is no basic difference, for example, between an arroyo with a stream in it and one without water as far as most of the "desert" plants and land animals are concerned; or whether such a stream winds to the sea or sinks from sight in an interior valley.

Biologically, the Great Basin has been variously delineated with key plants—a common practice has been to use the Creosote bush (*Larrea divaricata*) as an approximate indicator (exclusive of its fringe areas) of the southern limits of a "biologically" defined Great Basin, with equally unsatisfactory results.[1]

---

[1] Also, the term "Basin and Range Province" has been used by geomorphologists and structural geologists to include not only the Great Basin proper in the restricted sense of drainage, but additional areas in southern California and southwestern Utah, which have the characteristic north-south, parallel mountain blocks.

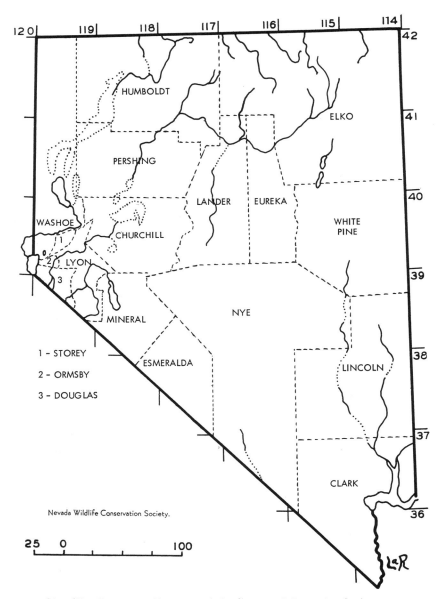

FIG. 17. County outline map of the State and its major drainages.

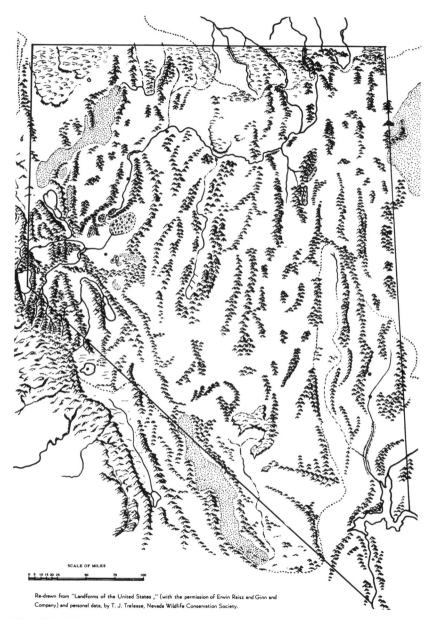

Re-drawn from "Landforms of the United States ," (with the permission of Erwin Raisz and Ginn and Company) and personal data, by T. J. Trelease, Nevada Wildlife Conservation Society.

FIG. 18. Diagrammatized geomorphologic pictorial of Nevada, showing the north-south strike of major mountain ranges and courses of the main rivers. Western Great Basin drainage limits are shown by dotted lines on north and south, and by the sharp eastern face of the Sierra Nevada escarpment on the west.

And so, not only because it is the most generally applicable designation, but the most widely used, the term "Great Basin" will here mean that part of the central intermontane West lying between the Rocky and Sierra Nevada Mountains which has no drainage to the exterior. This includes large parts of southern Oregon, fragments of southern Idaho and southwestern Wyoming, most of western Utah, most of Nevada and a large section of southeastern California—literally a little bit of everything in sight.

The typical physiography of the Great Basin can be better understood against the background of a summary of its formation.

The phenomenon of "block faulting," in which portions of the earth's crust move up or down, has been largely responsible for the present appearance of the Basin. As height differentials build up between various parts of these moving blocks, the depressions begin to fill with soft, unconsolidated debris derived from erosion of higher areas; the sediments so deposited in these troughs are mostly fanglomerates and lake beds, with local development of fluviatile or river sands and, even more restrictedly, of glacial materials. All of these are easily dissectible by later flowing waters, and even subject to much deflation by winds.

Interbedded with these sediments are great thicknesses, in many places, of intrusives and lava flows, basalt predominating over rhyolite in the latter case. Because of the unconsolidated nature of the sedimentaries, and the ease with which they are moved by water, even intermittent streams normally have little difficulty in carving out the troughs. However, in an area of enclosed basins, the opposite effects are achieved—the basin is rapidly filled with sediments, for there is no place for them to be carried.

When lakes with no outlets occupy these basins, as was usually the case, these sediments are finely sifted and become typical lake bottom silts over most of the basin. When the lakes disappear temporarily, fanglomerate materials, spreading out from the two ranges paralleling and enclosing the basin, sweep in and cover the lake beds so that over a long period of time the sediments of such a basin become grossly layered—fanglomerates, lake silts, fanglomerates, lake silts, etc.

As the mountain blocks on each side continue to move upward—or the basin sinks—sediments may accumulate to a remarkable depth. Southern Nevada Miocene beds are known to exceed 3,000 feet, or better than half a mile, in depth.

As a consequence of this structure, younger stream courses parallel the mountain ranges in a systematic north-south pattern. The older streams, however, cut across the ranges; being antecedent to the mountains, these ancient rivers have been able to downcut their channels at the same rate the land was rising.

Physically, the Great Basin presents a study in extremes. Its two highest peaks—Boundary Peak on the western border and Wheeler Peak near the Utah line—slightly exceed 13,000 feet in height, several thousand feet above timberline. Its lowest valleys range down to 500 feet above sea level along the Colorado River in the extreme southern tip of the State.

In the intervening country in central Nevada, a prominent block of mountain ranges—Shoshone, Toiyabe, Toquima and Monitor Ranges—

all have peaks rising more than 10,000 feet above sea level, the tallest being 11,807 ft. Mt. Jefferson in the Toquimas. To the northeast of this block, and northwest of Wheeler Peak, the massive Ruby Mountains scale up to 11,400 ft. in Ruby Dome. The Rubies have more than passing interest in being the site of more extensive glaciation than any of the other interior Nevada ranges.

Along Nevada's northern margins the Columbia Plateau sweeps briefly into the State—the country here is generally high, but peaks are not impressive.

The lofty Sierra Nevada Mountains invade west-central Nevada in the vicinity of Lake Tahoe. The Nevada spur here culminates in 10,800 ft. Mt. Rose near Reno. The main body of the Sierras is not much higher here, for the range becomes progressively lower in its northern latitudes. To the south, below Nevada's 13,145 ft. Boundary Peak, the southern Sierras exceed 14,000 feet in height, terminating in Mt. Whitney, the 14,500 ft. sovereign of all mainland United States crags.

The only other prominent eminence in the State is a great limestone block called Mt. Charleston in the Spring Mountains of southern Nevada west of Las Vegas (11,910 ft.). Here, in the south, where bottomlands along the Colorado River are the lowest elevations in the State (500 ft.), Charleston Peak towers far above all others in Nevada, rising 9,500 feet above its 2,500-foot plain. The high Toquima Mountains are not as impressive, for their valley floors are already 5,000 feet above sea level.

Larger expanses of west-central Nevada have bottomlands of not much under 4,000 feet, this rising progressively toward the State's northern border, where the valley lows average better than 5,000 feet.

In summary, the lowest section of the State is its southern tip, where levels of around 2,500 feet prevail in bottomlands around Las Vegas. Land swells upward toward the center of the State until the Toiyabe block is reached where major valleys may have a base elevation of more than 6,000 feet. They drop off gradually on all sides from this highland, most prominently to the west and southwest, where basins of less than 4,000 feet are common. Northward, the dip down to 4,500 and 5,000-foot valleys gently begins to ascend again toward the Columbia Plateau country.

(2) DYNAMICS

In order to understand the origin of the native fish fauna of Nevada, one must not only have a knowledge of external landscape features, but know how these have moved and isolated or brought together drainage systems in the past to produce the distinctive associations of fishes we have today.

Geomorphology, which deals with the visible form and shape of the land, is the final but never completed expression of dynamic forces which are continually altering the face of the earth. There is a cyclic balance, going first one way and then another, between those agencies which elevate and add to the heights of various portions of the earth's crust, and those which counter these movements by wearing away rocks and soil.

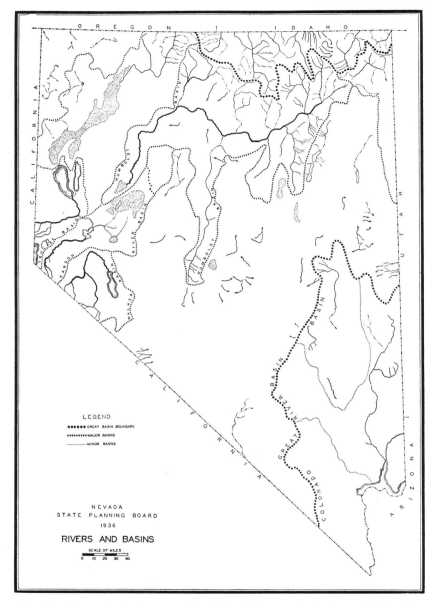

FIG. 19. The major drainages of the State. The oceanic Snake and Colorado tributaries and systems are seen at the northeastern (north-facing) and southeastern edges of the State, respectively. The Bonneville System of the eastern Great Basin, which also had a connection to the Snake System, appears at intervals along the eastern face of the State, particularly in the east-facing northeastern sector. Major aspects of the Lahontan System can be clearly seen in the central Humboldt, and the western Truckee, Carson and Walker Rivers. The minor Amargosa System shows up in southern Nevada west of the Colorado System limits.

We understand that birds and mammals readily migrate or move from one place to another, and that the larger of these, at least, easily cross mountain, desert and river barriers. We know that entire communities of plants move about also, but at a much slower rate—so slowly, indeed, that changes in plant communities are not generally detectable in a lifetime unless these changes are man-caused, such as fire, over-grazing or agricultural land-uses.

Yet if we could ascend in a balloon, moor it motionless several miles above the earth and speed time up so that each year becomes a second, we would see vegetation creeping actively below us much as a large herd of cattle or sheep would move in grazing.

In a similar very slow and visually undetectable manner, streams are moving about also. A stream of water is always actively cutting its channel headward to its own source, deepening its valley near its headwaters and cutting less slowly as it loses its power of flow over lowlands, where it may actually be aggrading or building up land surfaces rather than tearing them down. Its entire picture is one of constant, tireless, inexorable change.

In mountains, where waters are swifter and have more cutting power, we see the typical picture of streams heading on both sides of a divide and cutting back towards the crest so that headwaters of opposing streams continually come closer together. Eventually one cuts into the other and "captures" its source waters, which may entirely divert the captured stream or markedly decrease its volume of flow.

The fact that drainage systems existing on opposite sides of a mountain range have usually been isolated for long perids of time before the first stream captures occur means that fish faunas of the two sides will differ from each other in many respects, depending upon the length of time involved as well as upon the origin of each fauna.

B—Past Times

(1) PHYSIOGRAPHIC HISTORY OF NEVADA

The first affinities of Nevada fish faunas were oceanic ones, and can be disposed of briefly.

For all but a small part of the 300 million plus years of the Paleozoic Era, most of the area now called Nevada was constantly under warm, shallow seas which connected in various directions and at various times with oceans to the west, south and occasionally the east.

During the 115 million years of the following Mesozoic Era, the picture was one of decreasing marine connections. The two oldest periods of this era, the Triassic and Jurassic, represent a combined time lapse of 55 million years in which at least parts of the State were covered by marine waters. In the oldest, the Triassic, for some 30 million years half, or more than half, of Nevada was under water, with seaway connections to larger bodies of marine water both north and south.

During the Jurassic Period, the next 25 million years saw a gradual and continual decline of surface marine waters, caused by a correspondingly gradual encroachment of land from the east until, by the end of the period, nothing remained under ocean waters except a very small angle of the State which now ends bluntly in Lake Tahoe.

At no time during the 65 million years of the following Cretaceous Period does the Nevada region appear to have been covered by salt water.

Certainly, a substantial fish fauna must have been present on the Nevada land mass during these times, for the bony fishes were well developed and dispersed by Middle Cretaceous in freshwater habitats, but we have little record of them here, principally because sedimentary rocks of this age are found only in small areas of eastern and southeastern Nevada.

The North American continental outline was much the same at the beginning of the 70 million year Cenozoic Era as it is today—the inland sea arms which had been so conspicuously a part of the landscape in prior times, bringing wide, shoal ocean waters over the land, became things of the past and Nevada was thereafter undisturbed by marine waters.

From this point on we can logically infer that the freshwater fish fauna developed rapidly and took on modern aspects.

We can also expect to find more-and-more evidences of the vicissitudes of both drainage systems and their fish inhabitants as erosional and orogenic forces changed the shape of things. The most important erosional agency was the drainage systems with their grinding water mills—rivulets, streams and rivers moving about, carrying their loads of silt and fishes with them, now wearing down the surface, now building it up.

The first of the Cenozoic epochs, the Paleocene, is not detectable in Nevada rocks. To all intents and purposes, we can begin with the Eocene, 60 million years ago, and dispose of it just as briefly. No fish fossils are known; in fact, few fossils of any kind have come to light, and the only Eocene sedimentary rocks are some that have been doubtfully assigned here from the Comstock Lode area of west-central Nevada and some from east-central Nevada.

The succeeding Oligocene Epoch—12 million years in duration—has not been found even provisionally in Nevada, the nearest locality being one in California near Death Valley.

This conspicuous absence of Eocene and Oligocene deposits in Nevada is related to the fact that during the Early Cenozoic Era, the Great Basin was largely a land of high, broken profile and drainage to the sea, so that such sediments were washed out of the area rather than being deposited in it.

The Miocene Epoch, of 16 million years, brings us a comparative wealth of sedimentary deposits and fossils.

The last major time interval of the Cenozoic, the Pliocene Epoch (12 million years) produced as many sedimentary remains as did the earlier Miocene, and these are equally fossiliferous. And we are near enough here to the present day to be able to see some of the remains of the Pliocene landforms in our modern landscapes. No vestiges of earlier geomorphologic features can be detected.

Technically, the last Cenozoic epoch, a trivial one of somewhat more than a million years in length, the Pleistocene, brings us up to currency, and we find innumerable evidences of freshwater deposits in the area as well as increasing numbers of fish remains—all recent in aspect and representing fishes now living in our lakes and streams.

It has been said that "geologic history is only a succession of geographies" (Blackwelder 1948: 5, quoting Salisbury); i.e., in any given geologic time interval, the landscape is always the final external expression of internal geologic and external geomorphic forces which operate to produce landforms, chemically alter rocks and wear away or build up features of the land. And it is these landscapes with which we theoretically come to be interested, and our task of deciphering the past history of drainage systems is one of trying to find evidences of these drainages in the vestiges that still remain of ancient landforms.

The quest is short if our plans involve any actual tracing of drainages against the evidence of older landscapes, since these are detectable, and then only with some question, no farther back than the Pliocene. But we can draw inferences, here and there, and work out a shotgun mosaic with far more blanks than answers—and end up with speculations which are often frustrating rather than satisfying but nevertheless more comprehensive than one would imagine offhand from the meagerness of the data.

To begin where beginnings are, we can not expect, for practical purposes, to carry our inquiries back farther than the last stages of the Jurassic Period, some 135 million years ago. At this point, the Mesozoic Era was nearly half finished, and the region of Nevada was permanently freed from ocean waters.

During a good portion of the next 65 million years succeeding the terminal phases of the Jurassic Period, Nevada evidently was the site of prominent mountain chains whose often lofty peaks occupied most of the State, with flatter lands flanking them east and west. It was as if the Sierra Nevadas had been displaced two hundred miles to the east and broadened into a north-south cordilleran belt reaching into western Utah.

Associated with these large mountain systems was a drainage network of sizeable dimensions and while we can reason only by inference, we are probably safe enough to say that the minor streams, as they do today, largely paralleled the uplands and were eventually led away into main rivers which flowed across broad outwash plains to the seas—west, east and southeast. The most important drainage systems must have centered on Nevada, since its extensive ranges formed a continental divide at this period.

The following Cretaceous time was a long one of slow wastage of Nevada mountains, with a gradual encroachment of a broad, shallow sea trough up from the south over a region where our Rocky Mountains now lie. Then this sea as gradually withdrew and the Cretaceous closed with a renewed period of mountain building, this time bringing the Rocky Mountains into existence.

The old massive range farther west in Nevada had long since been reduced to a gently sloping plain with consequent modifications in the character of its drainage systems—from turbulent, clear mountain streams, these must have become sluggish, silt- and sand-ridden waterways meandering across their flatlands. Corresponding changes in the fish fauna must have been many and marked.

During better than half of the following Cenozoic Era, for some 40

FIG. 20. Wabuska Hot Springs, Lyon County. Lying in the northwestern corner of Mason Valley, these springs are representative of many thermal areas common to the Great Basin. These particular springs were originally fishless, the source water being too hot. Now the cooler pools teem with the introduced Mosquitofish (*Gambusia affinis*), a little fish whose presence is, unfortunately, not in the best interests of the small, warm water fishes native to Nevada, with many of whom it has been planted. The Mosquitofish is greatly over-rated as a practical control of mosquitoes. Looking northeast. Photo by the author, 1940.

million years, minor mountain ranges which had persisted in Nevada were inexorably being leveled, probably mainly by streams moving across sloping lowlands to the west and emptying into the Pacific Ocean. Nevada was a region of much greater precipitation, higher humidities and correspondingly increased streamways. All-in-all, the general picture seems to have been a subtropical one, similar to that now prevailing in our eastern Gulf States.

Redwood forests covered wide areas from the Pacific Coast to the Rocky Mountains.

This leveling of the land was interrupted again toward the end of the Miocene Epoch by concomitant elevations of the Sierra Nevada Mountains for the first time and numerous block-faulted minor ranges between the Sierras and the Rockies, the latter changes beginning first.

This two-fold effect—the breaking up of the extensive, pre-existing drainage systems into smaller, often isolated basins and a cutting off of the moisture-laden winds from the Pacific—combined to produce semi-arid conditions in an area of desiccating valleys without exterior drainage. The Great Basin had come into existence essentially as we know it today—a land of evaporating pans.

For some 12 million years of following Pliocene times, crustal disturbances seemed to have been slight and were outmatched by erosional degradation with topography wearing down to the aspect of an ancient land surface—low and isolated hills standing weakly above a generally planed surface. As a consequence, Pacific Coast moisture again became available and the semi-arid climate ameliorated.

This lasted until the end of the Pliocene, when orogenic unrest rejuvenated the major features of the Great Basin, leaving it much as we find it today. And we are correspondingly in a better position to tell something of the drainage patterns, patterns which have changed little since the beginning of the Pleistocene, a rough million years ago.

In general, Great Basin climate of the Pleistocene—the "Ice Age"—was cooler and more humid than might be expected behind the massive barrier of a two-mile high Sierra Nevada chain. This was because of more widespread conditions which produced vast glaciers reaching down to the central United States, south of which were fluctuating areas of increased rainfall or pluviation. These, in effect, over-rode local tendencies toward aridity inherent in the Great Basin structure.

The Pleistocene was a succession of at least four glacial periods and their interglacial recessions. These were much better expressed to the northeast of the Great Basin than in that province proper—in our area, glaciers were not widespread, but formed only in the higher mountains; the Sierra Nevadas, the Rubies and the Rockies having the most prominent ones. From available evidence, interglacial periods considerably exceeded those of glaciation in time intervals, the latter appearing to represent only a relatively small portion of the Pleistocene.

The second glacial stage seems to have been the longest of all. Essentially, Great Basin conditions during the Pleistocene can be more properly spoken of as "Pluvial" and "Interpluvial" rather than "Glacial" and "Interglacial," for the dominant expression of the Ice Age here was that of increased and colder rains and not of extensive ice sheeting. Maximum precipitation or pluviation seems to have occurred during initial recessions of the glaciers.

Changes in Great Basin drainage systems during the Pleistocene seem to have been largely minor ones, those of advancing and retreating lakes—now inundating river courses, now exposing them—as rising and falling waters reflected fluctuations in precipitation received over the Basin. And only the most recent of these major lakes have left recognizable beaches and terraces along mountain slopes—at the same time that they were eradicating any evidences of former lakes. Of the hundreds of lakes which have come and gone in the multitudinous enclosed basins of the elevated, dissected country between the Sierra Nevada and Rocky Mountains, only these last have left certain traces.

Fortunately, even though only comparatively recent bodies of water have left their marks on the land, we have another source of evidence upon which to base generalizations concerning previous affinities of drainage systems—the fishes that live in those systems.

As Hubbs and Miller (1948B: 28) have pointed out, Great Basin fishes do not form a uniform faunal unit—they are rather the modern expression of past associations of elements of several different, isolated drainage complexes. By studying relationships of Nevada species with those fishes in drainage systems around the periphery of the Great Basin, it is possible to postulate probable general entry areas for many of these species into the Basin, and consequently deduce the existence of stream freeways which seem now to be hidden from view beneath extensive lava flows in the northwest, and which would otherwise be totally unsuspected.

(2) PLEISTOCENE LAKES OF NEVADA

The map in back of book gives a concise pictorial of the last, detectable, major lakes which occupied the still unchanged drainage basins of Nevada during the period of increased rains. With minor shifts, patterns shown by these Late Pleistocene lakes must have prevailed throughout most of the epoch, with water levels fluctuating according to the available precipitation in any given period.

LAKE LAHONTAN

The picture is most readily seen if we show the situations during maximum development of lake surfaces. The large, tortuous and much attenuated body of water occupying west-central and northwest Nevada in these times was called "Lake Lahontan."[2] At its greatest extent, the lake had a maximum depth of slightly more than 875 feet over Pyramid Lake, covering an area of some 8,500 square miles and feeding from a drainage or hydrographic basin of about 45,000 square miles. The hydrographic basin, and parts of the lake itself, overlap slightly into Oregon and more extensively into northeastern California.

The broadest expanse lay across the present Fallon area (Lahontan Valley), where the open sweep of water was better than 30 by 70 miles; at its greatest height, Lake Lahontan surrounded a central, jagged island of somewhat more than 3,000 square miles. The island consisted

---

[2] After Baron Louis Armand La Hontan (1666—c. 1713), an early French explorer in America. He got an uncertain and probably not a great distance west of the Mississippi River, but the Great Basin was unknown to him. Hague (1877: 818) first applied Lahontan's name to this Pleistocene area.

of most of western Pershing County, a long northern arm extending into south-central Humboldt County, and a southern tip ending in extreme northwestern Churchill County.

(a) History

(1.) Russell

Israel C. Russell, a Federal geologist who made the first—and, in many respects, the most comprehensive—study of the history of Lake Lahontan, had no reference points from which to gauge an estimate of the absolute age of the ancient sea. He could only comment that "The life of the old lake must have embraced at least several thousand years" (1896: 127), when writing of the time involved in the precipitation of the extensive tufa deposits which are everywhere conspicuous about the basins.

Whatever the age, we can give the following abbreviated sequential conclusions he arrived at:

(1) That 500 feet above the 1882 Pyramid Lake level (3867.6 ft. el.) was the maximum height of Lahontan during its first rise, reserving the Lahontan beach 30 feet higher for a later rise.

(2) That from this 500-foot maximum, the inland sea waters dropped during the following low period of "low water, and perhaps complete desiccation" (1885: 251) to a condition of greater depletion than existed in his time. After this, basins began filling again, with some unknown change taking place in the waters, perhaps because of the partial removal of salines by burial under valley alluvia and by wind deflation during an intervening dry period. This chemical change would explain the differing nature of calcareous deposits characteristic of this phase, which he called "Thinolite tufa." During this stage the lake rose to the thinolite terrace and lingered for a long time, with thinolite tufa being deposited from, Russell thought, a much more concentrated solution than in the "lithoid" period.

(3) That the lake continued to expand to the "dendritic" terrace, 210 feet above the thinolite terrace and 180 feet below the lithoid terrace—Russell's postulated maximum—depositing dendritic tufa as it rose.

(4) That, still on the upswing, water crept up to the highest beach, the "Lahontan terrace," 30 feet above the lithoid terrace, stayed there but a short while and began to pulse downward. No chemical precipitates mark this highest point, but the faint beach is unmistakable. Russell found two terraces, in some places indicated by two gravel embankments, superimposed on lithoid tufa which belong to this second highest rise.

(5) That finally, levels receded until the basins dried out completely a second time. Then, about 300 years ago, they gradually filled to their 1882 conditions.

(2.) Antevs

Ernst Antevs, a glacial chronologist of premier stature, has placed Pleistocene Lake Lahontan well into the final half of the last great glaciation in North America (1925)—the Wisconsin Stage, which may have begun 150,000 years ago and ended some 25,000 years ago. Antevs detects three main water levels of Lahontan:

(1) The maximum stage, 65,000 years ago, for which Russell (1885) determined its greatest depth as 886 feet in the basin now occupied by Pyramid Lake, 529 feet over the Black Rock Desert, 500 feet over Fallon, 435 feet in Walker Lake and 325 feet in Honey Lake near Susanville, California. To this highest stage, which occurred during the Iowan substage of glaciation (with which the Sierra Nevadan "Tahoe" substage correlates), Antevs would restrict the name "Lake Lahontan" proper.[3] A type of stony lime or calcium carbonate referred to as "lithoid tufa" was deposited by waters of this phase of the lake as a coating on rocks along shores up nearly to the highest water mark.

(2) There followed next an interpluvial stage during which glaciers retreated, less precipitation fell, and the waters of Lahontan receded some 420 feet and remained at this low ebb for a long time, as evidenced by wave terraces cut in rock on Anaho Island in Pyramid Lake. The dominant terrace of this stage, the "thinolite" terrace, is so named because of a distinctive kind of lime tufa which was precipitated over lithoid tufa up to the terrace. The average date for this low point seems to have been about 45,000 years ago, and Antevs uses the name "Thinolite Lake" for this phase, although rather than a single body of water, it had separated into a Fallon lake to the southeast and a long, narrow, tortuous Pyramid-Black Rock lake to the northwest—its once large central island became a continuous block of land cutting the old lake in half, and entirely isolating Walker Lake.

(3) An upswing occurred again with the onset of the Mankato (eastern U. S.)-Tioga (western U. S.) glacial substage which carried Lahontan basin waters 210 feet above the interpluvial low of the Thinolite Lakes to a point 210 feet below the all-time high of Lithoid Lake. This, the "Dendritic Lake" of Antevs, reached its climax about 25,000 years ago and deposited dendritic tufa over the thinolite tufa. While this body of water was more extensive than the Thinolite Lakes, its waters never again cut off the large central island so conspicuous in the lithoid tufa stage—a 25-mile wide neck of land connected the former island to the mainland just north of Lovelock.

The extreme northeast arm around Winnemucca and northward became a separate smaller lake and Walker basin was again isolated from the main body of water. The Honey Lake extension, however, remained intact.

From this point on, the waters of Lahontan basin declined until, according to Antevs' chronology, they reached the 1882 level of Pyramid Lake (used as a reference point because it was one of Russell's fieldwork years on Lahontan) about 10,000 years ago.

Presumably, the basin has remained much the same from then till now. Only fragmented, remnant waters persisted—the lakes of recent

---

[3]However, we suggest, to make it slightly less confusing, that the term "Lake Lahontan" be used for all waters occupying the Lahontan basin during the last half of the Wisconsin Glacial, and the terms "Lithoid Lake," "Thinolite Lake" and "Dendritic Lake," the last two by Antevs, designate the substages of Lake Lahontan.

times—Pyramid and now-dry Winnemucca; Humboldt; North and South Carson; and Walker. Of these, Pyramid Lake is the deepest and apparently the only permanent residual lake. Some of the others have come and gone several times, at least. The one possible exception is Walker Lake; here biological evidence indicates that this lake dried up at least once in the post-Lahontan period, for the peculiar lacustrine sucker, the Cui-ui (*Chasmistes cujus*), which must have been widespread during Lithoid Lake times and consequently present in Walker basin (if indeed that basin was part of Lahontan), does not now occur in Walker Lake, a body of water which is perfectly suited for it (see discussion of the Walker system).

Today (1962), Pyramid Lake, due mainly to artificial diversion of its main affluent (the Truckee River), and secondarily to semi-drought conditions, is some 70 feet below its level in Russell's time.

An interesting sidelight of Lahontan history is the probable deflection of both the Truckee and Walker Rivers from their old courses to the ones they now follow.

There is evidence that originally pre-Lahontan Truckee River flowed past Hazen into Lahontan Valley in which Fallon is situated. Then, so the geological story goes, undoubtedly because the three major western rivers of the system—Truckee, Carson and Walker—fed the growing lake in Lahontan Valley (which is the Carson Desert of pioneer days), the lake there rose much faster than the less fortunately equipped Pyramid-Black Rock system. Consequently, when Fallon waters topped their divide in the region of Fernley, they spilled northward toward Pyramid Lake, downcutting an outlet channel. At this point in time, the Truckee River was poised rather delicately on an apron between the two areas—degrading by the northward moving waters was enough to alter the Truckee's direction. Rapid filling of all basins flooded the lower Truckee, so it seemingly did not become firmly entrenched in its new pathway until the Thinolite Lakes stage, when its lower course was again freed from lacustrine waters.

Walker River, similarly balanced on a broad fan, appears once to have flowed northward past Wabuska and into the Carson River near the site of old Fort Churchill. Probably pretty much the same sequence of events engineered the deflection here as in the case of the Truckee River. Swiftly rising Fallon waters backed up the Wabuska channel, overflowed the divide, and cut a river channel toward Walker Lake. When the Lithoid phase merged into the lower Thinolite stage, this channel was freed and occupied by the now eastward-flowing Walker River.

Presumably, in pre-Lahontan days, without Walker River as its affluent or inflowing stream, Walker Lake must have been much smaller and more ephemeral than in post-Lahontan times; once the river became part of the Walker system, the lake became a major one.

(3.) Jones

In contrast to Antevs' results, the University of Nevada geologist J. Claude Jones, felt the entire history of Lake Lahontan could be accounted for over a period beginning about 3,000 years ago, and that its major decline began within the last 1,000 years.

His estimates were based largely on examinations of remnant waters of the Lahontan basin and their affluent streams with a view to determining how long it would take certain chemicals to accumulate in lakes at the rates at which streams were bringing them in. However, Jones' conclusions have not seemed tenable to his colleagues, since Lahontan waters appear to have been freshened anew several times during their existence, and by various mechanisms, and he made no allowances for this. Later, more conclusive evidence from carbon-14 studies (see Broecker et al) make it plain that the lake's history covers a much greater span than Jones thought.

It is neither expedient nor desirable to discuss the Russell-Antevs interpretations beyond a major discrepancy or two which later studies have brought to light.

(1) First, Russell's belief that Lahontan Basin very probably dried down completely, once during the inter-Lahontan recession (Antevs' "Thinolite stage") and again at the end of the post-Lahontan drop, is not compatible with ichthyological evidence. This has been discussed by Snyder (1917A: 39–40), Hutchinson (1937: 53–54) and Hubbs and Miller (1948B: 27). The endemic lakesucker, the Cui-ui (*Chasmistes cujus*), provides the most striking evidence against such desiccation. Today it is found only in Pyramid Lake; under normal conditions, it leaves the lake only during a brief spawning season and ascends the Truckee River but a comparatively short distance. When the river is too low for the Cui-ui to get out, it spawns in the lake.

During Lahontan days, it must have been widespread in the great lake, and seemingly could not have withstood lacustral drying by taking to the much reduced streams as other species might have done. And the type is so distinct that there is no basis for supposing it might have differentiated in the post-Lahontan time available to it after the basins began to fill again. The fact that the Cui-ui is not found in Walker Lake, the other supposed Lahontan remnant, can be due to two circumstances; either Walker Lake did dry up completely, or at least concentrated down to salinities which killed its fish fauna, or it was never connected to Lake Lahontan at any time, which we have some reason to believe. This will be referred to later under Walker Lake.

Additional data which Hubbs and Miller (1948B: 27) advanced to substantiate this point—that of the presence of coastal fishes, *Gasterosteus* and *Fundulus*, in Lahontan waters—are not valid, since we have no reason to believe now that these fishes survived the Pliocene in Nevada.

(2) Second, although there is no suitable solution at present, the conflicting views of Russell and Antevs as to whether tufa deposition occurred essentially during upward or downward pulsations, should be resolvable by additional studies of this aspect of the problem.

On the basis of mechanical precipitation of calcium carbonate in the basins, Antevs' ideas of a straight-line relationship between increasing concentration of solutes and consequent increasing precipitation of those solutes from their solvents are the first which would naturally occur to an investigator—and also have the value of offering the simplest solution.

However, Russell's apparently anomalous aquatic system which deposits solutes as it rises and as concentrations increase much more slowly, or even decrease, depending upon conditions, can be as simply explained if we assume that certain algae or minute, greenish, single-celled plants have been the calcium-removal agent. In other words, that certain types of tufa deposition are a function of the biological activities of these algae and not a mechanical matter of increasing concentrations of solutes. Therefore, as the lake deepened, calcium carbonate precipitation would keep pace with the rising waters.

Jones (1925) discussed this at length, comparing recent tufa-forming algae in Pyramid Lake with those in Salton Sea, and reviewed some of the literature. This will be referred to in more detail in the section on limnology under Pyramid Lake.

(4.) Broecker and Orr

Very recently, W. S. Broecker of Columbia University's Lamont Geological Observatory, and Phil C. Orr of the Santa Barbara Museum (California), have collected tufa specimens from various parts of the Lahontan basin. Coming chiefly from the vicinity of Pyramid Lake, these have been analyzed for their carbon-14 values at Columbia with some totally unexpected results in terms of what has been accepted from long use of the Russell-Antevs hypotheses.

"The results of the study suggest a high water period from 25,000 to about 14,000 years ago. This period was preceded by an interval of moderately low water level extending back to at least 34,000 years before present. Following a recession to a moderately low water level close to 13,000 years ago Lake Lahontan and possibly Lake Bonneville also rose to their maximum levels close to 11,700 years ago. This rapid rise was followed by an equally rapid fall close to 11,000 years ago. This latter decline is recorded by terrestrial deposits in many of the wave-cut caves on the shorelines of the ancient lakes. There is some evidence for another maximum close to 10,000 years ago. The lakes have almost certainly remained low since 9,000 years ago." (Broecker and Orr 1958).

The most striking variances of carbon-14 data from previous theory are:

(1) Age-reversal of the three major terraces—Lithoid, Thinolite and Dendritic. Whereas Russell and Antevs both agreed that they followed each other in the just listed sequence, Carbon-14 sampling indicates that the order was Thinolite, Dendritic and Lithoid. In personal correspondence, Broecker has written that the field evidence for Lithoid deposition antecedent to the other two types is very weak;

(2) That the life of Lahontan may be only half that ascribed to it by Antevs:

"Measurements indicate that nearly all the tufa found in the Lahontan area has formed during the past 30,000 years" (Broecker and Orr 1956: 29).

(b) Associations

Since none of its fishes are boldly unique, the now isolated Lahontan System obviously got its fish fauna from some surrounding drainage or drainages, in the near or distant past. All adjacent areas are possibilities, but the most reliable remaining evidence—that of fish relationships—points strongly to the Klamath Lakes region to the northwest. Lesser affinities appear to be with the Bonneville System to the east.

However, attempts to find even the remotest remains of physical connections between Lahontan and any of its neighboring systems have been failures. Russell (1885) was emphatic in interpreting modern geomorphic features as showing Lahontan to have an entirely unbroken rim, unlike the easily detectable northern overflow connection of Bonneville with the Snake River System.

During Russell's painstaking geologic fieldwork, Cope (1883) came to the same conclusion on biological evidence. After noting that Fortieth Parallel Survey geologists had outlined ancient seas in the Bonneville and Lahontan rimlines, Cope wrote:

"It is exceedingly probable that it will be shown that a third lake existed in Oregon, north of the supposed northern boundary of Lake Lahontan, which is now represented by the Warner Lakes, Abert's Lake, Summer Lake and Silver Lake, and probably by Harney's and Malheur Lakes on the eastern side of the Oregon desert. As will be shown later, the larger species of fishes found in such of these lakes as contain them, are identical, and different from those of the lakes of the Bonneville series. One species, the *Catostomus tahoensis*, is common to this area and that of the true Lahontan Lakes (Tahoe and Pyramid), and this Oregon lake may have been continuous with that of Nevada, at a point some distance east of the mountains" (pp. 134–135).

James Blake may have been the first to discuss the geologic possibility of a northwest outlet for Lahontan toward the Columbia River. He made a trip from Winnemucca (Nevada) to southeastern Oregon and published two papers on this problem in the California Academy of Sciences series in 1872.

This region between Lahontan and Klamath is today covered widely and deeply with lava flows which have obliterated whatever evidence might have existed as proof for these connections.

Subsequent writers who have expressed ideas on Lahontan affinities have been Gilbert (1898), Snyder (1917A) and Hubbs and Miller (1948B).

The great basins left barren by the disappearance of Lahontan waters have been tempting as unique ecological areas to other biologists. Ornithologists have suggested a relationship between the peculiarities inherent in this region and certain endemic western Great Basin races of birds. Specifically, Behle (1942: 305–306) felt that a race of the Horned Lark (*Otocoris alpestris lamprochroma*) originated in the recession basin of Lake Lahontan.

Hall (1946: 61) makes a similar case for several mammals:

Kit Fox (*Vulpes macrotis nevadensis*).
Long-tailed Pocket Mouse (*Perognathus formosus melanurus*).
Dark Kangaroo Mouse (*Microdipodops megacephalus ambiguus*).
Ord Kangaroo Rat (*Dipodomys ordi inaquosus*).
Botta Pocket Gopher (*Thomomys bottae canus*).
Montane Meadow Mouse (*Microtus montanus undosus*).

The present writer has previously pointed to a parallelism between the distribution of some cold-blooded land vertebrates and insects, and the now-dry Lahontan Basin (1946, 1948). The late Edwin C. Van Dyke, a talented and life-long student of insect geography, used the Lahontan northwest connection as an explanation for the presence of the distinctive carabid beetle *Nebria eschscholtzi* Menetries 1844 about Pyramid Lake (1943).

LAKE BONNEVILLE

In the eastern Great Basin, lying almost entirely in western Utah, a large inland sea to which the name Lake Bonneville[4] was given existed contemporaneously with Lake Lahontan and showed evidences of the same general series of fluctuations. In area and depth, Bonneville greatly exceeded Lahontan, being about 20,000 square miles in extent and slightly over 1,000 feet deep at its maximum.

Its three stages have been given the following names:

(1) *Bonneville Lake* (the "Lake Bonneville proper" of Antevs), the first stage, possibly the sea's maximum development;

(2) *Stansbury Lake*, the second stage during the interpluvial low, when the level dropped some 700 feet below the first stage; and

(3) *Provo Lake*, the third stage (Mankato-Tioga glacial re-advance), characterized by an increase in depth to a point some 400 feet below the first maximum, and representing an areal expanse of about 13,000 square miles.

Unlike Lahontan, Bonneville had an outlet to the north, discharging into the Columbia River via the Snake River. Great Salt Lake is its major remnant.

Only a very small arm of Lake Bonneville (Bonneville Lake stage) reached Nevada, extending up into the Pilot Creek area of extreme eastern Elko County. Thousand Springs Creek was a main tributary at the north end of this arm. An insignificant overlap of the Nevada line seems to have occurred farther south. Drainage systems supplying the western edge of Lake Bonneville were, naturally, more inclusive of Nevada territory, reaching from near the Idaho border along a narrow, fluctuating strip to a point roughly 250 miles to the south.

Lake Bonneville has more genera and species of fishes than Lake Lahontan according to Hubbs and Miller (1948B: 29, 32) but corroborative evidence for the longer isolation of the latter body of water from adjacent drainage systems is testified to by the larger number of endemic forms found in Nevada. Lahontan has a total of about 30 forms, including subspecies, and Bonneville some 28. Physiographic evidence of the former's longer isolation is found in the lack of an overflow connection.

---

[4]From Benjamin Louis Eulalie de Bonneville. In 1832 Captain Bonneville went into the Rocky Mountain fur trade on a two-year furlough from the United States Army with New York financier backing. Joe Walker was a member of his party. Bonneville was popularized by Washington Irving, but his exploratory accomplishments were inconsequential. In subsequent years, it was authentically determined that Bonneville's true mission was that of a spy for the United States government.

OTHER NEVADA LAKES

During highwater stages of Lakes Lahontan and Bonneville, the Great Basin was an expansive series of lakes, large and small. Almost every enclosed basin seems to have had its own body of water. Hubbs and Miller (1948B) have admirably presented the results of their several and collective years spent in analyzing these basins and their fish faunas, and we will mention here only a few of the larger or otherwise fish-important of these systems.

(a) Eastern Nevada

Lying along the west edge of the Bonneville drainage, but with their own separate drainages, were a series of rather large, isolated valley lakes. Five of these were roughly about the same size—

(1) Steptoe Lake

Long Steptoe Valley in which Ely lies was the site of a double lake, one discharging into the other, the entire drainage area extending for better than 150 miles along a north-south axis. Pluvial (Pleistocene) *Lake Steptoe* occupied the lower half of this trough and emptied into Pluvial *Lake Waring* to the north, ultimately draining itself and being replaced by a stream affluent to Lake Waring, Pluvial *Steptoe River*. The present Steptoe Slough or Duck (Willow) Creek represents the lower portion of this ancient river.

(2) Spring Lake

Southeast of Lake Steptoe, Pluvial *Spring Lake* filled the lower reaches of Spring Valley (Schell Creek Valley), a narrow system some 80 miles long, and was rather deep, as these lakes went.

(3) Clover Lake

Clover and Independence Valleys west of Lake Waring held the plump, "H"-shaped Pluvial *Clover Lake*, of which the present transient alkali pan Snow Water (Eagle) Lake is a remnant.

(4) Lake Franklin

Pluvial *Lake Franklin* lay along the east flank of the impressive Ruby Mountains, southwest of Clover Lake, covering present-day Ruby and Franklin Marshes. The Lake Franklin drainage system was about 70 miles in length. Butte Valley, east of Ruby Marsh, contained a smaller lake, Pluvial *Lake Gale*, which was tributary to Lake Franklin via a northeast bay of the latter.

These lakes, and a few other smaller ones, show evidence of relatively long isolation from both Lahontan and Bonneville drainages in their almost mutual possession of a modified type of dace (*Rhinichthys*) which Hubbs and Miller regard as probably generically new and distinct from *Rhinichthys*. It is still undescribed.

(b) Southern Nevada

This large area was not so thickly beset with lakes as was the eastern section just discussed, and most of the bodies of water here were small and ephemeral.

(1) Railroad Lake

The largest of these, extensive Pluvial *Railroad Lake,* lay adjacent to the eastern Nevada series, and to the southwest. Railroad Lake was not as narrow as the eastern series, however, being nearly 50 miles long and almost 20 miles wide, and had a drainage area not much less than those of the discussed eastern section; Hubbs and Miller (1948B: 90) estimated this as some 6,000 square miles.

Two native species of fishes are characteristic of the Railroad Lake system—the Railroad Valley Springfish (*Crenichthys nevadae*), which is found nowhere else, and a total of eight local forms of the common Tui Chub (*Siphateles bicolor*) (Hubbs and Miller 1948B: 91), none of which have yet been described. The Tui Chub occurs throughout the Lahontan system and other major drainages to the west, northwest and north, but the genus *Crenichthys* is confined to the Railroad and White River systems in southeastern Nevada.

FIG. 21. Pleistocene White River Channel, cut spectacularly through limestone at the north end of the Warm Springs is in extreme northern Clark County. Through such defiles, Pluvial White River provided a continuous passageway for Colorado River fishes to penetrate deeply into otherwise totally inaccessible country in what is now east-central Nevada. Looking northerly. Photo by the author, 1952.

(2) White River

An extended area east and southeast of the Railroad Valley section contains two prominent, mainly fluviatile systems which are separated except at their extreme southern ends. The westernmost of these, adjacent to the Railroad system on the east, is that of Pluvial *White River*, now represented by modern White River in the north, some surface flowage in Pahranagat Valley in the middle, and the Moapa or Muddy River at the south end.

Endemic species restricted to this system are:

White River Springfish (*Crenichthys baileyi*).
White River Mountainsucker (*Pantosteus intermedius*).
White River Colorado Gila (*Gila robusta jordani*).
Moapa Dace (*Moapa coriacea*).
White River Spinedace (*Lepidomeda albivallis*).
Pahranagat Spinedace (*Lepidomeda altivelis*).

(3) Carpenter River

The eastern partner, Pluvial *Carpenter River*, has a more extensive modern remnant, Meadow Valley Wash, which connects with Moapa River a short distance north of the Colorado River. Affinities of the fish fauna of both of these pluvial streams are with the Coloradan fauna, but sufficient isolation has been involved in their history to have produced a new genus and several distinctive new species.

Disposing of the many small, transient lakes in an ill-defined region southwest of the White River system, we pass to an interesting study of isolation farther to the southwest.

(4) Lake Manly

The Death Valley system is mainly Californian, but its Nevadan portions add four fish species to our State list, three of them restricted to Nevada. Pluvial *Lake Manly*, 90 miles long, filled Death Valley during the late Pleistocene, and had as a northern and western affluent Owens River, once a stream of considerable magnitude, which eventually discharged into the southwest edge of Lake Manly after passing through three other sizeable bodies of water—Pluvial *Lakes Owens, Searles* and *Panamint*, the latter a long, narrow water of remarkable depth (more than 900 feet). Lake Manly, about 600 feet deep at its maximum, was the terminal sump of the series.

(5) Amargosa River

On the Nevada side, Pluvial *Amargosa River* arose in the reaches north of what is now Beatty, in southwestern Nye County, flowed southward for some 120 miles, thence wheeled west and north to empty into the southern end of Lake Manly in Death Valley. A couple of small lakes lay on its course, and the larger Pluvial *Lake Tecopa* intervened only a few miles from Manly. A major side contributary was the warm spring discharges of Ash Meadows on the east side of the drainage about midway down the main streamcourse, where another small lake existed.

(6) Lake Pahrump

Lake Pahrump, a fairly large pluvial expanse, filled Pahrump Valley, adjacent to the Ash Meadows locality to the southeast, overflowing through a northwestern arm via Ash Meadows into the Amargosa. The following tiny desert fishes are found nowhere but in the Nevada part of the Amargosa drainage:

Amargosa Speckle Dace (*Rhinichthys osculus nevadensis*).
Ash Meadows Poolfish (*Empetrichthys merriami*).
Pahrump Poolfish (*Empetrichthys latos*).
Devil Pupfish (*Cyprinodon diabolis*).

These are all confined to the Ash Meadows-Pahrump area. Some fishes in the Death Valley basin may have come across divides northeast of Ash Meadows by transfers with the Colorado system; others seem to have been derived from Lahontan drainage much to the north via Mono Lake.

(7) Fish Lake

Pluvial *Fish Lake* lies in a small drainage basin just north of the Death Valley system near the Nevada-California line in what is now Esmeralda County. This basin is about 50 miles long, and harbored a thin, small lake 15 miles long. Hubbs and Miller (1948B: 44) found two forms of chub (*Siphateles*) in Fish Lake Valley, one of which they regard as a native subspecies, the other as probably introduced from the westerly adjacent Owens River.

(c) Central Nevada

Working northward from the vicinity of this Death Valley block of lakes across an intervening region which Hubbs and Miller have referred to as the "Area of sterile basins" (1948B: 45), we arrive at one other Pleistocene lake which deserves brief mention:

Lake Toiyabe

Pluvial *Lake Toiyabe*, in Big Smoky Valley, was a deep, 40-mile lake in a restricted drainage valley only some 75 miles long. Three native fishes are the sum and substance of the present fauna here; the Lahontan Speckle Dace (*Rhinichthys osculus robustus*)—found throughout the vast adjoining Lahontan system, and two undescribed subspecies of the Tui Chub (*Siphateles bicolor*). Affinities are clearly with the Lahontan basin.

(d) Northwestern Nevada

In an extreme corner of the State, narrowly separated from Lake Lahontan, lies a complex of Pleistocene lake beds which overlap into both California and Oregon.

(1) Lake Meinzer

At the California line, Pluvial *Lake Meinzer,* a 40-mile, rather wide and deep sheet of water in Long Valley, northwestern Washoe County, occupied a drainage basin not much larger than the lake. Only a speckledace (*Rhinichthys*) lives in the valley today.

(2) Summit Lake

Due east of Long Valley is a tiny depression which, in late Pleistocene times, bore a Pluvial *Summit Lake* much larger than that presently in the site but, nonetheless, still a very small body of water. It lies in what is now western Humboldt County, and is of interest in that it contains populations of Cutthroat Trout (*Salmo clarki*). However, there is good reason to suspect that these were planted from Pyramid Lake stock many years ago.

(3) Lake Alvord

Northeast of Summit Lake is the Alvord Basin, which held Pluvial *Lake Alvord* into which the Thousand Creek system of northwestern Humboldt County drained. The lake was rather large, some 30 miles of it lying in Nevada. Hubbs and Miller found a very strikingly different chub, possibly generically distinct from *Siphateles,* in the modern basin (1948B: 60).

CHAPTER IV—RIVERS

CONTENTS

	PAGE
A—The Lahontan System	93
(1) Historical and Descriptive	93
Humboldt River	93
Truckee River	97
Carson River	99
Walker River	100
(2) Characteristics of Lahontan Rivers	101
Physical and Chemical Conditions	101
Biological Conditions	102
B—The Snake System	104
(1) Historical and Descriptive	104
(2) Characteristics of the System	105
Physical and Chemical Conditions	105
Biological Conditions	106
C—The White River System	107
(1) Historical and Descriptive	107
(2) Characteristics of the System	108
Physical and Chemical Conditions	108
Biological Conditions	109
D—The Colorado River System	110
(1) Historical and Descriptive	110
(2) Characteristics of the System	112
Physical and Chemical Conditions	112
Biological Conditions	113
E—The Amargosa River System	114
(1) Historical and Descriptive	114
(2) Characteristics of the System	115
Physical and Chemical Conditions	115
Biological Conditions	115
F—The Nevada Bonneville System	116
(1) Descriptive	116
Elko County	116
White Pine County	116
(2) Characteristics of the System	117
Physical and Chemical Conditions	117
Biological Conditions	117

CHAPTER IV
RIVERS

There is small difference between the late Pleistocene landscape and its modern drainage pattern but one—little water is left in the picture. Streams and lakes are mere spectres of their former solid substances. Channels and basins are still there but water is a rarity today where it was once in abundance. The old minor streambeds swell with water only during spring runoffs and an occasional cloudburst. Main rivers run low except in the spring and the rare wet years.

A—The Lahontan System
(1) HISTORICAL AND DESCRIPTIVE
HUMBOLDT RIVER

This stream seems to have first been officially noted by Ogden in 1825. "Mary's," "Paul's," "Unknown," "Barren" and "Ogden's" Rivers were its names in the years following until Frémont made it an honorarium to the famous German explorer, Alexander von Humboldt, in 1849. As is common in such matters, von Humboldt never saw the river.

The Humboldt was the longest and most important eastern affluent of Pleistocene Lake Lahontan. After recession of the great lake, Humboldt River waters flowed a hundred new miles southwesterly to enter Humboldt Lake and disappear in the sink below—in dry years, not even to reach the sink.[1]

Today, in some 200 miles of its flattest length, the sluggish, silt-laden river drops only about 1,100 feet between Elko and Lovelock, or a bit

---

[1] The terms "Humboldt Sink" and "Carson Sink" have been variously used, and the earlier, pioneer applications are seldom interpreted today as they were then. Originally, the enlargement of the Humboldt River below Lovelock or Big Meadows was called both "Humboldt Lake" and "Humboldt Sink." This thin sheet of water, with its marshy areas, was formed behind an easily eroded gravel bar left by Lake Lahontan, a bar which was quickly breached by an outlet channel so that any lake which was backed up by the bar would soon disappear—it was in the process of doing so when the immigrant arrived on the scene and threw dirt into the break to impound more water.

The overflow from Humboldt Lake meandered south and east into a much larger *playa* or *pan* in the northern part of the Carson Desert. This pan was another "Humboldt Sink" of the early travelers; it is the "North Carson Lake" or even the "Carson Sink" of those who came later, suffering from historical apathy. Carson Desert has become Churchill, and later, Lahontan, Valley to the faithful who have seen part of it change from the wasteland it verily was— the dreaded *original* "40-Mile Desert" of the despairing wagon trains—to a garden spot of the State. Perhaps no single notion could have been more incredulous to the emigrants who struggled across this waste of sand and alkali than that part of it should one day be fertile ranch and farm land.

The original "Carson Sink" was the smaller, more southern playa of the Carson Desert, later called the "South Carson Lake" and today largely drained and supporting a luxuriant grassland locally referred to as "Government Meadow."

more than 5 feet per mile. Rapids are conspicuously rare over long stretches of this journey. In its eastern drainage, a series of tributaries descend from more mountainous elevations to contribute to the mainstem from both north and south. Travelling from east-to-west, as the river flows for much of its length, the following major affluents come into it from the north:

(1) Mary's River, draining out of the Jarbidge Mountains in extreme northern Elko County. Running parallel to it a few miles to the west is the Snake River's north-flowing tributary, Bruneau River.

(2) North Fork of the Humboldt River, draining an area south-southeast of Mountain City. The two major branches of the North Fork

FIG. 22. Quiet beaver pool on Martin Creek, northeast of Paradise Valley, Humboldt County, in the Santa Rosa Mountains. Photo by the author, 1940.

bracket the beginnings of the northward-flowing Owyhee River, another Snake system tributary.

(3) Susie and the larger Maggie Creeks come into the Humboldt at Carlin, bringing water from both sides of the Independence Mountains. The source of south-flowing Maggie Creek closely parallels the source of north-flowing South Fork of the Owyhee River, which latter enters the Snake River and thence the Columbia River.

(4) Rock Creek, draining the west side of the Tuscarora Mountains.

(5) Evans and Kelly Creeks, coming together into a common course and emptying at Red House.

(6) The Little Humboldt River, a complex chain of creeks draining an area north and east of Winnemucca. This includes the North and South Forks, with their tributary inlets, to the east, and Martin Creek to the north. The North Fork and Martin Creek originate high in the

Santa Rosa Mountains and the system enters the Humboldt River at Winnemucca.

No waters of importance come into the Humboldt from here to its sink. The main side streams to the south are:

(1) Lamoille and Secret Creeks, emptying into the Humboldt at Halleck through a common channel and draining an important northwest section of the massive Ruby Mountains in Elko County;

(2) South Fork of the Humboldt, coming from the Rubies farther south, with its main tributary, Dixie Creek;

(3) Pine Creek, entering the Humboldt near Palisade in Eureka County, and draining a long area to the south;

(4) Reese River, the longest of all the Humboldt River tributaries. This has its beginnings in northwestern Nye County from the high reaches of the southern Toiyabe Mountains, and drains the west side of this towering range with its many small streams. It comes into the Humboldt at Battle Mountain in Lander County. This is the last of the southern tributaries, and is on the order of 160 miles in length.

Many of the early pioneers who came down the Humboldt had flavored comments to make of the placid river; those of Frémont, not the earliest of the sagebrush travelers, are as apt as any—and certainly more often quoted:

"The Humboldt river rises in two streams in mountains west of the Great Salt Lake, which unite, after some fifty miles, and bears westwardly along the northern side of the basin. . . . The mountains in which it rises are round and handsome in their outline, capped with snow the greater part of the year, well clothed with grass and wood, and abundant in water. The stream is a narrow line, without affluents, losing by absorption and evaporation as it goes, and terminating in a marshy lake, with low shores, fringed with bulrushes, and whitened with saline incrustations. It has a moderate current, is from two to six feet deep in the dry season, and probably not fordable anywhere below the junction of the forks during the time of melting snows, when both lake and river are considerably enlarged. The country through which it passes (except its immediate valley) is a dry sandy plain, without grass, wood, or arable soil; from about 4,700 feet (at the forks) to 4,200 feet (at the lake), above the level of the sea, winding among broken ranges of mountains, and varying from a few miles to twenty in width. Its own immediate valley is a rich alluvion, beautifully covered with blue-grass, herd-grass, clover, and other nutritious grasses, and its course is marked through the plain by a line of willow . . . serving for fuel" (Frémont 1849: 9–10).

Lieutenant E. G. Beckwith, of the Pacific Railroad Surveys, added his own impressions, a few years later, of the middle part of the stream:

". . . The river-bottom is a mile wide, the stream, just level with its banks, winding from side to side, to where the second banks or bluffs, twenty feet high, rise to the level of the main plain of the valley. Willows line the stream in many parts, but trees are nowhere seen on the Humboldt. Its water, even at this season, is not superior [June 8] and becomes less so as you descend it, and as it subsides after the spring rise. It is now 40 yards wide when all collected in one channel, and eight feet deep, flowing with a moderate current. There are no fish in this part of it larger than minnows. . . . It is infested with mosquitoes and sand-flies" (p. 35).

"*June 11.*—Our last camp was in a large bottom of coarse grass—the last found on this river above its sink—known to emigrants as Lassen's Meadows [now partially covered by Rye Patch Reservoir], in which the

FIG. 23. Lower Truckee River, several miles west of Wadsworth, Washoe County, looking east. Courtesy Nevada Highway Department and Adrian Atwater.

river terminates its general western course, and turns south for 40 miles, where it reaches the marshy lake in which it disappears" (Beckwith 1855: 36).

The Humboldt did not always flow to its lake. In 1841, John Bidwell looked out over a tortuous lower channel and saw that it "was extremely dry that year, and as we approached the sink it ceased to run, and we were enabled to cross dry shod in several places as we descended it" (Bancroft 1890: 53).

These were the observations of men who saw the Humboldt in its historical prime, so to speak. Time, and the human being, have changed it since then. At this writing (1960), Rye Patch Reservoir has been recently dry for the first time in its more than 20 years of existence. Needless to say, the never magnificent Humboldt is a poorer stream now, and not quite the fitting memorial it once was to the great German explorer whose name it bears and who never saw it.

TRUCKEE RIVER

This river was discovered, for historical purposes, in January of 1844 by Frémont and Carson and named the "Salmon Trout River." In October of the same year, an "emigrator" party under the aegis of 81-year-old Caleb Greenwood ("Old Greenwood") approached the stream and called it the "Truckee" in gratitude to an Indian chief who gave them directions toward it.

The Truckee is a major western contributant to Lahontan Basin, rising in California's Lake Tahoe on the eastern face of the Sierra Nevada Mountains and flowing north, east and north to the valley of Pyramid Lake more than a hundred miles downstream.

The drop in elevation for that distance is one of about 2,500 feet, from 6,200 to 3,700 feet, or about 25 feet per mile, producing a much more turbulent and actively degrading stream than the Humboldt. More than two-thirds of this gradient lies in the upper half of the river, from Tahoe to Reno, where it is descending through broken ridges forming the eastern face to the Sierras—the last half of its course is through flatter and less precipitous channels.

The Truckee is an old stream—an "antecedent" one. As the Great Basin block through which its channel now cuts—Virginia Range—was thrust upward, grinding waters wore their present defile from Vista to Wadsworth, downcutting keeping pace with the upthrust.

After leaving the Virginia Range's eastern edge, the river turns abruptly north and meanders through Lahontan lake sediments between two mountain blocks to its depository in Pyramid Lake. During the highest stages of Lake Lahontan, the lower 35 miles of the Truckee River were under lake waters. Deflection of the pre-Lahontan Truckee eastward flow from the Fallon area to a northerly course into Pyramid Lake by rising Lahontan waters has already been mentioned.

Intensive studies by Hardman and Venstrom (1941) on past and present fluctuations in runoff in the Truckee River basin, which also indirectly reflect similar changes in other Lahontan basin streams, tend to show that some of the ills affecting trout streams are traceable back to the years before irrigation. Their conclusions:

"(1) That drouth-conditions prevailed on the Truckee River Watershed for many years prior to 1840.

"(2) That a period of greatly increased precipitation began about 1860 which, although broken with minor drouth-periods of short duration, lasted until about 1917.

"(3) That since 1917 a drouth-period, comparable in intensity but not in duration to the period prior to 1840, has existed.

"(4) That the period from 1860 to 1917, and particularly that portion of the period which began in 1890, was unusually moist for this area" (p. 89).

No reservoirs have been built athwart the Truckee River, either in its 35-mile California or 75-mile Nevada courses, but numerous power and

FIG. 24. The Truckee River just below Verdi, looking northwest. Here the river has emerged from its steep cañon between Verdi and Lake Tahoe and is cutting through old lakebeds of Mio-Pliocene age. Out of the picture to the left are the famous Verdi Pliocene leaf fossil beds.

irrigation diversion dams dot its channel. The power dams particularly have virtually wiped out sections of the river as trout habitats. At Derby Dam, several miles west of Wadsworth, most of the remnant river is diverted into the Fallon basin to the east for Lahontan Reservoir.

The list of tributaries entering the Truckee is basically small. The most important of these are, from source to end, on the north side:

(1) Donner Creek, running a couple of miles east from three-mile long Donner Lake into the Truckee River west of the town of Truckee, Sierra County, California.

(2) Prosser Creek, draining cañons and small lakes north of Donner Lake and flowing into the Truckee a mile or so upstream from the original outlet of the Little Truckee River.

(3) Little Truckee River. This stream has main sources in Independence (two miles long) and Webber (three-quarters of a mile long) Lakes northwest of Truckee town. Sagehen Creek is its chief contributant, entering the Little Truckee about four miles north of Boca Reservoir which now sits across the mouth of the Little Truckee.

(4) Dog Creek, coming out of Dog Valley just north of Verdi, Washoe County.

To the south, several creeks drain the area north of Lake Tahoe, such as Martis, Juniper, Gray and Bronco before Verdi is reached. Between Verdi and Reno the only sizeable stream is Hunter Creek. South of Reno, the main drainage channel is Steamboat, which originally received all of the creeks along the east face of the Carson Range from Washoe Valley north. These included—from south to north—Franktown and Ophir Creeks in Washoe Valley, and Brown, Galena, White, Thomas and Evans Creeks in Pleasant Valley and the Truckee Meadows. Downstream from this area, nothing of importance accrues to the Truckee River.

CARSON RIVER

This is the next major stream system south of the Truckee, feeding northeasterly into Lahontan Basin. Probably first seen by Joe Walker in 1833, Frémont named it in 1849 for his guide. Its greatest length is about 150 miles, measured from headquarters of the longest fork—East Carson—to its farthest point downstream, the Sink of the Carson (or North Carson Lake). About 120 miles of this is in Nevada.

The long East Fork originates in Alpine County, California, under the north shadow of towering Sonora Peak—a Sierran crest nearly 11,500 feet high—flows north through Silver King Valley and down into Carson Valley, Douglas County, Nevada, where it joins the shorter West Fork just southeast of Genoa. The main channel then meanders northeast out of the valley, cutting easterly through the southern part of the Virginia Range as does its sister stream, the Truckee, farther north. It then wends northeast again through narrow valleys out onto the flat bed of the original Carson Desert (now fertile Lahontan or Churchill Valley).

West Fork of the Carson begins its mainstem some 20 airline miles northwest of the East Fork in a Sierran bowl backed up to the south by the 9,400-foot Nipple. Some of the small cirque lakes from which the West Fork begins are but a shade—geomorphologically speaking— from being captured by beginning headwaters of California's North Fork of the Mokelumne River.

Carson River's upper reaches are turbulent and their rate of descent is great, coming as they do off the sheer eastern facade of the Sierra Nevadas, but not until waters converge on the foot of the range can they be called rivers in our sense. Unlike the Truckee, there is no convenient beginning lake surface from which to compute average gradient, and, as a consequence, attempts to draw comparisons between the two are arbitrary. Carson River's upper stems descend the range more precipitously than those of the Truckee, while gradient from the Valley of the Carson to the plains of Carson Sink is more mature and gentle. From convergence of the two forks between Genoa and Minden to

Carson Sink is about 110 miles which is negotiated with a total drop of about 800 feet, or a bit more than 7 feet per mile.

Away from the Sierra Nevadas, its waters take on aspects typical of Great Basin meanders—placid and brown.

In its natural state, the channel bifurcated just west of present-day Fallon (Churchill County). A short southeastern branch fed smaller Carson Lake proper (or South Carson Lake) and sometimes Carson Sink. South Carson Lake is now the "Government Meadow" or "Pasture." A longer feeder arm flowed northerly to Carson Sink (or North Carson Lake, also Humboldt Sink). The Sink was a common meeting ground for Humboldt and Carson Rivers in overflow years, both contributing water to this, the lowest point in the vicinity (3,850 feet elevation).

But one important holding dam has been established across Carson channel—that of Lahontan Reservoir some 15 miles west of Fallon, and several smaller irrigation diversion dams exist, such as the Mexican and Quilici Dams.

WALKER RIVER

The first recorded discovery of this 135-mile stream is generally attributed to Joe Walker, for whom Frémont named the river and the lake into which it flows in 1849. Walker, heading the Pacific Coast section of Bonneville's party in 1833, came down the Humboldt to its sink, then struck south to Walker Lake and worked westward over the Sierra Nevada Mountains. He apparently passed Mono Lake, traversed the south border of Yosemite and descended into the Great Valley in the vicinity of the Tuolumne and Merced Rivers of California.

Physically, Walker River exists as two separate and nearly equal forks for better than 50 percent of its length. Both branches flow generally north, with some eastering, their headwaters lying high on the eastern face of the Sierra Nevadas in Mono County, California. That spectacular mountain mass known as Sweetwater Range is a northward projecting peninsula of the Sierras which splits the two forks.

Heading in Twin Lakes Basin below the northeast rim of Yosemite, the East Fork of the Walker River meanders for some 85 miles northeast, east and north, receiving tributaries from the east flank of the Sweetwaters and from the west face of Walker Range. In southern Mason Valley, south of Yerington (Lyon County), the East Fork fuses with its companion branch to the west.

The 75-mile long West Fork, from its origin on the northeast slopes of 11,700-foot Tower Peak on the north Yosemite rim, works downhill and to the north, turning northeast near Topaz Lake and entering Smith Valley. East across Smith Valley, the antecedent West Fork pierces the Singatze Range, as both the Truckee and Carson Rivers cut through their blocking hills farther north, and enters Mason Valley to meet the east branch.

From this point of union, the Walker River—now a slow-moving, silt-laden stream downcutting through fine-grained Lahontan beds—flows for 50 miles north, east and south to feed into the north end of Walker Lake. The great bend of the Walker River north of Yerington has already been mentioned as a delicate perch where the river was deflected from a tributary of the Carson basin to its present position as the major stream of Walker Lake.

In Nevada, the East Walker is superior to the West Fork as a trout stream, having a more rocky bottom.

(2) CHARACTERISTICS OF LAHONTAN RIVERS

Because major affluents of the Lahontan Basin—Humboldt, Truckee, Carson and Walker Rivers—are all essentially similar in originating as small, clear, swift streams high in the surrounding mountains and ending as larger, turbid, and sluggish waterways evaporating in saline-alkaline enclosed basins, we can safely consider them, for our purposes, as one limnological unit. And because the author knows the Truckee River better than the others, it can be used as typical of the unit.

PHYSICAL AND CHEMICAL CONDITIONS
WATER MOVEMENTS

In its 2,500-foot drop from Lake Tahoe (6,200 feet) to Pyramid Lake (3,800 feet)—a distance of some 110 miles—the Truckee River runs a gamut of velocity extremes. In its upper, rapid reaches it is an actively downcutting or degrading stream, rapidly widening and deepening its valley. After leaving the mountains, it becomes slower, meandering and its cutting action is greatly reduced. At Pyramid Lake, where its movement disappears, it is aggrading or building up its delta into the lake as the silt it carries settles out. Carson and Walker Rivers show the same pattern.

TEMPERATURE

Unlike lakes, stream temperatures more closely reflect adjacent air temperatures and streams also show a uniform increase in temperature from headwaters to flatlands. Warming up is not pronounced in all cases, however, since increased evaporation in lower, sluggish reaches of these streams keeps them cooler in warm weather than might be expected.

TURBIDITY

Streams such as these show marked extremes in opacity. They begin as clear water and gradually pick up solutes and visible mechanical sediments as they course downward, until they assume a light gray-brown color in their lower levels and become opaque. Light fails to penetrate to significant useful depths for plants in opaque waters. After runoffs, usually in the spring, lower stretches generally clear up considerably and, barring disturbances, may be quite transparent.

OXYGEN

This gas decreases in concentration as stream waters move downward and slow their pace. The Truckee River is not typical in this respect (with regard to the other Lahontan streams) since its oxygen peak does not occur as it leaves Lake Tahoe but farther down stream where increased turbulence brings its waters closer to equality with the oxygen content of the air. My data have shown early spring oxygen highs of nearly 12 parts per million in its waters between Lake Tahoe and the town of Truckee, and late summer lows of around 8 ppm near its Pyramid Lake terminus.

DISSOLVED SOLIDS

These increase downstream as water dissolves more material from the rocks and soils it flows over. The Humboldt River shows maximum dissolved solids for the Lahontan system, and these increase from its sources to about 1,000 parts per million near the Sink.

HYDROGEN-ION CONCENTRATION OR pH

These streams are mildly alkaline to begin with, and characteristically show an increase in alkalinity from headwaters to lowlands. Again, the Humboldt can be used as the example of extremes here. Its alkalinity fluctuates around the 8.5 pH point, depending upon volume, distance from source, flushing effects of runoffs, etc.

BIOLOGICAL CONDITIONS

PLANKTON

While practically no systematic work has been done with the plankton flora and fauna of Nevada streams, there is no reason to believe that Lahontan rivers differ basically from other streams of similar type in this respect. Flowing (lotic) waters are notoriously poor in plankton compared to standing (lentic) water bodies such as lakes and ponds— and since plankters form such an important beginning link in the food chain cycle of lakes, it is to be expected that there would be significant differences between the two types of waters.

The author has collected enough quantitative plankton samples in the Truckee River to note that, as would be expected, plankters are scarce in the upper, clearer waters and increase in variety and numbers as river waters are followed downward to their slower, wider reaches. Here several factors operate to promote growth of these microorganisms—warmer temperatures, increased dissolved solid materials, decreased velocity and turbulence and a proportionately greater volume of quiet backwaters where plankton can accumulate.

In the samples I have collected, phytoplankton (plants) have been overwhelmingly dominant, with little zoöplankton (animals) being evident. Diatoms[2] formed the bulk of the phytoplankton.

OTHER BIOTA

Smaller aquatic animals, such as insects, show a marked segregation of species depending upon the stream conditions they live under. In upper, swifter portions of the river the flat, streamlined types with

---

[2] Single-celled algae encased in remarkably sculptured shells of silica.

higher oxygen requirements and greater tolerances for low temperatures are characteristic. They are replaced by other species in the more sluggish parts of the stream.

Such a succession is not so conspicuous among fishes, since there are much fewer species involved, but it exists nevertheless. Belding Sculpin and Mountain Whitefish—among native species—and the introduced Brook Trout, are examples of types favoring higher, colder headwaters. More at home in warmer, slower stretches are such indigenous species as the Tahoe Sucker, Lahontan Redshiner and Lahontan Tui Chub, as well as the introduced Asiatic Carp.

Original Fish Fauna of the Lahontan River System Within Historical Times

(A) SALMON AND TROUT FAMILY, Family *Salmonidae*

1. Lahontan Cutthroat Trout (*Salmo clarki henshawi*)

(B) WHITEFISH FAMILY, Family *Coregonidae*

2. Mountain Whitefish (*Prosopium williamsoni*)

(C) SUCKER FAMILY, Family *Catostomidae*

3. Tahoe Sucker (*Catostomus tahoensis*)
4. Lahontan Mountainsucker (*Pantosteus lahontan*)
5. Cui-ui Lakesucker (*Chasmistes cujus*)—in the lower Truckee River only during the spring spawning run

(D) CARP AND MINNOW FAMILY, Family *Cyprinidae*

6. Lahontan Redshiner (*Richardsonius egregius*)
7. Lahontan Tui Chub (*Siphateles bicolor obesus*)
8. Lahontan Speckle Dace (*Rhinichthys osculus robustus*)
9. Soldier Meadows Desertfish (*Eremichthys acros*)

(E) SCULPIN FAMILY, Family *Cottidae*

10. Belding Sculpin (*Cottus beldingi*)

Fishes Successfully Introduced Into the Lahontan River System

(A) SALMON AND TROUT FAMILY, Family *Salmonidae*

1. Brown Trout (*Salmo trutta*)
2. Rainbow Trout (*Salmo gairdneri*)
3. Brook Trout (*Salvelinus fontinalis*)

(B) CARP AND MINNOW FAMILY, Family *Cyprinidae*

4. Sacramento Blackfish (*Orthodon microlepidotus*)
5. Asiatic Carp (*Cyprinus carpio*)

(C) CATFISH FAMILY, Family *Ictaluridae*

6. Channel Catfish (*Ictalurus punctatus*)
7. White Catfish (*Ictalurus catus*)
8. Black Bullhead (*Ictalurus melas*)
9. Brown Bullhead (*Ictalurus nebulosus*)

(D) TOPMINNOW FAMILY, Family *Poeciliidae*
10. Mosquitofish (*Gambusia affinis*)

(E) PERCH FAMILY, Family *Percidae*
11. Yellow Perch (*Perca flavescens*)

(F) SUNFISH FAMILY, Family *Centrarchidae*
12. Largemouth Blackbass (*Micropterus salmoides*)
13. Smallmouth Blackbass (*Micropterus dolomieui*)[3]
14. Bluegill Sunfish (*Lepomis macrochirus*)
15. Green Sunfish (*Lepomis cyanellus*)
16. Sacramento Perch (*Archoplites interruptus*)
17. White Crappie (*Pomoxis annularis*)[4]

FIG. 25. Owyhee River, East Fork, looking north eight miles below Mountain City, Elko County. Photo by the author, July 21, 1939.

B—The Snake System

(1) HISTORICAL AND DESCRIPTIVE

The Snake system, a large network of streams centering on the Snake River, drains southern and western Idaho and some adjacent parts of

---

[3]Introduced in late 1956 into the upper Humboldt River between Elko and Battle Mountain. Plants were mainly made in Palisade and Carlin Cañons.
[4]Introduced very successfully into Rye Patch and Lahontan Reservoirs in early 1956.

Wyoming, Utah, Nevada, Oregon and Washington. Its main channel, the Snake proper, is in the neighborhood of 1,000 miles long.

The system drains northeastern and northcentral Nevada via seven northerly flowing tributaries which are, from west to east: Lake Creek, South Fork of the Owyhee River, East Fork of the Owyhee River, Bruneau River, West Fork of the Jarbidge River, East Fork of the Jarbidge River and Salmon Falls Creek. All head in Elko County, with the exception of a fork of Lake Creek originating in eastern Humboldt County.

While there exists some local variation among them, in general it can be said that they represent pretty much the same type of stream—by-and-large, their waters are clearer, swifter and contain less dissolved and suspended matter than do rivers of the Lahontan system.

In Nevada, Bruneau and Jarbidge Rivers are small but torrential streams rapidly downcutting their steep cañons through lava beds. Forks of the Owyhee and Salmon Falls Creek meander somewhat in places through wider bottomlands, but always with a good current.

Historically, the first record of discovery and use of this northeastern corner of the State seems to stem from an 1825 trip made up its streams by a trapping brigade under Peter Skene Ogden of Hudson's Bay Company. From here, Ogden's party crossed the low divide separating the Snake and Lahontan basins and explored upper waters of the Humboldt River in the same year.

From this time on there is good evidence that increasing numbers of trappers congregating in adjacent parts of Utah and Idaho rapidly over-ran the area in their search for beaver skins.

(2) CHARACTERISTICS OF THE SYSTEM

PHYSICAL AND CHEMICAL CONDITIONS

WATER MOVEMENTS

In relation to the Snake drainage in general, the Nevada stems are in the nature of high headwater streams. Their gradient is quite steep, averaging nearly 200 feet per mile. The Jarbidge and Bruneau Rivers are swift and strongly pitched, while the Owyhee and Salmon Falls streams are somewhat flatter in profile.

TEMPERATURE

These waters, in keeping with their headwaters characteristics, are cold, natural trout streams. The swifter flows are the coldest, and while diversions have further decreased velocities in the lower stretches, in the main even such slow and many-pooled sections as the lower Salmon Falls Creek will support trout.

TURBIDITY AND OXYGEN

Except for spring runoffs, turbidity is lacking in these streams. The coldness of the water, as well as its upstream turbulence, results in high oxygen contents, which range from better than 6.0 to some 10.0 parts per million.

DISSOLVED SOLIDS, pH AND SUMMARY

Dissolved solids are very low in the upper parts of the system, and increase as the waters cover more ground in their flow toward the Snake River. Hydrogen-ion concentration (pH) varies from alkaline for the majority of streams (8.5 pH or less) to acid (6.0) for a few—the Jarbidge River has been recorded at pH 5.0 by Durrant (1934). Carbon dioxide and bicarbonates are low (up to about 10 and 125 parts per million, respectively), while carbonates are lacking (this latter is typical of most Nevada streams).

BIOLOGICAL CONDITIONS

Plankton is scarce in the upper waters, and increases with lessening of velocity and greater amounts of dissolved solids in stretches farther downstream. In the faster portions, typically flattened, streamlined, rock-clinging immature stoneflies, mayflies and caddisflies make up a dominant part of the food chain for fish. In pool areas below, the insect fauna increases in number and diversity as the water warms up.

Original Fish Fauna of the Nevada Snake System Within Historical Times

(A) SALMON AND TROUT FAMILY, Family *Salmonidae*
1. King Salmon (*Oncorhynchus tshawytscha*)[5]
2. Cutthroat Trout (*Salmo clarki*)
3. Rainbow Trout (*Salmo gairdneri*)
4. Dolly Varden Trout (*Salvelinus malma*)

(B) WHITEFISH FAMILY, Family *Coregonidae*
5. Mountain Whitefish (*Prosopium williamsoni*)

(C) SUCKER FAMILY, Family *Catostomidae*
6. Biglip Sucker (*Catostomus macrocheilus*)
7. Bridgelip Sucker (*Catostomus columbianus*)

(D) CARP AND MINNOW FAMILY, Family *Cyprinidae*
8. Chiselmouth (*Acrocheilus alutaceum*)
9. Northern Squawfish (*Ptychocheilus oregonense*)
10. Typical Columbia Redshiner (*Richardsonius balteatus balteatus*)[6]
11. Bonneville Columbia Redshiner (*Richardsonius balteatus hydrophlox*)[6]
12. Snake River Speckle Dace (*Rhinichthys osculus carringtoni*)

(E) SCULPIN FAMILY, Family *Cottidae*
13. Belding Sculpin (*Cottus beldingi*)

Fishes Successfully Introduced Into the Nevada Snake System

(A) SALMON AND TROUT FAMILY, Family *Salmonidae*
1. Brown Trout (*Salmo trutta*)
2. Brook Trout (*Salvelinus fontinalis*)

---

[5] Undoubtedly extinct now.
[6] These two subspecies intergrade in this area.

(B) CARP AND MINNOW FAMILY, Family *Cyprinidae*
3. Asiatic Carp (*Cyprinus carpio*)

C—The White River System

(1) HISTORICAL AND DESCRIPTIVE

This is a relict or "fossil" system of southeastern Nevada, now greatly reduced and intermittent. While it might technically be considered a part of the Colorado River system, White River is now discontinuous over most of its length with only its very lowermost part flowing into the Colorado. This, plus its very interesting native warm-water fishes, makes it rather unique.

FIG. 26. Hiko Pool, a man-made extension of a natural warm spring in the northern end of Pahranagat Valley, Lincoln County. Habitat of the White River Springfish, *Crenichthys baileyi*. Looking southeast. Photo by the author, 1937.

Hubbs and Miller (1948B: 96–100) have reviewed the evidence—geological and ichthyological—that larger, continuous rivers at one time connected the two major waterways of the system with the Colorado River to the south.

The western fork—White River—was once a flowing stream nearly 200 miles long. Beginning on the east slopes of White Pine Range southwest of Ely, in White Pine County—and fed by springs along the way—the channel can be traced today southward for its entire length to the Colorado River. In its upper portion, in the vicinity of Preston and Lund, there is a restricted surface flowage for some 40 miles, more-or-less, then the channel is dry until springs of Pahranagat Valley in

Lincoln County re-activate another 30 miles of flow. Below Pahranagat, there is 35 miles of dry channel; at this point surface water again fills the old river bed from prolific springs in Warm Springs Valley, the source of the Muddy or Moapa River which discharges into the Overton Arm of Lake Mead, Clark County, 25 miles away.

Ichthyological evidence points to considerable antiquity for the disruption of this waterway, for its isolated springs contain peculiar little endemic fishes generically different from fishes elsewhere.

The eastern fork—Meadow Valley Wash—is some 50 miles shorter than its sister fork. In its upper reaches it is perenially dry (in fact, goes by the name of "Dry Valley Wash") but has a small and consistent flowage in the vicinity of Caliente. In all but the driest years it continues as a definite stream all the way to its junction with the Moapa River at Glendale, in northern Clark County, a distance of about 75 miles. Two tributaries of Meadow Valley Wash of importance are Eagle Valley Creek, east of Pioche, and Clover Creek, east of Caliente. However, water from these streams seldom reaches the Wash due to irrigation needs.

West of the White River channel, a series of isolated springs in Railroad Valley, Nye County, contain a species of the peculiar genus of thermal fishes typical of White River, and indicate that this area too once was part of the White River system.

(2) CHARACTERISTICS OF THE SYSTEM

PHYSICAL AND CHEMICAL CONDITIONS

WATER MOVEMENTS

In White River channel, which we know best, water activity is mainly a function of the springs, mostly warm, which appear at irregular geographic intervals. Flowage is not great and is confined to very narrow channels. Velocities are more than average for the Lahontan system, when water is present. In 200 miles between Preston-Lund and Lake Mead, elevation drops from 6,000 feet to 1,200 feet, or shows an average channel descent of 24 feet per mile. Prior to the coming of the settler, much more of the channel must have carried water, since most of the springs are now utilized completely for irrigation and seldom contribute anything to the main stem.

TEMPERATURE

The many warm or thermal springs along the White River course provide temperature patterns reversed from the usual. Instead of waters which gradually warm up from source to terminus due to insolation, many White River springs begin hot or warm and end cold, providing a greater temperature diversity to the habitat.

TURBIDITY

This is noticeable in the system, especially in the Moapa or Muddy River at the lower end where the stream is downgrading through old, fine lake sediments, or during flash floods.

OXYGEN

Since gases in general are less soluble in warmer liquids than they are in cooler ones, it is distinctive of thermal waters that they seem, at first test, to be deficient in oxygen as far as fish life is concerned. The oxygen they contain is only a portion of that held in colder waters, and much less than that commonly believed necessary for fish. However, considerable populations of fishes, insects and other animals are found everywhere that temperatures are not harmful, showing adaptation of these types to what would be considered less than the normal amount of oxygen.

DISSOLVED SOLIDS, pH AND SUMMARY

Dissolved solids are proportionally quite high, since these waters percolate up through limestone, which is easily soluble in cold waters, and increasingly so in warm waters, if the necessary carbon dioxide is present.

Hydrogen-ion concentration, or pH, has, everywhere tested, been found to be slightly on the alkaline side.

Sampling at Big Pool in the Warm Springs area of northern Clark County (headwaters of the Moapa River) gave the following limnologic characteristics:

Temperature—93.2° F.; pH—7.3; CO_2—zero parts per million; CO_3—zero ppm; HCO_3—223 ppm; chlorides-sulfates, present; sulfides-chlorine-iron, absent; dissolved oxygen—3.4 ppm; water clear and palatable.

BIOLOGICAL CONDITIONS

In place of the usual succession of habitats we find in the normal stream profile—as mentioned under "Biological Conditions" for the Lahontan River system—in these warm springs we see restrictive forces of temperature at work. Certain fishes of the White River system are thermal endemics; i.e., they originated in warm, isolated waters of the area and are adapted to the peculiar living conditions found there. As a consequence, it is not uncommon to discover that cold water is a barrier which prevents them from appearing downstream away from the warm springs in which they originated. Such species are the Moapa Dace (*Moapa coriacea*) and the White River Springfish (*Crenichthys baileyi*).

BIOTA IN THE FOOD CHAIN

The invertebrate fauna of the system is so sketchily known that it seems rather pointless, on the one hand, to mention any of its elements. However, the broader aspects of the picture are deducible, and the first details can be added here and there to make a beginning.

Plankton exists, some of it contributed by attached algae which often grow in profusion, forming a basic food for certain of the fishes. In many of the warm springs situations, nothing beyond this simple grass-to-herbivore, algae-to-fish food chain exists, since larger predatory fishes are absent and all sizes of those fish present feed on algae. Often, such predation as does occur involves an insect-to-fish chain.

The writer has been able to sample the White River invertebrates only in Pahranagat Valley and the Warm Springs region, and then only in a minor way. His first visit to the Pahranagat Valley warm springs in 1937 was primarily a plant-collecting one and secondarily a search for dragonflies. The usual large aggregation of water beetles, bugs, dragonflies, mayflies and caseflies is found here, all of which are good fish food, and vice versa.

Original Fish Fauna of the White River System Within Historical Times

(A) SUCKER FAMILY, Family *Catostomidae*
1. White River Mountainsucker (*Pantosteus intermedius*)

(B) CARP AND MINNOW FAMILY, Family *Cyprinidae*
2. White River Colorado Gila[7] (*Gila robusta jordani*)
3. White River Speckle Dace (*Rhinichthys osculus velifer*)
4. Moapa Dace (*Moapa coriacea*)
5. Big Spring Spinedace (*Lepidomeda mollispinis pratensis*)
6. White River Spinedace (*Lepidomeda albivallis*)
7. Pahranagat Spinedace (*Lepidomeda altivelis*)

(C) KILLIFISH FAMILY, Family *Cyprinodontidae*
8. White River Springfish (*Crenichthys baileyi*)
9. Railroad Valley Springfish (*Crenichthys nevadae*)

Fishes Successfully Introduced Into the White River System

(A) SALMON AND TROUT FAMILY, Family *Salmonidae*[8]
1. Rainbow Trout (*Salmo gairdneri*)
2. Cutthroat Trout (*Salmo clarki*)
3. Brook Trout (*Salvelinus fontinalis*)

(B) CARP AND MINNOW FAMILY, Family *Cyprinidae*
4. Asiatic Carp (*Cyprinus carpio*)

(C) TOPMINNOW FAMILY, Family *Poeciliidae*
5. Mosquitofish (*Gambusia affinis*)

(D) SUNFISH FAMILY, Family *Centrarchidae*
6. Largemouth Blackbass (*Micropterus salmoides*)
7. Bluegill Sunfish (*Lepomis macrochirus*)

D—The Colorado River System

(1) HISTORICAL AND DESCRIPTIVE

This great river, 1,400 miles long and draining some 8 percent of the United States, briefly forms the extreme southeastern border of Nevada for some 100 miles and so contributes its rather unique and specialized

---
[7] Pronounced "Hee-la."
[8] All these have been planted in the Caliente-Panaca area.

fish fauna to our area. Flowing in this vicinity through semi-arid regions of little but flash drainages, it has but two permanent Nevada tributaries—the Virgin and Moapa Rivers.

The largest of these two, the Virgin River, heads in Kane County, Utah and flows west and southwest, crossing the tip of northwestern Arizona and terminating its 150-mile course in Nevada. The Moapa River, tiny by comparison, begins to flow in Warm Springs, northern Clark County, Nevada (on the course of the Pleistocene White River), and originally entered the Virgin River about 30 miles downstream

FIG. 27. Colorado River, looking north just above present site of Davis Dam, Clark County. This was what the unobstructed river looked like in 1937. Photo by the author.

from Warm Springs. However, the last five miles of its course are now inundated by the Overton Arm of Lake Mead, into which both the Virgin and Moapa Rivers empty.

Limnologically, the Colorado no longer exists as a river along the Nevada border—almost its entire channel, which was cut through deep, narrow cañons, is now occupied by reservoirs backed up behind two large dams. The first of these—Hoover or Boulder Dam built at the entrance to Black Cañon in the 1930s—impounds Lake Mead (Boulder Lake) which backs some 25 miles into Arizona.

Davis Dam, farther south near the extreme tip of the State, forms Lake Mohave which reaches to the tailrace of Hoover Dam. Characteristics of these two vast reservoirs will be considered in the section on Nevada lakes.

The only uninundated part of the Colorado River in Nevada, that below Davis Dam, has lost many of its original features because it now represents drainage from a lake rather than unimpeded river flowage.

Historically, the Colorado River was discovered and explored quite early. Captain García Lopez de Cárdenas, of Coronado's force, is generally credited as the first white man to see the river, moving up to it with a small force from Cibola or Zuñi in 1540. His brief contact with the river was in a portion of its Utah traverse. Padre Francisco Garcés, the first man known to touch upon any portion of Nevada, saw the State somewhere in the vicinity of its Colorado River border in 1776.

(2) CHARACTERISTICS OF THE SYSTEM

Since river conditions typical of the original Colorado no longer exist in that portion of the channel impinging on Nevada, its one-time limnology is only of academic interest in showing how the water environment has changed and how this has been reflected in alterations in the fish fauna.

PHYSICAL AND CHEMICAL CONDITIONS
WATER MOVEMENTS

Strong, continuous water motion is a dominant property of the Colorado River as a habitat. As stated, the only river section left is the 14-mile stretch below Davis Dam, but originally the entire length of the river adjacent to Nevada represented water which contained species unique to it. The effects of this swiftly-moving water are reflected in the rather bizarre forms of some of the dominant native fishes—the Gila Chub, Colorado Squawfish and Razorback Sucker all have flat, sloping heads which tend to hold them against the bottom when pointed upstream. The chub and sucker in turn have conspicuous dorsal keels which have a stabilizing effect in the current. The most peculiar fish in this regard is Miller's *Gila "cypha,"* an extremely streamlined form of the Gila Chub.

While movement of water in the channel below Davis Dam has not changed particularly, other river peculiarities have altered considerably.

TEMPERATURE AND TURBIDITY

This, of course, still shows the familiar pattern of gradually increasing as the river flows southward, but is significantly colder in the lower levels and warmer in the upper reaches. In addition to cooling the river waters, Lake Mohave behind Davis Dam is also a silting basin and water emerging from the dam is clear for many miles downstream. Actually, most of the silt is taken out of the river by Lake Mead, so that the once conspicuously opaque river is now colder and clearer—a much changed environment.

OXYGEN—DISSOLVED SOLIDS—pH

As would be expected, these are variable. The two oxygen readings the writer took in the Fort Mohave area gave results of 6.8 and 7.3 parts per million, indicating the anticipated sufficiency of the gas.

Dissolved solids are high enough to substantially enrich the water as far as plankton production is concerned, but are not marked. The published figures of 596 to 766 parts per million (U. S. Geol. Surv. Water Supply Paper 970) can be compared with 30–40 ppm for rainwater, 300 ppm as a generally accepted freshwater lake maximum, and 4,700 ppm for Pyramid Lake.

Hydrogen-ion (pH) concentration is on the alkaline side—the writer obtained a reading of pH 8.4 at Fort Mohave in 1951. Moffett (1942: 81) noted a pH of 8.2 in the river channel below Hoover Dam before the impoundment of Lake Mohave.

SUMMARY

The writer's field tests in 1951 at Fort Mohave below Davis Dam showed the following figures: CO_2 from 2.5 to 5.0 parts per million; HCO_3 from 159 to 161 ppm; chlorides—sulfates present; chlorine—sulfides—iron lacking.

BIOLOGICAL CONDITIONS

BIOTA IN THE FISH FOOD CHAIN

While the river invertebrate and plant life could be expected to change at least quantitatively with construction of Hoover and Davis Dams, we know little about such changes. Surveys in the river below Lake Mead before Davis Dam was built, and below Davis Dam later on show that the Colorado River still contains supplies of fish foods. Algae (chiefly *Cladophora* sp.) is a very common attachment material on submerged objects and is often found in trout stomachs. Smaller planktonic crustacea such as the widespread waterfleas *Daphnia* and *Diaptomus* occur in locally protected spots while such of the larger aquatic insects as mayflies and true flies were plentiful. The author and others (Moffet, 1942, has summarized these earlier efforts) have collected these as well as many others of the common types of aquatic insects, all utilizable as fish food.

Original Fish Fauna of the Colorado River System Within Historical Times

(A) SALMON AND TROUT FAMILY, Family *Salmonidae*
1. Colorado Cutthroat Trout (*Salmo clarki pleuriticus*)

(B) SUCKER FAMILY, Family *Catostomidae*
2. Flannelmouth Sucker (*Catostomus latipinnis*)
3. Razorback Sucker (*Xyrauchen texanus*)

(C) CARP AND MINNOW FAMILY, Family *Cyprinidae*
4. Colorado Squawfish (*Ptychocheilus lucius*)
5. Colorado Gila (*Gila robusta* and subspecies)
6. White River Speckle Dace (*Rhinichthys osculus velifer*) (Las Vegas Wash)
7. Virgin River Spinedace (*Lepidomeda mollispinis mollispinis*)
8. Woundfin (*Plagopterus argentissimus*)

Fishes Successfully Introduced Into the Colorado River System

(A) SHAD FAMILY, Family *Clupeidae*

1. Mississippi Threadfin Shad (*Dorosoma petenense atchafalayae*)

(B) SALMON AND TROUT FAMILY, Family *Salmonidae*

2. Rainbow Trout (*Salmo gairdneri*)
3. Brown Trout (*Salmo trutta*)

(C) SUCKER FAMILY, Family *Catostomidae*

4. Dusky Mountainsucker (*Pantosteus* sp.)

(D) CARP AND MINNOW FAMILY, Family *Cyprinidae*

5. Asiatic Carp (*Cyprinus carpio*)
6. Golden Shiner (*Notemigonus crysoleucas*)

(E) CATFISH FAMILY, Family *Ictaluridae*

7. Channel Catfish (*Ictalurus punctatus*)
8. Black Bullhead (*Ictalurus melas*)

(F) TOPMINNOW FAMILY, Family *Poeciliidae*

9. Mosquitofish (*Gambusia affinis*)
10. Black Molly (*Mollienesia latipinna*)
11. Swordtail (*Xiphophorus helleri*)
12. Moonfish (*Xiphophorus maculatus*) (Platy)

(G) SUNFISH FAMILY, Family *Centrarchidae*

13. Largemouth Blackbass (*Micropterus salmoides*)
14. Bluegill Sunfish (*Lepomis macrochirus*)
15. Green Sunfish (*Lepomis cyanellus*)
16. Black Crappie (*Pomoxis nigromaculatus*)

E—The Amargosa River System[9]

(1) HISTORICAL AND DESCRIPTIVE

This abbreviated drainage system of extreme southwestern Nevada and adjacent California is perhaps the most shattered of the relict drainages of the State. Modern Amargosa River is little but a travesty on the name, its course now largely occupied by a perennially dry bed except at its headwaters and at certain intermittent points, far-spaced, down-channel.

About 175 miles in total length, the Amargosa heads west of 7,425-foot Timber Mountain above Beatty (southern Nye County) and courses southerly through Amargosa Desert and Valley, respectively, where it swings sharply west and enters the southern tip of Death Valley. In its northerly path up through Death Valley it can be traced

---

[9]The simplest pronunciation of this word is the Spanish "Uh-marg-uh-suh" rather than the cumbersome but typical Anglicization "Amar-go-suh," which usually degenerates to "Arma-gosa." Means "bitter water."

to its technical terminus in Bad Water, the lowest point in the United States—280 feet below sea level.

For ten to fifteen miles in the vicinity of Beatty, chiefly above the town, surface flowage is evident, but very sparse, coming mainly from warm springs. In the desert below Beatty, the channel is dry except during spring runoffs and occasional cloudbursts.

About 55 miles south of Beatty, the dry channel receives discharge from Ash Meadows to the east, where a large number of spectacular warm springs percolate to the surface through limestone rocks. At this point, the Amargosa channel crosses into California. As far as Nevada is concerned, the fish fauna of the Amargosa system is confined to the cooler Beatty region and the warm springs of Ash Meadows.

(2) CHARACTERISTICS OF THE SYSTEM

PHYSICAL AND CHEMICAL CONDITIONS

WATER MOVEMENTS—TEMPERATURE—TURBIDITY

The remarks made for these categories under the White River system heading apply generally here as well. In the Ash Meadows area, water movement is found only in the vicinity of springs, and mostly is slow, with marshy ponds and tracts as the termini of the system. About Beatty, where the Amargosa is actively flowing for short distances, the environment is that of a cool, small and slow stream, except briefly at some of the warm sources near Springdale.

Similarly, temperature reversals are the pattern at Ash Meadows, while the usual situation prevails in the Amargosa near Beatty.

Turbidity is variable in the Ash Meadows micro-environments; periodically, flood conditions produce silting in the Amargosa channel proper.

OXYGEN—DISSOLVED SOLIDS—pH

As in the White River system, oxygen content is lowered, dissolved solids are probably rather marked, and the pH values are slightly alkaline.

SUMMARY

Many gross field examples were run in the Ash Meadows springs. Temperatures ranged from 72° F to 92° F; dissolved oxygen from 1.8 to 4.2 parts per million; pH was constant about the 7.3 point; CO_2 from 16 to 26 ppm; CO_3 lacking; HCO_3 from 244 to 271 ppm; chlorides—sulfates present; chlorine—sulfides—iron lacking.

BIOLOGICAL CONDITIONS

These are still much in need of study, and except for minor notations on Devil's Hole, only the usual specific listings could be given.

Original Fish Fauna of the Amargosa River System Within Historical Times

(A) CARP AND MINNOW FAMILY, Family *Cyprinidae*
1. Amargosa Speckle Dace (*Rhinichthys osculus nevadensis*)

(B) KILLIFISH FAMILY, Family *Cyprinodontidae*
2. Ash Meadows Poolfish (*Empetrichthys merriami*)
3. Pahrump Poolfish (*Empetrichthys latos* and subspecies)
4. Amargosa Pupfish (*Cyprinodon nevadensis* and subspecies)
5. Devil Pupfish (*Cyprinodon diabolis*)

F—The Nevada Bonneville System

(1) DESCRIPTIVE

Only a minor fragment of the vast Bonneville drainage system of Utah overlaps into eastern Nevada. Like elements of the Snake system to the north, the Nevada Bonneville consists of a series of separate streams, all small, flowing into Utah chiefly from east slopes of Nevada ranges. Not all of the Bonneville fishes are represented in Nevada waters.

The most important streams are listed below by counties—mileages given are not totals, but only the Nevada lengths.

ELKO COUNTY

(1) *Goose Creek* is in the extreme northeastern corner. It heads in Idaho, picks up numerous minor tributaries in Nevada, and after 25 miles, flows northeasterly into Utah. It is not actually a part of the Bonneville system.

(2) *Thousand Springs Creek*, which rises on the east slope of Burnt Creek Mountains, southwest of Goose Creek, flows tortuously eastward for more than 60 miles before entering Utah.

WHITE PINE COUNTY

(3) *Lehman Creek* system comes off the east side of the Snake Range (as do all other White Pine County creeks listed below) and becomes a part of Utah after a Nevada length of some 11 miles. The system originates from the three Lehman Lakes (Stella, Black and Me-Me).

(4) *Baker Creek*, now a part of the Lehman system, originates in Baker Lake.

(5) *Smith Creek* system, 12 miles. This includes two important tributaries, *Deadman* and *Deep Cañon Creeks*.

(6) *Hendrys Creek*, 11 miles.

(7) *Silver Creek* system, 15 miles.

(8) *Snake Creek* and tributaries, 13 miles.

(9) *Lake Creek* or *Big Springs Creek*, with 7 miles in Nevada, originates from large springs a short distance east of the Snake Range, and ends in Pruess Reservoir in Utah.

None of these streams is in direct touch with the main body of the Bonneville farther east, and in that sense they represent an isolated system. However, their naturally dry or diverted lower channels would make such contact if a sufficient volume of water flowed in them and their native fish fauna is entirely derived from the Bonneville system.

(2) CHARACTERISTICS OF THE SYSTEM

PHYSICAL AND CHEMICAL CONDITIONS

WATER MOVEMENTS—TEMPERATURE—TURBIDITY

In the main, these streams are those typical of mountain ranges of the Great Basin—small, swift and clear in their upper reaches and slower on the flats. Since they generally do not have enough volume to maintain a flow for any great distance after leaving their mountain flanks, they may be smaller there than near their headwaters. While clear most of the time, they are heavily silted during spring runoffs.

Temperatures vary from the near zero of melting snows to the much warmed lower stretches on valley floors, and temperature changes can occur in comparatively short distances. Ted Frantz, State Fishery Biologist who surveyed the streams of this area, found that "providing water is present [lower down] benches or fans often afford the better trout habitat than the higher cañon areas with their colder water temperatures, etc."

Oxygen values are high, *dissolved solids* are low and *pH* is around neutral.

BIOLOGICAL CONDITIONS

Original Fish Fauna of the Nevada Bonneville System Within Historical Times

(A) SALMON AND TROUT FAMILY, Family *Salmonidae*
1. Utah Cutthroat Trout (*Salmo clarki utah*)

(B) SUCKER FAMILY, Family *Catostomidae*
2. Utah Sucker (*Catostomus ardens*)
3. Bonneville Mountainsucker (*Pantosteus platyrhynchus*)

(C) CARP AND MINNOW FAMILY, Family *Cyprinidae*
4. Bonneville Columbia Redshiner (*Richardsonius balteatus hydrophlox*)
5. Bonneville Speckle Dace (*Rhinichthys* sp.)
6. Leatherside Chub (*Snyderichthys aliciae*)

(D) SCULPIN FAMILY, Family *Cottidae*
7. Bonneville Baird Sculpin (*Cottus bairdi semiscaber*)

Fishes Successfully Introduced Into the Nevada Bonneville System

(A) SALMON AND TROUT FAMILY, Family *Salmonidae*
1. Rainbow Trout (*Salmo gairdneri*)
2. Brown Trout (*Salmo trutta*)
3. Brook Trout (*Salvelinus fontinalis*)

(B) CARP AND MINNOW FAMILY, Family *Cyprinidae*
4. Asiatic Carp (*Cyprinus carpio*)

(C) SUNFISH FAMILY, Family *Centrarchidae*[10]

5. Largemouth Blackbass (*Micropterus salmoides*)
6. Sacramento Perch (*Archoplites interruptus*)

---

[10]These have been planted in Preuss Reservoir, adjacent Utah, which is fed by Nevada streams, but the species do not occur in the Nevada Bonneville proper.

CHAPTER V—LAKES AND MAJOR RESERVOIRS

CONTENTS

	PAGE
A—The Lahontan System	121
(1) Pyramid Lake	121
(a) Historical and Descriptive	121
(b) Characteristics of Pyramid Lake	124
Physical Conditions	125
Chemical Conditions	133
Biological Conditions	135
(c) Management of Pyramid Lake	138
(2) Walker Lake	141
(3) Lake Tahoe	144
(a) Historical and Descriptive	144
(b) Characteristics of Lake Tahoe	145
Physical and Chemical Conditions	145
Biological Conditions	147
(c) Management of Lake Tahoe	148
(4) Donner Lake	150
(5) Independence Lake	152
(6) Webber Lake	154
(7) Topaz Lake	156
(8) Lahontan Reservoir	157
(9) Stillwater Marsh	160
(10) Rye Patch Reservoir	163
(11) Washoe Lake	165
(12) Bridgeport Reservoir	168
(13) Twin Lakes	169
(14) Boca Reservoir	172
(15) Prosser Reservoir	172
(16) Summit Lake	172
(17) Eagle Lake	174
(18) Carson Lake	176
(19) Big Soda Lake	176
B—The Snake System	179
Wildhorse Reservoir	179
C—The White River System	180
Adams-McGill Reservoir (Sunnyside Reservoir)	180

CHAPTER V—CONTENTS—*Continued*

	PAGE
D—The Colorado River System	181
(1) Lake Mead	181
(a) Historical and Descriptive	181
(b) Characteristics of Lake Mead	182
Physical Conditions	182
Chemical Conditions	184
Biological Conditions	184
(c) Management of Lake Mead	185
(2) Lake Mohave	186
(a) Historical and Descriptive	186
(b) Characteristics of Lake Mohave	187
Physical Conditions	187
Chemical Conditions	188
Biological Conditions	188
(c) Management of Lake Mohave	188
E—The Amargosa River System	191

CHAPTER V

LAKES AND MAJOR RESERVOIRS

A—The Lahontan System

The remnant lakes left from the Pleistocene Nevada inland sea—Lake Lahontan—are now only two in number; Pyramid and Walker Lakes. Winnemucca Lake, basically an ephemeral overflow sump for Pyramid Lake, dried up in 1938. These lakes have one important thing in common—all are the evaporating pans for inflowing streams and as such are continually increasing in salinity and alkalinity.

(1) PYRAMID LAKE

(a) Historical and Descriptive

Pyramid Lake was first discovered by John Charles Frémont, in 1844:

"January 10.—We continued our reconnoissance ahead [Frémont and Carson], pursuing a south direction in the basin along the ridge; the camp following slowly after. On a large trail there is never any doubt of finding suitable places for encampments. We reached the end of the basin, where we found, in a hollow of the mountain which enclosed it, an abundance of good bunch grass. Leaving a signal for the party to encamp, we continued our way up the hollow, intending to see what lay beyond the mountain. The hollow was several miles long, forming a good pass, the snow deepening to about a foot as we neared the summit. Beyond, a defile between the mountains descended rapidly about two thousand feet; and, filling up all the lower space, was a sheet of green water, some twenty miles broad. It broke upon our eyes like the ocean. The neighboring peaks rose high above us, and we ascended one of them to obtain a better view. The waves were curling in the breeze, and their dark-green color showed it to be a body of deep water. For a long time we sat enjoying the view, for we had become fatigued with mountains, and the free expanse of moving waves was very grateful. It was set like a gem in the mountains, which, from our position, seemed to enclose it almost entirely. At the western [actually northern] end it communicated with the line of basins we had left a few days since; and on the opposite side it swept a ridge of snowy mountains, the foot of the great Sierra. Its position at first inclined us to believe it Mary's Lake, but the rugged mountains were so entirely discordant with descriptions of its low marshy shores and open country, that we concluded it some unknown body of water, which it afterwards proved to be" (Frémont 1845: 216).

Three days later, Frémont's party saw the peculiar rock which prompted the lake's name:

"The next morning [14th] the snow was rapidly melting under a warm sun. Part of the morning was occupied in bringing up the gun; and, making only nine miles, we encamped on the shore, opposite a very remarkable rock in the lake, which had attracted our attention [they were going south along the east side of the lake] for many miles. It rose, according to our estimate, 600 feet [actually it was less than 300 feet out of the water at the time] above the water; and, from the

FIG. 28. Pyramid Lake levels 94 years ago. In this famous 1868 photograph by T. H. O'Sullivan for the U. S. War Department's Fortieth Parallel Survey (Clarence King), the lake's waters rode at their highest post-Lahontan mark. This represents an elevation of about 3,876 feet above sea level (Hardman and Venstrom 1941), and shows the Pyramid essentially as Frémont saw it. Looking west down the chain of warm spring-formed tufa domes. Courtesy Office of U. S. Chief of Engineers, Photograph No. 77–KS–13 in the U. S. National Archives.

point we viewed it, presented a pretty exact outline of the great pyramid of Cheops. The accompanying drawing presents it as we saw it. Like other rocks along the shore, it seemed to be incrusted with calcareous cement. This striking feature suggested a name for the lake; and I called it Pyramid Lake; and though it may be deemed by some a fanciful resemblance, I can undertake to say that the future traveller will find a much more striking resemblance between this rock and the pyramids of Egypt, than there is between them and the object from which they take their name" (Frémont 1845: 217).

Physically, Pyramid Lake is a large 25-mile long, wedge-shaped body of water, with its long axis lying nearly north and south and its greatest width at the northern end. It is over 300 feet deep, and occupies the lowest of the old Lahontan basins. Chemically, its waters are noticeably saline and alkaline, with a high percentage of dissolved solids. Biologically it is very rich in plankton and certain aquatic animals, and in addition to supporting a fishery of prime importance in the original economy of the local Indians, it was, until comparatively recently, a valuable commercial fishery for Indians and immigrants

FIG. 29. Pyramid Lake levels today (1961), taken from the same spot as the O'Sullivan picture. Not only has water dropped more than 70 feet and exposed to their bases the tufa domes between the Pyramid and shore, but the Pyramid itself is now part of the expanding peninsula and can be reached dry-shod. This became possible for the first time in 1960. Photo by the author.

alike. Its expanse—10 miles at the northern end and 4 miles at the southern—was the home of an incredibly numerous but now extinct population of Cutthroat Trout.

Geographically, the lake lies northeasterly from Reno in southeastern Washoe County, entirely within the confines of the Pyramid Lake Pahute Indian Reservation and bordered west and east by the Virginia and Lake Ranges, respectively. The Truckee River, its only significant inlet stream, enters at the narrow southern end.

(b) Characteristics of Pyramid Lake

Because Pyramid Lake has been studied more intensively than any other Nevada lake, we can conveniently and profitably use it as the major example in the Lahontan system.

Limnological investigations of Pyramid Lake might be said to have begun with the work of Russell in the years 1881 to 1883—although primarily interested in the geology of the region, he collected samples of Lahontan waters for important chemical analyses. And while Russell noted something of the gross biota of the lake, this was more-or-less in

FIG. 30. Another O'Sullivan photo of the Pyramid and its associated tufa domes, from a point farther south than Fig. 28. 1868. Courtesy Office of U. S. Chief of Engineers, Photograph No. 77–KS–12 in the U. S. National Archives.

passing and his efforts to understand the chemistry of the Lahontan Basin waters constitute a major contribution to this aspect of the subject.

In 1883, Cope described a series of new fishes—only one of them, the Cui-ui, still being valid—from his recent collecting in the lake, and Call's 1884 report dealt with the molluscan fauna of the basin. Russell, publishing his results in 1885, used Cope's terminology for the fishes.

Years intervened, except for the occasional mention of some of the fishes by the State and Federal fisheries organizations. About 1911, Snyder began his initial investigations of the fishes of the Lahontan system, aided in this summer by the herpetologist C. H. Richardson, which culminated in his comprehensive report of 1917. This evidence of a revived interest in the history of Lake Lahontan showed further in the notices and publications of Gale (1915), Antevs (1925) and Jones (1925). Jones' work was a monograph on the subject, largely based on Pyramid Lake, in which new analyses were added to the chemical picture and some mention of the algal flora was made.

In 1935, a field report (unpublished) by United States Bureau of Fisheries biologist M. J. Madsen pointed out the uncertainties of keeping a fishery going in the lake and implied the pessimism which others had long been expressing—without water coming in, the lake could not long continue to survive.

Although brief, the single most important, strictly limnological consideration of the major Lahontan lakes was that of Hutchinson (1937) whose fieldwork was performed in a six-weeks period in 1933. For the first and, to date, only time, such limnological facets as morphometric constants, heat budgets, stability, etc., were given limited consideration.

In the summer of 1948, Thomas J. Trelease of the Nevada Fish and Game Commission and the writer, began occasional work with the lake. In 1954, the Fish and Game Commission initiated an official Dingell-Johnson project (No. F–4–R) which included Pyramid Lake in its area of study. Al Jonez, State Fishery Biologist, was the first technician assigned to the lake. In 1955 he was promoted to another assignment in southern Nevada and his place taken by V. Kay Johnson, who continues to work with the lake.

PHYSICAL CONDITIONS

WATER MOVEMENTS

Current effects of the inflowing, fluctuating Truckee River change considerably as the river flow varies from a trickle barely reaching the lake by late summer and early fall to the volume floods which come down in the late spring or, more rarely and precipitously, in the late fall—early winter. During the November floods of 1950, when off-season rains melted much of the snow pack in the eastern Sierras and the Truckee system found itself in the worst flood stage in many years, the muddy waters of the Truckee could be traced up the eastern, deeper side of Pyramid Lake to the north end—with a sharp line demarking green lake waters on the west and silt-laden river waters on the east. Water flowed into the lake at the maximum rate of 19,000 cubic feet per second. In non-flood times, current effects from the affluent must be unimportant except locally at the south end.

Another current factor of limited extent and importance is that supplied by the mixing waters of the numerous, but small warm and hot springs which enter the lake at various places.

The most important current-producing agency is the wind, whose efforts are modified locally by configuration of the shorelines. In the main, since the morphometry of the lake is essentially that of a smooth, elongate basin whose bottom and deeper sides, at least, offer few irregularities which would divert currents, the force of the wind is capable of producing pronounced current patterns over the broad sweeps of water which are available to it. The marked thermal stratification of

FIG. 31. Catch at the rocks at Sutcliffe, Pyramid Lake, circa 1920. Looking southeast. From the Nettie Sutcliffe Cooper collection.

late summer, as well as such chemical stratification as may exist, exerts its modifying effects on these wind-produced water movements.

Horizontal and returning (undertow) currents are often pronounced in Pyramid and are direct reflections of wind activity except at the inflowing river mouth. The swings or oscillations called "seiches,"[1] which result, usually, when winds momentarily pile water up on one shore and which can be compared to swings of a pendulum, are a distinctive part of Pyramid, since the lake is exposed to brief, irregular and often violent wind blows during most of the year. However, their characteristics have never been accurately established for the lake.

TEMPERATURES

Of the many temperature problems of lakes, only two are of general enough interest to the fish specialist to be outlined here—thermal stratification or heat layering, and heat budget.

Thermal stratification, in deeper lakes such as Pyramid, is the result of two factors—availability of solar heat or energy at various times of the year, and the distribution of this energy by the wind. A description of the annual heat cycle is probably the best approach to an understanding of the development of thermal stratification in a lake.

In lakes where this problem was first studied, it soon became apparent that four overturn phases in the annual cycle could be detected, correlated with the seasons, and beginning with the layerless fall waters and ending with the pronouncedly heat-stratified waters of late summer. We will consider these first as they were developed in the typical lakes of the north-central United States, and then point out how Pyramid deviates from this "classical" pattern.

Fall Overturn—during this cooling period, when lake waters are losing heat gained in spring and summer, surface waters sink downward as they lose heat and consequently become heavier. This vertical motion sets up currents which tend to distribute this heat loss throughout the basin as well as to move warmer, lighter waters into surface position for cooling. The more uniform the lake waters become in temperature, the easier it is for winds to move the water about and set up strong mixing currents—as a consequence of these forces, by the end of the fall overturn, the typical picture is one of a thoroughly mixed lake with the same temperatures from top-to-bottom.

Winter Stagnation—the classical picture here is obtained only in those lakes whose surfaces freeze to an ice cover in winter. Water is a peculiar substance in that, above 4° C. (about 39° F.) it becomes heavier as it cools, but between 4° C. and 0° C. (32° F.), the situation reverses, and water becomes lighter as it becomes colder (the reason ice floats rather than sinks). So, when waters cool below 4° C., they are lighter than the slightly warmer waters about them, and rise to the surface. If this were not so, ice would sink and lakes would freeze solid from bottom-to-top and life as we know it in water would be radically altered.

During this period there is an increasing temperature gradient from the ice cover at the top (0° C.) to 4° C. at the bottom. The fact that

---

[1] Pronounced "say-sh-es."

the water does not circulate, being protected from wind action, and may become oxygen-stagnant as well, is responsible for the name given to this phase.

Spring Overturn—this is a warming period, a reversal of the fall overturn, during which the waters begin to take on heat. Lighter, colder water (0° C. to 2° C.) was lying on top of heavier, warmer (4° C.) water during winter; in spring, as the surface waters rise to 4° C., the maximum density of water, they sink downward through colder but

FIG. 32. Pyramid Lake, looking southwest. The chain of tufa domes leading from shore to the Pyramid were probably formed beneath the mud of ancient Lake Lahontan when the Pyramid was under several hundred feet of water. Behind the Pyramid rises Anaho Island, site of a famous pelican-cormorant-gull rookery. At the present rate of water loss, Anaho Island will become a part of the mainland by 1975. This will spell the end of water bird life on the island with the extinction of one of the major rookeries in the West. Courtesy Nevada Highway Department and Adrian Atwater. Photo taken circa 1948.

lighter waters (2° C.), which latter are displaced upward to be warmed and sink in turn. Soon the lake waters, aided by wind mixing which becomes more effective as temperature extremes disappear, are again uniform in temperature from top-to-bottom as they were in the fall overturn.

Summer Stagnation—as temperatures continue to rise above the critical 4° C. point, water acts as we think it normally should, becoming progressively lighter with each degree of added heat, and so more buoyant. As temperature differences increase between surface and underlying waters, resistance to wind-mixing becomes greater and eventually only the relatively narrow surface layer can be circulated by the wind which moves it—much as a tractor track circulates over its rollers—over the colder waters below. When this situation is well-developed, markedly different temperature strata exist in the lake. The uppermost zone of summer circulation is called the *epilimnion,* and has a nearly uniform temperature. An intermediate stratum is called the *thermocline,* characterized by a rapid drop in temperature.[2] It is the thermocline upon which the epilimnion rotates.

FIG. 33. Infra-red photo showing a confluence of clouds, water and desert difficult to match anywhere but Pyramid Lake. Looking northwest to the Pyramid, 1940. Courtesy Ernest Mack.

The lowermost layer is the *hypolimnion,* which is often the thickest of the three layers and is characterized, like the epilimnion, by more uniformity of temperature. Because of its effective isolation by the thermocline from the top circulating waters, the hypolimnion usually becomes greatly depleted, or even exhausted, of oxygen by late summer. Organic matter is constantly raining down from upper waters, and

---

[2] By definition, the top of the thermocline is that region in which the temperature decrease attains 1° C. or better per each meter of descent (0.55° F. per foot). The bottom of the thermocline, conversely, would be that region where the per meter descent again becomes less than 1° C.

most of the hypolimnionic oxygen supplies are used up in the oxidation of this material.

The above pattern is drawn from a typical temperate lake of the "second order." The desiccating lake in the Pyramid basin presents a somewhat different picture, although more fieldwork is necessary before all its variable details will be apparent.

In the first place, such data as are now available indicate that the winter stagnation period is missing in Pyramid Lake—fall, winter and spring merge into one great overturn cycle during which the lake waters are essentially uniform in temperature by virtue of the lack of an ice cover and constant turmoil from the gusty, blustering wind blasts which are typical of the region.

Secondly, a thermocline may not develop during cooler summers when wind action over the lake is excessive. All-in-all, Pyramid Lake

FIG. 34. The Pyramid, from the southwest, in high water days. Photo from the Nettie Sutcliffe Cooper collection, taken in the mid-1920s.

seems to fit best into the conventional system of lake classification as a "tropical lake of the second order;" i.e., one in which bottom temperatures are variable but never reach 4° C. and there is generally but one circulation period in the year, rather than two.

The utility of some of these facts to the ichthyologist and the fishery biologist can be appreciated when we realize that fish will avoid the oxygen-low hypolimnion during summer and will tend to disappear into the somewhat warmer depths in winter.

Heat budgets show us the actual amounts of calories involved in heat gain and loss in a lake during the year and are computed as the heat increment differentials between minimal winter, and maximal summer, temperatures. To determine this, we must know the following mean figures; lake depth, summer and winter temperature extremes. From these data we can also determine the work of the wind in distributing heat throughout a lake.

Little data have been organized on the heat budget of Pyramid Lake. Hutchinson (1937) gave the first figures, calculated from his 1933 work and from Jones' 1914 figures published in 1925.

Other aspects of temperature regulation in the lake, such as heat gain through insolation and basin-water heat exchanges, have not been investigated for Pyramid.

TURBIDITY

Under this heading can be considered not only the mechanical sedimentaries of organic and inorganic origin suspended in water, but also the characteristics of light penetration attendant upon the particulate matter present.

Again, little work has been done on these factors in Pyramid Lake. Even a superficial visual examination shows the lake to be highly turbid

FIG. 35. Trout fishermen and their catch on the beach at Sutcliffe, Pyramid Lake (The Willows). From the Nettie Sutcliffe Cooper collection, mid-1920s. Looking southeast.

most of the year. Closer investigation reveals this to be due to large quantities of suspended organic matter in the form of living algae (single-celled plants) and microcrustaceans and small quantities of inorganic silt washed in from the river and from the shores. The wind also contributes a share.

Substantial amounts of dissolved solids are also present, and markedly affect the way that light penetrates water. Photoelectric or photodiode studies of light penetration are needed to determine these values, and could now easily be done with standard photographic and electronic equipment in the hands of aqualungers. From such limited observations as are available it is apparent that light penetrates to greater depths and in greater intensities—as far as human eye sensitivity is concerned—than would be suspected from a casual examination of surface turbidity.

Some of the biologists' problems relative to turbidity rates and concentrations involve the rate of filling of a lake basin with incoming sediments; the availability or lack of bottom areas to aquatic vegetation; and the zones of suspension between top and bottom where algae (phytoplankton) can carry on photosynthesis. The type and efficiency of food cycles often depend heavily on such factors as the last two mentioned. The former (gradual filling of the lake basin with sediments), while occurring in a short time, geologically speaking, usually

FIG. 36. "Old Woman and her Basket," looking southwest. An infra-red shot by Ernest Mack in 1941. Note the conspicuous white wavy beach lines. These are formed by the multitudinous shells of several different species of now extinct freshwater snails. Prior to the 1920 drop in water level, the lake was fresh enough to support a large, although even then, depauperate, snail fauna. Today—salinity-alkalinity levels are lethal, and the snails are gone. Tomorrow—the fish.

is so slow in terms of human generations that little attention is given to it except in artificial impoundments.

An illustration of the turbidity of Pyramid Lake is the average Secchi disk[3] reading of 12½ feet obtained by Tom Trelease and the writer in August of 1948. Although this will vary through the year, it can be compared with visibility of ocean waters where it is not uncommon to be able to see many times this distance off of coral reefs, or in clear waters carrying less dissolved materials such as Lake Tahoe.

---

[3]A white enameled disk, of variable size but commonly 8 inches in diameter, lowered into the water on a line and point at which it disappears from sight, both on the dropping and rising phases, averaged out.

CHEMICAL CONDITIONS

OXYGEN

Data on this gas in Pyramid Lake have come to hand from the fieldwork of Hutchinson in 1933 and that of the writer and Fish and Game Biologists since 1948. The former's limited fieldwork—six weeks—could not produce any overall picture, but we now know the pattern fairly well from other investigators.

Hutchinson's summer determinations show the expected lowered but adequate concentrations of the gas for this time of year. The writer has found extremes of 1.4 ppm (parts per million) at 275 feet in November and 9.5 ppm at the surface in February, showing the decrease of oxygen in the hypolimnion in the last stages of fall temperature stratification, and the maximum concentration of the gas in colder surface waters during the vigor of the overturn period, respectively.

During the long overturn period of winter and spring, oxygen values increase[4] in the colder waters and, like temperature values, become more uniform from top-to-bottom in response to wind-mixing of the waters.

The importance to the fishery biologist and the sportsman of knowing oxygen conditions of water lies in the relationships between the oxygen requirements of aquatic plants and animals and their bearing on productivity of the water. In areas of minimal or no oxygen, plants and animals cannot live, such as in the hypolimnion; in other cases, marginal amounts of oxygen may drastically reduce the attached or floating plants (large and small), decrease the foods available for animals at the beginning of the food chain and so result in lowered overall productivity of that particular body of water.

A general statement has been made, which is true with certain limitations, that such active, cool water game fishes as the trout require a rough minimum of 4.0 ppm of oxygen to prosper—many fishes, particularly those adapted to warm water environments, can do with much less oxygen (see White River-Amargosa River systems).

In connection with the importance of oxygen to the aquatic system, it might be well to point out here that without the work of the wind in mixing oxygen into water, there would be much less of the gas in solution since direct diffusion from the atmosphere accounts for very little of the water's oxygen.

DISSOLVED SOLIDS

These vary enormously over the world in the many different kinds of lake waters that exist. Pyramid Lake stands rather well above the middle point with its 4,700 ppms (parts per million) of dissolved solids. As comparative illustrations, Moosehead Lake in Maine contains 16 ppm—well below the usual 30–40 ppm for rain water—while Big Soda Lake, near Fallon, Nevada, represents one of the other extremes with 113,700 ppm. Lake Tahoe contains approximately 75 ppm of dissolved

---

[4] It is an universal property of gases that they dissolve more readily in cold, than in warm, water; hence, there is more oxygen in winter, than in summer, water, if it similarly fluctuates in temperature.

solids. A very general figure of 300 ppm has been given as the usual maximum for freshwater lakes.

The relatively high salt content of Pyramid is apparent when we note that sodium chloride makes up about 75 percent of the dissolved solids in the lake at the present time (1961).

HYDROGEN-ION CONCENTRATION OR "pH"

The pH of Pyramid Lake is noticeably on the alkaline side[5]—about 9.1—more alkaline than the average 6.5–8.5 range for lakes. The recorded extremes of all waters vary from pH 3.2 to 10.5, or from highly acid to highly alkaline.

Cyclic changes tend to be absent from Pyramid Lake, although this aspect of the hydrogen-ion concentration has not been studied and

FIG. 37. Looking southeast across the massive tumbled tufa block at Sutcliffe, Pyramid Lake, when the water was still high in the mid-1920s. Mrs. Cooper tells of seeing otter in the lake waters off these rocks. Photo from the Nettie Sutcliffe Cooper collection.

probably a slight lessening of alkalinity could be detected in the deeper waters during stagnation periods, among other things. It is too large a body of water, and lacks any protected coves or bays where plant activity could produce any easily manifested diurnal pH fluctuations such as are not uncommon in certain types of small ponds.

Permanent changes in pH are, however, taking place in Pyramid, but only very slowly. Because the lake is the desiccating end-pan of a river system, solutes are constantly accumulating in it and its salinity, in particular, is more marked with each yearly recession of the water level. That the pH changes very little over long periods of time is due to the buffering action of carbonates and bicarbonates in the hard

---

[5]On a scale of 0 to 14, 7 is the neutral point. Values less than 7 are increasingly acid as they approach zero, while above 7 they become correspondingly more alkaline.

water of Pyramid which resist increased changes in hydroxyl (alkaline) ion concentrations by replacing ions lost in the neutralization of the new increment of alkaline ions so that the original pH system is preserved. Only waters high in dissolved solids and having weak, rather than strong, acidic or basic elements, can have any substantial buffering action.

BIOLOGICAL CONDITIONS

As an *eutrophic* lake (one rich in nutrients), Pyramid could be expected to be highly productive in the biological sense, and indeed it

FIG. 38. Another view of the Sutcliffe rocks, looking north, and showing the lake water about 70 feet above the 1961 levels. The lake shore is now several hundred yards to the east. From the Nettie Sutcliffe Cooper collection, taken in the mid-1920s.

is. And this productivity is not confined, as it is in many lakes, to the upper summer circulating zone (epilimnion), but extends to all depths, as the writer's plankton sampling shows.

LIME–FORMING ORGANISMS

There is a large and well-known literature on marl formations concerning, especially, the lower plants, but except for the activities of J. Claude Jones (1925), no attention has been paid to the prominent Pyramid Lake algae which undoubtedly are lime producers.

The various kinds of *tufa* or calcium carbonate deposits of the Lahontan basin have been briefly named in the discussion of the age of Lake Lahontan. Apparently, such lime deposition occurs in at least five known ways in the basin—(1) through the agency of algae, (2) by mechanical precipitation against shores and headlands, largely by wave action, (3) by precipitation from supersaturated waters entering the cold lake waters from hot springs (which are numerous in the Pyramid basin), (4) by precipitation in bottom muds through the action of hot, supersaturated waters rising from below[6] and (5) by precipitation

FIG. 39. Pinnacles at north end of Pyramid Lake, looking south. These are chains of hot spring-formed tufa domes similar to those of the Pyramid area farther south on the east side of the lake. Courtesy Nevada Highway Department and Adrian Atwater.

throughout the lake waters as they concentrate during periods of general desiccation.

Professor Charles LeRoy Brown, formerly a member of the University of Nevada Biology Department (1918–1938),[7] determined algal samples for Jones in the latter's work on Pyramid Lake. Brown found that *Callothrix* algae, either *C. thermalis* or *C. parietina*, were growing in intimate association with newly formed tufa, and Jones felt that this alga, and possibly other blue-greens which Brown noted with *Cal-*

---

[6]In our 1948 fieldwork, Thomas Trelease and the writer determined this as the most likely way in which the symmetrical tufa "domes" were formed. In 1957, U.S.G.S. geologist Radbruch published the same conclusions.
[7]Deceased 13 March 1960.

lothrix (such as *Nostoc* and *Phormidium*) were all marl producers (Jones 1925: 7, 9).

Many tons of snail shells, of several different species recently extinct in the lake, form extensive beaches around the basin and represent the most prolific lime makers among the higher animals.

An interesting minor example of calcium carbonate precipitation is

FIG. 40. Derby Dam "fishway," lower Truckee River looking west and upstream. Inoperative virtually from the day it was built, it was in no better shape a generation later when this picture was taken (1941). Contemplating one of the chief factors in the decline of the Pyramid Lake fishery of the 1930s is Tom Trelease, then a student at the University of Nevada, now Chief of Fisheries for the Nevada Fish and Game Commission. Photo by the author.

that of the thin film which forms conspicuously on the exterior of the common Mormon Creeping Waterbug (*Ambrysus mormon*, Family Naucoridae). This predaceous species, which spends its time crawling about on the rough tufa fragments in very shallow water, apparently provides, by means of its minute hairs, nuclei about which contact crystallization of lime can begin from the saturated waters. Algae also frequently attach to the insects, and may account for at least some of the precipitation in their withdrawal of carbon dioxide from the water during their photosynthetic activities.

BIOLOGICAL PRODUCTIVITY

This all-important phase of the limnology of Pyramid Lake has yet to be quantitatively investigated, although its grosser qualitative characteristics are known. About all that can be done at present is the listing of the lake biota, and many additions remain to be made here.

The general configuration and content of the classical productivity pyramid, when adequately known for Pyramid Lake, will undoubtedly have its local peculiarities. Among other things, it will have a greatly reduced "bottom flora" component and probably an increased percentile of phytoplanktonic content.

BIOTA IN THE FISH FOOD CHAIN

As yet, no adequate listing of the plants and animals which call Pyramid Lake home has ever been made; that offered in the Appendix is reasonably complete for the macroscopic forms.

Lahontan drainage fishes are abundant in Pyramid except for those types which prefer colder waters, such as the Belding Sculpin and the Mountain Whitefish. The fishing picture has changed so much in the last two decades, and is still changing, that much of it is still a question mark.

Originally, of course, native Lahontan Cutthroat Trout and Cui-ui Lakesuckers were the only food fishes to be found in the lake and they existed in almost unbelievable numbers. With the extinction of the trout in the 1930s, only the Cui-ui and previously introduced Sacramento Perch were left and, because of cessation of angling, the latter was entirely unutilized.

Of recent years, stocking with Kokanee Salmon has been unsuccessful, but reasonably good fishing has been achieved in the stocking of Rainbow, and the restocking of Cutthroat, Trout.

(c) Management of Pyramid Lake

The two most important management problems at Pyramid Lake concern (1) the long-range one of keeping useful amounts of water in the lake and (2) a more immediate one of working under decreased inflow conditions to provide adequate access of trout to the necessary spawning grounds in lower reaches of the Truckee River.

As long as these two sets of conditions can be satisfactorily met there exists no problem so far as keeping fishable populations in the lake is concerned. Unlike Tahoe, Pyramid is literally a culture medium for all organisms in the food chain—temperatures are warm, nutrients plentiful and fish growth rapid—and so is capable of producing the maximum yields of game fish that could possibly be expected anywhere outside a fish hatchery.

The first problem is an extremely knotty one to which a great many people have given time and energy (and some thought)—probably no economic decisions have been more difficult to arrive at than those dealing with water rights and allocations. Since Pyramid Lake is the catch basin for the Truckee-Tahoe drainage system to the west, it was to be expected that agricultural use of these waters by increasing populations would decrease flow into the lake. The solution now is to attempt

FIG. 41. New Truckee River channel, reaching from Marble Bluff to Pyramid Lake, some two and a half miles away. This was dug from 1941 to 1945 and the diversion dam was apparently completed in 1945 (records are incomplete). It operated until 1950, when the diversion dam in the nearby old channel gave way during the November flood, and the river returned to its original course. The purpose of the new channel was to provide water deep enough for trout to spawn. Photo by the author, May 16, 1945.

an agreement whereby a minimal amount of water—enough to keep the lake from drying up—can be annually guaranteed, subject to climatic conditions, such as drought.

Unfortunately, like most water problems, this has gruesome legal complexities that do not appear currently solvable to the extent that any long-term existence for Pyramid Lake can be reasonably expected. It seems that the lake is destined to desiccate slowly, fluctuating locally with runoff conditions, until it either disappears except for some water in winter and spring (as in the case of other playas around it such as Winnemucca "Lake"), or becomes too much of a brine and alkali concentrate for fish to live in.

FIG. 42. View of part of the Pyramid-Winnemucca Lakes complex looking ENE from Grandview Peak, Pah Rah Mountains, Virginia Range, Washoe County, April 8, 1951. (Photo by author.) The lower half shows the southern end of Pyramid Lake, present and past maximum water lines, while above, in the middle, the dry valley of Winnemucca (with prominent white areas) shows the site of Winnemucca Lake, which dried completely in 1938. Mud Lake shows in the lower right hand corner. The straight dark line running diagonally across below center is the two and one-half mile artificial channel built to confine Truckee River waters for fish spawning, while below it meanders the wider natural channel. The original bifurcation of the Truckee River, feeding both Pyramid and Winnemucca Lakes, occurs about on a straight line drawn between the white-footed end of Marble Hills and the lower right hand corner of the photograph, close to the hills. At high water, Pyramid Lake backed up and inundated this point of bifurcation.

However, until such conditions prevail—and many conservationists and conservation agencies, particularly the State Fish and Game Commission and local sportsmen's groups, are actively trying to find better solutions—there will be many years of good fishing in Pyramid if (a) suitable stocking programs can be maintained for its waters and (b) enough development of the lower Truckee, particularly the river mouth, can be effected to allow lake fish access to the natural spawning areas. This latter would greatly cut down on the costs of maintaining fishable populations in the lake.

(2) WALKER LAKE

The lake's discovery is credited to the Joseph Reddiford Walker segment of the Bonneville party, which crossed Nevada in 1833, and it was named by Frémont in 1849 (see Walker River); the latter had his first sight of Walker Lake in 1845.

FIG. 43. Walker Lake, Mineral County, looking north along the more precipitous west side when the lake was at a much higher level than now (1962). Walker River enters at the north end. At the present time the lake is some 140 feet deep. Courtesy Nevada Highway Department and Adrian Atwater.

This extensive body of water has not been studied by the variety of investigators who have looked, many only momentarily, at Pyramid Lake, but the State Fish and Game Commission has recently (1954) embarked on a systematic survey of the lake. For our limited purposes here, we will emphasize its similarities to Pyramid—from which it does not differ greatly in generalities—rather than its minute differences.

Physically, Walker Lake is not as long as Pyramid—17 miles long, and narrower along a general north-south axis, with its greatest width lying south of its center. At the Narrows toward the north end, the lake slims to 1½ miles, while its greatest width of 6½ miles is farther south. A shallow lake compared to Pyramid, it now contains about 140 feet of water in its deepest part (1960). Chemically it resembles its larger Lahontan companion to the north in its marked salinity and alkalinity and has a greater percentage of dissolved solids. Biologically, both are similar in the richness of their biota, and in their importance as a fishery resource, although fishing for the great Lahontan trout has persisted longer in the southern lake.

Geographically, Walker Lake occupies the valley north of Hawthorne in northwestern Mineral County, its northern end in the Walker River Pahute Indian Reservation, its south one-third a part of the United States Naval Ammunition Depot. Towering Walker (originally Wassuck) Range borders the lake to the west, rising to 11,300-foot Mt.

FIG. 44. A few hours' fishing, Walker Lake, in the 1920s.
Courtesy Tom Trelease.

Grant. The eastern ranges are low, desert hills. The only major ingress stream, Walker River, enters at the north end.

Within historical times, the level of the lake has dropped some 80 feet due to decreased inflow from the river—at its modern high water mark, the lake had a maximum depth of about 225 feet.

The major management problem on Walker Lake revolves around the fact that the lake's alkalinity and/or salinity is now so high that only a Cutthroat Trout fishery can be maintained there. Plants of other trout are unsuccessful. Also, the lack of sufficient inflowing water produces the same situation at both Walker and Pyramid Lakes—virtually total elimination of natural spawning by trout.

Currently (1962), a research project is under way by the author and some of his students at the University of Nevada to determine the chief factor or factors responsible for the peculiar trout environment of Walker Lake. From the purely academic viewpoint such a problem has much of interest—from the practical standpoint, it will have to be solved before any extensive planting can be undertaken for management of the lake's fisheries.

Cal Allan, State Fishery Biologist, has been studying the lake since 1954 for the State Fish and Game Commission.

Original Fish Fauna of Pyramid and Walker Lakes Within Historical Times

(A) SALMON AND TROUT FAMILY, Family *Salmonidae*

1. Lahontan Cutthroat Trout (*Salmo clarki henshawi*)
2. Pyramid Rainbow Trout (*Salmo gairdneri smaragdus*), Pyramid Lake only[8]

(B) SUCKER FAMILY, Family *Catostomidae*

3. Tahoe Sucker (*Catostomus tahoensis*)
4. Cui-ui Lakesucker (*Chasmistes cujus*) Pyramid Lake only

(C) CARP AND MINNOW FAMILY, Family *Cyprinidae*

5. Lahontan Redshiner (*Richardsonius egregius*)
6. Lahontan Tui Chub (*Siphateles bicolor obesus*)
7. Lahontan Speckle Dace (*Rhinichthys osculus robustus*)

Fishes Introduced Into Pyramid and Walker Lakes

(A) SALMON AND TROUT FAMILY, Family *Salmonidae*

1. Kokanee Red Salmon (*Oncorhynchus nerka kennerlyi*)[9]
2. Brown Trout (*Salmo trutta*)[10]
3. Rainbow Trout (*Salmo gairdneri*)[10]

(B) CARP AND MINNOW FAMILY, Family *Cyprinidae*

4. Asiatic Carp (*Cyprinus carpio*)

(C) CATFISH FAMILY, Family *Ictaluridae*

5. Channel Catfish (*Ictalurus punctatus*)[11]

---

[8]This is an aberrant, unplaceable form described from a single specimen in Pyramid Lake. If it is a member of the Rainbow Trout series, as has been generally held, it probably represents an early, unrecorded introduction; however, it could as likely be an atypical deep-water Lahontan Cutthroat—in any event it is not a good taxonomic unit.

[9]Recent implantations of these fish into both lakes have definitely failed in Walker, and while momentarily successful in Pyramid, did not produce any fishing, and seems doomed.

[10]These have done much better in Pyramid than in Walker Lake, in which latter they do not seem to be able to maintain themselves.

[11]A plant of these was made in Pyramid Lake in 1953, with as yet unknown results.

6. Brown Bullhead (*Ictalurus nebulosus*) Pyramid Lake only[12]

(D) SUNFISH FAMILY, Family *Centrarchidae*[13]

7. Sacramento Perch (*Archoplites interruptus*)

(3) LAKE TAHOE

(a) Historical and Descriptive

The largest lake tributary to the Lahontan system (and the largest in the entire region at least from the standpoint of volume), Lake Tahoe is also one of the highest, having a surface elevation of about 6,225 feet above sea level, a length of some 22 miles, width of about

FIG. 45. Lake Tahoe, looking north along the east shore. Note the heavily-wooded contrast to the open aspect of most bodies of water in the State. Courtesy Nevada Highway Department and Adrian Atwater.

13 miles and a maximum depth of nearly 1,650 feet. The lake has a much smaller ratio of width-to-length than other large Lahontan bodies of water, with its more rounded outline, and presents a surface area of about 193 square miles or 123,328 acres. Situated as it is high in the coniferous forests of the Sierra Nevada Mountains, Lake Tahoe stands scenically in contrast to Pyramid Lake with its desert setting. Both are

---

[12]Miller and Alcorn list this as common in Pyramid Lake, apparently on the authority of Bond (1940: 246), but neither the Fish and Game personnel nor the author has ever found the species in the lake.

[13]Bluegill Sunfish (*Lepomis macrochirus*) have been introduced, but apparently not in survival quantities, to Pyramid Lake. It is not unlikely that Largemouth Blackbass (*Micropterus salmoides*) from sloughs upriver in the Truckee get into Pyramid, but they are so far undetected.

connected by the same water line, the Truckee River, rising in Tahoe and ending in Pyramid more than 100 miles away.

The same explorer officially discovered both lakes—Frémont.

Lake Tahoe has attracted the attention of several competent investigators, both geological and biological. Among the most prominent of the former was John Le Conte who, in the years 1883 and 1884 published the results of some "physical studies of Lake Tahoe" in the Overland Monthly, based on work done ten years earlier.

In 1907, Chancey Juday, a longtime member of the famous limnological pioneers of Birge and Juday, published a short survey of Lake Tahoe made in the summer of 1904. John O. Snyder's comprehensive work on the Lahontan Basin of western Nevada culminated in his "Fishes of the Lahontan system of Nevada and northeastern California" in 1917, and dealt in part with Lake Tahoe.

The last published limnological information on Tahoe was that by Kemmerer, Bovard and Boorman in 1923, when they reported on work done in the summer of 1913. This was very brief.

The number of people who have published information of biological and geological significance on Lake Tahoe cannot all be mentioned. Among some of the collectors of ichthyological material were Yarrow, Henshaw, Jordan, Cooper and Eigenmann.

At the present time, the State Fish and Game organizations of California and Nevada and the University of Nevada are doing routine limnological work on Tahoe, especially in areas having special significance in fishery management practices.

(b) Characteristics of Lake Tahoe

The ideas which make it easier to understand a body of water as an environment for fishes have already been discussed in some detail in the section dealing with Pyramid Lake. Here I would like merely to introduce some of the meager data which have accumulated regarding the broad limnological patterns of Tahoe.

PHYSICAL AND CHEMICAL CONDITIONS

TEMPERATURE

While temperatures may drop extremely low in the wintertime and the deep and extensive snowpacks last well into summer, the lake itself never has an ice cover except in small and shallow marginal areas. As Le Conte (1883) suspected, the peculiar characteristic of water whereby it takes up heat slowly and loses it slowly (referred to as the "high specific heat" of water) plays a part here, as undoubtedly does the agency of the wind. In both Tahoe and Pyramid the unceasing play of the wind against the surface waters is given considerable credit for their iceless winter condition.

The few temperature data that have appeared are not extensive enough to do much with. Hutchinson (1937) in comparing the heat budget of Tahoe with his work on Pyramid Lake, seems to have reliable evidence that Le Conte's earlier temperature readings were inaccurate.

Those of Kemmerer et al seem to be more indicative of the actual state of affairs and heat budgets worked out on these last figures give more reasonable results.

Le Conte's 1873 temperature series included all depths, those of Juday (1904) were run only to the 425 foot level, while Kemmerer et al covered the entire range but not in the detail of Le Conte's study.

FIG. 46. Marlette Lake, Carson Range, east of Lake Tahoe, looking northerly. Natural drainage is into Lake Tahoe, but an inverted siphon line has provided Virginia City with Marlette water since the early days of the Comstock Lode. Photo July 19, 1959, by the author.

TURBIDITY

Tahoe is a remarkably clear body of water and it is not unusual to be able to see the bottom easily in over 100 feet of water. Secchi disc readings that the author has made have given results of 76 feet in late July, while extreme visibility of over 100 feet was reported by Le Conte for August and restricted visibility of 65 feet by Juday in June when turbid incoming waters were clouding the lake. Compare these figures with those for Pyramid Lake.

OXYGEN

As would be expected in a lake which shows only weak thermal stratification, and hence is almost always well mixed, the oxygen readings that have been made indicate that a condition approaching uniformity in concentration from top-to-bottom is the usual one. Kemmerer et al made the most extensive oxygen measurements, finding, in July, only slight variation between surface and deep waters (7.0 to 9.0 parts per million). The author obtained similar results for the same month down to the 400-foot level. Earlier in the same year (May), our readings ran as high as 10.5 ppm at the 100-foot level.

The cold waters of the lake are well-suited to hold sizeable amounts of the gas in solution and constant wind activity keeps it well mixed so that extreme lake depths are available to fish populations all year round. This is not usually the case in shallower, warmer lakes whose bottom levels may stagnate late in the summer and be depleted of their oxygen content.

HYDROGEN-ION CONCENTRATION OR pH

Unlike Pyramid Lake with its relatively high alkalinity, Tahoe waters are slightly acid. The author's figure of 6.8 pH was obtained from zero to 100 feet of water at the north end of the lake in May and June. The chemical picture is quite different between the two bodies of water: Tahoe is the originating source of water in an acid, granitic basin and Pyramid is the alkaline evaporating pan of that same water after it has travelled a hundred miles.

BIOLOGICAL CONDITIONS

BIOLOGICAL PRODUCTIVITY

In contrast to the eutrophic condition of Pyramid, Tahoe must be classed as an *oligotrophic* lake (one poor in nutrients) and so biologically less productive of the two.

Original Fish Fauna of Lake Tahoe Within Historical Times

(A) SALMON AND TROUT FAMILY, Family *Salmonidae*
1. Lahontan Cutthroat Trout (*Salmo clarki henshawi*)
2. Tahoe Rainbow Trout (*Salmo gairdneri regalis*)[14]

(B) WHITEFISH FAMILY, Family *Coregonidae*
3. Mountain Whitefish (*Prosopium williamsoni*)

(C) SUCKER FAMILY, Family *Catostomidae*
4. Tahoe Sucker (*Catostomus tahoensis*)

(D) CARP AND MINNOW FAMILY, Family *Cyprinidae*
5. Lahontan Redshiner (*Richardsonius egregius*)
6. Lahontan Tui Chub (*Siphateles bicolor obesus*)
7. Lahontan Speckle Dace (*Rhinichthys osculus robustus*)

---

[14]The same remarks made previously about the Pyramid Lake "Emerald Trout" (Pyramid Rainbow Trout) apply equally well here—also see the discussion under this trout in the systematic section.

(E) SCULPIN FAMILY, Family *Cottidae*
8. Belding Sculpin (*Cottus beldingi*)

Fishes Introduced Into Lake Tahoe

(A) SALMON AND TROUT FAMILY, Family *Salmonidae*
1. Kokanee Red Salmon (*Oncorhynchus nerka kennerlyi*)
2. Brown Trout (*Salmo trutta*)
3. Rainbow Trout (*Salmo gairdneri*)
4. Brook Trout (*Salvelinus fontinalis*)
5. Lake Trout (*Salvelinus namaycush*)[15]

Introduced Lake or Mackinaw Trout provides the bulk of fishing in Tahoe, while its native trout, Cutthroat, has disappeared from the scene except for some sporadic artificial stocking. Rainbow Trout follow next, with both spring- and fall-spawning strains in the lake with the former being the more important of the two.

Brown Trout, Brook Trout and Kokanee Red Salmon are more limited spawners.

Tahoe is a Lake Trout-Rainbow Trout water from the standpoint of its fishery aspect, and has substantial habitat areas for both types. There is no significant interplay between the two species since the Lake Trout is partial to deeper waters during most of the year while the Rainbows generally live above them in the upper water levels.

(c) Management of Lake Tahoe

Major problems of a general nature facing the fishery specialist at Lake Tahoe revolve around the two facts that—

(1) the lake with its colder waters, is basically not as productive as is a warmer body of water, and

(2) even so, Tahoe is not currently producing what it should.

With respect to the first point, which is of least importance from the standpoint of practical management since little can be done to rectify it, the waters of Tahoe are cold, not warming up significantly even by late summer and they hold only small amounts of solids and nutrients, either dissolved or colloidal. These conditions combine to slow down general growth of many important food chain organisms. Small lakes and ponds can be artificially fertilized following much the same principles as the agriculturist uses in increasing plant yields. However, size of the body of water has a direct bearing on success or failure of this method and this plus the fact that the types of fishes which are adapted to Tahoe's cold waters are not those which seem to profit from this treatment, militates against any success for the enrichment method for the lake.

For the second point, three approaches can, and are, being explored —those of (a) finding some cold-adapted species of game fish which can be introduced with reasonable expectation of providing better yield than now exists at Tahoe; (b) attempting to bring back the original

---

[15] By far the most successful.

game fish, Cutthroat Trout, and re-establishing it in its former great abundance; and (c) increasing the numbers of fish now living in the lake.

Alternative (a) has been explored, both systematically and unsystematically, over a good many years at Tahoe. The most notable example would be the Lake Trout which was introduced many years ago and which has proven a satisfactory addition to the lake in that it produces most of the current yield and lives in deeper waters not desired by other game fishes.

The Kokanee Red Salmon planting program of recent years has also shown success of sorts—at least their establishment seems a certainty although they do not add much to the annual catch.

Alternative (b) is in process of experimental sampling by the Nevada Fish and Game Commission at this writing (1962). There seems no good reason to suppose that Cutthroat Trout cannot exist in Tahoe today, in spite of their unexplained dieoff of the Twenties. If their disappearance was related to disease carried by the Lake Trout, then certainly resistant Cutthroat strains can be found or produced which will be relatively immune. Problems here are the same as for

Alternative (c), which involves the Rainbow Trout. For both these fishes, the qualified consensus is that more control will be needed on spawning streams at Tahoe in order to prevent yearly hatches from being caught out of such streams, where fishing pressure is very great, before they can return to the lake. This will involve heavier planting, at least temporarily, and needed restrictions on fishable streams tributary to the lake. This latter is a difficult point to put over to the angler who likes to fish these streams, particularly if he doesn't care about fishing the lake.

FISHING

Fishing methods are varied on Lake Tahoe due to different habitat preferences of the several major game fishes. Two chief approaches exist—(1) "Deep lining" for the top producer, Lake Trout, which lives in bottom waters offshore except for the spawning season, and (2) "Top lining" for the more shallow-living Rainbow, Brook and Brown Trouts and the Kokanee Red Salmon.

Deep lining calls for specific skills and tackle needed to fish a suitable lure on or near the bottom in deep water. In addition to heavier equipment—rod, reel and weighted monel metal, copper or nylon line—a boat is necessary to get the baited-lure combination[16] into the right depths for the bottom-loving Lake Trout.

While more Lake Trout are taken than are other kinds, many fishermen consider them inferior in sporting characteristics to such as Rainbow, etc., since there is often little fight to them and it is sometimes difficult to determine whether the heavy fishing tackle is merely bouncing over the bottom or whether a Lake Trout is on the hook. Lack of fight is commonly due to expansion of the gas-filled swim bladder which paralyzes the fish as it is hauled up from high-pressure depths into

---

[16]An attractor followed by several feet of heavy leader holding the hooked bait or additional lure.

shallower, lower-pressure water. The fish is, however, a very good eating one.

Top lining is the commonest technique and may be either shore or boat fishing in shallower waters. The gamiest fishes are taken by this method and include Rainbow, Brook, Brown and even Lake Trout (in the colder months) as well as Kokanee.

Current regulations allow only forage fish native to the lake to be used as bait, to eliminate the possibility of undesirable rough fish becoming established in Tahoe.

(4) DONNER LAKE

This body of water is a medium-sized natural lake 2.7 miles long with an average, nearly uniform width of one-half mile (0.7 mile at the widest point, China Cove). The lake's long axis is almost east and west, and east of China Cove, the remaining 0.7 mile of the lake narrows markedly to an average width of some 0.3 mile. It is located in Nevada County, California, about 12 airline miles west of the Nevada State line and some 35 miles by United States Highway 40 from Reno.

Maximum depth is 220 feet, the last 12 feet of which (representing its active capacity of 9,500 acre-feet) are controlled by a concrete dam completed in 1927. The dam is built on Donner Creek several hundred yards east of the lake. Maximum elevation of the lake is around 5,935 feet, surface area is 840 acres and volume is approximately 25,000 acre-feet.

The storage rights were purchased by the Truckee-Carson Irrigation District and the Sierra Pacific Power Company in 1943 and this water is withdrawn via Donner Creek to augment supplies in the Truckee River for use downstream. The water, by agreement, is subject to only a limited drawdown before September 15, and is used during the winter months for power production on the Truckee River. Ultimate storage is in Lahontan Reservoir for irrigation and power purposes.

(a) Limnology

Donner Lake is biologically moderately rich, but little investigative work has been done on it. Unpublished records of work done by California Division of Fish and Game biologists show some data principally instigated for Kokanee research. Available California records plus those of the author indicate the usual thermocline development during the warm summer period with oxygen exhaustion in the hypolimnial depths. In September a moderate thermocline was present.

Through the succeeding three months the Fall overturn slowly evened out the temperature differential from top-to-bottom, until by January the water was thoroughly mixed and showed the same temperature and oxygen content throughout its column. In a typical winter, ice and snow cover the lake, when it is to be expected that the deeper waters again stagnate with respect to oxygen.

The water varies from pH 6.6 near the bottom to 7.5 at the top, lacks carbonates and shows a bicarbonate content varying from 22 to 32 parts per million. Considerable water turbidity was indicated by a Secchi disc reading of 34 feet.

(b) Biota

Our limited sampling of the food elements in Donner Lake showed an abundance of basically important algae, particularly several genera of diatoms. Of these, the commonest were *Asterionella formosa, Melosira granulata,* and *Surirella splendida.* Undoubtedly the summer flora is richer.

The freshwater mussel, *Margaritifera margaritifera,* was abundant

FIG. 47. Donner Lake, California, looking east.
Infra-red photo by Ernest Mack.

and particularly noticeable in the shallow muddy bottoms of the east end. The introduced Pacific Crawfish, *Pacifastacus leniusculus,* was also common. Remains of rotifers and copepods were found in our plankton net samples as well as monaxon spicules of a freshwater sponge. An extensive aquatic insect fauna is undoubtedly present, but was not too apparent at this time of the year (Fall).

Among vertebrates, all typical Lahontan fishes are found, as well as several introduced salmonids. The earliest game fish planted was the Lake or Mackinaw Trout which was brought in to establish a fishable population in 1895. Later, Brook, Brown and Rainbow Trout were successfully imported. The latest addition was the Kokanee Red Salmon which was first brought in in 1944. Planting for these and the rainbow has continued up to the present writing.

(c) Management

It might be said that interest in Donner Lake by professional biologists during the past decade (1950–1960) has been mainly focused on experiments dealing with kokanee possibilities for these waters (Curtis and Fraser 1948, Kimsey 1951 and 1955).

The major management problems concern the periodic drawdowns which have, on occasion, so exposed the shallows where spawning takes place as to virtually eliminate natural reproduction of kokanee for the year. Also, certain proportions of kokanee drop down into Donner Creek and then cannot return to the lake because of the dam. In the winter of 1948, Thomas Trelease and the writer found this creek below the dam choked with spawning kokanee trying unsuccessfully to get upstream into the lake.

As in other similarly sized bodies of water, a continued Rainbow Trout stocking program seems to be standard procedure in order to maintain fishable populations under pressure.

(5) INDEPENDENCE LAKE

Two-and-a-half miles long and one-half mile wide, Independence Lake lies barely east of the Sierran divide (like Donner and Webber Lakes). Its maximum elevation is 6,949 feet, and it contributes its overflow waters eastward to the Lahontan system via Independence Creek, the Little Truckee River and the Truckee River. Its general strike or longitudinal axis is southwest-northeast with Independence Creek flowing almost due north to the Little Truckee.

This beautiful High Sierra lake is almost cut in two during low water by a rock reef extending from both shores at almost the exact center of the lake. The slightly narrower southwestern portion is the deepest of the two sub-basins, with a maximum sounding of 130 feet. The northeastern section is 115 feet deep. Maximum surface area is 710 acres, greatest volume is 17,300 acre-feet and total watershed area is something like 8.5 square miles.

Independence Lake is located askew of the Nevada County-Sierra County line in California, 15 airline miles west of Nevada. From Reno it is most conveniently reached via United States Highway 40 west to Truckee (32 miles), California State Highway 89 northwest for 14 miles to the turnoff. The lake is 4 miles farther south over a mediocre dirt road.

The lake became the property of the Sierra Pacific Power Company in 1930, acquired from the Hobart Mills Lumber Company. Ten years later the power company renovated the dirt fill outlet structure so that the water level could be dropped 28 feet. We could not determine how much these and earlier efforts to control outflow had raised the water level, but from large, waterworn tree trunks protruding from 60 feet of water on the southeast edge it was quite evident that the lake had been much lower for long periods of time. When these trees were alive, Independence was obviously two small, adjacent lakes divided by a thin, rock reef.

(a) Limnology

Our limited investigation of the lake showed it to be about neutral in pH, varying from 7.5 in the upper areas to 6.7 near the bottom. Shallower than Donner, Independence possibly has an ephemeral thermocline development during some summers. This is demonstrated by comparative measurements made between the two bodies of water in the Fall of 1959, during which Independence Lake showed complete temperature homogeneity by December (42° F.). This was a drop, on the surface, of ten degrees, from October, and occurred a month earlier than in Donner Lake.

However, thermocline development in the preceding summer months is indicated by surface and bottom figures of 7.6 and 0.6 parts per million of oxygen in November. The fact that a thermocline may not be expected in average years is suggested by testimony of caretakers that 1959 was an exceptionally calm and windless season and that usually the lake basin is quite windy the year around.

Carbonates are lacking, and bicarbonates low, the latter varying slightly around 25 parts per million. The everpresent ice and snow cover during winter undoubtedly produces bottom stagnancy comparable to that in calm summers. A Secchi disc reading of 36 feet in October indicates moderate turbidity and plankton development.

(b) Biota

In spite of the cold water, larger plants such as *Potamogeton, Ranunculus* and *Chara* were found, and, in most cases, plentifully along the margins. Planktonic flora included twelve genera of algae, of which *Melosira distans* was the commonest species. *Daphnia* was the only identifiable microcrustacean, while the introduced crawfish, *Pacifastacus leniusculus*, was plentiful in the shallow waters of the northeast end. It was too late in the season to sample for the common aquatic insects which must occupy the lake.

No fishes were collected, but it is presumed that the common Lahontan basin species are undoubtedly present. California Division of Fish and Game stocking records for the preceding decade indicate that substantial numbers of Rainbow and Cutthroat Trout have been planted, as well as Kokanee Red Salmon. The latter were spawning in the outlet creek at the time of our last visit, 12 December 1959. Brook Trout and Lahontan Redshiners are also known to be a part of the fauna. The original popularity of the lake as a summer fishing spot was due entirely to the native Lahontan Cutthroat Trout which it harbored.

(c) Management

While Independence Lake is locally considered to be a rather heavily-fished water, its private ownership by the power company and its moderate isolation militate against any unrestricted usage in the sense of a drain on its game fishes. The fishes it contains appear to be in excellent condition and while the considerable drawdown of its winter waters may occasionally adversely affect the kokanee spawning areas, these latter seem to be maintaining themselves satisfactorily.

The only practical management procedures which are seemingly necessary here would be those of occasional stocking of the lake with the most generally popular game fish—the Rainbow Trout. However, it would be more desirable from the viewpoint of a biologist, to attempt to maintain and increase any populations of Lahontan Cutthroat Trout which are still surviving there. This is particularly important in the face of the threatened extinction of the native Cutthroat Trout elsewhere in the Great Basin. Needham and Gard's studies of Cutthroat Trout (1959) have indicated to them that the population of these fishes in Independence Lake seem indisputably to be the Lahontan strain.

(6) WEBBER LAKE

Webber Lake is 30 airline miles due west of Reno and 22 such miles west of the Nevada State line. By road, the distance is 55 miles, following the same route noted for Independence Lake to the turnoff from Highway 89. From this point, Webber Lake is 9 miles due west by good dirt road.

This is the smallest and most symmetrical of the three middle High Sierra lakes we have included in this report (Donner, Independence and Webber). This very striking body of water is about 0.7 mile in diameter, nearly circular and has a maximum high water depth of 64 feet. We found an apparent maximum annual fluctuation in water level of four feet.

Located in Sierra County, California, at a maximum elevation of 6,773 feet, Webber Lake has a surface area of 250 acres and a volume of 2,250 acre-feet. Its main affluent is small Lacey Creek, while its intermittent effluent—Webber Creek—flows eastward one-and-a-half miles to a confluence with the upper Little Truckee River. In Fall, only seepage leaves the lake.

Beginning as Little Truckee Lake, its name changed about 1860 after it had passed into the hands of Dr. David G. Webber in 1852. For the past 50 years or so it has been the property of Mr. W. H. Johnson, and for about the last quarter century was operated by his son and daughter-in-law, the Vernon Johnsons. Mrs. Johnson died during our Fall 1959 investigations of the lake, and its future is uncertain at the present time.

(a) Limnology

This rather shallow high altitude lake varies in temperature from its ice and snow cover in the winter to probably upwards of 60° F. for its surface waters in the summer. Our first temperature reading in October 1959 showed a surface figure of 52° F. and a bottom reading of 45° F. Our last findings in December showed 38° F. and 37° F., respectively, or an almost homogeneous condition with respect to both temperature and oxygen. In October, bottom stagnation prevailed and oxygen there was essentially zero parts per million while 7.0 ppm existed in the surface waters.

Carbonates are absent and bicarbonates sparse, varying from 18 to 33 parts per million during our investigation. Hydrogen-ion concentration (pH) seems slightly on the alkaline side, reading from 6.8 to 7.2, with the latter figure prevailing from top to bottom in the late fall and winter. It is doubtful if anything but a brief thermocline develops and then only during very calm summers. Turbidity is high, our Secchi disc reading 10 feet in October. The bottom, in shoal and deeps, is thick mud.

(b) Biota

Webber Lake, in spite of its high altitude and long winters, has many of the aspects of a shallow pond and is correspondingly fertile. Our limited samplings showed that basic food chain elements were plentiful. Several types of algae, chiefly diatoms, were identified of which *Fragilaria* was the commonest, with *Staurastrum* and *Pandorina* both abundant. Rotifers and microcrustacea (*Daphnia pulex*) were similarly present as were such higher plants as *Potamogeton epihydrus* in deeper waters and *Elodea canadensis* about the shallows.

Aquatic beetles and mayflies were much in evidence, as were freshwater Malacostraca (*Gammarus*). Tendipedid fly larvae were common in the bottom muds. These undoubtedly represent but a part of the potential fish foods of the lake.

Because of the steep cañon with its falls between Webber Lake and the Little Truckee River there undoubtedly were no fishes native to the lake. Webber early stocked his lake with various trout (1860), and Lahontan Redshiners and Lahontan Speckle Dace are now common. At one time or another, Brook, Brown, Cutthroat and Rainbow Trout have been planted here. For a period of some 25 years (1930 to 1955), the California Division of Fish and Game regularly stocked the lake with Brook, Brown and Rainbow Trout of which the first and last are most commonly caught now. They discontinued plantings because of problems involving free access by fishermen.

(c) Management

Webber Lake is an ideal fishing area in some respects. Certainly trout proliferate here in the nutrient-rich and cold waters. However, because of its high altitude, the lake is open for fishing only for about 4 months during the year. It is too small to have any value as a source of irrigation or power water, so only ineffectual attempts have been made to control its output. It has been exploited entirely as a resort and fishing spot and in the last couple of decades the only real emphasis has been on fishing. The owners did not or could not develop it, probably due both to its relative isolation and the short profit season.

The owners charge boating and beaching fees, and controversy between them and the California Division of Fish and Game finally resulted in cessation of trout planting by that agency in 1955. Rainbow and Brook Trout seem to be maintaining themselves satisfactorily at the present time, although it is difficult to find spots around the lake margins that would be suitable for spawning. Farther up Lacey Creek, however, there seems to be good spawning areas.

This, with Independence Lake, is another body of water the writer would like to see set aside for the vanishing Lahontan Cutthroat Trout. Other popular species of trout, particularly the rainbow, can be sought in any number of more accessible places and the usual suitable streams and lakes will always be stocked with them. A strong and, I think, compelling attraction could be built around these lakes by promoting them as strictly Cutthroat Trout waters and as such they would serve as a guarantee for future generations of sportsmen and biologists, teachers and conservationists, that the very distinctive Lahontan Cutthroat Trout would not go the way of the Passenger Pigeon.

(7) TOPAZ LAKE
(Walker River Irrigation District)

This moderate-sized reservoir is located in southern Douglas County, projecting slightly across the California line into northern Mono County, some 65 miles south of Reno on United States Highway 395. Formerly, a smaller amount of water, referred to locally as "Alkali Lake," occupied this basin. The larger, present reservoir was formed by leading water in from the West Walker River in 1921–1922. The resultant 45,000 acre-feet capacity was raised to nearly 60,000 acre-feet in 1937 with an earthfill levee at the south end. Altitude at maximum elevation is about 5,000 feet. Our greatest depth is in the neighborhood of 80 feet.

(a) Limnology

Topaz Lake is, rather unusually for a reservoir, quite rich in nutrients and plankton, in spite of its fluctuations—perhaps this is a consequence of overlaying reservoir waters on smaller volumes of natural waters. Its limnology, which should prove to be of more than passing interest when fully known, has only been partially investigated. In recent years the Nevada Fish and Game Commission has been working Topaz along with its Lake Tahoe project,[16*] and accumulating valuable information for the biologist and fisherman.

As with all reservoirs, the standard major limnological characteristics are quite variable. Temperatures range from an ice cover in colder winters to surface readings of around 70° F. in summer. Oxygen is well mixed and in good supply, running as high as 11.0 parts per million and over in the upper waters. Water reaction is weakly to moderately alkaline with pH readings varying from 7.8 to 8.9.

(b) Biota

The usual Lahontan system fishes are present in Topaz, and the lake has been further rather unique in still having the original game fish of the system—Cutthroat Trout—as the most important fishery, at least up until 1956, when heavy plantings of Rainbow Trout by both Nevada

---

[16*]Ray Corlett, Fishery Biologist, Dingell-Johnson Project No. F-4-R—"Lake Tahoe" project, 1954–1957.
Norman Wood, Fishery Biologist, 1957–1958.

and California seem to have reversed the situation. Something like 75 percent of the fisherman's catch in previous years has been cutthroats. Rainbow and Brown Trout made up the remainder of the catch, in that order. Kokanee Red Salmon have been recently introduced (Pister 1961), but the success of the operation is not yet apparent.

The commonest rough fish are the introduced carp and the native Lahontan Tui Chub and Tahoe Sucker—these exist in much greater populations than do the game fishes, which is usually the case.

The smaller forms of life—insects and their relatives and plankton—are abundant, forming a substantial food source for the heavy fish populations. External parasitization of trout by the "Anchor Worm" (*Lernia* sp.) is very common during warm summer months.

(c) Management

Since nothing like a complete drawdown is currently possible for Topaz,[17] the heavy rough fish population is an increasing problem as more fishing demands are made on the lake. Also, natural spawning areas for trout are decidedly limited and natural reproduction does not seem to be able to keep pace with heavy fishing. Although Cutthroat and Rainbow Trout hybridize in the lake (producing sterile, unreproductive offspring), the former seems best suited for its waters, but artificial stocking must be continued in order to maintain high fishing yields.

At first thought it might seem feasible to poison Topaz at one of its low stages to remove rough fish so that re-planted trout could proliferate with less crowding and competition. This would undoubtedly result in better temporary fishing in the few following years (as it did in Twin Lakes, Mono County, California some years ago) but undesirable fish come back in a short while and the situation rapidly returns to "normal." Besides, the question of how undesirable these rough fish are has never been satisfactorily settled. Certainly large populations of trout should have other sources of food than themselves to maintain suitable numbers. Trout are natural predators among fish, and the best balance in nature is obtained when the old classical picture of a food chain can be realized—vegetation (basic food) to herbivore (sucker-chub) to predator (trout).[18]

Stocking management presents no problems, merely the routine of rearing sufficient quantities of desired fishes and getting them into the lake in proportions which will maintain good fishing.

(8) LAHONTAN RESERVOIR
(Truckee-Carson Irrigation District)

This sizeable reservoir is part of the "Newlands Project," the first Federal reclamation effort of its kind in the country, and is located some 15 miles west of Fallon. Carson River water flows directly into

---
[17] Although pumping stations to do this have been seriously considered.
[18] Trout are more versatile, however, than this might indicate, since they are capable of living off the microscopic plant and animal life (plankton) to a considerable extent.

the reservoir, and Truckee River water is diverted into it from an adjoining northwesterly basin. Reservoir waters are used for power and irrigation.

Work began on the project around the turn of the century (1903), the Derby diversion dam was completed in 1905, and Lahontan Reservoir began to operate in 1915. Situated in Churchill County, the 1,325-foot long and 120-foot high concrete and fill dam was designed to back up some 293,000 acre-feet of water with a surface acreage of nearly 11,000. At maximum levels, the reservoir is about 17 miles long, over 100 feet deep at its lower end and backs up tortuously with more than

FIG. 48. Lahontan Dam and Reservoir, Churchill and Lyon Counties looking southwest. The dark ribbon is Derby Canal, leading Truckee River watershed runoff into the reservoir for use in the Carson River watershed. Courtesy Nevada Highway Department and Adrian Atwater.

half its length (but not half its volume) in adjacent Lyon County to the west. Maximum elevation is a little over 4,100 feet.

Lahontan waters drain easterly and divide variously, as they follow the original Carson River channels, into South Carson and North Carson areas of Lahontan Valley, that in North Carson containing the Stillwater Marshes.

(a) Limnology

As would be anticipated in a fluctuating reservoir, all of the key limnologic characteristics change with water level and season.[19] The

---

[19]Data mostly from Nevada Fish and Game Reports, Lahontan Project, and the writer.

waters seem always to be alkaline, varying from slightly so (pH 7.6) to strongly (pH 9.4) alkaline, although upper extremes appear to be more generally in the 8.0 range. Carbonates vary widely around a 12 parts per million average, and bicarbonates around a 125 ppm point. Oxygen concentrations also show much variation from good aeration during the inclement winters (8.0 to 9.0 ppm) to 1.5 ppm in late summer when stagnation effects are maximum. Lahontan has sufficient depth to show thermal and possibly chemical stratification when left relatively undisturbed by drawdowns and winds. Turbidity increases from summer to winter, clearing progressively during the remainder of the year. Runoff sediments from its affluents plus cyclic plankton growth appear to be most important here as causative agents.

(b) Biota

In addition to the expected Lahontan system fishes, the reservoir has populations of the introduced warm-water species which have established themselves so well in west-central Nevada and which account for a high percentage of successful angling in this area. Angling in Lahontan, however, has only been fair.

These warm-water fishes include Yellow Perch, Sacramento Perch, Largemouth Blackbass, White Crappie, Black Crappie, White Catfish and Brown Bullhead. Channel Catfish have been stocked (1956) but to date there is no indication of successful spawning for this species. There is reason to believe, however, that they will succeed here. The two trout—Brown and Rainbow—do moderately well in Lahontan.

The shore aspect is the bleak one of typical reservoirs, with little vegetative cover. Willows and cocklebur seem to offer some fingerling protection, largely because of the exclusion of other plants although Smartweed (*Polygonum amphibium stipulaceum*) grows submerged in the upper reaches of the reservoir.

Production of larger invertebrates (arthropods) is characteristically not great with the exception of the crawfish (*Pacifastacus leniusculus*), an introduced species, although the usual kinds of aquatic insects common to this area can be found there.

(c) Management

Work by Nevada Fish and Game personnel[20] on Lahontan has been intensive enough to show the problems of the reservoir and suggest possible solutions. In the main, these waters seem to be somewhat more suitable for certain of the so-called "warm water" or "pan" fishes of the Mid-West than they are for such cold-water types as trout, although there is some habitat overlap. The major problems seem to be:

"Have all possibilities regarding the most suitable species of fishes for Lahontan been reasonably exhausted?"

At the present time, these are being explored systematically and a several-year experimental program for different species of trout (the "cold-water" fishery) will undoubtedly be initiated to answer this

---

[20]Fishery Biologists Robert C. Sumner and V. Kay Johnson, Dingell-Johnson project No. F–3–R, 1954 to present.

phase of the question. If this fails, the Commission plans to investigate certain "cool-water" species such as the Mid-Western Walleye Pike (*Stizostedion vitreum* (Mitchill) 1818 (Percidae), which might team up with the Yellow Perch already established in the reservoir to produce something like the very good fishery which exists in the Mid-West for these species.

Finally, failing this, attention will be turned to improving to a maximum the "warm-water" fishery already in the reservoir.

The basic reason for such a program, which may seem like hitching the horse backwards to the buggy, is founded on the strong feeling for a trout fishery constantly expressed by sportsmen of the area.

The usual minor problem exists here, principally that of the almost

FIG. 49. Stillwater Marsh, Churchill County. A representative view.
Courtesy Robert Sumner, 1961.

routine one of control of rough or undesirable fishes. Because of periodic drawdowns, it will be possible to occasionally eliminate undesirables and restock with the most preferred or successful game fish. In the summer of 1959, White Crappie fishing rapidly rose to a position of dominance.

(9) STILLWATER MARSH
(Stillwater Wildlife Management Area)

The Stillwater Wildlife Management Area came into being in 1948 when—through the agency of the Nevada Fish and Game Commission —three conservation groups banded together to save the marshes from extinction as a waterfowl habitat. The State Commission, the United

States Fish and Wildlife Service (now the Bureau of Sport Fisheries and Wildlife) and the Truckee-Carson Irrigation District reached an agreement in that year which insured an adequate amount of water to maintain the area, and began work on it the following year.

Much of the 200,000 acres of the Area is alkali desert, its two main water sections—Stillwater Marsh and Indian Lakes—comprising but a minor part. Due to the type of land involved, and its pulsating water supply, it is not easy to arrive at a general figure. However, nearly 25,000 acres of water seem to exist currently in the Marsh, which has an average depth of slightly less than two feet. Overall management plans for the Area call for continued development of marshes so the just-quoted figures can be expected to increase.

Water for Stillwater comes from excess irrigation flows. As a consequence, although it remains today, as it always has been, part of the sump for Carson River, Stillwater receives most of its inflow in summer rather than spring, with corresponding habitat alterations.

Stillwater lies in the northeastern part of Churchill County's impressive Lahontan Valley, adjacent to Fallon. Its extreme southeastern section is a Federal Wildlife Refuge, while around it are shooting and fishing grounds, public and private.

(a) Limnology

Water temperatures in Stillwater Marsh show marked fluctuations from an ice cover in winter to highs in excess of 90° F. in shallow summer waters. Generally speaking, there does not seem to be much top-to-bottom variation, as would be expected in shallow waters. Alkalinity is not too pronounced, varying around an 8.0 pH average.

Carbonates are low (0 to a rare 40 parts per million), with bicarbonates fluctuating widely around a 300 ppm point. Oxygen figures also change greatly with the seasons, running in excess of 10 ppm down to zero under the ice cover in certain sections. Little thermal stratification is evident.

Turbidity likewise varies, from lacking in colder winter waters to being quite pronounced in spring and summer inflow and plankton-growing seasons.

(b) Biota

Stillwater Marsh has common native Lahontan drainage fishes such as redshiners and chubs, as well as large numbers of introduced carp and mosquitofish. However, it is attracting warm-water fishermen as one of the comparatively few waters in the State which are ideally suited to blackbass and their relatives.

Currently, fishing is variably good in different parts of the Marsh for most of these fish—Largemouth Blackbass, Black and Brown Bullheads, Bluegill Sunfish, Yellow Perch, Sacramento Perch and White Catfish. Commercial carp seining seems to be sporadically worthwhile— Nevada Fish and Game records show nearly 300,000 pounds of carp as coming recently from the activities of one seiner here in a little more than a year's time.

Stillwater's warm environment is a rich one, from the standpoint of organisms in the fish food chain, and these are prolific. Plankton are noticeably common during warmer months, showing the abundance of dissolved and colloidal nutrients in the water. Insects and some other

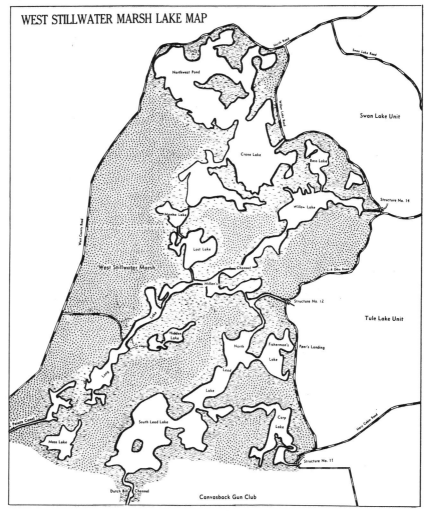

Fig. 50.

aquatic invertebrates are characteristically a dominant part of the fauna.

A wide variety of plants grow here, of which the commonest are: Sago Pondweed (*Potamogeton pectinatus*), Narrowleaf Cattail (*Typha angustifolia*), Common Spikerush (*Heleocharis palustris*), Common Bulrush (*Scirpus americanus*), Hardstem Bulrush (*Scirpus acutus*) and Coontail (*Ceratophyllum demersum*).

(c) Management

Productivity problems at Stillwater Marsh are rather complex, both administratively and biologically, but have been effectively tackled by all concerned and the results have been gratifying.

The fishery began more-or-less incidentally to the waterfowl program, for which latter the Area largely owes its existence and present status. However, this does not mean that fishery problems are neglected. Actually, the State Fish and Game Commission has been interested for the last decade in improving fishing yields from Stillwater and has had one-to-several biologists working on these problems.[21]

The main management problem revolves around the oftentimes uncertain water supply and consequent undesirable drying of parts of the Marsh. The best solution for such a situation calls for maximum co-operation between water users and Area managers. Within uncontrollable climatic limits, the outlook is promising enough that considerable expansion of the Marsh area can be contemplated as part of future plans. If some areas dry down temporarily, they can be rather easily replenished when filled again.

Minor problems such as need to control over-abundant rough fish like carp are a usual part of the picture.

(10) RYE PATCH RESERVOIR
(Pershing County Water Conservation District)

This is an extensive irrigation reservoir some 23 miles north of Lovelock on the Humboldt River, Pershing County. United States Bureau of Reclamation work began on the 800-foot long earthfill dam in 1934 and was completed in 1936. Eighty feet in maximum height, the dam was built to back up a total of 179,000 acre-feet of water with nearly 11,000 surface acres, but it has held more than this on occasion. The reservoir, at full capacity, extends some 22 miles up the Humboldt Valley from its dam and presents an exceptionally narrow aspect, with numerous little fingers flooding the many draws and coulees in this dissected area.

Maximum depth is around 60 feet and is relatively rapidly decreasing due to silting. The Humboldt River is the only affluent, and the reservoir provides badly needed storage water on the lower Humboldt, where it feeds the rich Lovelock farming region. The entire potential of the Humboldt drainage system can be focused on this reservoir—parts of Humboldt, Elko, Eureka and Lander Counties contribute to it. Elevation in the vicinity of the reservoir is approximately 4,100 feet.

(a) Limnology

Surface temperatures vary from nearly 80° F. in the summer to firm ice covers in the colder winters. Alkalinity is moderate (pH 8.4), with some variation when water levels move up and down. Averages for

---
[21] Robert C. Sumner, 1954 to present.
Donald King, 1954–1956.
V. Kay Johnson, 1954–1955.

carbonates seem to be somewhat less than 20 parts per million, and for bicarbonates around 250 ppm. Turbidity is high, as mentioned, because of pronounced silting, and oxygen concentrations are generally high.

Since Rye Patch is primarily an irrigation reservoir, its drawdown is often considerable. In recent years—1954, 1955 and 1961—the reservoir has been completely emptied. The effects were disastrous on fish life, but could be regarded as beneficial in another sense because it removed the large populations of rough fish and made it possible for the Fish and Game Commission to replant with desired game fishes.

FIG. 51. Rye Patch Dam and Reservoir, Pershing County, looking north. Lake Lahontan sediments, through which the Humboldt River is downcutting, are clearly visible. Courtesy Nevada Highway Department and Adrian Atwater.

(b) Biota

Since the essentially total kills of 1954–1955, the Commission has restocked the reservoir with Rainbow Trout, Channel Catfish, Brown Bullheads, Largemouth Blackbass and White Crappie. Bluegill, Yellow Perch, Sacramento Perch and Brown Trout were all fishes common to Rye Patch prior to 1954. At the present time they are definitely rebuilding their populations again.

These waters are relatively rich in microscopic animals and plants which make up the basic food chains for fishes—plankton—and have substantial amounts of larger intermediate food types such as insects. The common lower Lahontan area fishes are present.

(c) Management

The usual reservoir management problems exist in Rye Patch, with those attendant on complete drawdowns being most important at the present time. Very likely, those of 1954–1955–1961 will not be the last and in order to realize to the fullest the fishery potential of the reservoir, the State will have to be prepared to continue heavy stockings after drawdowns. The suitability of Rye Patch waters for warm-water fishes of the type which do so well there now is unquestioned, although sportsmen pressure will probably always exist to some extent for trout plantings. It is unlikely that any spectacularly adapted "miracle" fish can be found for such waters beyond those which already occupy it—although there are one or two other possibilities worth future consideration. In the summer of 1959, the recently introduced White Crappie furnished the bulk of the fishermen's take.

(11) WASHOE LAKE

This natural body of water occupies most of the bottomland of Washoe Valley today, 20 miles south of Reno on United States Highway 395. In years past, it has dried up completely.[22] When restricted in size, it is divided into "Big Washoe" in the south and middle, and "Little Washoe" at the extreme north end. Drainage is to the north, from Big to Little Washoe, thence through Steamboat Creek to the Truckee River at the east edge of the Truckee Meadows (Reno). While Washoe is a natural lake, it is in the nature of a semi-permanent playa, and may be pulled down to some extent for irrigational purposes.

At its greatest extent, Washoe Lake contains nearly 5,000 surface acres of water, has a maximum depth of about 10 feet and a very symmetrical, sand-ridged shoreline on the southeast. That it is much shallower than this over most of its expanse is shown by the continually roiled and opaque appearance from fine silts and organic matter kept in suspension by the blustery winds of the region.

---

[22]The last periods of total dryness for Washoe Lake were the years 1931 and 1933. The Reno Evening Gazette for September 22, 1933 carried the following item:

"WASHOE LAKE IS DRY SECOND TIME IN YEARS

"For the second time in a quarter of a century, Washoe Lake is dry. Nothing but fine sand and silt cover the lake bed and when one of the famous Washoe zephyrs sweeps across the valley, it carries huge clouds of dust from the lake bed.

"The lake, which is nearly four miles long and two miles wide, started shrinking early in the season, as the snowfall in the surrounding mountains was light, and the creeks began to dwindle early this summer. At the present time, there is not even a muddy patch anywhere on the lake bed.

"Two years ago, when there was a light winter, the lake dried up, but it was for the first time in at least 25 years, according to Leo Sauer, Washoe Valley rancher.

"Despite the drying up of the lake two years ago, there were plenty of catfish in the lake when it filled up during winter and spring. These fish burrow into the soft mud as the water begins to recede, and when a hard crust forms over the mud, this prevents the mud beneath from drying out."

(a) Limnology

Certain of the limnologic features of the lake would be expected to change with its depth and extent, since it receives fresh waters from mountains to the west (principally Ophir and Franktown Creeks) and freshens by discharge to the north. However, as would be apparent from the heavy sediments carried by the water, dissolved solids must be relatively high although these have never been investigated. Hutchinson (1937: 85) lists the results of a cursory chemical investigation, comparing Big and Little Washoe Lakes, there being less water in the valley in his day. Alkalinity varies, Hutchinson recording a pH of 8.9 while the writer has found it as low as 8.3, when the lake was higher.

FIG. 52. Washoe Lake in winter, from Slide Mountain, looking southeast. Photo by Ernest Mack.

Carbonates and bicarbonates also oscillate, averaging around 25 and 200 parts per million, respectively. Surface temperatures go from an ice cover in extremely cold winters to 80° F. in the summer. Oxygen content also changes considerably—usually wind action keeps ample amounts of oxygen in solution, but occasionally circumstances operate to reduce concentrations below those which will sustain fish, and large numbers die off. The writer has noted this on several occasions in the past 25 years, and State Fish and Game Commission studies in recent years have shown oxygen levels to be down during these events.

Several factors could operate here in differing combinations, to produce the effect. Warming of the water during spring and summer drives out oxygen; "blooming" of microscopic plants and animals (plankton) during this period further reduces oxygen; accumulation of organic debris on the bottom from small and large plants and animals takes oxygen out of solution. Long and tight snow-covered ice sheets would lower oxygen even in cold winter waters; and particularly balmy summers, with little wind action, would not mix the usual amounts of atmospheric gas with the water, etc.

(b) Biota

Fishes in the lake are the common Lahontan types, with the usual additions. There are substantial populations of Lahontan Tui Chub, Sacramento Perch, Yellow Perch, Asiatic Carp and Black Bullheads with carp far outnumbering everything else. Tahoe Suckers and Brown Trout were presumed to be present also, but a recent poisoning operation (see below) failed to produce them.

By far the commonest catch from Washoe Lake is the Black Bullhead. Although the original plants were supposedly Brown Bullheads, these appear not to have done too well, possibly because of the subsequent drying down of the lake.

(c) Management

The State Fish and Game Commission has been able to spend some time in recent years on the problems of Washoe Lake as part of the "Lahontan Project."[23]

At this writing (1962) information is available to show trends and suggest some management procedures. These would indicate that: (1) an attempt should be made to improve Brown Bullhead fishing since this species grows larger than the Black Bullhead; (2) some control should be practiced on the large numbers of undesirable or uncatchable fish, particularly chub and Sacramento Perch which utilize food resources but do not contribute much to the anglers' creels—other species might also be included here; and (3) establishment of some system of fish salvage.

Specifically the above could be accomplished by—

(1) A Brown Bullhead planting program, carried out to the point where it would be apparent which of the two species of bullhead, Brown or Black, is definitely best adapted to Washoe Lake waters. Obviously, if Blacks are going to be the overwhelmingly dominant species in any natural competition between the two, it would not be feasible under present conditions to indefinitely plant Browns unless Blacks could be completely excluded—not a very likely possibility.

(2) Since the lake teems with fish of little value to the sportsmen, the average person feels that these may be a considerable drain on the natural food supplies present. While the two mentioned above (Sacramento Perch and Lahontan Tui Chub) are not direct competitors of bullheads—feeding as they do above the bottom—the carp certainly is,

---

[23]Dingell-Johnson Project No. F-3-R, with Fishery Biologists Robert C. Sumner and V. Kay Johnson, 1955 to present.

and any of the carnivorous types of course would consume certain quantities of bullheads although such studies as have been made indicate that predation is not as great on catfish as we would ordinarily surmise. However, it must be pointed out here (as has been done elsewhere) that the food competition argument, although convenient and quick in its analysis and solution, is often fallacious.

Actually, the more fish living in a given body of water, the richer and more productive is that water or such numbers would not be present. Even carp, in direct competition with bullheads, represent large amounts of nutrient materials which break down, decay and dissolve upon the death of individuals and maintain the needed enrichment of the water for others. Only when competition is so severe that one species can virtually eliminate another does it generally need management. This is not the case here since Black Bullheads are abundant which they would not be if carp were restrictive.

To take a theoretical example, if nothing but bottom-feeding bullheads were present in the lake, very likely their population would not increase—in fact there is good experimental reason to believe they would decrease in numbers or become stunted.

(3) A fish salvage program is desirable since the lake is occasionally drawn down by irrigation needs, especially in dry periods. When this is necessary, the value of having a catch basin for fish at the outlet system is obvious—a great many specimens could be conveniently obtained for planting where needed elsewhere.

Poisoning of the lake was recently completed by the Fish and Game Commission (19–20 November 1960) preparatory to re-stocking. By rough estimate, carp outnumbered all other species by at least 10-to-1.

(12) BRIDGEPORT RESERVOIR
(Walker River Irrigation District)[24]

This impoundment is smaller than Topaz Lake, holding a maximum of around 53,500 acre-feet of water behind an earthfill dam, which was built in 1923–1924. The dam is 4 miles north of Bridgeport, Mono County, California, on the East Walker River and is located just off United States Highway 395 north of Mono Lake. The elevation is a little over 6,450 feet.

Surface area is 2,500 acres with about 30 feet as maximum depth. The reservoir is some 4.5 miles long and 2 miles wide and is fed by the outflow water from Twin Lakes about 15 miles upstream.

(a) Limnology

Like Topaz, Bridgeport Reservoir warms up considerably during the summer and usually supports heavy blooms of plankton at this season.

---

[24] I am indebted to Mr. E. P. Pister, California Department of Fish and Game, Fishery Biologist III, stationed at Bishop, Mono County, California, for much of the following information on Bridgeport Reservoir and Twin Lakes.

California biologists have noted occasional mass fish mortality in the summer, presumably indicating an oxygen stagnation period. Annual temperatures vary from the winter ice cover to more than 75° F., with the development of a thermocline in late summer.

Hydrogen-ion concentration (pH) has been recorded from 6.8 to 7.0, showing slight acidic tendencies. Surface oxygen readings have gone as high as nearly 10 parts per million in early spring.

(b) Biota

The fish fauna is that of the Lahontan system, with the usual modifications. Asiatic Carp and the three trout—Rainbow, Brook and Brown—are well established as introductions. See the listing for Twin Lakes.

(c) Management

Bridgeport Reservoir is essentially a Rainbow Trout water, and the management program is geared largely to this species. Pister (1958) writes:

"Present fisheries are sustained by our catchable rainbow trout program and, in the case of Bridgeport Reservoir, large numbers of brown trout fingerlings. Regular creel checks of these waters indicate that the catch comprises almost entirely rainbow trout, with an occasional eastern brook and brown trout being taken. The catch in Bridgeport Reservoir also is predominantly rainbow trout. However, owing to the planting of brown trout fingerlings, the percentage of brown trout in the catch exceeds that for Twin Lakes."

Speaking now of current planting allotments:

"Bridgeport Reservoir receives about 80,000 rainbow trout of the same general size annually [five per pound average] plus the aforementioned brown trout fingerlings. Angling . . . ranges from good to excellent. Since the last chemical treatment of Bridgeport Reservoir in 1955, growth of planted rainbow trout has been excellent, and the reservoir now ranks among the finest fishing waters in California."

(13) TWIN LAKES

Upper Twin Lake can be taken as an example of both lakes. Vestal (1950) reported on this lake in some detail in the course of chemically treating it to remove the large rough fish populations—most of the following information is from that report.

Physically, Upper Twin Lake is a little over one and a half miles long and a bit less than half a mile wide. Its greatest depth is 112 feet, surface area is 265 acres and volume is 12,455 acre-feet. At maximum level, elevation is 7,096 feet. Lower Twin Lake is somewhat the larger of the two on all points, and both lakes have been increased in size by impoundments. They lie in Robinson Cañon, about 14 miles southwest of Bridgeport, Mono County, California.

(a) Limnology

Vestal's late summer (August) data of 1949 gave the following results from top-to-bottom in 95 feet of water:

(1) Oxygen varied from nearly 8 to less than 4 parts per million;

(2) Temperatures dropped from 65° F. to less than 50° F.;

(3) pH fluctuated about the 7.0 or neutral point, showing marked acidic readings (less than 7.0) in the depths;

(4) Bicarbonate alkalinity (methyl orange) is slight and variable, on this particular survey showing 17 parts per million at the surface and 24 ppm at the bottom.

FIG. 53. Upper Twin Lake, looking westerly. Twin Lakes lie at the headwaters of the East Fork of the Walker River. Photo by courtesy of E. P. Pister.

A thermocline was noted from about 25 to 65 feet in depth in midsummer and flushing action from Robinson Creek and other tributaries seems to keep the hypolimnion rather shallow. As in Bridgeport Reservoir, there are the two textbook overturns, fall and spring, and strong ice covers in the winter. Plankton development is relatively poor although algae are a common sight.

(b) Biota

Fishes are Lahontan drainage species with the addition of the sporting three—Rainbow, Brown and Brook Trout—to augment the original Lahontan Cutthroat. The complete listing is:

(A) SALMON AND TROUT FAMILY, Family *Salmonidae*
1. Kokanee Red Salmon (*Oncorhynchus nerka kennerlyi*)
2. Lahontan Cutthroat Trout (*Salmo clarki henshawi*)
3. Rainbow Trout (*Salmo gairdneri irideus*)
4. Brown Trout (*Salmo trutta*)
5. Brook Trout (*Salvelinus fontinalis*)

(B) WHITEFISH FAMILY, Family *Coregonidae*
6. Mountain Whitefish (*Prosopium williamsoni*)

(C) SUCKER FAMILY, Family *Catostomidae*
7. Tahoe Sucker (*Catostomus tahoensis*)
8. Lahontan Mountainsucker (*Pantosteus lahontan*)

(D) CARP AND MINNOW FAMILY, Family *Cyprinidae*
9. Lahontan Redshiner (*Richardsonius egregius*)
10. Lahontan Tui Chub (*Siphateles bicolor obesus*)
11. Lahontan Speckle Dace (*Rhinichthys osculus robustus*)

(E) SCULPIN FAMILY, Family *Cottidae*
12. Belding Sculpin (*Cottus beldingi*)

A considerable variety of aquatic invertebrates exists in the lakes, forming the basis for a substantial food chain.

(c) Management

Pister's comments on the present situation are much the same as for Bridgeport Reservoir, which see:

"Our present planting allotments for the Twin Lakes are about 55,000 rainbow trout each (averaging about five per pound) on an annual basis . . . Angling in the three waters ranges from good to excellent."

Vestal (1950) gives some of the historical highlights of the Twin Lakes fishery. There seems general agreement that the building of Bridgeport Dam in 1923 cut off the original sizeable runs of native Cutthroat Trout to Twin Lakes. With the introduction of other species of trout later on, the decline of the cutthroat was well on its way.

By the early 1930s, the native rough fish (chubs and suckers) dominated the fish fauna to such an extent that sport fish angling had become negligible. Stocking with rainbow produced only a momentary improvement in the situation, and eradication of the rough fish by treating the Twin Lakes with rotenone was decided upon. Although this treatment kills all the fish, game fish are then re-introduced.

Such a program almost always yields good results for a variable number of years afterward, but eventually the rough fish again build up in numbers and the process must be repeated, so it is no permanent cure.

In 1958, California initiated a kokanee program (Pister 1961) for Twin Lakes which has been highly successful. From that date to 1961, almost an even half million kokanee have been introduced to the system (497,355). By 1961, creel censuses showed "39 percent of the catch in the upper lake and 51 percent of the catch in the lower lake were kokanee. These fish average about ten inches in length.

"Thus far in 1961, one spawning run of kokanee has already occurred, and a second appears to be starting at this time" (October) (Pister 1961).

(14) BOCA RESERVOIR
(Washoe County Water Conservation District)

This is an impoundment on the Little Truckee River at its junction with the Truckee River in Nevada County, California, a short distance west of the Nevada State line. Work started on the earthfill dam in 1937 and was completed in 1942, when the district took over maintenance of the unit. The reservoir is a little more than two miles long and about three-quarters of a mile in maximum width, with its main axis north-and-south. At maximum load it contains slightly more than 40,000 acre-feet of water.

The level of Boca changes considerably, and its waters present the typical muddy appearance of reservoirs. Trout fishermen generally have but poor success here. The area is reached via a 27-mile drive westward on United States Highway 40 from Reno. Since Boca is capable of a complete drawdown, management can consist only of fish planting whenever feasible.

(15) PROSSER RESERVOIR
(United States Government)

Work began on this unit of the 52 million dollar Washoe Project in 1960. Located astride Prosser Creek, which contributes to the Truckee River about two miles west of Boca, the dam will be an earthfill one similar to that at Boca. It is planned to hold back 30,000 acre-feet, of which 10,000 will be storage with the remainder allocated to flood control and bolstering the water system below Prosser during drier parts of the year. At complete drawdown, 1,200 acre-feet will remain as a fish reservoir. Area of the reservoir at capacity will be 870 acres. An earlier concrete dam of limited size at the mouth of Prosser Creek has long since vanished.

(16) SUMMIT LAKE

Lying at an elevation of about 6,000 feet in high, broken country at the northwest edge of the Lahontan drainage system is mile-long, tortuous Summit Lake of western Humboldt County. To the north is the Alvord drainage, reaching into southern Oregon. The tiny pocket which comprises the drainage area of Summit Lake can be little more than 10 by 15 miles, with the lake occupying a relatively small depression toward the south edge.

The basin is largely set aside for Shoshone, Pahute and other Indians as the "Summit Lake Indian Reservation" which was established in 1913 and enlarged in 1928. An old military establishment, Camp McGarry, existed at Summit Lake until its abandonment in 1890. From this was derived the name for Soldier Meadows just south of Summit Lake, in the Lahontan drainage.

The main affluents are Mahogany and Snow Creeks flowing off the precipitous west slopes of Mahogany Mountain, more than 8,000 feet high. The morphology of Summit Lake is rather peculiar. It superficially appears to drain into Soldier Meadows to the south, until the effluent gorge at the southeast corner the investigator is following turns upward toward Mahogany Mountain. An incomplete examination of

FIG. 54. Summit Lake, Humboldt County, looking northerly. Feeder creeks come off the steep slopes to the right and the supposed drainage connections would be at the extreme right. Fort McGarry was at the extreme middle left. Photo by the author, 1950.

the terrain seems to indicate lava blockage of an original southerly outlet.

The Indians living here have some interesting stories, not only about the lake, but about the single species of fish it contains—the Cutthroat Trout. During a visit to the lake with Tom Trelease in 1950, it was evident from the reservoir-like, whitish, high-water line several feet above the surface that the water had been higher in the recent past. Our visit was early in the season—May 13—when the lake, if not rising, should at least have evidenced stability. We were told by an Indian informant that water had been steadily receding for the past few years and that the older Indians knew of the lake going dry in the past.

Mr. Richard B. Millin, formerly United States Indian Service Range Conservationist stationed at the Nevada Indian Agency, Stewart, Nevada, several years ago sent the author the following letter on this isolated little lake:

"In January 1952 at Ft. McDermitt, Nevada, Pete Snapp aged about 80, stated that when he was a young boy the older people told him that in earlier days there had been no lake there and that the waters from the creeks ran to a clump of willows at the bottom of the depression and disappeared into the ground. There were native trout of modest size in these streams. The road from Ft. McGarry ran westward across the bottom of the depression on dry ground.

"When he first remembers the area the road was covered with water but the lake was not nearly so high as it has been in recent decades. This was probably about 1885.

"He does not know what caused the basin to flood instead of to drain out through its former underground channel."

The Indians also say that the Cutthroat Trout now common in the lake (and which may be the only existing population of the original Lahontan Cutthroat Trout, *Salmo clarki henshawi*) was introduced many years ago, apparently around the turn of the century, from Pyramid Lake. Hubbs and Miller (1948B: 61) quote—from J. Ray Alcorn's data—about the same figure we were told, circa 40 years. This may even have coincided with the formal 1913 establishment of the reservation. At any rate, there is now a good run of Cutthroat Trout up Mahogany Creek each spring, which serves as a source of food and revenue for the Indians.

Our field tests on the 25-foot deep lake yielded the following limnologic information: pH 8.15; carbonates 11 parts per million; bicarbonates 242 ppm; chlorides present; sulfates, sulfides and iron absent; temperature—surface 52° F., bottom 45.5° F.; dissolved oxygen—surface 9.2 ppm, bottom 9.0 ppm; floating matter, *Myriophylum spicatum* and filamentous algae. The•water is reservoir-like in its opaque muddiness.

Management of this small but very interesting body of water is a function of the Indians and the Indian Service, but both groups have been highly co-operative in allowing the State Fish and Game Commission to get cutthroat spawn here every spring for the past several years—the first agreement to let the State take spawn was reached with the Summit Lake Indians in 1950. Years ago an abortive attempt was made to tunnel through the south ridge confining the lake in order to tap its waters for the irrigation of Soldier Meadows but fortunately this project was abandoned—Summit Lake does not represent enough stored water to be of importance in irrigating even a small parcel of land and would have quickly been ruined as a trout habitat. Summit Lake is particularly important to the Fishery Biologist and the sportsman in view of the fact that its Cutthroat Trout population probably represents the last definitely known remnant of the once more widespread and famous Lahontan Cutthroat—as such, it can furnish the nucleus for restocking prominent fishing spots like Pyramid and Walker Lakes.

(17) EAGLE LAKE

Eagle Lake, a desiccating, smaller counterpart of Pyramid Lake, lies about 35 airline, and some 80 road, miles northwest of Pyramid Lake in Lassen County, California. It is a narrow lake with winding outline, its long axis striking SSW–NNE, the northern portion being characterized by extensive shoal areas only a few feet in depth. Its total length is about 13 miles and widths vary from a maximum of some 4 miles to a minimum of ⅓ mile in Pelican Channel about midway of the lake.

It occupies a somewhat isolated position within the Lahontan drainage system proper in that there have been no recent, natural connections between Eagle Lake and the Lahontan system of which Pyramid Lake is a part. Eagle Lake lies in its own minor basin which in the Pleistocene contained a larger body of water which presumably, from its fish evidence, had intermittent surface flowage contact with Lahontan waters around it.

Feeder streams are sporadic at the present time, with considerable amounts of water being diverted for irrigation before reaching the lake. Pine Creek, the main affluent, enters on the west side of the tortuous shoreline, with smaller streams coming in at three other points. The drainage basin of the lake is some 500 square miles in area, and the lake lies at an elevation of about 5,100 feet with a surface area of some 15,000 acres. An irrigation tunnel which lowered the lake level some years ago is no longer in operation.

According to Kimsey's data of 1948 (published in 1954), the lake does not quite reach 60 feet in depth in its deeper, southern portion. It is alkaline in reaction, having a pH of from 8.4 to 9.6. The average total alkalinity, expressed as parts per million, is given as about 709.

Unlike Pyramid Lake, which never freezes in the winter, Eagle Lake regularly is covered with ice which may reach a foot and a half in thickness so that a pronounced winter stagnation period develops. During the summer, temperatures may rise as high as 70° F., with oxygen concentrations thinning beyond fish requirements below the 35-foot level. The long sweep of the winds usually effectively prevents the formation of a definite thermocline (Kimsey 1954).

In 1917, Snyder recorded the following fishes as native to the lake:

(A) SALMON AND TROUT FAMILY, Family *Salmonidae*
1. Eagle Lake Trout (*Salmo aquilarum*)

(B) SUCKER FAMILY, Family *Catostomidae*
2. Tahoe Sucker (*Catostomus tahoensis*)
3. Lahontan Mountainsucker (*Pantosteus lahontan*)

(C) CARP AND MINNOW FAMILY, Family *Cyprinidae*
4. Lahontan Tui Chub (*Siphateles bicolor obesus*)

The Eagle Lake Trout, usually cited now as *Salmo gairdneri aquilarum* to indicate the fact that it has obvious rainbow features, was described by Snyder as new at the time of his report. The taxonomic status of this trout has varied with the opinions of different investigators since Snyder's time. Hubbs and Miller (1948B) declared it most likely a hybrid between a native cutthroat (*Salmo clarki*) and an introduced rainbow (*Salmo gairdneri*). Miller later (1950B) expressed some doubt about this interpretation and left the matter in the air. From the few specimens the author has examined, he is inclined to accept the hybridization view from the physical evidence as well as from the standpoint of the geographics of cutthroat and rainbow distribution (see discussion concerning the "Royal Silver" and "Emerald" Trout under the rainbow series farther on).

Snyder had an additional fish on the above list, the cyprinid *"Leucidius pectinifer,"* which he had described in the same publication as a new genus and species from Pyramid Lake, but which has been known for some time to be a synonym of the Lahontan Tui Chub, *Siphateles bicolor obesus*.

At any rate, the trout in Eagle Lake has been close to extermination for many years, and Kimsey's survey of the lake—while primarily dealing with the life-history of the Tui Chub—was also aimed at determining possible ways to increase the numbers of the waning trout through a better overall understanding of its habitat.

An additional cyprinid species, the Lahontan Redshiner (*Richardsonius egregius*) apparently was not in Eagle Lake during Snyder's visit there, but was collected later by Hubbs and Miller (1948B). Other species have also been introduced.

(18) CARSON LAKE

This, the Big or North Carson Lake and the Humboldt Sink of earlier times, was investigated briefly by Hutchinson in the summer of 1933, but with noteworthy results as far as additions to our aquatic microfauna is concerned. Essentially, the "lake," when it has water in it over any extent, consists of a thin sheet over a finely muddy, alkaline bottom interspersed with slough-like aggregations of tules and rushes here-and-there. Hutchinson's stations were in 10–12 inches of water, which seemed to be the maximum depths. In Russell's time (circa 1882), the usual winter and spring inflow was sufficient to produce a lake about 25 by 15 miles in extent and only a few feet deep, which disappeared completely in drier years.

(19) BIG SODA LAKE

This is the most distinctive and unique body of water in the State—and although it contains no fishes, is of interest on other points. Big Soda Lake occupies the low, symmetrical crater of an extinct volcano and is some mile by three-quarters of a mile in size, with a shallow bay now breaking the symmetry to the south. At this writing (1962) the greatest depth is slightly over 200 feet.

Formerly, the lake was the site of a "soda" industry where sulfates and carbonates of sodium were recovered for the commercial market by selective evaporation. In addition to these valuable salts, sodium borate and magnesium carbonate existed in the waters in minute amounts; chlorides made up nearly three-fourths of the dissolved solids, that of sodium (common salt) far outweighing anything else. At the present time, the concentrations of salts in the lake have dropped considerably over what they were in Russell's day (1882)—and for many years afterward, up to slightly beyond the turn of the Twentieth Century—because of the constant rise of water level and consequent freshening of

FIG. 55. Big Soda Lake today, at its highest known level. The soda works shown in Fig. 56 are under 50–60 feet of water, and wide areas of the south basin are inundated. While the water has freshened considerably, it is still too alkaline and saline to support fish life, but is a favorite with waterfowl and certain invertebrates, among which are insects specialized to highly concentrated waters. Photo by the author, December 1961.

FIG. 56. Big Soda Lake yesterday. Same view as Fig. 55. This photograph was taken by the U. S. Bureau of Reclamation January 12, 1905, and shows the old soda works intact on the south-southeast shore. Photo No. 115–JQ–110B in the collection of the U. S. National Archives, by whose courtesy it is used here.

the lake from increased irrigation works in the vicinity. Since 1882 the level of Big Soda has risen about 60 feet, nearly to the point of overflow, most of this rise occurring after 1900.

A much smaller crater adjacent to the southwest—Little Soda Lake —has freshened sufficiently in the same length of time to support fish life. When the first emigrants passed through the Carson Desert, Little Soda seems to have been dry, or largely so, and it was later converted into a series of evaporation basins for the recovery of "soda."

Big Soda Lake exhibits many peculiarities to the limnologist, particularly in its chemical dynamics. It does not overflow, yet large amounts of electrolytes seem to be disappearing from the basin, presumably through the same underground channels which bring water into it. Hutchinson's calculations (1937), which provide only rough approximations due to the nature of his data, indicate that from about 90,000 to 175,000 short tons of salt (sodium chloride) may have been lost by underground water replacement between 1882 and 1933. That there might be sufficient flow along this hydraulic gradient to accomplish this is indicated by his conservative inflow figures of some 265,000 tons per day between 1905 and 1933.

The hydrogen-ion concentration appears to have been decreasing through the recent years, at least. Hutchinson's 1933 figure of pH 9.3, which is not in itself a lethal alkalinity for at least some fishes of our area, had increased to 9.7—10.0 by the writer's field observations with standard Hellige discs in 1951. Little Soda Lake was not investigated by Hutchinson, but in 1951 it had, in its surface waters, an alkalinity of pH 9.1; 6 percent of the carbonates of Big Soda; 60 percent of the bicarbonates; over 7 times the dissolved oxygen; and fish.

In addition to its high alkalinity, Big Soda also showed another peculiarity which would greatly affect the suitability of its waters as a fish environment—complete lack of oxygen in the depths. Zero oxygen readings were obtained from both the 100- and 200-foot samples, and only 1.5 parts per million of dissolved oxygen at the surface (April). Whatever total factors account for this, the bottom is a thick deposit of sulfuretted hydrogen mud, and much invertebrate organic material constantly sifts down from the surface layers; this plus the even more important fact that the lake is short and deep and semi-protected and so resistant to wind-mixing, operates to produce the oxygen deficiency.

Pronounced chemical stratification was evident from both Hutchinsons' data and that of the writer. A freshening of the water and an increase in oxygen was noted in the shallow, somewhat swampy water of the south bay, where most of the life in the lake was seen. The biota of such a lake, where shiny brass water sampling equipment turns black before the investigator's eyes, is of more than passing interest in itself, whether or not it can be tied into a food chain pattern involving fishes.

Soda Lakes lie but a few miles west of Fallon.

B—The Snake System

The northeastern corner of Nevada—tributary to the Snake River—can be disposed of briefly. There are but two important fishable bodies of water, of which the one mentioned below can be taken as an example.

WILDHORSE RESERVOIR
(United States Indian Reservation, Duck Valley)

Built in 1936–1937 in a cañon of the Owyhee River's East Fork about 15 miles south of Mountain City, northern Elko County, this largest lake in the region is some three miles long and, in places, one mile wide. The 80-foot concrete dam backs up water to an elevation of 6,190 feet above sea level and impounds a maximum of 32,500 acre-feet of water having a surface area of 1,830 acres. Greatest depth is about 60 feet. This is all usable irrigation water since a complete drawdown can be effected.

FIG. 57. Wildhorse Dam and Reservoir, northern Elko County, looking south upstream. Courtesy Nevada Highway Department and Adrian Atwater.

The only striking recent management problem, beyond those inherent in all reservoirs, especially those which can be lowered completely, has been one of dealing with an abundant rough fish population. On 13–17 December 1955, the reservoir was poisoned to remove rough fish and replanted soon afterward with Rainbow Trout and Kokanee Red Salmon. A year or two later, rainbow fishing was "redhot," and kokanee were also contributing substantially to the yield. Typically, however, the rough fish population has been slowly rebuilding and the game fish will correspondingly decrease in numbers.

C—The White River System

ADAMS–McGILL RESERVOIR
(Sunnyside Wildlife Management Area)

This reservoir is impounded by an earthfill dam a little more than a third of a mile long. At the time Ted Frantz, State Fishery Biologist, made his survey (1956)—from which the following data are taken—the reservoir was three miles long and had a maximum width of 1,900 feet. The entire unit was shallow, with a greatest depth of about four feet. However, the dam has a maximum height of 15 feet so that much more water could be stored than was present in 1956.

FIG. 58. Adams-McGill Reservoir, Sunnyside, Nye County.
Photo courtesy Ted Frantz.

The location is some 70 miles south of Ely, in extreme northeastern Nye County and the reservoir lies along the White River streambed, with most of its inflow coming from numerous large warm and cold water springs nearby.

Limnologically, Frantz determined that the reservoir, at the points measured, had pH values varying from 8.2 to 9.6; dissolved oxygen of 4.0 to 6.1 parts per million; carbonates from 16 to 332 ppm; and bicarbonates from 193 to 209 ppm. Water temperature was 73° F. As would be expected in such shallow waters, vegetation was dense in suitable locations, bulrushes being common at the north end and both sides of the south end. Pondweed and *Chara* algae "were dense in the reservoir particularly at the north end and central section." In addition to the common White River system coldwater fishes (see White River), the reservoir appears to have a sizable population of Bluegill Sunfish, and had blackbass in past years—whether any of the latter still existed could not be ascertained.

D—The Colorado River System

No sizable natural lakes exist in that part of Nevada drained by the Colorado River (excluding the White River system) and its tributaries, but man-made Lake Mead is the largest and one of the most spectacular bodies of water in the State. Below it on the Colorado River, the newer Lake Mohave is smaller and shallower, but with the same reservoir characteristics in most ways.

These two reservoirs occupy the entire Colorado River channel from a point well into Arizona to the site of Davis Dam, a distance of some 180 miles. The only "natural" river channel remaining to the State is the short 20-mile stretch below Davis Dam, and because it is now the outlet stream from an impoundment, rather than the free-wheeling river, it is no longer characteristic of the original stream.

Just prior to this writing, the State Fish and Game Commission had terminated two intensive fishery projects in this area, one on Lake Mead[25] and the other on Lake Mohave, so that we have a substantial body of data for these two waters.

(1) LAKE MEAD
(United States Government)
(a) Historical and Descriptive

The magnificent backwater that is Lake Mead extends up the river basin some 115 miles behind Hoover Dam and, at capacity, contains 32 million acre-feet of water under a surface area of 163,000 acres. Constructed under the auspices of the United States Bureau of Reclamation, the dam was begun in 1931, dedicated in 1935, started power production in 1936 and was completely filled by 1941. Its three and a quarter million cubic yards of cement—located about 30 miles southeast of Las Vegas in Black Cañon—is arranged into a massive structure 726 feet high, 1,244 feet in crest length, 45 feet in crest thickness and 660 feet in bottom thickness.

Water rises behind the dam to a maximum height of 590 feet and spreads out upstream to the north into the 4-mile wide Boulder Basin. Above this, an eastering constriction of some 9 miles extent connects Boulder Basin with the slightly larger, 8-mile wide Virgin Basin into which flows the wide Overton Arm from the north.

The reservoir then irregularly and gradually narrows upstream to its nominal point of disappearance some 25 miles into northwestern Arizona. The total shoreline is something like 550 miles, and upper portions of the lake are depositing silt in the amount of about 137,000

---

[25]Dingell-Johnson Project F–1–R, Lakes Mead and Mohave, 1951 to 1954, Fishery Biologists Al Jonez and Robert C. Sumner.

acre-feet a year, thereby reducing storage capacity by that amount. Originally, both dam and lake were called "Boulder" since Congress had first authorized the project for Boulder Cañon. The present names are for a former president and a commissioner of reclamation.

(b) Characteristics of Lake Mead[26]

Reservoirs, such as Lakes Mead and Mohave, differ in certain ways from most natural waters. In the main, these differences are of two types: (1) greater annual fluctuation in water level and (2) more

FIG. 59. Lake Mead, before it had backed up to maximum elevation behind newly built Hoover (then Boulder) Dam. Looking east-northeast. Photo by the author, May 1937.

"flushing" action through the body of water due to constant and considerable inflow and outflow. These affect such basic lake conditions as temperature, density, and nutrient and chemical stratification, which determine essential mechanisms like food chains by controlling distribution and abundance of plankton organisms upon which all other aquatic animals live.

PHYSICAL CONDITIONS
TEMPERATURE

The establishment of a lentic or standing body of water astride the cold Colorado River has had the expected effect of warming these waters. Inflowing river water showed range extremes from 36° F. for

---

[26]Most of the following limnologic data on both Lakes Mead and Mohave are from Jonez and Sumner 1954.

winter to 80° F. in summer. The warmest part of the lake seems to be the Overton Arm, where surface waters reached maxima of nearly 90° F. in summer, and minima at 50 feet depth of about 51° F. Depths of down to 300 feet, farther south in the Overton Arm did not quite achieve these minima.

The often sharp differences between lake and river waters produce striking effects where they come together. If river water is colder than that of the lake, it may slide under the warmer lake waters for considerable distances without appreciable mixing. If river water is the

FIG. 60. Hoover (Boulder) Dam, Clark County, looking north at the mouth of Black Cañon. Courtesy Nevada Highway Department and Adrian Atwater.

warmest of the two, the reverse effect occurs. While, from a physical point of view, the interrelations between such water bodies with differing temperatures is quite complex, these seem to have little or no recognizable effects on the fishery.

In the Forel-Whipple system of lake classification, Mead would resemble a tropical lake of the third order—with surface waters never dropping to 4° C. and with circulation more extensive than the usual winter period. As in the case of Pyramid and other Nevada lakes, Forel's original attempt to categorize lakes—later modified by Whipple —has to be altered further before it can be applied to most of our major bodies of water.

Several investigators have noted the occurrence of a thermocline in Lake Mead, among them Moffett (1943: 180) and Sumner (Jonez and

Sumner 1954: 23), although thermal stratification has not been pronounced enough to cause summer bottom stagnation with its consequent loss of oxygen.

TURBIDITY

As would be expected, maximum areas of turbidity were those about the river mouths, where large volumes of silt-laden waters were being added to the lake. Water opacity further increased during flood periods. Generally, however, visibility was good, with Secchi disk readings of better than 40 feet being common. Occasionally, algal "blooms" produce temporary restriction of light penetration.

CHEMICAL CONDITIONS

OXYGEN

This vital gas is everywhere abundant in Lake Mead, Sumner noting extremes from 4.4 to 11.5 parts per million, with more average figures of 7.0 to 9.5 ppm prevailing (1954: 17). This is the expected situation in a reservoir of this type.

DISSOLVED SOLIDS

These seem not to exceed 800 parts per million. While this is more than twice the average figure generally given for freshwater lakes, it is not unexpected. All Lake Mead tributaries—the Rivers Colorado, Virgin and Moapa—carry large amounts of suspended and dissolved solids, and considerable deposits of such easily soluble substances as gypsum (calcium sulfate) and common salt (sodium chloride) lie in the Mead basin. Compared with the 4,700 ppm figure for Pyramid Lake, the southern Nevada reservoir is still quite fresh water.

HYDROGEN–ION CONCENTRATION OR pH

The alkaline nature of Lake Mead waters is evident from published figures ranging between pHs of 7.5 and 8.4, and indicates a considerable nutritive potential.

BIOLOGICAL CONDITIONS

BIOLOGICAL PRODUCTIVITY

Lake Mead and such richer bodies of water as Pyramid Lake have the point in common that both have sparse bottom faunas and better developed plankton resources—in the case of Pyramid, plankton are phenomenally developed, but they also form the dominant food element near the base of the chain in Mead.

Some algae have been collected and identified from Lake Mead— common species—as well as several microcrustaceans, and it is upon these and some insects that small fish depend largely for food. The more sizeable game fish eat each other and rough fish, while the latter subsist on algae growing on rocks or bottom and on plankton in the open water. Crawfish may be present in small numbers, although neither Moffett (1943) nor Sumner (1954) could demonstrate them.

BIOTA IN THE FISH FOOD CHAIN

In addition to the organisms alluded to in the preceding section, it might be pointed out that all fishes native to the Colorado River seem to be abundant today with two exceptions: (1) the native Colorado Cutthroat Trout (*Salmo clarki pleuriticus*) is deemed extinct in this area if, indeed, it ever existed this far down the Colorado; and (2) the Colorado Squawfish (*Ptychocheilus lucius*), once common in the original river, seems to have disappeared from the two large impoundments now straddling the Nevada portion of the Colorado.

The fish fauna of Lake Mead has changed greatly in composition with the introduction of several Mid-West species of warm-water pan fish of the Family Centrarchidae (blackbass, sunfish and crappie) as well as ictalurids (catfish) and forage fish. The overpowering carp does well in the reservoir, as always, and Rainbow Trout have been extensively planted also.

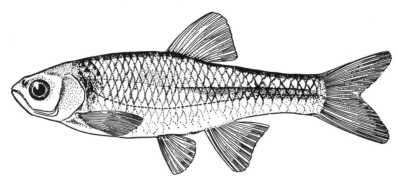

FIG. 61. Plains Red Shiner *(Notropis lutrensis lutrensis* (Baird and Girard)). This central United States minnow has been used as a bait fish in the Colorado River below Nevada, and has been propagated for that purpose by bait fish dealers in Arizona. Miller feels that there is little danger of the species becoming established in the river (1952: 34). In its original habitat it seems to prefer muddy, slower waters such as one finds in the Great Plains region.

(c) Management of Lake Mead

Fishery management of Lake Mead is the usual complex one of trying to make a reservoir as productive as a natural water of the same area and position would be. A great many specialists and interested laymen have tackled the problem and so far, without marked success. Certain factors strongly militate against a reservoir attaining any great degree of productivity, which, in turn, would be reflected in a greater fishable population.

The 1951–1954 Nevada Fish and Game Commission study of Lake Mead by Robert C. Sumner, Fishery Biologist, had many objectives, prominent among which were attempts to find answers to the productivity problem. Out of this came:

(1) Analyses of pertinent environmental conditions. These included applying the well-known fluctuation effects of reservoirs to Lake Mead

with the usual and expected results. Water usage here is too complex to allow of any regulation which would counteract the biologically detrimental raising and lowering of the surface level. Artificial enrichment is out of the question because of the great amount of water involved. In any event, this has not been demonstrated as workable in other large reservoirs where it has been tried.

(2) The possibility of adding forage fish to improve feeding conditions for the dominant game fish—Largemouth Blackbass. This was considered and the Mississippi Threadfin Shad (*Dorosoma petenense atchafalayae*) was brought in in 1953 as the best possibility. The shad has quickly and successfully become a prominent part of the blackbass food supply in Lake Mead and, to that extent, has improved the lake's productivity.

(3) Artificial stocking of game species to maintain them in the face of mounting fishing pressure. While Pyramid Lake has shown that this method can give results well beyond what might be theoretically expected, it is also true that the larger the body of water the more impractical it becomes to maintain fishable populations by stocking alone. Whether or not Mead is too voluminous for stocking to be effective remains to be seen.

(2) LAKE MOHAVE
(United States Government)
(a) Historical and Descriptive

Lake Mohave is a more recent addition to the Colorado River family of reservoirs than is Lake Mead. In 1942, some 65 miles downriver from Hoover Dam, the United States Bureau of Reclamation began construction of Davis Dam[27] on the old "Bullshead" damsite. The war intervened and construction was not resumed until 1946. A small amount of water storage began in 1948; by 1950, the unit was essentially complete and water started backing up behind the dam. By 1951, rising waters had advanced over 65 miles upstream to encroach on the tailrace of Hoover Dam.

Davis Dam is longer than Hoover but much lower, having a crestlength of 1,600 feet and a maximum height of 200 feet. Unlike its upriver concrete companion piece, Davis is a rock- and earth-fill embankment with about the same crestwidth (50 feet).

The reservoir itself—a bit more than 65 miles long—is extremely narrow for most of its length and is confined between high cañon walls. Only in the southern portion does it widen, where it is some 4 miles across in places. Total storage capacity is about 1,820,000 acre-feet of water, under an approximate surface area of nearly 30,000 acres. The shoreline totals up to about 200 miles. Silt deposition is not a problem since water issuing from Hoover Dam is clear and there is little suspended material in the scoured Colorado River bed that now forms Lake Mohave.

---

[27]The Bureau has a fondness for eulogizing itself—the name in this case is for another past commissioner of reclamation.

(b) Characteristics of Lake Mohave

Remarks about reservoir patterns under Lake Mead apply here, except that flushing action is much more direct and pronounced in Mohave due to its straighter, narrower aspect.

PHYSICAL CONDITIONS
TEMPERATURE

Temperature relationships in Lake Mohave are quite complex. In summer months, surface waters warm enough so that strong density

FIG. 62. Davis Dam, Clark County, looking north. Lake Mohave backs up behind the dam to the tailrace of Hoover Dam. The only part of the Colorado River left intact in Nevada issues from the spillways of Davis Dam. Courtesy Nevada Highway Department and Adrian Atwater.

differences exist between them and the incoming cold waters from Lake Mead. Some 20 miles below Hoover Dam, cold and warm waters come together, with the former sliding conspicuously underneath the latter. "Surface temperatures vary as much as 20° F. from the upstream side of the debris [point-of-contact] to the downstream side, a distance of approximately 10 feet" (Jonez 1954: 97–105).

The warm water lens increases in thickness downstream. In this same summer period, the bottom-flowing cold water emerges from Davis Dam at about 60° F. During the course of the Fish and Game study, the coldest recorded waters were 50° F., and the warmest, 80° F. Thermal stratification may be very pronounced in the summer months.

TURBIDITY

Jonez (1954) recorded visibility averaging about 20 feet for most of the year, with increased cloudiness on such occasions as algal blooms and inflows from side dry washes during intermittent storms.

CHEMICAL CONDITIONS

OXYGEN

This gas, as would be expected, is always more than sufficient in quantity for fish life, and was recorded variously as ranging from 5.0 to 10.0 parts per million.

DISSOLVED SOLIDS AND pH

Dissolved solids seem to average around 650 parts per million, with a high of 766 ppm, while pH is about 8.0 with some fluctuation.

BIOLOGICAL CONDITIONS

BIOLOGICAL PRODUCTIVITY AND BIOTA IN THE FISH FOOD CHAIN

The formation of a reservoir considerably reduced productivity of the Colorado River in this stretch. Certain aquatic insects such as Mayflies (Ephemeroptera) and Caseflies (Trichoptera) were common in the river and are now seldom seen in the reservoir. These are basic and popular fish foods.

Plankton, the fundamental substance in the food chain, is abundant in the warmer months. Freshwater shrimp (*Gammarus fasciatus*) was introduced to this part of the river in 1941 (Moffett 1942: 82) and is now firmly established as a valuable fish food. Jonez (1954: 107) found an occasional crawfish (*Pacifastacus* sp.) which generally seemed traceable to recent releases by California anglers.

Changing the environment from a river to a reservoir has had the expected effects on the fish fauna. Trout, mainly rainbow, have decreased except in the upper, colder reaches, while the widened and warmer southern part is now a better warm-water fish habitat with Largemouth Blackbass predominating. The fish list is much the same as for Lake Mead.

(c) Management of Lake Mohave

One important fishery management aspect revolves about the method of water release from Hoover Dam above Mohave. Temperature is suitable to trout if water is brought into the turbine penstocks through the lowermost (900 feet above sea level) of the two intake gates of Hoover Dam. This keeps the temperature of water released through the turbines at about 55° F., a good trout figure. Water warms up above desirable trout temperatures, sometimes for considerable periods of time, when the upper intake gates of Hoover (1,050 feet above sea level) are used. Ordinarily, this would be a management problem with no practical solution, but the 900-foot intake system is being modified so that it can be used regardless of the fluctuation of Lake Mead—this will continue to maintain a trout habitat in the upper part of Lake Mohave.

Other, more biological, management problems in Lake Mohave are

FIG. 63. Johnson Lake, above Snake Creek, east side of the Snake Range, White Pine County. Looking east. Elevation of the lake, 10,800 feet. Photo August 5, 1959, by author.

a bit more complex than in Lake Mead since Mohave provides a living area in its northern section for trout and in the southern part for warm-water fishes such as blackbass, sunfish and catfish. Consequently, there is a large overlap zone where such voracious fishes as the Largemouth Blackbass take considerable numbers of fingerling trout for food. This naturally makes the trout fishermen mad.

Some kind of forage fish seems to be the reasonable answer here, if one can be found with the proper habits—small, non-aggressive and

FIG. 63B. Liberty Lake, Ruby Mountains, Elko County. Elevation 9,700 feet. This lake drains into Favre Lake, and thence down the west side of the Rubies. Its surface area is 21 acres and its greatest depth 108 feet. It is currently overpopulated with Brook Trout, and presents a special management problem in attempts to maintain fishing in an environment basically poor in nutrients but excellent in sportsman appeal (data from Donald B. Thurston and Donald J. King of the Nevada Fish and Game Commission). The above picture was taken by T. H. O'Sullivan in 1868 while on the King Fortieth Parallel Survey for the U. S. War Department. At that time it was called Lake Marian. Courtesy Office of U. S. Chief of Engineers, Photograph No. 77–KS–15 in the U. S. National Archives.

non-carnivorous so that it will not consume eggs and young of game fish, yet energetic enough to make a good living for itself under the conditions that prevail and so produce large numbers of its own kind as food fish for the game species. As in Lake Mead, the newly-arrived Mississippi Threadfin Shad may do the trick. There, it is providing a good food source for blackbass and others in an environment relatively poor in nutrients—in Lake Mohave, it would not only do this but might act as a buffer between blackbass and trout.

E—The Amargosa River System

No prominent lakes, natural or artificial, exist in the Amargosa drainage, either in the Beatty or the Ash Meadows subdivisions. Springs—warm and cold—are the sources of most of the water, and while these, particularly in Ash Meadows, wander off into sloughs and marshes, they do not constitute any but mediocre bodies of water.

CHAPTER VI

THE FISHES

1—CHECKLIST OF THE SPECIES AND SUBSPECIES OF FISHES WHICH OCCUR IN NEVADA

Class *Osteichthyes*
    Subclass Actinopterygii
        Superorder *Teleostomi*
            Order ISOSPONDYLIFORMES
                Suborder Clupeoidei

SHAD AND HERRING FAMILY, Family *CLUPEIDAE**
(1) Mississippi Threadfin Shad*
    Dorosoma petenense atchafalayae 240

                Suborder Salmonoidei

SALMON AND TROUT FAMILY, Family *SALMONIDAE*
(2) King Salmon
    Oncorhynchus tshawytscha 247
(3) Kokanee Red Salmon*
    Oncorhynchus nerka kennerlyi 251
(4) Lake Trout*
    Salvelinus namaycush 257
(5) Brook Trout*
    Salvelinus fontinalis 264
(6) Dolly Varden Trout
    Salvelinus malma 272
(7) Cutthroat Trout
    Salmo clarki 275
    (7a) Lahontan Cutthroat Trout
        (*Salmo clarki henshawi*) 281
    (7b) Yellowstone Cutthroat Trout*
        (*Salmo clarki lewisi*) 295
    (7c) Utah Cutthroat Trout
        (*Salmo clarki utah*) 297
    (7d) Colorado Cutthroat Trout
        (*Salmo clarki pleuriticus*) 299

*Introduced species.

- (8) Rainbow Trout*
  Salmo gairdneri 300
  - (8a) Southcoast Rainbow Trout*
    (*Salmo gairdneri irideus*) 306
  - (8b) Kamloops Rainbow Trout*
    (*Salmo gairdneri kamloops*) 308
  - (8c) Tahoe Rainbow Trout*
    (*Salmo gairdneri regalis*) 311
  - (8d) Pyramid Rainbow Trout*
    (*Salmo gairdneri smaragdus*) 313
- (9) Golden Trout*
  Salmo aguabonita 317
- (10) Brown Trout*
  Salmo trutta 318

WHITEFISH FAMILY, Family *COREGONIDAE*

- (11) Mountain Whitefish
  Prosopium williamsoni 325

Order CYPRINIFORMES
Suborder Cyprinoidei

SUCKER FAMILY, Family *CATOSTOMIDAE*

- (12) Lahontan Mountainsucker
  Pantosteus lahontan 334
- (13) White River Mountainsucker
  Pantosteus intermedius 337
- (14) Biglip Sucker
  Catostomus macrocheilus 340
- (15) Bridgelip Sucker
  Catostomus columbianus 342
- (16) Utah Sucker
  Catostomus ardens 344
- (17) Flannelmouth Sucker
  Catostomus latipinnis 350
- (18) Tahoe Sucker
  Catostomus tahoensis 352
- (19) Razorback Sucker
  Xyrauchen texanus 357
- (20) Cui-ui Lakesucker
  Chasmistes cujus 363

CARP AND MINNOW FAMILY, Family *CYPRINIDAE*

- (21) Northern Squawfish
  Ptychocheilus oregonensis 376

*Introduced species.

(22)	Colorado Squawfish	
	Ptychocheilus lucius	381
(23)	Chiselmouth	
	Acrocheilus alutaceus	384
(24)	Colorado Gila	
	Gila robusta	388
	(24a) Swiftwater Colorado Gila	
	(*Gila robusta elegans*)	391
	(24b) White River Colorado Gila	
	(*Gila robusta jordani*)	393
(25)	Utah Gila	
	Gila atraria	395
(26)	Lahontan Redshiner	
	Richardsonius egregius	399
(27)	Columbia Redshiner	
	Richardsonius balteatus	403
	(27a) Bonneville Columbia Redshiner	
	(*Richardsonius balteatus hydrophlox*)	408
(28)	Tui Chub	
	Siphateles bicolor	410
	(28a) Lahontan Tui Chub	
	(*Siphateles bicolor obesus*)	412
(29)	Golden Shiner*	
	Notemigonus crysoleucas	421
(30)	Speckle Dace	
	Rhinichthys osculus	424
	(30a) Snake River Speckle Dace	
	(*Rhinichthys osculus carringtoni*)	428
	(30b) Lahontan Speckle Dace	
	(*Rhinichthys osculus robustus*)	430
	(30c) Amargosa Speckle Dace	
	(*Rhinichthys osculus nevadensis*)	433
	(30d) White River Speckle Dace	
	(*Rhinichthys osculus velifer*)	434
(31)	Moapa Dace	
	Moapa coriacea	436
(32)	Soldier Meadows Desertfish	
	Eremichthys acros	441
(33)	Asiatic Carp*	
	Cyprinus carpio	448
(34)	Goldfish*	
	Carassius auratus	454
(35)	Sacramento Blackfish*	
	Orthodon microlepidotus	457

*Introduced species.

- (36) White River Spinedace
  Lepidomeda albivallis 462
- (37) Colorado River Spinedace
  Lepidomeda mollispinis 465
  - (37a) Virgin River Spinedace
    (*Lepidomeda mollispinis mollispinis*) 465
  - (37b) Big Spring Spinedace
    (*Lepidomeda mollispinis pratensis*) 468
- (38) Pahranagat Spinedace
  Lepidomeda altivelis 472
- (39) Woundfin
  Plagopterus argentissimus 475

Suborder Siluroidei
Superfamily Siluroidea

NORTH AMERICAN CATFISH FAMILY, Family *ICTALURIDAE**

- (40) Channel Catfish*
  Ictalurus punctatus 481
- (41) White Catfish*
  Ictalurus catus 485
- (42) Brown Bullhead*
  Ictalurus nebulosus 489
- (43) Black Bullhead*
  Ictalurus melas 494
  - (43a) Northern Black Bullhead
    (*Ictalurus melas melas*) 497
  - (43b) Southern Black Bullhead
    (*Ictalurus melas catulus*) 497

Order CYPRINODONTIFORMES
Suborder Cyprinodontoidei

KILLIFISH FAMILY, Family *CYPRINODONTIDAE*

- (44) Amargosa Pupfish
  Cyprinodon nevadensis 500
  - (44a) Big Spring Amargosa Pupfish
    (*Cyprinodon nevadensis mionectes*) 504
  - (44b) Lovell Spring Amargosa Pupfish
    (*Cyprinodon nevadensis pectoralis*) 505
- (45) Devil Pupfish
  Cyprinodon diabolis 507
- (46) White River Springfish
  Crenichthys baileyi 512

*Introduced species.

FISHES AND FISHERIES OF NEVADA 197

(47) Railroad Valley Springfish
   Crenichthys nevadae ___ 517
(48) Ash Meadows Poolfish
   Empetrichthys merriami ___ 520
(49) Pahrump Poolfish
   Empetrichthys latos ___ 523
   (49a) Manse Ranch Pahrump Poolfish
      (*Empetrichthys latos latos*) ___ 527
   (49b) Pahrump Ranch Pahrump Poolfish
      (*Empetrichthys latos pahrump*) ___ 527
   (49c) Raycraft Ranch Pahrump Poolfish
      (*Empetrichthys latos concavus*) ___ 528

TOPMINNOW FAMILY, Family *POECILIIDAE**

(50) Mosquitofish*
   Gambusia affinis ___ 532
(51) Black Molly*
   Mollienesia latipinna ___ 536
(52) Swordtail*
   Xiphophorus helleri ___ 536
(53) Moonfish (Platy)*
   Xiphophorus maculatus ___ 536

Order PERCIFORMES*
Suborder Percoidei

PERCH FAMILY, Family *PERCIDAE**

(54) Yellow Perch*
   Perca flavescens ___ 538

SUNFISH FAMILY, Family *CENTRARCHIDAE**

(55) Sacramento Perch*
   Archoplites interruptus ___ 547
(56) Largemouth Blackbass*
   Micropterus salmoides ___ 554
(57) Smallmouth Blackbass*
   Micropterus dolomieui ___ 560
(58) Bluegill Sunfish*
   Lepomis macrochirus ___ 566
(59) Green Sunfish*
   Lepomis cyanellus ___ 571
(60) Black Crappie*
   Pomoxis nigromaculatus ___ 576
(61) White Crappie*
   Pomoxis annularis ___ 580

*Introduced species.

Suborder Cottoidei
SCULPIN FAMILY, Family *COTTIDAE*

(62) Baird Sculpin
    Cottus bairdi 585
  (62a) Bonneville Baird Sculpin
    (*Cottus bairdi semiscaber*) 586
(63) Belding Sculpin
    Cottus beldingi 588

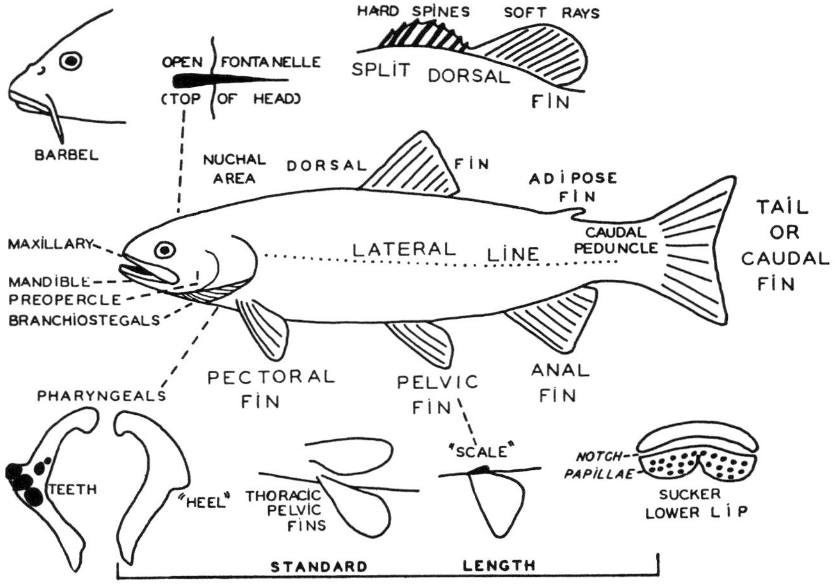

FIG. 64. An artificial composite, showing morphologic and anatomical terms commonly used in fish identification.

SUPPLEMENTARY ANNOTATED LIST OF UNSUCCESSFULLY INTRODUCED SPECIES[1]

The following list is concerned with those species which have been planted without known success in the State. Miller and Alcorn (1946) have previously summarized most of these data.

SHAD AND HERRING FAMILY, Family *CLUPEIDAE*

(1) American Shad .... *Alosa sapidissima* (Wilson) 1812?
    An early plant made in the mid-1880s in the Colorado River near Nevada was not successful.

---
[1]From La Rivers and Trelease, 1952.

SALMON AND TROUT FAMILY, Family *SALMONIDAE*

(2) Chum Salmon .... *Oncorhynchus keta* (Walbaum) 1792
Widely planted in west-central Nevada in 1939, but none are known to have survived (as would be expected).

(3) Silver Salmon .... *Oncorhynchus kisutch* (Walbaum) 1792
Planted in the Carson and Truckee Rivers in 1913.

(4) Red Salmon .... *Oncorhynchus nerka*
According to the Nevada Fish and Game Commission Biennial Reports for the years 1936, 1938 and 1941, several attempts were made to establish "*Oncorhynchus nerka*," the "Sockeye Salmon," in west-central Nevada. It is presumed that these were the anadromous form.

(5) Atlantic Salmon .... *Salmo salar* Linnaeus 1758
(5a) Landlocked Atlantic Salmon
(*Salmo salar sebago* Girard 1853)

An early introduction into Lake Tahoe and the Carson and Truckee Rivers (1880s), where they were reported a temporary source of good fishing.

(6) Cutthroat Trout .... *Salmo clarki*
(6a) Greenback Cutthroat Trout
(*Salmo clarki stomias* Cope 1871)

Planted in 1892–1893 in the Humboldt River near Elko, presumably from a Federal hatchery in Colorado. If any of the strain survived, mixture with other subspecies of cutthroats soon obliterated its characteristics.

(7) Rainbow Trout .... *Salmo gairdneri*
(7a) Northcoast Rainbow Trout
(*Salmo gairdneri gairdneri*)

Presumably brought into the State in the early 1900s, as "steelhead" trout, but unknown here at the present time.

(8) Golden Trout .... *Salmo aguabonita*
1918–1919 plants were made in the Lake Tahoe region, but no survivals are known. In 1960 specimens were introduced into some Ruby Mountain lakes in Elko County, with apparently good results.

WHITEFISH FAMILY, Family *COREGONIDAE*

(9) Lake Whitefish .... *Coregonus clupeaformis* (Mitchill) 1818
(9a) Great Lakes Whitefish
(*Coregonus clupeaformis clupeaformis*)

First brought into Lake Tahoe in the 1870s from Lake Michigan spawn, and later introduced into the Eureka area of central Nevada.

GRAYLING FAMILY, Family *THYMALLIDAE*

(10) Arctic Grayling............*Thymallus signifer* (Richardson) 1823

    (10a) American Arctic Grayling
        (*Thymallus signifer tricolor* Cope 1865)

        Introduced into Ruby Valley, Elko County, in the 1940s, and probably previously into the Lake Tahoe region. Thomas J. Trelease planted some 500 fingerlings in Kings Cañon near Carson City, Ormsby County, in 1948, but none are known to have survived.

CARP AND MINNOW FAMILY, Family *CYPRINIDAE*

(11) Tench........................*Tinca tinca* (Linnaeus) 1758

    A shipment was sent to a Virginia City applicant in 1885 (Storey County).

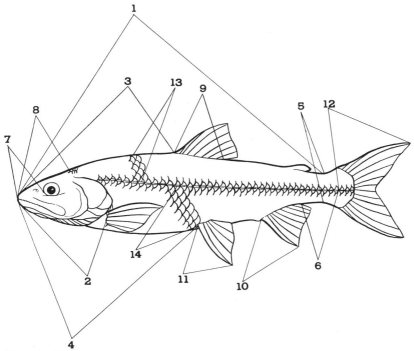

FIG. 65. Diagram showing method of taking proportional measurements. (1) *Body length*, recorded in millimeters. (2) *Length Head*, measured from tip of snout to posterior edge of opercle, the opercular flap, which is likely to shrink, not being considered. *Depth Head*, measured at occiput. *Depth Body*, the greatest depth. (3) *Snout to Dorsal*. (4) *Snout to Ventral*, tip of snout to anterior end of base of fin. (5) *Depth Caudal Peduncle*, measured at the narrowest place. (6) *Length of Caudal Peduncle*, base of posterior anal ray to end of last vertebra; not to base of lowermost caudal ray, as the latter point is often indefinite. (7) *Length Snout*, tip of snout to anterior border of eye. *Diameter Eye*, varies greatly with age. In poorly preserved specimens the tissue surrounding the eye is often shrunken, leaving the opening abnormally large. Only well-preserved examples nearly equal in size ought to be compared. *Interorbital Width*, measured on skull, the points of the dividers being closed as nearly as possible between the eyes. (8) *Snout to Occiput*, tip of snout to the point on occiput where scales of body first appear. (9) *Length Base of Dorsal, Length Base of Anal*, from base of anterior to base of posterior ray, the extent of the membrane posteriorly not being taken into account. *Height Dorsal*, (10) *Height Anal, Length Pectoral*, (11) *Length Ventral*, the length of longest ray in fin. (12) *Length Caudal*, measured from end of last vertebra to tip of upper caudal lobe. *Scales Lateral Line*, counted to end of last vertebra; not on base of caudal, where they frequently become densely crowded and difficult to make out. (13) *Scales Above Lateral Line*, from lateral line upward and forward to a point about midway between occiput and insertion of dorsal; not between lateral line and base of dorsal, as in the latter region the scales are sometimes minute, densely crowded, and indistinct. (14) *Scales Below Lateral Line*, from upper edge of base of ventral upward and forward to the lateral line. The series in the lateral line is not enumerated in this or the previous count. *Scales Before Dorsal*, the number of rows or series between occiput and base of dorsal. *Dorsal Rays, Anal Rays*, when the posterior ray is cleft to the base it is still counted as a single ray. The anterior ray is often simple and preceded by one or two short, spine-like rays closely united to it. The spine-like rays are not enumerated (Snyder 1908: 79).

2—ILLUSTRATED KEY TO THE FISHES OF NEVADA

(1) Pelvic fins thoracic or jugular in position (i.e., moved forward from their usual position near the anal fin), always present, typically with 1 spinous and 5 soft rays each (reduced to 1 spinous and from 3 to 4 soft rays in Cottidae); anterior rays of dorsal fin spinous _____ 2

 Pelvic fins abdominal in position, lacking entirely in some of the Cyprinodontidae; spinous rays usually lacking _____ 11

(2) Body lacking true scales, often covered with prickles; head large, broad, flattened dorso-ventrally; body noticeably small and tapering behind the large head (Family COTTIDAE) _____ SCULPINS __ 3

 Body uniformly covered with conspicuous scales; head much smaller in proportion to the body, typically narrow and flattened laterally _____ 4

(3) No spines (at most, only a tubercle or two) below single spine at preopercular angle; Lahontan and Columbia River systems.

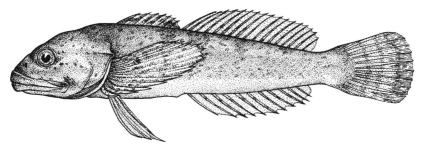

FIG. 66. Belding Sculpin, *Cottus beldingi*. Drawn by Silvio Santina.

 One or two spines below the single spine at preopercular angle; Bonneville system.

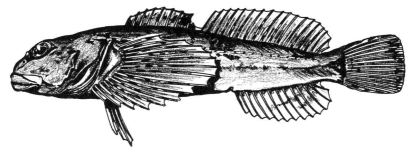

FIG. 67. Bonneville Baird Sculpin, *Cottus bairdi semiscaber*. Drawn by William L. Brudon for Miller 1952. Courtesy California Department of Fish and Game and Leo Shapovalov.

(4) Two distinct dorsal fins present, the anterior with spinous rays, the posterior one soft-rayed; less than 3 anal spines (Family PERCIDAE)_____PERCHES

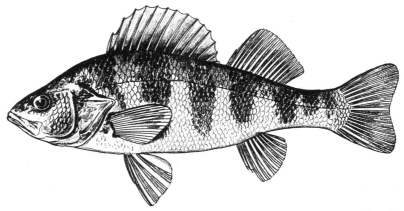

FIG. 68. Yellow Perch, *Perca flavescens*. Drawn by William L. Brudon for Miller 1952. Courtesy California Department of Fish and Game and Leo Shapovalov.

_____ Only one dorsal fin present (in blackbasses, the dorsal fin has a moderate-to-deep notch between the anterior spinous and the posterior soft-rayed portions, but these two parts are always connected); 3 or more anal spines (Family CENTRARCHIDAE)_____ SUNFISHES____ 5

(5) Anal fin spines 5 or more in number_____ 6

_____ Anal fin spines 3 in number_____ 8

(6) Dorsal fin spines 12 to 13 in number; dorsal fin base much longer than anal fin base; gillrakers short.

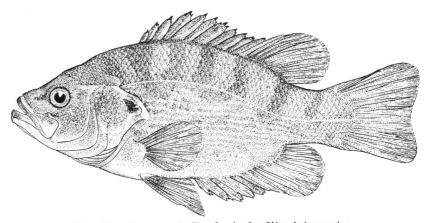

FIG. 69. Sacramento Perch, *Archoplites interruptus*. Drawn by Silvio Santina.

------- Dorsal fin spines 6 to 8 in number; dorsal fin base subequal to anal fin base (i.e., both about the same length); gill-rakers long and slender (Genus *Pomoxis*)__CRAPPIES____ 7

(7) Dorsal fin larger, its base length about equal to distance from origin of dorsal fin (front end) to eye; dorsal fin spines usually 7 to 8 in number; mandible shorter than pectoral fin; maxillary (including supramaxillary), shorter, reaching posteriorly to about a vertical line from posterior edge of eye pupil; body speckled.

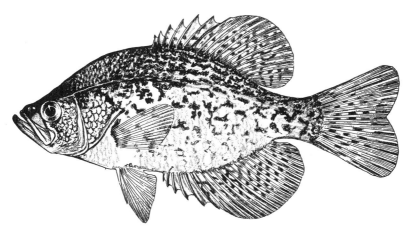

FIG. 70. Black Crappie, *Pomoxis nigromaculatus*. Courtesy California Department of Fish and Game and Leo Shapovalov.

------- Dorsal fin smaller, its base length much less than distance from origin of dorsal fin to eye; dorsal fin spines usually 6 in number; mandible about equal in length to pectoral fin; maxillary (including supramaxillary) longer, extending posteriorly to about a vertical line which is as near posterior edge of eye as it is to posterior edge of eye pupil; body banded.

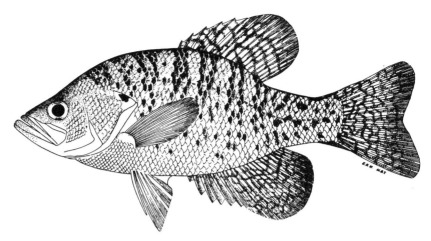

FIG. 71. White Crappie, *Pomoxis annularis*. Reproduced from *The Fishes of Ohio*, by Milton B. Trautman (Columbus, Ohio: Ohio State University Press, 1957), p. 477. Used by permission of the publisher.

(8) Scales small, 58 or more in the lateral line; body depth about one-third of the standard length (Genus *Micropterus*) .... BLACKBASSES____ 9

_____ Scales large, 53 or less in the lateral line; body depth usually about one-half the standard length (Genus *Lepomis*) .... SUNFISHES____ 10

(9) Dorsal fin deeply notched, almost divided into two fins, the shortest spine of the notch less than half as long as the longest spine of the fin; upper jaw (maxillary) extending behind hind margin of eye in adults (measured with mouth closed); from 58 to 69 scales in lateral line; cheek scales in 9 to 12 rows.

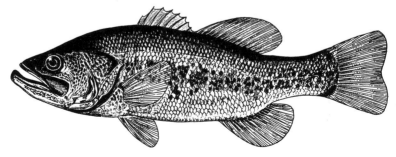

FIG. 72. Largemouth Blackbass, *Micropterus salmoides*. Courtesy California Department of Fish and Game and Leo Shapovalov.

_____ Dorsal fin less deeply notched, the shortest spine at the emargination being more than half the length of the longest fin spine; upper jaw extending beyond middle of eye pupil but not to hind margin of eye; from 68 to 81 scales in the lateral line; cheek scales in 14 to 18 rows.

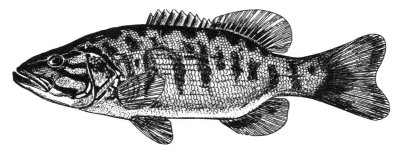

FIG. 73. Smallmouth Blackbass, *Micropterus dolomieui*. Courtesy California Department of Fish and Game and Leo Shapovalov.

(10) Pectoral fins short and rounded, their length about one-fourth that of the standard length; mouth large, the upper jaw or maxillary reaching behind front edge of eye (in adults, the upper jaw is two or more times the width of the eye); usually lacking any definite, isolated black spot on dorsal fin rays, except for a darkening at ray bases.

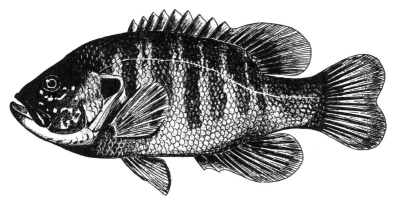

FIG. 74. Adult Green Sunfish, *Lepomis cyanellus*. Courtesy California Department of Fish and Game and Leo Shapovalov.

------- Pectoral fins long and pointed, their length about one-third that of the standard length; mouth small, the maxillary not reaching the front edge of eye (in adults, the maxillary is but slightly longer than the width of the eye); possesses a prominent black, isolated spot on dorsal fin rays.

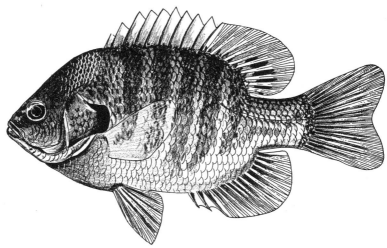

Fig. 75. Adult Bluegill Sunfish, *Lepomis macrochirus*. Courtesy California Department of Fish and Game and Leo Shapovalov.

(11) Dorsal fin with a greatly elongated, filamentous posterior ray; venter coming to a narrow, sharp, saw-toothed margin along the mid-line (Family CLUPEIDAE) .... SHAD AND HERRING

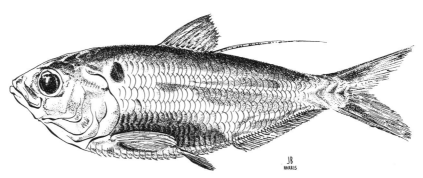

Fig. 76. Mississippi Threadfin Shad, *Dorosoma petenensis atchafalayae*. Drawn by John B. Harris.

_____ Dorsal fin usual, with no such elongated rays; venter rounded, unkeeled, except in *Notemigonus*, where the keel is fleshy and scaleless 12
(12) With an adipose fin (a short, fleshy process between dorsal fin and tail) 13
_____ Without an adipose fin 25
(13) Conspicuous barbels or tentacles about and adjacent to mouth; sharp, stout spines at leading edges of dorsal and pectoral fins (Family ICTALURIDAE) .... CATFISHES.... 14
_____ No barbels; no dorsal or pectoral fin spines 17

(14) Tail or caudal fin distinctly forked, the 2 lobes well-defined and often sharply pointed; animal predominantly colored silver-blue through silver to white, breeding individuals showing more bluish; supra-occipital bone extending rearward and connecting with a forward-projecting process of the dorsal fin; lower jaw distinctly shorter than upper jaw_____(Catfishes proper)____ 15

_____ Tail or caudal fin square, convex or weakly concave, not forked; species predominantly colored in yellows, browns and blacks; supra-occipital bone extending rearward but distinctly falling short of union with the dorsal fin anterior process; jaws about equal in length _____ (Bullheads)____ 16

(15) Anal fin rays from 24 to 30 in number, counting all rudiments; pelvic fins, when extended rearward, overlapping origin of long anal fin; black-spotting conspicuous on all except larger, older fishes—from 5 pounds and above, spotting may be lacking; young usually have dark edges along top of tail fin, leading edge of dorsal and distal edge of caudal fins.

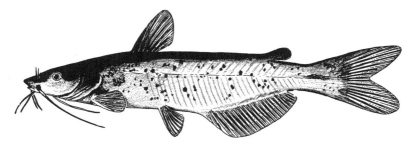

FIG. 77. Channel Catfish, *Ictalurus punctatus*. Courtesy California Department of Fish and Game and Leo Shapovalov.

_____ Anal fin rays from 19 to 23, counting rudiments (in some cases, there are 24); pelvic fins, when extended rearward, not reaching origin of shorter anal fin (i.e., shorter than in the Channel Catfish); never any black-spotting on body, although breeding specimens may be blue-black as in the Channel Cat; young without the darker fin edges as described for *Ictalurus punctatus*.

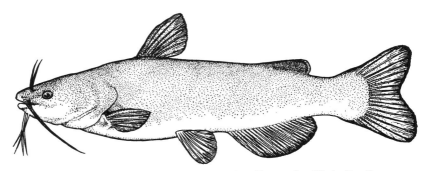

FIG. 78. White Catfish, *Ictalurus catus*. Drawn by Silvio Santina.

(16) Anal rays generally 18–19 (rarely 16–22), counting rudiments; pectoral fin spine without strong, definite barbs or teeth on posterior edge, although occasionally some individuals will show 5–10 poorly developed teeth on this posterior edge, teeth which are little more than bumps (see test for teeth of Brown Bullhead in next couplet); inter-radial membranes of anal fin jet black—fin is never base-striped nor -spotted nor uniformly pigmented on both rays and membranes as in the following species—outer two-thirds of these membranes always darker than the rays themselves; ventral color in adults contrasting with darker sides, giving the individual a bicolored look with a lightening of color vertically in front of the caudal fin base.

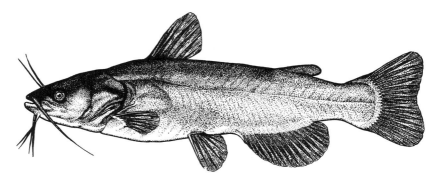

FIG. 79. Black Bullhead, *Ictalurus melas*. Drawn by Silvio Santina.

------ Anal rays generally 22–23 (rarely 19–24), counting rudiments; pectoral fin spine with strong barbs on posterior edge (in the young these number from 5–10 prominent teeth whose lengths average more than half the pectoral spine diameter—in adults and older fish, these teeth or barbs become smaller and more numerous, from 10–25 (a practical test for this is to grasp the spine in the plane of the fin between thumb and forefinger, hold tightly, and pull outward—if your grasp holds, or the spines prick, it is this species); inter-radial membranes of anal fin not black, the spotting characteristically densest near free margins of fin, or producing a vague longitudinal stripe near fin base, or showing as faint mottlings scattered unevenly on both rays and membranes—if individuals are washed out in color, pigmentation is about equally distributed on both rays and membranes; ventral color in adults usually fusing gradually into the mottled sides with no lightening of color vertically in front of the caudal fin base.

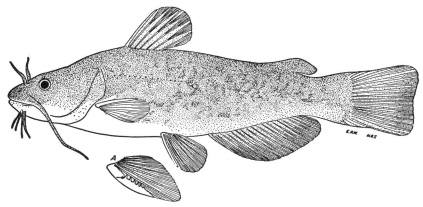

FIG. 80. Brown Bullhead, *Ictalurus nebulosus*. (A) Pectoral fin, showing serrations on posterior or inner edge of pectoral spine. Reproduced from *The Fishes of Ohio*, by Milton B. Trautman (Columbus, Ohio: Ohio State University Press, 1957), p. 424. Used by permission of the publisher.

(17) Mouth small, the maxillary process not extending behind the eye in adults, usually ending before or just at anterior edge of eye; jaw teeth weak; scales in the lateral line fewer than 105 (Family COREGONIDAE)------------
--- WHITEFISHES

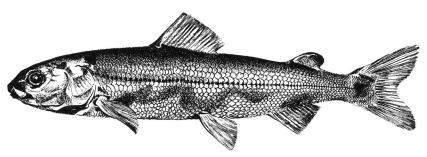

FIG. 81. Mountain Whitefish, *Prosopium williamsoni*. Drawn by John B. Harris.

------- Mouth large, the maxillary process extending at least to hind edge of eye in adults, usually considerably posterior to the eye; jaws usually bearing strong teeth; mouth deeply cleft; lateral line scales more than 115 (Family SALMONIDAE)_____ SALMON AND TROUT___ 18

(18) Anal fin elongate, from 13 to 19-rayed (rarely with 12–18 rays); vomer narrow, long, flat, with weak teeth; gillrakers 19–40 on first gill arch; branchiostegals 13–19; species with or without black spots, the adults with anal and dorsal fins seldom spotted (Genus *Oncorhynchus*) _____ PACIFIC SALMON___ 19

------- Anal fin short, 9 to 12-rayed (rarely 13); gillrakers 20 or less on first gill arch; branchiostegals 10–20; dorsal fin black-spotted_____ 20

(19) Gillrakers comparatively short and few in number, 19–28 on the first gill arch (rarely 29); gillraker formula 7–9 + 11–13[1]; anal rays usually 15–17; distinct black spots on back and caudal fin.

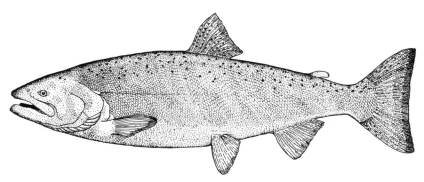

FIG. 82. Female King Salmon, *Oncorhynchus tshawytscha*.
Drawn by Silvio Santina.

------- Gillrakers comparatively long and numerous, 30–50 on the first gill arch; gillraker formula 11–24 + 20–26; anal rays usually 14–15; spotting not evident on back and/or caudal fin, but fine black speckling may be present.

---

[1]Interpreted as 7 to 9 gillrakers on the upper limb of the first gill arch, **and** 11 to 13 gillrakers on the lower limb.

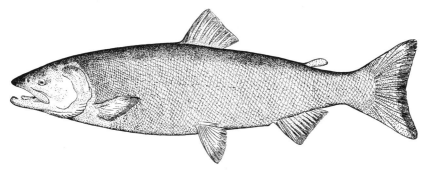

Fig. 83. Red Salmon, *Oncorhynchus nerka*. Drawn by Silvio Santina.

(20) Species with dark spots on a lighter background; fewer than 190 scale rows crossing the lateral line; vomer flat, its toothed surface plane, teeth on vomerian shaft in alternating rows or in one staggered row, those on the shaft placed directly on surface of bone, not on a free crest (Genus *Salmo*)_____ TROUT____ 21

‑‑‑‑‑‑ Species with light spots, white or gray, on a darker background; often with red spots on sides; with over 190 scale rows crossing the lateral line; vomer boat-shaped, the shaft depressed and without teeth (Genus *Salvelinus*)_____ CHARR____ 23

(21) Red dash of color on the dentary (between lower jaw and isthmus) usually evident in life; dorsal fin rays usually 10 (9–11); maxillary process in adults extending behind the eye, measuring about 1.6 to 2.25 into head; hyoid teeth (those behind the patch of teeth on the tongue) usually present, but few and scattered and easily broken off; vertebrae usually 60–61 (58–62); no red-spotting on flanks.

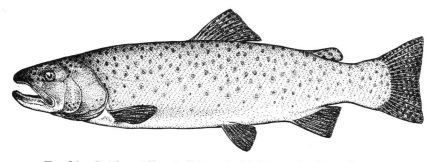

Fig. 84. Cutthroat Trout, *Salmo clarki*. Drawn by Silvio Santina.

‑‑‑‑‑‑ No red dash of color on dentary; dorsal fin rays usually 11–12 (10–13); maxillary process in adults shorter, not extending behind the eye; hyoid teeth always absent; vertebrae usually more or less than 60–61_____ 22

(22) Usually with a few red spots on the sides, these surrounded by a whitish halo; black-spotting large, often sparse, and many spots white-haloed; body color brownish-yellow to greenish; spotting absent or only slightly developed on tail (caudal) fin; vertebrae usually 57–58 (56–59); adipose fin of young individuals orange, without dark margining or spotting.

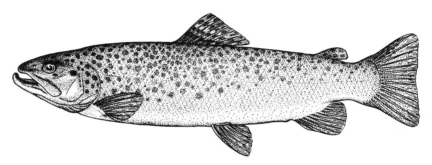

FIG. 85. Brown Trout, *Salmo trutta*. Drawn by Silvio Santina.

_____ No red-spotting on the sides; black-spotting small, numerous, no spots haloed; body color gray-to-blue above, the reddish lateral band usually but slightly interrupted by faint parr marks on adults; spotting well-developed on tail fin; vertebrae usually 63 (59–65); adipose fin of young specimens olive, with black margining or spotting.

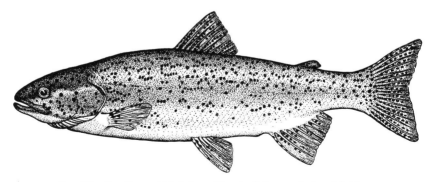

FIG. 86 Southcoast Rainbow Trout, *Salmo gairdneri irideus*. Drawn by Silvio Santina.

(23) Tail prominently forked in adults; body gray-spotted without red spots; fins not strongly bright-edged; vomer with a raised crest extending backward from the head of the bone, this crest armed with strong teeth.

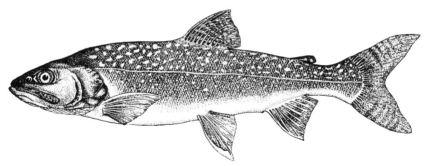

FIG. 87. Lake Trout, *Salvelinus namaycush*. Drawn by Silvio Santina.

- Tail weakly forked in adults; body yellow- or red-spotted in life; fins strongly bright-edged; vomer without a raised crest which extends backward, the head of the bone toothed 24
(24) Conspicuous dark-green vermiculations ("worm tracks") on back and dorsal fin; reddish spotting on body sides enclosed in blue; dorsal and caudal fins finely mottled; body stouter, the head heavy.

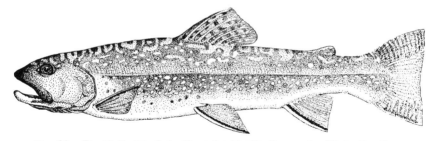

FIG. 88. Brook Trout, *Salvelinus fontinalis*. Drawn by Silvio Santina.

- Vermiculations lacking, the back spotted similarly to the sides; reddish or yellowish spotting on sides not surrounded by blue; dorsal and caudal fins unmarked and clear; body slimmer, head smaller.

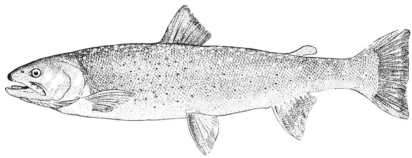

FIG. 89. Dolly Varden Trout, *Salvelinus malma*. Drawn by Silvio Santina.

(25) Head with scales _____ 26
_____ Head without scales _____ 35
(26) Origin of dorsal fin behind origin of anal fin, about on a perpendicular with hind base of anal fin; dorsal fin generally 6-rayed (counted at bases); pelvic fins conspicuous, nearly as large as pectoral fins; third anal ray unbranched (counting rudiments); anal fin of male unlike that of female, modified into a long narrow reproductive organ (Family POECILIIDAE) _____ TOPMINNOWS

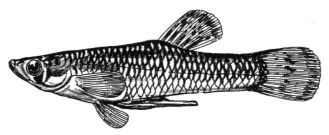

FIG. 90A. Male Mosquitofish, *Gambusia affinis*. Drawn by William L. Brudon for Miller 1952. Courtesy California Department of Fish and Game and Leo Shapovalov.

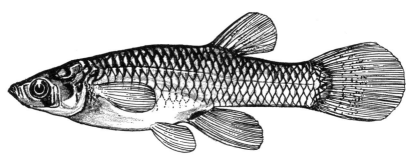

FIG. 90B. Female Mosquitofish, *Gambusia affinis*. Drawn by William L. Brudon for Miller 1952. Courtesy California Department of Fish and Game and Leo Shapovalov.

_____ Origin of dorsal fin about on a perpendicular with origin of anal fin or anterior to it; dorsal fin generally 8-rayed or more (counted at bases); pelvic fins inconspicuous, much smaller than the pectorals, or entirely absent; third anal ray branched, except in some immatures; anal fins of male and female similar (Family CYPRINODONTIDAE) _____ KILLIFISHES _____ 27
(27) Origin of dorsal fin markedly anterior of origin of anal fin; jaw teeth tricuspid; pelvic fins usually present except in *Cyprinodon diabolis* (Genus *Cyprinodon*) _____
_____ PUPFISH _____ 28

------ Origin of dorsal fin about on a perpendicular with origin of anal fin; jaw teeth not tricuspid; pelvic fins never present_____ 30
(28) Found only in Devil's Hole, Ash Meadows, Nye County; pelvic fins absent (tail fin often bilobed, body dwarfed).

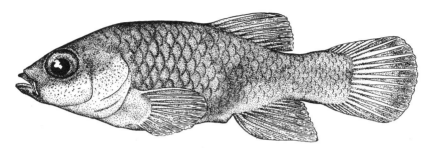

FIG. 91. Devil Pupfish, *Cyprinodon diabolis*. Drawn by Silvio Santina.

------ Found in the Amargosa system surrounding Devil's Hole; pelvic fins usually present, occasionally lacking on one side, rarely lacking on both sides (tail fin rounded or truncate; body generally normal-sized for the genus, rarely slightly dwarfed).

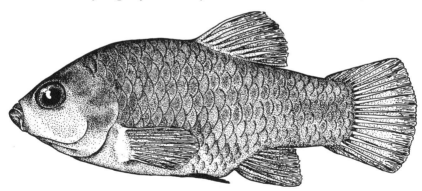

FIG. 92. Amargosa Pupfish, *Cyprinodon nevadensis*. Drawn by Silvio Santina.

(29) Pectoral fin rays usually 17; scales average 9.79 between dorsal and pelvic fins; scales average 15.44 around peduncle; scales average 24.75 around the body; preopercular pores average 13.36 (*Cyprinodon nevadensis pectoralis*)
_____LOVELL SPRING AMARGOSA PUPFISH
------ Pectoral fin rays usually 16; scales average 9.25 between dorsal and pelvic fins; scales average 14.17 around peduncle; scales average 22.67 around the body; preopercular pores average 12.48 (*Cyprinodon nevadensis mionectes*)_____ BIG SPRING AMARGOSA PUPFISH

(30) Lower pharyngeal teeth conical; jaw teeth bicuspid; intestine markedly coiled; jaws equal in length (i.e., the lower jaw not projecting forward beyond the upper jaw) (*Genus Crenichthys*) .... SPRINGFISH.... 31

.... Lower pharyngeal teeth molar-like; jaw teeth conical; intestine merely S-curved; jaws unequal in length, the lower jaw projecting forward beyond the upper jaw (Genus *Empetrichthys*) .... POOLFISH.... 32

(31) Lateral dark spots in one series; Railroad Valley, Nye County.

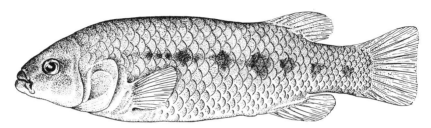

FIG. 93. Railroad Valley Springfish, *Crenichthys nevadae*.
Drawn by Silvio Santina.

.... Lateral dark spots in two series; White River system.

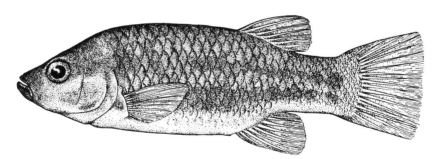

FIG. 94. White River Springfish, *Crenichthys baileyi*.
Drawn by Silvio Santina.

(32) Found only in Ash Meadows, Nye County; lateral line scales usually 30 or less in number; downward slope of snout quite marked so that mouth lies below the plane of longitudinal section.

FIG. 95. Ash Meadows Poolfish, *Empetrichthys merriami*.
Drawn by Grace Eager.

- Found only in Pahrump Valley, Clark County; lateral line scales usually 31 or more; downward slope of snout distinctly less abrupt so that mouth lies approximately on the plane of longitudinal section.

FIG. 96. Pahrump Poolfish, *Empetrichthys latos*. Drawn by Grace Eager.

(33) Distance between anal origin and caudal base shorter, averaging 346[2] in males and 328 in females; head depth less, averaging 267 in males and 281 in females; caudal peduncle length less, averaging 199 in males and 206 in females (*Empetrichthys latos latos*)
.... MANSE RANCH PAHRUMP POOLFISH
- Not as above 34
(34) Length of middle caudal rays shorter, averaging 221 in males and 214 in females (*Empetrichthys latos pahrump*) .... PAHRUMP RANCH PAHRUMP POOLFISH
- Length of middle caudal rays longer, averaging 232 in males and 228 in females (*Empetrichthys latos concavus*)
.... RAYCRAFT RANCH PAHRUMP POOLFISH
(35) Origin (front end) of anal fin from 1½ to 2½ times as far from tip of snout as from base of tail fin; tail fin rays usually 18 in number (16-branched); dorsal fin rays usually more than 10; pharyngeal teeth numerous, more than 10 in number, arranged in a single row like teeth in a comb; mouth usually directed downward, distinctly protractile, sucker-like, generally with papillose lips (Family CATOSTOMIDAE) .... SUCKERS 36

[2]Measured in thousandths of the standard length. Data from Miller 1948.

.... Origin of anal fin from 1 to 1½ times as far from tip of snout as from base of tail fin; tail fin rays usually 19 (17-branched); dorsal fin rays usually 10 or less; pharyngeal teeth fewer, less than 10, arranged in from 1 to 3 rows, not comb-like; mouth not especially directed downward, lacking papillose lips (Family CYPRINIDAE) .... MINNOWS AND CARP.... 44

(36) Nuchal (neck) region with a high, sharp-edged hump, formed by the greatly enlarged and expanded interneural vertebral spines; hump is detectable on all but very small individuals.

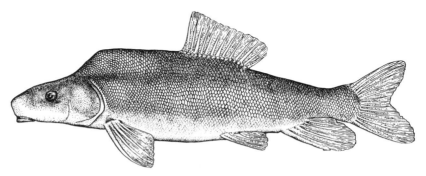

FIG. 97. Razorback Sucker, *Xyrauchen texanus*. Drawn by Silvio Santina.

.... No such hump, at most merely with an upslope on the otherwise smoothly curving dorsal outline.. 37

(37) Lips thin, lacking papillae; mouth semi-terminal, the lower jaw oblique.

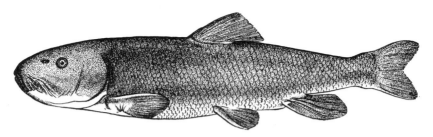

FIG. 98. Cui-ui Lakesucker, *Chasmistes cujus*. Drawn by Silvio Santina.

.... Lips thick with many papillae; mouth inferior (i.e., opening directly downward) 38

(38) With a distinct notch at each corner of the mouth; edge of jaw inside lower lip with a hard cartilaginous sheath; upper lip recurved; a small flap or "scale" of skin present at base of each pelvic fin, in axil (Genus *Pantosteus*) .... MOUNTAINSUCKERS.... 39

------- Without such a notch between upper and lower lips, although occasionally a very slight indentation occurs in some individuals; cartilaginous sheath, if present, flexible and not hard; upper lip nearly flat, not recurved; no flaps or "scales" in pelvic fin axils (Genus *Catostomus*) --- COMMON SUCKERS---- 40

(39) More than 35 scales in front of dorsal fin; fontanelle (space between parietal and frontal head bones) open; Lahontan drainage system of western Nevada.

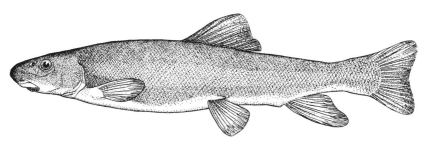

FIG. 99. Lahontan Mountainsucker, *Pantosteus lahontan*.
Drawn by Silvio Santina.

------- Less than 35 scales in front of dorsal fin; fontanelle closed; White River drainage system of eastern Nevada.

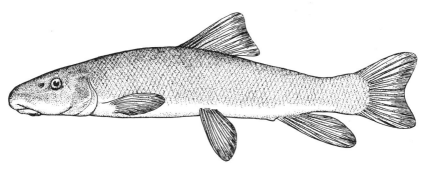

FIG. 100. White River Mountainsucker, *Pantosteus intermedius*.
Drawn by Silvio Santina.

(40) Lower lips not deeply incised medially, allowing several transverse rows (2–4) of papillae to cross the midline between the incision and the forward edge of the lower jaw (body scales small-to-moderately large, about 92 to 114 along the lateral line); Columbia River system.

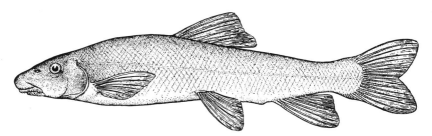

FIG. 101. Bridgelip Sucker, *Catostomus columbianus*. Drawn by Silvio Santina.

―――― Lower lips deeply incised medially, allowing one or no transverse rows of papillae to cross the midline between the incisures (body scales variable) ―――――――――――――――― 41

(41) Body slimmer and more attenuated than is usual for the genus, reflected quantitatively in the very slim caudal peduncle, whose least depth goes into body length at least 16 times; lips massively enlarged, hindermost edge of lower lip reaching posteriorly beyond nostrils and often to the eye (body scales more than 90 along the lateral line); Colorado River system.

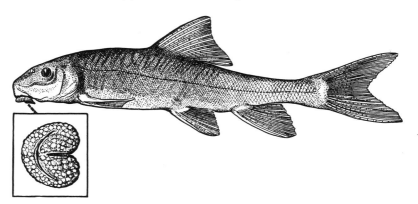

FIG. 102. Flannelmouth Sucker, *Catostomus latipinnis*. Drawn by William L. Brudon for Miller 1952. Courtesy California Department of Fish and Game and Leo Shapovalov.

―――― Body stouter and average for the genus, with caudal peduncle averaging about 12 into body length; lips smaller, hindlips not reaching behind nostrils (body scales variable) ―――――――――――――――――――――――――― 42

(42) Lips large, lower lips reaching posteriorly about to nostrils (body scales along lateral line larger, less than 80 in number); Columbia River system.

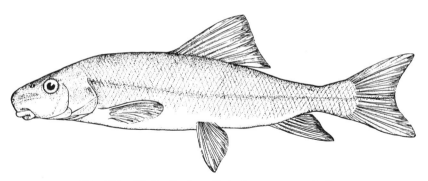

Fig. 103. Biglip Sucker, *Catostomus macrocheilus*. Drawn by Silvio Santina.

.... Lips smaller, lower lips falling distinctly short of reaching to the nostrils (body scales variable) 43

(43) Body scales larger, less than 80 in the lateral line; Bonneville system of Utah and eastern Nevada.

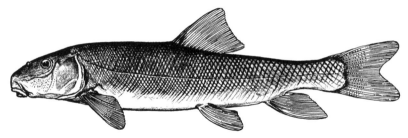

Fig. 104. Utah Sucker, *Catostomus ardens*. Drawn by William L. Brudon. Courtesy California Department of Fish and Game and Leo Shapovalov.

.... Body scales smaller, more than 80 in the lateral line; Lahontan system of Nevada and adjacent California.

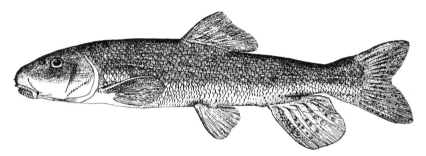

Fig. 105. Tahoe Sucker, *Catostomus tahoensis*. Drawn by Silvio Santina.

(44) Abdomen with a distinct, fleshy, scaleless keel between pelvic and anal fins.

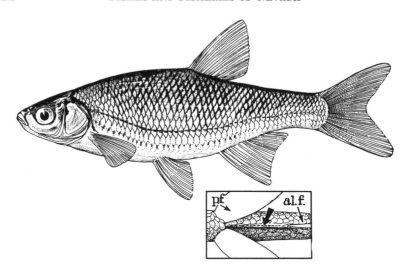

FIG. 106. Golden Shiner, *Notemigonus crysoleucas*. Drawn by William L. Brudon for Miller 1952. Courtesy California Department of Fish and Game and Leo Shapovalov.

| Abdomen without such modification _____ 45
(45) Distinct and strong spines developed at the anterior edge of the dorsal fin _____ 46
| No such spines (certain species occasionally have the first simple ray of dorsal fin hardened in very old individuals, but this is not a sharp, bony spine) _____ 52
(46) Dorsal fin very long, with more than 12 soft rays extending nearly to base of caudal fin; posterior edges of dorsal and anal fins about same distance from base of caudal fin; dorsal fin spine usually serrated; inner border of pelvic fins not adhering to the body at any point _____ 47
| Dorsal fin much shorter, usually with less than 10 soft rays, at least its own length removed from the base of caudal fin; posterior edge of dorsal fin at most only reaching to middle of anal fin; dorsal fin spines smooth; inner border of pelvic fins adhering to body for at least half their lengths _____ 48
(47) Barbels in two pairs on upper jaw; more than 32 scales in the lateral line (except in the "mirror" or "leather" varieties which have lost much of their scales).

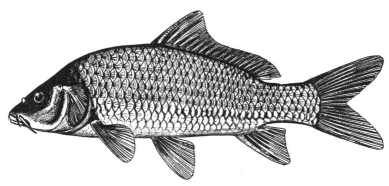

FIG. 107. Asiatic Carp, *Cyprinus carpio*. Drawn by William L. Brudon for Miller 1952. Courtesy California Department of Fish and Game and Leo Shapovalov.

------- Barbels lacking; less than 30 scales in the lateral line.

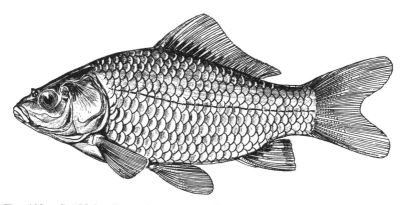

FIG. 108. Goldfish, *Carassius auratus*. Drawn by William L. Brudon for Miller 1952. Courtesy California Department of Fish and Game and Leo Shapovalov.

(48) Maxillary barbels present; body scaleless (an occasional individual has a few scales on the back and elsewhere in patches); of the two spines at the leading edge of dorsal fin, the anteriormost is the largest; body color a brilliant burnished silver.

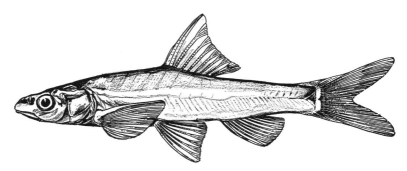

FIG. 109. Woundfin, *Plagopterus argentissimus*. Drawn by William L. Brudon for Miller 1952. Courtesy California Department of Fish and Game and Leo Shapovalov.

-------- Maxillary barbels absent; body covered with small scales (occasionally lacking under some of the fins); anteriormost of the two dorsal spines is the smallest; body color much less brilliant (Genus *Lepidomeda*) SPINEDACE---- 49

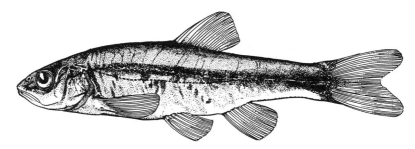

FIG. 110. White River Spinedace, *Lepidomeda* sp. Drawn by William L. Brudon for Miller 1952. Courtesy California Department of Fish and Game and Leo Shapovalov.

(49) Mouth[3] more oblique (a line from uppermost tip of premaxillary to middle of caudal peduncle passes above middle of pupil); snout sharper; dorsal spines stronger; dorsal fin higher (depressed length 1.8 to 2.1 in predorsal length), and more sharply pointed than in any other form; head more compressed (width 1.9 to 2.0 in its length); almost no pigment on shoulder girdle in advance of scapular bar. Additional characters—pigment on opercles about as in *L. mollispinis;* a band of coarse pigment crossing chin behind upperlip; pelvic fin length 1.4 to 1.5 in head; upper jaw length 1.15 to 1.3 in postorbital; (formerly) cooled, swift outflow from Ash Spring and Upper Pahranagat Lake, Pahranagat Valley, in course of Pluvial White River.

---

[3]This is the recently published key of Miller and Hubbs (1960: 15–16), with slight rearrangement.

FIG. 111. Holotype of *Lepidomeda altivelis*, UMMZ 125004, 56 mm. S. L. Photograph by William L. Brudon for Miller and Hubbs 1960. Courtesy University Michigan Press and Dr. Robert R. Miller.

Mouth less oblique (a line from uppermost tip of premaxillary to middle of caudal peduncle passes below middle of pupil); snout more rounded; dorsal spines variably weaker; dorsal fin low to moderate (depressed length 2.2 to 2.4 in predorsal length, except in *L. mollispinis pratensis*), and varyingly less pointed; head broader (width 1.5 to 1.85 in its length); pigment on shoulder girdle extending variably forward beyond scapular bar.... 50

(50) Melanophores typically extending across opercle and subopercle, and to angle of preopercle; lower half of outer face of shoulder girdle in adults with considerable pigment in front of vertical from pectoral insertion; size larger (commonly 80–90 mm., largest 103 mm. in standard length) and colors particularly bright; dorsal spines stronger; pharyngeal arch and teeth much more massive (in an average adult about 0.25 percent weight of fish), the whole arch thicker (less flattened dorsoventrally), anterior angle usually less conspicuous, more evenly rounded, edge of dorsal surface usually broadly rounded. Additional characters—pelvic fin length 1.4 to 1.8 in head length; upper jaw length 1.1 to 1.5 in postorbital; springs and spring-fed creeks in Recent White River Valley (upper headwaters of Pluvial White River system).

FIG. 112. Holotype of *Lepidomeda albivallis*, UMMZ 173781, 69.5 mm. S. L. Photograph by William L. Brudon for Miller and Hubbs 1960. Courtesy University of Michigan Press and Dr. Robert R. Miller.

▭ Melanophores confined to upper half of opercle and to upper part of upper limb of preopercle; lower half of outer face of shoulder girdle in adults lacking pigment in front of vertical from pectoral insertion; size smaller (only rarely more than 80 mm. in standard length) and colors less bright; dorsal spines weaker; pharyngeal arch and teeth much smaller and more delicate (in an average adult about 0.10 percent weight of fish), the whole arch flatter (more compressed dorsoventrally), anterior angle usually sharp and conspicuous, edge of dorsal surface medially and on anterior limb usually rather sharply ridged (the pharyneal arch and tooth distinctions are not sharp in the localized form *L. mollispinis pratensis*, of which no large adults are known); Virgin River system in Utah, Nevada, and Arizona, and (formerly) Big Spring in Meadow Valley, Nevada.

FIG. 113. Holotype of *Lepidomeda mollispinis mollispinis*, UMMZ 141673, 88 mm. S. L. Photograph by William L. Brudon for Miller and Hubbs 1960. Courtesy University Michigan Press and Dr. Robert R. Miller.

(51) Dorsal fin lower and more rounded (its depressed length 1.45 to 1.65 in distance from dorsal origin to occiput), when the fin is erected at about 45° the outer edge in the adult usually slopes downward and backward; pelvic fin shorter (length 1.5 to 1.85, usually 1.65 to 1.8, in head length); mouth smaller (upper jaw length 1.4 to 1.8, usually 1.45 to 1.6, in postorbital), and less oblique (a line from uppermost tip of premaxillary to middle of caudal peduncle passes below pupil); Virgin River system in Utah, Nevada and Arizona (*Lepidomeda mollispinis mollispinis*)▭▭▭▭▭ VIRGIN RIVER SPINEDACE

▭ Dorsal fin higher and more pointed (its depressed length 1.2 to 1.45 in distance from dorsal origin to occiput), when the fin is erected at about 45° the outer edge in the adult is usually about vertical; pelvic fin longer (length 1.35

to 1.6, usually about 1.5, in head length); mouth larger (upper jaw length 1.25 to 1.4 in postorbital), and more oblique (a line from uppermost tip of premaxillary to middle of caudal peduncle passes through lower part of pupil); (formerly) outflow of Big Spring, in meadow adjacent to Meadow Valley Wash, in the course of Pluvial Carpenter River, Nevada (*Lepidomeda mollispinis pratensis*)_____ BIG SPRING SPINEDACE

(52) A horny sheath covering the lip of the lower jaw (this can be lifted free with a needle)_____ 53

------ No such sheath present on the lower lip_____ 54

(53) A horny sheath covering only the lip of the lower jaw (this sheath is an external covering not to be confused with the small cartilaginous plate on the upper jaw of this species, a plate which is not visible externally, but is covered by the fleshy upper lip); Columbia River system.

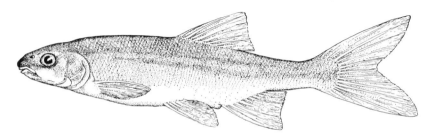

FIG. 114. Chiselmouth, *Acrocheilus alutaceus*. Drawn by Silvio Santina.

------ Horny sheaths covering the lips of both jaws; found only in Soldier Meadows, Humboldt County.

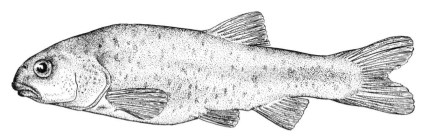

FIG. 115. Soldier Meadows Desertfish, *Eremichthys acros*. Drawn by Silvio Santina.

(54) Scales in the lateral line numbering 100 or more.

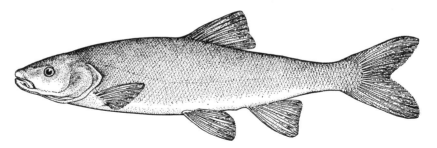

Fig. 116. Sacramento Blackfish, *Orthodon microlepidotus*.
Drawn by Silvio Santina.

‑‑‑‑‑‑‑‑ Scales in the lateral line less than 100‑‑‑‑‑‑‑‑‑‑‑‑‑‑‑‑‑‑‑‑‑‑‑‑‑‑‑‑‑‑‑‑‑‑‑‑‑‑ 55
(55) Species with only a single row of pharyngeal teeth, the lesser
or outer row never developed[4]‑‑ 56
‑‑‑‑‑‑‑‑ Species with two rows of pharyngeal teeth, the lesser or outer
row occasionally lacking on one side ‑‑‑‑‑‑‑‑‑‑‑‑‑‑‑‑‑‑‑‑‑‑‑‑‑‑‑‑‑‑‑‑‑‑‑‑‑ 57
(56) Scales in the lateral line larger, less than 65; maxillary
smaller, not reaching anterior edge of eye; intestine
longer, about equal in length to the standard body
length; pharyngeal arch uniformly and smoothly
rounded in the vicinity of the "heel."

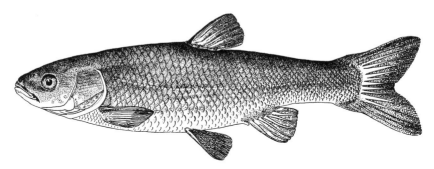

Fig. 117. Lahontan Tui Chub, *Siphateles bicolor obesus*.
Drawn by Silvio Santina.

‑‑‑‑‑‑‑‑ Scales in the lateral line smaller, more than 65; maxillary
larger, reaching posteriorly to about the anterior edge
of eye; intestine shorter, only about one-half the standard length; pharyngeal arch with a quite prominently

---

[4] It is unfortunate that pharyngeal teeth characteristics must be used in a key designed as much as possible so that anyone can use it. However, it is unavoidable in this instance, since no better character is available. The serious student of fishes will learn to dissect and clean the pharyngeal teeth in this family since much of the important classification is based on these teeth. With a little experience, the individual can learn to extract these "throat teeth" borne on the modified fifth gill arches by working inward under the posterior edge of the gill cover or opercle.

developed "heel" which breaks the otherwise smooth contour of the pharyngeal bone.

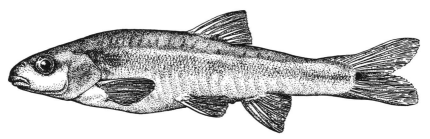

Fig. 118.   Moapa Dace, *Moapa coriacea*. Drawn by Silvio Santina.

(57) Barbels always absent from the posterior angles of the maxillary processes 59

      Barbels usually present on the posterior angles of the maxillary processes (since up to 50 percent of specimens in any given population may lack barbels, it is necessary to make extensive collections in determining these forms) 58

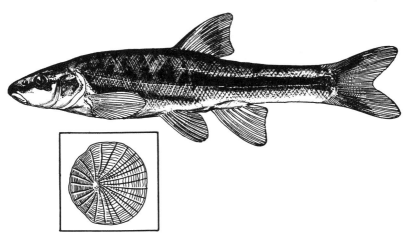

Fig. 119.   Speckle Dace, *Rhinichthys osculus*. Drawn by William L. Brudon for Miller 1952. Courtesy California Department of Fish and Game and Leo Shapovalov.

(58) Snake River drainage of northeastern Nevada; Rivers Owyhee, Bruneau, Jarbidge, Salmon and tributaries[5] (*Rhinichthys osculus carringtoni*) ....
.... SNAKE RIVER SPECKLE DACE

      Lahontan drainage of western, central and northern Nevada; Rivers Humboldt, Truckee, Carson, Susan, Walker and

---

[5]This is the only currently practicable way to separate the known Nevada forms of this species.

associated tributaries and lakes (*Rhinichthys osculus robustus*) .... LAHONTAN SPECKLE DACE

―――― Amargosa River drainage of southwestern Nevada-southeastern California (*Rhinichthys osculus nevadensis*) ____
.... AMARGOSA SPECKLE DACE

―――― White River drainage of eastern and southeastern Nevada (*Rhinichthys osculus velifer*) _____
.... WHITE RIVER SPECKLE DACE

(59) An accessory "scale" or flap of skin present in the axil (base) of the pelvic fin; sides with a deep orange-to-red broad band in adults of both sexes, heightened to brilliancy during the breeding season (Genus *Richardsonius*) .... REDSHINERS 60

―――― No such scale or flap of skin present; without the above described color pattern _____ 62

(60) Rays of anal fin usually numbering 8–9, rarely 10; body comparatively slender; Lahontan drainage system.

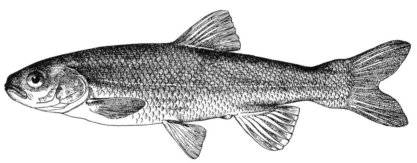

FIG. 120. Lahontan Redshiner, *Richardsonius egregius*.
Drawn by Silvio Santina.

―――― Rays of anal fin from 10–22; body comparatively robust; Snake River tributaries of northern Nevada _____ 61

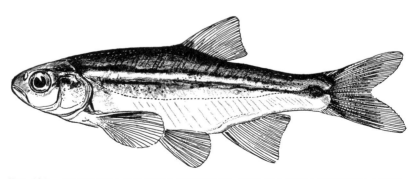

FIG. 121. Bonneville Columbia Redshiner, *Richardsonius balteatus hydrophlox*. Drawn by William L. Brudon for Miller 1952. Courtesy California Department of Fish and Game and Leo Shapovalov.

(61) Anal fin rays usually 14–18, rarely 13–22; Columbia-Snake system below the falls of the Snake River (intergrades with the Bonneville form as indicated below) (*Richardsonius balteatus balteatus*)_____
_____ TYPICAL COLUMBIA REDSHINER

_____ Anal fin rays usually 11–12, rarely 10–13; Bonneville system and the upper Snake River drainage (*Richardsonius balteatus hydrophlox*)_____
_____ BONNEVILLE COLUMBIA REDSHINER

_____ Anal fin rays usually 13–14; northeastern Nevada, in Snake River tributaries (*Richardsonius balteatus : : balteatus X hydrophlox*)_____ INTERGRADES

(62) Head long and pike-like, the mouth deeply cleft with maxillary extending backward beneath eye at least to leading edge of pupil; pharyngeal teeth subconical, scarcely hooked, sharp-edged, the lower limb of the pharyngeal bone greatly elongated (Genus *Ptychocheilus*)_____
_____ SQUAWFISH____ 63

_____ Head not as above, the mouth less deeply cleft and maxillary not reaching to leading edge of eye; pharyngeal teeth compressed, close-set, strongly-hooked, pharyngeal bone not as described above (Genus *Gila*)____ GILA CHUBS____ 64

(63) Scales large, numbering 67 to 80 in the lateral line; anal rays averaging 8; Columbia River system.

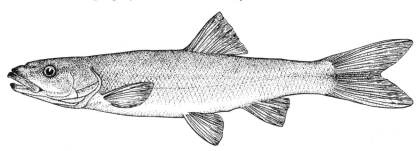

FIG. 122. Northern Squawfish, *Ptychocheilus oregonensis*.
Drawn by Silvio Santina.

_____ Scales smaller, numbering 83–93 in the lateral line; anal rays averaging 9; Colorado River system.

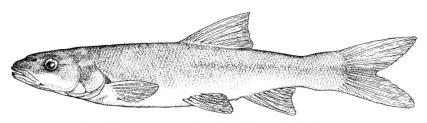

FIG. 123. Colorado Squawfish, *Ptychocheilus lucius*.
Drawn by Silvio Santina.

(64) Origin (front end) of dorsal fin immediately over the origin of the pelvic fins; lateral line scales numbering 70 or less; caudal peduncle relatively deep, resembling that of a *Siphateles*.

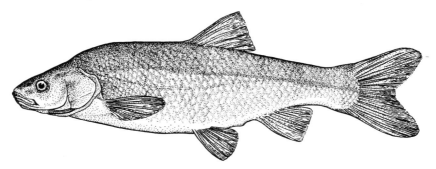

Fig. 124. Utah Gila, *Gila atraria*. Drawn by Silvio Santina.

_____ Origin of dorsal fin behind the origins of the pelvic fins; lateral line scales approximately 80 or more; caudal peduncle moderately robust to extremely slender and pencil-shaped *(Gila robusta)* _____COLORADO GILA____ 65

(65) Dorsal fin rays 8–10, usually 9; anal fin rays 7–10, usually 9; pelvic fin rays usually 9/9, rarely 10/10; body fully scaled, scales numbering from 79 to 96 in the lateral line; basal radii of scales poorly developed; nuchal hump detectable in older specimens; caudal peduncle least depth goes into head length from 3.3 to 4.3 times; small rivers *(Gila robusta robusta)* _____
_____ TRIBUTARY COLORADO GILA

_____ Not as above_____ 66

(66) Dorsal fin rays 10–11; anal fin rays 10–11; pelvic fin rays usually 9/9, rarely 10/10; body scales often lacking over dorsum, venter, and caudal peduncle, or consisting there of minute embedded scales; lateral line scales 75–88; basal radii completely lacking in scales; nuchal hump prominent; caudal peduncle least depth from 5.0 to 6.5 into head length; large rivers.

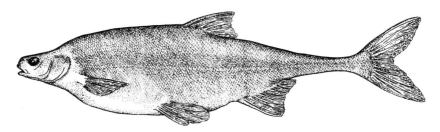

Fig. 125. Bonytail Colorado Gila, *Gila robusta elegans*. Drawn by Silvio Santina.

------ Dorsal fin rays 9; anal fin rays 9; pelvic fin rays 9/9; body fully scaled, scales numbering from 89 to 94 in the lateral line; basal radii of scales conspicuous; nuchal hump absent; caudal peduncle least depth from 3.3 to 4.1 into head length; White River system.

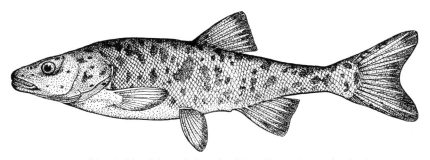

FIG. 126. White River Colorado Gila, *Gila robusta jordani*. Drawn by Silvio Santina.

KEY TO FAMILIES OF NEVADA FISHES

(1) Pelvic fins thoracic or jugular in position (i.e., moved forward from their usual position near the anal fin), always present, typically with 1 spinous and 5 soft rays each (reduced to 1 spinous and from 3 to 4 soft rays in Cottidae); anterior rays of dorsal fin spinous_____ 2

------ Pelvic fins abdominal in position, lacking entirely in some of the Cyprinodontidae; spinous rays usually lacking_____ 4

(2) Body lacking true scales, often covered with prickles; head large, broad, flattened dorso-ventrally; body noticeably small and tapering behind the large head (Family COTTIDAE)_____ SCULPINS

------ Body uniformly covered with conspicuous scales; head much smaller in proportion to the body, typically narrow and flattened laterally_____ 3

(3) Two distinct dorsal fins present, the anterior with spinous rays, the posterior one soft-rayed; less than 3 anal spines (Family PERCIDAE)_____ PERCHES

------ Only one dorsal fin present (in blackbasses, the dorsal fin has a moderate-to-deep notch between the anterior spinous and the posterior soft-rayed portions, but these two parts are always connected); 3 or more anal spines (Family CENTRARCHIDAE)_____ SUNFISHES

(4) Dorsal fin with a greatly elongated, filamentous posterior ray; venter coming to a narrow, sharp, saw-toothed margin along the mid-line (Family CLUPEIDAE) HERRINGS

------ Dorsal fin usual, with no such elongated rays; venter rounded, unkeeled, except in *Notemigonus*, where the keel is fleshy and scaleless _____ 5

(5) With adipose fin, a short, fleshy process between dorsal fin and tail 6
 .... Without adipose fin 8
(6) Conspicuous barbels or tentacles about and adjacent to mouth; sharp, stout spines at leading edges of dorsal and pectoral fins (Family ICTALURIDAE) .... CATFISHES
 .... No barbels; no dorsal or pectoral fin spines 7
(7) Mouth small, the maxillary process not extending behind the eye in adults, usually ending before or just at anterior

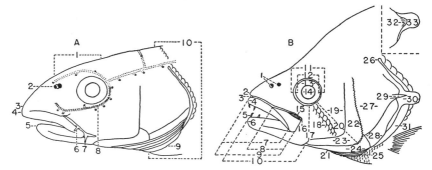

FIG. 127.   (A) Head of a soft-rayed fish—(1) forehead. (2) nostrils. (3) frenum, the membrane which binds together the tips of upper jaw and snout, resulting in a nonprotractile upper lip. (4) tip of upper jaw, the symphysis of premaxillaries. (5) tip of lower jaw, the chin, the symphysis of dentaries; since tip of lower jaw is posterior to tip of upper jaw, lower jaw is inferior or included and mouth is inferior, subterminal and ventral. (6) a barbel which is in advance of posterior end of upper jaw and above maxillary and in the groove between maxillary and preorbital bones. (7) barbel at posterior end of upper jaw. (8) part of anterior portion of lateral canal system, the shaded portion of which is the infraorbital canal; if all pores are connected to canal, the infraorbital canal is said to be complete, but if one or more pores are isolated, it is incomplete. (9) branchiostegals. (10) head depth.

(B) Head of a spiny-rayed fish—(1) nostrils. (2) tip of terminal snout. (3) groove between tips of snout and symphysis of upper jaw; since there is no binding frenum, the upper jaw may be moved forward, resulting in a protractile upper jaw. (4) tip of upper jaw and symphysis of premaxillaries. (5) canine tooth. (6) tip of lower jaw and symphysis of dentaries (the chin); since tip of lower jaw extends as far anteriorly as does tip of upper jaw, the mouth is terminal. (7) premaxillary length. (8) maxillary length. (9) upper jaw length, sometimes called "maxillary length." (10) lower jaw or mandibular length. (11) eye or orbital length. (12) eyeball length. (13) iris. (14) pupil. (15) suborbital width. (16) supramaxillary bone (shaded). (17) cheek depth, or height. (18) row of cheek scales (these are the ones that are counted). (19) cheek. (20) length from eye to preopercular angle. (21) isthmus. (22) serrated posterior edge of preopercle. (23) preopercle. (24) interopercle. (25) branchiostegals. (26) upper angle of gill cleft. (27) opercle. (28) subopercle. (29) opercular spine, absent in many species. (30) membranous opercular flap, the "earflap." (31) gill cleft, which in most species extends from its upper angle to isthmus. (32) posterior extension of opercular bone without a spine. (33) membranous opercular flap, the "earflap." Reproduced from *The Fishes of Ohio*, by Milton B. Trautman (Columbus, Ohio: Ohio State University Press, 1957), p. 59. Used by permission of the publisher.

edge of eye; jaw teeth weak; scales in the lateral line fewer than 105 (Family COREGONIDAE) .... WHITEFISHES

— Mouth large, the maxillary process extending at least to hind edge of eye in adults, usually considerably posterior to the eye; jaws usually bearing strong teeth; mouth deeply cleft; lateral line scales more than 115 (Family SALMONIDAE) .... TROUT

(8) Head with scales 9

— Head without scales 10

(9) Origin of dorsal fin behind origin of anal fin, about on a perpendicular with hind base of anal fin; dorsal fin generally 6-rayed (counted at bases); pelvic fins conspicuous, nearly as large as pectoral fins; third anal ray unbranched (counting rudiments); anal fin of male unlike that of female, modified into a long narrow reproductive organ (Family POECILIIDAE) .... TOPMINNOWS

— Origin of dorsal fin about on a perpendicular with origin of anal fin; dorsal fin generally 8-rayed or more (counted at bases); pelvic fins inconspicuous, much smaller than the pectorals, or entirely absent; third anal ray branched, except in some immatures; anal fins of male and female similar (Family CYPRINODONTIDAE) .... KILLIFISHES

(10) Origin (front end) of anal fin from 1½ to 2½ times as far from tip of snout as from base of tail fin; tail fin rays usually 18 in number (16-branched); dorsal fin rays usually more than 10; pharyngeal teeth numerous, more than 10 in number, arranged in a single row like teeth in a comb; mouth usually directed downward, distinctly protractile, sucker-like, generally with papillose lips (Family CATOSTOMIDAE) .... SUCKERS

— Origin of anal fin from 1 to 1½ times as far from tip of snout as from base of tail fin; tail fin rays usually 19 (17-branched); dorsal fin rays usually 10 or less; pharyngeal teeth fewer, less than 10, arranged in from 1 to 3 rows, not comb-like; mouth not especially directed downward, lacking papillose lips (Family CYPRINIDAE) .... MINNOWS

3—ACCOUNTS OF SPECIES

THE BONY FISHES, Class *OSTEICHTHYES*

THE RAY–FINNED BONY FISHES
Subclass ACTINOPTERYGII

THE TRUE BONY FISHES
Superorder TELEOSTOMI

THE ISOSPONDYL FISHES
Order *ISOSPONDYLIFORMES*

THE HERRING AND SHAD–LIKE FISHES
Suborder CLUPEOIDEI

HERRING AND SHAD
Family *CLUPEIDAE*

This extensive family contains many species of great economic importance over the world—herring, shad, anchovy, etc. Distribution is worldwide, from the colder northern oceans to the tropics. Most of them inhabit oceans, but some are river-dwellers and a few show the same anadromicity as the Salmonidae. The only Nevada representative is a fresh-water form recently introduced into Lake Mead as a forage fish.

DIAGNOSIS

Body elongate to short and deep, noticeably compressed laterally, scales cycloid or pectinate; head bare, often much shortened; mouth variable in size, inferior or terminal, without barbels; teeth small or lacking; no pharyngeal teeth; branchiostegals variable, about 6 to 15 in number; pseudobranchiae present and often large; gills 4 in number, the fourth followed by a slit; gill membranes unconnected, unattached to the isthmus; fins—adipose lacking, dorsal median or somewhat posterior, occasionally lacking, its last ray sometimes greatly elongated, pectorals and ventrals modest in size, with or without accessory scales, the ventrals rarely lacking, anal long and narrow, caudal strongly forked; lateral line absent; some with short, muscular, fowl-like gizzard stomach; air bladder connected to the ear; belly often compressed and in some species edged with bony serratures.

This highly successful family goes back quite a way in the fossil record, some clupeids being known from the Cretaceous of various parts of the world.

SHAD
Genus *Dorosoma* Rafinesque 1820
("Lance body," referring to the shape of the young)

MISSISSIPPI THREADFIN SHAD

Dorosoma petenense atchafalayae (Evermann and Kendall)
(Atchafalaya Gizzard Shad)

(*Meletta petenensis* Günther 1866: 603)
Signalosa atchafalayae Evermann and Kendall 1898 (1897): 127–128
Signalosa atchafalayae, Jordan and Evermann, 1898B: 2809–2810
Signalosa atchafalayae, Jordan, Evermann and Clark, 1930: 46–47
Signalosa atchafalayae, Schrenkeisen, 1938: 33
Signalosa petenensis, Miller, 1950C: 390
Signalosa petenensis atchafalayae, Hubbs, 1951: 297
Signalosa petenensis atchafalayae, Parsons and Kimsey, 1954
Dorosoma petenense, Berry, Huish and Moody, 1956: 92
Dorosoma petenense, Eddy, 1957: 43
Dorosoma petenensis atchafaylae (sic), Kimsey, Hagy and McCammon, 1957
Signalosa petenensis, Moore, 1957: 60
Dorosoma petenense, Shapovalov, Dill and Cardone, 1959: 166, 171
Dorosoma petenense, Miller, 1960: 373

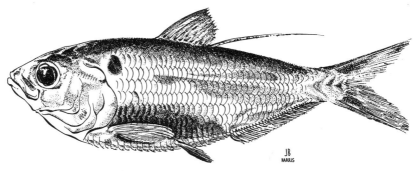

FIG. 128. Mississippi Threadfin Shad, *Dorosoma petenensis atchafalayae*. Drawn by John B. Harris.

ORIGINAL DESCRIPTION

"Type locality, Atchafalaya River at Melville, Louisiana. Type, No. 48790, U. S. N. M. Cotypes, No. 48791, U. S. N. M.; No. 532, U. S. F. C.; and No. 5775, L. S. Jr. Univ. Mus. Collector, F. M. Chamberlain, May 5, 1897. Associate type localities, Grand Plains Bayou and Black Bayou, Mississippi. Collectors, F. M. Chamberlain and H. R. Center.

"Length of type, 4¾ inches to base of caudal, or 5¾ inches to tips of caudal rays.

"Head 3⅗; depth 3⅐; eye 3½ in head; snout 5½; maxillary 3⅕; D. I, 12; A. I, 24; scales 42–15; scutes 17 + 10. Body oblong-elliptical, compressed, the back in front of dorsal narrow; ventral edge sharp, serrate; head small, mouth terminal, oblique, lower jaw slightly included; snout rather pointed, not blunt, as in *Dorosoma cepedianum;* maxillary in 3 pieces, long and curved, reaching vertical at front of pupil, the outer edge not notched; no teeth. Caudal peduncle short, compressed, and deep. Origin of dorsal fin over base of ventrals, much nearer tip of snout than base of caudal, the last ray filamentous, about one-fourth longer than head and nearly reaching base of caudal; the first dorsal ray about 2 in the last one; pectoral 1¼ in head, reaching base of ventrals; ventrals short, reaching only half way to vent, their length 1⅘ in pectorals; anal rays short, base of fin 1⅙ in head; scutes moderate; caudal widely forked, the lower lobe the longer; scales large, thin, deciduous, somewhat crowded anteriorly; accessory scales at bases of pectorals and ventrals; base of caudal with small scales. Color, bluish-black or dark olivaceous on back and sides to level of the jet-black humeral spot; rest of sides and under parts bright silvery; dorsal and caudal dusky; other fins plain.

"The cotypes from Grand Plains Bayou are 2 females with ripe roe. They are 4½ and 5½ inches long, respectively, and differ from the types only in the deeper body and the much darker coloration of the upper parts.

"This species appears to be rather common in the larger lowland streams and bayous of Louisiana and Mississippi. It probably does not reach a large size, adult examples being less than 6 inches long. It is not used as food, but is of considerable value as bait in the catfish fishery of the Atchafalaya River and its connecting lakes and bayous" (Evermann and Kendall 1898: 127–128).

DIAGNOSIS

Head about 3½ times, depth about 3 times, in body length; head moderate, snout semi-sharp, mouth oblique; 3-piece maxillary un-notched on outer curved margin, extending posteriorly to a vertical line dropped from front edge of pupil; branchiostegals 5; pseudobranchiae large; gillrakers short and numerous, around 350 in number; teeth lacking; stomach like a bird gizzard; scales 40–43/14–15; scutes 16–17 + 9–11; fins—dorsal origin over base of ventrals and set closer to anterior end of animal than to base of caudal fin, last dorsal ray filamentous, long, almost reaching to base of caudal fin, I, 12 rayed—anal long, narrow, I, 24 rayed; caudal strongly forked, the lower fork largest; ventral edge of body serrate and sharp; color—blackish to deep olivaceous dorsally, and laterally to humeral spot below this, color changes to bright silver, with dorsal and tail fins smoky; up to about 6 inches in length.

TYPE LOCALITY

Melville on the Atchafalaya River, Louisiana.

RANGE

Arkansas and Tennessee to Gulf of Mexico.

The Threadfin Shad was introduced into the Colorado River by tri-State agreement—Arizona, California and Nevada—in 1953. California stock was obtained from Watts Bar, Tennessee River, Tennessee, and flown to its western destination, the Colorado River and southern California ponds.

Fish for Nevada were brought in by the State Commission from Kentucky Lake, Tennessee River, Tennessee in 1953. Some of these were placed in the Overton arm of Lake Mead and others held in ponds nearby at Overton. The latter did not survive, but the initial plant in Lake Mead prospered and was added to from California stock next year.

By 1956, the threadfin seemed to be distributed the entire length of the Colorado mainstem from Lake Mead to Mexico.

TAXONOMY

Described originally as a separate species, the Mississippi Threadfin has not developed any complex synonymy. Its placement as a subspecies of the more widespread *Dorosoma petenense,* named from Lake Peten in Yucatan, is a result of a better understanding of the variability of the species. There has also been some tendency to elevate the subgenus *Signalosa* to generic status, but the differences between *Signalosa* and *Dorosoma* do not seem important.

Etymology—named for the type localities, Lake Peten in Yucatan and the Atchafalaya River in Louisiana.

LIFE HISTORY

The Threadfin Shad is a small active fish with a pronounced schooling instinct and a double-spawning period—spring and fall—each year. According to California Fish and Game investigations along the Colorado River, crustacea of several kinds, and algae, are their main food sources, with insects utilized when available. From this same source comes the following spawning information:

"During the early spawning, groups of two or three fish break from the school and swim rapidly to the surface. The female struggles and actually splashes at the surface, forced up by the presence of several males below her. The eggs are extruded and float down slowly.

"The eggs do not seem to be initially adhesive, but shortly after extrusion are observed sticking to submerged and floating objects.

"The small muscular 'gizzard' of the Threadfin shad usually has bits of sand in it. The stomach contents that have passed through its 'gizzard' have the appearance of having been ground. It therefore appears that the sand may be actually sought out, in addition to being taken incidentally during feeding" (Kimsey, Hagy and McCammon 1957: 3–4).

ECONOMICS

Threadfin Shad rate high as forage for warm water fish such as blackbass, crappie, etc. Their introduction to the Colorado River was premised chiefly on the need for a good food fish for Largemouth Blackbass. With changing of long stretches of the Colorado River into impoundments suitable for the development of a blackbass fishery, native forage fish have disappeared or become rarities. A good deal of thought and planning went into selection of a fish which would have the following characteristics: (a) small size, (b) prolificness, (c) non-competitive feeding habits, (d) habitat needs similar to those of blackbass and (e) ability to do well in the newly-formed environments of the Colorado River.

Threadfin were known to meet the first four requirements very satisfactorily—the fifth could only be determined by trial, and present indications are that the shad has come through with flying colors here also. Furthermore, threadfin have specific mannerisms which make them additionally attractive as a bass bait—they school in the open where they are easily seen and move erratically and flashily, calling attention to their presence.

SALMON-LIKE FISHES, Suborder SALMONOIDEI
SALMON AND TROUT
Family *SALMONIDAE*

This well-known family is a large and variable group, some species of which are familiar to almost everyone. Typically a northern hemisphere freshwater assemblage, it has many species which exhibit pronounced anadromicity, i.e., which live in the sea but run up freshwater coastal streams to spawn. The Nevada fauna of salmonids is composed almost entirely of the freshwater segment of the family; its only anadromous representatives are the salmon and trout living in the upper reaches of the Snake River tributaries, some individuals of which belong to coastal populations normally living in the ocean. Probably these species were able to ascend the major Nevada Snake tributaries (Owyhee, Bruneau, Jarbidge, and Salmon Rivers) for some distance originally, but dam construction is in process of eliminating anadromous salmonids from these waters, if it hasn't already done so.

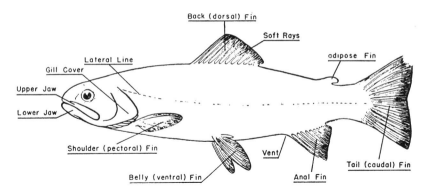

Typical Trout or Salmonoid fish with descriptive terms.

FIG. 129. Courtesy Oregon State Game Commission and Harold P. Smith.

From the economic point of view, Salmonidae are the most important fishes in Nevada, and provide most of the sport fishing in the State. Of the ten forms now occuring here, only three are native, the remainder representing introductions as a result of previous fisheries management programs carried on by State and Federal agencies.

DIAGNOSIS

Body elongate, scales cycloid; head bare; mouth terminal, variable in size, barbels lacking; jaws toothed; pharyngeal teeth lacking; branchiostegals 10 to 20 in number; pseudobranchiae present; gills 4 in number, the fourth followed by a slit; gill membranes unconnected, unattached to the isthmus; fins—adipose present, pectorals ventral, pelvics abdominal, caudal usually forked; lateral line present; stomach siphonal; air bladder large; pyloric caeca numerous; ova large, free in abdominal cavity before deposition.

Characteristic of waters generally north of the Fortieth Parallel, and very successful as a rule, having achieved considerable diversity in habitat preferences and tolerance ranges. Many of the larger species are anadromous; others live in freshwater lakes and run up tributary streams to spawn, while the remainder constitute the resident stream populations. As a group, salmonids are young, being unrepresented below the Pliocene Epoch. The multitudinous numbers of intergrading forms which exist in western United States waters are good indicators of the youth of the group and its consequent plasticity.

KEY TO GENERA OF NEVADA SALMONIDAE

1. Anal fin elongate, from 13- to 19-rayed (rarely with 12 to 18 rays); vomer narrow, long, flat, with weak teeth; gillrakers 19–40 on first gill arch; branchiostegals 13–19; species with or without black spots, the adults with anal and dorsal fins seldom spotted (*Oncorhynchus*) _____ PACIFIC SALMON

   ____ Anal fin short, 9- to 12-rayed (rarely 13); gillrakers 20 or less on first gill arch; branchiostegals 10–20; dorsal fin black-spotted _____ 2

2. Species with dark spots on a lighter background; fewer than 190 scale rows crossing the lateral line; vomer flat, its toothed surface plane, teeth on vomerian shaft in alternating rows or in one staggered row, those on the shaft placed directly on surface of bone, not on a free crest (*Salmo*) _____ TROUT

   ____ Species with light spots, white or gray, on a darker background; often with red spots on sides; with over 190 scale rows crossing the lateral line; vomer boat-shaped, the shaft depressed and without teeth (*Salvelinus*) _____ CHARR

PACIFIC SALMON

Genus *Oncorhynchus* Suckley 1861

("Hook snout")

KEY TO SPECIES

Gillrakers comparatively short and few in number, 19–28 on the first gill arch (rarely 29); gillraker formula 7 to 9 plus 11 to 13;[1] anal rays usually 15–17; distinct black spots on back and caudal fin (*Oncorhynchus tshawytscha*) _____ KING SALMON

Gillrakers comparatively long and numerous, 30–50 on the first gill arch; gillraker formula 11 to 24 plus 20 to 26; anal rays usually 14–15; spotting not evident on back and/or caudal fin, but fine black speckling may be present (*Oncorhynchus nerka kennerlyi*) _____ KOKANEE RED SALMON

---

[1]Interpreted as 7 to 9 gillrakers on the upper limb of the first gill arch, and 11 to 13 gillrakers on the lower limb.

KING SALMON

Oncorhynchus tshawytscha (Walbaum)

(Spring Salmon, Chinook Salmon, Quinnat Salmon, Tyee Salmon)

Salmo tshawytscha Walbaum 1792: 71
Salmo orientalis Pallas 1811: 367
Salmo quinnat Richardson 1836: 219–220
Fario argyreus Girard 1856: 218 / 1859: 212–213
Salmo confluentus Suckley 1858
Salmo quinnat, Girard, 1859: 306–307
Salmo cooperi Suckley 1861
Salmo richardi Suckley 1861
Salmo warreni Suckley 1861
Oncorhynchus quinnat, Günther, 1866: 158
Oncorhynchus chouicha, Jordan and Gilbert, 1882: 306–307
Oncorhynchus tschawytscha, Gilbert and Evermann, 1895: 198–200
Oncorhynchus tschawytscha, Jordan and Starks, 1896: 791
Oncorhynchus tschawytscha, Jordan and Evermann, 1896B: 479–480 / 1902: 161–164
Oncorhynchus tschawytscha, Snyder, 1917A: 79, 85 (Nevada)
Oncorhynchus tschawytscha, Pratt, 1923: 43
Oncorhynchus tschawytscha, Locke, 1929: 181–182
Oncorhynchus tschawytscha, Jordan, Evermann and Clark, 1930: 55
Oncorhynchus tschawytscha, Davidson and Hutchison, 1938: 667, 669–670
Oncorhynchus tschawytscha, Schrenkeisen, 1938: 38–39
Oncorhynchus tshawytscha, Dimick and Merryfield, 1945: 26–29
Oncorhynchus tschawytscha, La Monte, 1945: 106–107
Oncorhynchus tschawytscha, Miller and Alcorn, 1946: 190
Oncorhynchus tschawytscha, Eddy and Surber, 1947: 116
Oncorhynchus tschawytscha, Carl and Clemens, 1948: 41, 47–48
Oncorhynchus tschawytscha, Miller and Miller, 1948: 183–184
Oncorhynchus tschawytscha, Clemens and Wilby, 1949: 85–86
Oncorhynchus tschawytscha, Shapovalov and Dill, 1950: 385
Oncorhynchus tschawytscha, La Rivers, 1952: 92–93
Oncorhynchus tschawytscha, La Rivers and Trelease, 1952: 113
Oncorhynchus tschawytscha, Gabrielson and La Monte, 1954: 93–94
Oncorhynchus tschawytscha, Eddy, 1957: 48
Oncorhynchus tschawytscha, Moore, 1957: 62
Oncorhynchus tschawytscha, Slastenenko, 1958A: 68–69
Oncorhynchus tshawytscha, Carl, Clemens and Lindsey, 1959: 78–79
Oncorhynchus tshawytscha, Shapovalov, Dill and Cordone, 1959: 171

ORIGINAL DESCRIPTION

"37. Salmo, *Tshawytscha. Pennant arct.—zool. introd.* 124. D. 12. P. 16. V. 10. A. 15.

"Corpus grandiflumum latum, pondere 50 vel 60 circiter librarum immaculatum superne obscure cinereum; a lateribus argenteum pinnis e caeruleo albescentibus. Maxillae rectae & nunquam curvatae. Dentes magni in seriebus diversis. Squamae majores sunt, quam in Salmone

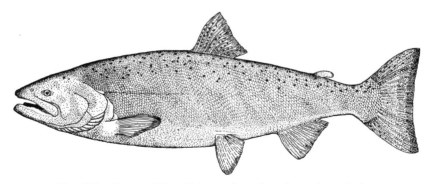

FIG. 130. Female King Salmon, *Oncorhynchus tshawytscha*.
Drawn by Silvio Santina.

communi. Cauda lunata. Caro ejus, tamdiu in mari reflat, rubra, quae in aquis dulcibus albescit. Habitat in fluminibus Kamtschatkae.
"*Tschawitscha. Krascheninnikow descr. Kamtschatkae* p. 178.
"Salmoni commune similis verum multo latio est. Rostrum acutum. Maxilla supera longior infera. Opercula longa, angusta. Cauda aequalis. Dorsum subcaeruleum maculis nigris minimis conspersum. Latera argentea. Abdomen album. Caro & cruda & cocta rubicunda" (Walbaum 1792: 71–72).

DIAGNOSIS

Head and greatest body depth each about 4 times in body length; branchiostegals from 15 to 16 on one side and from 18 to 19 on the other; gillrakers on the first arch from 7 to 9 above the angle and from 11 to 13 below the angle (written 7–9 + 11–13 in following diagnoses); pyloric caeca from 140 to 185; vertebrae 66; scales from 23 to 34 above the lateral line, from 135 to 155 along the lateral line, and from 19 to 39 below the lateral line (written 23–34/135–155/19–39 in following diagnoses); fins—dorsal 11-rayed, anal 16-rayed; color

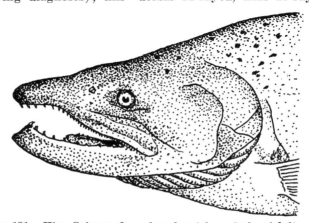

FIG. 131. King Salmon, *Oncorhynchus tshawytscha*. Adult male.
Drawn by Silvio Santina.

dark above (with blue or green lights), silvery on sides—back, dorsal and caudal fins profusely black-spotted—breeding males become red-splotched; the largest of the Pacific Salmon, attaining lengths of over 5 feet; average weights are around 20 pounds, with one 125-pounder recorded.

TYPE LOCALITY

Rivers of the Kamchatka Peninsula, Siberia.

RANGE

North Pacific coastal waters and tributary freshwater streams from northern China to northern California. Although formerly known to occur in certain headwaters of the Snake River tributaries in Nevada, their status in some of these streams is now uncertain due to the construction of impassable dams which have cut the upper Snake drainage at several points. Gilbert and Evermann's comments are timely, since they made some study of the situation in this region at a time when upstream conditions were in process of change:

"The principal tributaries of that portion of Snake River which is accessible to salmon are the following; Salmon Falls River or Salmon Creek, Malade River, Bruneau River, Owyhee River, Boise River, Payette River, Salmon River, Grande Ronde River, Clearwater River, and Palouse River . . .

"Bruneau River was formerly an important stream for spawning salmon, which reached its head waters in October, according to the statement of Mr. J. L. Fuller, of Bliss, Idaho. We are informed that a dam recently constructed in the Lower Bruneau now wholly prevents the ascent of fish.

"The Owyhee River is still open to salmon, so far as could be learned from reports. Mr. J. L. Fuller has seen them in the extreme headwaters of the Owyhee in Nevada" (1895).

Another obscure testimonial for salmon in northern Nevada, of more historical than scientific interest, is provided by a note which appeared in the Scientific American in 1874. The author, a resident of Elko, Nevada, signed himself only as "G. A. F."

"About 80 miles north of this place (Elko) on the north slope of Bull Run Mountain, which never loses its massive banks of snow, rises a small stream, formed by springs that furnish the purest and coldest water I ever drank. The stream, after running a distance of half a mile, is about 2 feet deep and about 6 feet wide on an average; at this point a succession of hot springs rise on the banks, and flow into the stream, increasing the volume of water about one third. The water of these springs is so intensely hot that less than three seconds are consumed in boiling eggs in it. The creek above and below this point swarms with fine brook trout; . . .

"This stream is one of the many that form the headwaters of the Columbia River; and to this point, over eighteen hundred miles from its mouth, in the spring and fall, the salt water salmon come in hundreds to spawn."

Miller and Miller (1948) could not say definitely whether this salmon still ascended the Owyhee.

FIG. 132. Courtesy Oregon State Game Commission and Harold P. Smith.

TAXONOMY

Although formally described 168 years ago, the specific name *tshawytscha* has been in consistent use for only about the last 65 years. For most of the interim, Richardson's *quinnat* was the accepted designation. The spelling "tschawytscha" has only relatively recently been changed to *tshawytscha*.

Etymology—this is the native Siberian and Alaskan name, and has also been spelled "tchaviche."

LIFE HISTORY

A carnivorous, fall-spawning species ascending West Coast rivers farther than other salmon—running nearly 2,000 miles upstream during the breeding season. Maturity is reached in from three to seven years, the young returning to the ocean during their first year.

ECONOMICS

The most important species of Pacific Salmon in the main United States, furnishing an extensive commercial and sport fishery. Its value in northern Nevada was originally that of providing local Indians with a readily utilizable food supply, a food supply which no longer exists due to manmade changes in the streams.

Several attempts were made to introduce the species into west-central Nevada to provide temporary fishing; in the years 1879, 1908, 1909, and 1910, about half-a-million spawn were imported, hatched and planted in the lakes and major rivers of the Lahontan basin. Snyder's notation apparently records about the last of these fish to be caught: "An example of this species 10½ inches long was caught by an angler while trolling in Lake Tallac June 20, 1911" (1917A: 85), which is part of the Lahontan drainage lying southwest of Lake Tahoe. Obviously, any such plants could not endure.

RED SALMON
Oncorhynchus nerka (Walbaum)
(Sockeye Salmon, Blueback Salmon)

Salmo nerka Walbaum 1792: 71
Salmo lycaodon Pallas 1811: 370
Salmo paucidens Richardson 1836: 222
Fario aurora Girard 1856: 218 / 1859: 308–310
Salmo kennerlyi Suckley 1861: 307 / 1874: 145–147
Hypsifario kennerlyi, Gill, 1862: 330
Salmo kennerlyi, Günther, 1866: 120
Oncorhynchus lycaodon, Günther, 1866: 155
Oncorhynchus nerka, Jordan and Copeland, 1878: 431
Oncorhynchus nerka kennerlyi, ibid
Oncorhynchus nerka, Jordan and Gilbert, 1882: 308
Oncorhynchus nerka, Jordan and Evermann, 1896B: 481 / 1902: 155–159
Oncorhynchus nerka, Pratt, 1923: 43
Oncorhynchus nerka, Foerster, 1925
Oncorhynchus nerka, Locke, 1929: 182–183
Oncorhynchus nerka, Jordan, Evermann and Clark, 1930: 55
Oncorhynchus nerka, Davidson and Hutchison, 1938: 667, 670
Oncorhynchus nerka, Schrenkeisen, 1938: 39–40
Oncorhynchus nerka kennerleyi, ibid
Oncorhynchus nerka, La Monte, 1945: 106
Oncorhynchus nerka, Miller and Alcorn, 1946: 190
Oncorhynchus nerka, Carl and Clemens, 1948: 42, 50–51
Oncorhynchus nerka kennerlyi, ibid: 52–53
Oncorhynchus nerka kennerlyi, Curtis and Fraser, 1948: 111
Oncorhynchus nerka, Clemens and Wilby, 1949: 88–89
Oncorhynchus nerka, Shapovalov and Dill, 1950: 385
Oncorhynchus nerka kennerlyi, Simon, 1951: 43
Oncorhynchus nerka kennerlyi, Beckman, 1952: 22–23
Oncorhynchus nerka kennerlyi, La Rivers, 1952: 93
Oncorhynchus nerka kennerlyi, La Rivers and Trelease, 1952: 113–114
Oncorhynchus nerka kennerlyi, Gabrielson and La Monte, 1954: 94
Oncorhynchus nerka, Eddy, 1957: 48
Oncorhynchus nerka, Moore, 1957: 62
Oncorhynchus nerka, Slastenenko, 1958A: 73–75
Oncorhynchus nerka, Carl, Clemens and Lindsey, 1959: 81–83
Oncorhynchus nerka, Nelson, 1959
Oncorhynchus nerka, Shapovalov, Dill and Cordone, 1959: 171

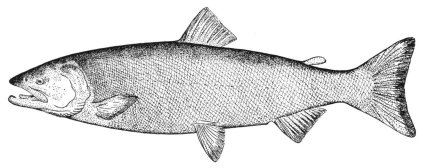

FIG. 133. Red Salmon, *Oncorhynchus nerka*. Drawn by Silvio Santina.

ORIGINAL DESCRIPTION

"36. SALMO, *Nerka*. Ruffis *Krasnaya ryba*, i.e., piscis ruber. *Pennant arct. zool. introd.* 125. D. 11. P. 16. V. 10. A. 15.

"Salmoni Salari forma similis, sed minor. Pondus 16 libras non superat. Quando flumina intrat colore argenteo splendet, dorso & pinnis caerulescentibus: Dentes tunc sunt parvi & maxillae rectae. Postea, dum in aqua dulci moratur, dentes crescunt & maxillae praefertim in maribus incurvantur. E mari adscendit aeflate flumina Kamtschatkae. Caro intense purpurea.

"*Narka. Krascheninnikow descr. Kamtschatkae*, c. 9. p. 181.

"Caput per parvum. Rostrum breve acutum. Lingua subcaerulea. Corpus parum compressum longitudine circiter 21 unciarum & latitudine 4 unc. macrolepidotum. Dorsum subcaeruleum nigro maculatum; abdomen & latera alba. Cauda bifurca. Caro ruberrima" (Walbaum 1792: 71).

DIAGNOSIS

Head and greatest body depth each about 4 in body length; branchiostegals 13–15; gillrakers 14–15 + 22–23; pyloric caeca 75–95; vertebrae 64; scales 18–26/125–145/17–27; fins—dorsal 11-rayed, anal 14–16 rayed; color, back bright blue, sides silvery—bottom fins pale, upper fins dark, adults without spotting—breeding males become red-blotched; attains lengths of about 2 feet and weights of from 3–8 pounds.

TYPE LOCALITY

Rivers of the Kamchatka Peninsula, Siberia.

RANGE

North Pacific coastal waters and tributary freshwater streams from Japan to Oregon. Introduced into adjoining parts of central California and Nevada.

Kokanee Salmon have been introduced into Nevada in large numbers in recent years as part of a concerted effort by the State Fish and Game Commission to definitely establish whether or not these types are suitable for our waters.

These first plantings occurred in 1950 in Lake Tahoe in cooperation with California and continued until 1954, when the Nevada program was terminated and judged a success. California had begun kokanee planting in the area (Donner Lake) in 1944, terminating in 1947.

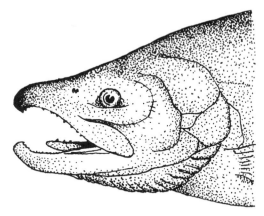

FIG. 134A. Red Salmon, *Oncorhynchus nerka*, adult male. Drawn by Silvio Santina.

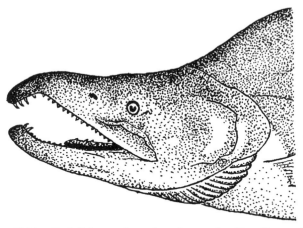

FIG. 134B. Red Salmon, *Oncorhynchus nerka*. Breeding male. Drawn by Silvio Santina.

The following tabulation summarizes plantings and their known results to date:

State involved	Locality	Years	Numbers	Results
California	Donner Lake	1944–1947	?	Successful
California-Nevada	Lake Tahoe	1950–1954	?	Successful
Nevada	Pyramid Lake	1951–1953	1,471,178	Failure
Nevada	Walker Lake	1951–1954	853,609	Failure
Nevada	Lahontan Reservoir	1951	5,000 fry	Failure
Nevada	Virginia Lake, Reno	?	?	Successful
Nevada	Wildhorse Reservoir	1954–1960	275,528	Successful

FIG. 135. Courtesy Oregon State Game Commission and Harold P. Smith.

TAXONOMY

Suckley's *kennerlyi* was described in 1861, 69 years after Walbaum characterized the species *nerka*. It has only been in the last 20 years that the most biologically logical status of *kennerlyi*—that of a landlocked subspecies of *nerka*—has been generally accepted.

Etymology—"nerka" is a Russian name, while Dr. C. B. R. Kennerly was naturalist to the Northwest Boundary Commission under Archibald Campbell.

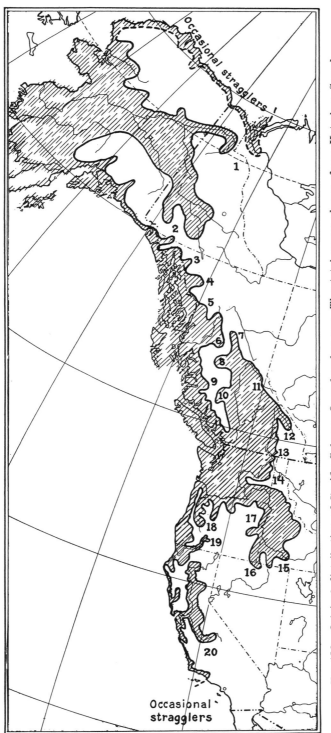

FIG. 136. Original distribution of Pacific Salmon, Oncorhynchus spp. The total range shown, from Kotzebue Sound southward, is that of the King Salmon, Oncorhynchus tshawytscha. Red Salmon, Oncorhynchus nerka, does not occur in numbers much south of the Columbia River. Numbers indicate literature references for specific localities. Courtesy University of California Press and Erhard Rostlund.

LIFE HISTORY

Similar to the King Salmon (*O. tshawytscha*), most individuals maturing, spawning and dying usually at the end of their fourth year. Unlike the former, however, Red Salmon are essentially plankton feeders during their growing period.

ECONOMICS

The most important salmon of the Alaskan canning industry. Both anadromous and landlocked populations seem to have been introduced into Nevada at one time or another, only the latter surviving. It seems safest, as Miller and Alcorn (1946) have done, to consider all introductions prior to 1940 to be anadromous stocks. These plants were made between 1936 and 1940 in various lakes, reservoirs and streams of westcentral Nevada, and were all unsuccessful.

At this writing (1960), the California-Nevada cooperative kokanee planting program for Lake Tahoe is completed, as is the Nevada Fish and Game Commission's kokanee stocking of Pyramid and Walker Lakes.

As plankton feeders, kokanee are an intermediate link in the food chain of larger fishes of economic value. An instance of this is its use as food by the famed Kamloops Rainbow Trout of Washington and British Columbia. According to Dymond (1936), kokanee are the staple food of Kamloops Trout larger than 16 inches in length, and the latter seldom exceed this length in lakes from which the kokanee is absent.

CHARRS

Genus *Salvelinus* Richardson 1836

(An old European name for the charr)

KEY TO SPECIES

1. Tail prominently forked in adults; body gray-spotted without red spots; fins not strongly bright-edged; vomer with a raised crest extending backward from the head of the bone, this crest armed with strong teeth (*Salvelinus namaycush*) LAKE TROUT
   - Tail weakly forked in adults; body yellow- or red-spotted in life; fins strongly bright-edged; vomer without a raised crest which extends backward, the head of the bone toothed 2
2. Conspicuous dark-green vermiculations ("worm tracks") on back and dorsal fin; reddish spotting on body sides enclosed in blue; dorsal and caudal fins finely mottled; body stouter, the head heavy (*Salvelinus fontinalis*) .... BROOK TROUT
   - Vermiculations lacking, the back spotted similarly to the sides; reddish or yellowish spotting on sides not surrounded by blue; dorsal and caudal fins unmarked and clear; body slimmer, head smaller (*Salvelinus malma*) .... DOLLY VARDEN TROUT

LAKE TROUT

Salvelinus namaycush (Walbaum)

(Mackinaw Trout, Gray Trout)

Salmo namaycush Walbaum 1792: 68
Salmo pallidus Rafinesque 1817: 120
Salmo amethystinus Mitchill 1818: 410
Salmo symmetricus Prescott 1851: 340
Salmo adirondacus Norris 1865: 255
Salmo namaycush, Günther, 1866: 123
Salmo namaycush, Suckley, 1874: 151–153
Cristivomer namaycush, Gill and Jordan, 1878: 356
Cristivomer namaycush, Jordan, 1878F: 794–795
Cristivomer namaycush, Jordan and Copeland, 1878: 430
Salvelinus namaycush, Jordan and Gilbert, 1882: 317
Cristivomer namaycush, Jordan and Evermann, 1896B: 504 / 1902: 203–205
Cristivomer namaycush, Juday, 1907: 141
Cristivomer namaycush, Snyder, 1917A: 85 (Nevada) / 1940: 138
Cristivomer namaycush, Forbes and Richardson, 1920: 56–57
Cristivomer namaycush, Pratt, 1923: 48
Cristivomer namaycush, Locke, 1929: 188
Cristivomer namaycush, Jordan, Evermann and Clark, 1930: 59
Cristivomer namaycush, Fish, 1932: 315–316
Cristivomer namaycush, Schultz, 1936: 138
Cristivomer namaycush, Schrenkeisen, 1938: 52–53
Cristivomer namaycush namaycush, Hubbs and Lagler, 1941: 36–37 / 1947: 36–37
Cristivomer namaycush, MacLulich, 1943
Cristivomer namaycush namaycush, Royce, 1943
Cristivomer namaycush, La Monte, 1945: 115
Cristivomer namaycush, Miller and Alcorn, 1946: 180 (Nevada)
Cristivomer namaycush, Wales, 1946: 115, 118
Cristivomer namaycush namaycush, Eddy and Surber, 1947: 112–114
Cristivomer namaycush, Carl and Clemens, 1948: 43, 64–65
Cristivomer namaycush, Fry and Chapman, 1948: 19–35
Cristivomer namaycush, Miller and Kennedy, 1948: 176–189
Salvelinus namaycush, Miller, 1950B: 3
Cristivomer namaycush, Shapovalov and Dill, 1950: 385
Cristivomer namaycush namaycush, Simon, 1951: 46–47
Salvelinus namaycush, Beckman, 1952: 25–26
Salvelinus namaycush, La Rivers, 1952: 93 (Nevada)
Salvelinus namaycush, La Rivers and Trelease, 1952: 115 (Nevada)
Salvelinus namaycush, Fry and Gibson, 1953: 56–57
Cristivomer namaycush, Gabrielson and La Monte, 1954: 95–96
Salvelinus namaycush, Gibson and Fry, 1954: 252–260
Salvelinus namaycush, Morton and Miller, 1954
Cristivomer namaycush, Slastenenko, 1954: 652–659 / 1958A: 86–90

Salvelinus namaycush, Van Oosten, 1956
Salvelinus namaycush, Eddy, 1957: 51
Salvelinus namaycush, Moore, 1957: 65
Salvelinus namaycush, Trautman, 1957: 194–195
Salvelinus namaycush, Carl, Clemens and Lindsey, 1959: 58–61
Salvelinus namaycush, Shapovalov, Dill and Cordone, 1959: 171
Salvelinus namaycush, Kimsey, 1960: 229–230

ORIGINAL DESCRIPTION

"28. SALMO, *Namaycush. Pennant. arct. zool. introd.* 191.
"Caput, dorsum, pinna dorsalis obscure caerulea. Latera subsulca, maculis albis & rubicundulis notata. Abdomen argenteum. Caro alba egregii saporis. Habitat in freto Hudsonis" (Walbaum 1792: 68).

DIAGNOSIS

Head and greatest body depth each about 4 in body length, head a bit more; branchiostegals 11–13; gillrakers 5–6 + 9–11; pyloric caeca 116–138; vertebrae 61–64; scales 30–36/184–213/24–25 in the lateral line; fins—dorsal 11 to 14-rayed, anal 11 to 13-rayed; color pale-to-dark gray, conspicuously dotted with paler rounded spots (often red-tinged), head generally vermiculate on top; reached lengths of more than 5 feet and weights in excess of 100 pounds, but these giants have been effectively disposed of. Commercial catches, when good, would average something like 8 pounds.

TYPE LOCALITY

Hudson's Bay, Canada.

RANGE

Northeastern half of the United States west and north to the Arctic Circle, predominantly a lake species. Introduced into parts of the western United States. In Nevada, it is known only from interstate Lake Tahoe where it had been planted on three known separate occasions before 1900.[1]

TAXONOMY

As can be seen by examining the accompanying synonymicon, the Lake Trout has been replaced in the genus *Salvelinus* by Miller (1950) after having had over 50 years stability in the genus *Cristivomer.* Jordan (with Gilbert) in 1882 listed the species in *Salvelinus* after having originally placed it in *Cristivomer* in 1878. However, in 1896, Jordan and Evermann returned it to *Cristivomer,* where it has

---

[1]Smith (1896: 433), in a summary of plantings in the Western United States, records an 1885 shipment of 100,000 Lake Trout eggs to Parker, who terminated his Nevada State Fish Commissionership with his report of this year. The subsequent fate of these is unknown. The years 1889 and 1895 both saw Lake Tahoe plantings of Lake Trout and State Fish Commissioner Mills noted that 48,000 Lake Trout were planted in the Douglas County portion of Lake Tahoe in 1896 (1897: 22). A bit later (1907–1908) the species was introduced into the Truckee River and Walker Lake, but did not survive (27,245 fry).

Miller and Alcorn (1946) have previously summarized these data.

remained until recently. It is too early to judge the extent to which this change will be generally accepted, but if analysis of the relationships here is correct, the change will be welcome in the interests of nomenclatural simplification.

Etymology—"Namaycush" is an Indian name.

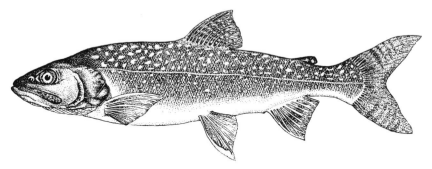

FIG. 137. Lake Trout, *Salvelinus namaycush*. Drawn by Silvio Santina.

LIFE HISTORY

A large, carnivorous, fall-spawning species without marked migratory instincts. It prefers deep waters, living in the depths of lakes except during the breeding season, when it enters shoal water to spawn over rocky or gravelly bottoms; spawning is at greater depths in large bodies of water in which winds disturb the shallow areas. Fish (1932) records spawning at depths of nearly 100 feet in Lake Erie, and the species is known to range into at least the 800-ft. levels in the deeper Great Lakes. As would be surmised from their range, they are essentially cold water fishes, not doing well in waters warmer than 65° F. (18.5° C.). Consequently they have prospered in Lake Tahoe, a cold lake with a maximum depth slightly less than 1,650 feet.

Royce's data (1943) indicate the situation with respect to Lake Trout spawning in New York can be rather complex. He noted three different races each with different spawning habits: (1) a race which goes upstream in September and spawns, (2) a shallow-spawning lake population depositing eggs in October, and (3) a group of trout spawning in the depths from October into November. There is no data which at present would indicate more than one strain in Lake Tahoe, but there is much to be learned yet of these trout in our waters.

Royce showed that some variation existed in several New York lakes in the minimum spawning ages and sizes of Lake Trout, with the seventh year for females and the sixth for males being generally the rule. Spawning time occurs on the wane of temperature and daylight during the fall, with many other factors such as cyclic weather changes, physiological characteristics of certain populations or races, and physical and biological peculiarities of individual lakes playing roles resulting in variations in spawning times in different areas (Royce, p. 24).

"Larger differences in the time of lake trout spawning occur among different lakes. In New York State the trout spawn earliest in the smaller, shallower lakes at higher altitudes and latest in the larger, deeper lakes at lower altitudes . . . In this state the most important factor influencing the time of lake trout spawning is the depth of the lake . . . It appears that in New York State the time of lake trout spawning may vary from late September to early December" (Royce, pp. 32, 34).

He also found it to be the rule that Lake Trout spawned in the shallower lakes at about the time of fall overturn in the lakes.

The three races of Lake Trout that have been described[2] all appear to be further distinguishable in that they each have somewhat different spawning periods, a distinction noticed by Milner as far back as the 1870s, although all are fall and winter spawners. Our Lake Tahoe population is unquestionably derived from stock of the Common Lake Trout, with a general spawning period including October and November.

FIG. 138. Courtesy Oregon State Game Commission and Harold P. Smith.

From available information, spawning goes on for about a month and all investigators agree that the preferred type of bottom is one composed of "clean" gravel or rubble which is free of sand or mud.

"The location of these suitable areas of bottom in the lake is primarily determined by currents which keep that bottom swept clean. The lake trout will roll the smaller stones around and fan off the silt but they cannot remove sand or mud from the crevices" (Royce 1943: 39, 41). Unlike many other trout, the Lake Trout make no nest or "redd," the work of the earlier-arriving males in sweeping the area clean apparently being the only activity of this type. The males of this species are also rather distinctive in lacking any indication of the misshapen breeding jaws or "kype" of other salmonids; as several authors have pointed out, the kype is most prominent on those members

---

[2]Common Lake Trout (*Salvelinus namaycush namaycush*), Siscowet Lake Trout (*S. n. siscowet* (Agassiz) 1850) and Rush Lake Trout (*S. n. huronicus* (Hubbs) 1930).

of the Salmon family that travel the greatest distances to their spawning grounds with the Pacific Salmon (*Oncorhynchus*) showing it most prominently.

Authors have noted some sexual dimorphism in the Lake Trout during their courtship and mating—the males became light-colored on the back which then contrasted markedly with the dark sides—females were typically dark all over (Merriman 1935, Royce 1943).

Royce has observed spawning closely and reported it in some detail:

"The males began their courtship upon the appearance of the females on the area. The usual procedure was for the male to nudge the female in the side with his snout . . . and then to attempt the spawning act. It was not unusual for two or more males to court and attempt to spawn with the female at the same time. During the courtship the males displayed the characteristic coloration . . . and commonly held the dorsal fin erect . . .

"The spawning act or attempts at it usually consisted of one or two males approaching a female, pressing against the female's sides with their vents in close proximity and then quivering all over . . . usually the mouths of both sexes were open and the dorsal fin of the males held erect.

"The spawning act was not attempted solely by two or three trout but often by several. As many as seven males and three females were seen at one time all pressing together in one large group and quivering in unison. No spawning act was seen to last for more than a few seconds and it seems that a female must accomplish many unions to empty completely her ovaries" (1943: 50, 52).

Because of the nature of the gravel bottom over which these fish spawn, it is not necessary to bury the eggs to protect them—and no attempts are made to do so—for they fall readily in among the stones and seem amply protected. The efficiency of fertilization in this seemingly rather haphazard process is high—on the order of 80 percent and better.

As would be expected in a fall-winter spawning fish, Lake Trout eggs require lower temperatures for optimum development—several authors (Cook 1929, Embody 1934, Royce 1943) mention satisfactory incubation of Lake Trout eggs at temperatures just slightly above freezing, while Royce (p. 56) noted that "At the New York State hatchery at Rome high mortality occurred in Lake Trout eggs developing at water temperatures of over 50° when other trout eggs developed normally. Lake Trout eggs from the same source developed normally in other hatcheries at lower temperatures."

Royce's work indicated an incubation period of about four months which would, however, vary with the temperature. Fry of this species appear to keep well hidden among the rubble and gravel of the spawning bottom, and very little appears to be known of these earliest stages. They seem to stay close to the bottom and move into deeper waaters quite early.

Scale readings to determine age is difficult in all trout, but has been particularly bothersome to the biologist in the case of the Lake Trout. Not many investigators have tackled the problem but some of what has been done shows the following results:

Investigator	Area	Length (in ½ yrs., i.e., 5½, etc.)	1	2	3	3½	4	4½	5	5½	6	6½	7	7½	8	8½	9
Smith and Van Oosten 1940	Lake Michigan		...	...	...	16.3	...	19.2	...	22.2	...	23.8	...	22.9	...	29.1	26.6
Royce 1943	Keuka Lake, N. Y.		7	9.5	11	...	17	...	20	...	24	...	...	...	...	...	...
Royce 1943	Lake Simcoe, Ontario		...	...	...	...	16	22.5	...	...	25.5	...	28	...	29.5	...	...
Royce 1943	Seneca Lake, N. Y.		...	...	...	...	...	25.5	...	...	29	...	...	...	...	...	...

The feeding habits of Lake Trout are well known, at least in their generalities; individuals 15–16 inches long and up are almost entirely fish eaters, while the younger specimens consume greater and greater amounts of invertebrates, particularly arthropods, the smaller the fish are. Several investigators (Corlett; La Rivers; Miller 1951) have noted the following three associated and abundant nongame fishes as articles of diet for the Lake Trout in Tahoe: Belding Sculpin, Tahoe Sucker and Lahontan Tui Chub.

Representatives of the other fish in the lake (see Lake Tahoe section, p. 147, for a complete listing of the major associated animals) are also occasionally found in Lake Trout stomachs. The author has found the introduced crawfish, *Pacifastacus leniusculus* (Dana) 1852, in one stomach. This crawfish, which grows to a substantial size (8 inches or more) invaded the lake from the Truckee River in which it was planted many years ago from the Pacific Northwest, and is found at considerable depths. It undoubtedly provides the best nonvertebrate source of food for the larger Lake Trout.

Recently (1960), Kimsey reported on food habits of the Lake Trout in Donner Lake. He found that Lahontan Tui Chub constituted the great bulk of their fish food (about 90 percent) and added the Lahontan Redshiner and Lahontan Mountainsucker which were sparingly used. In this study of 31 stomachs, fish made up more than 98 percent of the total Lake Trout diet.

ECONOMICS

A valuable species in the Great Lakes, where it is second only to the Common Whitefish (*Coregonus clupeaformis*) as a commercial fishery. It was introduced and well established in Lake Tahoe before the turn of the present century (plantings are known in Nevada for the years 1885, and specifically for Lake Tahoe in 1887, 1895, 1896, 1907 and 1908) and has contributed to sport fishing there ever since. Small plants made in the Truckee River and Walker Lake in 1907–1908 failed.

In determining the suitability of waters where stocking with this species is contemplated, depth, bottom type, temperature and oxygen cycles need to be considered. The Lake Trout requires deep waters which will stay cool, and rocky bottoms with little or no decaying organic matter to deplete the oxygen content of the water. It is not always considered the best trout, either from the standpoint of gaminess or food qualities, and hence its introduction into western waters already supporting native trout should be discouraged, as a general rule. However, the species has probably been unjustifiably condemned in Lake Tahoe as the immediate factor chiefly responsible for the decline of the native Cutthroat Trout there. While the Lake Trout undoubtedly would utilize some cutthroat as food, the former's predilection for deeper waters would remove it from too intimate contact with other trout; if the spawning streams tributary to Tahoe had been properly managed, it is possible that the cutthroat might have held its own. There is some speculation that the 1922–1928 wholesale demise of cutthroat in Lake Tahoe was caused by an epizoötic carried by the Lake Trout.

Snyder summarized the local situation as regards the Lake Trout as it existed in the decade between 1910 and 1920:

"Specimens of this long-headed, ravenous-looking trout were frequently seen at Tallac. It is found in Fallen Leaf, Cascade, and other smaller lakes as well as Lake Tahoe, and it is generally regarded as inferior to both the rainbow and the cutthroat trout. It is difficult to say anything in defense of the introduction of the species. Mr. Mills reports one weighing 26 pounds" (1917A: 85).

Regardless of the pros and cons on the merits of the Lake Trout, points which have been, and probably always will be, argued lustily, the fact remains that it has, for many years, provided by far the dominant sport fishery of Tahoe and this appears to be what can be expected for at least the near future. And while it requires special tackle and equipment, and so may be argued as a fish for the few, it is still taken in considerable numbers from the lake waters. As a lake spawning species, it has proven that it can adequately maintain its fishable populations while stream spawners such as the late lamented cutthroat have, we will all agree with profound regret, followed the dinosaur into extinction, and must be artificially replenished.

Most people now consider the Lake Trout an excellent eating fish, an interesting change of opinion from 30 years ago. Perhaps this is due in some measure to the disappearance of the better tasting cutthroat.

BROOK TROUT
Salvelinus fontinalis (Mitchill)

(Eastern Brook Trout, Speckled Trout, Speckled Char(r))

Salmo fontinalis Mitchill 1815: 435
Salmo fontinalis, Richardson, 1836: 176–178 (color plate 83, fig. 1)
Salmo hudsonicus Suckley 1861: 310 / 1874: 119–120
Salmo fontinalis, Günther, 1866: 152
Salmo fontinalis, Suckley, 1874: 123–129
Salvelinus fontinalis, Jordan and Copeland, 1878: 430
Salvelinus fontinalis, Jordan and Gilbert, 1882: 320
Salvelinus fontinalis, Jordan, 1891: 6, 16
Salvelinus fontinalis, Jordan and Evermann, 1896B: 506 / 1902: 207–209
Salvelinus fontinalis, Rutter, 1903: 148
Salvelinus fontinalis, Snyder, 1917A: 85 (Nevada)
Salvelinus fontinalis, Pratt, 1923: 48
Salvelinus fontinalis, Locke, 1929: 188–189
Salvelinus fontinalis, Jordan, Evermann and Clark, 1930: 60
Salvelinus fontinalis, Hazzard, 1932
Salvelinus fontinalis, Ricker, 1932A: 69–110 / 1932B / 1934: 76–87
Salvelinus fontinalis, Jenkins, 1936: 231–233
Salvelinus fontinalis, Schultz, 1936: 138
Salvelinus fontinalis, Schrenkeisen, 1938: 54
Salvelinus fontinalis, Hoover, 1939: 81–91
Salvelinus fontinalis, Kuhne, 1939: 28–29

Salvelinus fontinalis, Snyder, 1940: 137
Salvelinus fontinalis, White, 1940: 176–186 / 1941: 258–264 / 1942: 471–473
Salvelinus fontinalis fontinalis, Hubbs and Lagler, 1941: 36–37 / 1947: 36–37
Salvelinus fontinalis, Murphy, 1941: 167
Salvelinus fontinalis, Smith, 1941: 461, 468–471 / 1947: 281, 287–296
Salvelinus fontinalis, Hinks, 1943: 25–26
Salvelinus fontinalis, MacLulich, 1943
Salvelinus fontinalis, La Monte, 1945: 118
Salvelinus fontinalis, Miller and Alcorn, 1946: 179 (Nevada)
Salvelinus fontinalis, Wales, 1946: 115, 117
Salvelinus fontinalis fontinalis, Eddy and Surber, 1947: 110–112
Salvelinus fontinalis, Carl and Clemens, 1948: 62–63
Salvelinus fontinalis, Clemens and Wilby, 1949: 95–96
Salvelinus fontinalis, Graham, 1949
Salvelinus fontinalis, Elliott, 1950
Salvelinus fontinalis, Everhart, 1950: 25
Salvelinus fontinalis, Shapovalov and Dill, 1950: 385
Salvelinus fontinalis, Cooper, 1951
Salvelinus fontinalis, Fry, 1951
Salvelinus fontinalis fontinalis, Simon, 1951: 44–46
Salvelinus fontinalis, Smith, 1951A: 1–3 / 1951B: 1–6
Salvelinus fontinalis, Beckman, 1952: 24–25
Salvelinus fontinalis, Cooper, 1952
Salvelinus fontinalis, La Rivers, 1952: 93 (Nevada)
Salvelinus fontinalis, La Rivers and Trelease, 1952: 114 (Nevada)
Salvelinus fontinalis, Wilder, 1952: 169–198
Salvelinus fontinalis, Fry and Gibson, 1953: 56–57
Salvelinus fontinalis, Black, Fry and Black, 1954
Salvelinus fontinalis, Gabrielson and La Monte, 1954: 98–99
Salvelinus fontinalis, Scott, 1954: 32–33
Salvelinus fontinalis, Shomon, 1954: 10–11
Salvelinus fontinalis, Slastenenko, 1954
Salvelinus fontinalis, Dymond, 1955: 549–550
Salvelinus fontinalis, Job, 1955
Salvelinus fontinalis, Rupp, 1955
Salvelinus fontinalis, Shepard, 1955
Salvelinus fontinalis, Stokell, 1955: 135–137
Salvelinus fontinalis, Eddy, 1957: 51
Salvelinus fontinalis, Koster, 1957: 33–34
Salvelinus fontinalis, Moore, 1957: 65
Salvelinus fontinalis, Trautman, 1957: 191–193
Salvelinus fontinalis fontinalis, Brasch, McFadden and Kmiotek, 1958
Salvelinus fontinalis, Bridges and Mullan, 1958
Salvelinus fontinalis, Mullan, 1958
Baione fontinalis, Slastenenko, 1958A: 82–85
Salvelinus fontinals, Carl, Clemens and Lindsey, 1959: 63–64
Salvelinus fontinalis, Shapovalov, Dill and Cordone, 1959: 171
Salvelinus fontinalis, Needham, 1961

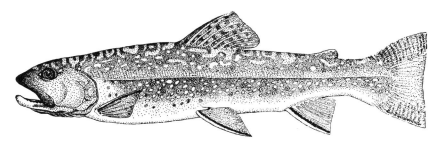

FIG. 139. Brook Trout, *Salvelinus fontinalis*. Drawn by Silvio Santina.

ORIGINAL DESCRIPTION

"3. *Common Trout.* (*Salmo fontinalis*) With yellow and red spots on both sides of the lateral line, concave tail, and sides of the belly orange red.

"Back mottled pale and brown. Sides dark brown, with yellow and red spots; the yellow larger than the red, and surrounding them. The latter appear like scarlet dots. Lateral line straight. The yellow spots and red dots both above and below that line. Sides of the belly orange red. Lowest part of the abdomen whitish, with a smutty tinge. First rays of the pectoral, ventral and anal fins white, the second black, the rest purplish red. Dorsal fin mottled of yellowish and black.

"Tail rather concave, but not amounting to a fork; and of a reddish purple, with blackish spots above and below.

"Eyes large and pale. Mouth wide. Teeth sharp. Tongue distinct. Skin scaleless.

"Is reckoned a most dainty fish. They travel away to Hempstead and Islip, for the pleasure of catching and eating him. He is bought at the extravagant price of a quarter of a dollar for a single fish not more than ten or twelve inches long. He lives in running waters only, and not in stagnant ponds; and, therefore, the lively streams, descending north and south from their sources on Long-Island, exactly suit the constitution of this fish. The heaviest Long-Island trout that I have heard of, weighed four pounds and a half" (Mitchill 1815: 435–436).

DIAGNOSIS

Head and greatest body depth each about $4\frac{1}{2}$ in body length; branchiostegals about 12, gillrakers about 6 + 11; pyloric caeca about 34, scales about 37/200–230/30; fins—dorsal 10–12 rayed, anal 9–10 rayed; color—sides red-spotted, the spots smaller than eye pupil—back lacking such spotting, barred or mottled with olivaceous or black—dorsal and caudal fins spotted with blackish, ventral fins each with a conspicuous white, cream or orange border along leading edge, followed by a bordering dark stripe; record lengths exceeding 30 inches and weights of about 15 pounds are known, but the averages are about 12 inches and two pounds.

TYPE LOCALITY

Vicinity of New York City, N. Y.

RANGE

East coast from Labrador to Georgia, west to Saskatchewan. Has been extensively and successfully introduced into many western United States streams. It is now one of the commonest trouts in Nevada, having been stocked into the State's water on innumerable occasions since the 1880s. A good many of the early planted specimens seem to have come from Marlette Lake, in the Carson Range bordering Lake Tahoe to the east, Ormsby and Washoe Counties. According to Smith (1896), Brook Trout were well established in Marlette by 1892. He also mentioned the 1875 implantation of Brookies into Prosser Creek, California—a tributary of the Truckee River from whence they could have easily entered Nevada a few miles downstream. Fish Commissioner records tell of an 1883 implantation for the west-central rivers (Truckee, Carson and Walker), the Humboldt River and Washoe Lake, with supposed good results. In 1889, spawners were removed from Marlette for the Carson City Hatchery (Mills 1891: 4), and other river plants were made in the Humboldt, Truckee and Carson in 1891 to 1893.

Since 1900, importations of eggs and fish have run into the millions.

TAXONOMY

Etymology—"fontinalis" refers to living in springs.

FIG. 140. Courtesy Oregon State Game Commission and Harold P. Smith.

LIFE HISTORY

A carnivorous, fall-spawning, cold-stream species preferring water from 55° F. to 65° F. in temperature, spawning on gravel beds in shallow water of the smaller tributaries. The eggs hatch the following spring when the water begins to warm up.

The staple article of diet is the insect, aquatic and terrestrial, even for the larger specimens. Fry begin with the smallest suitable items, such as the microcrustacea (Cladocera, Ostracoda and Copepoda), some 25–30 days after hatching. In larger trout, only a small percentage ever appear to take other fish for food, and then only occasionally. The only type of Brook Trout that consistently feeds on other fish is the "salter" or sea-run form of the Atlantic coast, which subsists in the ocean on various suitable species.

No selectivity has been demonstrated on the part of this trout for its food, but it seems to take in greatest quantity that type of food which is most abundant. Seasonal variation in foods appears to be due simply to what is available at different times of the year. There is evidence that Brook Trout will feed at temperatures too low for digestion to take place, so that food will be moved through the digestive tract without much if any benefit to the animal (Leonard 1941). Benson (1953) determined that when temperatures were between 55° F. and 66° F., these trout seemed to show maximum feeding capacities.

In the matter of temperatures generally, Brook Trout tolerances are considered to range from just above the freezing point of water to about 75° F. They show varying degrees of discomfort at temperatures above this, and also seem to have the greatest cold tolerance capacities of any trout.

Growth rates have been extensively studied—in fact, with the possible exception of the Rainbow Trout, the brook is the most investigated of all our trout—and indicate that there is a sharp natural dropoff after the fourth year of life, with some individuals attaining twice this age.

Average growth rates, in inches, below are taken from Bridges (1958):

	1 yr.	2 yrs.	3 yrs.	4 yrs.	5 yrs.	6 yrs.
Squannacock River drainage, Mass.	3.1	4.7	6.3	8.5	12.0	
Sunkhaze Stream, Maine		6.5	7.5	9.4	11.8	
Moser River, Nova Scotia	1.8	5.1	7.3	8.4	10.5	12.9
Pigeon River, Michigan		3.9	6.0	7.9	8.9	
West Gallatin River, Montana		4.7	8.5	11.8		

During spawning, males take on the typical exaggerations of hooked lower jaw, deeper body and increased coloration, markedly distinguishing them from the females, who show, in addition to these differences, the characteristic prominent genital papillae. Courtship actions as described by several observers, take place while the female is scooping out the redd or egg-nest in the gravel bottom. The male takes up a position somewhat downstream or behind the female and darts at her, occasionally nudging her and then dropping back to his original position.

Intruding fish are driven away during this period, by either one or both of the fish.

During actual spawning, the female positions herself at the bottom of the redd, flattened out against the gravel with bottom fins spread out. The male moves quickly into the redd, curves his body to press the female against the bottom and releases milt while vibrating strongly. This stimulates the female to vibratory movements and the release of her eggs at the same time. Eggs and milt immediately settle to the bottom of the redd and adhere to the gravel or stones and fall into the crevices.

After spawning is completed—and more than one male may supply milt to a redd—the female sweeps gravel into the redd, and, by working upstream and stirring up finer materials, in a couple of hours time produces a low mound where once was a hollowed-out redd. The male takes no part in this.

These redds are usually situated in riffle-area gravel bars or similar bottoms kept clean and loose by incoming ground water.

The spawning season is generally from October nearly to January, with recorded temperatures from 40° F. to 50° F. Light plays some part in this (Hoover and Hubbard 1937) and has been used to trigger spawning in fish cultural activities. By subjecting potential spawners to controlled light conditions in the winter and spring, it is possible to begin collecting eggs in July and August rather than waiting for October and November, the usual time.

Several investigators (White 1930, Hazzard 1932) have reported a nearly 80 percent hatch of eggs from redds, which is similar to results achieved in hatcheries. The number of eggs produced by females varies with the size of the fish, Vladykov (1956) noting 4,800 eggs for a 22-inch fish.

Bridges (1958) gives the following revised incubation-period after Embody (1934):

° F.	Hatching days
35.5	143
38.5	109
40.5	90
43.0	75
46.0	62
50.0	45
52.0	40
54.0	37

It is not known to what extent hybridization occurs in nature with other trout and chars, but several rather distinctive crosses have been accomplished in the hatchery, where Brook Trout have been successfully crossed with Brown Trout, Lake Trout and Arctic Char (*Salvelinus alpinus* (Linnaeus)). Some of these may even be suitable for rearing and release, according to the Swedish investigator Alm (1955) and Stenton (1952); the latter has reported on the introductions of the "Splake" (Brook x Lake) into several Canadian lakes. The cross with Brown Trout is called a "Tiger," because of its distinctive striping. See the section on Hybridization in the Appendix.

Brook Trout do not seem to move about much in the stream environment, being most active in early morning and late afternoon, quiescent at night and during the brightest part of the day. During spawning, there apparently is an overall short upstream movement, while during the colder months a reverse trend occurs. Cooper (1952) and Newell (1957) have pointed out something of interest to the hatcheryman—if water temperatures are less than 50° F., stocked Brook Trout drop downstream rapidly, but this is checked if planting occurs when temperatures are above 50° F. The only brooks with any pronounced migratory tendency are the "salters" or sea-run populations, and no one has explained why some stream individuals become salters while others live permanently in freshwater.

Mortality, as in wild populations generally, is high, several studies indicating this to range from 2.5 percent to 5.1 percent survival of fingerlings from eggs. From fingerlings on, rates become much better, although there is wide disagreement on actual figures, from Newell's 8.69 percent to Cooper's 30 percent (1953) (Bridges 1958).

Brook Trout are subject to numerous parasites, most of which show

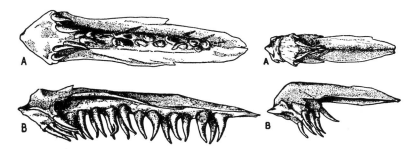

FIG. 141. Left side, vomerian bone of the Brown Trout (*Salmo trutta*)—(A) ventral surface, (B) lateral surface.
Right side, vomerian bone of the Brook Trout (*Salvelinus fontinalis*)—(A) ventral surface, (B) lateral surface.
Courtesy Dr. Carl L. Hubbs and the Cranbrook Institute of Science.

up as important only under crowded hatchery conditions. Bridges (1958) summarizes the 25 species as: 2 bacteria, 2 fungi, 5 protozoa, 1 copepod, 2 spiny-headed worms, 4 roundworms, 4 tapeworms, and 5 flukeworms.

ECONOMICS

For North America as a whole, the Brook Trout rates as one of the most important sport fishes, its value in this respect decreasing for the western United States where it is overshadowed by the Rainbow Trout. Both are similar in that they furnish much of the early angling returns, not being difficult fish to catch. At least this is the reputation of the Brook Trout in its native northeastern area where it is plentiful.

In Nevada, the Brook Trout is generally the most adaptable game fish for the high, cold, small mountain streams and lakes, where it is often the only fish. However, because of its reputation in this respect, it has been planted in many mountain waters which should not, in their present, natural condition, be considered as suitable for fish. This results in the big-headed, slender-bodied, semi-starved "Slim Jim" of the fisherman. Such poorly developed specimens also occur at lower elevations where stream conditions are poor.

Ted Frantz found, in his stream survey below the 40th Parallel of latitude, that it was the rule rather than the exception to find Brook Trout in poor shape. However, some Nevada streams do produce good, deep-bodied brookies.

With respect to the "Slim Jims," while of little use to the sportsman, they are fascinating to the aquatic biologist for their ability, in places, to exist in a seemingly totally inadequate environment. I recall a population in a tiny perpetually snow-fed lake at about 11,000 feet elevation in the Convict Creek Basin of the southern Sierras, which I first saw several years ago. While some insects and minute amounts of algae are present, standard biological assay methods fail to convince the investigator that fish can even stay alive there, yet at last reports the lake still holds its brooky population—even though big-headed, stunted and starving, they were managing to exist.

And not all Brook Trout are pygmies. Ted Frantz has told me of a 32-inch specimen reported to him by Ralph Kaufman of Baker, White Pine County. This fish was caught by Mr. Kaufman in Baker Lake, Snake Range.

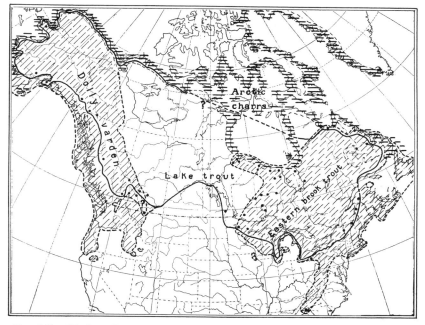

FIG. 142. Native distribution of the charrs, *Salvelinus* spp. The Dolly Varden, *Salvelinus malma*, ranges farther south, into northeastern Nevada, than the map indicates. The Lake Trout, *Salvelinus namaycush*, occurs within the limits of the solid line. Courtesy University of California Press and Erhard Rostlund.

DOLLY VARDEN TROUT
Salvelinus malma (Walbaum)
(Dolly Varden Char(r), Bull Trout)

Salmo malma Walbaum 1792: 66
Salmo curilus Pallas 1811: 251
Salmo spectabilis Girard 1856: 218
Salmo bairdii Suckley 1861: 309
Salmo parkei ibid
Salmo bairdii, Günther, 1866: 121
Salmo parkii, ibid
Salmo tudes Cope 1873: 24
Salmo pluvius Hilgendorf 1876: 25
Salvelinus spectabilis, Jordan, 1878G: 79
Salvelinus bairdii, ibid: 82
Salvelinus malma, Bendire, 1882: 86–87
Salvelinus malma, Jordan and Gilbert, 1881D: 38 / 1882: 319
Salvelinus malma, Turner, 1886: 104–105
Salvelinus malma, Evermann, 1891: 50
Salvelinus malma, Jordan and Evermann, 1896B: 507–508
Salvelinus parkei, ibid, 1902: 210–212
Salvelinus parkei, Pratt, 1923: 49
Salvelinus malma spectabilis, Jordan, Evermann and Clark, 1930: 61
Salmo parkei, ibid
Salvelinus malma, Dymond, 1936: 63–64
Salvelinus malma spectabilis, Schultz, 1936: 139
Salvelinus malma spectabilis, Schrenkeisen, 1938: 57–58
Salvelinus malma, DeLacy and Morton, 1942
Salvelinus malma spectabilis, Dimick and Merryfield, 1945: 33–34
Salvelinus malma, La Monte, 1945: 118
Salvelinus alpinus malma, Carl and Clemens, 1948: 63–64
Salvelinus malma, Clemens and Wilby, 1949: 94–95
Salvelinus malma, Shapovalov and Dill, 1950: 385
Salvelinus malma, La Rivers, 1952: 93
Salvelinus malma, La Rivers and Trelease, 1952: 115
Salvelinus malma spectabilis, Gabrielson and La Monte, 1954: 99
Salvelinus alpinus malma, Eddy, 1957: 52
Salvelinus malma, Moore, 1957: 65
Salvelinus alpinus malma, Slastenenko, 1958A: 79–80
Salvelinus malma, Carl, Clemens and Lindsey, 1959: 61–62
Salvelinus malma, Shapovalov, Dill and Cordone, 1959: 171

ORIGINAL DESCRIPTION

"22. SALMO, *Malma.* Ruffis *Golet. Pennant. arct. zool. introd.* 126. D. 12. P. 14. V. 8. A. 10.

"Corpus bipedale, perangustum, cylindraceum. Caput illi Farionis simile. Squamae minutae. Dorsum & latera caerulescentia, maculis coccineis conspersa. Abdomen album. Pinnae ventrales & analis rubrae. Cauda parum bifurca. Adscendit fluvios Kamschatkae.

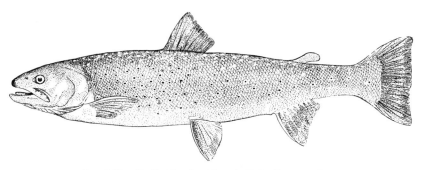

Fig. 143. Dolly Varden Trout, *Salvelinus malma*.
Drawn by Silvio Santina.

"*Malma*. Kamtschatkis, *Goltra. Kratscheninnikow descr. Kamtschatkae* 183.

"Peritonaeum coccineo maculatum. Aetate progrediente magis in latitudinem, quam in longitudinem crescit" (Walbaum 1792: 66).

DIAGNOSIS

Head and greatest body depth about 3½ and 4 in body length, respectively; branchiostegals 10–12; gillrakers 5–10 + 8–10; pyloric caeca large, 45–50; scales about 39/190–250/36; fins—dorsal 11-rayed, anal 9-rayed, caudal truncate, adipose large; color—olivaceous above and on sides, white on belly—laterally with large, conspicuous, round spots nearly the size of the eye, dorsally with somewhat smaller, pale spots—lower fins colored much as in *Salvelinus fontinalis*, with white along the leading edges contrasting sharply with darker stripe behind. Maximum lengths up to 3 feet and weights up to 20 pounds are known, but averages only about 2 to 3 pounds.

TYPE LOCALITY

Rivers of the Kamchatka Peninsula, Siberia.

RANGE

Freshwater streams entering the ocean and bordering the North Pacific from Kamchatka to central California; in the United States, occurs eastward to Montana and northern Nevada in the Snake River System.

TAXONOMY

Etymology—"Malma" is a Kamchatka Peninsula native name.

LIFE HISTORY

A fall-spawning, carnivorous species preferring cooler waters, and entering the ocean in many parts of its range. Essentially a "Western" Brook Trout, the Dolly Varden has a life history much like that of the "Eastern" Brook Trout, *Salvelinus fontinalis*.

FIG. 144. Courtesy Oregon State Game Commission and Harold P. Smith.

ECONOMICS

A species of char(r) which is generally considered inferior to the brook and other trout as a game fish, although, as with other species, its gaminess depends largely upon the waters in which it is found. The Dolly Varden is, strangely enough to most anglers, considered an undesirable rough fish in Alaskan streams because of its competition with the commercially valuable Red Salmon (*Oncorhynchus nerka*), and bounties are offered for its local extermination.

TROUT

Genus *Salmo* Linnaeus 1758

(The old Latin name, from *salio*, to leap)

KEY TO SPECIES

1. Red dash of color on the dentary (between lower jaw and isthmus) usually evident in life; dorsal fin rays usually 10 (9–11); maxillary process in adults extending behind the eye, measuring about 1.6 to 2.25 into head; hyoid teeth (those behind the patch of teeth on the tongue) usually present, but few and scattered and easily broken off; vertebrae usually 60–61 (58–62); (no red spotting on flanks) (*Salmo clarki*) .... CUTTHROAT TROUT

    ---- No red dash of color on dentary; dorsal fin rays usually 11–12 (10–13); maxillary process in adults shorter, not extending behind the eye; hyoid teeth always absent; vertebrae usually more or less than 60–61 2

2. Usually with a few red spots on the sides, these surrounded by a whitish halo; black-spotting large, often sparse, and

many spots white-haloed; body color brownish-yellow to greenish; spotting absent or only slightly developed on tail (caudal) fin; vertebrae usually 57-58 (56-59); adipose fin of young individuals orange, without dark margining or spotting (*Salmo trutta*) .... BROWN TROUT

---- No red-spotting on the sides; black-spotting small, numerous, no spots haloed; body color gray-to-blue above, the reddish lateral band usually but slightly interrupted by faint parr marks on adults; spotting well-developed on tail fin; vertebrae usually 63 (59-65); adipose fin of young specimens olive, with black margining or spotting (*Salmo gairdneri*) ....
.... RAINBOW TROUT

CUTTHROAT TROUT

Salmo clarki Richardson

(Blackspotted Trout)

Salmo clarkii Richardson 1836: 225
Fario stellatus Girard 1856: 219 / 1859: 316–318
Fario clarkii, Girard, 1856: 219 / 1859: 314–315
Salmo brevicauda Suckley 1861: 308
Salmo purpuratus, Günther, 1866: 116
Salmo stellatus, Cooper, in Cronise, 1868: 494
Salmo clarkii, Suckley, 1874: 112–113
Salar clarkii, Jordan, 1878F: 796–797
Salar clarkii, Jordan and Copeland, 1878: 430
Salmo purpuratus, Jordan and Gilbert, 1881B: 460 / 1881D: 39 / 1882: 314–315
Salmo purpuratus, Cope, 1883: 141
Salmo mykiss clarki, Jordan, 1891: 14
Salmo mykiss clarkii, Jordan and Evermann, 1896B: 492
Salmo clarkii, ibid, 1898B: 2819
Salmo clarkii, ibid, 1902: 176–179
Salmo clarki, Pratt, 1923: 44
Salmo clarkii, Jordan, Evermann and Clark, 1930: 56
Salmo clarkii clarkii, Schultz, 1936: 136
Salmo clarkii, Schrenkeisen, 1938: 43–46
Salmo clarki, Snyder, 1940: 130–131
Salmo clarkii, Smith, 1941: 1947
Salmo clarkii, Shapovalov, 1941: 442
Salmo clarkii, Hubbs and Miller, 1943: 353
Salmo clarkii, La Monte, 1945: 114
Salmo clarkii, Turner, 1946
Salmo clarkii clarkii, Carl and Clemens, 1948: 42, 53–54
Salmo clarkii, Miller and Miller, 1948: 183
Salmo clarkii clarkii, Clemens and Wilby, 1949: 89–90
Salmo clarki, Miller, 1950B: 4
Salmo clarkii, Shapovalov and Dill, 1950: 385
Salmo clarki, Fleener, 1952: 235–248

Salmo clarki, La Rivers, 1952: 94
Salmo clarki, La Rivers and Trelease, 1952: 114
Salmo clarkii, Gabrielson and La Monte, 1954: 97–98
Salmo clarki, Irving, 1955
Salmo clarki, Eddy, 1957: 49
Salmo clarki, Koster, 1957: 30–31
Salmo clarki, Moore, 1957: 63–64
Salmo clarki, Slastenenko, 1958A: 55–56
Salmo clarki clarki, Carl, Clemens and Lindsey, 1959: 66–68
Salmo clarki, Needham and Gard, 1959
Salmo clarkii, Shapovalov, Dill and Cordone, 1959: 171

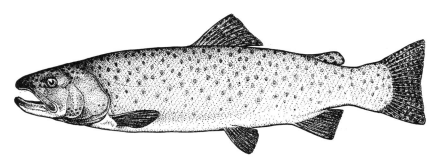

Fig. 145 Cutthroat Trout, *Salmo clarki*. Drawn by Silvio Santina.

ORIGINAL DESCRIPTION

"[Dr. Gairdner does not mention the Indian name of this trout, which was caught in the Katpootl, a small tributary of the Columbia, on its right bank. I have therefore named it as a tribute to the memory of Captain Clarke, who notices it in the narrative prepared by him of the proceedings of the Expedition to the Pacific, of which he and Captain Lewis had a joint command, as a dark variety of Salmon-trout (see p. 163). In colour this species resembles the *Mykiss* of Kamtschatka, and there is no very material discrepancy in the number of rays in the fins. Vide Arct. Zool., Intro., p. cxxvi.—R].

" 'Colour.—Back generally brownish purple-red, passing on the sides into ash-grey, and into reddish-white on the belly. Large patches of dark purplish-red on the back. Dorsals and base of the caudal ash-grey, end of caudal pansy-purple. Back, dorsal, and caudal studded with small semilunar spots. A large patch of arterial-red on the opercule and margin of the preopercule. Pectorals, ventrals and anal greyish-white, tinged with rose-red. Teeth.—Both jaws armed with strong hooked teeth, a single row on each palate-bone, a double row on the anterior half of the vomer and on the tongue. Dorsal profile nearly straight. Ventrals opposite to the middle of the first dorsal. Fissure of mouth oblique. Extremity of caudal nearly even. Fins.—Br. 11; P. 12; V. 8; A. 13; D. 11—0.

FIG. 146. "Cutthroat trout at moment of spawning. The white 'cloud' of milt is filling the pit and has spread beyond the female (nearer the camera)" (Osgood R. Smith 1941). Courtesy of Dr. Smith.

" 'Dimensions.

	Inches	Lines
Extreme length	14	0
Greatest height of body	2	10⅔
Greatest thickness	1	6
Length from end of snout to eye	1	1⅙
Length from end of snout to angle of opercule	3	2⅓
Length from end of snout to pectorals	1	9½
Length from end of snout to dorsal	6	4⅔
Length from end of snout to ventrals	7	6
Length from end of snout to anal	10	0
Length from end of snout to adipose	10	1½
Length of ventrals	1	6
Length of attachment of dorsal	1	4⅔
Length of attachment of anal	1	3½
Length of caudal	2	0
Greatest breadth of ditto	2	10⅔

Gairdner, *in lit.*

"[There appear to have been two specimens of this species sent to me by Dr. Gairdner. In both the spinal column contains sixty-two vertebrae. The teeth, which are closely set, rather long, slender and acute, and, in the older specimen, considerably curved, are in number as follows: *Intermax.* lost; *labials* 28–30; *palate-bones* 15–17; *vomer* 13, two in front and the others in a single flexuose series, as long as the dental surface of the palate-bones; *lower jaw* 13–13; *tongue* 6–6, in two almost parallel rows. The lingual teeth are the largest and most curved, those of the lower jaw are next in size, then follow the vomerine, palatine, and labial teeth, which are equal to each other. The pharyngeal teeth are also proportionally long, and there is an oblong plate, rough with very minute ones, on the isthmus which unites the lower ends of the branchial arches. This space is quite smooth in *S. salar,* in several, if not all the English trouts, and in *S. quinnat, Gairdneri,* and in the imperfect specimen which I have referred to *S. Scouleri.* In the latter the surface of the arches is also quite smooth, but in the *quinnat* and *Gairdneri* minute rough points become visible with a good eye-glass. In all the trouts the compressed rakers have their thin inner edges more or less strongly toothed. In one of the specimens of *S. Clarkii* the spinal column is nine inches long, in the other six.—R.]" (Richardson 1836: 225–226). The *Salmo quinnat* and *S. scouleri* to which Richardson refers are synonyms of, respectively, *Oncorhynchus tshawytscha* and *O. gorbuscha.* As may be surmised from the form of the above, the actual description is a quotation from a letter of Gairdner's to Richardson, with the latter's comments preceding and following this in brackets.

FIG. 147. "Cutthroat trout separating immediately after spawning" (Osgood R. Smith 1941). Courtesy of Dr. Smith.

DIAGNOSIS

Head and depth about 4 in body length, respectively; branchiostegals 9–12, usually 10–11; gillrakers 22 + 27, usually 23 + 25; maxillary somewhat abbreviated, extending only a short distance behind the eyes; hyoid or basibranchial teeth almost always present; pyloric caeca 50–85; scales 27–40/140–190/30–37; fins—dorsal 9–12 rayed, usually 10–11—anal 10–12 rayed—pectoral 12–15 rayed, usually 13–14; color —exhibits great diversity, from the deep olivaceous tones of specimens from the higher, colder waters; the multicolored yellow, pink and red of the land-locked breeding populations of lower, warmer waters; to the uniform silvery sea-run or "steelhead" individuals. The conspicuous black-spotting and red slash mark below the mandible are usually typical of the animal wherever found;[1] weights and lengths in excess of 40 pounds and 40 inches, respectively, are known, but the average is much less.

TYPE LOCALITY

Cathlapootl River, a tributary of the Columbia River, Washington.

RANGE

Western North America from Alaska to California and east to the Rocky Mountains; absent from many local areas in the intermontane region.

TAXONOMY

This is quite involved, since, like the rainbow series, the cutthroats occupy a wide range and show the diversity of their habitats in their great size and color variation. This has resulted in the description of many "species" which later years proved to be at best only geographical races of a single widespread, young and variable species.

For many years, Jordan and his disciples considered our cutthroats as variants of the Kamchatka (southeastern Siberia) trout *Salmo mykiss* of Walbaum, which does not occur in North America.

Etymology—Named for William Clark, of the 1804–1805 Lewis and Clark expedition to the Pacific Northwest.

LIFE HISTORY

Since this information is given in considerable detail under the Lahontan Cutthroat, it will not be repeated here except to say something about the sea-run or "steelhead" cycle which is not a conspicuous part of the habit pattern of the Lahontan form.

Of the cutthroats living in streams tributary to the ocean, some are permanent stream residents, others occasionally enter the ocean and the remaining portion of the population represents individuals who are found in streams only during their spawning and immature periods. These latter are the "steelheads" or "harvest trout." They begin life in freshwater and mature in salt water. In the sea they are typical salmon-

---

[1] Cal Allan informs me that the red slash is often lacking in individual specimens from Walker Lake, particularly the immatures, while older fish may show only a faint orange or pink tinge in this area (personal correspondence 1958).

Fig. 148. "Female cutthroat trout with tail flat against the gravel at start of digging action. She is covering eggs, which are exposed just back of her tail, beside the male. This picture was taken one minute after that shown in Fig. 147" (Osgood R. Smith 1941). Courtesy of Dr. Smith.

looking, silvery-sided, blue-backed steelheads with none of the better developed and often brilliant colors found in stream populations. Several factors may be responsible for this pattern of color response to the different water types, but probably exposure, or lack of it, to light is the governing one. In the sea, trout obviously live much deeper than in shallower streams, and the same effects are noticed in deep, freshwater lakes.

In spring-spawning cutthroats, these steelheads appear at the mouths of the rivers and creeks in the fall and slowly work upstream during the winter until they arrive at their spawning grounds often hundreds of miles inland—as they go they change in color to the more resplendent freshwater hues and are indistinguishable from the resident stream cutthroats by spawning time. After spawning, they return to the sea to repeat the cycle next year. In one to three years their progeny will follow them. For the freshwater variation of the steelhead pattern, see the Lahontan Cutthroat. Partial cutthroat life history work has been done or mentioned by Snyder (1917A), Muttkowski (1925), Hazzard and Madsen (1933), Smith (1941 and 1947), Calhoun (1944) and Fleener (1952), among others.

ECONOMICS

In addition to the value of the cutthroat as a sport fishery, large numbers are taken each year as part of the commercial catch particularly when they are entering rivers from the ocean. In years past a part of the commercial yield was provided by isolated fishermen in inland areas, but the advent of modern icing methods has reduced such activity almost entirely to those interested in sport fishing.

FIG. 149. Courtesy Oregon State Game Commission and Harold P. Smith.

LAHONTAN CUTTHROAT TROUT
Salmo clarki henshawi Gill and Jordan
(Lahontan Blackspotted Trout)

Salmon-trout, Frémont, 1845: 218
Salmo henshawi Gill and Jordan 1878: 258 (in Jordan 1878G)
Salmo tsuppitch, Jordan, 1878G: 358
Salar henshawi, Jordan and Copeland, 1878: 431
Salmo tsuppitch, Jordan and Henshaw, 1878: 196–197
Salmo henshawi, ibid: 197–198
Salmo purpuratus henshawi, Jordan and Gilbert, 1882: 316
Salmo purpuratus henshawi, Cope, 1883: 141, 152
Salmo purpuratus Henshavi, Russell, 1885: 62
Salmo mykiss henshawi, Eigenmann and Eigenmann, 1891: 1132
Salmo mykiss henshawi, Jordan, 1891: 14
Salmo mykiss henshawi, Jordan and Evermann, 1896B: 493
Salmo clarkii henshawi, ibid, 1898: 2819
Salmo clarkii tahoensis, ibid: 2870
Salmo henshawi, ibid, 1902: 180
Salmo tahoensis, ibid: 181
Salmo henshawi, Rutter, 1903: 148
Salmo henshawi, Juday, 1907: 141–146
Salmo tahoensis, ibid: 141
Salmo clarkii tahoensis, Snyder, 1917A: 71
Salmo henshawi, ibid: 70 / 1940: 131–134
Salmo henshawi, Pratt, 1923: 45
Salmo tahoensis, ibid
Salmo tahoensis, Jordan, Evermann and Clark, 1930: 57
Salmo clarkii henshawi, Schrenkeisen, 1938: 45
Salmo clarkii tahoensis, ibid

Cut-throat of Pyramid Lake, Snyder, 1940: 104
Salmo henshawi, Murphy, 1941: 167
Salmo henshawi, Shapovalov, 1941: 442
Salmo henshawi, Smith, 1941: 461–468
Salmo clarkii henshawi, Calhoun, 1944: 80–85
Salmo clarkii henshawi, La Monte, 1945: 115
Salmo clarkii henshawi, Miller and Alcorn, 1946: 175
Salmo clarkii henshawi, Eddy and Surber, 1947: 108
Salmo clarkii henshawi, Smith, 1947: 281–287
Salmo clarkii henshawi, Miller, 1950B: 4
Salmo clarkii henshawi, Shapovalov and Dill, 1950: 385
Salmo clarki henshawi, La Rivers, 1952: 94
Salmo clarki henshawi, La Rivers and Trelease, 1952: 114
Salmo clarkii tahoensis, Böhlke, 1953: 14
Salmo clarkii henshawi, Shapovalov, Dill and Cardone, 1959: 171
Salmo clarkii henshawi, Needham and Gard, 1959: 69–70

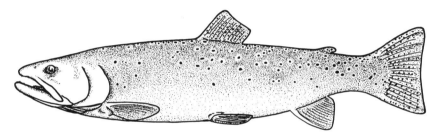

FIG. 150. Lahontan Cutthroat Trout, *Salmo clarki henshawi*.
Courtesy Thomas J. Trelease.

ORIGINAL DESCRIPTION

"Silver Trout of Lake Tahoe. Head little carinate; body elongate, not much compressed; scales in 160–184 rows. Streams of Cal." (Gill and Jordan in Jordan 1878G: 358).

This inadequate description is somewhat improved by portions of two key couplets appearing just above it in the text—

"†Hyoid bone with an elongate band of small teeth (easily scraped off by careless observers).

"*c*. Head large and long—acuminate; hyoid teeth weak; caudal somewhat forked."

This was elaborated upon in a description published later in the same year as part of the Wheeler geographical surveys west of the 100th meridian:

"The following description is taken from Mr. Henshaw's specimens, of which the one numbered 17,086 in the Museum Register is the one measured, and which may be considered as the type of *Salmo henshawi*:

"Body elongate, not greatly compressed; head comparatively slender and long-acuminate, its upper surface very slightly carinated; muzzle somewhat pointed, but bluntish at the tip; head not convex above in either direction; maxillary rather short, about as in *S. clarki*, not reaching much beyond eye; teeth on vomer usual, a rather small, narrow, but distinct patch of teeth on the hyoid bone, the patch narrower and the teeth smaller and more closely set than in *S. clarki;* these teeth appear to be sometimes deciduous. Dorsal fin small, its last rays ⅔ the height of the first, the outer margin even, as in *S. irideus.* Caudal fin short, moderately forked; scales medium, 27—160—27.

"Color rather dark, the sides silvery; back about equally spotted before and behind; sides with rather distant spots. In some, the belly is also spotted; in others, the spots are quite sparse and very round.

"Top and sides of head spotted, even to the end of snout. Dorsal and caudal spotted.

"From *S. clarki* it differs in the form of the head, in the forked tail, in the smaller patch of hyoid teeth, and somewhat in color" (Jordan and Henshaw 1878: 197).

DIAGNOSIS

As in *S. clarki*—being good as a subspecies largely only because of its geographical isolation from other cutthroat stocks. After all the peering at scales is over—the measuring of ratios, the counting of black spots and the description of vivid colors is done—there is little of substance the taxonomist can grasp other than the fact that the Lahontan Cutthroat occupied an area by itself. Also, very probably because of the great diversity of habitat within the Lahontan basin and its affluents, the cutthroat exhibited almost every color combination and size range known for all other cutthroat stocks combined. This can be said to be characteristic of the subspecies, although it is one of little use to the taxonomist. This diversity has given rise to some synonyms, and Snyder (1940: 131) even suggested that it might be convenient to separate the Pyramid Lake cutthroat proper from the rest of the Lahontan basin population because of its color qualities.

TYPE LOCALITY

Lake Tahoe, California-Nevada.

RANGE

The Lahontan drainage system of west-central Nevada; Pyramid, Tahoe and Walker Lakes; Humboldt, Truckee, Carson and Walker Rivers and their tributary lakes and streams. Unlike the Rainbow Trout, *S. clarki henshawi* has seldom been successfully transplanted out of its native basin; extensive stockings were made in eastern United States and California without known success. Eddy and Surber (1943: 108) list an introduction into Minnesota in the early 1920's, which did not survive. Jordan (Days of a Man) recorded a San Joaquin system stocking.

In addition, the Lahontan Cutthroat has been planted in Nevada streams not originally part of its range. Miller and Alcorn (1946: 175-176) give an account, taken from the 1938 field notes of Hubbs, detailing the stocking of the Toiyabe and Toquima Ranges (Lander and Nye Counties), Grass Valley north of Austin (Lander County), and Roberts Creek northwest of Eureka (Eureka County). The Toiyabe specimens were originally brought in from the Reese River system (Washington Creek) to Kingston Creek (south of Austin on the east side of the range) in 1873 by George and Henry Schmidtlein. After establishment there, plantings were made over the remainder of the range.

Subsequent Nevada Fish Commission reports show that large numbers of "blackspotted" spawn were hatched and reared and sent to about every county in the state through the years up to about 1930 when the waning cutthroat population of Pyramid Lake ended this activity.

In his stream survey of part of the State (1950), Ted Frantz found what appear to be Lahontan Cutthroats in several *streams* in the Toiyabe (*Tierney Creek*), Toquima (*Pine, Moore*) and Desatoya (*Edwards*) Ranges of central Nevada. Whether these are remnants of the early Reese River stock brought in by the Schmidtleins or later plantings, no one seems to be sure. Cutts from Heenan Lake,[1] Alpine County, California, have also been rather widely distributed in Nevada and could have been the source of these specimens.

FIG. 151. Heenan Lake. This view is looking westerly and spawning station is at the east end of the narrow lake. Photo May 1957, by Robert Sumner.

Lahontan Cutthroats were also transported many years ago from the Reese River system (Italian Creek on the west side of the Toiyabes) to Callahan Creek on the east side of the Toiyabes by James Callahan, according to information given Mr. Frantz by the son, Mr. John Callahan, present owner of the ranch (north of Austin). No cutthroats now

---

[1]Heenan Lake, originally a tiny natural pond, was enlarged to a reservoir and stocked with Cutthroat Trout from nearby Blue Lake. Heenan drains into the East Carson River, Blue Lake into the West. Heenan Lake lies in the northeast corner of Alpine County, California near the Nevada State line south of Markleeville and southwest of Topaz Lake. The original source of cutthroats for Blue Lake is unknown.

exist in Callahan Creek, but an inventory of Italian Creek showed cutts and hybrids with rainbows in the upper elevations.

In 1947–1948, Bruce Clogston, a former Washoe County and State Game Warden, planted cutts from Heenan Lake into Catnip Reservoir in northern Washoe County.

In 1961, the Wyoming Game and Fish Commission stocked Lahontan Cutts into Seven Mile Lake, Albany County and Eight Mile Lake, Carbon County, Wyoming in a successful attempt to find a trout capable of performing satisfactorily in highly alkaline lakes. These two Wyoming lakes have total solids of 8,816 and 11,744 parts per million, respectively. Their pH, however, is not as high as those of Pyramid and Walker Lakes, being distinctly less than 9.0 in both cases. Brook Trout did not survive in Eight Mile Lake, where both Brook and Cutthroat Trout were initially kept in livecars to determine adaptability to the water. These cutthroat were reared from eggs originally obtained from the California Department of Fish and Game, and information concerning these plants was generously furnished us by Mr. Jack Kanaly and Mr. Fred Beal of the Wyoming Commission.

As conditions change over the original range of the subspecies, populations of *S. clarki henshawi* change with them—usually this means fewer and smaller individuals in any given area, since the changes are invariably detrimental, consisting largely, as they do, of diversionary structures which obstruct up-stream spawning runs, lessen flow in stream channels (resulting in warmer, less-oxygenated water), and lower the lake levels. In combination with pollution, which is the great oxygen-removal factor, these forces produce a general deterioration of the fishes' environment which has meant a consistent restriction of range for many once-teeming species. The Lahontan Cutthroat, as a consequence of some of the above-mentioned factors, has lost ground over most of its natural area.

Its original populations are extinct in Pyramid Lake, which produced the largest specimens on record. It has been virtually eliminated from its type locality, Lake Tahoe, where it was once the dominant species. It has waned in Walker Lake so alarmingly in the last ten years that its extermination could be forseen, but for restocking action by the fish and game authorities; and the one-time teeming river populations are restricted, where they exist at all, to a fraction of their former abundance. Localities in which the cutthroat is dominant, or even plentiful, are rarities. In Nevada, the last such reservoir of the subspecies is Summit Lake in southwestern Humboldt County—here it is the only species of fish present, and exists in relatively large numbers, protected by its inaccessibility and the fact that the lake is on an Indian reservation.

TAXONOMY

It was many years after the publication of the original description of *Salmo clarki henshawi* that the full extent of variability of its component populations was recognized. In addition to fluctuating between specific and subspecific status in the 75 years of its existence as a formal taxonomic unit, variant segments of the *S. c. henshawi* population have each in turn received names which have similarly enjoyed both subspecific and specific rank. The early day practice of describing fishes from one or a few specimens from widely separated localities without any knowledge of the characteristics of intervening populations led to the creation of numerous species reasonably distinct from each other. It was a good many years, in some cases, before the intergrading characteristics of adjoining populations became evident—in the meantime, formidable and badly tangled synonymies grew up involving practically all of the entities we now recognize as reasonably valid species and races. At this date (1960) some of these synonymies have been satisfactorily resolved, while others are only in process of resolution. One of the most confusing synonymicons involved *S. c. henshawi*.

The entity *henshawi*, named in honor of Henry W. Henshaw, naturalist with the Wheeler survey of the 100th meridian, who is credited with its discovery, began its existence in a state of confusion. The same year it was described as *"Salmo henshawi"* by Gill and Jordan in the second edition of Jordan's Manual of the Vertebrates, etc. (1878G), it was referred to as *"Salar henshawi"* by Jordan in a separate paper and the name *"Salmo tsuppitch"* of Richardson (1836: 224), a synonym of *Oncorhynchus kisutch*, was applied by Jordan and Henshaw to a color variant of the Lake Tahoe cutthroat population; *"tsuppitch"* they called the "Black Trout of Lake Tahoe," and *"henshawi"* the "Silver Trout of Lake Tahoe." Their statement that "The species *tsuppitch* as thus defined is closely related to *Salmo clarkii* and to *Salmo henshawi*, from both of which it may be distinguished by the want of hyoid teeth and by the smaller and more conical head" (p. 196) opens the possibility that they had on hand specimens of the Rainbow Trout to which Snyder many years later gave the name *Salmo regalis* (1912: 26).

Speaking further of *Salmo tsuppitch*:

"With *henshawi* it agrees most closely, and if old specimens of the latter, as may perhaps be possible, shed or absorb their hyoid teeth, there may be some difficulty in distinguishing them. The color and general appearance of the two (in spirits) are remarkably similar; but Mr. Henshaw informs me that they are readily distinguishable in life, *S. henshawi* being known to the fishermen as Silver Trout; *S. tsuppitch* as Black Trout.

"It [*S. tsuppitch*] attains a large size both in this stream [North Fork of the Kern River] and in Lake Tahoe, where it is also very abundant. In Lake Tahoe this species is distinguished from the succeeding [*S. henshawi*] by the popular name of 'Black Trout' on account of its very dark colors. Its method of capture and habits while in the lake appear to be identical with those of the following species [*S. henshawi*], and under that will be detailed. It is the belief of all the fishermen on the lake that the 'Black Trout' spawn only in the streams making into the lake, notably the Truckee River, their passage up for this purpose occurring in May, June, and July—June being the special month. At this season many are speared during the run, the Indians especially taking advantage of the opportunity. Such is not the case with the Silver Trout (*S. henshawi*), which is said never to be found in the streams. During the last part of May, while on a visit to Pyramid Lake, which with Winnemucca Lake receives the waters of Tahoe through the Truckee River, we caught in a short time a couple dozen of these trout, averaging about two pounds each [*S. tsuppitch*]. They appear not to attain so great a size here as in Tahoe [!]. The waters of Pyramid Lake fairly teem with these fine fish, and as the lake is within the limits of the Indian reservation, they have been protected, fishing for market, except by the Indians, being prohibited by the superintendent" (Jordan and Henshaw 1878: 196–197).

By 1882, Jordan was listing *henshawi* as a subspecies of *Salmo purpuratus* (Jordan and Gilbert, 1882B), in 1891 as *S. mykiss henshawi* and in 1896 Jordan and Evermann had abandoned the name "*tsuppitch*" for any Tahoe trout, recognizing the misidentification, and were using *henshawi* for both Black and Silver Tahoe trout, regarding the two as color variants unworthy of formal separation. They felt the black type to be the typical *henshawi*, the silver type to be the variant.

However, in 1898, Jordan and Evermann reconsidered their earlier stand and described the silver type as the new subspecies *Salmo clarkii tahoensis*, reapplying to it their old name of "Silver Trout of Lake Tahoe." *Henshawi* was again thus restricted to the smaller, darker fish which spawned in the streams while *tahoensis* designated the deeper water silvery lake populations spawning in the lake itself; distinctions of no value, as we now know. Snyder's 1917A elucidation of the different color, size and age groups of Lahontan Cutthroat trout running up the Truckee River from Pyramid Lake clearly showed the integrity of the name *henshawi* for the populations as a single taxonomic unit. The apparently great variation of sections of the *S. c. henshawi* population within the Lahontan drainage basin is not at all remarkable when the variety of habitats occupied by them is considered; furthermore, there was little discontinuity in the population as a whole, and what there was appears to be relatively recent.

The color variants of *Salmo clarki henshawi* were briefly summed up by Snyder (1917A), who considered *henshawi* as a specific unit:

LAKE TAHOE	PYRAMID AND WINNEMUCCA LAKES AND LOWER TRUCKEE RIVER
(1) "Tahoe Trout," large-spotted, dark; larger.	"Redfish," brilliantly colored with prominent red cheeks; large, with some silvery examples.
(2) "Silver Trout," small elongate spots, chunky body, silvery color; smaller.	"Tommy," relatively large-spotted; smaller, with some silvery examples.

The name *tahoensis* lingered in the literature until the late 1930s, and it was somewhat later, within the last ten years, that *henshawi* became stabilized in its present subspecific status.

Etymology—Named for Henry W. Henshaw, naturalist of early government western surveys.

LIFE HISTORY

S. c. henshawi is a typically carnivorous, spring-spawning trout showing great diversity in color types over its range. Its feeding habits do not diverge from the typical trout patterns—invertebrates, while

FIG. 152. Picture taken at Sutcliffe, Pyramid Lake. From the Nettie Sutcliffe Cooper collection.

small; becoming increasingly predaceous on other fish with greater size.

In lakes such as Pyramid and Walker, which are or were ideal habitats in most ways for trout, insects are relatively scarce so that the teeming microcrustacea (ostracods, copepods, and Cladocera) are utilized by developing fish not only in their small stages but apparently after they have achieved sizes where they would normally easily handle insect food.

Unfortunately, despite the one-time importance of the cutthroat as a commercial fishery in Pyramid and Winnemucca Lakes and as a sport fish in the west-central Nevada streams and Lake Tahoe, no satisfactory growth-rate studies seem to have been published on it. Growth rates were maximal in Pyramid, Winnemucca and Walker Lakes, decreasing proportionately at higher and colder altitudes.

John Otterbein Snyder, working for the U. S. Bureau of Fisheries on the fishes of the Lahontan basin, has left us the best overall account of the Lahontan Cutthroat Trout, particularly as regards descriptions and life history—were it not for his accurate and extensive notes on the color variations of such conspicuously colorful populations as those occurring in Pyramid Lake and running up the Truckee River, we would know much less about them than we do, since they are now extinct.

The stock living in Pyramid and Winnemucca Lakes (the latter was dry by 1938), which ascended the Truckee River to spawn, developed two distinctive runs, separated by time and composed of quite different appearing individuals. The first run, of the largest and most colorful fish, began with the rise of the river during the late fall and winter rains, and was usually over in March. These "Redfish" apparently did not go up river much beyond Verdi, where the cañon steepens and narrows and the water becomes swifter and more turbulent. These specimens, called "Tomoo-agaih" ("Winter Trout") by the Pahute Indians, had the entire body suffused with red. Snyder described living specimens as (1917A: 71):

"A large male redfish dipped from the pool below the dam at Thisbe (March 26) [between Reno and Wadsworth] was colored as follows: Whole body suffused with pink and yellow, the color approaching vermilion in some lights or darker red in others, the yellow with metallic reflections. The yellow color is more intense above and below, the pink brightest in the region of the lateral line, but not distinctly outlined as a stripe. Opercle bright, livid red; subopercle like body; preopercle reddish yellow, much brighter than body, but duller than opercle. An indefinite, small pinkish spot midway between eye and opercle. A few deep-orange, coin-shaped spots somewhat smaller than eye,[2] scattered here and there on the body near bases of pectorals and on the breast. The caudal fin is yellow, the dorsal suffused with yellow, the paired fins with purplish red. The area beneath the mandible is strongly marked with bright red, the color confined to the side next to the branchiostegals. More brilliantly colored examples are often seen

---

[2] These spots are also seen on Heenan Lake Trout, as well as occasionally in Walker Lake.

where the red of the opercle is more livid and spreads to the shoulder girdle, the lateral stripe better defined and more intense in color, and the entire head and body of a brighter hue. The females are similar though paler in color. The small orange spots referred to are remarkable in that they appear, without regularity, on any part of the body except in the region of the lateral line. One was seen on the lower jaw, another on the base of the caudal, and one on the adipose fin. In addition to the bright colors noted above, the head and body are sprinkled with black spots which are smaller than in the Tahoe trout. The red color beneath the jaw is apparently always present in the redfish."

As the redfish migrations tapered off, large schools of smaller and darker fish began their up-stream run from Pyramid Lake. The Pahute name "Tama-agaih" ("Spring Trout") was generally translated by the white angler into "Tommy." This second cutthroat surge up-river began about April, overlapping with the waning redfish runs, and usually terminated early in May. Around the first part of July, the lower Truckee River was generally cleared of trout for the year. Silvery individuals of both redfish and tommy migrations were often seen in the lower river.

At the time of Snyder's studies, 9 to 10 pound fish were common in Pyramid Lake, while 20-pounders were not rare. The variation in color, partially delineated in the above description of breeding individuals, was striking not only in different parts of the fishes' range, as might be expected between waters with such different limnologic characteristics as Tahoe and Pyramid, but was also remarkably pronounced even among the individuals of a given body of water, such as Pyramid Lake. Snyder gives the following description of a typical specimen of this extinct population (1917A: 72):

"An example 2 feet long, taken in Pyramid Lake May 29, in life was clear greenish olive on the upper surface, this color becoming diffuse on the sides, where it gave place to silvery, strongly tinged with pink, the ventral surface of the body, throat, and chin being white. The whole dorsal surface when seen from the side appeared silvery with a suffusion of pink, instead of greenish olive, as when viewed from above. Cheeks pinkish, the upper edge of opercle brassy. Paired fins and anal tinted with orange red. Dorsal and adipose fins like the back. A red gash present on the under mandible. Small black spots scattered rather evenly over head and body, varying in size from about a third the diameter of pupil to mere specks. They were largest on the upper surface and on the caudal fin. Many examples similar in color were seen. The red was never entirely absent from the throat, and small, round, brassy spots were frequently found on the head and body. Large, fat examples are sometimes taken, the general color of which is very pale and silvery, the spots being few and small, exact counterparts of the silver trout of Lake Tahoe. Several specimens taken on the western side of Pyramid Lake late in May were brilliantly colored and dark like the migrating trout. Their fins were frayed, they had spawned, and it was quite apparent that they had but recently returned from the river. Rarely an individual is caught which has no spots below the lateral line except on the caudal fin and only a few above. Trout from Winnemucca Lake are like those from Pyramid Lake. An examination of many from both localities gave the impression that those from the former were somewhat lighter in color."

ECONOMICS

It is difficult to do more than generalize about the value of the Pyramid Lake fishery, where the Lahontan Cutthroat was most exploited commercially. For nearly three-quarters of a century, from the 1860s to the 1930s, the lake was a bountiful source of easily obtainable market and sport fish. For a good many of those years, trout were shipped to market by the railroad carload, and hundreds of individual Indian peddlers fanned out to ranches and towns east, south and west of the lake, driving their wagons and cars as far as the

FIG. 153. A catch of Pyramid Lake Cutthroat Trout, in the early 1920s, at Sutcliffe. Courtesy Tom Trelease.

spoiling time of fish allowed, selling the cordwood-stacked large trout for a few cents or a dollar each. The value of the fish to the sporting angler was also a considerable part of the value of the fishery as a whole—Pyramid Lake was famous over the country for the size of the trout it produced.

According to Sumner (1940), who studied the decline of the Pyramid fishing industry, 1920 was reportedly the last year that angling was considered good in the Truckee River above the lake, and 1934 saw the last of good boat fishing on Pyramid itself. The period from 1935 to 1938 brought a reduction of 85 percent in the numbers of fish stamped to be sold by the Indians, and average weight of the 1938 fish was almost double that of the 1935 fish. The increase in size of individual catches and the fact that small fish were seldom seen indicated the spawning cycle had been interrupted and reproduction was not taking place. As a consequence, the "Tommy" runs of former years (spawning migrations of younger mature fish following the older redfish runs) disappeared. The 1938 fish seemed to be about the same age (8 to 9 years), indicating them to be the progeny of the 1929 or 1930 spawning run. Other evidence also pointed to the 1929 up-stream migration as being the last breeding run of any consequence.

The most important factor in this decline was the diversion of water from the Truckee River to the adjacent Fallon basin, a diversion which could be forseen in 1903 with the construction of Derby Dam (completed in 1905) on the river six miles west of Wadsworth. From this point, a large canal carried much of the Truckee River into Lahontan Valley in Churchill County to provide irrigation and power for the newly created Newlands Project, the first Federal reclamation project of its type in the country. There had been other minor dams placed across the river at many points prior to this, but without serious effects upon the spawning runs.

Derby, in addition to being large enough to carry all the Truckee River flow during the latter's low periods in late summer and fall, presented an insurmountable obstacle to upstream movements of trout. However, the restriction of the migratory run from its original theoretical distance of over 100 miles (the length of the Truckee River between Tahoe and Pyramid) for some of the smaller, more active fish, to 30 miles and less between Pyramid and Derby, could not have conceivably been important in exterminating the Pyramid Lake cutthroat, for the many spawning beds available in the restricted lower river would have supported a sizeable amount of reproduction. Higher average water temperatures during the spawning season in these lower Truckee River reaches, however, now could become restrictive.

The Derby diversion, aided by the onset of somewhat drier years, initiated a drop in the level of Pyramid Lake which has continued steadily since about 1920. As the lake dropped, the river mouth became increasingly sand-choked, until a point was reached where it was impossible for trout to enter the river (about 1930). During the decade of the 1920s, each year was apparently a touch-and-go proposition as far as trout getting out of the lake was concerned.

From then on it was a matter of the remainder of the trout population living out their span and dying of old age without being able to to reproduce. During this last phase of their existence, the average size of the fish progressively increased, giving the false impression to the uninitiated for the first few years that fishing conditions were better than they had ever been and that the often-mentioned danger of the Lahontan Cutthroat becoming extinct in Pyramid Lake was only a myth.

During this same period, changing river conditions due to up-stream diversions and pollutions, and the continually heavy planting of Rainbow Trout, effectively eliminated most of the thriving river populations of cutthroat. The smaller volume of water in the river channel became warmer and held less oxygen in solution; pollution further depleted the oxygen, virtually eliminating it in local stretches of the river over considerable parts of the year; the large introduced populations of rainbow not only competed for food with the cutthroat in an ever-changing environment unfavorable to the latter, but also hybridized with it, producing sterile offspring and further reducing the cutthroat population by "diluting" it and rendering it incapable of reproducing.

Conditions have changed so markedly in many stretches of the Truckee River that even the rainbows could not prosper if they were not constantly and heavily stocked into the stream by fish and game authorities—and in this retrograding environment, the European Brown Trout has become a most successful species, maintaining itself naturally in what are probably maximum populations for the existing stream conditions. It is reasonable, however, to suppose that, regardless of changes in the Truckee River, the river cutthroats might well have become depleted because of the greatly increased angling pressure of recent decades.

The taking of spawn at the river mouth; depredations of the White Pelican (*Pelecanus erythrorhynchos*), Farallon Cormorant (*Phalacrocorax auritus albociliatus*) and other piscivorous birds; introduction of foreign fish (Sacramento Perch and Asiatic Carp); and the increasing salinity and alkalinity of the lake waters have all appealed to the uninformed at various times in the past and present as scapegoats to be blamed for the decline and extinction of the Lahontan Cutthroat. All these factors together could not have conceivably reduced the cutthroat populations had the fish been able to spawn in even greatly reduced numbers. Certain of these arguments eliminate themselves for obvious reasons to the thinking person—Indians and pelicans had been taking fish at the river mouth during spawning runs for millenia with no effects on the cutthroat population. Other points have been easily disproven by simple example and experiment. Sacramento Perch and Asiatic Carp have both lived in great numbers in the lake for many years, and the cutthroats there continued to thrive until upstream spawning conditions became unsatisfactory.

In 1948, Thomas Trelease and the writer showed to their own satisfaction that trout could be kept in Pyramid Lake water with no ill-effects, even though the water was then admittedly more saline and alkaline than it was when the level stood 50–60 feet higher. This was done at a time when public opinion denied that trout could live in the lake—and public convictions had been strengthened by professional biologists who had studied the situation hastily and concluded unwisely. As far as mere existence in its waters was concerned, immense numbers of chub, carp, Sacramento Perch, cui-ui and suckers have been there all the time as living testimonials of the illogic of such reasoning.

This optimism has been amply justified—in the years that have followed the lowering of these crude, trout-filled cages into the lake, the Nevada Fish and Game Commission has embarked on an increasingly successful stocking program with the best developments (at this writing, 1960) yet to come. Since 1950, Mr. Trelease has set up planting schedules which have placed better than one and a half million kokanee fingerling (1950–1953), over three-quarters of a million fingerling and semi-reared rainbow (1954–1955) and more than half a million fingerling cutthroat (1953)—total, more than three million—in the lake and adjacent lower Truckee River, with more to follow. Even a few Brook Trout were released in 1955 (4,752). Fishing has returned to Pyramid Lake.

The original value of the Pyramid and Walker Lake cutthroat fishery to the aboriginal economy should not go unmentioned. The Indian occupants of prehistoric Lovelock Cave netted and hooked fish in Humboldt Lake and the lower Humboldt River several thousand years ago, as attested by artifacts dug from the cave floor. While they undoubtedly took any and all fish that came to hand, trout must have constituted a majority of these, particularly during the spawning runs.

Frémont (1845) was impressed with the fish he found at the mouth of the Truckee River after coming down along the eastern edge of Pyramid Lake: "An Indian brought in a large fish to trade, which we had the inexpressable satisfaction to find was a salmon trout; we

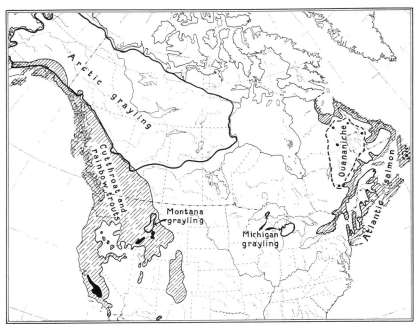

FIG. 154. Original distribution of trout, *Salmo*, spp., and graylings, *Thymallus* spp. Black areas in the West indicate absence of trout. Courtesy University of California Press and Erhard Rostlund.

gathered around him eagerly. The Indians were amused with our delight, and immediately brought in numbers; so that the camp was soon stocked. Their flavor was excellent—superior, in fact, to that of any fish I have ever known. They were of extraordinary size—about as large as the Columbia river salmon—generally from two to four feet in length. From the information of Mr. Walker, who passed among some lakes lying more to the eastward, this fish is common to the streams of the inland lakes. He subsequently informed me that he obtained them weighing six pounds when cleaned and the head taken off; which corresponds very well with the size of those obtained at this place. They doubtless formed the subsistence of these people, who hold the fishery in exclusive possession.

"I remarked that one of them gave a fish to the Indian we had first seen [along the east shore of the lake], which he carried off to his family. To them it was probably a feast; being of the Digger tribe, and having no share in the fishery, living generally on seeds and roots. Although this was a time of the year when the fish have not yet become fat [January 15, 1844], they were excellent, and we could only imagine what they are at the proper season . . .

"In the mean time, such a salmon-trout feast as is seldom seen was going on in our camp; and every variety of manner in which fish could be prepared—boiled, fried, and roasted in the ashes—was put into requisition; and every few minutes an Indian would be seen running off to spear a fresh one."

YELLOWSTONE CUTTHROAT TROUT
Salmo clarki lewisi (Girard)
(Yellowstone Blackspotted Trout)

Salar lewisi Girard 1856C: 219 / 1859: 318–320
Salmo lewisii, Günther, 1866: 122–123
Salmo carinatus Cope 1872C: 471
Salmo lewisi, Suckley, 1874: 139–140
Salmo purpuratus, Jordan and Gilbert, 1882: 315
Salmo mykiss lewisi, Jordan and Evermann, 1896B: 493
Salmo clarkii lewisi, ibid, 1898B: 2819
Salmo lewisi, ibid, 1902: 179
Salmo lewisi, Pratt, 1923: 44
Salmo lewisi, Locke, 1929: 188
Salmo lewisi, Jordan, Evermann and Clark, 1930: 56
Salmo clarkii lewisi, Schultz, 1936: 136
Salmo clarkii lewisi, Schrenkeisen, 1938: 44
Salmo clarkii lewisi, Hubbs and Lagler, 1941: 35–36 / 1947: 36–37
Salmo clarkii lewisi, Miller and Alcorn, 1946: 176–177 (Nevada)
Salmo clarkii lewisi, Eddy and Surber, 1947: 108
Salmo clarkii lewisi, Carl and Clemens, 1948: 54–55
Salmo clarkii lewisii, Clemens and Wilby, 1949: 90
Salmo clarkii lewisi, Miller, 1950B: 4
Salmo clarkii lewisi, Simon, 1951: 35–38
Salmo clarki lewisi, Beckman, 1952: 18–19
Salmo clarki lewisi, La Rivers and Trelease, 1952: 114
Salmo clarki lewisi, Slastenenko, 1958A: 57
Salmo clarki lewisi, Carl, Clemens and Lindsey, 1959: 69–70
Salmo clarkii lewisi, Needham and Gard, 1959: 68

ORIGINAL DESCRIPTION

"Body rather thickish upon its middle region; head moderate, constituting a little less than the fifth of the total length. Lower jaw longest. Maxillary gently curved, its posterior extremity reaching a vertical line drawn immediately behind the orbit. Anterior margin of dorsal fin a little nearer the extremity of the snout than the base of the caudal fin. Ground color of the upper region bluish grey, of the inferior region yellow or orange. The back, peduncle of tail, dorsal, adipose, and caudal fins, spotted with black. The belly and lower fins are unicolor, a deep orange hue existing along the rays and also in the shape of a dot upon the abdominal scales, and which disappears by long standing in alcohol.

"This is the trout alluded to in Lewis and Clarke's 'Travels.' They 'caught (at the Falls of the Missouri) half a dozen trout from sixteen to twenty-three inches long, precisely resembling in form and the position of the fins, the mountain or speckled trout of the United States, except that the specks of the former are of a deep black, while those of the latter are of a red or gold color: they have long sharp teeth on the palate and tongue, and generally a small speck of red on each side behind the front ventral fins; the flesh is of a pale yellowish red, or when in good order, of a rose colored red.'—London edition of 1814, p. 192, 4to. And further on, page 487, we read: 'The mountain or speckled trout are found in the waters of the Columbia within the mountains; they are the same with those found in the upper part of the Missouri, but are not so abundant in the Columbia as in that river. We never saw this fish below the mountains, but from the transparency and coldness of the Kooskoskee, we should not doubt of its existence in that stream as low as its junction with the south east branch of the Columbia.'

"It would be an interesting point to compare, side by side, specimens caught in the Columbia, with those of the Missouri River. We should not be surprised if the result of such a comparison should refer the specimens from the basin of the Columbia to *Fario gairdneri.*

"Specimens of this species were collected at the Falls of the Missouri River, Rocky Mountains, by Dr. Geo. Suckley, U.S.A., under Gov. I. I. Stevens" (Girard 1856C: 219–220).

DIAGNOSIS

A slightly smaller-scaled form than the Lahontan Cutthroat, having recorded extremes of from 156 to 190 scales in the lateral line, with 165–170 being usual (Schultz 1936: 136). The black-spotting also apparently tends to be absent from the lower parts of the body and to be nonuniform; i.e., more widely spaced anteriorly whereas, although subject to wide variation in size and density, the spotting of the Lahontan type was more uniform.

TYPE LOCALITY

Falls of Missouri River (= Great Falls, Montana).

RANGE

Both sides of the northern United States Rocky Mountains, in headwaters of both the Snake and Missouri systems.

TAXONOMY

Etymology—Named for Meriwether Lewis, of the 1804–1805 Lewis and Clark expedition to the Columbia River.

ECONOMICS

The Yellowstone Cutthroat is of interest to us only in that it has been extensively introduced into various waters in the State in times past and has undoubtedly furnished considerable fishing as a result. A perusal of various sources, such as the U. S. Fish Commissioner reports, as well as those of the California and Nevada Fish and Game organizations, shows that better than 3 million eggs and fish supposedly of the Yellowstone Cutthroat have been distributed and planted in Nevada.

However, all this is largely of historical importance only since the persistent rainbow planting program—which was going on simultaneously and has continued unabated to the present time—has overwhelmed the Yellowstone introductions and as far as the writer is aware, no cutthroat recognizable as a Yellowstone can now be found in Nevada waters with the possible exception of Pyramid Lake.

The Yellowstone Cutthroat has been a popular fish with culturists for many years, and importations into Nevada have been numerous, beginning with the first known one of 1918 when O'Malley (1919: 9, 29) records a shipment of 50,000 eggs to the Nevada Commission at Ely, presumably for hatching there. The first plants into western Nevada seem to have been made in 1925 according to Fish Commission Reports when 150,000 eggs were hatched and planted with Lahontan Cutthroats into Walker Lake, and Washoe, Lander and Elko Counties. Several million eggs have been received by the State since about 1930 for hatching and distribution.

Walker Lake received more Yellowstones in 1941 and 1942 (Miller and Alcorn 1946: 177) and has been particularly plentifully supplied with them during the past few years (prior to 1958). Cal Allan informs me that survival has been poor and growth much slower at Walker Lake than in the case of our native cutthroats.

Pyramid Lake has received several quotas of Yellowstones in the last few years, apparently with indications of good survival for some of them (Kay Johnson).

In streams where they have been more-or-less recently planted, Ted Frantz found, as in Walker Lake, that they did not do well.

UTAH CUTTHROAT TROUT
Salmo clarki utah Suckley
(Utah Lake Blackspotted Trout)

Salar virginalis Girard 1856C: 220 (in part)
Salmo utah Suckley 1874: 136
Salmo virginalis, ibid: 135
Salmo virginalis, Yarrow, 1874: 363–368
Salmo purpuratus, Jordan and Gilbert, 1882: 315
Salmo mykiss virginalis, Jordan and Evermann, 1896B: 495
Salmo virginalis, ibid, 1902: 182
Salmo utah, Jordan, Evermann and Clark, 1930: 56
Salmo utah, Hatton, 1932: 18–23
Salmo utah, Tanner, 1936: 161–163
Salmo clarkii utah, Schrenkeisen, 1938: 44
Salmo clarkii utah, Miller and Alcorn, 1946: 177–178
Salmo clarkii utah, Miller, 1950B: 4
Salmo clarki utah, La Rivers and Trelease, 1952: 114

ORIGINAL DESCRIPTION

"For this variety or kind we will, for the present, apply the provisional name of *Salmo utah*.

"CHARACTERS—Highest point of convexity of dorsal profile rather anterior to the same on *S. virginalis;* scales appear somewhat larger, (but this may be more apparent than real, owing to the insufficient material for comparison;) appearances of fish more silvery, spots much smaller in size and more irregular in shape; in other respects resembling *S. virginalis*" (Suckley 1874: 136).

DIAGNOSIS

Supposedly a rather large-scaled form and one—like the Lahontan Cutthroat—subspecifically definable as well on the basis of its geographic isolation as for any other characteristic.

TYPE LOCALITY

Utah Lake, Utah County, Utah.

RANGE

The Utah Cutthroat is the native trout in the Bonneville Basin of Utah and eastern Nevada, but now seems to be extinct over much of its former range and rare elsewhere. In fact, Miller seems to think it is entirely extinct.

Since the early days, Utah Cutts have been distributed into various streams in the State beyond their original range. Miller and Alcorn (1946: 177) record "old-timer" statements to the effect that Cleve Creek, Spring Valley, White Pine County, received Utah Cutthroats before 1881 from Utah (supposedly Trout Creek, Juab County), and that eventually most of the streams of that area were stocked.

The earliest record seems to be that of Hubbs, from local testimony (Miller and Alcorn 1946: 177), that cutthroats were planted in 1876 near the Moorman Ranch (Illipah) in Jakes Valley, White Pine County. However, there is no certainty these were from Utah—they may have been the Lahontan strain. In the same year, the waterworks tanks of Hamilton were planted to trout, and these later were turned into Illipah Creek. Other species of trout have also been planted in Illipah, so that, if the original stockings here were Utah Cutts, they would have been rapidly inundated.

Ted Frantz informs me that there is a good possibility of Utah Cutthroats still being in some *creeks* of the Snake (*Pine* and *Ridge*) and Schell Creek (*Muncy*) Ranges, White Pine County. In 1953, 44 cutts from Pine Creek were taken to the Spring Creek Rearing Station and a few days later were planted near the headwaters of Hampton Creek in the Snake Range, with apparent success. Pine Creek was also kept closed to fishing for many years, first by Robert Dickey, one-time Fish and Game Commissioner from White Pine County, and later by the Fish and Game Commission because it was recognized as desirable that this particular strain of cutthroat be kept as pure as possible.

ECONOMICS

Of little importance in Nevada.

COLORADO CUTTHROAT TROUT
Salmo clarki pleuriticus Cope
(Colorado River Blackspotted Trout)

Salmo pleuriticus Cope 1872C: 471
Salmo spilurus pleuriticus, Jordan and Gilbert, 1882: 314
Salmo mykiss pleuriticus, Jordan, 1891: 14, 28
Salmo mykiss pleuriticus, Jordan and Evermann, 1896B: 496
Salmo clarkii pleuriticus, ibid, 1898B: 2819
Salmo pleuriticus, ibid, 1902: 186
Salmo pleuriticus, Pratt, 1923: 45
Salmo pleuriticus, Jordan, Evermann and Clark, 1930: 57
Salmo clarkii pleuriticus, Schrenkeisen, 1938: 44
Salmo clarkii pleuriticus, Dill, 1944: 149
Salmo clarkii pleuriticus, Miller, 1950B: 4, 27, 29–30, 34
Salmo clarkii pleuriticus, Shapovalov and Dill, 1950: 385
Salmo clarkii pleuriticus, Simon, 1951: 39
Salmo clarki pleuriticus, Beckman, 1952: 18

ORIGINAL DESCRIPTION

"This is the abundant mountain trout of the head-waters of the Green and Platte Rivers, and even of the Yellowstone. It is rather a stout species, with obtusely descending muzzle, and large eye entering the head only four times. The cranial keel is a marked character; its elevation is greater between the orbits than on the posterior part of the frontal bones. The interorbital width is 1.33 times the long diameter of the interpalpebral opening of the eye. The dorsal fin is nearer the origin of the marginal rays of the caudal fin than to the end of the muzzle, but is midway between the latter and the termination of the scales on the sides of the fin. Radii, D. II. 11–12 and 13; A. II. 11. Br. XI. The scales range from 40 to 45 below the first dorsal ray to the lateral line. The maxillary bone extends to a little beyond the orbit, and is not expanded.

"This is a spotted species, and the spots are chiefly found above the lateral line and on the whole caudal peduncle, and on the dorsal and caudal fins. They are usually rather scattered, less numerous on the peduncle than in *S. spilurus,* and more so anteriorly; those on the fins are smaller and less numerous. There is, however, variation in the size and number of the spots. The sides are ornamented with short, broad longitudinal bars of crimson; a band of the same color occupies the fissure within each ramus of the mandible and skin on the median side of it. The fins are all more or less crimson; but none of these are black-bordered. The largest specimens are 10–12 inches long.

"Seven specimens of this species are in the collections from the heads of Green River; from Medicine Lodge Creek, Idaho, (two specimens;) four from the Junction, Montana. A specimen each from Yellow Creek and the Gallatin Fork of the Missouri, Montana, represent at least a color variety of this fish. The spots are much smaller and much more numerous, though distributed over the same regions; they are less numerous on the caudal fin. In the Gallatin specimen there are 51 scales above the lateral line; in the other 44. Another variety from the

Yellowstone Basin is only represented by young specimens. They have no spots on the caudal fin.

"A number of dried specimens from the Yellowstone Lake, of larger size than the specimens above described, probably belong to this species. They are rather more closely spotted on the caudal peduncle and fin, but are similar in all important respects. The only discrepancy which I find is the relatively smaller eye, (not orbit,) which enters the head five times, and the greater prolongation of the maxillary bone. These characters are, perhaps, due to the larger size attained by the individuals. They are from a foot to eighteen inches in length" (Cope 1872C: 471).

DIAGNOSIS

A very fine-scaled form, having from 180 to 205 scales in the lateral line (Miller 1950: 29), with most of the black-spotting concentrated in the posterior part of the body, around the tail. Like all cutthroats, the Colorado Cutthroat was variable, and an orange or reddish lateral coloring was reportedly the usual case, as the specific name *pleuriticus* indicates.

TYPE LOCALITY

Head-waters of Green River, Wyoming.

RANGE

Head-waters of the Colorado River basin in Utah, Wyoming, Colorado, and New Mexico.

TAXONOMY

Etymology—"pleuriticus" alludes to the side, calling attention to the lateral red coloration.

ECONOMICS

This form of the cutthroat is included only because there have been literature records of the Colorado Cutthroat for the lower Colorado River and as such the fish could be presumed to have occupied that part of the river which forms the extreme southern boundary of Nevada. However, there is reason for suspecting that these lower Colorado River records are erroneous and that the Colorado Cutthroat was regularly found only in the upper reaches of the system (Miller 1950). As such, it probably then could not be considered a native of Nevada. To write it off still further (and make us wonder why we ever brought it up in the first place) there is some possibility that, like its many cousins in the cutthroat group, this variety may now, like the swan-tailed butterfly fish, be extinct.

RAINBOW TROUT

Salmo gairdneri Richardson

(Steelhead Trout)

Salmo gairdnerii Richardson 1836: 221–222
Fario gairdneri, Girard, 1859: 313–314
Salmo masoni Suckley 1860: 345

Salmo truncatus Suckley 1862: 3 / 1874: 115–116
Salmo gairdnerii, Günther, 1866: 118–119
Salmo gairdneri, Suckley, 1874: 114
Salmo gairdneri, Jordan and Gilbert, 1882: 313
Salmo gairdneri, Jordan and Evermann, 1896B: 498 / 1902: 190–191
Salmo gairdneri, Pratt, 1923: 45–46
Salmo gairdnerii, Locke, 1929: 186
Salmo gairdnerii, Jordan, Evermann and Clark, 1930: 58
Salmo gairdnerii gairdnerii, Schultz, 1936: 137
Salmo gairdnerii, Schrenkeisen, 1938: 49–51
Salmo gairdnerii, Murphy, 1941: 167
Salmo gairdnerii, Shapovalov, 1941: 442–443
Salmo gairdnerii, Hubbs and Miller, 1943: 353
Salmo gairdnerii, La Monte, 1945: 110
Salmo gairdnerii gairdnerii, Carl and Clemens, 1948: 42, 57–59
Salmo gairdnerii, Miller and Miller, 1948: 183
Salmo gairdnerii gairdnerii, Clemens and Wilby, 1949: 92–93
Salmo gairdnerii, Everhart, 1950: 23
Salmo gairdnerii, Miller, 1950B: 4
Salmo gairdnerii, Shapovalov and Dill, 1950: 385
Salmo gairdneri, Beckman, 1952: 19–20
Salmo gairdneri, La Rivers, 1952: 94
Salmo gairdneri, La Rivers and Trelease, 1952: 114
Salmo gairdnerii, Gabrielson and La Monte, 1954: 96–97
Salmo gairdnerii, Scott, 1954: 29–30
Salmo gairdneri, Dymond, 1955: 546–548
Salmo gairdnerii, Shomon, 1955: 12–13
Salmo gairdnerii, Stokell, 1955: 97–104
Salmo gairdneri, Eddy, 1957: 50
Salmo gairdneri, Koster, 1957: 31–32
Salmo gairdneri, Moore, 1957: 64
Salmo gairdneri, Trautman, 1957: 188–190
Salmo gairdneri, Garside and Tait, 1958
Salmo gairdneri, Slastenenko, 1958A: 58–60
Salmo gairdneri, Carl, Clemens and Lindsey, 1959: 71–74
Salmo gairdnerii, Needham and Gard, 1959
Salmo gairdnerii, Shapovalov, Dill and Cordone, 1959: 171

ORIGINAL DESCRIPTION

"[The specific name which I have given to this salmon is intended as a tribute to the merits of a young though able naturalist, from whom science may expect many important acquisitions, and especially in the history of the Zoology of the Northwest coast of America, should his engagements with the Hudson's Bay Company permit him to cultivate that hitherto neglected field of observation.—R]

"This species ascends the river in the month of June, in much smaller numbers than the *quinnat*, in whose company it is taken. Its average weight is between six and seven pounds.

"'Colour.—Back of head and body bluish-grey; sides ash-grey. Belly white. The only traces of variegated marking are a few faint spots at the root of the caudal. Form.—Profile of dorsal line nearly straight, tail terminating in a slightly semilunar outline. Ventrals correspond to commencement of dorsal and adipose to end of anal. Teeth.—Jaws fully armed with strong hooked teeth, except a small space in centre of upper jaw. Vomer armed with a double row for two-thirds of its anterior portion. Palate-bones also armed with strong teeth. Fins.— Br. 11–12; P. 13; V. 11; A. 12.

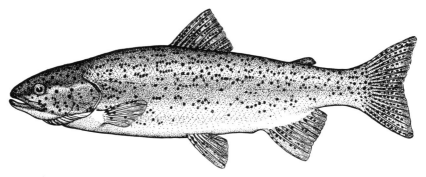

FIG. 155. Southcoast Rainbow Trout, *Salmo gairdneri irideus*. Drawn by Silvio Santina.

" 'Dimensions.

	Inches	Lines
Extreme length	31	0
Greatest height of body	5	9½
Circumference of ditto	14	0
Breadth between the eyes	2	0
Breadth between the nostrils	1	2⅓
Length from end of snout to nostrils	1	2⅓
Length from end of snout to eyes	1	9½
Length from end of snout to angle of opercule	5	2⅓
Length from end of snout to pectorals	6	3½
Length from end of snout to dorsal	12	0
Length from end of snout to ventrals	12	3½
Length from end of snout to anal	21	0
Length from end of snout to adipose*	21	0
Length of pectorals	3	4⅔
Length of ventrals	3	0
Length of attachment of dorsal	3	0
Height of dorsal	2	4⅔
Height of adipose	1	2⅓
Length of caudal	4	8⅓
Its greatest breadth	4	0
Length of attachment of anal	2	4⅔

Gairdner, *in lit.*

"[In this species the gill-cover resembles that of *S. salar* still more strongly than that of the *quinnat* does, the shape of the suboperculum in particular being precisely the same with that of *salar*. The teeth stand in bony sockets like those of the *quinnat,* but are scarcely so long.

Those of the lower jaw and intermaxillaries are a little smaller than the lingual ones, and somewhat larger than the palatine or labial ones. The tongue contains six teeth on each side, the rows not parallel as in the *quinnat*, but diverging a little posteriorly. The pharyngeals are armed with small sharp teeth. The numbers of the teeth, excluding the small ones which fall off with the gums, are as follows: *Intermax.* 4-4; *labials* 21-21; *lower jaw* 11-11; *palate bones* 12-12; *vomer* lost; *tongue* 6-6. When the soft parts are entirely removed, the projecting under edge of the articular piece of the lower jaw is acutely serrated, in which respect this species differs from all the others received from Dr. Gairdner. There are sixty-four vertebrae in the spine.—R.]" (Richardson 1836: 221-222). In the footnote accompanying this description (see asterisk after "adipose" above), Richardson comments:

"*Dr. Gairdner must have accidentally put down wrong figures here

FIG. 156. Courtesy Oregon State Game Commission and Harold P. Smith.

in transcribing his notes, as the adipose is not opposite to the commencement of the anal, but to its end.—R."

DIAGNOSIS

Head 4.0 to 4.5 in body length, depth 3.75 to 5.0 in body length; branchiostegals 11-13; gillrakers 8 + 12; maxillary rather narrow, eye small and almost above middle of maxillary; hyoid or basibranchial teeth absent; pyloric caeca about 42; vertebrae 58-65; scales 23-32/120-160/20-30; fins—dorsal 11-rayed, anal 10-12 rayed; color—bluish or greenish above, silvery on sides—dorsal areas everywhere profusely, irregularly black-spotted—in spring both sexes have a prominent broad pinkish lateral stripe—no red slash mark on the throat; known to reach lengths in excess of 40 inches and weights over 50 pounds (as in the Kamloops variety).

TYPE LOCALITY

Fort Vancouver, Columbia River, Washington.

RANGE

Coastal streams from southern Alaska to Baja California. Has been widely introduced throughout the Western Hemisphere as well as into Europe, Asia and Australia.

TAXONOMY

The very variable Rainbow Trout series consists of a large number of described entities, many of which have fluctuated from full specific status to complete synonymy and back. It has become evident, of late years, with the accumulation of sufficient specimens over the range of the series, that the whole represents a widely variant species consisting of many intergrading geographic races, nearly each one of which was originally described from one or a few specimens as a distinct species. The name *gairdneri* is the first valid specific term available for this complex, a stabilization which has crystallized only in the last decade.

As a species, *S. gairdneri* is related to the similarly variable *S. clarki*, the former occupying the coastal streams, in general, and the latter more typical of the interior streams, although also occurring with the rainbow in some of the coastal areas. The two species hybridize readily when artificially mixed in regions where only one occurred originally, the resulting progeny being sterile. If rainbow are being heavily and continually stocked into native cutthroat streams, this hybridization tends to eliminate the cutthroat by dilution, although very probably changing stream conditions have played a more important part in the disappearance of cutthroat from many of their native streams than have the rainbow. Certainly such is obviously true of the Truckee River, fifty years ago one of the finest cutthroat streams in the West; were conditions the same on the Truckee now as then, there would undoubtedly be a substantial cutthroat population in residence, regardless of rainbow stocking.

Etymology—Meredith Gairdner was a young Hudson's Bay Company doctor stationed for a while on the Columbia River.

LIFE HISTORY

The Rainbow Trout is somewhat more migratory than other freshwater species, as people have discovered who have tried to stock limited stretches of stream for their own fishing. This strong urge to move downstream from a planting site tends to spread them out in areas of irrigation ditches where many are lost if such ditches are dried up. It also is one of the apparent reasons they have reportedly not done well in places such as England. Malloch (1910: 234) commented.

"These trout [speaking of *Salmo g. irideus*] were introduced into this country [England] from America over twenty years ago. As time went on they were thought to be much superior to the *fario* [Brown Trout] as sporting fish. They may be so in their own country, but our climate does not seem to suit them. In a few English rivers they have done fairly well, as also in Blagdon Lake. If left to themselves for a year or two, however, I think they would all disappear. I have had ten years' experience of them in Scotland, and have introduced them into many lochs. They did well for three or four years and weighed about 1½ lbs.; but after that we saw few of them. Some of them spawned, but never in sufficient numbers to establish themselves . . . In New Zealand, where they have rich feeding, they grow to over 20 lbs. and rise to fly. It is strange, however, that in this country when they become 2 lbs. or more in weight they seldom rise to a fly. When hooked they jump out of the water like sea-trout [Brown Trout], and fight well to the last."

Jenkins later noted that

"A great drawback to the acclimatisation of the Rainbow Trout [in England] lies in the fact that it is a migratory species, and after the owner of a certain stretch of water has gone to the expense of hatching, rearing, and planting out this species, he finds that it suddenly disappears" (1925: 233).

The Rainbow is usually listed as a spring-spawner by most authors, but the spawning period extends over several months, from late fall to spring, depending upon the locality and elevation, and there have been definite fall-spawning strains developed and used by hatcherymen. Food consists of the usual insects and other aquatic arthropods in the smaller stages, with fish added to the diet of larger rainbows. The Kamloops has produced the record sizes for rainbows, some being known to exceed 50 pounds in weight and this large size seems to be the result of feeding upon the abundant and smaller Kokanee Salmon which live with them.

The rainbow is usually a surface or shoal water feeder, and its spawning habits are somewhat unique among freshwater trout in that two males generally accompany each spawning female. Nestbuilding and egg-coverage are similar to that of other species. Egg-hatching depends upon temperatures, but by and large can be said to take about five to six weeks. Like the Brown Trout, the rainbow is capable of thriving in warmer waters than most other species and as a consequence is a good species to place in deteriorating streams.

Numerous parasites are known to infest trout of all species, and it is presumed that the majority of these will be found, on occasion, in the usual percentages on Nevada specimens. The highest incidences of recorded infections are from immatures crowded in hatcheries. However, Jonez and Sumner (1954) have reported the common *Lernaea*[1] (an external tiny crustacean) and an internal nematode worm[2] from rainbow in Lake Mohave, Clark County.

Cal Allan informs me that immature trout in Walker Lake are commonly infested with an adult tapeworm in the vicinity of the pylorus, but that these seem to be lacking in adult trout. He has found nematodes only infrequently in trout, but heavy infections are common in the chubs and Sacramento Perch.

Kay Johnson reports that trout in the warmer lower Truckee River in summer bear considerable concentrations of *Lernaea* and fungous, and that nematodes are common in trout in both the lower Truckee and in Pyramid Lake.

---

[1] Family Lernaeidae, Order Copepoda, Class Arthropoda.
[2] *Contracoecum multipapillatum* (Drasche), Family Heterocheilidae, Class Nematoda.

ECONOMICS

In the western United States, at least, and particularly in Nevada, the rainbow is the most important hatchery reared trout, the most widely planted fish in streams and lakes and, usually, the species currently most popular with anglers, although it is not always the gamiest or best adapted trout in many of the streams into which it is continually being stocked in response to pressure from the sportsman.

SOUTHCOAST RAINBOW TROUT

Salmo gairdneri irideus Gibbons

(Southern Steelhead Trout)

Salmo iridia Gibbons 1855: 35
Salar iridea, Girard, 1856C: 220 / 1859: 321–323
Salmo irideus, Günther, 1866: 119–120
Salmo iridea, Cooper, in Cronise, 1868: 494
Salmo irideus, Cope, 1872C: 470
Salmo iridea, Suckley, 1874: 129–134
Salmo irideus, Cope and Yarrow, 1875: 684
Salar irideus, Jordan and Copeland, 1878: 431
Salmo irideus, Jordan and Gilbert, 1881D: 38 / 1882: 312
Salmo irideus, Jordan, 1891: 6, 16
Salmo mykiss irideus, Jordan and Gilbert, 1894: 139
Salmo irideus, Jordan and Evermann, 1896B: 500 / 1902: 295–297
Salmo irideus, Rutter, 1903: 148 / 1908: 141–142
Salmo irideus, Malloch, 1910: 234–237
Salmo irideus, Snyder, 1916: 585 / 1940: 123–124
Salmo irideus shasta, Snyder, 1917A: 85
Salmo irideus, Pratt, 1923: 46
Salmo irideus, Locke, 1929: 186–187
Salmo irideus, Jordan, Evermann and Clark, 1930: 58
Salmo irideus, Jenkins, 1936: 233–234
Salmo irideus, Schrenkeisen, 1938: 46–48
Salmo gairdnerii irideus, Kuhne, 1939: 30
Salmo gairdnerii irideus, Hubbs and Lagler, 1941: 35–36 / 1947: 36–37
Salmo gairdnerii irideus, Miller and Alcorn, 1946: 178
Salmo gairdnerii irideus, Eddy and Surber, 1947: 108–110
Salmo gairdnerii irideus, Miller, 1950B: 4
Salmo gairdnerii stonei, Shapovalov and Dill, 1950: 385
Salmo gairdnerii irideus, Simon, 1951: 41–42
Salmo gairdneri irideus, Beckman, 1952: 19
Salmo gairdneri irideus, La Rivers, 1952: 94
Salmo gairdneri irideus, La Rivers and Trelease, 1952: 114
Salmo gairdneri irideus, Weinrub and Bilstad, 1955
Salmo gairdneri irideus, Slastenenko, 1958A: 63–64

ORIGINAL DESCRIPTION

"Salmo iridia.—Gibbons. Body elongated, sub-compressed; head about one-fourth of total length.

"Eyes large, circular, horizontal diameter nearly one-third the length of the head. Facial outline elliptically rounded. Vertical line from the posterior extremity of the upper maxillary will graze the posterior edge of the iris.

"Teeth minute, numerous, regular, incurved. A series of from three to five incurved teeth in each margin of the tongue. Those on the edges of the palatines and on the vomer, numerous.

"Length of body to its greatest depth, 9 to 2. First dorsal rises from a point midway between the extremity of the snout and the end of the lateral line. The adipose and anal terminate opposite to each other. Ventrals under the first fourth or half of the first dorsal. Caudal forked. First dorsal with five irregular, interrupted black horizontal bands. Other fins black punctate, ventrals tipped with orange, caudal and adipose with black margin.

"Scales small. Back cineritious, with light purple tint. Sides along

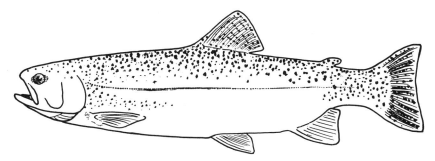

FIG. 157. Rainbow Trout, *Salmo gairdneri irideus.*
Courtesy Thomas J. Trelease.

the lateral line light vermillion, interrupted by rounded dark patches, which become nearly or quite obsolete in older specimens. Sides and belly below these, silver-tinted, finely black punctate.

"D.14; P.13; A.12; V.11; C.19; with accessories. Length five inches.

"The three specimens from which this description was taken were obtained by Mr. Nevins from the San Leandro Creek. They are evidently young fish" (Gibbons 1855: 35).

DIAGNOSIS

As in *S. gairdneri*, being distinctive as a subspecies chiefly in its consistently larger scales (about 21/135/20), smaller size and brighter colors, the side stripe in both sexes being a conspicuous bright pink-to-red color.

TYPE LOCALITY

San Leandro Creek, Alameda County, California.

RANGE

Originally confined to coastal streams from Washington to southern California. Now widely planted in inland waters, being very common in the Great Basin.

TAXONOMY

The long-standing practice of mixing various rainbow varieties indiscriminately in the hatchery and field has resulted in such a complete suppression of subspecific characteristics that in many localities, the varietal assignment of the rainbow in question must be entirely an arbitrary one. While it is known that several races of rainbow have been mixed in Nevada streams, most of the identifiable populations seem to fit the description of *S. g. irideus* better than that of any other strain.

Etymology—"irideus," like a rainbow.

FIG. 158 not available at press time.

ECONOMICS

Since about 1880 Rainbow Trout have been brought into Nevada, and since these seem to have been largely the subspecies *irideus*, they will be discussed here. Federal hatchery statistics for about a 35-year period in the beginning of the present century show nearly 12 million eggs and fish allocated for Nevada, in addition to those independently obtained and reared by State installations. Most sportsmen still prefer rainbow, and the relative ease with which they are handled in hatcheries makes them the most economical fish to raise and plant. Satisfying the demand of sportsmen for more fishing in an era of deteriorating fishing waters has resulted in fish and game programs whose main objectives are often to dump fish into the streams ahead of the angler who can, if he wishes, immediately catch them out. This has forced the situation into a direct one of supply and demand, with the degenerating streams and lakes acting as temporary distributing stations in the fish's journey from Federal and State hatchery to the fisherman's creel. To put it drastically, until stream conditions can be bettered one of the best alternative programs would be that in which the fish and game authorities reared and processed the fish and the sportsmen bought them frozen over the counter. As long as things remain so, the Rainbow Trout will be the most satisfactory fish to handle.

KAMLOOPS RAINBOW TROUT
Salmo gairdneri kamloops (Jordan)
(Kamloops Trout)

Oncorhynchus kamloops Jordan 1892A: 405
Salmo kamloops, Jordan, 1892B: 60–61 / 1925A: 323
Salmo gairdneri kamloops, Jordan and Evermann, 1896B: 499–500[1]
Salmo kamloops, ibid, 1902: 192–193
Salmo kamloops, Pratt, 1923: 46
Salmo gairdnerii, Jordan, Evermann and Clark, 1930: 58
Salmo kamloops, Dymond, 1936: 62
Salmo gairdnerii kamloops, Schultz, 1936: 138
Salmo gairdnerii kamloops, Schrenkeisen, 1938: 50

---

[1]They cite "*Salmo kamloops*, Jordan, Proc. U. S. Nat. Mus., 1892, Kamloops Lake, British Columbia. (Type, No. 44238. Coll. A. C. Bassett.)" as the original description, but according to Miller (1950B: 7) this was never published.

Salmo kamloops, La Monte, 1945: 110–111
Salmo gairdnerii kamloops, Carl and Clemens, 1948: 59–60
Salmo gairdnerii kamloops, Miller, 1950B: 7–8
Salmo gairdneri kamloops, Beckman, 1952: 19
Salmo gairdnerii kamloops, Scott, 1954: 30
Salmo gairdneri kamloops, Slastenenko, 1958A: 60–62
Salmo gairdnerii kamloops, Shapovalov, Dill and Cordone, 1959: 171

ORIGINAL DESCRIPTION
"Description of a New Species of Salmon
"*Oncorhynchus kamloops*, from the lakes of British Columbia
"By David Starr Jordan

"(Copy of MSS. sent to Smithsonian Institution for publication).

"*Oncorhynchus kamloops*, species Nova:
"Head, $4\frac{1}{2}$ in length to base of caudal; depth, $4\frac{1}{3}$; dorsal rays, 11, not counting the rudiments; anal rays, 11 in one specimen, 12 in the other, besides 3 rudiments; scales, 30, 145, 26 (in second specimen 135 scales); about 120 pores; length of body, largest specimen, $16\frac{1}{4}$ in.; smaller specimen, $15\frac{3}{4}$ in.

"Body moderately elongated, somewhat compressed, the general form resembling that of the silver salmon (*Oncorhynchus kisutch*); jaws in the typical specimens not prolonged, the maxillary extending beyond the eye, its length not quite half the head; snout slightly rounded in profile, the profile regularly ascending; eye large, about as long as snout, four and a half times in head; teeth moderate, some of those in the outer rows in each jaw moderately enlarged; teeth on tongue and vomor [sic], as usual in *Oncorhynchus;* opercles striate, not much produced backward; branchiostegal rays, eleven on each side; dorsal fin rather low, its longest ray slightly greater than the base of the fin, $1\frac{3}{5}$ in head; anal fin lower and smaller than usual in *Oncorhynchus*, its outline slightly concave, its longest ray greater than the base of the fin and a little more than half head; adipose fin moderate; caudal fin rather broad, distinctly forked, its outer rays about twice inner; pectoral fins rather long, $1\frac{1}{3}$ in head; ventrals moderate, $1\frac{3}{4}$ in head; gill-rakers comparatively short and few in number, about 6 plus 12 or 13.

"Coloration, dark olive above, brightly silvery below, the silvery color extending for some distance below the lateral line where it ends abruptly; when fresh the middle of the sides in both specimens was occupied by a broad band of bright, light rose pink, covering about one-third of the total depth of the fish; back above with small black spots about the size of pin heads irregularly scattered and somewhat more numerous posteriorly; a very few faint spots on upper part of head; dorsal and caudal fins rather closely covered with small black spots similar to those on back but more distinct; a few spots on the adipose fin, which is edged with black; lower fins plain; the upper border of the pectoral dusky; a vague dusky blotch on the upper middle rays of the anal; ventrals entirely plain.

"The intestines had been removed and so no account can be given of the pyloric caeca.

". . ." (Jordan 1892A: 405–406).

DIAGNOSIS

Features of *S. gairdneri* in general, being subspecifically separable mainly because of finer scalation (about 30/145/28), larger size and more contrast between the dark greenish back and prominent rosy sides.

TYPE LOCALITY

Kamloops Lake, British Columbia.

RANGE

Originally the interior of British Columbia south to and including northeastern Washington. It has been widely introduced elsewhere, however, south to California, Nevada and Utah, and east to Colorado and southeastern Canada.

In Nevada, one small plant was made in Walker Lake in 1958 without success, and several varying sized plants have gone into Pyramid Lake between 1956 and 1958 (some 155,000 fish, up to 5 inches in length). Most of these were planted in the Truckee River just above the lake, the remainder in the lake proper.

TAXONOMY

Beyond the taxonomic tangle created by Jordan when he first described the Kamloops, no changes have occurred in its nomenclature that could be considered important. As Miller pointed out in 1950 (pp. 7–8), Jordan sent copies of his original description to three sources for publication—the United States National Museum, Forest and Stream and the California Board of Fish Commissioners. There is no record that the National Museum ever published its copy, but Forest and Stream and the California Commissioners did.

Jordan proposed the Kamloops as a species of Pacific Salmon (*Oncorhynchus*) but decided, while his material was still in press, that it more properly belonged to the trout genus *Salmo*. However, he was able to make the change only in the copy sent to the California Commissioners, so his name appeared as *Oncorhynchus kamloops* in "Forest and Stream" and as *Salmo kamloops* in the California board's publication. Since "Forest and Stream" appears to have been issued about a month earlier than the Board's report, this is the prior form of the name.

Rather oddly, Jordan (in Jordan and Evermann 1896B: 500) cites the apparently never printed U. S. National Museum paper as the original description and gives it a date of 1892 also.

LIFE HISTORY

Beyond its propensity to attain larger sizes (supposedly), the Kamloops Rainbow Trout has essentially the same life history pattern as other rainbows. And in the matter of size, differences here seem to be entirely nutritional. The Kamloops is a notorious feeder on Kokanee Red Salmon where these two occur together.

ECONOMICS

While an abundant and popular fish in its native range, in Nevada the Kamloops seems to hold no special possibilities. The Walker Lake plant was unsuccessful (as are other trout except the cutthroat), and those introduced into Pyramid Lake have not shown any unique capabilities, the return being small, but commensurate with the modest size of the plants.

TAHOE RAINBOW TROUT

Salmo gairdneri regalis Snyder

(Royal Silver Trout)

Salmo regalis Snyder 1912: 26 / 1917A: 79–80 / 1940: 130
Salmo regalis, Pratt, 1923: 47
Salmo regalis, Jordan, Evermann and Clark, 1930: 56
Salmo gairdneri regalis, Schrenkeisen, 1938: 50
Salmo regalis, Shapovalov, 1941: 442–443
Salmo regalis, Hubbs and Miller, 1948B: 37
Salmo regalis, Miller, 1950B: 4
Salmo gairdneri regalis, Shapovalov and Dill, 1950: 385
Salmo gairdneri regalis, La Rivers and Trelease, 1952: 114
Salmo regalis, Böhlke, 1953: 15
Salmo regalis, Moore, 1957: 64
Salmo gairdnerii regalis, Needham and Gard, 1959: 17–18, 37
Salmo gairdnerii regalis, Shapovalov, Dill and Cordone, 1959: 171

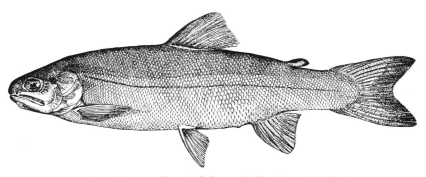

FIG. 159. Tahoe Rainbow Trout, *Salmo gairdneri regalis*. Re-drawn from Snyder (1917A) by Silvio Santina.

ORIGINAL DESCRIPTION

"Head, 4.4 in length to base of caudal; depth, 4; depth caudal peduncle, 9.5; snout, 4.5 in head; length of maxillary, 2.1; vertical diameter of eye, 5.8; dorsal rays, 11; anal, 11; longitudinal series of scales above lateral line, 144; in series between lateral line and middle of back, 29; pores in lateral line, 120.

"Body shaped as usual among trout, perhaps somewhat flatter than ordinary, the width contained 2¾ time in the depth near middle of body. Snout short and rounded; maxillary weak, narrow, and short, scarcely extending beyond a vertical through posterior edge of orbit. Opercles and branchiostegals rather thin and papery. Jaws weak; their teeth small and sharp; teeth in bands on vomer and palatines; 2 rows of 5 teeth each on the tongue; basibranchials smooth and without teeth. Branchiostegals, 11. Gillrakers on first arch, 7 + 12, slender and pointed. Lateral line almost straight from opercle to caudal. Scales very thin, not deeply embedded, moderate in size on the sides, extremely small on middle of back anterior to dorsal fin and on throat and abdomen. Fins all comparatively thin and frail, not like those of trout from mountain streams; pectorals and ventrals sharply pointed; dorsal and anal with slightly concave edges; caudal deeply cleft, the lobes pointed; adipose fin thin, narrow, and elongate.

"Color in alcohol, dusky above, silvery on the sides, white beneath; dorsal and caudal fins dusky; anal somewhat dusky toward the border; pectorals and ventrals immaculate; no spots or bars on the head, body, or fins. On close examination, the scales are observed to be silvery, those on the dorsal region closely speckled with black, those beneath with but little luster.

"The type, which will be deposited in the United States National Museum, is a specimen measuring 323 millimeters in length, collected near Brockway, Lake Tahoe, Cal., August 23, 1912, by Mr. W. P. Lyon" (Snyder 1912: 26–27).

DIAGNOSIS

A finer scaled rainbow than *S. g. irideus* (144–150 scales in the lateral line), lacking the typical rainbow black-spotting and lateral red stripe, being silvery on the sides. Snyder commented (1917A: 79) that freshly caught specimens readily lost their scales when handled.

TYPE LOCALITY

Brockway, east side of Lake Tahoe, California, from deep water.

RANGE

Known only from Lake Tahoe, and presumed extinct.

TAXONOMY

This was the first of the two Lahontan system rainbows that Snyder described (see *S. g. smaragdus* below), and he had only *S. clarki henshawi* to compare the Tahoe Rainbow with in 1912. Although, as Snyder commented, *S. g. smaragdus* was usually confused with *S. c. henshawi*, technical differentiation was not difficult since the former lacked the hyoid teeth of the latter. If the possibility be ignored that Snyder's Lahontan rainbow represented stock introduced in the 1860s or 1870s, they were the only members of the variable rainbow series to occur anywhere east of the crest of the Sierra Nevada mountains, except the similarly suspect Eagle Lake Rainbow (see discussion under *S. g. smaragdus*). In the light of the facts that rainbow had been artificially stocked into Lake Tahoe for some years prior to Snyder's investigation and that he made no mention of them, it is interesting to note his statement:

"It seems probable that the relationships of *S. regalis* and also of *S. smaragdus*, a representative species found in Pyramid Lake, are with *S. irideus* of the western slope of the Sierras, *S. smaragdus* having apparently departed farther from the parent form than has *S. regalis*" (1917A: 80).

Etymology—"regalis," royal.

LIFE HISTORY

A presumably spring-spawning, extremely rare fish. Snyder had four specimens in his possession when describing it in 1912, and apparently never saw another one, since he made no mention of additional specimens in his more extensive 1917 resumé of Lahontan fishes. No individuals referrable to his description have been officially recognized since his early work, and the type is presumed extinct. Accordingly, the only life history notations we have are those of Snyder:

"The market fishermen and most of the anglers who visit Lake Tahoe confuse examples of this species with silvery individuals of *S. henshawi*, and as it is next to impossible to get taintless information regarding *S. regalis*, almost nothing definitely is known of its habits. It seems to be very rare, or at least specimens are seldom caught. It does not appear to enter the creeks which are tributary to the lake, and its time of spawning is unknown.

"An examination of the stomach contents revealed the remains of many small insects—ants, bugs, beetles, etc.—which usually may be seen floating on the surface of tributary streams and are carried out into the lake. This would indicate surface feeding, which may possibly occur at night. As this trout may be taken with a spinner, it may be inferred that it also feeds upon minnows" (1917A: 79–80).

PYRAMID RAINBOW TROUT
Salmo gairdneri smaragdus Snyder
(Emerald Trout)

Salmo smaragdus Snyder 1917A: 80–81
Salmo smaragdus, Pratt, 1923: 47
Salmo smaragdus, Jordan, Evermann and Clark, 1930: 58
Salmo smaragdus, Hutchinson, 1937: 54–55
Salmo irideus smaragdus, Schrenkeisen, 1938: 47–48
Salmo smaragdus, Hubbs and Miller, 1948B: 37
Salmo smaragdus, Miller, 1950B: 4
Salmo gairdneri smaragdus, La Rivers and Trelease, 1952: 114
Salmo smaragdus, Moore, 1957: 64
Salmo gairdnerii smaragdus, Needham and Gard, 1959: 17–18, 37

ORIGINAL DESCRIPTION

"Description of type no. 75596, United States National Museum, a specimen measuring 480 mm. in length, from Pyramid Lake, Nevada, May 22, 1913. Sex, male.

"Head 4.5 in length to base of caudal; depth 5; depth caudal peduncle 10.5; snout 3.8 in head; vertical diameter of eye, 5.6; length maxillary 1.9; dorsal rays 11; anal 11; scales in lateral series 124; above lateral line 26; between lateral line and base of ventral 23; between occiput and dorsal about 70.

"Body slender, the head and snout elongate when compared with *S. regalis*. Maxillary comparatively broad, rather thin and weak, not extending far beyond orbit. Gill cover obtusely rounded behind; the opercle wide and with the upper edge broadly rounded, imparting to the gill cover a pronounced and characteristic shoulder. Branchiostegals 11. Gillrakers 9–11, long and acutely pointed. Vomerine teeth in 2 distinct rows, the toothed area somewhat longer than in *S. henshawi*. Strong teeth on anterior part of tongue; none on basibranchials; edge of maxillary except extreme posterior part with teeth.

"Scales notably large and loosely embedded; those immediately behind occiput and on ventral surface minute. Fins naked. Lateral line with 123 pores, slightly decurved beyond its origin, and then straight to middle of caudal.

"Longest (first) dorsal rays contained 1.8 times in head; pectoral 1.6; pectorals and ventrals acutely pointed; ventral inserted on a vertical through base of third dorsal ray. Caudal deeply forked, the lobes pointed. Adipose fin small and rather narrow" (Snyder 1917A: 80–81).

DIAGNOSIS

The largest-scaled variant of the three types of *S. gairdneri* attributed to Nevada, having about 124 scales in the lateral line; otherwise colored much as in *S. g. regalis*, with a rich emerald back, silver sides and white belly.

TYPE LOCALITY AND RANGE

Pyramid Lake, Nevada; known only from the type specimen, and now extinct.

TAXONOMY

S. g. smaragdus was the Pyramid Lake counterpart of the Lake Tahoe *S. g. regalis*; from what meager information we have, both were lacustrine, with no contact at any point in the hundred-odd miles of Truckee River connecting the two bodies of water, which lie at elevations of 6,225 and 3,800 feet, respectively. Utilizing statistics primarily from Snyder (1917A), it is possible to briefly compare the three Nevada rainbows as follows:

	Salmo gairdneri irideus	*Salmo gairdneri regalis*	*Salmo gairdneri smaragdus*
Scales	Intermediate in size, averaging 135 above lateral line	Small in size, averaging 147 above lateral line	Large in size, averaging 124 above lateral line
Color	Bright pink to red side stripe	Sides silvery	Sides silvery
Black spotting	Conspicuous and irregular over back	Entirely lacking	Slight traces
Head size	Largest, 4.0 times in body length	Smaller, 4.4	Smallest, 4.5
Body proportions	Heavier, depth 3.8 times in body length	Intermediate, 4.0	Slimmer, 5.0
Eye	Larger, 5 times in head length	Smallest, 5.8	Smaller, 5.6

These points are not as clear cut as they appear in tabulated form; so few specimens of *S. g. regalis* and *S. g. smaragdus* are known that such comparisons as the above have only historical interest.

At this point it may be profitable to organize some of the significant facts revolving about the presence of native rainbow stock in an otherwise cutthroat region. Five items stand out in bold relief as factors to be evaluated in considering the zoögeographic significance of the rainbow here: (1) The Cutthroat Trout was the overwhelmingly dominant salmonid of the Lahontan system; (2) the Rainbow Trout was similarly the dominant coastal stream species (of the two, where they occurred together); (3) the Great Basin provides the type of habitat to which cutthroat stock is, in general, the best adapted of the two species (note the present situation in Walker Lake); (4) modern rainbow strains, when introduced into cutthroat streams heavily and consistently, as is done in current propagation practices of fish and game authorities, aid in eliminating the cutthroats by producing sterile hybrids which dilute out the cutthroat strain; and (5) the Lahontan rainbows, *S. g. regalis* and *S. g. smaragdus*, were evidently ill-adapted forms, to judge from their well-authenticated scarcity.

From these facts, two stand out as wholly incongruous to an acceptance of *regalis* and *smaragdus* as Great Basin endemics: (1) their lack of success in competition with the formerly large cutthroat populations and (2) nowhere else, with the exception of the debatable *Salmo aquilarum* of Eagle Lake, are rainbows found as natives east of the Sierras. The evidence is admittedly conjectural, but that is all we have, and as such, suggests that these "species," like the Pacific Terrapin, *Clemmys marmorata*, were not pre-Caucasian members of the Great Basin fauna (La Rivers 1942: 67).

Considering the two above points in order, this lack of success as indicated by their rarity is what would be expected of a species newly introduced in small numbers into a drainage system completely dominated by a closely related competing species. Not only would it be extremely difficult for such an introduction to maintain themselves against competition for food, but hybridization with its production of sterile offspring would "dilute" them out of the habitat. Not until such a time as environmental circumstances might change, and so produce unsatisfactory living conditions for the cutthroat, would there be any chance of a rainbow introduction building up sizeable populations.

It is true, with the coming of the white man, that cutthroat populations declined and rainbow populations prospered, but this was not due to naturally changing conditions which allowed the rainbow to utilize a superior trait of adaptability to the disadvantage of the cutthroat. It was caused by (a.) deterioration of streams (and usually, as a consequence, their affluent lakes) as habitable environments by diversions and pollution, which lessened resident populations of both cutthroat and rainbow, and (b.) dominance of rainbow by virtue of heavy and steady artificial stocking of this type by fish and game authorities, with the results mentioned above.

As for the second point, the mere finding of a supposedly resident rainbow east of the Sierran divide should be examined suspiciously as a discordant note in the known geographic range of the species in the

Western U.S. Snyder's *S. aquilarum*, of rainbow stock and supposedly the counterpart of the Lahontan *regalis-smaragdus* combination in Eagle Lake, in the northwest corner of the Lahontan drainage system, if valid, would be an argument in favor of rainbow endemicity east of the Sierras. However, its status is doubtful and has little value in resolving the problem. Hubbs and Miller (1948B: 38) considered *aquilarum* a hybrid between the native cutthroat of Eagle Lake (*S. c. henshawi*) and the introduced rainbow. Miller (1950B: 6) retrenched somewhat and noted that their interpretation may have been erroneous. The problem is being investigated further, but no conclusions have been reached as yet, to my knowledge.

As far as the actual opportunity for rainbow stock to be introduced into the Lahontan system is concerned, early day activities involving the Tahoe region of the Sierra Nevada afforded ample incentive and circumstances for the transplanting of the popular rainbow of California into adjacent Nevada. It is difficult now to imagine the amount of traffic some of the Sierran passes carried in the late 1850s and early 1860s when the the original California gold-seekers were crowding the Henness Pass and Placerville roads to get to the silver fields of Washoe —wagons, riders and pedestrians were literally "bumper-to-bumper." Some of the early-day immigrants brought catfish across the plains wrapped in wet burlap sacking, wetting them down at some springs, and keeping them in small temporary pools in others.

By 1877, the office of Fish Commissioner was functioning in Nevada, filled by a succession of men energetically engaged in introducing everything from fish to terrapins to crawfish and salmon flies into the waters of the Western Great Basin. California and Federal agencies were in action even earlier. There can be little doubt that the period between 1860 and 1880, to choose a suitably representative time, was opportune for the many private and unrecorded translocations of fish which have occurred across the Sierras—and which still continue sporadically even though frowned upon by the authorities.

By the time Snyder began his fieldwork in the Lahontan basin (1911), rainbow stock could have enjoyed a good many years of occupancy in the Lahontan system. From Lake Tahoe, rainbow had direct and relatively easy access to Pyramid Lake via the Truckee River, and it is to be expected that the quite different environments of the two lakes would result in rainbow populations with minor characteristics of their own, differing slightly from each other and from the parent stock. In conclusion, it might be mentioned that the introduced *S. g. irideus* occasionally produces silvery individuals in Lake Tahoe and the Upper Truckee River of a superficial "steelhead" appearance much like that described for *regalis* and *smaragdus* so that the coloration of these latter, once thought to be distinctive, is of no value as a segregative criterion.

Etymology—"smaragdus," a precious stone of green color.

LIFE HISTORY

Snyder never saw but the type specimen, and, as far as the writer knows, this is the only individual of *S. g. smaragdus* in existence, dead or alive. Snyder examined the stomach contents of the type, which he caught trolling with a spinner, and found "parts of a small minnow together with the remains of insects (beetles and bugs); but whether the latter had been taken directly or as part of the captured fish could not be determined" (1917A: 81). He presumed it to spawn in the lake, lacking information to the contrary, and, as an index to its scarcity, mentioned that "it is said by all to be extremely rare, so that the most persistent fisherman sees but one or two during a season" (p. 80). The Pahute Indians supposedly differentiated between this fish and the cutthroat, calling it "A-gaih."

GOLDEN TROUT

Salmo aguabonita Jordan

Salmo mykiss agua-bonita Jordan 1892D: 62–65[1]

This beautiful trout has been planted in Lake Tahoe on two previous occasions, both failures (1918 and 1919). Shebley wrote of the second plant, which consisted of 250,000 eggs hatched out in the Lake Tahoe Hatchery: "The resulting fry were carefully reared and planted in the streams flowing into the lake, where conditions appeared to be most favorable for them" (1921: 33).

Recently (1960), the Nevada Fish and Game Commission attempted to establish Golden Trout in two lakes of the Ruby Mountains, northern Nevada:

"ELKO—Earl A. Frantzen, Nevada Fish and Game Commissioner from Elko, reported a poisoning program at Cold Lakes in preparation for the planting of golden trout started here Aug. 2. The two lakes being treated are located in the Ruby Mountains above the Sorenson Ranch, between Lamoille and Ft. Halleck, at an elevation of about 10,000 feet.

"The golden trout, known for their adaptability to high mountain lakes, are scheduled to be planted in Cold Lakes this fall. A shipment of 39,000 eggs was received last month and the eggs are being hatched at the Ruby Fish Hatchery.

"One of these lakes contains 3.7 surface acres and the other measures 1.8 surface acres. They are situated at the head of the middle fork of Cold Creek and are accessible only by horseback or foot.

---

[1] See Jordan 1892D in the bibliography for the various sources in which the original description appeared. Although that in the California State Board of Fish Commissioners' Biennial Report appeared first (1892), Jordan later listed the 1893 description in the United States National Museum Proceedings as the original citation.

"Brook trout presently in the lake have displayed a tendency to overpopulate—partly as a result of a lack of fishing pressure—and fish experts expect the goldens to maintain a better balance in the lakes. The Cold Lakes were selected because of ideal spawning conditions and F-G officials hope to utilize the site as a stocking source for other waters in the mountains of Elko County" (Reno Evening Gazette, 11 August 1960).[2]

BROWN TROUT

Salmo trutta Linnaeus

(German Brown Trout, Loch Leven Trout)

Salmo trutta Linnaeus 1758: 308–309
Salmo fario ibid: 309
Salmo trutta, Günther, 1866: 22–34
Salmo fario, ibid: 59–76
Salmo fario, Snyder, 1917A: 85
Salmo fario, Locke, 1929: 185–186
Salmo trutta, Jenkins, 1936: 227–231
Salmo trutta, Schultz, 1936: 137
Salmo eriox, Schrenkeisen, 1938: 51–52
Salmo trutta, Kuhne, 1939: 32
Salmo levenensis, Snyder, 1940: 136
Salmo trutta fario, Hubbs and Lagler, 1941: 35–36 / 1947: 34–37
Salmo trutta, Murphy, 1941: 167
Salmo trutta, Shapovalov, 1941: 445
Salmo trutta, La Monte, 1945: 114
Salmo trutta fario, Miller and Alcorn, 1946
Salmo trutta, Wales, 1946: 115–116
Salmo trutta fario, Eddy and Surber, 1947: 106–108
Salmo trutta, Everhart, 1950: 22
Salmo trutta, Shapovalov and Dill, 1950: 385
Salmo trutta fario, Simon, 1951: 33–35
Salmo trutta, Beckman, 1952: 21–22
Salmo trutta, La Rivers, 1952: 94
Salmo trutta fario, La Rivers and Trelease, 1952: 114
Salmo trutta fario, Sigler, 1952
Salmo trutta, Gabrielson and La Monte, 1954: 97

---

[2]As this goes to press (1962), the following information comes to hand:
"Last fall, following the eradication project, Cold Lakes were replanted with goldens. Approximately 3,100 of the fish, fingerling size or a little less than an inch in length, were transported into the lakes via horseback and planted. They were left then to the winter. Fisheries personnel of the Owyhee District were anxious to see how the trout were doing or whether they had made it at all. In latter July of this year [1961], personnel rode back into the two lakes and reported that the goldens had made it. About 300 or more were spotted by walking the shores, and from less than an inch, many had grown to around four inches in size. The goldens are still too small for fishing but the summer after next, Cold Lakes ought to provide some real good golden fishing. By that time, they will have grown large enough." (Anonymous, 1961. Nevada Wildlife, Nevada Fish and Game Commission, Vol. 2, No. 3: 6, Sept.)

Salmo trutta, Scott, 1954: 27–28
Salmo trutta, Dymond, 1955: 544–545
Salmo trutta, Stokell, 1955: 67–80
Salmo trutta fario, Eddy, 1957: 50
Salmo trutta, Koster, 1957: 32–33
Salmo trutta, Moore, 1957: 63
Salmo trutta, Trautman, 1957: 185–187
Salmo trutta, Slastenenko, 1958A: 53–55
Salmo trutta, Carl, Clemens and Lindsey, 1959: 64–66
Salmo trutta, Shapovalov, Dill and Cordone, 1959: 171

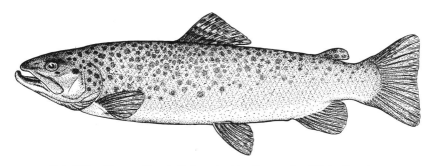

FIG. 160. Brown Trout, *Salmo trutta*. Drawn by Silvio Santina.

ORIGINAL DESCRIPTION

"Trutta. 3. S. ocellis nigris iridibus brunneis, pinna pectorali punctis 6. *Fn. Suec.* 308.
                                            B.—D.12. P.13. V.10. A.9. C.20
Art. gen. 12. *Syn.* 14. Salmo latus, maculis rubris nigrisque, cauda aequali. *Gron. mus.* 2. *n.* 164. Salmo latus, cauda subrecta maxillis aequalibus, maculis nigris annulo albido.   D.14. P.12. V.12. A.10. C—
Habitat in fluviis Europae."
                                                      (Linnaeus 1758: 308–309).

"Fario. 4.S. maculis rubris, maxilla inferiore sublongiore.
Art. gen. 12. *Syn.* 23. *Spec.* 51.
                                      B.—D.13. P.10. V.10. A.10. C.18.
Fn. Suec. 309. idem. B.10. D.14. P.14. V.9. A.11. C.20.
Habitat in Sueciae, Helvetiae *fluviis*."
                                                              (Linnaeus 1758: 309).

DIAGNOSIS

Head about 4.0 in body length, depth 3.0 to 5.0 in body length; branchiostegals 8–11; gillrakers 6–9 + 9–10; maxillary narrow and extending beyond posterior orbital margin in older specimens, eye medium in size; pyloric caeca very variable, ranging from 33 to 80; vertebrae 58–60; scales [1]26–30/100–120/20–22;[2] fins—dorsal 10 or 11

---
[1] Counted obliquely rearward from origin of dorsal fin to lateral line.
[2] Counted obliquely forward from origin of anal fin to lateral line.

rayed;[3] anal 8 or 9 rayed;[3] color—with large, conspicuous spots on sides and back, against a generally brownish color, often with an olive cast, especially on the back—spots about size of eye pupil, or somewhat smaller, reddish or brown on sides, brown or black on back, generally conspicuously ringed with white—black spots, in particular, are often cross or "X"-shaped—the dorsal and anal fins have noticeable yellowish leading margins; a 40-pounder from Scotland seems to be the record catch, while specimens exceeding 20 pounds are not uncommon in this country. The Truckee River yielded a record Brown Trout of more than 20 pounds in 1949 and one of 20 pounds in 1948.[4] When an angler catches one of these monsters he can be sure he has a fish that has spent most if not all its life in those waters successfully eluding the hundreds of hooks and lures that have been tossed at it—in contrast to the thousands of rainbows which are dumped into the streams one day and literally hooked out the next week.

TYPE LOCALITY

Sweden.

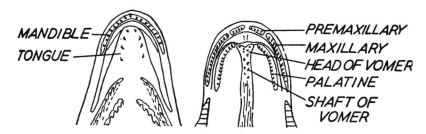

FIG. 161. Mouth dentition of the Brown Trout (*Salmo trutta*). Left, floor of mouth; right, roof of mouth. Courtesy Dr. Carl L. Hubbs and the Cranbrook Institute of Science.

RANGE

Originally confined to Europe (southeast nearly to India, south to include a portion of northwestern Africa and northwest to Iceland), but now known from many other parts of the world, including points as far as New Zealand—Australia, India, Japan, South Africa and South America. With the exception of the Japanese plants—which came from the United States—parent stock for these introductions came from England.

The United States importation of this famous sport fish was by way of Germany; von Behr sent Brown Trout eggs to New York in the years 1883, 1884 and 1885, some of which were hatched in Michigan.

Intergrading Loch Leven Trout eggs came into New York from Scotland in 1885 and were distributed to several states, including Michigan. Following this and other introductions, spread of the Brown Trout over other parts of the United States was swift.

---

[3]Since first 2–3 rays are vestigial and skin-covered, and the last ray may be split nearly to its base, the apparent count results in 2–4 less than the actual number of rays. These are the apparent counts.

[4]Reno Evening Gazette, June 5, 1949.

In only a few areas were implantations unsuccessful—Texas, Kentucky, Missouri and Virginia. Additional attempts would undoubtedly be profitable, since the Brown Trout has been established in neighboring states. The species also occurs abundantly in southeastern Canada and in certain portions of western Canada (Wiggins 1950). Sea-run populations are also known from the eastern Canadian coast.

In 1895, California planted some 250 three-inch Loch Leven (*S. t. levenensis* Walker) in Webber Lake, Sierra County, a lake which drains via the Little Truckee River into its mainstem, the Truckee, which is one of Nevada's main western affluents to the Lahontan Basin. This is the first known possible date of entry of the Brown Trout into Nevada.

An early biennial report of the California Fish Commissioners covering the years 1905–1906 mentions *S. trutta* as being in the Truckee

FIG. 162. Courtesy Oregon State Game Commission and Harold P. Smith.

River (1907). In 1917A, Snyder reported *S. fario* from Lake Tahoe, some small surrounding lakes and in the Truckee River during his field seasons of 1911–1912.

In the years from 1929 to 1941, better than a quarter of a million eggs and fish are known to have been brought into Nevada for planting. The species is now distributed in all the major and minor streams of the Lahontan Basin, as well as in many small streams and lakes in the neighboring Sierra Nevada Mountains.

TAXONOMY

The great variety of common and technical names applied to the Brown Trout, while serving as an indication of the local diversity of the stock, have also been the source of much confusion. European ichthyologists have varied in approach from considering all native European trout to be the one species *Salmo trutta* to the other extreme of breaking it into more than a dozen species. Linnaeus, the original namer of names, added somewhat to the complexity of the situation by apparently describing a variety other than what he originally intended as *S. trutta*.

Currently, both European and American specialists generally agree that there is only one Brown or European Trout in spite of the formidable synonymy that this species has accumulated over the years. Earlier, Widegren (1862), from the mass of names applied to the animal, united *eriox, trutta* and *fario* into one and called it *trutta*, which the cataloguer Günther noted in 1866.

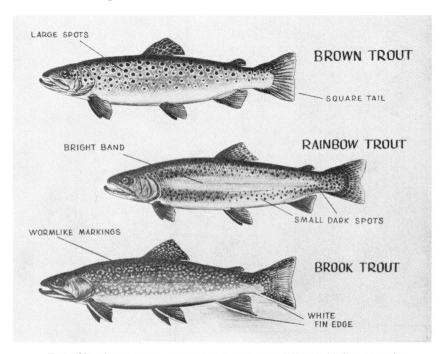

FIG. 163. Some trout comparisons. Courtesy Wisconsin Conservation Department, Jens von Sivers and N. H. Hoveland.

Such things do not always settle the problem, however. Collett (1875) considered the same three names and selected *eriox* as the prior or most suitable name. In 1930, Hubbs called attention to the earlier use of *trutta* by Widegren, and since then this has been the most consistently used designation.

In Europe, the name *Salmo trutta trutta* is used for the sea-run populations, much as we distinguish our Steelhead Cutthroats as *Salmo clarki clarki*. *Salmo trutta fario* applies to the permanent stream fish, as *Salmo clarki henshawi* does to our landlocked Lahontan Cutthroat. And *Salmo trutta levenensis* refers to a variety which is native to the Loch Leven area of Scotland and adjacent England.

Both these last two types of browns (*fario* and *levenensis*) have been introduced to the United States, but have lost their identities through subsequent thorough mixing in hatcheries and streams. We can now refer with certainty to our freshwater browns as *Salmo trutta fario* since what we have in our streams is apparently comparable to the freshwater browns of Europe.

LIFE HISTORY

The Brown Trout seems to be predominantly an insect feeder, although, like all trout, graduates to fish in its larger stages. In general, it can be considered a highly carnivorous and wary species. It is usually referred to as a fall spawner, but locality must be taken into account here and spawning covers the entire period from fall to early spring. Spawning is similar to that of other trout, the female working against the gravel bottom until she has a redd or nest about as long as she is and in the form of a slight hollow. After egg-and-milt deposition she moves upstream and covers the eggs by stirring up more gravel. Hatching occurs in about a month, with temperature a dominant controlling factor. Hybrids are known between browns and the native brooks which seem to be superior fish with zebra markings and sturdy constitutions but are, unfortunately, sterile and not capable of perpetuating themselves.

ECONOMICS

Because of the changing nature of most of our streams, mentioned above and elsewhere, the Brown Trout has and is rapidly replacing the Brook Trout (*Salvelinus fontinalis*) in many parts of the eastern United States. In Nevada, it provides relatively good fishing in such one-time superlative cutthroat streams as the Truckee River, in spite of the fact that it was only meagerly and intermittently planted there. Being able to better tolerate the existing stream conditions, the Brown Trout reproduces itself in maximum numbers for its given set of environmental factors. Because it is a more wary fish than the rainbow and not as easily caught, it is not so generally popular with anglers, although the ones caught usually average much better in size than rainbows from the same stream. While the Brown Trout does not have the delicate flavoring of the brookie, it is gamey and sizeable and in many parts of the country still provides fishing where native species have declined or disappeared because of increasing stream pollution. For such reasons, the one-time antagonism toward the brown has largely dissipated.

WHITEFISHES
Family *COREGONIDAE*

These salmonoids do not bear too much superficial resemblance to their relatives, salmon and trout, but share the same technical characteristics of the group as a whole. Several economically important species belong to this family, none of which occur in Nevada.

DIAGNOSIS
Mouth moderate-to-small, mandible shorter and uniting with the quadrate bone beneath or anterior to the eyes; maxillary broad; teeth inconspicuous or partially lacking; scales moderate-to-large.

MOUNTAIN WHITEFISH
Genus *Prosopium* Milner 1878[1]
("Masked," from the large preorbitals)

MOUNTAIN WHITEFISH
Prosopium williamsoni (Girard)
(Rocky Mountain Whitefish)

Coregonus williamsoni Girard 1856A: 136 / 1859: 326–327
Coregonus williamsoni, Günther, 1866: 187
Coregonus Williamsonii, Cooper, in Cronise, 1868: 494
Coregonus couesii, Milner, 1874: 88 (Rpt. U. S. Fish Comm. 1872–73)
Coregonus villiamsonii, Cope and Yarrow, 1875: 682–683
Coregonus couesii, Jordan, 1878G: 362
Coregonus williamsoni, Jordan and Copeland, 1878: 429
Coregonus williamsoni, Jordan and Henshaw, 1878: 194
Coregonus williamsoni, Jordan and Gilbert, 1881B: 460 / 1882B: 297
Coregonus couesii, ibid, 1882B: 297
Coregonus williamsoni, Jordan, 1890C: 46, 49 / 1891: 34–35
Coregonus williamsoni, Eigenmann and Eigenmann, 1891: 1132
Coregonus williamsoni, Evermann and Rutter, 1895: 486
Coregonus williamsoni, Evermann and Smith, 1896: 291–293
Coregonus williamsoni, Jordan and Evermann, 1896B: 463
Coregonus williamsoni, Jordan and Snyder, 1909: 428–430
Coregonus williamsoni, Snyder, 1917A: 69 / 1919: 3
Coregonus williamsoni, Pratt, 1923: 38–39
Coregonus williamsoni, Locke, 1929: 180–181
Prosopium williamsoni, Jordan, Evermann and Clark, 1930: 65
Prosopium williamsoni, Schultz, 1936: 139
Prosopium williamsoni, Tanner, 1936: 158, 164
Prosopium williamsoni, Schrenkeisen, 1938: 67

---
[1]*Coregonus,* also used for this species, is a combination referring to the "eye pupil" and "angle."

Prosopium williamsoni, Dill and Shapovalov, 1939A: 226–227
Prosopium williamsoni, McHugh, 1939: 39–50 / 1940: 131–137 / 1941: 337–343
Prosopium williamsoni, Carl and Clemens, 1948: 32–34
Prosopium williamsoni, Miller and Miller, 1948: 184
Coregonus williamsoni, Shapovalov and Dill, 1950: 385
Prosopium williamsoni, Sigler, 1951: 1–21
Prosopium williamsoni williamsoni, Simon, 1951: 28–29
Prosopium williamsoni, Beckman, 1952: 15–16
Coregonus williamsoni, La Rivers, 1952: 92
Coregonus williamsoni, La Rivers and Trelease, 1952: 115
Coregonus williamsoni, Eddy, 1957: 56
Prosopium williamsoni, Moore, 1957: 67
Prosopium williamsoni, Slastenenko, 1958A: 95–96
Prosopium williamsoni, Carl, Clemens and Lindsey, 1959: 40–41
Coregonus williamsoni, Shapovalov, Dill and Cordone, 1959: 171

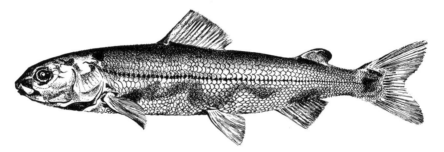

FIG. 164. Mountain Whitefish, *Prosopium williamsoni*.
Drawn by John B. Harris.

ORIGINAL DESCRIPTION

"A most important (I was almost going to say unexpected) discovery of a white fish was made by the party on the R.R. Survey of California and Oregon, commanded by Lt. R. S. Williamson. And since it is different from its hitherto known congener, we will call it Coregonus williamsoni as commemorative of that Survey. Its head is rather small, being contained about five times in the total length, which measures eleven inches. The mouth is very small and the posterior extremity of the maxillary does not extend as far back as the anterior rim of the orbit. The scales are large; eighteen rows of them may be counted between the anterior margin of the dorsal and the insertion of the ventrals: nine above the lateral line, and eight below it. The lateral line, itself, is perfectly straight. The caudal fin forked as usual. The pectorals are rather small. I have alluded to the color in saying it was a white fish; add to it a bluish grey hue along the back. It was collected by Dr. Newberry in the Des Chutes river, a tributary of the Columbia" (Girard 1856A: 136).

DIAGNOSIS

Head 4.5 to 5.0 in body length, depth 4.0 to 5.0 in body length; branchiostegals about 9; gillrakers short and thick, 8–10 + 13–15;

maxillary short and broad, anterior edge of eye just about reaching posterior edge of maxillary; pyloric caeca very numerous, often 100 or more; vertebrae 56–60; scales 8–10/80–95/6–10; fins—dorsal 11–14 rayed, anal 11–14 rayed; color not striking, white below, more silvery on sides and darkening to bluish above, fins black-tipped; in breeding males the lateral scales become bumpy or tuberculate; a small fish, seldom exceeding about a foot in length.

TYPE LOCALITY

Des Chutes River, Oregon.

RANGE

Northern half of the western United States, chiefly in the mountain streams; northward into Canada. Lacking in the interior of the Great Basin.

FIG. 165. Courtesy Oregon State Game Commission and Harold P. Smith.

TAXONOMY

During most of the last two decades, the species has been assigned to the genus *Prosopium*, which was erected to differentiate more clearly between eastern and western whitefish. These differences seem to be good.

Etymology—named for Lieut. R. S. Williamson, in charge of one of the U. S. Pacific Railroad surveys.

LIFE HISTORY

This species is a late fall-spawning insectivorous salmonoid less voracious than trout and salmon. It has a strong penchant for spawn of its own and other species, and, according to stomachs the author has examined and to the published results of other studies, insects make up practically its entire diet under usual conditions.

No systematic study has been made of the whitefish in our area, but

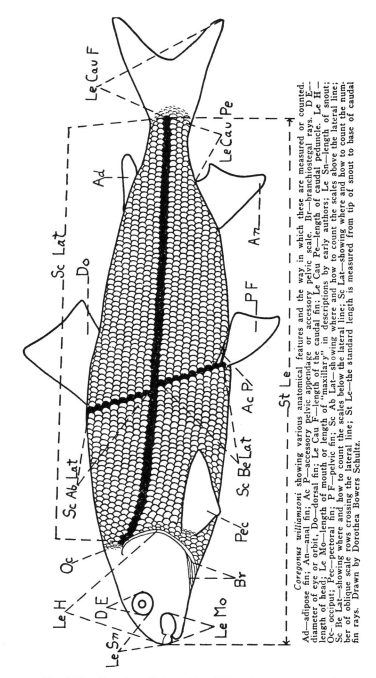

FIG. 166. Courtesy University of Washington Press and Dr. Leonard P. Schultz.

Coregonus williamsoni showing various anatomical features and the way in which these are measured or counted. Ad—adipose fin; An—anal fin; Ac P—accessory pelvic appendage or accessory pelvic scale. Br—branchiostegal rays. D E—diameter of eye or orbit, Do—dorsal fin; Le Cau F—length of the caudal fin; Le Cau Pe—length of caudal peduncle. Le H—length of head; Le Mo—length of mouth or length of "maxillary," in descriptions by early authors; Le Sn—length of snout; Oc—occiput; Pec—pectoral fin; P F—pelvic fin; Sc Ab Lat—showing where and how to count the scales above the lateral line; Sc Be Lat—showing where and how to count the scales below the lateral line; Sc Lat—showing where and how to count the number of oblique scale rows crossing the lateral line; St Le—the standard length is measured from tip of snout to base of caudal fin rays. Drawn by Dorothea Bowers Schultz.

the population of Logan River—a Utah stream with many resemblances to the Truckee River—has been intensively examined by Sigler (1951). He found very little movement of the species upstream during the spawning period, although lake populations seem to provide migrations into affluent streams in impressive numbers. Snyder (1917A: 69) reported this for Tahoe:

"It spawns in October, large numbers then moving up the tributaries of Lake Tahoe. The migration is said to last about two weeks, being at its height near the middle of the month. Males collected at that time had tubercles on the posterior part of the body and tail, both above and below the lateral line." Jordan and Henshaw (1878) had previously noted the same spawning time for whitefish in Tahoe.

Sigler's studies indicated that breeding began in the third year, none of the younger fishes showing any maturity. He found up to 14,000 eggs in 24-ounce females and noted that the egg diameter increased with the size of the specimen. In over-all size, Sigler reports maxima of some 19 inches for lengths and only slightly less than 3 pounds for weights. Others have found Mountain Whitefish weighing better than 3 pounds (C. J. D. Brown, as quoted by Sigler) and there are angler reports of weights in excess of 4 pounds.

Strongly insectivorous as the species is, it finds satisfactory feeding areas where such aquatic insects as mayflies (Ephemeroptera), case- or caddis-flies (Trichoptera), true flies (Diptera) and stoneflies (Plecoptera) are dominant parts of the fauna. The author has found these orders of insects, plus dragonflies (Odonata), snails (Gastropoda) and crawfish (Decapoda) in the stomachs of Truckee River specimens. Sigler (1951) noted the first four mentioned to be important in the Logan River studies in this order (by volume): caseflies, true flies, mayflies and stoneflies.

The type of food found in whitefish stomachs indicates that this species is, like the Brown Trout and the Belding Sculpin, a bottom feeder by preference, but it frequently feeds at the surface when bottom foods are scarce, particularly in lakes. The data of Rawson and Elsey (1950) from Pyramid Lake, Alberta, Canada, indicate heavy feeding on such minute floating planktonic crustacea as Cladocera as a usual thing.

Since the whitefish is an inhabitant of the upper and cooler waters, and has certain stream-size requirements as well, it is not generally found much beyond the points where the Lahontan streams leave the flanks of the Sierras and meander eastward, warming and slowing as they go. Sigler found that:

"The upstream movement of the whitefish apparently stops where the pools have less than a maximum width of 16 feet and a maximum depth of 4 feet, at the season of least flow. It is believed that less water than this does not provide acceptable cover for whitefish" (1951: 7). In the Logan River, whitefish were found only between elevations of 7,300 and 4,500 feet, which would roughly correspond to conditions in the Lahontan basin.

ECONOMICS

Although generally spurned by Nevada anglers, this species is gamier on light tackle than many trout, and is a good food fish. It is apparently decreasing in western Nevada streams, like the trout, in the face of changing stream conditions. Its utilization in earlier days when it existed more abundantly is related by Jordan and Henshaw (1878: 194):

"At Lake Tahoe it was found very abundantly in October, being met with at that season in all of the few small streams that issue into the lake from the adjoining mountains, as Trout Creek and Truckee River. This month and later is their spawning season, and as they pass up many are intercepted by the Indians, who find a market for considerable numbers in the settlements and logging camps about the lake. Having constructed a suitable net of mosquito-netting, which is affixed

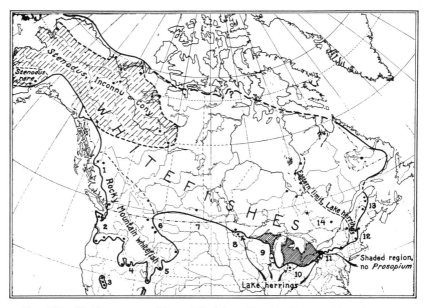

FIG. 167. Native range of the whitefishes, Family Coregonidae. Courtesy of University of California Press and Erhard Rostlund.

to a long pole, the Indian, accompanied by one or two squaws, proceeds to the stream where it is sufficiently narrow for his purpose. Placing the net at the head of one of the deep sandy-bottomed pools which are found at every turn of the streams, he awaits quietly until all the fish near by have been frightened into it by the squaws, who advance from below and beat the water with sticks. With a sudden scoop he usually empties the pool, taking perhaps six to a dozen fish from each. All that we saw caught in this manner were quite small, averaging perhaps ten inches in length, but they attain a much larger size."

Snyder's brief comment is: "It rises to the fly at times, is as game as a trout, and by some is preferred as a food fish" (1917A: 69).

The possible competition between other fishes and trout is always foremost in the anglers' and fishery management biologists' minds, and, as is generally true, some aspects have been over-emphasized in the past. As bottom feeders, whitefish certainly compete with Brown Trout, where they occur together. Also, as mentioned, the sculpin is involved in bottom food competition with the other two. However, there is no evidence that this competition is a matter of any great importance to either the angler or the biologist since it probably has little over-all effect on the trout populations any given area will support from a practical standpoint.

Probably, as Sigler suggests, more good management could be accomplished by modifying the creel limit on whitefish so that more could be taken and they would not be thrown away in preference to trout.

THE CYPRINIFORM FISHES
Order *CYPRINIFORMES*
Division CYPRINI
THE SUCKERS AND CARP-LIKE FISHES
Suborder CYPRINOIDEI

SUCKERS
Family *CATOSTOMIDAE*

This large and common family comprises a considerable part of the North American freshwater fish fauna. Of the approximately 14 genera and 80 species in the group, all are North American except two species. The idea that catostomids are ancestral to the next family, the Cyprinidae, is not a logical one to this writer, either on the basis of the fossil record, present distribution or comparative morphology. However, most current opinion leans to the concept that the Asiatic-originating catostomids gave rise to cyprinids and have been almost entirely superseded by the latter in the ancestral home.

As food fishes, suckers are not generally popular, usually because of their boniness, although many are well-flavored. All are spring spawners, and our species are relatively numerous, both specifically and individually. They are a prominent converter of plant materials and organic debris to proteins which serve as more concentrated foods for the fishes that prey upon them.

DIAGNOSIS

Body elongate-to-stout, rounded; head bare, conical, mouth ventral, occasionally ventro-terminal, medium in size, generally protractile with fleshy lips, barbels lacking; jaws without teeth; pharyngeal teeth present; branchiostegals 3; pseudobranchiae present; gills 4, the fourth followed by a slit; gill membranes applied to the isthmus, limiting the side openings; fins—adipose lacking, pectorals ventral, pelvics abdominal, caudal usually forked; lateral line usually present; stomach simple; air bladder large, 2–3 compartmented, with no surrounding bony capsule; pyloric caeca lacking.

Somewhat carp-like, bottom-feeding species with sucking mouths, all living in fresh water.

KEY TO GENERA OF NEVADA CATOSTOMIDAE

1. Nuchal (neck) region with a high, sharp-edged hump, formed by the greatly enlarged and expanded inter-neural vertebral spines; hump is most pronounced in adult specimens (*Xyrauchen*)_____ RAZORBACK SUCKERS

    No such hump, at most merely with an upslope on the otherwise smoothly curving dorsal outline _____ 2

2 (1). Lips thin, lacking papillae; mouth semi-terminal, the lower jaw oblique (*Chasmistes*) .... LAKE SUCKERS

.... Lips thick with many papillae; mouth inferior (i.e., opening directly downward) 3

3 (2). With a distinct notch at each corner of the mouth; edge of jaw inside lower lip with a hard cartilaginous sheath; upper lip recurved; a small flap or "scale" of skin present at base of each pelvic fin (in axil) (*Pantosteus*) .... MOUNTAIN SUCKERS

.... Without such a notch between upper and lower lips, although occasionally a very slight indentation occurs in some individuals; cartilaginous sheath, if present, flexible and not hard; upper lip nearly flat, not recurved; no flaps or "scales" in pelvic fin axils (*Catostomus*) .... COMMON SUCKERS

MOUNTAINSUCKERS

Genus *Pantosteus* Cope 1875

("All bone," probably referring to the thick head bones and obliterated fontanelle)

More than 35 scales in front of dorsal fin; fontanelle (space between parietal and frontal head bones) open; Lahontan drainage system of western Nevada (*Pantosteus lahontan*) .... LAHONTAN MOUNTAINSUCKER

Less than 35 scales in front of dorsal fin; fontanelle closed; White River drainage system of eastern Nevada (*Pantosteus intermedius*) .... WHITE RIVER MOUNTAINSUCKER

LAHONTAN MOUNTAINSUCKER

Pantosteus lahontan Rutter

(Lahontan Sucker)

Pantosteus lahontan Rutter 1903: 146 / 1908: 118, 120
Pantosteus lahontan, Snyder, 1915: 578 / 1917A: 49–50 / 1927: 4
Pantosteus lahontan, Jordan, Evermann and Clark, 1930: 105
Pantosteus lahontan, Schrenkeisen, 1938: 83
Pantosteus lahontan, Murphy, 1941: 167
Pantosteus lahontan, Hubbs, Hubbs and Johnson, 1943: 54–58
Pantosteus lahontan, Shapovalov and Dill, 1950: 385
Pantosteus lahontan, La Rivers, 1952: 95–96
Pantosteus lahontan, La Rivers and Trelease, 1952: 115
Pantosteus lahontan, Eddy, 1957: 75
Pantosteus lahontan, Moore, 1957: 93
Pantosteus lahontan, Shapovalov, Dill and Cordone, 1959: 171

ORIGINAL DESCRIPTION AND DIAGNOSIS

"Head 4.5 in length, depth 5.5; eye 6 in head; D. 10 or 11; A. 7; scales 17–81 to 96–12, 47 to 50 before dorsal. Body terete, caudal peduncle but little compressed; interorbital slightly convex, or flat, width of bone 2.8 in head; eye posterior, 3 in snout, 2.5 in interorbital space, 1.5 in distance between eye and upper end of gill-opening; snout equal to half of head, broadly rounded both vertically and horizontally, projecting beyond the large mouth; 4 rows of papillae on upper lip, 4 rows across symphysis of lower lip, 10 papillae in an oblique row from corner of mouth to inner corner of lobe of lower lip; isthmus broader than interorbital, equal in distance between pupils; fontanelle present, but less than half width of pupil in a 6-inch specimen; dorsal inserted from 49 to 52 hundredths of body from tip of snout; ventrals inserted under ninth ray of dorsal, halfway between tip of snout and

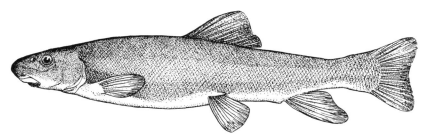

FIG. 168. Lahontan Mountainsucker, *Pantosteus lahontan*.
Drawn by Silvio Santina.

tip of middle caudal rays; caudal 1.5 in head, deeply emarginate, not forked; pectoral 1.3 in head; height of dorsal about 1.4 in head, the base equal to snout, margin slightly emarginate; ventral 1.7 to 1.8 in head. Very dark, almost black above, abruptly paler below, lower fins slightly dusky. Maximum length, about 6 inches.

"Found in abundance in Susan River, and also in Little Truckee River and Prosser Creek.

"Types (No. 50587, U. S. Nat. Mus.) from Susan River, collected by Rutter and Chamberlain" (Rutter 1903: 146).

TYPE LOCALITY
Susan River, Lassen County, California.

RANGE
Confined to the Lahontan drainage system of western Nevada and adjacent California; Walker, Carson, Truckee and Susan Rivers on the west side of the Lahontan system; Quinn River on the north of the system; and Humboldt and Reese Rivers on the east side; and their tributary streams, such as the mentioned Prosser Creek.

TAXONOMY
The Mountainsuckers are closely related, as a group, to the genus *Catostomus*, the Common Suckers, and seem to be a segment which has

become isolated and adapted to higher and smaller streams in the intermontane West. The genus *Pantosteus* was established by Cope in 1875 to separate certain suckers of the intermontane West from *Catostomus* by virtue of having thicker skulls and cutting edges on both jaws.

LIFE HISTORY

A small, summer-spawning sucker apparently confined to streams and occurring in much smaller numbers than the associated *Catostomi*. They feed much like other suckers, on rock algae and decaying organic material.

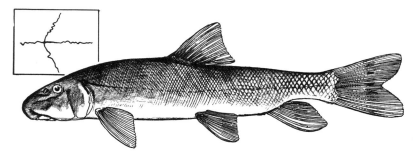

FIG. 169A. Utah Bluehead Mountainsucker, *Pantosteus delphinus utahensis*, showing closed skull fontanelle. Found in adjacent areas of Utah, it has been used as bait fish on the Colorado River. Drawn by William L. Brudon for Miller 1952. Courtesy California Department of Fish and Game and Leo Shapovalov.

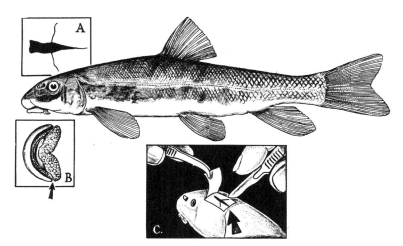

FIG. 169B. Bonneville Mountainsucker, *Pantosteus platyrhynchus* (Cope) 1874. A common species in the Bonneville Basin, it has been used as a bait fish in the Colorado system of southern Nevada (Miller 1952: 27–28). (A) Showing open fontanelle. (B) The notch at lip corners. (C) Method of demonstrating the fontanelle. Drawn by William L. Brudon for Miller 1952. Courtesy California Department of Fish and Game and Leo Shapovalov.

"Females with nearly ripe eggs were observed in Long Valley Creek, July 13. Ripe eggs were found in examples in Carson River, July 20, and in Quinn River, July 30. Both males and females appeared to be migrating up the Humboldt River early in July, for they were congregating in large numbers below obstructions" (Snyder 1917A: 49).

Hubbs, Hubbs and Johnson (1943: 54) have recorded hybridization between this species and *Catostomus tahoensis* (which see).

ECONOMICS

Unimportant, other than the part they must play in the food chains of associated carnivorous game fish, such as trout. However, because of their scarcity compared with *Catostomi*, their role here is certainly insignificant.

WHITE RIVER MOUNTAINSUCKER
Pantosteus intermedius (Tanner)
(White River Sucker)

Notolepidomyzon intermedius Tanner 1942: 29–32
Pantosteus intermedius, La Rivers, 1952: 96
Pantosteus intermedius, La Rivers and Trelease, 1952: 115
Pantosteus intermedius, Moore, 1957: 93

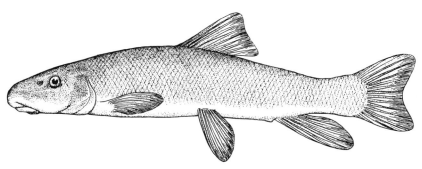

FIG. 170. White River Mountainsucker, *Pantosteus intermedius*. Drawn by Silvio Santina.

ORIGINAL DESCRIPTION AND DIAGNOSIS

"Description of the Type No. 4252: Head 4.0 times in length to base of caudal; depth 5.1; depth of caudal peduncle 11; dorsal rays 11; anal rays 7; scales before the dorsal 29; scales above lateral line 11; scales below the dorsal line 12; scales on the lateral line 85; snout to dorsal in proportion to total length 2; scales on the caudal peduncle 11 to 12.

"The head is long and slender, depth about $\frac{2}{3}$ the length, interorbital width 2.1 of length; top of head flat; fontanelle completely closed; width of mouth contained four times in length of head; width of mouth including lip 10 mm; papillae in 4 rows on upper lip and 8 to 9 rows on lower lip, the outer and inner rows smaller; cleft on lower lip moderate; 9 rows of papillae cleft to inner cutting edge of mouth.

"The dorsal is about ¼ greater in height than length of base; anal 1.3 in head; the anal just reaches base of caudal; the ventrals just reach the anus; 5 scales above the lateral line on the caudal peduncle; lateral line straight except for a short upward curve at its origin where it passes above the operculum.

"In life this species is greyish to olive green above becoming light yellow to whitish on the venter. In alcohol the specimens are dark to blackish above; mottled above, along and irregularly below the lateral line and whitish on the belly region; with a dark circular area on the operculum" (Tanner 1942: 29-31).

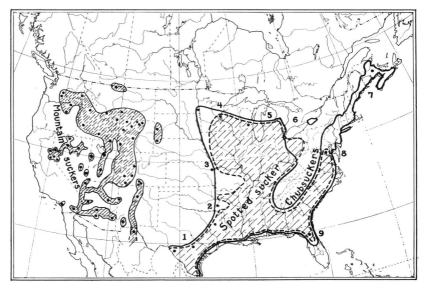

FIG. 171. Original range of mountainsuckers, *Pantosteus* spp. Courtesy of University of California Press and Erhard Rostlund.

TYPE LOCALITY

"White River, streams and springs at Lund and Preston, White River Valley, White Pine County, Nevada" (Tanner 1942: 31). The type is deposited in the Brigham Young University collection at Provo, Utah. Topotypes are at the University of Nevada in Reno, and elsewhere.

RANGE

Known only from the type locality and that portion of the relict White River system immediately to the south.

TAXONOMY

As an official entity, the White River Mountainsucker has led a short and uneventful life. The genus in which it was originally described, *Notolepidomyzon*, was first proposed by Fowler as a subgenus in 1913 and defined somewhat better by Snyder (1915) although it is difficult to determine if the latter was using the term generically or subgenerically from his text. At best, the characters upon which *Notolepidomyzon* is based seem to have no more than subgeneric value.

LIFE HISTORY

No data has been published on the species' life cycle, but presumably it is not dissimilar from that of other members of its group. Its food probably consists characteristically of algae and the organic debris in muds.

ECONOMICS

Unimportant except the part it plays, as the only sucker in its waters, in converting plant materials into animal proteins for utilization by larger species in the food chain. This is insignificant, from the standpoint of human economics, since the only larger carnivorous fish in the same waters is a nongame race of *Gila robusta*. They are used as bait fish in the Colorado River, Lakes Mead and Mohave quite extensively.

COMMON SUCKERS
Genus *Catostomus* Le Sueur 1817
("Inferior mouth")

KEY TO SPECIES

1. Body scales very small to moderate in size, about 80 to 115 along the lateral line_____ 2

   ____ Body scales large, about 60–80 in the lateral line_____ 4

2 (1). Lower lips not deeply incised, allowing several transverse rows of papillae to cross the midline between the incision and the forward edge of the lower jaw (lateral scales 92–114; dorsal fin rays 11–14; Columbia River system) (*Catostomus columbianus*)_____ BRIDGELIP SUCKER

   ____ Lower lips deeply incised, usually allowing room for only one row of papillae to cross the midline_____ 3

3 (2). Dorsal fin rays from 10 to 12 in number; caudal peduncle thick and broad, its least depth about 12 times into body length; lips moderate-sized for the genus (lateral scales 82–95; Lahontan system) (*Catostomus tahoensis*)_____
   _____ TAHOE SUCKER

   ____ Dorsal fin rays 13–15; caudal peduncle very slim and narrow, its least depth about 16 times into body length; lips conspicuously enlarged (lateral scales 98–105; Colorado River system) (*Catostomus latipinnis*)_____
   _____ FLANNELMOUTH SUCKER

4 (1). Dorsal fin rays from 12 to 15 in number, usually more than 12; lobes of lower lip long and full, extending back to below the nostrils, their length (from tip of upper lip) more than one-half the depth of the caudal peduncle

(lateral scales 65–79; Columbia River system) (*Catostomus macrocheilus*) .... BIGLIP SUCKER

Dorsal fin rays 11–13; lobes of lower lip short, not extending back to below nostrils, their length (from tip of upper lip) about one-half, or less, the depth of the caudal peduncle (lateral scales 61–79; Bonneville system) (*Catostomus ardens*) .... UTAH SUCKER

BIGLIP SUCKER

Catostomus macrocheilus Girard

(Columbia Coarsescale Sucker, Columbia Basin Sucker)

Catostomus macrocheilus Girard 1856B: 175 / 1859: 225
Catostomus macrochilus, Günther, 1868: 20
Catostomus macrochilus, Jordan, 1878A: 171
Catostomus macrochilus, Jordan and Copeland, 1878: 416
Catostomus macrochilus, Jordan and Gilbert, 1882: 128
Catostomus macrocheilus, Jordan and Starks, 1895: 852
Catostomus macrocheilus, Jordan and Evermann, 1896B: 178 / 1902: 50–51
Catostomus macrocheilus, Snyder, 1908: 81, 165–169 / 1916: 580 / 1917A: 47
Catostomus macrocheilus, Pratt, 1923: 57
Catostomus macrocheilus, Jordan, Evermann and Clark, 1930: 106
Catostomus macrocheilus, Hubbs and Schultz, 1932: 5
Catostomus macrocheilus, Carl, 1936
Catostomus macrocheilus, Schultz, 1936: 145
Catostomus macrocheilus, Schrenkeisen, 1938: 85
Catostomus macrocheilus, Hubbs, Hubbs and Johnson, 1943: 19–33
Catostomus macrocheilus, Dimick and Merryfield, 1945: 34–35
Catostomus macrocheilus, Carl and Clemens, 1948: 66–68
Catostomus macrocheilus, Miller and Miller, 1948: 176–177
Catostomus macrocheilus, La Rivers, 1952: 96
Catostomus macrocheilus, La Rivers and Trelease, 1952: 115
Catostomus macrocheilus, Eddy, 1957: 76
Catostomus macrocheilus, Moore, 1957: 88
Catostomus macrocheilus, Slastenenko, 1958A: 161–162
Catostomus macrocheilus, Carl, Clemens and Lindsey, 1959: 86–88
Catostomus macrocheilus, Macphee, 1960: 119–125
Catostomus macrocheilus, Weisel, 1960: 109–129

ORIGINAL DESCRIPTION

"This species is very different from both of the preceding ones [*C. occidentalis* and *C. labiatus*] by a larger and more elongated head, a larger mouth, and hence much larger lips, covered with large papillae. The scales which cover the body are larger than in *C. occidentalis*, and smaller than in *C. labiatus*. The head constitutes the fifth of the total length; the horizontal diameter of the eye is contained nearly six times in the length of the side of the head. The head itself is subquadrangularly pyramidal, truncated anteriorly with the upper edge of the snout projecting. The anterior margin of the dorsal is a little nearer the end of the snout than the insertion of the caudal fin. Its upper margin is concave. The anal is well developed, for its tip extends beyond the base of the caudal. The ventrals are inserted opposite the middle of the dorsal. The pectorals are large and long.

"D.17; A.9; C.5, 1, 8, 8, 1, 6; V.10; P.18.

"Bluish black above; yellowish golden on the sides and whitish beneath.

"Collected at Astoria, O. T., by Lieut. W. P. Trowbridge, U. S. A." (Girard 1856B: 175).

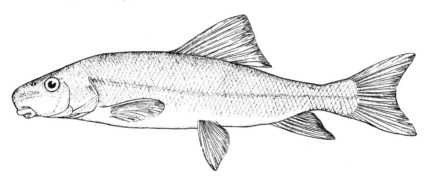

FIG. 172. Biglip Sucker, *Catostomus macrocheilus*.
Drawn by Silvio Santina.

DIAGNOSIS

Head about 4 in body length, depth about 5; head large, quadrangular, with projecting snout; fontanelle present (open); mouth large, ventral, with very large lips, the lower lip deeply incised along the median line with 6–8 papillous rows, none of which are prominently complete across the median line; a coarse-scaled sucker, scales 12–16/65–79/8–10, 30–40 before the dorsal fin; fins—dorsal much longer than high, 12–15 rayed—pectorals elongated—caudal conspicuously forked; caudal peduncle much accentuated in young specimens; peritoneum light-to-dark, not visible through body wall in young individuals; coloration dark, with a darker side stripe, belly contrastingly pale.

TYPE LOCALITY

Columbia River at Astoria, Oregon.

RANGE

Northwestern United States: Puget Sound area of Washington south to Six Rivers region of Oregon, and eastward up the Columbia River tributaries to western Montana and northern Nevada; northward into Canada. Does not occur in the upper waters of its drainage systems, being replaced there by a subspecies of the common eastern sucker, *Catostomus catostomus griseus* Girard 1856, a fine-scaled form.

TAXONOMY

There has been no confusion surrounding this particular species as has been the case with so many others. *Catostomus macrocheilus* is the coarse-scaled sucker of the Columbia River, differing from the finer-scaled *Catostomus columbianus* not only in the mentioned matter of scales, but also in the structure of the mouth—the incisure of the lower lip being much more pronounced in the coarse-scaled species—and in the color of the peritoneum.

Etymology—"macrocheilus," meaning large lip.

LIFE HISTORY

Little is known of this species. Miller and Miller (1948) have given fragmentary data, obtained incidentally to collecting, which offer some clues:

"The habitat of these suckers was typical of the usual environmental predilections of species of *Catostomus*. On the South Fork of the Owyhee River, which was 20 feet wide where seined, there were many abandoned beaver dams where the water was up to 4 feet deep and the current was moderate. The water temperature was 68° F. [17 August 47]. On the Salmon River, pools were 8 x 16 feet in major dimensions, the water was up to 3 feet deep, the current was moderate, and the temperature was 74° F." [18 August 47] (p. 177). The species seems to prefer slower water than its fine-scaled associate, *Catostomus columbianus*. Probably their habits do not differ from those known for suckers generally.

Dimick and Merryfield, in pollution studies of the Willamette River in Oregon noted that

"During April and May, adult coarse-scaled suckers, some measuring 16 inches or more, may be seen moving upstream in large schools to spawning areas. The yellowish colored eggs are deposited in riffle areas of creeks where they adhere to gravel. It is thought that the young hatch in about two weeks time and then drift downstream into quieter water" (1945: 36).

Hubbs, Hubbs and Johnson have recorded hybridization between this species and *Catostomus columbianus* (which see), which they called *C. syncheilus*.

ECONOMICS

Undoubtedly plays the usual part in the food chain, converting plant materials to animal protein and being in turn food for larger carnivorous fishes such as the large salmon, trout, squawfish, etc., populations with which it is associated. The species was much used as a food staple in the original Indian economy.

BRIDGELIP SUCKER

Catostomus columbianus (Eigenmann and Eigenmann)
(Columbia River Sucker)

Pantosteus columbianus Eigenmann and Eigenmann 1893: 151
Pantosteus jordani, Jordan and Evermann, 1896B: 171–172 / 1902: 45–46
Pantosteus jordani, Pratt, 1923: 55
Pantosteus columbianus, Jordan, Evermann and Clark, 1930: 105

Pantosteus jordani, ibid: 104
Catostomus syncheilus Hubbs and Schultz 1932: 5–13
Catostomus syncheilus, Schultz, 1936: 143
Pantosteus columbianus, Schrenkeisen, 1938: 82–83
Pantosteus columbianus, Tanner, 1942: 28
Catostomus columbianus, Miller and Miller, 1948: 177–178
Catostomus syncheilus, ibid: 178
Catostomus columbianus, La Rivers, 1952: 96
Catostomus columbianus, La Rivers and Trelease, 1952: 115
Catostomus columbianus, Eddy, 1957: 78
Catostomus columbianus, Moore, 1957: 89
Catostomus columbianus, Slastenenko, 1958A: 160–161
Catostomus columbianus, Carl, Clemens and Lindsey, 1959: 92–93

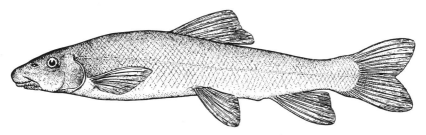

FIG. 173. Bridgelip Sucker, *Catostomus columbianus*.
Drawn by Silvio Santina.

ORIGINAL DESCRIPTION

"1. *Pantosteus columbianus* E. & E.

"Three specimens, 92–100 mm. Boise River, Caldwell, Oregon. Related to *P. generosus*, the eye larger, the caudal longer. Head $4\frac{2}{5}$–$4\frac{3}{5}$; D. II, $11\frac{1}{2}$ or $12\frac{1}{2}$; A. I, $8\frac{1}{2}$; scales 16–19–80–100–15; eye $1\frac{1}{2}$–2 in snout, $1\frac{1}{5}$ to $1\frac{3}{5}$ in interorbital, $3\frac{3}{4}$ to little more than 4 in head. All the fins pointed, the caudal lobes considerably longer than the head. Light brown, with indistinct clouds of darker" (Eigenmann and Eigenmann 1893: 151–152).

DIAGNOSIS

Body rather plump, head and depth about 4 in body length; fontanelle present (open); mouth moderately large, ventral, the lips prominent and coarsely papillose, the lower not strongly incised, 2–4 rows of papillae crossing the median line between the lateral lobes; a fine-scaled sucker, scales 20–22/97–111/19–22, 53–58 before the dorsal fin; fins—dorsal 11–13 rayed, usually 12; caudal peduncle abruptly constricted, thick, least depth about $12\frac{1}{2}$ times into body length; peritoneum black, showing prominently through the body wall of young individuals; coloration dark above, light below—yearlings typically mottled on back, with 3 dark areas on sides.

TYPE LOCALITY

Boise River at Caldwell, Oregon.

RANGE

Streams of the lower and middle Columbia River basin, entering northern Nevada through the Snake River tributaries.

TAXONOMY

This species has led a rather checkered taxonomic existence, which Miller and Miller (1948) finally untangled. When first described it was placed in the Mountainsucker genus *Pantosteus* and the name later synonymized with *Pantosteus jordani* where it remained for more than 50 years. Meanwhile it was redescribed as *Catostomus syncheilus* by Hubbs and Schultz in 1932 from Washington specimens and was so considered until Miller and Miller (1948).

Etymology—named for its major habitat, the Columbia River.

LIFE HISTORY

The species has not been the subject of any particular life history study, so far as I am aware, but, as in the case of *Catostomus macrocheilus*, this can be assumed not to deviate from the generally known sucker pattern. Miller and Miller make the following comments for northern Nevada collected material (1948: 180):

"In contrast to *C. macrocheilus*, the usual habitat preference of *columbianus* is for the more swiftly flowing portions of streams. At each of the four collecting stations where the junior author secured this species, but not *macrocheilus*, he noted that the current was swift, and that the bottom was composed of boulders, rocks or gravel, except in the head-waters of the Bruneau River where the current was moderate and the bottom consisted of mud and sand. This station, however, was at a high elevation, only 4 miles north of the pass separating this portion of the Columbia system from that of the Lahontan basin, and the water was cold, only 60° F. (air 74° F.)" [20 August 47].

UTAH SUCKER

Catostomus ardens Jordan and Gilbert

(Utah Lake Mullet)

Catostomus ardens Jordan and Gilbert 1881B: 464 / 1882: 128–129
Catostomus ardens, Jordan, 1890C: 46, 47–48 / 1891: 31
Catostomus ardens, Jordan and Evermann, 1896B: 179–180 / 1902: 52–53
Catostomus ardens, Pratt, 1923: 57
Catostomus ardens, Snyder, 1927: 3
Catostomus ardens, Jordan, Evermann and Clark, 1930: 105
Catostomus ardens, Hatton, 1932: 28–30
Catostomus ardens, Tanner, 1936: 166–167
Catostomus fecundus, ibid
Chasmistes liorus, ibid
Catostomus ardens, Schrenkeisen, 1938: 84
Catostomus ardens, Lowder, 1951: 19–20
Catostomus fecundus, Simon, 1951: 56–57

Catostomus ardens, La Rivers, 1952: 96
Catostomus ardens, La Rivers and Trelease, 1952: 115
Catostomus ardens, Miller, 1952: 14, 25
Catostomus ardens, Eddy, 1957: 76
Catostomus ardens, Moore, 1957: 88

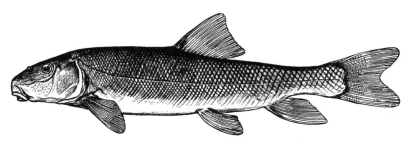

FIG. 174. Utah Sucker, *Catostomus ardens*. Drawn by William L. Brudon. Courtesy California Department of Fish and Game and Leo Shapovalov.

ORIGINAL DESCRIPTION

"13. Catostomus ardens, sp. nov.
(?*Catostomus guzmaniensis* Cope and Yarrow; not of Girard)
"A large, thick-lipped species, allied to *C. macrochilus*, etc.
"Body rather elongate, subfusiform, little compressed, the back broad and somewhat elevated. Head conical, broad and convex above, the front regularly sloping from the nape to the snout. Mouth entirely inferior, the mandible quite horizontal, the premaxillaries scarcely raised above the level of the base of the mandible. Upper lip very wide, full, pendant, with about eight rows of coarse, irregular papillae, of which the second and third rows from the inside are much larger than the others; upper lip continuous with the lower at the angle of the mouth, the lower lip cut to the base in the middle by a deep, abrupt incision. Front of eye midway in head. Eye very small, 7 in head, 3½ in the convex interorbital space. Isthmus broad, half broader than the eye. Fontanelle large, as in the other species noticed in this paper. Scales crowded anteriorly, 9–65–9. Breast with evident imbedded scales. Dorsal fin inserted a little behind the middle of the body, long and low, its anterior rays but three-fourths the length of the base of the fin, 1½ the length of the last rays; the free edge of the fin straight. Caudal fin short and broad, about equally forked, its upper lobe two-thirds the length of the head. Pectoral short and broad, their length three-fourths that of the head. Ventrals short, not quite reaching vent. Anal very high, reaching caudal. Dorsal rays 13; anal 7. Length of head 3⅔ in body to base of caudal; greatest depth 4½. Teeth essentially as in the others.
"Color blackish above, blotched with darker, the whole back and sides obscurely spotted; belly white; a narrow, bright, rosy, lateral band on the anterior part of the body, overlying the blackish; fins mostly dusky mottled; top and sides of head rendered dusky by the presence of many dark specks.

"This species is described from a large adult male nearly 18 inches in length, besides which we have a single young specimen" (Jordan and Gilbert 1881B: 464).

DIAGNOSIS

Head about 4, depth about 4.5, times in body length; head broad and conical; fontanelle present (open); mouth ventral, the lips full, pendulous and papillose, the lower lip completely divided along the median line into two rounded lobes, each bearing about 6 rows of papillae; a coarse-scaled sucker, scales 9–14/61–79/8–14, 29–36 before the dorsal fin; fins—dorsal 12–13 rayed, rather long—anal usually 7-rayed, long —caudal rather short, moderately forked; caudal peduncle thick, least depth about 10–12 in body length; coloration blackish above, pale below, breeding males with reddish sides; reaches maximum lengths and weights of better than two feet and 12 pounds, respectively.

TYPE LOCALITY

Utah Lake, Utah County, Utah.

RANGE

Utah and the upper Snake River drainage of Idaho, Wyoming, Nevada.

TAXONOMY

The names applied to this animal since its description have led to great confusion over the intervening years. Until the unpublished study of Lowder in 1951, *Catostomus ardens* was generally relegated to the synonymy of *C. fecundus*, particularly since Tanner's 1936 report. To reiterate the uncertainties surrounding this species:

In 1875, Cope and Yarrow described *C. fecundus* from Utah Lake, their specimens having been collected by Yarrow and H. W. Henshaw, members of the Wheeler Surveys West of the 100th Meridian. Three years after its description, Jordan treated the species in several publications within the space of a year, doing much to create the confusion which has persisted to the present day. Earlier in the year (1878B: 417), he very briefly characterized the new genus *Chasmistes* with *Catostomus fecundus* as the type. Later in the year (1878D: 149–151), he more fully and adequately described *Chasmistes*, still with *fecundus* as the only species. However, in an addendum to the same paper (pp. 219–220), Jordan decided that he had in reality been dealing with two different genera under the specific name of *fecundus*. His examination of the Cope and Yarrow type in the U. S. National Museum collection convinced him that *fecundus* was a true *Catostomus*, and that the specimens upon which he had based the new genus *Chasmistes*, instead of representing *fecundus*, were an undescribed species which he proceeded to name *Chasmistes liorus* (p. 219), thereby returning *fecundus* to the genus *Catostomus*.

He first applied the common name of "Big-mouthed Sucker of Utah

Lake" to *Chasmistes liorus,* in later years referring to it as the "June Sucker of Utah lake." *Catostomus fecundus* he called the "Webug Sucker."

In 1881B, Jordan and Gilbert described the new species *Catostomus ardens* from Utah Lake, allying it to *C. macrocheilus. Chasmistes liorus* was discussed in some detail in the same publication, as was *Catostomus fecundus*—reference was made to the latter's occurrence in the lake in enormous numbers, making Utah Lake the "greatest sucker pond in the universe," but no comparisons were drawn between *fecundus* and *ardens.* In 1891, Jordan commented further upon the sucker population of Utah Lake, this time listing *ardens* as the commonest sucker. An 1896 (Jordan and Evermann) notation is: "Utah Lake, rather scarce, and not yet seen elsewhere. This species resembles *Chasmistes liorus. (fecundus,* fertile, in allusion to its supposed abundance, but the 'fecund' species which has made Utah Lake the 'greatest sucker pond in the world' is really *C. ardens.*)" (p. 181).

In 1917A, Snyder (p. 50) included a footnote mentioning Jordan's original mistake in referring *fecundus* to *Chasmistes.* Tanner (1936: 166–167), to my knowledge, was the first to discuss comparatively the two entities *fecundus* and *ardens;* after examining hundreds of suckers from Utah Lake, he came to the conclusion that the two species were the same, in which case *fecundus* would have priority. Later, a student of Tanner's, Lowder, undertook the specific problem of unravelling the Utah Lake suckers and presented his results in a master's thesis in 1951. It appears, from his work, that the two are distinct,[1] although the extent to which hybrids between the two might be a confusing element has still to be determined. It would also seem that the Utah Lake *Chasmistes liorus,* is valid, resurrecting it from the synonymy of *Catostomus fecundus* where Tanner placed it in 1936.

Etymology—"ardens," burning; in reference to the bright red colors of the breeding male.

---

[1] In addition to differences which seem valid but are of such a nature that comparisons are difficult, the characters Lowder found to separate the three Utah suckers may be summarized as:
1. Lower lip thin, half or less as thick as in *Catostomus*, with median, "V"-shaped incision separating the two lobes of the lower lip except at their anterior border; upper lip very thin, overhung by snout so as to be virtually invisible in lateral or front view; upper lip with or without papillae, but if they are present, papillae are not arranged in rows; mouth typically subterminal as in other *Chasmistes*.. *CHASMISTES LIORUS*
...... Lower lip twice or more as thick as in *Chasmistes* above the median incision appearing as a line between the lobes which are not pulled apart; upper lip thick, not overhung by snout and always conspicuous as an external ridge of varying size; upper lip prominently papillose, papillae arranged in rows; mouth terminal, typically *Catostomus*-like .... *CATOSTOMUS*....2
2. Upper lip very wide, its papillae large and arranged in 7–9 rows, rows 2–3 enlarged.. *CATOSTOMUS ARDENS*
...... Upper lip wide, but only about half the width of *C. ardens*, its papillae small and arranged in 4–6 rows, only the 3rd row slightly enlarged .... *CATOSTOMUS FECUNDUS*

LIFE HISTORY

The original describers of *Catostomus fecundus* said that:[2]

"This species is abundant in Utah Lake, and is called 'sucker' by the settlers. They run well up the rivers to spawn in June; feed on the bottom and *eat spawn* of better fish; spawning beds on gravel;" (Cope and Yarrow 1875: 678).

Simon says of it in Wyoming:

"The author on June 11, 1941, observed rosyside suckers spawning in almost exactly the same manner as western white suckers [*Catostomus commersonni sucklii* Girard 1856]. That is, one female was followed for a short distance by two males, one on each side and slightly behind her. Suddenly the male came parallel with the female, and closed in on her. As the contact was made between the bodies of the three fish (tuberculate sides of the males probably help in this activity), all quivered rapidly and seemed to arch their backs, spreading out the dorsal fin to its fullest extent; eggs and milt were emitted at this time. The tails of the males vibrated violently, stirring up gravel and sand from the bottom of the stream . . ."

"With body held vertically, this sucker often feeds from the surface where small whirls and eddies collect flotsam; with body half exposed, it also feeds on surface drift along the shoreline" (1946: 56–57).

Simon also records the species as living in water "well above 80° F."

Miller summarizes some highlights for the species as: "It is an adaptable species, living in lakes, rivers or creeks at warm to very cold temperatures, in slow to rapid current, in silty to clear water where the bottom varies from soft mud to clay, gravel and stones, and where there is usually some algae or submerged plants or both. In Bear lake [Utah], it lives in water at least as deep as 76 to 80 feet" (1952: 25). The life cycle of *Catostomus ardens* does not differ in its essential details from that of *C. tahoensis* in Pyramid Lake, Nevada.

ECONOMICS

Cope and Yarrow (1875: 678–679) continued their remarks as: "bite at hook sometimes; are extremely numerous, and are considered a nuisance by the fishermen, but they meet with a steady sale in winter at an average price of 2½ cents per pound."

Jordan and Gilbert (1881B: 463) commented:

"This species occurs in Utah Lake in numbers which are simply enormous, justifying Mr. Madsen's [an early commercial fisherman] assertion that the lake is the 'greatest sucker pond in the universe.' It is very destructive to the trout. It ascends the rivers in the spring to spawn at the same time as the latter species, on the eggs of which it feeds. In the interests of the food supply of Salt Lake City an organized attempt at the reduction or extirpation of this species may become necessary. The old trout feed largely on the young of the species, but

---

[2] Since *C. fecundus* and *C. ardens* were not separately recognized at the time Cope and Yarrow described the former, their remarks will apply equally well to both—and it is not known at present that their life cycles differ in any significant respects.

the 'suckers eat the trout first.' No full description of *Catostomus fecundus* has yet been published. It will be seen from the following account that it is well separated from all its congeners, and that in many respects it approaches *Chasmistes liorus*. It is, in fact, probably the parent stock of the genus *Chasmistes*.

"This species seems to reach a smaller size than the other lake suckers."

Jordan in 1891: 34—

"In a single haul of the large seine made in a channel on the south side of the lake [Utah], fifty trout ranging from two to two and one-half pounds were taken. With these were taken six June suckers *Chasmistes liorus* weighing three pounds each, two hundred 'Mullet' *Catostomus ardens* weighing about two pounds each, one webug *Catostomus fecundus* weighing one pound, and about two hundred chubs *Leuciscus atarius*, the largest weighing one and one-fourth pounds. This list gives a fair index to the relative abundance of the larger fishes of the lake. The 'Sucker' and 'Webug' are, however, at times proportionately more abundant."

Tanner, who exhaustively studied the sucker population of Utah Lake in later years, noted the much changed environmental and taxonomic picture of his day:

"At this writing Jan. 1936 practically all the Suckers as well as other fish in Utah Lake have been killed by the severe drought of the past four years. The surface of Utah Lake has been reduced from a normal surface area of 93,000 acres to about 50,000 acres. During the winter of 1934-35 the water was so shallow that hundreds of tons of suckers and carp were killed due to freezing and crowding in the few deep holes. They are so completely depleted that the commercial fishermen have had to abandon all fishing. In the spring of 1935 there were no suckers to run up Provo River, something that has never happened before in the history of Utah Lake. Some fishermen have proposed bringing in suckers from Idaho and restock the lake. It is hoped that our State Game Department will not permit this" (1936: 167).

As shown by some of the above comments, this sucker was considered a problem to early commercial trout seiners in Utah Lake, but probably proportionately no more so than in similar situations elsewhere. The trout (*Salmo clarki utah*) also existed in tremendous numbers at least before the advent of commercial fishing which is clear proof that without man to upset the natural balance, the eating of trout eggs by suckers was of no consequence. This sucker, evidently like many other kinds of rough fish, was also utilized as food by the early settlers and certainly to some extent by the Indians as well. Otherwise it seems to occupy the usual place in the food chain for its kind.

By Lowder's time (1951), the Utah Lake sucker population had increased considerably from its drought low of nearly 20 years before and was the basis for commercial seining in spite of increasing human pollution of its waters.

This sucker has appeared in the bait fish industry of southern Nevada, these specimens apparently all coming from Utah. (Miller 1952: 25).

FLANNELMOUTH SUCKER

Catostomus latipinnis Baird and Girard

Catostomus latipinnis Baird and Girard 1853B: 388
Catostomus latipinnis, Günther, 1868: 14–15
Catostomus discobolus, Cope 1872B: 435
Catostomus latipinnis, Jordan, 1878A: 178–179 / 1891: 26
Catostomus discobolus, Jordan, 1878A: 179–180
Catostomus latipinnis, Jordan and Copeland, 1878: 416
Catostomus latipinnis, Jordan and Gilbert, 1882: 125–126
Catostomus latipinnis, Evermann and Rutter, 1895: 481
Catostomus latipinnis, Jordan and Evermann, 1896B: 174–175 / 1902: 47–48
Catostomus latipinnis, Gilbert and Schofield, 1898: 489–490
Catostomus latipinnis, Pratt, 1923: 56
Catostomus latipinnis, Jordan, Evermann and Clark, 1930: 107
Catostomus latipinnis, Schrenkeisen, 1938: 86
Catostomus latipinnis discobolus, Hubbs, Hubbs and Johnson, 1943: 60–62
Catostomus latipinnis discobolus, Simon, 1951: 58–59
Catostomus latipinnis, Beckman, 1952: 33–34
Catostomus latipinnis, La Rivers, 1952: 96
Catostomus latipinnis, La Rivers and Trelease, 1952: 115
Catostomus latipinnis, Miller, 1952: 13, 26
Catostomus latipinnis, Koster, 1957: 43
Catostomus latipinnis, Eddy, 1957: 77
Catostomus latipinnis, Moore, 1957: 89
Catostomus latipinnis, Shapovalov, Dill and Cordone, 1959: 171

ORIGINAL DESCRIPTION

"General shape subfusiform; head proportionally small, contained five times and a half in the total length. Eyes small, situated near the upper surface of the head; the mouth is small, the lips large and fleshy. All the fins are very much developed and constitute a very prominent feature. The upper margin of the dorsal is slightly concave; the posterior margin of the caudal, crescent shaped; the anal, ventrals and pectorals are posteriorly rounded or subconical.

"D I. 14. A II. 8. C 5. I. i. i. I. 6. V 10. P 18.

"The scales are of medium size, considerably smaller on the back than on the sides and belly. The lateral line runs through the middle of the sides from head to tail.

"The upper part of the body is reddish brown; the upper part of tail and sides, greenish brown; the belly, yellowish orange; the caudal is olive; the anal, ventrals, and pectorals, show traces of deep orange, especially on their outer margin.

"Rio San Pedro, of the Rio Gila" (Baird and Girard 1853B: 388).

DIAGNOSIS

Body slim, streamlined; head about 4.75 times, depth about 5.25 times, in body length; head medium in size, slender, snout prominent;

fontanelle present; mouth medium, with greatly thickened, strongly developed lips, the lower lip incised completely, only one row at most, of papillae, crossing the median line—hind edge of lower lip reaching posteriorly to a point opposite the eye; a fine-scaled sucker, scales 17/98–105/17; fins remarkably elongated—dorsal rays 11–13, usually the former—caudal very prominent, its rudimentary base rays markedly developed; strongly forked; caudal peduncle excessively slender, least depth about 16½ times in body length; color—deep olive on back, contrastingly paler beneath, with flanks and fins orange-tinted; maximum lengths and weights to about 2 feet and 8 pounds, respectively.

TYPE LOCALITY

San Pedro River in the Gila Basin of Arizona.

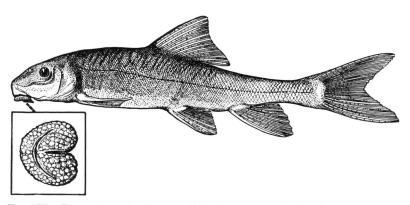

FIG. 175. Flannelmouth Sucker, *Catostomus latipinnis*. Drawn by William L. Brudon for Miller 1952. Courtesy California Department of Fish and Game and Leo Shapovalov.

RANGE

The Colorado River and its tributaries and the Gila River drainage of southern Arizona.

TAXONOMY

The unique distinctiveness of this sucker, with its slim body, enlarged fins and very prominent lips has probably contributed to its nomenclatural stability—it would be difficult to mistake this species for any other sucker.

Etymology—"latipinnis," or broad fin.

LIFE HISTORY

This is a streamlined, swift-water sucker well adapted to the turbulence of the Colorado River system. The species runs upstream to spawn in the spring and is characteristically herbivorous in food habits. Hubbs, Hubbs and Johnson have recorded hybridization between this sucker and a Mountainsucker, *Pantosteus delphinus utahensis* in the Virgin River of Utah (1943: 60–62).

ECONOMICS

Because of its large size, the sucker was originally an important fish food for Indians living along the Colorado River and some of its tributaries. The perennial question of rough fish damage on trout spawning grounds has also involved the Flannelmouth Sucker, particularly since the species ascends swift water on its spawning runs and very often spawns on the same beds as do the trout. However, its typically strong and sucker-like herbivorous feeding patterns preclude anything but accidental or minor local feeding on trout eggs and then undoubtedly only during the concomittant excitement of spawning.

TAHOE SUCKER

Catostomus tahoensis Gill and Jordan

(Nevada Sucker)

Acomus generosus, Cooper, in Cronise, 1868: 495 (misidentification) (Lake Tahoe)
Catostomus tahoensis Gill and Jordan 1878: 173–174 (in Jordan 1878A) (Lake Tahoe)
Catostomus araeopus Jordan and Henshaw 1878: 188–189 (Carson River)
Catostomus tahoensis, ibid: 188 (Lake Tahoe)
Catostomus araeopus, Jordan, 1878A: 173
Catostomus tahoensis, Jordan and Copeland, 1878: 416
Catostomus tahoensis, Jordan and Gilbert, 1882: 127
Catostomus tahoensis, Cope, 1883: 134, 152 (Pyramid Lake)
Catostomus Tahoensis, Russell, 1885: 62 (Pyramid Lake)
Catostomus tahoensis, Eigenmann and Eigenmann, 1891: 1132
Catostomus araeopus, Gilbert, 1893: 229 (Reese River)
Catostomus tahoensis, Jordan and Evermann, 1896B: 177
Pantosteus araeopus, Jordan and Evermann, 1902: 46
Catostomus tahoensis, Rutter, 1903: 147 / 1908: 118, 121–122
Chasmistes chamberlaini Rutter 1903: 147
Catostomus tahoensis, Snyder, 1908: 83 / 1916: 580 / 1917A: 42–47
Catostomus arenarius, Snyder, 1917A: 47–49 / 1917B: 202–203
Catostomus tahoensis, Pratt, 1923: 56
Catostomus tahoensis, Jordan, Evermann and Clark, 1930: 107
Catostomus tahoensis, Hubbs and Schultz, 1932: 7
Catostomus tahoensis, Schrenkeisen, 1938: 86
Catostomus tahoensis, Murphy, 1941: 167
Catostomus arenarius, Shapovalov, 1941: 443
Catostomus tahoensis, Hubbs, Hubbs and Johnson, 1943: 54–58
Catostomus tahoensis, Miller 1946A: 49
Catostomus tahoensis, Hubbs and Miller, 1948A: 22, 24 / 1948B: various mentions / 1951: 299–300
Catostomus tahoensis, Kimsey, 1950: 438 / 1954: 399, 408
Catostomus tahoensis, Shapovalov and Dill, 1950: 386
Catostomus arenarius, ibid
Catostomus arenarius, Hubbs and Miller, 1951: 299–300

Catostomus tahoensis, Lowder, 1951: 14–15
Catostomus tahoensis, La Rivers, 1952: 96
Catostomus tahoensis, La Rivers and Trelease, 1952: 115
Catostomus arenarius, ibid
Catostomus arenarius, Böhlke, 1953: 27
Catostomus tahoensis, Eddy, 1957: 78
Catostomus tahoensis, Moore, 1957: 89
Catostomus tahoensis, Shapovalov, Dill and Cordone, 1959: 171

ORIGINAL DESCRIPTION

"*g.* Body shorter than in the next [*longirostris*], but still elongated, its greatest depth 4½ to 5 in length; head very large and long-acuminate, the muzzle nearly one-half its length, overhanging the rather large mouth: lips moderate; the upper pendent, with about 3 rows of small papillae; the lower rather full, similarly papillose: eye nearly median, rather small, 8½ in head: scales small and crowded forwards, closely imbricated, 83 to 87 in the course of the lateral line and about

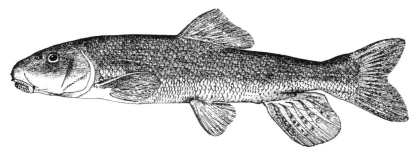

FIG. 176. Tahoe Sucker, *Catostomus tahoensis*. Drawn by Silvio Santina.

28 in a cross-series from dorsal to ventrals: coloration very dark; fins dusky; scales everywhere punctate. Size large . . . TAHOENSIS, 32." (Jordan 1878A: 160–161, in a key) . . .

"Habitat.—Lake Tahoe, Nevada.

"The Sucker of Lake Tahoe is closely related to *Catostomus longirostris*, but seems to differ constantly in the shorter head and more contracted body. It is said to be very abundant in Lake Tahoe. 'They are caught in nets and sometimes with hook, but like all this family are rather poor as food' (Cooper). *Acomus generosus* of Girard, with which this species has been identified, is a very different species, belonging to a different genus" (Jordan 1878A: 173–174).

DIAGNOSIS

Head about 4 times, depth about 5 times, in body length; head large, snout prominent, about half the head length; fontanelle well-developed; mouth large, with medium-sized lips, the lower lip deeply incised so that only one row of papillae, at most, crosses the median line; a fine-scaled sucker, scales 16–19/82–95/12–15, 40–50 before dorsal fin; fins —dorsal 10–11 rayed—anal 7 rayed, rarely 8—pectoral 14–16 rayed— caudal moderately forked; caudal peduncle thick, least depth about 12 times into body length; color—dark above, light below, breeding males with a bright red lateral line; grow to about 2 feet in length in the larger lakes.

TYPE LOCALITY

Lake Tahoe, California (Cooper).

RANGE

The Lahontan drainage system of west-central Nevada and adjacent California—the Walker, Carson, Truckee, Susan and Humboldt Rivers; Tahoe, Pyramid and Walker Lakes; and tributary streams.

TAXONOMY

The first detectable mention of this species was a notation by Cooper in 1868 in a chapter he contributed to Cronise's book, "The Natural Wealth of California." Cooper collected specimens in Lake Tahoe and erroneously attributed them to the *Catostomus generosus* of Girard. Theodore Gill had occasion to compare the Cooper specimens with Girard's U. S. National Museum types shortly afterward, noted the differences, and suggested the name of *Catostomus tahoensis* in an unpublished manuscript. Gill and Jordan formally characterized the species in 1878 in a bulletin of which Jordan was the author, the description being attributed to "Gill and Jordan." In the same publication, Jordan described *C. araeopus* from Kern River, California also listing "young specimens" of *araeopus* from the Carson River, Nevada; the young specimens were typical *C. tahoensis*, while *C. araeopus* later proved to be a synonym of the more widespread *C. occidentalis* of Ayres.

The *Chasmistes chamberlaini* of Rutter was shown by Snyder (1917A: 43) to be merely a dried head of *Catostomus tahoensis*. At the same time, Snyder described *Catostomus arenarius* from Pyramid Lake, Nevada (1917A: 47–49), allying it with such coarse-scaled species as *C. macrocheilus* and *C. occidentalis*, thus providing the Lahontan system with a coarse-scaled and a fine-scaled form (*C. tahoensis*) such as certain of the adjacent drainages were supposed to have (Klamath, Columbia and Sacramento). In 1951, Hubbs and Miller decided, after examining the types, that the species was a synonym of *C. tahoensis*, as had long been suspected.

There seems to have been some uncertainty in Jordan's mind about the original citation of *C. tahoensis* for in the synonymicon given for the new species on p. 173 (1878A), he lists a supposedly earlier description which, as far as I have been able to determine, is only imaginary—

"1878—*Catostomus tahoensis* Gill and Jordan, Bull. U. S. Nat. Mus. xi, p. ___."

Jordan and subsequent authors have designated the pagination of the original description as 173–174, but the inadequacy of this as a characterization is evident from the quotation above. Actually, the only acceptable description was that given as part of the key to Catostomi, also fully quoted above, on pages 160–161.

LIFE HISTORY

The best known of any of the Lahontan catostomids. A late spring-early summer spawner, and the commonest herbivorous fish in the streams; in the larger lakes, the Tahoe Sucker is almost as abundant as Chub (*Siphateles*). Spawning occurs progressively later in the season with increase in altitude and occupancy of colder waters. The following comments are those of Snyder (1917A: 43), who made the most extensive observations on this species of any investigator, either before or since his time:

"The males appear first on the spawning beds and are always represented there in large numbers, each female being attended by from two to eight or more. Twenty-five males were seen attending one female in a pool. Occasionally another female would enter the pool from below, when she would be met and inspected by a school of males and then allowed to pass on without further notice. Several of these passing females proved on examination not to be ripe. On account of the presence of so many males nothing definite can be observed of the spawning act, more than that the eggs are extruded and shaken down in the gravel by the female while the males struggle over and under her, churning the water to foam by their activities. Eggs artificially removed from a ripe female and quietly cast into the water upstream from the males attracted no more attention than did so much coarse sand. However, they were immediately gobbled up by numbers of *Richardsonius egregius*, which also attended the females, plunging into the melée of spawning fishes for eggs at every opportunity.

"The bed, or nest, is a somewhat concave depression in the coarse sand or gravel, measuring from 1½ to more than 3 feet in diameter. The nest is located in shallow water, usually less than 12 inches deep, which often proves fatal to the young, for the falling water of the river uncovers the beds at times, and the eggs quickly perish in the hot sun. The eggs are found in large numbers deep among the pebbles. In spawning there is no opportunity for the female to make any selection among the males. Large and small males appear to have an equal opportunity in fertilizing the eggs, for no fighting occurs.

"Spawning was in progress in the lower Truckee River April 22; Pyramid Lake May 20; ceased May 24; Eagle Lake May 25, the period about ended there; in tributaries of Lake Tahoe June 11.

"Individuals of this species are very shy, the females being more difficult to catch than the males. When spawning they may be closely approached if one moves very slowly without producing any crunching of the gravel underfoot or allowing a shadow to fall on the water.

"In the lakes this species attains a large size, one specimen measuring a little over 2 feet. The Indians call them 'auwa-go,' or 'a-wuh,' and occasionally catch them along with *Chasmistes*, but reject them as being undesirable for food. The flesh is sweet and palatable."

In our investigations in Pyramid Lake and vicinity, Thomas J. Trelease and the writer have found large schools of smaller spawning individuals (12 inches and less) ascending the river at its mouth during night time—on such occasions, at various times in April, migrating specimens were so numerous as to be easily scooped out of the water by hand. The fact that such runs were composed of small fish can be attributed to low water in the river—not enough of an inflow to allow larger fish to get out of the lake. These schools also run during overcast days. Our meager temperature data of 54° F. to 57° F. (12.2 C.

to 13.8 C.) give some indication of the spawning temperatures. About a week after migratory runs become common at the mouth of the Truckee River, spawning is in process in the vicinity of Reno, about 50 miles up-river. These river populations use the same gravel beds that Rainbow Trout had spawned on several months previously.

In the latter part of May, 1952, the author found moderate numbers of suckers 8 to 14 inches long working their way up a tiny stream coming into the southwest edge of Pyramid Lake north of Camp Foster.

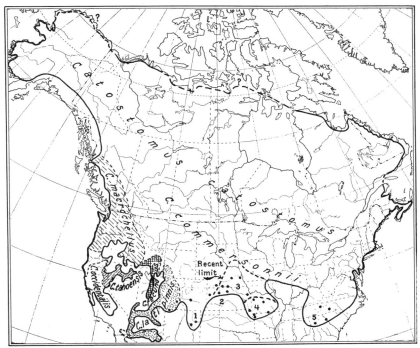

FIG. 177. Distribution of suckers, *Catostomus* spp. The range of *Catostomus columbianus* would be included in that shown for *C. macrocheilus*, and *C. ardens* would lie within the distribution shown for *C. fecundus*. Courtesy of University of California Press and Erhard Rostlund.

The stream gradient was quite steep with a small boulder bottom—looking more like a lamprey environment than a sucker spawning bed, and some individuals had achieved a distance of about a hundred yards from the lake and 30 feet above its level.

While predominantly herbivorous, the Tahoe Sucker takes in considerable amounts of animal protein in its feeding, as attested by examinations of stomach contents. Some stomachs contain as many fragments of Gammaridae (Crustacea: Amphipoda) and Tendipedidae (Insecta: Diptera) as they do algae remains. In brackish bodies of water such as Pyramid and Walker Lakes (where insects, as a group, are conspicuous by their absence), algae growing on rocks support large populations of micro-crustacea (Ostracoda, Copepoda and Cladocera), examples of which commonly appear in sucker stomachs; in streams such as the Truckee River the large insect fauna no doubt replaces the micro-crustacea as the most significant protein element in fish diets.

Cal Allan has observed spawning in Walker Lake during June. No redds were constructed, the fish spawning indiscriminately over coarse gravel.

Hubbs, Hubbs and Johnson (1943: 54–58) have recorded hybridization between *Catostomus tahoensis* and *Pantosteus lahontan* in west-central Nevada and adjacent California waters. In a relative numbers chart, they list 3,282 *C. tahoensis*, 2,007 *P. lahontan,* and 28 hybrids between the two; one-third of the *C. tahoensis* were collected in habitats from which *P. lahontan* was absent. Specimens were taken mainly in the Carson, Humboldt and Little Truckee Rivers, and showed typical intermediate characteristics in nearly a dozen measured categories.

ECONOMICS

Because of its abundance in the Lahontan system generally, this species is one of the basic elements in the aquatic food chain, representing, probably, the greatest amount of transformation of plant materials into animal proteins that occurs among our native fishes. Originally, this must have been important to the large resident populations of stream Cutthroat Trout, as well as lacustrine trout, both in the lakes and returning there after exhausting spawning runs.

While the Pahute Indians may not, from such testimony as we have today, have originally paid much attention to the Tahoe Sucker as a food fish, this could only have been because of the tremendous abundance of both Cutthroat Trout and the Cui-ui Sucker, for the Tahoe Sucker is a good-tasting and easily prepared fish, especially the larger individuals. At the present time, the Tahoe Sucker plays no direct part in man's economy.

RAZORBACK SUCKERS
Genus *Xyrauchen* Eigenmann and Kirsch 1888
("Razor nape")

RAZORBACK SUCKER[1]
Xyrauchen texanus (Abbott)
(Humpback Sucker)

Catostomus texanus Abbott 1860: 473
Catostomus texanus, Günther, 1868: 12
Catostomus cypho Lockington 1881: 237–240

---

[1] As good an argument can be made for use of the old name "Razorback," as for "Humpback." "Razorback" is more illusory, but certainly more distinctive, and avoids the fallacy that this is a deformed species. Where a physical characteristic is genetically fixed in a naturally-occurring population, it cannot be spoken of biologically as a malformation, which the term "humpback" strongly infers. In addition, "razor" is a more accurate translation of the generic name than is "hump."

Catostomus cypho, Jordan and Gilbert, 1882: 129
Xyrauchen cypho, Jordan, 1891: 26
Xyrauchen uncompahgre, ibid: 26–27
Xyrauchen cypho, Evermann and Rutter, 1895: 482
Xyrauchen cypho, Jordan and Evermann, 1896B: 184 / 1902: 57–58
Xyrauchen uncompahgre, ibid, 1896B: 184–185
Xyrauchen cypho, Gilbert and Schofield, 1898: 491–492
Xyrauchen texanus, Snyder, 1915: 579–580
Xyrauchen cypho, Pratt, 1923: 59
Xyrauchen texanus, Jordan, Evermann and Clark, 1930: 108
Xyrauchen texanus, Tanner, 1936: 159, 167–168
Xyrauchen texanus, Schrenkeisen, 1938: 89–90
Xyrauchen texanus, Moffett, 1942: 82 / 1943: 182
Xyrauchen texanus, Dill, 1944: 150–151
Xyrauchen texanus, Miller, 1946B: 410
Xyrauchen texanus, Shapovalov and Dill, 1950: 386
Xyrauchen texanus, Simon, 1951: 62
Xyrauchen texanus, Wallis, 1951: 89
Xyrauchen texanus, Beckman, 1952: 36–37
Xyrauchen texanus, Douglas, 1952
Xyrauchen texanus, La Rivers, 1952: 95
Xyrauchen texanus, La Rivers and Trelease, 1952: 116
Xyrauchen texanus, Eddy, 1957: 69
Xyrauchen texanus, Moore, 1957: 90
Xyrauchen texanus, Shapovalov, Dill and Cordone, 1959: 171

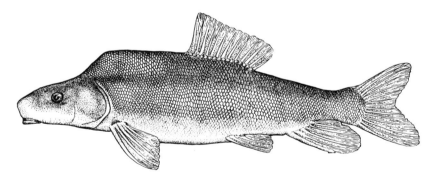

FIG. 178. Razorback Sucker, *Xyrauchen texanus*.
Drawn by Silvio Santina.

ORIGINAL DESCRIPTION

"*Spec. char.* Head somewhat compressed, large, constituting somewhat more than one-fourth of the total length. Eye small, longitudinally oval; its longitudinal diameter constituting one-twelfth of the length of the side of the head. Mouth large, with the labial papillae moderately developed. Body moderately compressed; a dorsal gibbosity extends from the occiput, attaining its greatest height an inch from the occiput, and disappearing at the anterior insertion of the dorsal fin; it is carinated throughout its whole extent. Dorsal fin one-third longer than high; its base enters five and a half times in the total length; its anterior margin equidistant between the base of the caudal and the extremity of the snout. The insertion of the ventrals is opposite the centre of the dorsal fin, and much nearer the base of the caudal than the extremity of the snout. The posterior extremity of the anal fin extends beyond the rudimentary rays of the caudal. The scales are of medium size, with a subcentric nucleus near the anterior margins of their free portions, from which radiate numerous striae, and around which are numerous well defined ridges. The lateral line is nearly straight throughout its course.

"The numbers of the fin-rays are D, 15. P, 16. V, 10. A, 7. C, 18 5/5.

"*Color.* Upper surface of the head, back, and sides, a dull slate color; belly white (not silvery). Throat yellow.

"Total length, 14 inches.

"*Habitat.* Colorado and New rivers.

"I am indebted to Dr. John L. LeConte, for a note containing a description of this fish, noticing many peculiarities which the specimen (a stuffed one) does not now exhibit" (Abbott 1860: 473).

DIAGNOSIS

Head and body depth each about 4 times in body length; fontanelle well-developed; mouth rather large, ventral; lips small—upper lip thin and only weakly papillose—lower lip smooth and nonpapillose (at least in adults), completely divided medianly into two smooth, rounded flaps separated by a "V"-shaped incision which is widest at the forward or leading edge of the mandible; a distinctive nuchal hump or elevation immediately behind the head, formed by the greatly developed interneural bones—this enlargement of the nape bone structure produces the hump which is so distinctive of the older specimens; a medium-scaled sucker, scales about 13–16/73–81/13, 60 before dorsal fin; fins—dorsal 12–15 rayed, low and long—anal 7-rayed—caudal large and powerful; caudal peduncle stout, its least depth about 11 times into body length, slimmer in young specimens; peritoneum black; color—dusky above and laterally, yellowish-white below; reaches lengths of over 2 feet and weights of about 10 pounds.

TYPE LOCALITY

Colorado and New Rivers, Arizona.

RANGE

Colorado River and Gila River basins.

TAXONOMY

The original description by Abbott in 1860 was overlooked for nearly 70 years, the Razorback Sucker bearing the synonym *X. cypho* of Lockington, 1881, for most of that period. The synonymy has never been complex, Jordan and Evermann's *X. uncompahgre*, 1891, being the only other name proposed, and not used for long. The origin of the specific name *texanus* is obscure, since the species does not range into that state.

LIFE HISTORY

The Razorback Sucker, like the Flannelmouth, is built to stem river waters. In addition to the streamlining of form, the peculiar formation of the vertebrae behind the head is such that this part of the back is lifted and knife-edged to form a dorsal crest from which the common name is derived. This, with the long, flat, sloping head, undoubtedly steadies the fish against the bottom in currents where the water has a tendency to push down on the anterior part of the body while the dorsal keel provides increased stability when faced into the current. It also does well in Lakes Mead and Mohave.

This sucker is a spring spawner, and, as would be expected, attains greater lengths and weights in the lower, warmer and more sluggish parts of the Colorado River system than in the headwaters. The food of the species consists of vegetable matter and material sifted from bottom ooze.

Hybrids have been noted between *Xyrauchen texanus* and both *Catostomus insignis* and *C. latipinnis* by Hubbs and Miller (1953).

Douglas (1952) has given us the most extensive observations of which I am aware on the spawning behavior of this species:

"SPAWNING LOCALE

"The general area where the humpback sucker was observed in spawning activities is at the Needles Boat Landing, located in the northwest corner of Lake Havasu [on the Colorado River, San Bernardino County, California, below Davis Dam]. Observations were made from the shore of a small bay south of the landing and its adjacent point.

"The bottom of the bay is silt over sand intermixed with a smattering of boulders and some gravel spotted throughout the shallow areas. Numerous depressions over a foot in diameter were seen and appeared to be largemouth black bass (*Micropterus salmoides*) nests. Aquatic vegetation consists of small clumps of blue-green algae growing on the boulders and bottom, and bullrush (*Scirpus* sp.) patches along the marginal areas at the head of the bay. Tamarisk (*Tamarix gallica*) adds considerable shade to the littoral zone in the apex of the bay.

"The bottom adjacent to the point to the south of this bay consists of a three-foot wide silt strip, located about 25 feet offshore, bordered by rubble both lakeward and shoreward. No aquatic vegetation was found in this area and the beach is open.

"SPAWNING OBSERVATIONS

"On March 2, 1950, Warden Leo Rossier and the writer noted spawning humpback suckers and carp (*Cyprinus carpio*) in the littoral area of the small bay and point south of Needles Boat Landing. Two suckers, both ripe males exuding milt, were seined. Their lengths were 21.2 and 21.3 inches and their weights 4.8 and 5.5 pounds, respectively.[2] They were in brilliant breeding coloration. The dorsal and lateral integument

---

"[2]All length measurements are to fork of caudal fin."

was black to a point about one inch below the lateral line, with a brilliant orange coloration extending ventrad from this point. Limited time prevented further observations on this date.

"On March 15th at 3:30 p.m. two groups of suckers were seen working their way along the marginal areas in the apex of the bay south of the landing approximately 15 feet from shore, in about 18 inches of water. In group one, six fish were noted revolving clockwise at a slow rate in four-foot diameter circles. Suddenly the caudal fins of all fish began to vibrate violently and one fish jumped clear of the water. It appeared that five of the fish, presumably males, were exerting pressure on one fish, presumably a female; three were crowding together on the female's right side and two on her left. The two most proximate males were pressing against the sides of the female with their heads just behind her hump. Ensuing violent motion raised bottom silt so that further observations were prohibited. Occasional water agitation was noted during this one and one-half minute period of activity and the location of the fish could easily be determined by the cloud of suspended matter raised from their violent motions. Following the spawning act the fish separated and moved away singly, two into deeper water and the other four parallel to shore.

"The second group consisted of three fish lying in a depressed pocket (possibly a largemouth black bass nest) about three feet from shore in 10 inches of water. Caudal fins of the two outer suckers, presumably males, began vibrating and a column of water about five inches high by seven inches in diameter was thrown into the air. Again bottom silt occluded the actual spawning act. After about one and three-quarters minutes of activity, the fish separated. A marine plankton net was towed through the immediate area, but no spawn was recovered.

"Within one-half hour 48 suckers between 20 and 30 inches long were counted passing through this area. Carp were intermingling with the suckers and ranged in length between 10 and 15 inches. There was a slight wave action in the bay area, due to an easterly breeze.

"On March 16th a strong east wind created wave action, preventing detailed observations of fish movements. However, a small school of suckers was located in the same area as of the previous afternoon, but no tight grouping was noted. A set was made with a standard sampling gill net at 10:15 a.m. and run at 12 m. Four suckers were captured, and extruded milt when removed from the net. They ranged from 4.8 to 5.6 pounds and 21.2 to 23.1 inches.

"The surface was calm by noon but no spawning groups were seen. A few fish were jumping in the center of the bay. The gill net was run several times during the afternoon in an attempt to capture a gravid female sucker. Five suckers ranging from 4.3 to 5.5 pounds and 19.5 to 21.3 inches were caught, but all were ripe males.

"At 3:05 p.m. some grouping of the suckers in the apex of the bay was noted, but no actual spawning appeared to be taking place. A slight surface ripple made observations somewhat difficult at this time. Two plankton tows were taken with the marine net. One complete tow consisted of a 40-foot drag each way. An area over the spawning grounds 15 feet by 40 feet was covered. All depths were sampled. No spawn was recovered, but one larval humpback sucker, 0.4 inch in total length, was captured (identification by Dr. Carl L. Hubbs).

"On the morning of March 17th the bay area was watched from the previous location, but no suckers were in evidence. An overnight net set was run and only seven ripe carp and three bluegill (*Lepomis macrochirus*) were caught. By 10 a.m. it appeared that no suckers were moving into this bay area, so the observation station was moved to the first point south of the boat landing, where suckers had been collected on March 2d. The water surface was calm and 27 suckers were clearly visible milling about just offshore.

"One group, consisting of two males and one female (presumably) between them, was noted moving in a counter-clockwise circle about three feet in diameter. The males appeared to be 'herding' the female with their heads and humps, nudging the female in the genital region. After about 30 seconds of this circular movement all three lay on the bottom on their ventral surfaces. The males' caudal fins vibrated rapidly and appeared to be slapping the female in the genital region. At this time this group was in two feet of water approximately 25 feet from the shore, over silt bottom. Silt erupted in this activity and further observation was impaired. The silt cloud continued in suspension for three minutes before the fish were seen to separate and move away. This action was noted between 10:05 and 10:13 a.m.

"A second group, consisting of three males and one female (presumably), was observed moving along the silt-bottomed strip between 10:15 and 10:17 a.m. The female escaped to deeper water and no spawning activity was noted. The female appeared lighter in color, being more brown than black on the dorsal side, but of the same size as the males. At 10:20 a.m. wave action from passing motor boats dispersed the groups and only single fish were seen.

"By 10:45 a.m. only seven suckers were observed in the shallow area, though several silt 'eruptions' were noted 50 feet offshore in 4.5 to 6 feet of water, and apparently some spawning activities were taking place there. A silt cloud was observed about 50 yards south of the observation station and on moving to this point one female (presumably) and three males were seen resting on the bottom, after the silt had settled and the bubbles had cleared. In about four minutes, all had dispersed and gone to deeper water" (pp. 150–154).

ECONOMICS

In addition to its usual importance as a basic converter of plant tissues to animal protein (see *Catostomus tahoensis*), this species seems to have been one of the staple fish foods of Indian tribes living along the Colorado River.

LAKESUCKERS

Genus *Chasmistes* Jordan 1878H

("One who yawns")

CUI-UI[1] LAKESUCKER
Chasmistes cujus Cope

Chasmistes cujus Cope 1883: 149
Chasmistes cujus, Russell, 1885: 62
Chasmistes cujus, Jordan and Evermann, 1896B: 183
Chasmistes, Gilbert, 1896: 1
Chasmistes cujus, Seale, 1897: 269
Lipomyzon (Pithecomyzon) cujus, Fowler, 1913: 54
Chasmistes cujus, Snyder, 1917A: 50-54
Chasmistes cujus, Pratt, 1923: 58
Chasmistes cujus, Jordan, Evermann and Clark, 1930: 108
Chasmistes cujus, Hutchinson, 1937: 54-55
Chasmistes cujus, Schrenkeisen, 1938: 88
Chasmistes cujus, Sumner, 1940: 222
Chasmistes cujus, Alcorn, 1943: 35
Chasmistes cujus, Hubbs and Miller, 1948B: 27
Chasmistes cujus, Trelease, 1949: 9-10, 23, 31
Chasmistes cujus, Lowder, 1951: 11, 13
Chasmistes cujus, La Rivers, 1952: 95
Chasmistes cujus, La Rivers and Trelease, 1952: 115
Chasmistes cujus, Eddy, 1957: 75
Chasmistes cujus, Moore, 1957: 87
Chasmistes cujus, Miller, 1961: 385-386

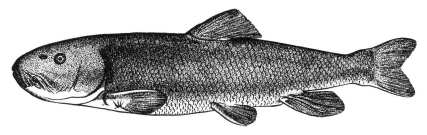

FIG. 179. Cui-ui Lakesucker, *Chasmistes cujus*. Drawn by Silvio Santina.

ORIGINAL DESCRIPTION

"I procured but one specimen of this fish from Pyramid Lake, where it is difficult to obtain. The size is large; the specimen I procured measured eighteen inches in length. The head is wide and flat, the width of the interorbital space being more than half the length. The upper lip is very thin; the lower lip is represented by folds on each side, which do not connect round the symphysis. Scales 13-65-11. Dorsal rays 12; anal I. 8. The eye enters the length of the head 8.5 times, and the interorbital width 4.5 times. The swim-bladder has but two cells. The colors are pale olive.

"The pharyngeal teeth of this species are much like those of the

---

[1]Most commonly rendered as "kwee-wee," but the Pahute pronunciation is closest to "Koo-ee-wee." Also spelled "Couia."

C. liorus in their triangular section; they are, nevertheless, of delicate construction. The head of this species is relatively larger and wider than in any of the others, which gives it a heavy and clumsy appearance.

"This fish is said by the fishermen to inhabit the deepest water, and to be seen in numbers only at the time of breeding. Its habits in this respect agree with what is said of the *C. luxatus* of the Klamath Lake. The Indian name of the *Chasmistes cujus* is 'Couia'." (Cope 1883: 149).

DIAGNOSIS

A plump, robust species; head from 3.4 to 4.0 times in body length, depth from 4 to 5 times in body length; fontanelle well-developed; head very large, blunt, eye very small proportionately; mouth ventro-terminal, un-sucker like; lips thin, obscurely papillose, the lower lip somewhat pendant; a coarse-scaled sucker, scales 13–14/59–66/10–12, 28–35 before the dorsal fin; fins—dorsal 10–12 rayed, small—anal 7-rayed, rarely 8—caudal weak, moderately forked; caudal peduncle thick, least depth about 12 times in body length; peritoneum nearly black; color—black-brown above, more broken laterally, fading into flat white on the venter—breeding males develop a reddish color on the sides—fresh specimens, especially females, with a decided bluish-gray cast; attain lengths of about 2 feet and weights of about 6 pounds.

TYPE LOCALITY AND RANGE

Pyramid Lake, Nevada, ascending the affluent Truckee River only during the spring spawning season (April-June).[2]

TAXONOMY

Completely stable, as indicated by the synonymicon. With its restricted distribution and distinctive appearance, it has not been possible for anyone to confuse this species with any other fish, or to attempt its subdivision on trivial characters.

The genus *Chasmistes* itself is limited in distribution, possessing species in three drainage systems which, while adjacent to one another, have disrupted the continuity of the genus by their loss of interconnecting waterways. Three species have been described in the Klamath Lakes of southeastern Oregon (which badly need modern re-evaluation), one in Utah Lake in the Bonneville system, and *C. cujus* of the Lahontan system's Pyramid Lake. All are lacustrine and, in most cases, confined to one lake. Because of the importance of the genus in Great Basin chronologies, particularly in the Lahontan system, some of its taxonomic characteristics will be pertinent here.

Jordan first characterized the genus in the briefest fashion in 1878H: 417 as:

---

[2]There is a fantasy, persistently current, that the cui-ui has only one other close relative in the world. Depending upon the temperament of the individual responsible at the moment, this takes the form of a species in far off romantic Peru, or one in far off romantic India, or one in far off romantic Egypt. Some of these people also believe there is a sea monster in Pyramid Lake, or that it is 1,200 feet deep, or that it has an underground drain to the Pacific Ocean.

"21. ³CHASMISTES Jordan, *gen. nov.* 1878. Big-mouthed Suckers."
All of the species currently assigned to *Chasmistes* were described between 1878 and 1898: *C. liorus* Jordan 1878 (Utah Lake), *C. brevirostris* Cope 1879 (Klamath Lake), *C. cujus* Cope 1883 (Pyramid Lake), *C. stomias* Gilbert 1898 (Upper Klamath Lake), and *C. copei* Evermann and Meek 1898 (Upper Klamath Lake). The *Chasmistes luxatus* of Cope, 1879, also described from Klamath Lake, was transferred to the new genus *Deltistes* by Seale in 1897, an entity still recognized by Jordan, Evermann and Clark in 1930. Not all of the above-listed *Chasmistes* are in good taxonomic repute, *C. liorus* in particular having a very uncertain status (see *Catostomus ardens*, p. 346).

FIG. 180. The author surrounded by stranded and dead cui-ui. These had started an upstream run at the mouth of the Truckee River and were almost immediately left high and dry by a drop in the river water. Fortunately for the survival of this unique species, when the schools cannot get upriver, they spawn in the inflowing fresh water at the mouth of the river. Photo taken by Thomas Trelease in May, 1940.

Two untenable names, the generic *Lipomyzon* Cope 1881 for *C. brevirostris*, and the subgeneric *Pithecomyzon* Fowler 1913 for *C. cujus*, have also been contributed to the synonymicon of the genus *Chasmistes*. The mistakenly referred *Chasmistes fecundus* of Jordan

---

"³This genus is distinguished from *Catostomus* by the very large, terminal mouth, the lower jaw being very strong, oblique, its length about one-third that of the head. The lips in *Chasmistes* are little developed, and are very nearly smooth. The type of the genus is *C. fecundus* Cope & Yarrow. It will be elsewhere fully characterized." (See comments under *Catostomus ardens*, p. 346, for the confusion surrounding the inception of the name *Chasmistes*.)

and *Chasmistes chamberlaini* of Rutter have already been noted (pp. 346 and 354).

The importance of *Chasmistes* in an understanding of drainage system affiliations within the Great Basin has been mentioned above; the specific contribution which a knowledge of the zoögeography of the genus makes toward a resolution of these problems is discussed in an introductory section (pp. 81, 82).

Etymology—"cujus" is assumed to be the latinization of the Indian name which Cope rendered as "Couia."

LIFE HISTORY

A large, omnivorous, spring-spawning lacustrine sucker with strong anadromous instincts, known only until recently from adult specimens. In spite of the local importance of the species as a food fish for the Indians, and its somewhat wider naive popularization as a "living fossil," Snyder (1917A) was the first to make public any pertinent information on its life cycle. In the 34 years that had elapsed between its description and Snyder's work, nothing had accumulated in print concerning its life history; and in the 43 years that have followed Snyder, little else has come to light. As it has been some years since the species has been able to ascend the Truckee River where spawning could be observed, the writer has no personal data on this phase of their cycle, so recourse must be had to Snyder's competent notations:

"Nothing seems to have been added to the brief original description of this species given by Cope, and until now but one specimen, the type, was preserved in any museum. Its distribution is restricted to Pyramid and Winnemucca Lakes [the latter now dry], where it lives in deep water beyond the reach of ocular observation, except during the brief spawning period, when a migration is made for a short distance up the Truckee River.

"The annual run begins about April 15, varying somewhat of late years with the condition of the river. The season of 1913 afforded an unusually good opportunity for observation, as the water was comparatively low and clear, while during the entire spring a reasonably steady flow into both Pyramid and Winnemucca Lakes was maintained.

"The first 'cui-ui' appeared in the river April 13, when several schools passed up rather hastily and lodged in pools below an impassable irrigation dam. This preliminary wave having passed, none was seen again until on the morning of April 22, when schools of 20, 30, or even 50 or more individuals were observed moving slowly and steadily upstream. It was customary for them to congregate and lie for a while below a rapid place, then suddenly and speedily shoot up, singly or in pairs or in small straggling schools, their brilliant red and brassy sides flashing in the bright sunshine. None of these stopped to spawn. Some which were dissected did not appear to be ripe. They were very shy and fled at once on the approach of a shadow, the jar of crunching gravel, or a heavy footfall; but the observer could come close if the move was steadily made. The passage of large numbers continued intermittently, until about May 16, when it became evident that the migration was waning rapidly. After May 11 none was seen moving upstream.

"In the meantime spawning had begun and was progressing with great activity. On April 24 the first females were seen depositing eggs. However, several ripe males and females were secured a little earlier. By May 5 every suitable bar or gravel bed was occupied by spawning fishes, whose activities entirely ceased before the 16th.

"The spawning is entirely suckerlike; it occurs in relatively shallow water where the flow is rapid, often at the head of a bar which turns or parts the current. At times the dorsal fins project above the surface, and in very shallow places where there is much crowding the whole back is exposed. Two, three, or even five or more males attend a single female during the spawning act. They wriggle over, alongside of, and around her, thrashing the water with such violence that close observation is impossible. Spawning fishes are easily alarmed, but if the observer approaches in the water he may occasionally get close enough to pick up specimens without difficulty. Eggs may be stripped and fertilized with ease.

"The ovaries are large, the eggs small and very abundant. No enemy appeared on the spawning beds, but the habit of depositing the eggs in shallow water often exacts an enormous toll from the young of the species, for a sudden fall in the volume of the river may leave many nests high and dry in a single day.

"No doubt the migration and spawning activities here described are fairly typical. Usually the water is so high, swift, and roily that very little of what is going on beneath its surface can be seen. Of late years irrigation projects and power plants have at times seriously interfered with the flow of the river and consequently disturbed the normal life of some of its native species. During the winter of 1911–12 the snow was very light in the mountains and there were no heavy rains. The dam at Tahoe was closed early, and a large amount of water was at the same time diverted from the channel of the Truckee above Derby. The lower part of the river then became so reduced that water began to flow back from Pyramid Lake (where it was higher than usual), up the river, down the slough, and into Winnemucca Lake, the surface of which is lower than that of Pyramid Lake. This flow continued until the water of the channel between the lakes was practically as brackish as that of the lakes themselves. No 'cui-ui' appeared in the river until a full month after the usual time, and then not until high water suddenly forced back the brackish flow and sent a fresh stream out into the lake. On the advent of this directing current the usual rush of 'cui-ui' from the lakes began; large schools passed up the river (May 17) and spawned at once. During the earlier back flow sufficient depth was maintained for easy passage of the fish, but it seems probable that there being no inflow of fresh water the waiting migrants were unable to find the mouth of the river.

"The time of departure of the fish from the river could not be determined because of high water, as no 'cui-ui' were seen at any time going down stream. A few individuals were seen in the river June 14, when the water suddenly cleared. On June 5, and for many days thereafter, large numbers of dying, dead, and decaying specimens were found at the mouth of Winnemucca Slough. This mortality among the 'cui-ui' is said to be a regular feature of the season at this place. If a similar death rate prevails in the lower Truckee, it was not evident at the time. However, the river was deep, the current strong, and the lake was stormy when the examination was made. A few dead individuals are always found along the river after the breeding season. It is possible and quite probable that the death rate is high just after the breeding season, but there is nothing to indicate that all the fish die after spawning. The dead and dying examples bore no evident scars.

"Diligent inquiry brought forth no account of the species spawning in the lakes. No one was found who had even seen one there. Hours of observation from tufa domes failed to detect any among the myriads of fishes which could be easily identified. Yet on May 11, 1913, large numbers of 'cui-ui' were found depositing eggs along the shallows near some springs on the southwest shore. Both ripe males and females were examined. The Indians were after them almost immediately, and they declared that these were the first that they had seen in the lake. None was observed here May 16 or later. A few individuals were found spawning in Winnemucca Slough.

"During the breeding season the males differ from the females in color and there is some variation in both sexes. The males have a dense black stripe 5 to 6 scales wide extending 10 scales below the dorsal fin from the opercle to the base of the caudal. The borders are interrupted here and there by brassy or silvery scales. Above the black stripe are numerous reddish-bronze scales with dark spots. The middle of the back is dusky, and this dark surface, together with the black stripe below, causes the red area to stand out boldly, especially when the fish turning in the water flashes the metallic red in the sunshine. Below the black stripe the body is silvery, many scales having a brassy sheen. The ventral surface is clean, dead white. The upper part of the head is blackish, the lower part whitish. The fins are slate blue, the paired ones lighter than the others. The tubercles are yellowish white. When the fishes are observed in the water from directly overhead, the stripes are very prominent, converging posteriorly and meeting over the caudal peduncle. The head appears lighter than the body, and the fins are very distinct. Some individuals are much duller, but in every specimen there is at least a strong trace of the red and dark stripes.

"In the female the whole upper surface is dark brownish-black, not the olive color usually seen in suckers. The sides are brassy, often more or less dull, and frequently the darker color is broken up into clouds on the sides. Occasionally the entire dorsal region of a female is tinged with a reddish coppery hue, the edges of the scales having a decided metallic luster. Sometimes the females are called black suckers. At times whole schools of both sexes were seen, apparently ready to spawn, but without a single individual with fully developed colors.

"Some Indians assert that they can distinguish between 'cui-ui' from the different lakes. Those from Winnemucca, said to be lighter in color and inclined to be spotted, are known as Izhi-'cui-ui.' The writer after examining many specimens from both lakes was unable to detect any difference.

"Observers differ somewhat as to the most distant point reached by *C. cujus* during the nuptial migration. It appears in large numbers at the great bend of the Truckee [Wadsworth], and it certainly ascends the river somewhat beyond the confines of the ancient Lake Lahontan. It never quite approaches the swift water above Reno. It would no doubt be a physical impossibility for the species to stem the turbulent water of the river canyon. The great blunt head and huge body, loaded down with eggs and fat, and the relatively small and weak caudal fin are not calculated to lend speed or endurance to a fish entering the current of a river for perhaps the first time. If Pyramid and Winnemucca Lakes contract and become too salty for fresh-water species, as

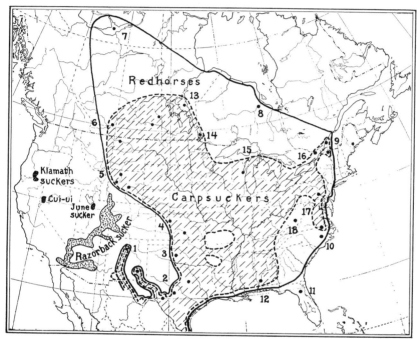

FIG. 181. Range of the lakesuckers, *Chasmistes* spp. and the Razorback Sucker, *Xyrauchen texanus*. Courtesy of University of California Press and Erhard Rostlund.

may possibly transpire if much water is withdrawn from the Truckee River for irrigation purposes, this species no doubt will disappear.

"Spawning appears to be more active at night than in the daytime, and so, also, is migration. This became evident from direct observation and from the fact that early morning usually revealed greatly changed conditions in the river population.

"At times 'cui-ui' appeared in such large and densely packed schools that considerable numbers were crowded out of the water in shallow places, especially on the gently sloping river bars. Once several hundred were observed stranded near the mouth of the river. In some places they were jammed together in masses two or three deep. Some were crowded entirely out and dead, while others were in water a foot deep, yet pushing close to the main group in a perfectly demoralized condition. When one such conditionally free individual was carried some distance away and headed upstream, it passed on its way with great speed, but if removed a short distance only it returned to the mass like an iron to the magnet. It was impossible to separate any number and get them started away from the stranded school. Cormorants, gulls, and pelicans in great numbers were attacking them, and many of the still wriggling fishes had lost their eyes and strips of flesh had been torn from their sides.

"The stomachs of all specimens examined were devoid of food.

"The largest specimen seen measured 670 millimeters, the smallest 410" (pp. 50–52). The heaviest male that Snyder recorded was a 3½ pound specimen 21¼ inches long, while the largest female was a 6-pounder 24¾ inches in length.

The deteriorating condition of the Truckee River has considerably changed the picture for the cui-ui since Snyder's time. The drastically reduced inflow into the lake due to increased diversions, together with over four decades of somewhat drier weather, has resulted in a drop of about 70 feet in the level of Pyramid Lake. The same factors which exterminated the Lahontan Cutthroat Trout (see *Salmo clarki henshawi*, p. 291) in the lake have steadily encroached upon the spawning success of the cui-ui. Fortunately for the latter species, it can spawn in the lake when necessity demands, although it makes every attempt each year to get into the river. Since at least 1948 it has been unable to spawn in the river, the inflow of water being so sparse that only the smaller Tahoe Suckers could get across the long stretches of sand-choked river mouth.

During their annual instinctive congregation at the river mouth, schools of cui-ui move back and forth along the shore, thrashing the surface of the water occasionally in what is evidently pre-nuptial behavior, for specimens caught from such schools when they first appear are invariably "green." While such display is not extensive, there is no way of telling just how much this inability to run upriver has reduced the total cui-ui population of the lake—but there is a strong suspicion that such is the case. Changed river conditions seem also to have advanced the time of their appearance at the river mouth by about a month since Snyder's studies—they now become apparent about the middle of May, and can usually be found into early June, after which they disappear again into the depths of the lake.

The first concerted attempts to study the life history of the species since Snyder's field observations, have been made by Thomas J. Trelease, of the Nevada Fish and Game Commission. By examining the stomachs of active, non-breeding specimens taken in commercial net hauls, he established the fact that the species is essentially a zoöplankton feeder; from the admixture of algal filaments with the plankton fragments, it seems probable that most of the feeding is done about rocks where thick algae coatings are heavily populated with microcrustacea. The gillrakers of the cui-ui are very fine and numerous, and the strong possibility exists that they can extract useable quantities of micro-crustacea from the open lake waters, where such species as *Daphnia pulex, Moina hutchinsoni, Diaptomus sicilis, Cyclops vernalis* and *Marshia albuquerquensis* exist, some of them so abundantly as to contribute substantially to the opacity of the water. Possibly some of the very numerous nannoplankton (diatoms and algae) can also be extracted from open waters.

In addition to studying the animal's feeding habits, which had been unknown, Mr. Trelease has attempted, with somewhat less success to date, to rear individuals from the eggs. Until the last year or two, no one had seen anything but an adult *C. cujus*, at least to anyone's knowledge, and these only when they were spawning. The occasional dead or dying individual which floats to the surface of the lake is so rarely seen as to be disregarded as a source of information. Presumably the cui-ui fingerlings dropped downriver into the lake soon after hatching, and lived there in relatively deep water until maturity. Mr. Trelease has succeeded in developing embryos to the yolk-sac stage, but not beyond, and the author has subsequently had similar results. Within the last year or two, V. Kay Johnson, State Fish and Game biologist for Pyramid Lake and vicinity, has taken sub-adult cui-ui in gill nets set at various parts of the lake, and these appear to be the first such specimens to be recognized.

ECONOMICS

A palatable fish esteemed by Indians and whites alike. They can be hooked only during the spawning season, and whereas they could formerly be taken in the river in large numbers, they must now be sought in the lake waters at the river mouth due to their inability to get into the river. They will not take bait or lures of any kind, but can be readily snagged when the schools are concentrated. The technique consists of throwing a large, weighted 3-cornered hook, attached to a stout line, offshore and allowing it to settle to the bottom. It is retrieved with a series of sudden jerks and, in a substantial percentage of cases, a cui-ui is snagged. Tahoe Suckers (*Catostomus tahoensis*) and occasionally Lahontan Tui Chub (*Siphateles bicolor obesus*), sometimes appear on the snag line. While there is no limit on the number an Indian can take, whites who favor the cui-ui are restricted to five per day, for which they currently pay a tribal fee of one dollar per day or $3 per calendar year.

The cui-ui has long been a favorite food of the Pahute Indians, sharing popularity with the now extinct original Cutthroat Trout. However, after the loss of the latter as a source of food and livelihood, the reservation Indian became a reluctant cattleman, paying little attention to the remaining cui-ui. Snyder, who made his observations during the era of actual use of the species as a staple food, gives us a clear picture of the one-time importance of *C. cujus* in the Indian economy:

"The flesh of this species is highly prized by the Indians. In former times the coming of the 'cui-ui' was a great event, not only for the Pyramid Lake tribe but also for other Piutes from far to the south, who sometimes reached the fishing grounds in such a starved condition that many were unable to survive the first feast. At present numerous little camps may be seen along the river during the spawning period.

The fishes are caught in large numbers and tons of them are dried for later use. They are taken most easily when the river is roily, the fishermen hooking them with an improvised gaff which is drawn quickly through the muddy water. Knowing the 'cui-ui' habit of resting in schools in quiet water, the Indian establishes his camp accordingly, and the willows, wire fence, or hastily constructed racks are soon covered with unsalted drying fish, which attract numbers of flies and send characteristic odors a long distance down the wind.

"When properly cooked, the flesh is sweet and palatable, equal to that of some fishes which bring a fair price in the city markets. The uncleanly methods of preservation employed by the Indians have caused the 'cui-ui' to be regarded with prejudice, and white people of the region will not eat them" (1917A: 53).

CARP AND MINNOWS
Family *CYPRINIDAE*
(Carp, Chub, Dace, Minnows, Shiners, etc.)

This is a very large family of small-to-medium sized freshwater fishes containing a total of over 200 genera and some two thousand species, about a quarter of which occur in the United States. In both numbers of species and individuals, cyprinids are the dominant group of freshwater fishes. Eastern United States species rarely exceed a foot in length, but in the West some Cyprinidae grow to lengths of several feet, and all of these larger species are of some value as food, either to local Indian tribes, or to whites.

In the matter of food, it can generally be said that most cyprinids are good tasting, only their marked boniness making them generally unsuitable for human consumption.

Although otherwise worldwide in distribution, Cyprinidae are absent from South America and Australia. Their almost complete withdrawal from the vast region encompassing northern Canada, Alaska and adjacent Siberia—which they once inhabited—is of more than passing interest to the ichthyogeographer, and shows that this area has become uninhabitable for the family fairly recently in terms of the geological time scale. At one time, presumably, when climates were warmer and the topography suitable, freshwater stream captures in the Bering Straits region must have moved fishes of this family back and forth between the New World and the Old. From the variety of species that exist there today, southern Asia is generally felt to be the area of cyprinid origin, and they appear not to have reached North America until the Cenozoic was well started (possibly in Oligocene times).

DIAGNOSIS

Body generally scaled; head bare; mouth terminal to inferior, only a few genera with barbels; jaws without teeth; pharyngeal teeth present; branchiostegals 3; pseudobranchiæ usually present; gills 4, the fourth followed by a slit; gill membranes applied to the isthmus, limiting the side openings; fins—adipose lacking, pectorals ventral, pelvics abdominal, caudal usually forked; lateral line nearly always present; stomach simple, lacking appendages; air bladder generally large, often divided into anterior and posterior portions, but occasionally lacking. Closely related to the suckers.

KEY TO GENERA OF NEVADA CYPRINIDAE

Abdomen with a distinct fleshy, scaleless keel between pelvic and anal fins (*Notemigonus crysoleucas*)_____ GOLDEN SHINER
Abdomen without such modification_____ 1
1. Distinct and strong spines developed at the anterior edge of the dorsal fin_____ 2
   No such spines (certain species occasionally have the first simple ray of dorsal fin hardened in very old individuals, but this is not a sharp spine)_____ 5

2 (1). Dorsal fin very long, with more than 12 soft rays extending nearly to base of caudal fin; posterior edges of dorsal and anal fins about same distance from base of caudal fin; dorsal fin spine usually serrated; inner border of pelvic fins not adhering to the body at any point _____ 3

―――― Dorsal fin much shorter, usually with less than 10 soft rays, at least its own length removed from the base of caudal fin; posterior edge of dorsal fin at most only reaching to middle of anal fin; dorsal fin spines smooth; inner border of pelvic fins adhering to body for at least half their lengths _____ 4

3 (2). Barbels in two pairs on upper jaw; more than 32 scales in the lateral line (except in the "mirror" or "leather" varieties which have lost much of their scales) (*Cyprinus carpio*) _____ ASIATIC CARP

―――― Barbels lacking; less than 30 scales in the lateral line (*Carassius auratus*) _____ GOLDFISH

4 (2). Maxillary barbels present; body scaleless (an occasional individual has a few scales on the back and elsewhere in patches); of the two spines at the leading edge of dorsal fin, the anteriormost is the largest; body color a brilliant burnished silver (*Plagopterus argentissimus*) _____ WOUNDFIN

―――― Maxillary barbels absent; body covered with small scales (lacking occasionally only under some of the fins); anteriormost of the two dorsal spines the smallest; body color much less brilliant (*Lepidomeda*) _____ SPINEDACE

5 (1). A horny sheath covering the lip of the lower jaw (this can be lifted free with a needle) _____ 6

―――― No such sheath present _____ 7

6 (5). A horny sheath covering only the lip of the lower jaw (this sheath is an external covering not to be confused with the small cartilaginous plate on the upper jaw of this species, a plate which is not visible externally, but is covered by the fleshy upper lip) (*Acrocheilus alutaceus*) _____ CHISELMOUTH

―――― Horny sheaths covering the lips of both jaws (*Eremichthys acros*) _____ SOLDIER MEADOWS DESERTFISH

7 (5). Scales in the lateral line numbering 100 or more (*Orthodon microlepidotus*) _____ SACRAMENTO BLACKFISH

―――― Scales in the lateral line less than 100 _____ 8

8 (7). Species with only a single row of pharyngeal teeth, the lesser or outer row never developed[1] _____ 9

―――――

[1] It is unfortunate that pharyngeal teeth characteristics must be used in a key designed as much as possible so that anyone can use it. However, it is unavoidable in this instance, since no better character is available. The serious student of fishes will learn to dissect and clean the pharyngeal teeth in this family since much of the important classification is based on these teeth. With a little experience, the individual can learn to extract these "throat teeth" borne on the modified fifth gill arches by working inward under the posterior edge of the gill cover or opercle.

.... Species with two rows of pharyngeal teeth, the lesser or outer row occasionally lacking on one side............................ 10

9 (8). Scales in the lateral line larger, less than 65; maxillary smaller, not reaching anterior edge of eye; intestine longer, about equal in length to the standard body length; pharyngeal arch uniformly and smoothly rounded in the vicinity of the "heel" (*Siphateles bicolor obesus*) ....
.... LAHONTAN TUI CHUB

.... Scales in the lateral line smaller, more than 65; maxillary larger, reaching posteriorly to about the anterior edge of eye; intestine shorter, only about one-half the standard length; pharyngeal arch with a quite prominently developed "heel" which breaks the otherwise smooth contour of the pharyngeal bone (*Moapa coriacea*) .... MOAPA DACE

10 (8). A barbel usually present on the posterior angle of the maxillary process, small but seldom obsolescent except in very young individuals (up to 50 percent of specimens in any given population may lack barbels, hence the necessity of extensive collections in determining these forms) (*Rhinichthys osculus*).................................. SPECKLE DACE

.... Barbels always lacking completely... 11

11 (10). An accessory "scale" or flap of skin present in the axil (base) of the pelvic fin; sides with a deep orange-to-red broad band in adults of both sexes, heightened to brilliancy during the breeding season (*Richardsonius*)..... REDSHINERS

.... No such scale or flap of skin present; without the above described color pattern.. 12

12 (11). Head long and pike-like, the mouth deeply cleft with maxillary extending backward beneath eye at least to leading edge of pupil; pharyngeal teeth subconical, scarcely hooked, sharp-edged, the lower limb of the pharyngeal bone greatly elongated (*Ptychocheilus*)..................... SQUAWFISH

.... Head not as above, the mouth less deeply cleft and maxillary not reaching to leading edge of eye; pharyngeal teeth compressed, close-set, strongly-hooked, pharyngeal bone not as described above (*Gila*)............................ GILA CHUBS

ALTERNATIVE KEY CHARACTERS

As an additional aid in determining genera in this difficult family, the following variations of the above key are offered:

(A) For 5(1), substitute—

5 (1). Species with only a single row of pharyngeal teeth, the lesser or outer row never developed (*Orthodon, Eremichthys, Acrocheilus, Siphateles* and *Moapa*)............................ 6

.... Species with two rows of pharyngeal teeth, the lesser or outer row occasionally lacking on one side (*Rhinichthys, Ptychocheilus, Richardsonius, Gila*)............................

6 (5). Lateral line scales 100 or more (*Orthodon*)............................
.... SACRAMENTO BLACKFISH
.... Lateral line scales 90 or less.. 7
7 (6). Horny sheaths on lips (*Eremichthys* and *Acrocheilus*)........ 8
.... No such horny sheaths (*Siphateles* and *Moapa*)........................ 8
8 (7). These four genera can be separated by the same characters given in the main key.

(B) For 11 (10), substitute—

11 (10). Pharyngeal teeth subconical, scarcely hooked, sharp-edged, the lower limb of the pharyngeal bone greatly elongated; head long and pike-like, etc. (same characters as in main key) (*Ptychocheilus*)............................ SQUAWFISH
.... Pharyngeal teeth compressed, close-set, strongly hooked, etc. (same characters as in main key for *Gila*)............................ 12

12 (11). Separate *Gila* and *Richardsonius* by pelvic axillary scales and color

SQUAWFISH

Genus *Ptychocheilus* Agassiz 1855

("Folded lip," in reference to the folding of the mouth skin behind the jaws)

KEY TO SPECIES

Scales larger, numbering 67 to 80 in the lateral line; anal rays averaging 8; Columbia River system (*Ptychocheilus oregonensis*).. NORTHERN SQUAWFISH

Scales smaller, numbering 83 to 93 in the lateral line; anal rays averaging 9; Colorado River system (*Ptychocheilus lucius*)......
.... COLORADO SQUAWFISH

NORTHERN SQUAWFISH

Ptychocheilus oregonensis (Richardson)

(Columbia River Squawfish)

Cyprinus oregonensis Richardson 1836: 305
Ptychocheilus gracilis Agassiz and Pickering 1855: 229
Ptychocheilus oregonensis, Girard, 1856B: 209 / 1859: 298–299
Luciscus oregonensis, Günther, 1868: 239–240
Leuciscus gracilis, ibid: 240–241
Gila oregonensis, Jordan and Copeland, 1878: 424
Ptychochilus oregonensis, Jordan and Gilbert, 1882: 226
Ptychocheilus oregonensis, ibid, 1894: 139
Ptychocheilus oregonensis, Jordan and Evermann, 1896B: 224–225 / 1902: 68–69

Ptychocheilus oregonensis, Jordan and Starke, 1896: 853
Ptychocheilus oregonensis, Snyder, 1908: 84–85, 170
Ptychocheilus oregonensis, Pratt, 1923: 69
Ptychocheilus oregonensis, Jordan, Evermann and Clark, 1930: 114
Ptychocheilus oregonensis, Hubbs and Schultz, 1931: 1–6
Ptychocheilus oregonensis, Clemens and Munro, 1934
Ptychocheilus oregonensis, Schultz, 1936: 150
Ptychocheilus oregonensis, Chapman and Quistorff, 1938
Ptychocheilus oregonensis, Schrenkeisen, 1938: 103–104
Ptychocheilus oregonensis, Foerster and Ricker, 1941
Ptychocheilus oregonensis, Ricker, 1941
Ptychocheilus oregonensis, Dimick and Merryfield, 1945: 37
Ptychocheilus oregonensis, Carl and Clemens, 1948: 71, 82–83
Ptychocheilus oregonensis, Miller and Miller, 1948: 181
Ptychocheilus oregonensis, Taft and Murphy, 1950: 149
Ptychocheilus oregonense, La Rivers, 1952: 99
Ptychocheilus oregonense, La Rivers and Trelease, 1952: 116
Ptychocheilus oregonense, Eddy, 1957: 94
Ptychocheilus oregonense, Hasselman and Garrison, 1957
Ptychocheilus oregonense, Moore, 1957: 102
Ptychocheilus oregonense, Slastenenko, 1958A: 209–210
Ptychocheilus oregonense, Carl, Clemens and Lindsey, 1959: 107–
Ptychocheilus oregonensis, Maxfield, Liscom and Lander, 1959
Ptychocheilus oregonensis, Thompson, 1959

ORIGINAL DESCRIPTION

"This species is also an inhabitant of the Oregon, or Columbia River, and is so similar in general appearance to the last, that it may be readily confounded with it, though it is certainly specifically distinct, as may be seen by the following differences.

"Form more tapering forwards, the shoulder not being so high; *head* longer, forming one-fourth part of the length of the fish, including the middle caudal rays: *snout* obtuse and even with the margins of the upper and lower jaw when the mouth is closed; mouth considerably larger, being cleft as far back as the edge of the orbit; anterior suborbital more oblong and perforated by a greater number of foramina; the gill-cover less widely rounded, and the edge of the operculum concave, though not so much so as in *leuciscus gracilis*, pl. 78. The *dorsal* also stands farther back, being nearer to the tip of the tail than to the point of the snout, while the *ventrals* stand under the first dorsal ray, and midway between the orbit and base of the central caudal rays. The distance from the gill-openings to the ventrals reaches from the latter to half way between the anal and caudal. The size of the *scales*, generally, and their number on the lateral line, is the same as in *leuciscus caurinus*, but their form is more perfectly orbicular, and those on the belly are proportionally smaller.

"Fins.—*Br.* 3–3; *P.* 15; *D.* 10; *V.* 9; *A.* 9; *C.* 19 9/9.

" 'Colour of the back and top of the head intermediate between yellowish grey and brocoli-brown, passing gradually on the sides, below the lateral line, into sulphur-yellow, the latter colour prevailing also on the cheeks, gill-covers, and bases of the fins. The *belly* is silvery white' (Dr. Gairdner).

"Dimensions.

	Inches	Lines
Length from tip of snout to extremity of caudal	13	0
Length from end of central caudal rays	11	11
Length from base of ditto	10	10
Length from end of anal	8	7½
Length from beginning of ditto	7	8½
Length from anus	7	7
Length from end of dorsal	7	2
Length from ventrals	6	1
Length from beginning of dorsal	6	1
Length from pectorals	3	2
Length from tip of gill-cover	3	1
Length from tip of bony operculum	2	11
Length from nape	2	4
Length from posterior edge of orbit	1	5
Length from anterior ditto	0	11
Length from anterior nostril	0	7
Length from tip of labial	1	2½
Length of margin of upper jaw one side	1	2½
Length of ditto under jaw	0	10
Length of under jaw to articulation	1	4½
Height of gape of mouth	1	0
Length of ditto transversely	0	10½
Length of pectorals	1	8½
Length of ventrals	1	6
Height of dorsal anteriorly	1	8
Height of ditto posteriorly	0	9
Length of attachment of dorsal	1	2
Length of attachment of anal	0	11
Height of anal anteriorly	1	6
Height of ditto posteriorly	0	8½
Length of caudal lobes	2	4½
Length of its middle rays	1	1½
Depth of caudal fork	1	0
Height of body before dorsal	2	3
Greatest thickness	1	2½
Greatest circumference	5	3
Thickness at nostrils	0	9½
Thickness between orbits	0	10¾
Thickness of nape	1	3

Weight 9 oz. 2 dr." (Richardson 1836: 305–306).

DIAGNOSIS

Head about 4 times, depth from 4.5 to 5.0 times, in body length; head long and slender, rather pike-like; mouth almost horizontal, large, the maxillary reaching backward to a point below the eye; lips thick, lacking barbels; eye small; pharyngeal teeth in 2 rows, 2/5–4/2, the pharyngeal bone with its straight portion very slim and attenuated, the teeth far apart—teeth little flattened, nearly conical, weakly curved at ends, the curve being toward the bone angle, grinding surfaces lacking; lateral line noticeably decurved, nearer venter than back; scales 13–20/65–75/11–18, 46–56 before the dorsal fin; fins—dorsal about 10-rayed, set well back—anal about 8-rayed—caudal strong, deeply forked; caudal peduncle thick, least depth about 10 times in body length; intestinal canal short, less than the body length, and with only one main loop; color olivaceous with lighter venter, young specimens with a black spot at mid-base of caudal; reportedly reached lengths up to 4 feet and weights of about 60 pounds, our second largest North American "minnow."

TYPE LOCALITY
Columbia River, Washington.

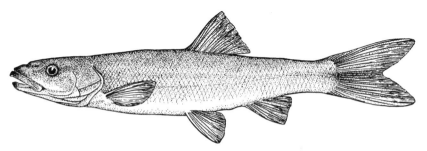

FIG. 182. Northern Squawfish, *Ptychocheilus oregonensis*.
Drawn by Silvio Santina.

RANGE
Columbia River basin of the Pacific Northwest, east to Montana, north into Canada. Reaches Nevada up the Salmon, Jarbidge, Bruneau and Owyhee Rivers, tributaries of the Snake River.

TAXONOMY
A stable species with only one synonym. The type is so distinctive in appearance and technical characteristics that Agassiz unhesitatingly founded his new genus *Ptychocheilus* on a specimen lacking both its pharyngeal teeth and its intestines, two important diagnostic structures among cyprinids. Three related species constitute the remainder of the genus; the Sacramento Squawfish (*P. grandis* (Ayres) 1854) in the Sacramento River drainage to the west, the Colorado Squawfish (*P. lucius* Girard 1856) southward in the Colorado River system and *P. umpquae* Snyder 1907, in certain Oregon coastal streams.

LIFE HISTORY
The Northern Squawfish is a rather markedly carnivorous, spring-spawning species, abundant in both lake and stream habitats within its range, although preferring the former where it seems to have much choice. It is quite trout-like in its manner of feeding and insects seem to constitute a good portion of the diet of younger individuals. Lake

populations run up affluent streams to spawn and seem to prefer the lake as a place to live during the rest of the season.

Taft and Murphy's studies of the biology of the Sacramento Squawfish (*Ptychocheilus grandis* (Ayres) 1854) do not contain much information on spawning behavior, but certainly what they observed can be applied generally to the northern species. They noted that the eggs were adhesive and could be found sticking to rocks or gravel. The females in spawning were accompanied by one or more males.

Studies on the life history and ecology of the Northern Squawfish in the Pacific Northwest by Paul D. Zimmer[1] of the United States Fish and Wildlife Service have contributed much valuable information on this hitherto little known fish. From these we have the following:

"With regard to water temperatures at time of spawning, it appears that in the lower Columbia River area the principal spawning takes place at about 65° F. Through artificial incubation of squawfish eggs with water temperature controls, it was learned that it took approximately four days for the eggs to hatch at 65° F. Spawning takes place during the months of July and August, although some ripe males were found as early as April . . . Ripe and recently spent adults have been captured in mud bottom sloughs and bays, and others have been taken in the same condition over gravel areas of certain lakes. We could find no evidence to indicate that the squawfish prepared a nest in the gravel, nor could we detect any movement of squawfish out of lakes or main rivers at spawning time.

". . . recent food habits study . . . indicates that the squawfish is quite omnivorous. The diet includes insects, crayfish, mussels, wheat, young salmon, cottids, and other fishes.

". . . preliminary analysis [of scales] shows that of the group of fish that was sampled, those in their third year had an average length of 9.3" while those in their eighth year had an average length of 16.8″″" (Laythe 1957).

Thompson's abstract of recently published work on the food of the Northern Squawfish is:

"A study is presented of 3,546 stomachs of the squawfish, *Ptychocheilus oregonensis* (Richardson), collected from April 1955 to April 1956 in the lower Columbia River. The basic food table of the squawfish, based on a modified point system with emphasis on salmon predation is presented. Sixty-three percent of the stomachs examined were empty. Size of squawfish, season of the year, and geographical distribution within the river affect the occurrence and importance of food items.

"Major food items were fishes, crayfish, and insects, and, in much lesser amounts, plant materials, mollusks, and miscellaneous items. Squawfish from 3 to 8 inches long subsisted on a diet of insects; above that length fishes and crayfishes attained importance. At 11 inches, fishes and crayfishes were dominant and insects were only 5 percent of the stomach content" (1959).

ECONOMICS

To complete Thompson's abstract—

"All occurrences of juvenile salmon in squawfish stomachs were

---
[1] See Hutchinson, 1957.

related to releases of young salmon from hatcheries. The role of the squawfish as a predator on salmon was limited to time and place where juvenile salmon concentrations were high following release" (1959).

Because of its size and feeding habits, the Northern Squawfish has been a source of sport for anglers and it was, in pre-Whites time, an important food fish for the Indians. In a past day—"fishing through the ice for squawfish is one of the popular winter amusements at the Idaho lakes" (Jordan and Evermann 1902: 69).

The impact of the squawfish on game fishes can be of some importance in certain areas, and has been the prime mover for a presently active study of the problem in the Columbia River:

". . . we are confronted with a very serious problem of squawfish predation at time of our hatchery releases from both Federal and State stations located below McNary Dam. Sampling in the main river throughout a thirteen-month period indicates that our main problem of predation appears only at time of hatchery releases. We are now engaged in a program of removing as many squawfish as possible from the immediate vicinity of any programmed release. Until we devise a more satisfactory method of collecting the predators, we will continue to use nylon gill nets for this purpose. The nets are fished for about one week prior to, during, and for as long a period after release as is necessary.

"We have done some limited investigation to determine whether squawfish have a commercial value. Dr. Joseph Stearn of the University of Washington, Technology Department, has smoked and pickled some of them for us. His report concerning the pickled squawfish states that in his opinion it is equal to, or superior to, any pickled product on the market. The flesh of the squawfish is very tasty when smoked, but the small, forked bones are of considerable bother. Several years ago we provided a fish buyer with several hundred pounds of squawfish which he shipped to a dealer in California. He found a ready market for them in the oriental trade" (Laythe 1957).

COLORADO SQUAWFISH

Ptychocheilus lucius Girard

(Colorado River Squawfish, White Salmon)

Ptychocheilus lucius Girard 1856B: 209
Ptychocheilus lucius, Cooper (in Cronise), 1868: 496
Leuciscus grandis, Günther, 1868: 239
Gila lucius, Jordan and Copeland, 1878: 424
Ptychochilus lucius, Jordan and Gilbert, 1882: 227
Ptychocheilus lucius, Jordan, 1891: 28
Ptychocheilus lucius, Evermann and Rutter, 1895: 482
Ptychocheilus lucius, Jordan and Evermann, 1896B: 225 / 1902: 69–70
Pytchocheilus lucius, Gilbert and Schofield, 1898: 492
Ptychocheilus lucius, Snyder, 1915: 581
Ptychocheilus lucius, Pratt, 1923: 69
Ptychocheilus lucius, Jordan, Evermann and Clark, 1930: 114
Ptychocheilus lucius, Hayes, 1935: 35–36

Ptychocheilus lucius, Schrenkeisen, 1938: 104
Ptychocheilus lucius, Moffett, 1942: 82 / 1943: 182
Ptychocheilus lucius, Dill, 1944: 154–155
Ptychocheilus lucius, Miller, 1946B: 410, 415 / 1952: 29
Ptychocheilus lucius, Shapovalov and Dill, 1950: 386
Ptychocheilus lucius, Simon, 1951: 80–81
Ptychocheilus lucius, Wallis, 1951: 89–90
Ptychocheilus lucius, Beckman, 1952: 43
Ptychocheilus lucius, La Rivers, 1952: 99
Ptychocheilus lucius, La Rivers and Trelease, 1952: 116
Ptychocheilus lucius, Eddy, 1957: 95
Ptychocheilus lucius, Moore, 1957: 103
Ptychocheilus lucius, Shapovalov, Dill and Cordone, 1959: 172

ORIGINAL DESCRIPTION

"A very characteristic species. The body is compressed, but the head is flattened or depressed and very much developed, constituting nearly one fourth of the entire length. The dorsal and ventrals are situated quite posteriorly. The scales are below the median in size, and the lateral line is bent downwards upon the abdomen. The pharyngeal bones are very slender; the inferior limb is almost exiguous and proportionally as long as in *P. grandis*. There are, however, but four teeth

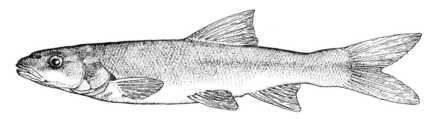

FIG. 183   Colorado Squawfish, *Ptychocheilus lucius*.
Drawn by Silvio Santina.

upon the main row, instead of five, as in the case of *P. grandis*. Color bluish grey above; silvery golden beneath.

"Collected in the Rio Colorado, by A. Schott, under Major W. H. Emory, Commissioner U. S. and Mex. Boundary" (Girard 1856B: 209).

DIAGNOSIS

General appearance as in *P. oregonensis;* head about 3.5 times in body length, body depth about 5.5 times in body length; pharyngeal teeth in two rows, 2/5–4/2; scales 20–29/83–93/13–20; fins—dorsal usually 9-rayed, set well back—anal usually 9-rayed—caudal strong, deeply forked; caudal peduncle thick, least depth about 13 times in body length; color as in *P. oregonensis,* the young with the same black spot at middle of caudal base, and two weak lateral zones, an upper dark one and a lower pale line; the largest of our American Cyprinidae, a "minnow" reaching lengths of 5 feet and weights of more than 80 pounds.

TYPE LOCALITY

Colorado River.

RANGE

Colorado River basin, southwestern United States. Occurred, at least originally, in extreme southern Nevada where the Rio Colorado forms the boundary between Nevada and Arizona.

TAXONOMY

Quite stable, as shown by the synonymicon. Like the Northern Squawfish, this fish is so distinctive it cannot be confused with any

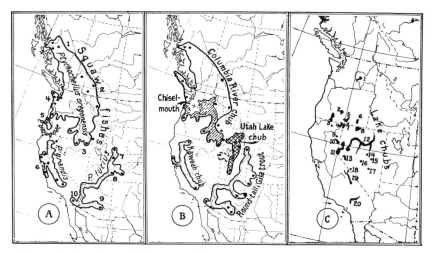

FIG. 184. A. Range of the squawfishes, *Ptychocheilus* spp.
B. Range of the Chiselmouth Chub, *Acrocheilus alutaceus*, and the Gila Chubs, *Gila* spp.
C. Range of the Lake Chubs, *Siphateles* spp.
Courtesy of University of California Press and Erhard Rostlund.

other species. Fortunately for the taxonomist, the four species of squawfish have no intergrading borderline populations since each occupies a different drainage system with no recent connections.

Etymology—"lucius," pike.

LIFE HISTORY

Beyond the fact that the Colorado Squawfish is obviously, from its design, an avid carnivore, nothing of much substance seems to be known of its life history. Comments on the species within the past 20 years in southern Nevada have amounted only to reporting its presence or absence from a given area under study (Moffett 1942, 1943; Wallis 1951; Jonez and Sumner 1954). In the case of the Jonez and Sumner report, they wrote that no occurrences of the species had been noted in southern Nevada since Moffett's work.

This animal has apparently been very sensitive to its changing environment and seemingly is a river form which decreases in reservoir waters. Just what the critical factor is in its case is unknown.

ECONOMICS

Beyond the fact that the Colorado Squawfish has long been recorded as a one-time important food source for Indians within its range, little can be said of its economic position. That the species was also often heavily utilized by the newly arrived white man is evident from the common names he has applied to it: "White Salmon of the Colorado River," "pike," etc. Undoubtedly such a sizeable and strong carnivore as the squawfish would be regarded suspiciously by modern fishery investigators as a threat to stocked trout populations in its waters, and there should be no reason to believe the large fish would not sample its share of other fishes with which it is associated.

However, changing the Colorado River into the two reservoirs Lakes Mead and Mohave along the Nevada border seems to have effectively eliminated this interesting fish as an active element in our fauna, although it is not extinct farther up the river.

CHISELMOUTH
Genus *Acrocheilus* Agassiz 1855
("Sharp lip")

CHISELMOUTH
Acrocheilus alutaceus Agassiz and Pickering
(Hardmouth, Squaremouth)

Acrocheilus alutaceus Agassiz and Pickering (in Agassiz) 1855: 99
Lavinia alutacea, Girard, 1856B: 184 / 1859: 240
Acrochilus alutaceus, Günther, 1868: 276
Acrochilus alutaceus, Jordan and Copeland, 1878: 418
Acrochilus alutaceus, Jordan and Gilbert, 1882: 150–151
Acrocheilus alutaceus, Jordan and Evermann, 1896B: 208
Acrocheilus alutaceus, Snyder, 1908: 84, 169
Acrocheilus alutaceus, Jordan and Snyder, 1909: 427
Acrocheilus alutaceus, Pratt, 1923: 65
Acrocheilus alutaceus, Jordan, Evermann and Clark, 1930: 112
Acrocheilus alutaceus, Hubbs and Schultz, 1931: 1–6
Acrocheilus alutaceus, Schultz, 1936: 146
Acrocheilus alutaceus, Schrenkeisen, 1938: 99
Acrocheilus alutaceus, Dimick and Merryfield, 1945: 36
Acrocheilus alutaceus, Miller and Miller, 1948: 181
Acrocheilus alutaceum, La Rivers, 1952: 98
Acrocheilus alutaceum, La Rivers and Trelease, 1952: 116
Acrocheilus alutaceum, Eddy, 1957: 82
Acrocheilus alutaceum, Moore, 1957: 108
Acrocheilus alutaceum, Slastenenko, 1958A: 188
Acrocheilus alutaceum, Carl, Clemens and Lindsey, 1959: 117–119

ORIGINAL DESCRIPTION

"Caught at Willamet Falls, and in Wallawalla River. Nose prominent and rounded.

"Tail rather slender. Caudal large. Dorsal much larger than the anal. The color light brown above, (there being a white and very fine line on the edge of each scale,) blending into yellowish brown upon the sides, and passing into pure white upon the abdomen. Gill-cover golden brown. Dorsal and caudal of the same color as the sides of the body. Pectorals orange, gradually paler towards the base. Ventrals as the pectorals, but more uniformly orange. Anal also orange, but more bright and reddish. It occurs in the rapids and falls of the River. It is caught by the natives while fishing in the Falls for Salmon" (Agassiz and Pickering, in Agassiz, 1855: 99).

Agassiz' description of the genus *Acrocheilus,* in the same paper, contains much that is important to the above description of *A. alutaceus,* but is too long to reproduce here in its entirety. The most important elements may be abstracted as:

"As a genus I would characterise it by the peculiar structure of the edging of the mouth, which in the lower jaw constitutes a transverse

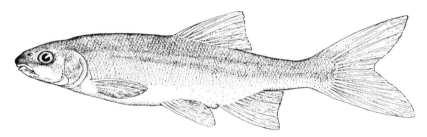

FIG. 185. Chiselmouth, *Acrocheilus alutaceus.* Drawn by Silvio Santina.

broad flat plate, very similar in appearance to the dental plates of *Myliobates,* being thicker along the outer edge and tapering gradually along the inner edge. This transverse plate is square and cut at right angles externally towards the symphysis of the two jaws. In consequence of this peculiar structure of the margin of the mouth and its armature, the lower jaw is as it were cut transversely, and has in no degree the rounded outline about the symphysis of its branches which is observed in most *Cyprinidae* . . . The scales . . . present a striking difference. They have not, as in Chondrostoma, the ordinary type of Leuciscus, but resemble rather the scales of Barbus in their elongated form, their small size, their many radiating furrows diverging in every direction, and their ornamental pigment cells which are especially numerous along the posterior margin . . .

"Unfortunately the two specimens collected in Columbia River are deprived of their intestines, and in one of them only, were the pharyngeal bones, with their teeth, preserved; but these afford further evidence of the correctness of my view in considering these fishes as a type of a distinct genus peculiar as far as is now known to the northwest coast of America. There is but a single row of teeth and only five teeth in that one row on the left and four on the right side. The isolated teeth stand on a cylindrical peduncle swelling into a sharp hook, but the inner surface is cut obliquely like the incisors of Rodents, and present a flat grinding surface resembling closely the dentition of Chondrostoma and Chondrochilus, differing however in the more club-shaped form of the teeth, and the sharp terminal hook, and also the smaller number of teeth in one row . . . the peritoneum is also black" (Agassiz 1855: 96–99). Agassiz presents an illustration of the pharyngeal teeth and arch from several angles, a helpful refinement of description too seldom followed by his successors.

DIAGNOSIS

Head and greatest body depth each about 4 times in body length; head rather blunt; mouth horizontal, wide, overlapped by broad snout, the maxillary not quite attaining the forward edge of eye; upper jaw protractile, bearing a fleshy lip surrounding a rather small cartilaginous plate which is completely hidden by the lip; the sharp cartilaginous sheath of the lower jaw covers the lip and is externally visible; lips lacking barbels; eye rather large; pharyngeal teeth in a single row, robust, hooked, tipped with grinding surfaces of varying extent, depending upon the age of the individual, usually 5–4, rarely 5–5, lateral line much decurved; scales about 20/85/16; fins—dorsal long, about 10-rayed—anal large, about 9-rayed—caudal broad and rather deeply forked; caudal peduncle moderately slim, least depth about 13 times into body length; intestine long, at least twice body length, and with more than one main loop; peritoneum jet-black; color—rather dark above, lighter below, usually with the body thickly spotted with small, irregular, conspicuous black dots—young with a more-or-less discernible blackish spot at mid-base of caudal; reaches a length of about a foot.

TYPE LOCALITY

Willamette Falls and Walla Walla River, Oregon.

RANGE

Columbia River and Malheur Lake systems.

TAXONOMY

Acrocheilus alutaceus is a singularly distinctive animal with no particularly close relatives, being the only species in the genus. In the more than 100 years of its existence, this distinctiveness has precluded the acquiring of any synonyms, slight differences in spelling being the only changes except for Girard's placement of the species in *Lavinia* a year after its initial description. As with *Ptychocheilus,* Jordan emended the diphthong "ei" in *Acrocheilus* to "i" for a few years, then returned to the original spelling.[1]

Etymology—"alutaceus," leathery.

LIFE HISTORY

Little, if any, published work has appeared on this species. Certain assumptions are safe to make, however; stomach contents are invariably

---

[1]Emendations of *alutaceus* to *alutaceum* and *Ptychocheilus oregonensis* to *oregonense* were originally at the suggestion of Robert R. Miller. The International Commission has now ruled a return to the original spellings.

mainly vegetation, with algae predominating. The long, convoluted intestine would, to the anatomist, also suggest an herbivorous diet as would the cartilaginous sheathing of the mouth. And beyond the fact that the chiselmouth has been noted spawning at the same times and in the same areas as the native trout, nothing can be said concerning spawning procedures, numbers of eggs, hatching times, etc.

ECONOMICS

Beyond the fact that the chiselmouth, because of its abundance and availability, was reportedly utilized as aboriginal food, not much economic importance can be attached to it. Its habits certainly preclude the greatest cross the fishery biologist has to bear—"does it eat trout eggs?" for this is probably one of those species who balance up the biologist's ledger by allowing themselves to be eaten by trout (for which they get scant credit and no praise).

THE GILA CHUBS

Genus *Gila* Baird and Girard 1853

(Named for the Gila River)

KEY TO SPECIES

1. Origin (front end) of dorsal fin immediately over the origin of the pelvic fins; lateral line scales numbering 70 or less; caudal peduncle relatively deep, resembling that of a *Siphateles* (*Gila atraria*) _____ UTAH GILA

   ------ Origin of dorsal fin behind the origins of the pelvic fins; lateral line scales approximately 80 or more; caudal peduncle moderately robust to extremely slender and pencil-shaped (*Gila robusta*) _____ COLORADO GILA ___ 2

2 (1). Dorsal fin rays 8–10, usually 9; anal fin rays 7–10, usually 9; pelvic fin rays usually 9/9, rarely 10/10; body fully scaled, scales numbering from 79 to 96 in the lateral line; basal radii of scales poorly developed; nuchal hump detectable in older fish; caudal peduncle least depth goes into head length from 3.3 to 4.3 times (small rivers) (*Gila robusta robusta*) _____ TRIBUTARY COLORADO GILA

   ------ Dorsal fin rays 10–11; anal fin rays 10–11; pelvic fin rays usually 9/9, rarely 10/10; body scales often lacking over dorsum, venter, and caudal peduncle, or consisting there of minute embedded scales; lateral line scales 75–88; basal radii completely lacking in scales; nuchal hump prominent; caudal peduncle least depth from 5.0 to 6.5 into head length (large rivers) (*Gila robusta elegans*) _____
   _____ BONYTAIL COLORADO GILA

   ------ Dorsal fin rays 9; anal fin rays 9; pelvic fin rays 9/9; body fully scaled, scales numbering from 89–94 in the lateral line; basal radii of scales conspicuous; nuchal hump

absent; caudal peduncle least depth from 3.3 to 4.1 into head length (small streams) (*Gila robusta jordani*) ....
.... WHITE RIVER COLORADO GILA

COLORADO GILA
Gila robusta Baird and Girard
(Bonytail, Roundtail)

Gila robusta Baird and Girard 1853A: 368
Gila grahami ibid 1853B: 389
Ptychocheilus vorax Girard 1856B: 209 / 1859: 285–286
Gila robusta, Cooper, in Cronise, 1868: 496
Leuciscus robustus, Günther, 1868: 241
Leuciscus zunnensis, ibid
Gila robusta, Cope and Yarrow, 1875: 663–664
Gila robusta, Jordan and Copeland, 1878: 424
Gila vorax, ibid
Gila robusta, Jordan and Gilbert, 1882: 228
Gila robusta, Jordan, 1891: 27
Gila robusta, Evermann and Rutter, 1895: 483
Gila robusta, Jordan and Evermann, 1896B: 227–228
Gila robusta, Gilbert and Schofield, 1898: 493
Gila robusta, Snyder, 1915: 581 / 1921: 25
Ptychocheilus vorax, ibid, 1921: 25, 28
Gila robusta, Pratt, 1923: 69
Gila robusta, Jordan, Evermann and Clark, 1930: 114
Gila robusta, Hayes, 1935: 37–39
Gila robusta, Tanner, 1936: 159, 168
Gila grahami, ibid
Gila robusta, Schrenkeisen, 1938: 104–105
Gila robusta, Dill, 1944: 154
Gila robusta, Miller, 1945A: 104 / 1946B: 410
Gila robusta, Hubbs and Miller, 1948A: 4, 10
Gila robusta, Kopec, 1949: 56
Gila robusta, Shapovalov and Dill, 1950: 386
Gila robusta robusta, Simon, 1951: 79
Gila robusta robusta, Wallis, 1951: 90
Gila robusta robusta, Beckman, 1952: 45
Gila robusta, La Rivers, 1952: 100
Gila robusta, La Rivers and Trelease, 1952: 116
Gila robusta, Eddy, 1957: 96–97
Gila robusta, Koster, 1957: 58–59
Gila robusta, Moore, 1957: 100
Gila robusta, Shapovalov, Dill and Cordone, 1959: 172

ORIGINAL DESCRIPTION

"Body very much swollen anteriorly, and tapering very suddenly from the dorsal fin to the insertion of the caudal. Head very much depressed above, sloping very rapidly from the nape to the snout, and forming one fourth of the entire length. Eyes proportionally small and subcircular. Mouth tolerably large; the posterior branch of the maxillary does not reach the vertical line of the pupil. Dorsal fin situated on the middle of the back, and a little higher than long. Caudal crescentic. Anal situated behind the dorsal. Insertion of ventrals in advance of the anterior margin of the dorsal. The posterior tip of the pectorals does not reach the insertion of the ventrals. All the soft rays are bifurcated. Lateral line composed of about ninety scales. Color greyish brown above, lighter below.

"Formula of the fins: D I. 9. C 8. I. 8. 8. I. 7. A I. 9. V I. 9. P 15" (Baird and Girard 1853A: 368).

Heading the section in which this species, as well as *G. elegans, G. gracilis* and the new genus *Gila* were characterized, is the statement:

"The species of fishes here described as new, were caught in the Zuni River, New Mexico, by Dr. S. W. Woodhouse, while attached as Surgeon and Naturalist to the expedition of Capt. Seetgreaves [sic], for the exploration of Zuni and its tributaries" (p. 368).

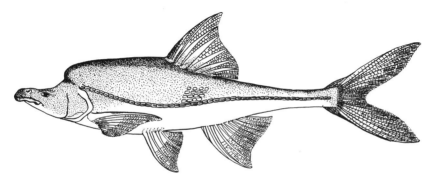

FIG. 186. *Gila cypha* Miller 1946, a rather bizarre form showing extreme adaptation to the swift and turbulent waters of the Colorado River. Miller described this species upriver from Nevada, the type coming from the Grand Cañon of the Colorado River near the mouth of Bright Angel Creek, Arizona. Re-drawn by Silvio Santina from Miller 1946.

DIAGNOSIS

Head 4–5 times in body length, body depth 5–6 times in body length; head wide and flattened, tending to be concave in profile; mouth large, horizontal, overhung by prominent snout; maxillary usually reaching to front part of orbit; lips lacking barbels; eye rather small; pharyngeal teeth in two rows, 5/2–2/4, close together, compressed, hooked, lacking grinding surfaces; lateral line markedly decurved; scales 17–23/65–98/10–12, quite variable in size, reduced and partially lacking in some forms; fins—well-developed, varying in the different forms from strong to very prominent, depending upon their habitats—dorsal 8–11 rayed, rarely 7, set behind middle of body, slightly back of pelvic fins—anal 8–11 rayed, rarely 7—caudal large and strongly forked; caudal peduncle varies from thick in the slow water forms to thin and pencil-shaped in the swift current types; intestines short, less than length of body; peritoneum dusky; color—brownish-black above, characteristically paler beneath, breeding males usually developing reddish tinges on the venter, and distinctive red marks on the cheeks; grow to lengths of about 18 inches.

Major differences between the several subspecies are summed up in Table I.

TYPE LOCALITY

Zuni River, New Mexico.

RANGE

The Colorado River and its tributaries. In Nevada the species occurs in the main river channel between Nevada and Arizona, as well as in the Moapa and Virgin Rivers and Pahranagat Valley.

TAXONOMY

This species, and the several varieties which are currently recognized, has developed a somewhat complex synonymicon in the past hundred years. About a third of the names applied to the *G. robusta* complex have been found valid in a subspecific sense to distinguish between some of the widely divergent ecotypes which have developed due to habitat differences; the remainder of the names have fallen into snyonymy. The distinctness of such a form as *intermedia,* with its complete squamation and thick caudal peduncle, from a type such as *cypha,* with reduced squamation and pencil-shaped peduncle and prominent nuchal hump, is quite marked until intergrading entities like *robusta* and *elegans* are compared with them. These distinctions are directly correlatable with the differences between the environments of the various forms— *intermedia* living in small, slow streams, *cypha* in the swift, roily difficult main stem, and *robusta* in situations intermediate between the two.

Etymology—"robustus," stout.

LIFE HISTORY

A spring-spawning species, the Colorado Gila does not seem to have been systematically studied—scattered references to it are found in the literature, the "passing mention" sort of thing. It is an omnivore, taking aquatic arthropods (as well as those that fall into water), smaller fishes and algae. It is still a common fish in most of its original areas, from the headwaters of the Colorado system (Simon 1951: 79) to the Nevada reservoirs of Lakes Mead and Mohave (Jonez and Sumner 1954: 70). The latter examined several ripe females, 12 to 14 inches long, at a gravel shoal in March, but never observed the actual spawning.

ECONOMICS

In addition to its expected utilization by Indians living along its streams, the Colorado Gila has, in company with many other "rough" fish, furnished a great deal of the fishing enjoyed by the "small fry"— it readily takes bait and the trout fisherman has often pulled in "roundtails" when after bigger game. The usual questions of its relations to trout could, of course, be raised by the biologist, and there would be no denying the gila must consume its fair share of trout eggs and young when conditions are favorable.

SWIFTWATER COLORADO GILA
Gila robusta elegans Baird and Girard

Gila elegans Baird and Girard 1853A: 369
Gila emoryi ibid: 1853B: 388
Gila elegans, Girard, 1856B: 205 / 1859: 286–287
Gila elegans, Cooper, in Cronise, 1868: 496
Leuciscus elegans, Günther, 1868: 241
Gila elegans, Cope and Yarrow, 1875: 664
Gila emorii, ibid: 667
Gila elegans, Jordan and Copeland, 1878: 424
Gila emorii, ibid
Gila elegans, Jordan and Gilbert, 1882: 227–228
Gila emorii, ibid: 229–230
Gila elegans, Jordan, 1891: 27–28
Gila elegans, Evermann and Rutter, 1895: 482
Gila elegans, Jordan and Evermann, 1896B: 226–227
Gila emoryi, ibid: 227
Gila elegans, Gilbert and Schofield, 1898: 492
Gila elegans, Snyder, 1915: 580–581 / 1921: 28
Gila elegans, Pratt, 1923: 69
Gila elegans, Jordan, Evermann and Clark, 1930: 114
Gila emoryi, ibid
Gila elegans, Schrenkeisen, 1938: 104
Gila elegans, Moffett, 1942: 82 / 1943: 182
Gila elegans, Dill, 1944: 153–154
Gila elegans, Miller, 1945A: 104–105
Gila robusta elegans, Miller, 1946B: 410–415
Gila robusta elegans, Wallis, 1951: 90
Gila robusta elegans, Beckman, 1952: 44–45
Gila robusta elegans, La Rivers, 1952: 101
Gila robusta elegans, La Rivers and Trelease, 1952: 116
Gila robusta elegans, Eddy, 1957: 97
Gila robusta elegans, Shapovalov, Dill and Cordone, 1959: 172

ORIGINAL DESCRIPTION

"Closely allied to the preceding species [*G. robusta*]. Its body, however, is more slender, and its tail proportionally more elongated. The caudal fin is more deeply emarginate and more developed, as indeed are all the fins. The head is very much depressed and flattened on the snout. Eyes elliptical. The scales are proportionally more elongated than in the preceding species and are broader anterior than posteriorly; the lateral line has about ninety of them. The number of rays in the fins affords also a distinctive mark between both species. The ventrals have no rudiment of spiny ray. Color light brown.

"Fin rays: D III. 9. C 9. I. 9. 9. I. 10. A III. 10. V 9. P 16" (Baird and Girard 1853A: 369).

DIAGNOSIS

See Table I.

TABLE I

The known subspecies of *Gila robusta*, three of which are of concern to the Nevada area, can be tabulated as follows (data mainly from Miller 1946B: 414):

	Dorsal rays	Anal rays	Pelvic rays	Squamation	Nuchal hump	Peduncle depth in head	Habitat	Range
Gila robusta intermedia	7–8*–9	7–8*–9	8/8–9/9	Complete; lateral line 65–87; basal radii usually present	Absent	2.6–3.4	Small streams	Gila River basin
Gila robusta robusta	8–9*–10	7–8–9*–10	9/9*–10/10	Complete; lateral line 79–96; basal radii poorly developed	Slight in older fish	3.3–4.3	Small rivers	Colorado River basin
Gila robusta elegans	10–11	10–11	9/9*–10/10	Often lacking on dorsum, venter and caudal peduncle, or being only tiny embedded scales; lateral line 75–88; no basal radii	Conspicuous	5.0–6.5	Large, swift rivers	Colorado River basin
Gila robusta jordani	9	9	9/9	Complete; lateral line 89–94; basal radii conspicuous	Absent	3.3–4.1	Small streams	White River

*Usual condition

TYPE LOCALITY

Zuni River, New Mexico.

RANGE

The larger channels of the Colorado River system, in swift water, southwestern United States. Occurs in Nevada along the main channel of the bordering Colorado River, as well as in the lower part of the Virgin River.

TAXONOMY

It has only been within the last twenty years or so that the subspecificity of *elegans* in the *robusta* complex has been recognized (Miller 1946B); prior to that, it had held the status of a species for over 90 years.

Etymology—"elegans," elegant.

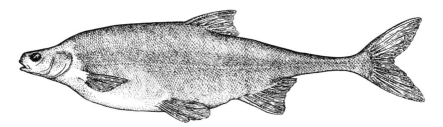

FIG. 187. Bonytail Colorado Gila, *Gila robusta elegans*.
Drawn by Silvio Santina.

LIFE HISTORY

The extent to which superficially distinct types can segregate out in specific populations encountering selective habitats is well-illustrated by this subspecies. The "Bonytail" represents environmental selection of those traits in the species which are of paramount importance to living in swift water—reduced squamation to decrease water resistance, greater streamlining and more powerful fins to make propulsion a more efficient process, and head dorsum concavity to aid the animal in steadying itself on the bottom against strong currents.

WHITE RIVER COLORADO GILA

Gila robusta jordani Tanner

(Nevada Bonytail Chub)

Gila jordani Tanner 1950
Gila robusta jordani, La Rivers, 1952: 101
Gila jordani, La Rivers and Trelease, 1952: 116
Gila jordani, Eddy, 1957: 98

ORIGINAL DESCRIPTION AND DIAGNOSIS

"Description of the type specimen BYU 9959: Head 3.6+; depth 4.3+; eye 6; D. 9; A. 9; teeth 2, 5-4, 2. Body somewhat elongate, more robust than *Gila elegans;* back only slightly elevated and head depressed; least depth of caudal peduncle about one-fifth of its length and 3¼ in length of head; mouth width 18 mm., the upper lip just below the level of the lower part of the orbit; eyes, small, anterior; fins about as in *G. robusta;* pectorals not reaching the ventrals in any specimens of this species; caudal fin forked; dorsal set a little back of the ventrals; middle caudal rays less than half the length of the upper caudal lobe; proximal portions of the fins in life whitish, in preservative an orange color; scales 26-91-14; in the series 89 to 94 on the lateral line; 26 to 27 above and 13 to 14 scales below the lateral line. In life the body is greenish with black blotches, while the preserved specimens are a lead gray with black blotches and white belly; young specimens have a silvery color when in spirits. Type about 6¼ inches long.

"Remarks: *Jordani* is a small scaled species belonging to the subgenus Gila, as proposed by Miller, 1945. It is related to *G. robusta.* It

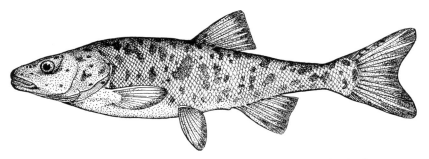

FIG. 188. White River Colorado Gila, *Gila robusta jordani.*
Drawn by Silvio Santina.

differs from *robusta* in body proportions such as head and depth and by being less elongate; by having more scales above, below, and on the lateral line; and in life by having a greenish color intermixed with black blotches" (Tanner 1950: 32-35). An illustration of the type accompanies the description.

TYPE LOCALITY
Crystal Spring, Pahranagat Valley, Lincoln County, Nevada.

RANGE
The subspecies is as yet unknown beyond the confines of Pahranagat Valley, but since it is obviously a relict of the Pleistocene White River system, the main stem of which connected Pahranagat Valley with the region to the north, and flowed south from the valley to ultimately enter the Colorado River through Moapa Valley, it may be found at any one of several isolated points north or south of the type locality.

TAXONOMY
While described as a species, *jordani* fits snugly into the *robusta* complex between *intermedia* and *robusta;* in actuality, *jordani* agrees so closely with typical *robusta* that were it not for the isolation of the

former so that it no longer has contact at any point with other populations of *robusta*, it would hardly be possible to regard it as a nameable entity. It is obviously an isolated segment of *G. robusta robusta* stock, from which it has differentiated less than has the more southern *G. r. intermedia*.

Etymology—Named for the famous ichthyologist, David Starr Jordan.

LIFE HISTORY

Nothing is known of the life cycle of this form. The writer first noticed it on a mid-winter trip (1948–1949) to Pahranagat Valley, where it was seen in Crystal Spring, but no specimens taken. It is a rapid swimmer, darting about so quickly that it is difficult to get a good look at it. A few passes with the seine netted us only some carp (*Cyprinus carpio*).

ECONOMICS

Small as it is, *Gila robusta jordani* is the largest carnivorous fish in the upper part of the White River system, probably feeding mainly on insects, but certainly taking its toll of the associated springfish, *Crenichthys baileyi*. Other than the part it plays in the local food chain, it is not involved in human economics.

UTAH GILA
Gila atraria (Girard)
(Utah Chub, Utah Lake Chub)

?*Tigoma lineata* Girard 1856B: 206 / 1859: 292–293
Siboma atraria ibid: 208 / 1859: 297–298
Tigoma squamata Gill 1861: 42
Leuciscus atrarius, Günther, 1868: 244
Protoporus domninus Cope 1872C: 473
Hybopsis bivittatus ibid: 474
Hybopsis timpanogensis Cope 1874A: 134
Siboma atraria, Cope and Yarrow, 1875: 667–668
Siboma atraria longiceps ibid: 668–669
Alburnops timpanogensis, Jordan and Copeland, 1878: 420
Protoporus domninus, ibid: 423
Siboma atraria, ibid: 424
Gila lineata, ibid
Myloleucus squamatus, ibid: 425
Squalius cruoreus Jordan and Gilbert 1881B: 460
Squalius rhomaleus ibid: 461
Squalius atrarius, ibid
Minnilus bivittatus, Jordan and Gilbert, 1882: 195–196
Minnilus timpanogensis, ibid: 196
Squalius cruoreus, ibid: 234
Squalius lineatus, ibid: 236–237
Squalius rhomaleus, ibid: 240
Squalius squamatus, ibid: 241
Squalius atrarius, ibid:

Leuciscus atrarius, Jordan, 1890C: 46, 48 / 1891: 33, 35
Squalius atrarius, ibid, 1891: 33
Leuciscus lineatus, Evermann and Rutter, 1895: 483
Leuciscus lineatus, Jordan and Evermann, 1896B: 232–233 / 1902: 70–71
Myloleucus pulverulentus, Fowler, 1913B: 71
Siboma atraria, Snyder, 1917A: 59–60 / 1921: 25–27
Tigoma lineata, ibid
Leuciscus lineatus, ibid: 59 / 25
Richardsonius atrarius, ibid: 59–60
Leuciscus lineatus, Pratt, 1923: 70
Tigoma atraria, Jordan, Evermann and Clark, 1930: 119
Tigoma atraria, Hatton, 1932: 34–37
Tigoma atraria, Hayes, 1935: 40–54
Siboma atraria, Tanner, 1936: 160, 169–170
Leuciscus lineatus, ibid: 170
Tigoma atraria, Schrenkeisen, 1938: 114
Gila atraria, Miller, 1945A: 105 / 1952: 23, 31–32
Gila atraria, Simon, 1951: 78–79
Gila atraria, La Rivers, 1952: 100
Gila atraria, La Rivers and Trelease, 1952: 116
Gila atraria, Eddy, 1957: 97
Gila atraria, Neuhold, 1957
Gila atraria, Moore, 1957: 100

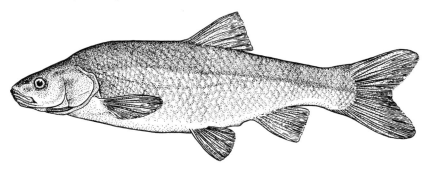

FIG. 189. Utah Gila, *Gila atraria*. Drawn by Silvio Santina.

ORIGINAL DESCRIPTION

"The largest specimen of this species which we have examined is about seven inches in length, and although small, compared to the specimens of *S. crassicauda*, to which we had to compare it, yet the distinctive features between the two species appear very striking. And first of all, the imbrication of the scales of *S. atraria*, is such as to expose more of their surface than in *S. crassicauda*, and moreover the lateral line in *S. atraria* runs along the seventh row of scales from the insertion of the ventrals upwards, leaving eleven rows above it, to the base of the dorsal fin, whilst in *S. crassicauda*, there are as many rows of scales below as above the lateral line. The absolute number of longitudinal rows of scales is the same in both species. The head is proportionally larger than in *S. crassicauda*, but the fins are much less developed. The ground color is olivaceous, the sides and the back being nearly black or brownish black, from the number of confluent maculae and dots. The fins themselves are blackish upon an olivaceous ground. The sides and upper part of the head, are likewise brownish black.

"Found in a spring, in Utah District, near the Desert, by Lt. E. G. Beckwith" (Girard 1856B: 208).

DIAGNOSIS

Head and body depth each about 3.5 times into body length; head wide, older specimens with a concave profile; mouth strongly oblique, lower jaw projecting beyond upper; maxillary attaining front edge of orbit; lips lacking barbels; eye small; pharyngeal teeth in two rows, usually 2/5–4/2, short, compact, one bearing a flat crushing surface; lateral line decurved; scales about 10/55–63/5; fins—dorsal usually 9-rayed, nearly median in position, inserted directly over pelvic fins—anal usually 8-rayed—caudal well-forked; caudal peduncle stout, least depth about 10 times in body length; intestine short, less than body length; color—olivaceous above, grading to lighter below; lengths up to about 16 inches.

TYPE LOCALITY

Fish Springs, Tooele County, Utah.

RANGE

Eastern section of the Great Basin; Bonneville system and the Snake River basin above Shoshone Falls and throughout all drainages of Utah; barely enters Nevada along its eastern border, in streams tributary to the Bonneville drainage system.

Apparently, Utah Gilas have been transplanted into various waters in eastern Nevada beyond their original range, which was the Bonneville system. Miller and Alcorn (1946: 182) record original testimony to the effect that early settlers brought this chub into Shoshone Springs, Spring Valley, White Pine County and to the Geyser Ranch in Duck (Lake) Valley, northern Lincoln County. They also found it on the Murphy Ranch, Steptoe Valley, White Pine County, assumed to be another introduction.

In his stream survey of the Snake Range (White Pine County), Ted Frantz found the gila only in Lake Creek (Big Spring).

TAXONOMY

The complex synonymicon is a good index of the amount of confusion which has surrounded *Gila atraria* during the more than 100 years it has been known. It is possible that Girard's *Tigoma lineata* was based on a specimen of *G. atraria*, but the type of *lineata* is lost, its type locality unknown, and the inadequacy of Girard's description is such that it could be applied with equal disquietude to several of his species. The fact that *G. atraria* has been placed in more than ten genera and redescribed as new time and again from various parts of its range by succeeding authors shows not only the lack of clear-cut characteristics among species of the group to which it belongs, but also the futility of such broad and vague generalizations which Girard and other early writers offered as descriptions of their species.

The species as currently accepted was chiefly stabilized in a recent resumé of the subgenera of *Gila* (Miller 1945A) where it was assigned to the subgenus *Siboma*, an early Girard genus. The Utah Gila is a *Siphateles*-appearing species which also presents many of the aspects of the redshiner (*Richardsonius*); in fact, Snyder (1917A: 59) considered it a *Richardsonius*.

Etymology—"atraria" blackish, from *ater* (*atra*-), black, and -*arius*, pertaining to.

LIFE HISTORY AND ECONOMICS

The author has been unable to find any suitable material published on the life history of this species. Neuhold's few comments of 1957 are mainly economic in nature:

"Among the non-game fishes that assume a role of importance in Utah is the Utah chub, *Gila atraria* (Girard), although it is not important economically. It has little value as a food fish, nor is it used in any commercial products. Recently, it has been used by fish hatchery people as a supplementary food material, but its utilization here is limited and has, as yet, reached no great proportions. It is of little value as a game fish since the sportsmen seldom attempt to catch them. Its real importance lies in the fact that the Utah chub reaches large numbers in areas in which more desirable game fish are present. Since the numbers of the chub are so large in some important trout waters, the competition for food and space with trout is suggested. Hazzard (1935) states that the Utah chub and trout are in direct competition for food in Fish Lake, Utah.

"The Utah chub inhabits a wide range in the state [Utah]. It has wide elevational and latitudinal distribution and is found in every drainage of the state. It is predominant at elevations between 5,000 and 9,000 feet. The lakes in which the Utah chub is present are also considered to be good habitat for trout and are being maintained and developed as such. Consequently when the Utah chub begins to outnumber the trout, cause for concern is justified. In the light of such developments a life-history study of the Utah chub is imperative.

"One of the most important phases of a life-history study is that concerned with age and growth. Collection of data was initiated in June of 1952 at Panguitch Lake and Navajo Lake in southern Utah" (217–218).

THE REDSHINERS

Genus *Richardsonius* Girard 1856

(Named after the prominent English zoölogist, Sir John Richardson)

KEY TO SPECIES

1. Rays of anal fin numbering from 8 to 9, rarely 10; body comparatively slender; Lahontan drainage system (*Richardsonius egregius*) _____ LAHONTAN REDSHINER

   Rays of anal fin from 10 to 22; body comparatively robust; Snake River tributaries of northern Nevada (*Richardsonius balteatus*) _____ COLUMBIA REDSHINER 2

2 (1.) Anal fin rays 10–13, usually 11–12; Bonneville basin and the upper Snake River drainage (*Richardsonius balteatus hydrophlox*)........ BONNEVILLE COLUMBIA REDSHINER

.... Anal fin rays 13–22, usually 14–18; Columbia-Snake system below the falls of Snake River; intergrades with the subspecies *R. b. hydrophlox* as below (*Richardsonius balteatus balteatus*)........... TYPICAL COLUMBIA REDSHINER

.... Anal fin rays usually 13–14; northeastern Nevada, in Snake River tributaries; intergrade between the two above subspecies ... *Richardsonius balteatus* : : *balteatus* X *hydrophlox*

LAHONTAN REDSHINER

Richardsonius egregius (Girard)

(Red-side Minnow, Red-side Bream, Lahontan Redside Shiner)

Tigoma egregia Girard 1859: 291–292 (Humboldt River)
Tigoma egregia, Günther, 1868: 211
Gila ardesiaca Cope 1875: 660
Gila egregia, Jordan and Copeland, 1878: 424
Gila ardesiaca, ibid
Squalius ardesiacus, Jordan and Gilbert, 1882: 235
Squalius egregius, ibid: 236
Squalius galtiae Cope 1883: 148 (Pyramid Lake)
Phoxinus clevelandi Eigenmann and Eigenmann 1889: 149
Phoxinus montanus, ibid, 1891: 1132
Leuciscus egregius, Jordan and Evermann, 1896B: 237
Richardsonius egregius, Snyder, 1917A: 54–58
Richardsonius microdon, ibid: 58–59
Tigoma egregia, ibid, 1921: 25
Leuciscus egregius, ibid: 25, 27–28
Richardsonius egregius, ibid: 28
Leuciscus egregius, Pratt, 1923: 70
Cheonda egregia, Jordan, Evermann and Clark, 1930: 118
Richardsonius egregius, Schrenkeisen, 1938: 113
Richardsonius egregius, Shapovalov, 1941: 444
Richardsonius microdon, ibid
Richardsonius egregius, Hubbs and Miller, 1943: 352, 356 / 1948A: 24
Richardsonius microdon, ibid
Richardsonius egregius, Miller, 1945A: 104
Richardsonius egregius, Kimsey, 1950: 438
Richardsonius egregius, Shapovalov and Dill, 1950: 386
Richardsonius egregius, La Rivers, 1952: 99–100
Richardsonius egregius, La Rivers and Trelease, 1952: 117
Richardsonius microdon, Böhlke, 1953: 38
Gila egregia, Eddy, 1957: 96
Richardsonius egregius, Moore, 1957: 101
Richardsonius egregius, Shapovalov, Dill and Cordone, 1959: 172

ORIGINAL DESCRIPTION

"Spec. Char.—Body rather elongated, sub-fusiform in profile. Head contained four times in the length, the caudal fin excluded. Snout sub-conical and thickish; gape of the mouth slightly oblique; jaws equal; posterior extremity of the maxillary extending to a vertical line drawn within the anterior rim of the orbit. Eye moderate sized, sub-circular; its diameter entering four times and a half in the length of the side of the head. Anterior margin of dorsal fin nearer the extremity of the snout than the fork of the caudal. Origin of ventrals equidistant between the insertion of the caudal fin and the extremity of the snout. Bluish black above; yellowish orange beneath.

"By its general aspect this species resembles *T. lineata,* whilst its structural characters are suggestive of *T. humboldti.* The dorsal and anal fin are nearly equal in size and shape; the upper edge of the former and the inferior edge of the latter are slightly sub-concave. The anterior margin of the anal is nearer the tip of the inferior lobe of the caudal than the isthmus; its base enters nine times in the total length, the caudal fin excluded. The ventrals are posteriorly sub-truncated; their tips not extending to the vent. The pectorals are rather slender, sub-ovate in their outline, their extremities not reaching the origin of the ventrals.

"D 1, 8 + 1; A 2, 9 + 1; C 6, 1, 9, 8, 1, 7; V 1, 8; P 13.

"The scales are about as large as in *T. humboldti,* and the lateral line is almost identical in its direction. Its system of coloration is likewise very similar to that of the species just alluded to. Dorsal region bluish black with a metallic reflect; middle of the flanks yellow, with numerous black dots constituting two dark zones, one above, the other below the lateral line, between which zones may be observed two streaks of red or golden orange. The belly exhibiting a uniform metallic yellow tint. The dorsal and caudal fins being greyish, whilst the remaining fins are more of a yellowish tint" (Girard 1859: 291–292).

DIAGNOSIS

Head and body depth each about 4 times in body length; head prominent; mouth rather oblique, jaws equal; maxillary nearly reaching front orbital edge; lips lacking barbels; eye large; pharyngeal teeth in two rows (rarely with 3 rows on one side), typically 2/4–5/2 (rarely 1/4–5/1), teeth of outer row prominently hooked, occasionally with a narrow crushing surface; lateral line weakly and smoothly decurved; scales 12–14/53–60/6–7, 27–33 before the dorsal fin; fins prominent—dorsal typically 8-rayed, rarely 7—anal 8–9 rayed, rarely 10, large—caudal moderately large, strongly forked; caudal peduncle stout, least depth 10–11 times in body length; intestine short, less than body length, S-shaped; peritoneum silvery, usually with some black spotting; color—in breeding season, one of our most brilliantly colored fishes, with brassy reflections and a bright red side stripe—ground color dark olivaceous above, lightening on sides and grading into silvery venter—fins yellowish-oranged; a small species, usually not much exceeding 4 inches in length.

TYPE LOCALITY

Girard's description was based on a single specimen credited to F. Creutzfeldt of Lt. E. G. Beckwith's party exploring for railroad routes in the West in 1853–1854. Creutzfeldt and others were killed by Indians in the Great Salt Lake basin before reaching Nevada (see Chronology, 1859, page _____ for details). Girard records no locality for the type, but, according to Snyder (1921: 28), the register of specimens at the U. S. National Museum bears an entry dated February 1857, listing the Humboldt River for this specimen.

RANGE

Confined to the extensive Lahontan drainage system of Nevada and adjacent California; Walker, Carson, Truckee, Susan, Quinn, Reese and Humboldt Rivers; Walker, Tahoe and Pyramid Lakes; and tributary streams and ponds. A very common species.

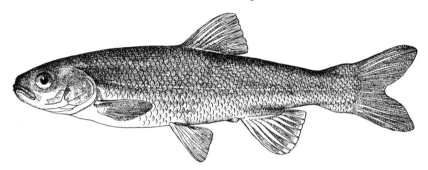

FIG. 190 Lahontan Redshiner, *Richardsonius egregius*.
Drawn by Silvio Santina.

Recent stream survey experiences of Nevada Fishery Biologists indicate that the redshiner appears now to be largely absent from the upper stream habitats that Snyder (1917A) recorded for it.

Ted Frantz tells me that shocking programs for the Carson Range (southern Washoe County) and the west side of the Toiyabe Range (Nye County) failed to produce any redshiners—and that none were found in the upper Reese River (Nye County), although they were present in the lower Reese (Lander County). Perhaps agricultural and other uses have depleted these upper reaches to the point where they are no longer available to the species.

While this species has been carried about by bait fishermen on many occasions, only two prominent translocations out of their native drainages are known to me.

In 1950, Kimsey recorded the presence of the Lahontan Redshiner and three other Lahontan species: Tahoe Sucker (*Catostomus tahoensis*), Lahontan Tui Chub (*Siphateles bicolor obesus*) and Lahontan Speckle Dace (*Rhinichthys osculus robustus*) in certain headwaters of the Sacramento River drainage. These may or may not have been from natural causes (see the Tahoe Sucker for the discussion of this).

In December 1956, Ted Frantz introduced Lahontan Redshiners into Railroad Valley Pond No. 2, Nye County, from the lower Reese River by Highway 50 bridge. On 21 July 1957, he repeated the plant with 20 redshiners.

TAXONOMY

Like many of the small western minnows, this species has been frequently misplaced and shifted through several genera (at least seven) and has accumulated several synonyms in a period of over 100 years. Girard established the genus *Richardsonius* in 1856 for a species described 20 years before by Richardson, but failed to recognize that his *Tigoma egregia* of three years later would also fit into *Richardsonius*. It was not until after the turn of the present century that *R. egregius* became stabilized as we currently know it, when Snyder published his valuable contribution on Lahontan fishes (1917A); at the same time, he established the name *Richardsonius microdon* which was later shown to be a hybrid between *R. egregius* and *Siphateles bicolor obesus* (Hubbs and Miller 1943: 353, 356). The first Lahontan synonym was the *Squalius galtiae* of Cope, 1883, from Pyramid Lake.

Only two species are presently recognized in *Richardsonius, egregius* and the following, *balteatus,* occupying adjacent but now unconnected drainages, and both part of the Nevada fauna.

Etymology—"egregius," surprising.

LIFE HISTORY

These are small, late spring-early summer spawning carnivorous cyprinids becoming one of our most resplendent fishes during their breeding period. Little is published of their life history beyond the extensive observations of Snyder (1917A), which we willingly resort to:

"This beautiful little fish is almost universally distributed throughout the brooks, rivers, and lakes of the region. It is found not only in the lower courses of the rivers where the water is deep and quiet, but it also stems the swift currents of the high mountain tributaries, following closely in the wake of the smallest trout. It is best known as a river species—the bait fish of the angler, the 'minny' of the small boy, and the food of the kingfisher and tern. It delights in the slow ripples and the quiet, shallow pools, where large numbers may be seen swimming lazily about over the submerged bars, occasionally turning their silvery sides to the bright sun. In the lakes it congregates in large schools, swimming about submerged logs, tops of fallen trees, wharves, and other sheltered places.

"The relationships of the species appear to be with *R. balteatus* and *R. hydrophlox,* in some characters approaching the latter more closely. It may not be out of place to remark, however, that speculation regarding this and other western species of the genus will remain more or less futile until something more definite is known of the fauna of the upper Columbia and the Bonneville Basins.

"This is one of the most brilliantly colored fishes of the West, not even excepting the trout. The brightest hues appear for a short time only during the breeding season and are seen at their best in the males.

"While examining Fallen Leaf Creek, a tributary of Lake Tahoe, on

June 10, large schools of this species were observed making their way up the smaller tributaries, progressing chiefly at night, when they might be seen or heard struggling over the small rapids and leaping the miniature falls. Both sexes were highly colored, the males especially so, where small whitish epidermal nodules appear on the surface of the head, on the opercles, throat, and body, each scale along the side having from one to three on its posterior edge. Later observation indicated that these fishes were at that time rapidly attaining the most advanced stages in their brilliant nuptial color.

"During the nuptial migration of *Catostomus*, before noted, large numbers of this species followed the female suckers, feeding on the eggs. Many eggs were found in the stomachs of the minnows. Neither male nor female spawning suckers objected to the presence of the minnows, the latter swarming about and at times darting over and under them.

"Many examples were caught with hook and line in Lake Tahoe. None of these exhibited bright colors, although taken at the time when others of the species were spawning in the streams near by. On dissection the ovaries of some of these were found to contain large eggs, although in most cases they were immature. Females here far outnumbered the males.

"On May 14 specimens in nuptial colors and nearly ripe eggs were secured in the Truckee River near Pyramid Lake. Examples taken at the same place April 24 had only traces of red color. Specimens from the Humboldt River collected after July 1 had spawned. These were generally lighter in color than those of the Tahoe region, as were also examples from Carson and Walker Rivers.

"A few specimens taken August 11 in Walker River near the outlet contained large numbers of almost fully developed eggs. Some rather brightly colored males were found there at the same time.

"The food seems to consist mostly of aquatic larvae and winged insects. No algae or other vegetable matter was found in the stomachs, except such as might have been swallowed by accident" (Snyder 1917A: 54-57).

ECONOMICS

A very important species in the food chain, converting portions of the voluminous aquatic insect populations into food-units convenient to the use of larger carnivorous game and sport fishes, such as trout. Its popularity as a bait fish in the catching of game species, while of some importance from the standpoint of human economics, is relatively inconsequential from the overall biological viewpoint.

COLUMBIA REDSHINER

Richardsonius balteatus (Richardson)

(Red-sided Minnow, Red-sided Bream, Northern Redside Shiner)

Cyprinus (Abramis) balteatus Richardson 1836: 301
Richardsonius lateralis Girard 1856B: 202 / 1859: 279-280
Tigoma humboldti Girard 1856B: 206 / 1859: 291
Richardsonius balteatus, Girard, 1859: 278-279

Abramis balteatus, Günther, 1868: 309
Abramis lateralis, ibid
Richardsonius balteatus, Jordan and Copeland, 1878: 425
Richardsonius lateralis, ibid
Squalius humboldti, Jordan and Gilbert, 1882: 234
Richardsonius balteatus, ibid: 251
Richardsonius lateralis, ibid
Leuciscus gilli Evermann 1891: 44
Leuciscus balteatus, Eigenmann, 1894: 112–113
Leuciscus balteatus lateralis, ibid
Leuciscus balteatus, Gilbert and Evermann, 1894: 46
Leuciscus balteatus, Jordan and Starks, 1895: 853
Leuciscus humboldti, Jordan and Evermann, 1896B: 236–237
Leuciscus balteatus, ibid: 238–239
Leuciscus balteatus, Snyder, 1908: 85, 174–175 / 1917A: 56
Richardsonius humboldti, ibid, 1917A: 60
Leuciscus humboldti, ibid, 1921: 25
Tigoma humboldti, ibid: 25, 28
Leuciscus balteatus, Pratt, 1923: 71
Richardsonius balteatus, Jordan, Evermann and Clark, 1930: 118
Cheonda humboldti, ibid: 119
Richardsonius balteatus balteatus, Schultz, 1936: 150
Richardsonius balteatus, Schrenkeisen, 1938: 112
Richardsonius humboldti, ibid: 113
Richardsonius balteatus, Miller, 1945A: 104, 109 / 1952: 31
Tigoma humboldti, ibid
Richardsonius balteatus, Carl and Clemens, 1948: 71, 84–85
Richardsonius balteatus, Miller and Miller, 1948: 183
Richardsonius balteatus, Weisel and Newman, 1951
Richardsonius balteatus, La Rivers, 1952: 100
Richardsonius balteatus, La Rivers and Trelease, 1952: 117
Gila balteata, Eddy, 1957: 96
Richardsonius balteatus, Moore, 1957: 101
Richardsonius balteatus, Slastenenko, 1958A: 208–209
Richardsonius balteatus, Carl, Clemens and Lindsey, 1959: 105–107

ORIGINAL DESCRIPTION

"This pretty little bream, which is an inhabitant of the Columbia, was sent to me by Dr. Gairdner.

"Description.

"Colour.—'Back of head and body mountain-green, with iridescent tints of yellow and blue. Belly silvery-white.—A bright gold-yellow band behind the eye on the margin of the preoperculum, and a broad scarlet-red stripe beneath the lateral line, extending from the gill-opening to the anal. Fins of an uniform greenish-grey colour without brilliancy' (Gairdner).

"Form much compressed, the depth of the body being equal to one-fourth of the distance between the tip of the snout and the caudal fork, while its thickness is only equal to a tenth of the same distance. The profile curves moderately from the snout to the dorsal, just before which the depth of the body is greatest, but it continues to be considerable at the insertion of the anal, the belly running as it were into an acute edge at that place: the short piece of the tail behind the anal is narrow. The *head*, forming exactly one-fourth of the length of the fish, excluding the caudal, has a conical profile when the mouth is shut, the apex being formed by the tip of the lower jaw, which projects a very little beyond the commissure of the mouth. The top of the head is comparatively broad and rounded, its thickness at the nape being equal to that of any part of the body, and the snout, when viewed from above, appearing obtuse. *Eyes* large, much nearer to the snout than to the gill-opening. *Nostrils* near the eyes. *Mouth* toothless, small, its commissure descending obliquely and not reaching farther back than the nostrils: the lower jaw, when depressed, projects considerably beyond the upper one. Gill-covers.—Bony *operculum* quadrangular, its slightly-convex under edge being equal to the anterior one, and fully one-third longer than the upper or posterior one: the latter is widely emarginated, or cut with a concave curve. The *suboperculum*, one-third of the height of the operculum, is rounded off posteriorly in the segment of a circle, forming an obtuse tip to the gill-cover: both these bones are edged with membrane. *Preoperculum* narrow.

"Scales thin and sub-orbicular, their transverse diameter being rather greater than their longitudinal one. A few crenatures may be obscurely seen on their basal edges with a lens, and very faint lines proceeding from them towards the centre. There are about fifty-seven scales on the lateral line, and the greatest diameter of one taken from the anterior part of the sides measures a line and a half. A linear inch includes sixteen or seventeen of them *in situ*. The *lateral line* is curved convexly downwards, just before the ventrals, rising so as to run straight through the tail. It is formed by a short tube on each scale.

"Fins.—Br. 3-3; P. 17; D. 11; V. 9; A. 19 to 22; C. 19-7/7.

"The *ventrals* are attached a little anterior to the middle, between the tip of the snout and the base of the caudal, or opposite to the eighth ray of the dorsal; their tips reach to the anal. The anal and dorsal are high anteriorly, and become considerably lower posteriorly, with a slight concave sweep; the articulations of the first ray of each are obsolete.

"The *air-bladder* is divided by a contraction into two portions, of which the lower one is the largest. There are forty vertebrae in the spine.

"Dimensions.

	Inches	Lines
Length from tip of snout to end of caudal lobes	5	9½
Length from end of central caudal rays	5	0
Length from base of ditto	4	7
Length from end of anal	4	0
Length from end of dorsal	3	2½
Length from beginning of anal	2	11
Length from anus	2	10½
Length from beginning of dorsal	2	6½
Length from ventrals	2	1
Length from tip of gill-cover	1	1
Length from nape	0	9
Length from posterior edge of orbit	0	6¾
Length from anterior ditto	0	3⅙
Length from nostrils	0	2⅓
Length from tip of labials	0	3⅔
Length of lower jaw	0	4¾
Length of pectorals	0	11½
Length of ventrals	0	8½
Length of attachment of anal	1	1½
Height of anal anteriorly	0	9
Height of anal posteriorly	0	2½
Height of dorsal anteriorly	0	10½
Height of dorsal posteriorly	0	3½
Length of attachment of dorsal	0	8
Length of lobes of caudal	1	3½
Length of its longest rays	1	2
Length of its central ditto	0	4¾
Depth of caudal fork	0	9½
Depth of body at ventrals	1	3½
Thickness of body where greatest	0	6"

(Richardson 1836: 301–302).

DIAGNOSIS

Head 4.0 to 4.5 times in body length, body depth 3 to 4 times in body length; head rather small; mouth terminal, somewhat oblique, short, jaws equal; maxillary about reaching front edge of orbit; lips lacking barbels; eye large; pharyngeal teeth in two rows, 2/5–4/2, lacking crushing surfaces; lateral line weakly decurved; scales 12–13/55–63/ 5–6, about 33 in front of the dorsal fin; fins—dorsal 9–10 rayed—anal 10–22 rayed, usually 11–18—caudal well-forked; caudal peduncle stout, least depth about 11 times into body length; intestine short, less than body length; peritoneum silvery; color—much like *Richardsonius egregius*, the males becoming brilliantly streaked with red and brassy reflections during breeding time; about 4 inches in length.

TYPE LOCALITY

Supposedly Ft. Vancouver or vicinity, Columbia River, Washington (Gilbert and Evermann 1894: 46).

RANGE

Columbia River basin and the Bonneville drainage system, entering Nevada along the north edge of the northeastern corner in the affluents to the Snake River: Salmon, Jarbidge, Bruneau and Owyhee Rivers, and along the eastern border from the Bonneville system of Utah.

TAXONOMY

The species *Richardsonius balteatus,* like other western minnows just discussed, has been variously placed in several genera and has accumulated several synonyms in its nearly 125 years of nomenclatural existence. Of most interest, from the Nevada standpoint, is the history of the synonym *R. humboldti.* Described as *Tigoma humboldti* by Girard in 1856, it led a varied existence; Snyder (1917A: 60) included a brief footnote on it, calling it a "nominal" species, and giving such of its history as was known—the major problem has always been the collection locality of the only specimen, the type, which is still in existence (U. S. National Museum). It is listed as having come from the Humboldt River, as its name obviously implies, supposedly taken by F. Creutzfeldt,[1] botanist on the Gunnison-Beckwith 1853–1854 railroad surveys through the Great Basin. However, J. S. Bowman is known to have collected two specimens also. It was provisionally carried on lists for many years, always with the notation that additional specimens had never been found in the Humboldt River, which was inhabited only by *R. egregius.* Finally, in 1945A, Miller (p. 104) decided, after an examination of the type, that it was synonymous with *R. balteatus;* Jordan, Evermann and Clark, in 1930, had off-handedly suggested this with the statement: it . . . "may be identical with . . . *hydrophlox,* but this seems unlikely" (p. 119). Even though structurally it seems an obvious *R. balteatus,* the erroneous type locality kept it lingering on in the literature until the weight of evidence was overwhelming enough to force the issue. There have been similar cases involving specimens of many plants and animals collected during these early government surveys where resolution of the problems was equally difficult due to lack of locality data, or confusion of labels.

Etymology—"balteatus," girdled.

LIFE HISTORY

Weisel and Newman (1951) have an excellent summary of certain phases of the life history of this species in western Montana. From this, the following account is largely drawn.

Spawning from April to July, depending upon the altitude, and at two years of age or older, the Columbia Redshiner takes on coloration surpassed by few fishes in brilliance.

"Ripe males are brassy on all fins, along the sides ventral to the dark lateral band, and in a narrow strip above this dark area. They also have a brassy half-moon below the eye. There is a dark, rosy wash just behind the operculum and at the base of the pectorals, which continues posteriorly in a narrower band to just above the origin of the anal fin.

---

[1]Actually, Creutzfeldt and Gunnison were killed by Indians near Sevier Lake, Utah, on Oct. 25, 1853, never reaching the Humboldt.

The top of the head, and back, and the lateral band are dark olive to black. The small but numerous tubercles are distributed most profusely over the top of the head and the back, from the nostrils to the origin of the dorsal fin. The pectoral, pelvic and dorsal fins also are frequently tuberculate, and a roughness can be felt on the scales. The gravid females show the same general coloration as the males but it is not so brilliant. Instead of having the brassy hue below the eye and on the sides, they are pale gold. Tubercles are generally lacking; if present, they are only weakly developed" (Weisel and Newman 1951: 188).

In spawning, courtship activities seem to be lacking as was nest-building or subsequent defensive patterns. The redshiners gathered in schools up to about 50 individuals, darting into the riffle and adjacent areas for the rather erratic spawning behavior which usually seemed to involve small groups of half-a-dozen or so individuals. They could discern no "pairing-off," and only a few eggs seemed to be deposited at any one time. Individuals appeared to take part in spawning several times. Their egg counts averaged somewhat less than 2,000 for 6 females and they estimated that it took each female several days to release this many eggs since no spawned clusters were found which exceeded 20 eggs.

The fish immediately ate any of the adhesive, demersal (sinking) eggs that did not fall into crevices. They also fed regularly during spawning on the usual algae and arthropods, etc., and though many were heavily parasitized, this did not seem to affect their general condition.

"Redside shiners are definitely cannibalistic and are probably their own worst egg predators" (p. 189).

"At laboratory temperatures of 21° to 23° C., the eggs hatch in 3 to 7 days. The young remain as prolarvae for 8 days. During this period they are quietly hidden in the rocks of the spawning area. The post-larval stage lasts until they are about 46 days old. They take food when 10 days old. Although the yolk sac is absorbed in these young, the pelvic fins are not developed and a portion of the dorsal and ventral fin folds persists. The postlarvae are active swimmers and apparently make no effort to remain hidden. Juveniles are essentially similar to adults" (p. 194).

ECONOMICS

See the Lahontan Redshiner (*Richardsonius egregius*).

BONNEVILLE COLUMBIA REDSHINER

Richardsonius balteatus hydrophlox (Cope)

(Silverside Minnow, Bonneville Redside Shiner)

Clinostomus hydrophlox Cope 1872C: 475–476
Clinostomus montanus ibid: 476 / 1874A: 136
Clinostomus taenia ibid 1874A: 133
Gila montana, Cope and Yarrow, 1875: 657–658
Gila hydrophlox, ibid: 658
Gila taenia, ibid: 658–660

Gila hydrophlox, Jordan and Copeland, 1878: 424
Gila montana, ibid
Gila taenia, ibid
Squalius hydrophlox, Jordan and Gilbert, 1882: 233
Squalius taenia, ibid: 234
Squalius montanus, ibid
Leuciscus montanus, Jordan, 1891: 32
Leuciscus hydrophlox, ibid: 48
Leuciscus hydrophlox, Jordan and Evermann, 1896B: 238
Richardsonius hydrophlox, Snyder, 1917A: 59–60 / 1921: 27
Leuciscus hydrophlox, Pratt, 1923: 70
Cheonda hydrophlox, Jordan, Evermann and Clark, 1930: 119
Richardsonius hydrophlox, Hayes, 1935: 55–66
Richardsonius balteatus hydrophlox, Schultz, 1936: 150
Richardsonius hydrophlox, Tanner, 1936: 160, 168–169
Richardsonius hydrophlox, Schrenkeisen, 1938: 113
Richardsonius balteatus hydrophlox, Miller and Miller, 1948: 176, 183
Richardsonius balteatus hydrophlox, Simon, 1951: 81–82
Richardsonius balteatus hydrophlox, La Rivers, 1952: 100
Richardsonius balteatus hydrophlox, La Rivers and Trelease, 1952: 117
Richardsonius balteatus hydrophlox, Miller, 1952: 31

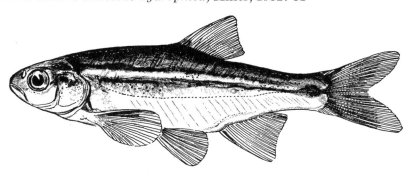

FIG. 191. Bonneville Columbia Redshiner, *Richardsonius balteatus hydrophlox*. Drawn by William L. Brudon for Miller 1952. Courtesy California Department of Fish and Game and Leo Shapovalov.

ORIGINAL DESCRIPTION

"This species and the following are typical forms of the genus, and interesting as the first that have been detected west of the Mississippi River. Length of head 4.75 times in total, exclusive of caudal fin; depth of body 4.5 times in same. Eye 5 times in head, one and a half times in interorbital width. Front straight; lower jaw projecting beyond upper; mouth descending; end of maxillary just reaching line of orbit. Isthmus narrow. Teeth, 5.2–2.4. Scales, 15–58–7. Radii, D. I. 8; A. I. 11. Ventrals not reaching anal. Length, 6 inches.

"Color above olive, with a blackish inferior border, extending from the superior margin of the orbit. Below this, a crimson band, and still lower, a blackish band, passing from the epiclavicular region above the lateral line to the basis of the caudal fin. Below this, crimson in front, silvery behind. Fins unspotted. Suborbital bones crimson; cheek golden." (Cope 1872C: 475–476).

DIAGNOSIS

Distinguished chiefly by its smaller number of anal fin rays, the usual number being 11 or 12, but varying from extremes of 10 to 13. Typically, *R. b. balteatus* has 14–18 as the anal fin ray number, varying from extremes of 13 to 22.

TYPE LOCALITY

Blackfoot Creek, Idaho.

RANGE

Occupies the Columbia River drainage basin at the higher altitudes and tributaries above the major falls, into Utah, Idaho and Nevada (basin of the upper Snake River, Salt Lake basin, and some of the rivers of eastern Washington).

TAXONOMY

Much that has been said concerning the variable and confused history of previously discussed western minnows applies to this subspecies. It appears to be the populations of *R. balteatus* of the more inland portions of the Columbia system, only slightly differentiated from typical *balteatus* which inhabits the lower, more western sections of the drainage.

Etymology—"hydrophlox," water flame, alluding to the brilliant breeding colors.

LIFE HISTORY AND ECONOMICS

See Northern and Lahontan Redshiners.

CHUBS

Genus *Siphateles* Cope 1883
("Incomplete siphon," meaning obscure)

TUI CHUB

Siphateles bicolor (Girard)
(Klamath Roach, Klamath Chub)

Algansea bicolor Girard 1856B: 183 / 1859: 238–239
Leuciscus bicolor, Günther, 1868: 245
Leucos bicolor, Jordan and Henshaw, 1878: 193–194
Leucus bicolor, Jordan and Gilbert, 1882B: 246
Leucus anticus, ibid
Myloleucus parovanus Cope 1883: 143
Myloleucus bicolor, Jordan, 1890B: 287
Rutilus bicolor, Jordan and Evermann, 1896B: 244–245
Rutilus bicolor, Rutter, 1903: 147 / 1908: 135–136
Rutilus bicolor, Snyder, 1908: 86, 178 / 1917A: 60
Myloleucus parovanus, ibid, 1917A: 60
Algansea obesa, ibid
Siphateles bicolor, ibid
Rutilus bicolor, Pratt, 1923: 72
Siphateles bicolor, Jordan, Evermann and Clark, 1930: 122

Siphateles bicolor bicolor, Schultz, 1936: 146
Siphateles bicolor, Schrenkeisen, 1938: 119
Siphateles bicolor, Shapovalov, 1941: 444
Siphateles bicolor, Hubbs and Miller, 1948B: various isolated mentions
Siphateles bicolor, Shapovalov and Dill, 1950: 386
Siphateles bicolor, Harry, 1951
Siphateles bicolor, La Rivers, 1952: 98
Siphateles bicolor, La Rivers and Trelease, 1952: 117
Siphateles bicolor, Eddy, 1957: 98
Siphateles bicolor, Moore, 1957: 106
Siphateles bicolor, Shapovalov, Dill and Cordone, 1959: 172

ORIGINAL DESCRIPTION

"Of all the species hitherto known of this genus, the one here referred to has the largest scales, five rows of which may be counted from the origin of the ventrals to the lateral line, and nine from the lateral line to the anterior margin of the dorsal, in all fifteen rows. The ventrals are inserted a little in advance of the anterior margin of the dorsal. The body is thickest anteriorly, and tapers backwards; the nape is slightly swollen. The head enters about four times and a half in the total length. The back and sides are of a metallic bluish black, intermingled on the lower half of the flank with a golden hue. The inferior surface is white, contrasting with the color of the back.

"Caught in Klamath Lake, O. T., by Dr. John S. Newberry, under Lieut. R. S. Williamson" (Girard 1856B: 183).

DIAGNOSIS

Head and body depth each from 3.5 to 4.0 times in body length; head rather pointed; mouth moderately oblique; maxillary small, not reaching anterior edge of orbit; lips lacking barbels; eye large; pharyngeal teeth in a single row, usually 4/5, but often 5/5 and rarely 4/4, large, somewhat hooked and with narrow crushing surfaces—arch with hind limb varying from larger than front limb, to smaller; lateral line very weakly decurved; scales 10–16/41–63/7–8, 22–35 before the dorsal fin; fins—dorsal 8–9 rayed, rarely 7, rather small, inserted directly over pelvic fins—anal 7–9 rayed, rarely 10–11, rather small; caudal rather prominent, moderately forked; caudal peduncle stout, least depth about 9 times in body length; peritoneum dusky; intestine short, about equal to body length; color—dark olive above, fading into the white venter—specimens from deep or winter or roily waters are generally silvery and pale overall—young individuals have a noticeable dark lateral stripe—breeding fish usually take on yellow or pink tints in the fins, and occasionally appear with a faint lateral pink stripe; seldom exceed a foot in length.

TYPE LOCALITY

Klamath Lakes, Oregon.

RANGE

Drainage systems on both sides of the Sierra Nevada Mountains, from the San Joaquin and Lahontan systems north to the lakes of southern Oregon and the Columbia River.

TAXONOMY

The situation is here somewhat the same as that involving *Gila robusta*. The several presently recognized subspecies of *Siphateles bicolor* were all originally described as species, and enjoyed specific standing until well after the turn of the present century. Even with this, and the allocation of these entities to several successively different genera, few synonyms have accumulated. The genus *Siphateles* was based by Cope on young examples of *S. b. obesus* from Pyramid Lake, which he described as *S. vittatus* in 1883.

LIFE HISTORY AND ECONOMICS

See *Siphateles bicolor obesus*.

LAHONTAN TUI CHUB

Siphateles bicolor obesus (Girard)

(Lake Chub, Lake Minnow, Pyramid Lake Roach)

Algansea obesa Girard 1856B: 183 (Humboldt River) / 1859: 239
Leuciscus obesus, Günther, 1868: 245
Leucos obesus, Jordan and Henshaw, 1878: 192–193 (Lahontan system)
Leucos formosus, ibid: 193 (Washoe Lake)
Leucus obesus, Jordan and Gilbert, 1882B: 245
Leucus formosus, ibid: 1615
Leucus olivaceus Cope 1883: 145 (Pyramid Lake)
Leucus dimidiatus ibid (Pyramid Lake)
Siphateles vittatus ibid: 146 (Pyramid Lake)
Leucus olivaceus, Russell, 1885: 62
Leucus dimidiatus ibid
Siphateles lineatus ibid
Myloleucus obesus, Jordan, 1890B: 287–288
Algansea obesa, Eigenmann and Eigenmann, 1891: 1132 (Lake Tahoe)
Algansea olivacea, ibid (Donner Lake)
Rutilus symmetricus, Gilbert, 1893: 231
Rutilus symmetricus (in part), Jordan and Evermann, 1896B: 245–246
Rutilus obesus, ibid
Rutilus olivaceus, Fowler, 1913B: 66
Leucus vittatus, ibid
Leucus dimidiatus, ibid
Richardsonius microdon (in part) Snyder, 1917A: 58–59 (Lake Tahoe)
Siphateles obesus, Snyder, 1917A: 60–63 (Pyramid Lake) / 1917B: 202–204 / 1918: 298 / 1921: 28
Leucidius pectinifer, Snyder, 1917A: 64–67 (Pyramid Lake)
Rutilus symmetricus (in part), Snyder, 1921: 25
Algansea obesa, Snyder, 1921: 28
Rutilus olivaceus, Pratt, 1923: 72
Hesperoleucus symmetricus (in part), ibid
Siphateles olivaceus, Jordan, Evermann and Clark, 1930: 122
Siphateles obesus, ibid
Hesperoleucus symmetricus (in part), ibid: 121

Leucidius pectinifer, ibid: 123
Siphateles obesus, Brues, 1932: 279
Siphateles bicolor obesus, Schultz, 1936: 147
Siphateles obesus, Hutchinson, 1937: 86, 92
Leucidius pectinifer, ibid: 68, 86
Siphateles olivaceus, Schrenkeisen, 1938: 119
Siphateles obesus, ibid
Siphateles obesus, Shapovalov, 1941: 444
Siphateles obesus, Alcorn, 1943: 35
Leucidius pectinifer, ibid
Siphateles obesus obesus, Hubbs and Miller, 1943: 352 / 1948A: 4, 24
Siphateles obesus pectinifer, ibid: 1948B: 37
Siphateles obesus pectinifer, Miller, 1949A: 448
Siphateles bicolor obesus, Kimsey, 1950: 438
Siphateles bicolor obesus, Shapovalov and Dill, 1950: 386
Siphateles bicolor pectinifer, ibid
Siphateles bicolor obesus, La Rivers, 1952: 98
Siphateles bicolor obesus, La Rivers and Trelease, 1952: 117
Siphateles bicolor pectinifer, ibid
Leucidius pectinifer, ibid
Leucidius pectinifer, Böhlke, 1953: 33
Richardsonius microdon, ibid: 38
Siphateles bicolor obesus, Shapovalov, Dill and Cordone, 1959: 172
Siphateles bicolor pectinifer, ibid

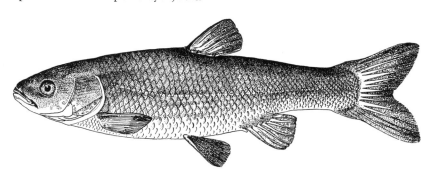

FIG. 192. Lahontan Tui Chub, *Siphateles bicolor obesus*.
Drawn by Silvio Santina.

ORIGINAL DESCRIPTION

"A very corpulent species covered with scales of moderate development, and so far, the smallest in the genus. The depth is contained about three times in the length, caudal fin excluded. There are eight longitudinal rows of scales between the origin of the ventrals and the lateral line, and fourteen rows above it to the anterior margin of the dorsal, in all twenty-three rows. Dorsal region bluish grey; sides greyish; belly yellowish.

"Specimens of this species were collected in the waters of the Humboldt River by the late J. Soulé Bowman and Lieut. E. G. Beckwith" (Girard 1856B: 183).

DIAGNOSIS

Differs chiefly from other subspecies of *Siphateles bicolor* in having a higher average scale count (smaller scales)—scales in the lateral line, for example, averaging about 55 for *S. b. obesus*, and about 45–50 for *S. b. formosus* (Sacramento Valley), and *S. b. bicolor* (Klamath Lakes). Geographical segregation is the most practical criterion to use for these subspecies.

TYPE LOCALITY

Humboldt River, Nevada.

RANGE

Streams and lakes of the Lahontan drainage system of western Nevada and adjacent California: Rivers Walker, Carson, Truckee, Susan, Quinn, Humboldt and Reese; Lakes Walker, Tahoe and Pyramid; and tributary ponds and streams. Chubs occur in several completely isolated interior valleys, such as Big Smoky and Little Fish Lake Valleys in Nye County and Dixie Valley in Churchill County, etc. Chubs of uncertain origin are found in Fish Lake Valley, Esmeralda County, and Ted Frantz informs me that there are some in Potash Well No. 7 of Railroad Valley, Nye County (these last were probably introduced).

TAXONOMY

The subspecies *Siphateles bicolor obesus* has developed perhaps the most complex synonymicon of any member of the Nevada fish fauna, as an examination of the synonymy readily shows. Originally described by Girard in 1856 with one of the briefest of Girard's inadequate descriptions of this period, *S. b. obesus* has been assigned to at least seven genera in nearly a hundred years and has acquired four synonyms, all of which, fortunately, have actually caused little trouble because of the geographic discontinuity of the various *obesus* populations.

The remarkable E. D. Cope, paleontologist, herpetologist and part-time fancier of fish, in an enthusiastic moment, contributed three of the four synonyms in an 1883 publication based on Pyramid Lake specimens. Cope had made a trip through the West just prior to this paper, and visited Pyramid Lake, where he was impressed with the abundance of fish life. The only lasting name in this family that Cope contributed in his paper was the generic term *Siphateles*, but it remained for Snyder, in 1917(A), to properly associate *obesus* as a *Siphateles*, and for Schultz (1936) as first reviser, to indicate the synonymy and to choose *bicolor* over *obesus*. Snyder's solution of one problem was somewhat counteracted by his posing of another; in the same paper, he proposed the new genus and species *Leucidius pectinifer* for a Pyramid Lake chub which differed from *Siphateles* only in having much finer and more numerous gillrakers. Snyder felt this to be a generic character, and in spite of the fact that *"Leucidius"* was completely indistinguishable from *Siphateles* on any other count, he was so impressed with the importance of the gillraker difference as to make a statement of affinities which completely excluded *Siphateles!*

"*Leucidius* is a lake minnow. It takes its place at once among an array of peculiar western genera, *Pogonichthys, Mylopharodon. Orthodon, Lavinia, Mylocheilus, Gila, Meda, Tiaroga,* and others" (1917A: 64).

In subsequent years, *pectinifer* was placed in the correct genus and considered as a subspecies of both *obesus* (Hubbs and Miller 1943: 352) and *bicolor* (Shapovalov and Dill 1950: 386). The former authors still maintain that *pectinifer* is a nameable taxonomic entity on the subspecific level, while the present writer feels it is a straight synonym (see Life History).

LIFE HISTORY

A large, fluviatile-lacustrine, omnivorous minnow, spawning in late spring-early summer, and extremely abundant in most of its varied habitats. Snyder's excellent observations have yet to be improved upon —he listed his notes under two species, *Siphateles obesus* and *Leucidius pectinifer*. Speaking of the first:

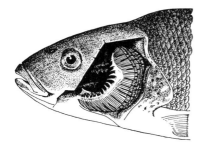 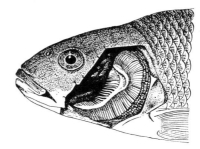

FIG. 193. Lahontan Tui Chub, *Siphateles bicolor obesus*, showing coarse- and fine-gill rakered forms. Re-drawn from Snyder (1917A) by Silvio Santina.

"In life the color is deep olive above, lighter on the sides, white below; upper parts and sides with a very pronounced brassy reflection. Some of the scales are darker than the others; some are pinkish. Fins olive, with a little red. In some examples the belly is suffused with yellow and the fins are strongly tinged with red. Others are more green than olive, and some have the dark lateral color extending almost to the ventral surface. The pink and yellow tints are more prominent during the breeding season. The green is more pronounced in lake specimens. In an occasional individual a faint pinkish stripe may be detected along the side, its outline and extent being very indefinite. In the rivers the general color is sometimes lighter. Young specimens have a narrow dark stripe along the lateral line.

"On May 24 this species had begun to spawn in Pyramid Lake, at least the eggs were then ripe, and the milt flowed from the males when they were touched. Small and medium sized examples caught in Lake Tahoe June 15 were not yet ready to spawn.

"Large individuals (12 to 14 inches) do not appear to come near shore, at least during the daytime. During the summer they may be secured with a small spinner trolled at a depth of 20 feet or more.

They bite readily when the hook is baited with angleworms. When still fishing, specimens were caught only at a long distance from the boat and at a depth of about 20 feet. A line so rigged that the bait would settle 20 feet below the surface, and cast out from shore where the bottom shelved off very rapidly, would generally secure specimens. Many smaller ones were taken in the seine and gill nets near shore along with *L. pectinifer*. Nets set at night were sure to be full of fishes in the morning.

"Large schools of lake chubs gather about the wharves, fallen trees, and other sheltered places. At the mouth of Fallen Leak Creek, Lake Tahoe, at times the rising water slowly spreads out over the meadows, and when covering the ground but a few inches is invaded by great schools of this species. After sundown they appear in countless numbers, thrashing about in the grass and rushes. When approached they either scurry off in great haste, sometimes diving into a bunch of grass, or settle down perfectly still. On the approach of daylight they return to deeper water. Dissection showed that they were probably feeding, certainly not spawning. Algae, bits of plants, and fragments of insects were found in their stomachs." (1917A: 62).

Concerning *Leucidius pectinifer*:

"This seems to be the most abundantly represented species in Pyramid Lake, approaching the shore at times in enormous schools. Perched on a high tufa crag near the shore one may observe countless numbers of these fish slowly passing through the clear water. From the cliffs above they resemble large purple clouds reflected from the green surface of the lake. They bite eagerly at a baited hook, a small spoon, or an artificial fly. It is said that during the winter very few are seen.

"On May 16 the writer began observations on the western side of Pyramid Lake at The Willows [Sutcliffe]. The weather was then cold and squally, the lake rough and forbidding. Relatively few fishes were seen, although large schools of suckers were spawning and a few minnows might be observed here and there among the suckers and in the algae, while an occasional trout was caught by the Indians. On May 20 the weather suddenly settled and became warm. As the sun went down, its last rays from over the mountains fell on the surface of the lake, which was as calm and placid as a great mirror. About 2 o'clock on the following morning there was heard a vigorous lapping of the water, which in the quiet air appeared entirely without cause until it was found to accompany the leaping of vast numbers of fishes. Far out and up and down the shores the surface of the water fairly boiled. Spring had come, and with it, in the dim light of the early morning, myriads of fishes from the depths of the lake. Daylight revealed them everywhere, along the shore, among the bowlders, and in the algae, hovering in enormous schools over the bars and moving about in the clear water of the sheltered bays. From this time the suckers rapidly disappeared, while the large trout approached the shore in their eagerness to feed on the luckless minnows.

"Most of the fishes engaged in this particular migration were *L. pectinifer*, although many examples of *S. obesus* were seen among them. At this time the latter were often taken in deep water with a troll at some distance from shore. Most of these fish (*L. pectinifer*) were not ready to spawn, and no ripe individuals were seen until May 24, after which date spawning soon began.

"Residents report that during the summer large numbers of minnows frequent the inshore waters, but that in September they disappear and are seen no more until the following spring. Late in the summer great numbers of small minnows swarm in the shallows near shore. Young specimens of this species were taken from several schools in protected, shallow bays in Winnemucca Lake, June 17, 1913. They were of two distinct sizes, measuring, respectively, 42 to 58 and 65 to 80 millimeters. There was a black lateral stripe extending along the side, broader, darker, and more definite in outline posteriorly and enlarging greatly near base of caudal fin, where there is a more or less distinct dark spot. The scales of the lateral line are very thin, the tube is large, and when the scales are rubbed off the line appears broken or incomplete.

"The species appear to be entirely lacustrine" (1917A: 66–67).

Although Pyramid Lake has changed in many respects since Snyder's observations nearly 50 years ago, the general chub picture remains much the same. The decreased inflow (see introduction) which has dropped the lake level some 70 feet and increased the concentration of solutes (particularly sodium chloride) has not as yet made the lake uninhabitable and it teems with chub as it did in Snyder's day. During the summer, an observer flying over the lake can see dense, blackish schools of chub everywhere, in shoal water and at the surface over deep water, each school made up, usually, of thousands of individuals, which dissolve into deeper water when the plane approaches within a couple hundred feet of them. In the winter they disappear into deeper water, but do not become dormant. Stomach analyses of specimens taken in a gill net in 70 feet of water by Thomas Trelease and the writer contained identifiable remains of diatoms, microcrustacea, hydracarina and tendipedid larvae to the practical exclusion of everything else.

The omnivorousness of the species is readily apparent in their summer feeding habits. It is common to see an algal-covered rock whose surface is just breaking water completely surrounded by chub feeding on the algae and its high protein population of aquatic micro-Arthropoda (chiefly Crustacea). Schools swimming within casting distance of shore rocks will avidly take small spinners as quickly as they can be tossed to them, until the school has passed. The obviously large number of progeny which result from the above-mentioned populations of adult chub are evident everywhere; it is possible to stand on rocky points and watch a three to four foot band of fingerlings move in one direction adjacent to the shore for several hours without interruption—and offshore the water will suddenly change color here and there from the quick rise of countless numbers of two to three inch young to the surface film, or their sudden change in direction due to some disturbance.

Kimsey (1954) extensively studied the Lahontan Tui Chub in Eagle Lake, where the genetic populations are, at least in the taxonomic sense, identical to those of Pyramid Lake, both fine and coarse-gill rakered forms being present. He interpreted his populations as hybrids between the two subspecies *Siphateles bicolor obesus* and *S. b. pectini-*

fer. Similar growths seem to be achieved by both Eagle Lake and Pyramid Lake Chubs, 16 inches being the largest example recorded in each case. Kimsey's data indicates that spawning takes place initially in the third spring of life when they average a bit over 6 inches in standard length. He obtained what he considered a low figure of 11,200 eggs from an 11-inch fish; apparently all eggs do not mature at the same time, so the total figure could be much higher. Spawning seems to occur between 55° F. and 60° F., much of it at least in aquatic plant beds such as *Myriophyllum, Ceratophyllum* and *Potamogeton.*

As would be expected, feeding habits are essentially the same in the two lakes—in both cases the available plankton has been found in stomachs of recently-hatched fishes; algae such as diatoms and desmids, etc. Kimsey found no definite evidence that chub were fish-eaters, but the author has found remains of what appeared to be fingerling chub in the stomachs of two larger specimens from Pyramid Lake. Cal Allan, working Walker Lake, has considerable evidence of the expected predation of chubs, including cannibalism (1958).

Harry (1951) has published an embryologic study on the Lahontan Tui Chub which is reproduced in the Appendix (p. 697).

The validity of *pectinifer* as a subspecies seems best considered here since the argument for the case is based upon supposed correlation between the gillraker difference and habitat segregation, an environmental distinction which the writer regards as inconsequential, if indeed it ever existed. In brief, the fine-gillrakered *pectinifer* is considered the lacustrine segment of the Lahontan *obesus* populations, while "typical" *obesus* is the fluviatile differentiate. The reason they are so completely mixed in the lake at the present time is because the river *obesus* has been forced into the lake by degeneration of the stream environment, and the two types are, as consequence, hybridizing freely. This is a rather neat argument, on the surface, particularly satisfying to the human need for pigeon-holing such stray bits of information as these gillraker differences, for while most biologists are quick to agree that theoretically an animal may have one or several peculiarities of structure which are essentially nonfunctional and evolutionarily inoperative, they are usually equally loath to admit of a concrete instance— when such apparent structural differences are noted, they are immediately rationalized.

Hubbs and Miller (1943) have completely analyzed mass hybridization between the lacustrine *Siphateles mohavensis* and the fluviatile *Gila orcutti* in the Mohave Desert, postulating that a disruption of the Pleistocene lakes of the region forced the *Siphateles* into the environment of the *Gila* with the resultant hybridization. They remark specifically upon the Pyramid Lake problem:

"The other case, that of the fluviatile *Siphateles obesus obesus* and the lacustrine *Siphateles obesus pectinifer*, of the Lahontan system, is almost exactly parallel in genetic as well as in historical and ecological respects. Almost the same difference in number of gill-rakers is involved. It is somewhat arbitrary that we regard the second case as one of subspecific intergradation rather than one of interspecific hybridization" (p. 352).

As superficially similar as these two cases may appear on the surface, there are, in actuality, no "historical and ecological" parallels. Historically, the *Siphateles-Gila* situation in the Mohave Desert may stem from a dissolution of the terminal lakes due to post-Pleistocene drought, whereas *obesus-pectinifer* are in supposed confluence because of a breakdown in the stream environment of one of them. Ecologically, the remnant populations and their hybrids in Mohave have nothing habitable left for them but the waning streams, while the Lahontan change is occurring in waning lakes.

Irrespective of the fact that a real doubt now seems to exist concerning the endemicity of *Gila orcutti* to the Mohave Desert (Miller 1945A: 105), the "parallel" between these cases dissolves further upon examination of other points:

Siphateles bicolor obesus seems to have been everywhere essentially a lacustrine type. It was the overwhelmingly dominant native cyprinid in at least the lower Lahontan remnant lakes such as Pyramid, Walker, Humboldt, Carson and Honey, and probably occupied a similar dominance in Lahontan itself, as attested by fossil material in Lahontan Valley. The few specimens found in rivers would not constitute any stable stream population, and these could be expected to have occurred only in the lower, sluggish portions of streams such as the Truckee River, or at considerable distances upriver only in rivers such as the Humboldt which were sluggish for many miles. The type of *obesus* came from the Humboldt River, but no one knows what part. The University of Nevada collection contains a specimen caught in the Reese River miles from its confluence with the Humboldt, but there is no way of telling how many of the chub in this heavily fished stream were brought in as bait fish and dumped when the angler was through (carp were being planted there in the 1880s). Snyder listed specimens from both the Little Humboldt and the Susan Rivers—the former is quite placid in its lower stretches, and the latter would contain such chub as could escape from the drying basin of Honey Lake.

As a lacustrine type, it is difficult to argue the point that a distinguishable part of the population (coarse gillrakers) is being forced back into the lacustrine reservoir (fine gillrakers) by changing stream conditions, particularly when Snyder found large populations of both types living side-by-side in Pyramid Lake, as they do today, at a time when conditions on the Truckee River were as favorable as they had ever been for the support of fluvial chub. Even at the present time, with the Truckee River much deteriorated as an environment, suckers manage to run upriver to spawn and carp and Brown Trout and some Rainbow Trout live in its lower reaches. The lake chub is as hardy a species as any of these, as attested by its survival in desiccating Winnemucca Lake to the last. At present, chub are rare in the lower Truckee River and there is no indication that Snyder found any there during his field work. It is probable that the Tahoe chub were continually being swept into the upper reaches of the Truckee River, and perhaps some of the Pyramid populations were ascending the lower Truckee for short distances during sluggish water, but it is quite apparent that chub "disappeared" from the Truckee River a long time before they could be forced out by a changing environment, in order to produce the substantial populations of the "fluviatile" type which now occur in Pyramid Lake. The present writer feels no different about the two types than he did in 1952:

"However, on the basis of recent data, we are convinced that *pectinifer* has no valid standing as a taxonomic unit. Genetically, *pectinifer* might be preserved to indicate tui chub with fine gill rakers (as Shapovalov and Dill, 1950, have done), just as it might be feasible, under some circumstances, to so distinguish between people with blue eyes and people with brown eyes. Chub with coarse gill rakers occur side-by-side with individuals with fine gill rakers, contrary to Snyder's supposition that he could observe them segregating in separate schools in the lake. Gill net sampling in Pyramid Lake by the writers has resulted in catches of chubs with both types of gill rakers at the same time and place, and there is no sexual correlation between raker differences. Whether there are any differences in feeding habits between the two types remain to be seen, but the above sampling, during the winter, showed both to be mixed in the same schools and to be feeding on the bottom on the same materials. It is possible that at other times of the year one form may have an advantage over the other in being able to obtain more plankton from the water, but at present, that is a dubious point" (La Rivers and Trelease 1952: 117).

ECONOMICS

The Lahontan Tui Chub played a much more important part in the food chain formerly than now. Prior to the extinction of the Lahontan Cutthroat Trout, chub formed a substantial part of the diet of these game fish, being one of the dominant species converting algae and arthropod protein into food units suitable for the larger, carnivorous fishes.

With the stepped up trout planting program in Pyramid and Walker Lakes usefulness of the chub as a forage food is again becoming apparent. Sacramento Perch utilize them also in both lakes.

"Lake chubs when prepared for the table were found to be sweet and very palatable. They have at times been salted and sold in the markets as 'whitefish.' They contribute to the food of the larger trout and also to that of the water birds" (Snyder 1917A: 62).

An earlier economic notation is that of Jordan and Henshaw (1878: 193), writing of this chub under the name *Leucos formosus:*

"This fish is extremely abundant in some localities, as at Washoe Lake, Nevada, where it forms the chief food for several species of diving birds that frequent its waters. As a table-fish it is considered of no value whatever, and is never captured for table use."

If our streams were in even fair condition as trout waters, there would be little point in writing about the game qualities of a fish like the Lahontan Tui Chub—however, as streams and lakes deteriorate, such rough fish as the chub will eventually begin to assume some importance to the sport angler, even though they are not at present considered desirable as food. On light tackle, the chub is moderately spirited for its size and weight, and many have been caught incidentally to other fishing.

Kay Johnson has found that there is some fishing at Pyramid Lake specifically for chub, which are utilized as food. Their boniness is their only drawback.

GOLDEN SHINERS
Genus *Notemigonus* Rafinesque 1819
("Back, half, angle," alluding to the almost carinate condition of the back)

GOLDEN SHINER
Notemigonus crysoleucas (Mitchill)
(American Roach, American Bream, Bitterhead)

Cyprinus crysoleucas Mitchill 1814: 23
Cyprinus hemiplus Rafinesque 1817: 121
Notemigonus auratus ibid 1819: 421
Hemiplus lacustris ibid 1820: 6
Abramis versicolor DeKay 1842: 191
Abramis americanus, Günther, 1868: 305
Notemigonus chrysoleucus, Jordan and Gilbert, 1882: 250
Abramis crysoleucas, Jordan and Evermann, 1896B: 250
Abramis crysoleucas, Forbes and Richardson, 1920: 126–128
Abramis crysoleucas, Pratt, 1923: 73
Notemigonus crysoleucas, Jordan, Evermann and Clark, 1930: 115
Golden Shiner, Madsen, 1935A: 21
Notemigonus crysoleucas, Schrenkeisen, 1938: 105–106
Notemigonus crysoleucas, Kuhne, 1939: 58
Notemigonus crysoleucas, Hubbs and Lagler, 1941: 48, 56 / 1947: 57, 65
Notemigonus crysoleucas, Hinks, 1943: 49–50
Notemigonus crysoleucas crysoleucas, Carpenter and Siegler, 1947: 42
Notemigonus crysoleucas, Eddy and Surber, 1947: 156
Notemigonus crysoleucas, Everhart, 1950: 36
Notemigonus crysoleucas, Shapovalov and Dill, 1950: 386
Notemigonus crysoleucas auratus, Beckman, 1952: 41–42
Notemigonus crysoleucas, Miller, 1952: 19, 32–33
Notemigonus crysoleucas, Scott, 1954: 51
Notemigonus crysoleucas, Harlan and Speaker, 1956: 88, 331, 355
Notemigonus crysoleucas, Eddy, 1957:
Notemigonus crysoleucas, Koster, 1957: 56
Notemigonus crysoleucas, Moore, 1957: 99
Notemigonus crysoleucas, Slastenenko, 1958: 215–216
Notemigonus crysoleucas, Shapovalov, Dill and Cordone, 1959: 172

ORIGINAL DESCRIPTION
"3. *C. Crysoleucas**—Shiner.
"Lives in fresh ponds; head small and smooth; rather depressed on the upper side; mouth small, even, and toothless; eyes large in proportion to the head; body deep in proportion to its length. Colour blackish,

with shining white scales; Eyes and gill-covers golden, with a tinge of the same along the belly; lateral line bends downward, to correspond with the curve of the abdomen; pectoral and ventral fins, especially the latter, yellowish brown; tail forked; belly whitish, with ruddy rays. "Rays, B. 3: P. 17: V. 9: A. 14: D. 9: C. 19" (Mitchill 1814: 23).

DIAGNOSIS

Head 4.0 to 4.5, body depth 3.0 to 3.5, in body length, respectively; head small, laterally well-flattened; mouth small, oblique, terminal; top of upper lip about on the same plane as top of pupil; lips lacking barbels; maxillary reaching only to anterior nostril; jaws subequal; eye large; pharyngeal teeth in a single row, 5-5 or 4-4, basally narrowed, often somewhat hooked; scales 9-11/45-52/3, dark triangular

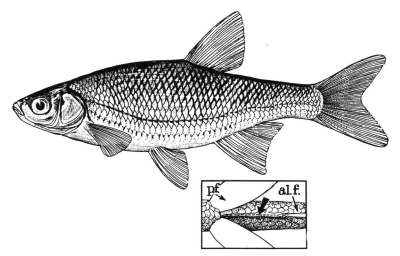

FIG. 194. Golden Shiner, *Notemigonus crysoleucas*. Drawn by William L. Brudon for Miller 1952. Courtesy California Department of Fish and Game and Leo Shapovalov.

spot generally visible at exposed base of each scale, lateral line complete, very prominently decurved; fins—strongly developed—dorsal 8-rayed, first ray about midway between beginning of caudal fin and posterior head margin, leading edge of dorsal set well behind leading edge of pelvic fins, a plumb line from base of first dorsal ray usually anywhere from middle to hind edge of belly edge of posteriorly stretched pelvic fins—anal 11–14 rayed—caudal large and strongly forked; caudal peduncle about 9 into body length; intestinal length varies from body length to nearly twice body length; peritoneum light, weakly dark-speckled; color—golden with marked greenish-olive tints dorsally, in certain lights appearing steel-bluish; flanks lighter yellow with silvery reflections, paling to whitish on the venter; dorsal and anal fins often darker-tipped, ventral fins more yellowish, becoming, particularly in the pelvics, tipped with brilliant orange in both sexes during breeding season; young individuals usually show a prominent dark stripe along sides; grows to maximum lengths of 12 inches, with averages from 6 to 8 inches. Its most distinctive feature, as far as the Nevada fauna is concerned, is the prominent, sharp, belly keel between the pelvic and anal fins.

TYPE LOCALITY

New York.

RANGE

Northern half of the Atlantic Coast west to Texas and the Dakotas. Has been used for bait fish in several western states, principally Arizona, Nevada and California, and is well established in the wild in many places in this area. Its importance as bait is mainly in the tri-state section of the Colorado River, where the three states touch.

TAXONOMY

Some three subspecies of the Golden Shiner seem to be currently recognized; an Eastern (*N. crysoleucas crysoleucas*), a Western (*N. c. auratus* Rafinesque 1819) and a Southwestern (*N. c. seco* (Girard) 1856). Western and Southwestern subspecies (*auratus* and *seco*) are both in use as bait fish along the Colorado. Although not easily differentiable, these are generally separated by the number of anal rays—averaging 11–13 (10–14) in the Western, 12–14 (11–16) in the Southwestern, Golden Shiner (Miller 1952: 33).

Although several species were at one time placed in the genus *Notemigonus*, it is currently considered monotypic, and the synonymicon has not been complex.

Etymology—"crysoleucas," gold white.

LIFE HISTORY

The Golden Shiner is a very abundant and common fish in its native range, a successful inhabitant of slower, weed-grown waters, whether lake or stream. It does so well in stagnant waters that often it is the only fish to live through the winter stagnation period in ponds where oxygen is completely depleted. University of Minnesota experiments show that the shiner, while making the usual oxygen demands on its environment at higher temperatures, is capable of existing on fractional oxygen percentages at low temperatures (Eddy and Surber, 1947).

As would be expected from its capabilities, the species is partial to muddy bottoms, and is versatile in food habits. Its long intestine and multipurpose pharyngeal teeth (with both hooks and grinding surfaces) plus the large number of thin, long gillrakers fit the Golden Shiner for most types of foods in its environment, even to working over bottom muds for its organic matter and consuming quantities of algae.

There is a sexual size difference, males being the smaller and generally showing a humping behind the head, particularly with age. The species is a spring-early summer spawner, from about April to July, depending upon the latitude and produces large numbers of strongly adhesive eggs.

ECONOMICS

The Golden Shiner has been of sole importance in the Nevada picture as an imported bait fish for the Colorado River area, where it seems to have been in use since about 1950. It has been established in nearby States for a much longer time—such as the Coconino National Forest of Arizona, where it has been known since at least 1934—and apparently its bait values have prompted all its western introductions.

The California Fish and Game Commission has actively cultivated the species in the Colorado River area—The Nevada Commission has not done this, but commercial bait dealers around Lake Mead have used the shiner extensively and it now seems well-established.

Its introductions have not been entirely without problems—in the Coconino area of Arizona the shiner has been a factor in decline of the trout fishery there (Borges 1949) but, as Miller states (1952: 33), its preferences for weed beds seem to make it unlikely the shiner will be a problem in largely weedless Lakes Mead and Mohave.

In the East, in addition to its natural position in the food chain as a good bass forage fish, the Golden Shiner is raised in ponds for hatchery fish food and large individuals even have some utilization as sport and pan fish. Like Yellow Perch, the shiner is a winter-active species and can be taken through the ice.

SPECKLE DACES

Genus *Rhinichthys* Agassiz 1850

("Snout fish," referring to the prominent snout)

SPECKLE DACE

Rhinichthys osculus (Girard)

Argyreus osculus Girard 1856B: 186 / 1859: 244
Argyreus nubilus ibid / 1859: 47
Argyreus notabilis ibid 1856B: 186
Argyreus nubilus, Günther, 1868: 188
Argyreus osculus, ibid
Argyreus notabilis, ibid
Ceratichthys ventricosus Cope 1874A: 136
Apocope oscula, Cope and Yarrow, 1875: 647
Apocope ventricosa, ibid: 648–649
Apocope nubila, Jordan and Gilbert, 1882: 210
Apocope oscula, ibid: 211
Agosia oscula, Evermann and Rutter, 1895: 484
Agosia oscula, Jordan and Evermann, 1896B: 309
Agosia nubila, ibid: 311
Agosia oscula, Gilbert and Schofield, 1898: 495
Agosia nubila, Snyder, 1917A: 68
Agosia oscula, Pratt, 1923: 86–87
Agosia nubila, ibid: 87
Apocope nubila, Jordan, Evermann and Clark, 1930: 140
Apocope oscula, ibid: 141
Apocope oscula oscula, Tanner, 1932: 135–136 / 1936: 160, 171
Apocope oscula nubila, Schultz, 1936: 148
Rhinichthys oscula, Murphy, 1941: 168

Apocope oscula nubila, Dimick and Merryfield, 1945: 40
Rhinichthys osculus, Wales, 1946: 115, 118
Rhinichthys osculus, Hubbs and Miller, 1948B: various mentions
Rhinichthys osculus, Shapovalov and Dill, 1950: 386
Rhinichthys osculus, Wallis, 1951: 87
Rhinichthys nubilus, La Rivers, 1952: 99
Rhinichthys nubilus, La Rivers and Trelease, 1952: 117
Rhinichthys nubilus, Miller, 1952: 30
Rhinichthys osculus, ibid
Rhinichthys osculus, La Rivers, 1956: 157
Rhinichthys osculus, Eddy, 1957: 84
Rhinichthys osculus, Koster, 1957: 64
Rhinichthys osculus, Moore, 1957: 113
Rhinichthys nubilus, Slastenenko, 1958A: 194–195
Rhinichthys osculus, Carl, Clemens and Lindsey, 1959: 123–124
Rhinichthys osculus, Shapovalov, Dill and Cordone, 1959: 172

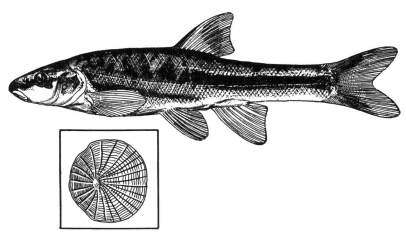

Fig. 195. Speckle Dace, *Rhinichthys osculus*. Drawn by William L. Brudon for Miller 1952. Courtesy California Department of Fish and Game and Leo Shapovalov.

ORIGINAL DESCRIPTION

(*Argyreus osculus*) "Has more the fascies of *A. atronasus* than of any other of its congeners both by the outline of its body and head, and the shape and position of the mouth. The head is comparatively small, forming the fifth of the length, with the exception of the lobes of the caudal. The eye is rather large and subcircular, its diameter entering about four times in the length of the side of the head. The dorsal and anal fins are well developed, the former being convex superiorly, and the latter subconvex exteriorly. The posterior margin of the caudal is crescentic. The posterior extremity of the ventrals extend as far as the vent, which is not the case in the two species described above" (*Argyreus nubilus* and *A. dulcis*).

"D 8 + 2; A 7 + 2; C 5, 1, 9, 8, 1, 6; V 8; P 14.

"The anterior two rays of both the dorsal and anal fins are mere rudiments, as already stated.

"The color is reddish brown above; olivaceous on the sides, with numerous dark blotches and dots. Beneath uniform yellowish white or silvery white.

"Many specimens, the largest of which measuring less than three inches, were collected by John H. Clark, under Col. J. D. Graham, U.S.A., in the Bobocomori, a tributary stream of the Rio San Pedro, itself flowing into the Rio Gila" (Girard 1856B: 186).

(*Argyreus nubilus*) "This is a very characteristic species. The head is very small, and the body, fusiform in shape and compressed, is thick and swollen upon its middle. The tail again is rather slender. The snout is subconical, but not more protruding than in the preceding species [*A. dulcis*]; the mouth is a great deal smaller than in the latter, with its barbels less conspicuous. The head constitutes about the fifth of the entire length. The eye is moderately developed and subcircular in shape; its horizontal diameter is contained about five times in the length of the side of the head. The dorsal, caudal and anal fins are of but moderate development, the pectorals and ventrals rather small.

"D 8 + 2; A 7 + 2; C 5, 1, 9, 8, 1, 6; V 8; P 12.

"The number of rays in the fins does not materially differ from that of the preceding species; in both, the anterior two rays of the dorsal and anal fins are mere rudiments.

"The color above is blackish brown, with a purplish hue along the middle of the flanks; the inferior regions are of a soiled white or yellowish brown. The upper surface of the head and upper half of the sides, including the eye, is deep black; inferiorly it is whitish or yellowish white.

"Specimens, four inches in total length, were collected at Fort Steilacoom, Puget Sound, W. T., by Dr. Geo. Suckley, U.S.A., under Gov. I. I. Stevens" (Girard 1856B: 186).

DIAGNOSIS

Head and body depth each about 4 times into body length; maxillary protractile or not; lips usually with one or two barbels, which are of varying lengths and which may be lacking on one or both sides; pharyngeal teeth in two rows, varying from 4/1–1/4 to 4/2–2/4, the lesser row occasionally lacking on one side—teeth strongly hooked, lacking crushing surfaces; lateral line complete or incomplete; scales very variable, 12–18/47–89/9–15; scales include radii on all fields; fins with edges square, rounded or falcate—dorsal 7–8 rayed, rarely 9—anal 7 rayed, rarely 8—pectorals variable in length, either too short to reach the ventral fins or long enough to reach well posterior to the bases of the ventrals—caudal rather large, broadly emarginate with well-rounded tips; caudal peduncle stout, least depth 10–12 times into body length; intestine short, less than the body length; color olivaceous above, lighter below, usually with a distinct blackish lateral stripe— breeding males with some pinkish at bases of ventral fins and around gill openings; a small species, seldom exceeding three inches in length.

TYPE LOCALITY

Bobocomori Creek, tributary to the San Pedro River in southeastern Arizona.

RANGE

Western United States, west of the Rocky Mountains, from Washington and Idaho to southern California and Arizona; breaks up into four geographic races in Nevada, where it is found in all the drainage systems, and is the most widespread species.

TAXONOMY

As would be expected of a species with its distribution, *Rhinichthys osculus* is a variable form which has been placed in several genera and has acquired a formidable synonymy. However, it suffered less at the

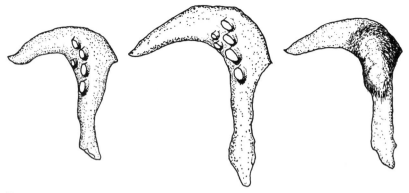

FIG. 196. Pharyngeal arches and teeth of *Rhinichthys*, showing conditions as they exist in the sampling of certain Nevada populations. From left to right: *R. osculus robustus*, *R. o. velifer* and *R. o. nevadensis* (Drawn by Bill O. Smith). *Robustus* averages 1–4/4–1 (Reno), *velifer* 2–4/4–2 (White River) and *nevadensis* varies from no teeth (0–0/0–0) to the following combinations—0–1/3–0, 0–3/1–0, 0–0/1–0 and 0–1/0–0 (Beatty) (counts made by Don Caldwell and Bill Smith in the author's laboratory).

hands of the taxonomic zealots of the last half of the Nineteenth Century than did, for example, the much more restricted *Gila robusta*.

At the time of its original description, Girard inadequately characterized not only *R. osculus* but also two of its synonyms immediately following it—*R. nubilus* and *R. notabilis*, the former from southern Arizona, the latter from northern Mexico. Cope and others described additional species in both *Rhinichthys* and *Apocope* (the latter now considered a subgenus of the former), and by the appearance time of the 1930 checklist of Jordan et al, a formidable series of untenable names existed officially under these two genera. In 1936 Schultz was listing some of these as subspecies of *osculus*, and in 1948, the current opinion was:

"The forms of *Rhinichthys* (subgenus *Apocope*) in the West exhibit so much overlapping in their characters that most of the nominal species are now regarded as comprising a single, wide-ranging species, *R. osculus* (Girard). This species inhabits all of the major drainage systems of Nevada" (Miller and Miller, p. 174).

Recently, for a brief period, the terms *osculus* and *nubilus* were confusingly interchanged in order to follow a line priority ruling of the International Commission on Zoological Nomenclature (1950)—in a reversal of this shortly thereafter, the Commission went back to the principle of first reviser.[1] As a consequence, the Great Basin Speckle Dace was listed in the two recent Nevada state lists as *nubilus* (La Rivers 1952: 99, La Rivers and Trelease 1952: 117).

Etymology—"osculus," small-mouthed.

LIFE HISTORY AND ECONOMICS

See the below-detailed subspecies.

GEOGRAPHIC KEY TO SUBSPECIES[2]

Snake River drainage of northeastern Nevada; Rivers Owyhee, Bruneau, Jarbidge, Salmon and tributaries (*Rhinichthys osculus carringtoni*)................ SNAKE RIVER SPECKLE DACE

Lahontan drainage of western, central and northern Nevada; Rivers Humboldt, Truckee, Carson, Susan, Walker and associated lakes (*Rhinichthys osculus robustus*)....................
.... LAHONTAN SPECKLE DACE

Amargosa River drainage of southwestern Nevada-southeastern California (*Rhinichthys osculus nevadensis*)....................
.... AMARGOSA RIVER SPECKLE DACE

White River drainage of eastern and southeastern Nevada (*Rhinichthys osculus velifer*)........ WHITE RIVER SPECKLE DACE

THE SNAKE RIVER SPECKLE DACE
Rhinichthys osculus carringtoni (Cope)

Apocope carringtonii Cope 1872C: 472–473
Apocope vulnerata ibid: 473
Tigoma rhinichthyoides ibid
Rhinichthys henshavii Cope 1874A: 133
Apocope carringtonii, Cope and Yarrow, 1875: 645
Apocope henshavii, ibid: 645–646
Apocope vulnerata, ibid: 646
Apocope carringtonii, Jordan and Copeland, 1878: 426
Apocope henshawi, ibid

---

[1] In the matter of *osculus* versus *nubilus*, "ii" endings versus "i" and some others, there is little to be said for the current International Commission for Zoological Nomenclature's policies except that they seem founded on indecisiveness, if not incompetence.

[2] This is the only currently practicable way to separate the known Nevada forms of this species.

Apocope vulnerata, ibid
Apocope carringtoni, Jordan and Henshaw, 1878: 191
Apocope vulnerata, ibid: 191–192
Apocope carringtoni, Jordan and Gilbert, 1882B: 209–210
Apocope vulnerata, ibid: 210
Apocope henshavii, ibid
Agosia novemradiata Cope 1883: 141
Agosia nubila carringtonii, Gilbert and Evermann, 1895: 41
Agosia nubila carringtonii, Jordan and Evermann, 1896B: 311–312
Agosia nubila carringtonii, Snyder, 1917A: 37, 67
Apocope carringtonii, Jordan, Evermann and Clark, 1930: 141
Apocope carringtonii, Hayes, 1935: 86–92
Apocope oscula carringtoni, Schultz, 1936: 148
Apocope carringtonii, Tanner, 1936: 160, 170–171
Agosia nubila carringtonii, Schrenkeisen, 1938: 146–147
Rhinichthys osculus carringtonii, Shapovalov and Dill, 1950: 386
Rhinichthys osculus carringtonii, Simon, 1951: 75–76
Rhinichthys nubilus carringtoni, La Rivers, 1952: 99
Rhinichthys nubilus nubilus, La Rivers and Trelease, 1952: 117
Rhinichthys osculus carringtonii, Shapovalov, Dill and Cordone, 1959: 172

ORIGINAL DESCRIPTION

"APOCOPE CARRINGTONII, Cope, *sp. nov.*

"This is a small species allied to the last [*Ceratichthys nubilus-Rhinichthys osculus*], but the muzzle is broader and less prominent, and the mouth larger. The muzzle is quite obtuse in profile and overhangs the mouth very little, and the end of the maxillary bone does not quite reach the line of the margin of the orbit. Barbels minute; teeth, 4.1–1.4; isthmus wide; eye a little smaller than one-fourth the length of the head, and 1.5 times in interorbital width. Scales, 10–60–11. Dorsal fin originating behind the point above the ventrals, and markedly nearer the basis of the caudal than the end of the muzzle. Caudal well forked. Radii, D. 8; A. 7; length of head a little more than four times in length to basis of caudal; depth five times in the same; length, 20 lines.

"Color olivaceous, with a dark lateral band from end of muzzle, and dark shades on the back.

"Four specimens from the Warm Springs, Utah. The species is named in honor of Campbell Carrington, zoologist of Dr. Hayden's expedition, to whose zeal in the cause of science we are indebted for the materials analyzed in this report, and that on the same subject in the Survey of Wyoming, 1871. Collection No. 9" (Cope 1872C: 472–473).

DIAGNOSIS

Since the varieties of the speckle dace are separable only with great difficulty, it is best to use the convenient criterion of geographical segregation to distinguish between the Nevada forms.

TYPE LOCALITY

Warm Springs, Box Elder County, Utah.

RANGE

Snake River drainage of southeastern Oregon, southern Idaho, northern Nevada and southwestern Wyoming; western Utah.

TAXONOMY

At the same time that Cope described this subspecies (as a species) he named as new two other Utah cyprinids which later became synonyms of *carringtoni*, one of which he diagnosed in a separate genus. In later years he added two more synonyms, each also in different genera. By the time of the Jordan-Evermann-Clark checklist of 1930 these matters were straightened out and it remained only for subsequent workers to orient it to its present genus as a subspecies of the much more widespread *Rhinichthys osculus*.

Etymology—Named for Campbell Carrington, an early government western survey naturalist.

LIFE HISTORY AND ECONOMICS

See the Lahontan Speckle Dace (*Rhinichthys osculus robustus*).

THE LAHONTAN SPECKLE DACE
Rhinichthys osculus robustus (Rutter)
(Black Minnow, Lahontan Speckled Dace)

Apocope ventricosa, Jordan and Henshaw, 1878: 192 (Lake Tahoe, Truckee River)
Agosia oscula, Eigenmann and Eigenmann, 1891: 1132 (Lake Tahoe)
Agosia nubila carringtonii, Jordan and Evermann, 1896B: 312 (Lake Tahoe, Lahontan Basin)
Agosia robusta Rutter 1903: 148 / 1908: 139–140 (Prosser Creek and other western Lahontan affluents)
Agosia robusta, Snyder, 1917A: 67–69 (Lahontan system) / 1917B: 202, 204–205
Apocope robusta, Jordan, Evermann and Clark, 1930 (erroneously listed as from "Upper Sacramento Basin")
Apocope robusta, Schrenkeisen, 1938: 147
Apocope robusta, Shapovalov, 1941: 444
Rhinichthys osculus robustus, Hubbs and Miller, 1948A: 17, 19, 22–24, 28 / 1948B: various isolated mentions
Rhinichthys osculus robustus, Kimsey, 1950: 438 / 1954: 408
Rhinichthys osculus robustus, Shapovalov and Dill, 1950: 386
Rhinichthys nubilus robustus, La Rivers, 1952: 99
Rhinichthys nubilus robustus, La Rivers and Trelease, 1952: 117
Agosia robusta, Böhlke, 1953: 29
Rhinichthys osculus robustus, Shapovalov, Dill and Cordone, 1959: 172

ORIGINAL DESCRIPTION

"Body heavy, highest above insertion of pectorals; the ventral outline curved almost as much as the dorsal. Head 3.8 to 4 in body; snout blunt, but little overlapping the premaxillary and never extending beyond it; mouth oblique, barbels usually absent, present on 10 to 50 percent of specimens from any one locality. Fins small; D. 8; A. 7;

pectoral about equal to head behind nostril, variable; caudal moderately forked, middle rays two-thirds length of longest; rudimentary caudal rays forming prominent keels along upper and lower edges of tail; margin of anal slightly rounded, the anterior rays not all produced, not extending beyond posterior rays when fin is depressed. Lateral line nearly always incomplete, but with scattered pores frequently extending to base of caudal; scales 56–77, varying about 12 in any one locality. Usually two dusky lateral stripes, the upper extending from snout to caudal, the lower branching off from the upper behind the head and ending along base of anal; cheek abruptly silvery below lateral stripe; tinged with orange about lower jaw, upper end of gill-opening, and at base of lower fins.

"Type (No. 50589 U. S. Nat. Mus.). Collected in Prosser Creek by Rutter and Atkinson.

"Taken in Spring Creek, Willow Creek, Susan River, Little Truckee River, and Prosser Creek" (Rutter 1903: 148).

DIAGNOSIS

Head and body depth each from 3.5 to 4.0 times into body length; maxillary protractile; barbels present or absent, when single occurring on either side, long or short; pharyngeal teeth in two rows, 4/1–1/4, sharp, hooked and without grinding surfaces—rarely the lesser row will be lacking; lateral line generally not complete; scales 12–14/56–77 /9–11; fins variable in shape—dorsal 7–9 rayed—anal usually 7 rayed, rarely 8—pectorals generally abbreviated, not extending to the ventrals except in the cases of long-finned individuals in which the pectorals may reach beyond the ventral bases—caudal prominent and moderately forked; caudal peduncle short, least depth 10–12 times into body length; intestine not as long as the body; color—olive-yellow dorsally, becoming brassy on the sides and yellowish below—spotting and striping black-to-olive—fins with red-yellow tips, and breeding individuals with some red or crimson spotted about the body; in size a small fish, averaging about two and one-half inches in length.

TYPE LOCALITY

Prosser Creek, eastern edge of Nevada County, California, a tributary of the Truckee River flowing into the river between the Nevada state line and Truckee, California.

RANGE

This is the speckle dace of the Lahontan drainage system of lakes (Tahoe, Pyramid and Walker) and rivers (Walker, Carson, Truckee, Susan, Quinn, Humboldt and Reese) and tributaries in Nevada and adjacent California.

Other speckle dace are found in isolated valleys between the Lahontan, White River and Amargosa systems which are as yet nondefinable. This includes such valleys as Big Smoky and Monitor in Nye County, and Vegas Wash in Clark County.

TAXONOMY

If it were not for the confinement of the subspecies *robustus* to the Lahontan system, it would be difficult to grace it with a separate name.

As Snyder aptly summed up its status: "It now appears that the name must be restricted to the *Agosia* of the Lahontan system or else be abandoned entirely, for if its range is extended to include even a part of the Sacramento system its variation becomes so great that it can not now be differentiated from the form or forms occupying the Oregon Lake region or the upper Columbia. As yet the present writer is unable to find any character or set of characters which will distinguish *Agosia robusta*, and the name is therefore only provisionally retained" (1917A: 67).

LIFE HISTORY

A small, omnivorous late spring-early summer spawning minnow. Snyder's observations, the first of any importance, are sketchy: ". . . A female caught at Tahoe City June 27 was full of nearly ripe eggs.

"This species inhabits both streams and lakes. In the rivers it is most often taken on the ripples. In lakes it frequents the shallow water, swimming near the bottom, or in crevices between rocks. From above, when seen in the water, the color is decidedly black. In the lakes it was taken with hook and line, a method which will often secure very small fishes where the net fails on account of deep, clear water or rough bottom.

"*Agosia nevadensis* and *A. velifer* are very distinct from this form and are easily recognized" (1917A: 68).

Lack of schooling behavior is a noticeable characteristic of this fish, especially when it can be compared with the schooling redshiners in such habitats as Pyramid Lake—most individuals will be found poking about the bottom in no more than very small groups, and often singly. Spawning very often takes place with the redshiners in Pyramid Lake, and individuals of both species were often observed feeding on any eggs available in the general melée. Speckle dace eggs are yellowish, and Pyramid Lake specimens showed an average diameter of 1.0 mm.

ECONOMICS

Biologically of varying importance as a food-chain species, and used to some extent by anglers as a bait fish. Probably because of its habits, it is afforded considerable protection from predatory fishes by the much more abundant, and masking Lahontan Redshiner, which seems to take the brunt of such predation. As such, the Lahontan Speckle Dace cannot generally be considered an important link in the food pyramid.

Like all the small "minnows," speckle dace have been widely used as bait fish and as such have been carried about by fishermen. To what extent this has occurred is indeterminable, but one instance of official planting of this type is known. In August of 1951, the Nevada Fish and Game Commission obtained some specimens from Sadler's Ranch in Diamond Valley, Eureka County, and planted them in Ruby Marsh, Elko County, as forage fish for the Largemouth Blackbass fishery there.

Prior to this, the Marsh suffered from lack of forage fishes, this lack being supplied by the smaller blackbass, to the general detriment of the fishery. The speckle dace seem to be well established in the Marsh, but apparently in sections more-or-less inaccessible to blackbass.

THE AMARGOSA SPECKLE DACE
Rhinichthys osculus nevadensis Gilbert

Rhinichthys (Apocope) nevadensis Gilbert 1893: 230–231
Agosia nevadensis, Jordan and Evermann, 1896B: 310–311
Agosia nevadensis, Snyder, 1917A: 68
Agosia nevadensis, Pratt, 1923: 87
Apocope nevadensis, Jordan, Evermann and Clark, 1930: 141
Agosia nevadensis, Schrenkeisen, 1938: 147
Rhinichthys (Apocope) nevadensis, Tanner, 1942: 31
Rhinichthys osculus nevadensis, Hubbs and Miller, 1948B: 102
Rhinichthys osculus nevadensis, Miller, 1948: 101
Rhinichthys osculus nevadensis, Shapovalov and Dill, 1950: 386
Rhinichthys nubilus nevadensis, La Rivers, 1952: 99
Rhinichthys nubilus nevadensis, La Rivers and Trelease, 1952: 117
Rhinichthys (Apocope) nevadensis, Böhlke, 1953: 38
Rhinichthys osculus nevadensis, La Rivers, 1956: 157
Rhinichthys osculus nevadensis, Shapovalov, Dill and Cordone, 1959: 172

ORIGINAL DESCRIPTION AND DIAGNOSIS

"Differing from other known species in the large head, the short deep body, very small eye, and in the reduction of the outer ventral ray to a mere rudiment.

"Head, $3\frac{2}{3}$ in length (varying from $3\frac{1}{2}$ to 4); depth, $3\frac{2}{3}$ (varying from $3\frac{1}{2}$ to 4). D., 8; A., 7. Lat. 1. 65. Ventral apparently with seven rays, the outer one rudimentary, and often to be detected with difficulty.

"Body robust, with broad heavy head, the least depth of caudal peduncle less than half the greatest height of body. Greatest depth of head at occiput 5 in length of body ($6\frac{1}{4}$ in nubila of equal size). Eye very small, half interorbital width, which equals distance from tip of snout to middle of eye, and is contained $2\frac{2}{3}$ times in head.

"Mouth terminal, very oblique, the lower jaw included, the premaxillaries not at all overlapped by the snout. The maxillary reaches the vertical from front of eye, and is one-third length of head. Maxillary barbel well developed.

"Scales very irregularly placed, and difficult to enumerate. The lateral line is incomplete in adults, and usually does not reach to opposite dorsal fin. In the young it is variously developed, often extending, though with many interruptions, to end of dorsal or base of caudal. Pores in lateral line (when complete) 58, about 66 oblique series, counted above lateral line.

"Fins small, the pectorals not reaching ventrals, the latter not to vent. Front of dorsal midway between base of caudal and middle of occiput.

"In spirits, the upper half of sides is speckled and marbled with brown; the belly and lower half of sides immaculate or sparsely spotted. A broad dark lateral stripe usually present, becoming more conspicuous posteriorly, and ending in an obscure black spot on base of tail. A dark stripe sometimes present along middle of lower half of sides.

"Numerous specimens were procured in the warm springs at Ash Meadows, Indian Creek, and Vegas Creek, Nevada" (Gilbert 1893: 230–231).

TYPE LOCALITY
Ash Meadows, Amargosa Desert, southeastern Nye County, Nevada.

RANGE
Confined to the Amargosa drainage system, Nevada and California. The Amargosa River heads near Springdale some ten miles north of Beatty, and disappears in sands of the upper part of the Amargosa Desert, rising at rare intervals to the surface. Its channel, dry for the most part, eventually enters the south end of Death Valley, after having received the very substantial discharge of the warm springs of Ash Meadows at the southeast edge of the desert, and those in the Shoshone area farther south.

TAXONOMY
This is one of several fishes Gilbert described from material taken by the famous early day naturalists, C. Hart Merriam and Vernon Bailey, in southern Nevada during the Death Valley expedition of the U. S. Department of Agriculture's Division of Ornithology and Mammalogy.[1] Headed by Merriam, the group did their field work in 1891.

This form owes its nomenclatural stability in good part to the fact that it occupies a restricted drainage system. Its only ups and downs have involved its generic assignments.

LIFE HISTORY
Has not been investigated but undoubtedly differs little if at all from that of the Lahontan Speckle Dace, which see. In its restricted environment, the Amargosa Speckle Dace has fewer associates than the Lahontan form—such common, competing and/or predatory species as the Lahontan Redshiner and Tui Chub are absent from the Amargosa system, as well as, originally, all game fish so that this aspect, at least, of the small fish's life-cycle has been simplified. Bullfrog is a big predator now.

ECONOMICS
None, other than its part in a very restricted food chain in which this tiny fish plays the unusual part of apex species; i.e., the last animal in the chain, whereas in more complex environments it usually in turn provides food for larger species.

WHITE RIVER SPECKLE DACE
Rhinichthys osculus velifer Gilbert

Rhinichthys (Apocope) velifer Gilbert 1893: 229 (Pahranagat Valley)
Agosia velifera, Jordan and Evermann, 1896B: 312
Agosia velifer, Snyder, 1917A: 68

---

[1] This later became the Bureau of Biological Survey and then the U. S. Fish and Wildlife Service.

Apocope velifera, Jordan, Evermann and Clark, 1930: 141
Apocope sp., Brues, 1932: 279–280
Agosia velifera, Schrenkeisen, 1938: 147
Apocope osculus, Sumner and Lanham, 1942: 319 (Preston)
Rhinichthys (Apocope) velifer, Tanner, 1942: 31
Rhinichthys nubilus velifer, La Rivers, 1952: 99
Rhinichthys nubilus velifer, La Rivers and Trelease, 1952: 117
Rhinichthys (Apocope) velifer, Böhlke, 1953: 38
Rhinichthys osculus velifer, La Rivers, 1956: 157

ORIGINAL DESCRIPTION AND DIAGNOSIS

"This species is closely related to *Rhinichthys yarrowi*, from which it differs in the much larger scales, the lateral line traversing 55 instead of 74 to 83 scales. Both species mark such perfect transition between *Apocope* and *Rhinichthys* that it seems best to reduce the former to the rank of a subgenus. About half the specimens of *yarrowi* have a narrow frenum, and this is present in each of the three type specimens of *velifer*. In both *yarrowi* and *velifer* the teeth are 2–4–4–2, as in typical *Rhinichthys*. The only character left to distinguish *Apocope* is the narrowness of the frenum when present, it being very wide in typical *Rhinichthys*.

"Head 4 in length; depth, 4¾. Snout narrow, but bluntly rounded, not projecting beyond the front of premaxillaries. Frenum joining premaxillaries to skin of forehead very narrow, varying in width in the three type specimens. It will probably be found that some specimens of this species, as of *yarrowi*, have protractile premaxillaries. Mouth small, horizontal, the maxillary reaching vertical from front of orbit, equalling diameter of eye, 3½ in length of head. Interorbital width, 3 in head.

"Teeth, 2, 4–4, 2, hooked, with sharp edges.

"Pectorals nearly reaching base of ventrals, the latter long, overlapping front of anal fin. Origin of dorsal fin midway between base of caudal and middle of eye.

"D., 8; A., 7. Lat. 1. 56 (pores). 10 scales in a series obliquely forward to lateral line from base of first dorsal ray.

"Color in spirits, brown along back, a black band from snout across cheeks and along middle of sides, with a narrow silvery streak above it. Lower half of sides and belly silvery; an ill-defined dark streak from base of pectorals back along sides to the end of the anal fin. A small black spot on base of caudal" (Gilbert 1893: 229–230).

TYPE LOCALITY

"Three specimens were taken in a hot spring in Pahranagat Valley, Nevada, May 25, 1891, by C. Hart Merriam and Vernon Bailey. Temperature of spring 36.11° C. (97° F.)" (Gilbert 1893: 230).

RANGE

Unknown beyond the confines of the relict White River system of southeastern Nevada.

The Amargosa Speckle Dace and the White River Speckle Dace are two examples of isolated populations on larger scales in the western

Great Basin. However, many more restricted speckle dace entities exist in Nevada which have not been adequately studied or characterized.

These are mainly in the completely broken interior section where almost all the individual valleys are separated, drainagewise, from each other. Prominent among these are:

Nye County—
Big Smoky Valley
Monitor Valley
Lander County—
Grass Valley
Eureka County—
Antelope Valley
White Pine County—
Spring Valley

Some of these are restricted to small springs or spring areas which, despite the limited amount of water, have been in existence as aquatic habitats since the disappearance of the Pleistocene lakes which occupied these valleys during the glacial periods some ten thousand years ago. Since Nevada is a relatively dry region, with few sizeable, permanent springs which can support fish life, it is remarkable that some of these isolated waters have persisted, with their fish, to the present day.

TAXONOMY

As with other Nevada *Rhinichthys, velifer* cannot be regarded as anything but a geographic expression of the great variation which exists in the widespread species *osculus*. Because of its restriction to this unique drainage system, *velifer* has been unencumbered with synonyms.

Etymology—"velifer," sail-bearing, referring to the dorsal fin.

LIFE HISTORY AND ECONOMICS

See the Lahontan Speckle Dace.

MOAPA DACE
Genus *Moapa* Hubbs and Miller 1948
(named from the type locality)

MOAPA DACE
Moapa coriacea Hubbs and Miller

Moapa coriacea Hubbs and Miller 1948A: 1–14, 28
Moapa coriacea, Kopec, 1949: 56
Moapa coriacea, La Rivers, 1949: 223 / 1950A: 19 / 1950B: 373 / 1952: 98
Moapa coriacea, Miller, 1949A: 449
Moapa coriacea, Wallis, 1951: 87–88
Moapa coriacea, La Rivers and Trelease, 1952: 117
Moapa coriacea, Eddy, 1957: 93
Moapa coriacea, Moore, 1957: 103

ORIGINAL DESCRIPTION

This is a combination of the generic and specific descriptions, and since the former contains the more standard information, and the latter is presented rather incidental to it, the generic characterizations are largely used here:

"The uniserial pharyngeal teeth typically number 5 on the left side and 4 on the right. The upper ones have a moderately strong hook and a faceted rather than excavated grinding surface. The hook and grinding surface grade toward obsolescence on the lower, rather stumpy teeth.

"The rather strong lower pharyngeal arch is hooked on the thin, morphologically inner edge. The 2 arms, measured from the teeth, are of subequal length. The lower arm is moderately slender, terete, and

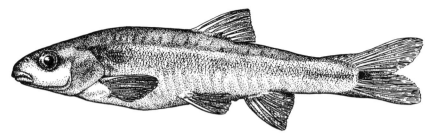

FIG. 197. Moapa dace, *Moapa coriacea*. Drawn by Silvio Santina.

curved. Below the main teeth the concave outer face is rather precipitous and lacks a shelf on which teeth of the outer (lesser) row might develop.

"No trace of a barbel can be found. The premaxillary groove although continuous is shallower on the median line than it is laterally. A hidden frenum, therefore, might be described as present. Both lips are rather thick. The upper lip is expanded at the front in a somewhat shield-shaped form. The snout and upper lip are about equal in forward projection, but the lower jaw is definitely included. When the mouth is tightly closed, the rather long maxillary fold is hidden between the rostral fold and the upper lip. The rostral-maxillary crease is scarcely continued onto the side of the snout. The fold of each lower lip is about twice as long as the width of the isthmus between the folds of either side. The weakly curved gape is subhorizontal. The maxillary extends nearly to or a little beyond the vertical from the front of the eye.

"As measured from the angle of the gill arch, the much restricted first gill slit is about as long as the eye. As the upper arm of the gill slit is scarcely developed, the pharyngeal roof shows almost no tendency to form a septum. The short, soft gill rakers number only 5 to 9, including all rudiments. There are no pseudobranchiae. The intestine is short (about one-half the standard length). It is straight posteriorly, but moderately coiled upon itself anteriorly, where it makes about three U-shaped convolutions. The peritoneum is blackish.

"The scales are very small and deeply embedded, giving the skin a distinctly leathery texture. In outline the scales are roughly subcircular. Radii are scattered on all fields. The scales along the lateral line number about 70 to 80. Especially on the trunk these scales are slightly larger than those immediately above or below the lateral line. This is particularly true of the breeding males. The lateral line is typically complete, rarely somewhat disrupted. It is very slightly decurved behind the head and is straight posteriorly.

"The origin of the dorsal fin lies directly over or slightly behind the pelvic insertion, closer to the caudal base than to the tip of the snout. This short-based fin has 8, occasionally 7, principal rays. None of the rays are spiny. The first ray is minute; the second, short and closely appressed to the third. The anal fin, also short-based, has a concave border, giving it a slightly falcate appearance. It has 7, or commonly 8, principal rays. The well-forked caudal is equilobate. There are usually 8 pelvic rays, but the number is occasionally reduced to 7, or even to 6" (Hubbs and Miller 1948A: 1-3).

FIG. 198. "Holotype of *Moapa coriacea* from Warm Springs near the source of Moapa River, southeastern Nevada; 74 mm. in standard length; U.M.M.Z. No. 143186." Photograph by Clarence M. Flaten (Hubbs and Miller 1948A). Courtesy Dr. Robert R. Miller.

DIAGNOSIS

Head about 3.5 times in body length, body depth about 4 times in body length; head rather short and semiconical; mouth nearly horizontal, lips somewhat thick, lacking barbels; maxillary extending posteriorly to front margin of eye; pharyngeal teeth in a single row, 5/4, rarely 4/5, teeth hooked and with crushing surfaces, grading from longest in the upper teeth to an almost blunted appearance in the lower teeth; pseudobranchiae lacking; lateral line complete, only weakly decurved; scales 17–22/69–79/12–15, 44–51 before the dorsal fin; fins—dorsal 8-rayed, rarely 7—anal usually 8-rayed, occasionally 7—pectoral 13–16 rayed, usually 14–15—pelvic 6–8 rayed, usually 8—caudal 17–20 rayed, usually 19, moderately forked, small; caudal peduncle stout, least depth about 10 times in body length; intestine very short, about half body length; peritoneum nearly black; color dark olivaceous above, brownish on sides, white on belly—along sides with a conspicuous blackish longitudinal line and vaguer longitudinal golden areas; a small species, about 3 inches in maximum length.

TYPE LOCALITY

"The holotype, a ripe female 74 mm. long to caudal base, was seined by Carl L. Hubbs, Robert R. Miller, and Alex Calhoun at Home Ranch in one of the sources of Moapa River (= Muddy River), 7.5 miles by road northwest of Moapa, Clark County, Nevada, on July 12–13, 1938. It is deposited in the University of Michigan Museum of Zoology (No. 143186)" (Hubbs and Miller 1948A: 8–9).

RANGE

Known only from the Warm springs area of northern Clark County, Nevada, inhabiting the numerous warm springs and their outlet streams which are the sources of the Moapa River, a modern occupant of the lower course of the old Pleistocene White River channel.

TAXONOMY

This newly-described genus and species is considered by its authors as probably an offshoot of *Agosia,* "or, more likely, that both were

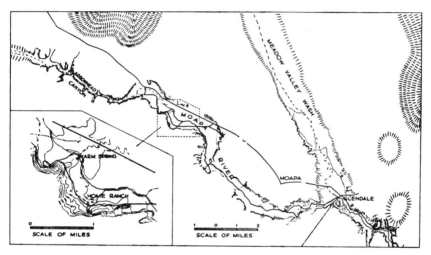

FIG. 199. The Moapa River of northern Clark County, with an enlargement of the Warm Springs area of *Moapa coriacea.* From Hubbs and Miller 1948A, courtesy Dr. Robert R. Miller.

derived from a common ancestral type" (p. 8). As they also point out, *Moapa* shows affinities with both *Gila* and *Rhinichthys:*

"In appearance the new genus approaches *Gila,* particularly the race of *G. robusta* that inhabits the same stream, but it differs from all species of that genus in the lack of teeth in the lesser (outer) row. In the relatively small, subinferior, nearly horizontal mouth, in the development of radii on all fields of the scale, in the leathery skin, and in the strong basicaudal spot, *Moapa* differs from *Gila* as a whole, but that genus, as at present constituted (Miller, 1945), is so diverse that none of these additional differences are consistently trenchant.

"*Moapa* resembles *Rhinichthys* (including *Apocope*) in the small, subinferior mouth, in the numerous scales, and in the development of radii on all fields of the scales. It differs from that genus in the consistently uniserial rather than usually biserial dentition and in having 5 rather than 4 teeth in the principal row on the left arch" (p. 5).

Comparisons between *Agosia* and *Moapa* are more in detail: "These genera have in common: (1) a small, subinferior and nearly horizontal mouth, with the lower jaw included; (2) pharyngeal arches of similar shape; (3) a hidden frenum; (4) a rather small eye; (5) small scales, with radii on all fields; (6) a broad, blackish mid-dorsal stripe; and (7) a conspicuous black spot at the base of the caudal fin. This spot is better developed in *Moapa coriacea*, however, than it is in any other western minnow. In many other features, as in coloration, in position of dorsal fin, and in number of gill rakers, *Moapa* and *Agosia* are very similar. *Moapa* differs from *Agosia* in the tooth formula, which is typically 5–4, rarely 4–5, instead of consistently 4–4; in the lack of a barbel (a small one is consistently evident in *Agosia*); and in the leathery skin" (pp. 5, 7).

Concerning the relationships of *Moapa* with *Siphateles*—"The dental formula of the new genus, 5–4, is characteristic of most forms of *Siphateles*, but the shape of the pharyngeal arch is very different, rather closely resembling that of *Gila*. In no other special way does *Moapa* resemble *Siphateles*, which genus, moreover, is unknown, either Recent or fossil, in the faunally distinctive Colorado River basin" (pp. 4–5).

Etymology—"Moapa" from the Moapa River, a southern Pahute word meaning "muddy" in allusion to the highly silted water. Residents of the area commonly substitute the term "Muddy River" for Moapa River. "Coriacea" from the Latin *coriaceus*, meaning "leathery."

LIFE HISTORY

The Moapa Dace is a thermal endemic, found only in water between 87° F. and 93° F. and occurring alike in streams and pools. Limnologic characteristics of its unique environment were grossly determined by the author in one of the major pools of the "Home Ranch" area as follows:

Water color—clear
pH—7.3 (varies up to 7.5 in the region—
Hubbs and Miller 1948A: 9)
Temperature—93° F.
Parts per million of oxygen—3.4 (ppm)
Methyl orange alkalinity—223 ppm
Phenolphthalein alkalinity—0 ppm
Free carbon dioxide—0 ppm
Chlorides-sulfates—present

(6 April 1950)

No studies have been published on the species' life history. As observed by the writer in pool environments where its general habits are easily seen, the Moapa Dace is a rather methodical "schooler," remaining conspicuously segregated from other similarly sized fishes as occur with it such as the White River Springfish (*Crenichthys baileyi*) and the introduced Mosquitofish (*Gambusia affinis*).

Stomach contents analyses made by the author show a preponderance of arthropod remains, principally insects, with some vegetal matter.

Their willingness to take a tiny baited hook attests to their carnivorousness and corollary anatomic evidence is found in their very short intestine.

ECONOMICS

Unimportant other than the food chain aspects. The Moapa Dace shared top spot in the aquatic food pyramid originally only with the similarly sized White River Springfish, which has matching habits, and certain of the larger aquatic bugs (Hemiptera)—a situation making for keen competitive possibilities; yet both species were, and are, common.

With introduction of the small but challenging Mosquitofish, and that burly bandit, the Bullfrog (*Rana catesbeiana*), the picture altered to an extent that others now crowd the energy exchange peaks—yet in spite of added predation and competition, both native fishes seem to be doing well.

DESERTFISHES
Genus *Eremichthys* Hubbs and Miller 1948
("Desert fish")

SOLDIER MEADOWS DESERTFISH
Eremichthys acros Hubbs and Miller
(Soldier Meadows Dace)

Eremichthys acros Hubbs and Miller 1948A: 3–4, 14–28 / 1948B: 41
Eremichthys acros, Miller, 1949A: 450–451
Eremichthys acros, La Rivers, 1952: 98 / 1954: 168
Eremichthys acros, La Rivers and Trelease, 1952: 117
Eremichthys acros, Eddy, 1957: 83
Eremichthys acros, Moore, 1957: 107

ORIGINAL DESCRIPTION

The initial remarks made here for the Moapa Dace, apply equally to the Soldier Meadows Desertfish.

"The rather heavy-set uniserial pharyngeal teeth number 5 on the left side and 4 on the right. On the left arch the uppermost tooth is somewhat longer and slenderer than the others. It has a weak to rather well-developed terminal hook. This tooth, like the 2 that follow, has a rather broad grinding surface, which varies from saucer-shaped in the half-grown to nearly flat or even somewhat raised in large adults. The grinding surface on the fourth tooth is variable, for it is either more marked, as well developed, or more reduced than those on the anterior teeth. The fifth tooth has neither grinding surface nor hooked tip and is rather small and stumpy, though commonly pointed. The basal parts of all teeth except the first one are heavy and somewhat swollen. The 4 teeth on the right arch are similar to those on the left. The upper 2 are moderately hooked, and the first is smaller and narrower than the second. The hook and grinding surfaces, flat to slightly saucer-like, tend toward obsolescence on the lower teeth.

"The heavy lower pharyngeal arch is very broad across the upper half. The 2 arms, measured from the teeth, are of nearly equal length. The lower arm is moderately broad, flat, and nearly straight. The outer face below the row of teeth descends rather abruptly, leaving no available space on which teeth of an outer (lesser) row might develop.

"There is no trace of a barbel or of a frenum. The upper lip is rather thick, and somewhat fleshy, especially medially, where it is expanded backward into a slight concavity of the rostral fold. Laterally, this lip

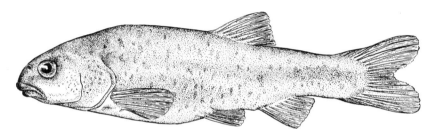

FIG. 200. Soldier Meadows Desertfish, *Eremichthys acros*.
Drawn by Silvio Santina.

FIG. 201. "Paratype of *Eremichthys acros* from Soldier Meadows, Humboldt County, Nevada; 58 mm. long; U.M.M.Z. No. 136874. Photograph by Clarence M. Flaten" (Hubbs and Miller 1948A), courtesy Dr. Robert R. Miller.

narrows evenly to the corners of the mouth, where it is about one-half as wide as it is at the apex. From the corners of the mouth a little more than half way to the apex of the lower jaw, the lower lip is expanded on each side into a somewhat fleshy lobe. In width each lobe about equals or somewhat exceeds the least distance between the lobes. The contour of the lower jaw is that of a broad U resting within the more rounded U of the upper jaw. The nearly straight gape is very slightly oblique. Even anteriorly it lies well below the level of the lower rim of the orbit.

"The prominent horny sheaths on the jaws provide the spectacularly distinctive generic character of *Eremichthys*. These sheaths are so loosely attached to the jaws that they may readily be plucked off with a needle. Each is a sharp-edged structure covering most of the arch of the jaw. The upper sheath is slightly narrower than the lower one and closes across the latter for a short distance at each side; elsewhere, when the mouth is closed, the two meet perfectly along their thin, sharp edges. Presumably, this specialized structure is an adaptation for feeding by grazing.

"The snout and upper lip project forward equally, and the lower jaw is only slightly included. The maxillary extends about to the vertical from the second nostril.

"As measured from the angle of the gill arch, the rather restricted first gill slit is about 1½ times the width of the orbit. A conspicuous,

FIG. 202. Head and mouth details of *Eremichthys acros*. On the right are the horny sheaths which come easily from the upper and lower jaws. From Hubbs and Miller 1948A, courtesy Dr. Robert R. Miller.

broad valve extends across the pharyngeal roof just behind the horny sheath of the upper jaw. The moderately short, fleshy gill rakers number 9 to 11, including all rudiments. There are no pseudobranchiae.

"The extensively coiled intestine is more than twice as long as the standard length of the fish. The peritoneum is brownish, with an admixture of black pigment.

"The irregularly arranged and loosely imbricated scales number about 70 to 80 along the lateral line. They are roughly oval in shape and bear numerous radii on all fields, though on the basal field these are fewer and more incomplete than they are elsewhere. The lateral line, decurved along its anterior half and nearly straight posteriorly, is variously developed. In 30 specimens analyzed for this character, it was complete in 13, interrupted in 10, and incomplete in 7 specimens.

The interruption usually occurs below the dorsal fin, where occasionally the lateral line is broken and displaced upward 1 to 3 scale rows. The scales along the lateral line are not enlarged.

"The origin of the dorsal fin is much nearer to the caudal base than to the tip of the snout, and in both sexes lies behind the vertical from the pelvic insertion. It is small, rounded, and short-based, with 8, or occasionally 7, principal rays, none of which are spiny. The anal fin is somewhat larger than the dorsal but is similarly shaped. It has 8, or

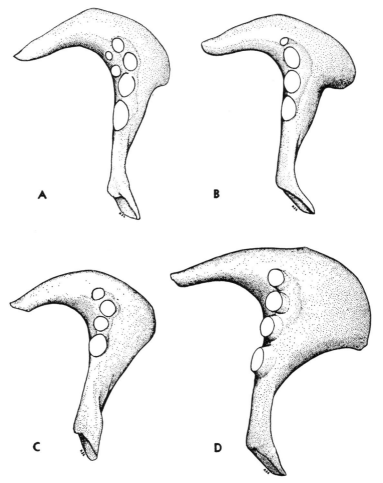

FIG. 203. Pharyngeal arches of four types of cyprinid fishes. Clear circles represent the position and number of the teeth. Drawn by Ann Green. Courtesy University of Michigan Press and Drs. Carl L. Hubbs and Robert R. Miller.
    (A) Colorado Gila (*Gila robusta*)
    (B) Moapa Dace (*Moapa coriacea*)
    (C) Lahontan Tui Chub (*Siphateles bicolor obesus*)
    (D) Soldier Meadows Desertfish (*Eremichthys acros*)

commonly 7, principal rays. The relatively large caudal fin is moderately forked. Its lower lobe is somewhat larger than the upper. The pelvic rays are usually 8, but vary from 6 to 9" (Hubbs and Miller 1948A: 14–16).

DIAGNOSIS

Head and body depth 3.0 to 3.5 times into body length; mouth subhorizontal, small; lips rather thick, lacking barbels; maxillary short, not reaching the eye, extending posteriorly to about the second nostril; both jaws equipped with conspicuous, loose, horny sheaths which are readily removable; pharyngeal teeth in a single row, seemingly constantly 5/4, uppers hooked and with crushing surfaces, lowers smaller and blunter; pseudobranchiae lacking; lateral line usually complete, weakly decurved; scales 15–18/68–78/10–13, 45–54 before the dorsal fin; fins—dorsal usually 8-rayed, occasionally 7—anal usually 8-rayed, not uncommonly 7 and very rarely 9–10—pectoral nearly always 16–17 rayed, but varying from 14 to 19—pelvic nearly always 8-rayed, but rarely 6–9—caudal generally 19-rayed, but varying from 17–21; caudal peduncle very thick, least depth about 7–8 times into body length; intestine long and much-coiled, more than twice as long as the body; peritoneum blackish-brown; color—olive-green on back, silvery below, and pronounced yellow reflections along the sides—sides weakly and vaguely mottled with blackish spotting, with a dorso-lateral deep greenish streak; small in size, 2½ inches being about maximum length.

TYPE LOCALITY

"The holotype (U.M.M.Z. No. 136873), a nuptial male 38 mm. long to caudal base, was collected by Robert R. and Ralph G. Miller in a spring-fed ditch near the northern edge of Soldier Meadows, Humboldt County, Nevada. It was seined on July 5, 1939 . . ." (Hubbs and Miller 1948A: 20).

RANGE

Restricted to the warm springs and creeks of Soldier Meadows, a high basin in the mountains of western Humboldt County, at an elevation slightly under 5,000 feet. The somewhat circular valley of Soldier Meadows lies west of the Black Rock Desert, and several hundred feet above it. Its drainage is southerly into the desert, which was a large segment of Pleistocene Lake Lahontan in Pluvial times. At the present, Soldier Meadows is effectively isolated from other remnants of Lahontan, although it was apparently partially encroached upon by Lahontan waters during their highest stage.

TAXONOMY

The describers' generic comparisons represent competent current opinion not only on this genus, but also on closely related genera.

Etymology—"acros," summit, from nearby Summit Lake.

TABLE II

	Eremichthys	Rhinichthys	Siphateles
Lip sheathing	Both jaws	Lower jaw occasionally with simple sheath	Never with lip sheathing
Scale radii	Greater extent (similar)	Greater extent (similar)	Lesser extent (dissimilar)
Fin position and size	Similar (dorsal more posterior)	Similar (dorsal more posterior)	Dissimilar (dorsal more anterior)
Dentition	5–4	1/4–4/1, rarely 5–4	5–4 (often 5–5, rarely 4–4)
Pharyngeal arch	Similar	Dissimilar	Similar
Barbels	Always absent	Present or absent	Always absent
Gill rakers	9–11	6–9, usually 6–7	8–36, usually 10–33
Anal rays	7–10, usually 8	7–8, usually 7	7–11, usually 8
Intestine	More than 2 times body length	Body length	Body length
Size	Small	Small	Large
Shape	Similar	Similar	Dissimilar
Lateral line	Often incomplete	Often incomplete	Complete

LIFE HISTORY

The Soldier Meadows Desertfish lives in warm spring-fed waters; Hubbs and Miller (1948A: 21) record a temperature as high as 100.4° F. (38° C.) for one spring source where a few specimens were seen. *E. acros* ranges, by their data (and as the author subsequently found), from water of 100 degrees down to at least 67 degrees; after the water cools, nothing is found but the associated species—*Rhinichthys osculus robustus* and *Catostomus tahoensis*, which also occur in water over 90 degrees. One of their comparative listings of this association in varying degrees of water shows the following:

"97°: *Eremichthys*, plentiful.

93°: *Eremichthys*, plentiful.

93°: *Eremichthys*, plentiful; 1 *Rhinichthys;* 1 young *Catostomus*.

92°: *Eremichthys*, plentiful; increased numbers of *Rhinichthys* and young of *Catostomus*.

82°: *Eremichthys*, plentiful; *Catostomus* young, many; *Rhinichthys*, not many.

"Below this point the stream broke up in the marshy meadow" (p. 23).

In one area, in waters from 79 to 86 degrees, they found the Soldier Meadows Desertfish associated with small numbers of speckle dace, most of the latter having what they considered as disease-caused fleshy swellings.

In the cooled waters of Soldier Creek below the Meadows proper, they found the Lahontan Redshiner (*Richardsonius egregius*), speckle dace and the sucker (*Catostomus*), but no *Eremichthys*. Farther down, the Lahontan Tui Chub (*Siphateles bicolor obesus*) appeared in the stream.

Studies in the author's laboratory by a graduate student, David Nyquist, indicated that large specimens had a decided tendency to eat the smaller fishes.

ECONOMICS

The Soldier Meadows Desertfish seems to be the top species in its food pyramid, at least in so far as vertebrates are concerned, since its predilection for warm waters removes it effectively from contention with the several typical Lahontan fishes living in colder waters just below it. In its smaller stages, it is prey to the usual array of predaceous aquatic insects, as are all thermal endemic fishes. So, as a basic converter of vegetable matter to animal protein, the species peaks a very short food chain in its restricted environment.

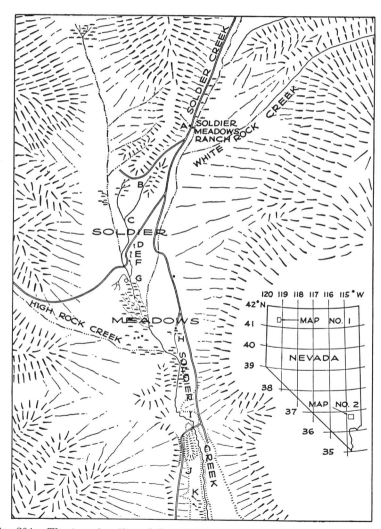

FIG. 204. The type locality of *Eremichthys acros*. From Hubbs and Miller 1948A, courtesy Dr. Robert R. Miller.

CARP

Genus *Cyprinus* Linnaeus 1758
(Latinization of the ancient Greek name for the carp)

THE ASIATIC CARP

Cyprinus carpio Linnaeus
(European Carp, Carp)

Cyprinus carpio Linnaeus 1758: 320
Cyprinus carpio, Günther, 1868: 25–31
Cyprinus carpio, Hessel, 1878: 865–900
Cyprinus carpio, Poppe, 1880: 661
Cyprinus carpio, Jordan and Gilbert, 1882: 254 / 1894: 139
Cyprinus carpio, Gilbert, 1893: 231
Cyprinus carpio, Jordan and Evermann, 1896B: 201
Cyprinus carpio, Forbes and Richardson, 1908: 104–110 / 1920: 104–110
Cyprinus carpio, Snyder, 1917A: 85
Cyprinus carpio, Pratt, 1923: 64
Cyprinus carpio, Coker, 1930: 195–200
Cyprinus carpio, Fish, 1932: 322–323
Cyprinus carpio, Hatton, 1932: 53–58
Cyprinus carpio, Hayes, 1935: 103–104
Cyprinus carpio, Jenkins, 1936: 282–284
Cyprinus carpio, Schultz, 1936: 145
Cyprinus carpio, Tanner, 1936: 172
Cyprinus carpio, Schrenkeisen, 1938: 96–97
Cyprinus carpio, Kuhne, 1939: 50–54
Cyprinus carpio, Frey, 1940
Cyprinus carpio, Sumner, 1940: 220
Cyprinus carpio, Hubbs and Lagler, 1941: 44, 53 / 1947: 53, 62
Cyprinus carpio, Murphy, 1941: 167
Cyprinus carpio, Moffett, 1942: 82 / 1943: 182
Cyprinus carpio, Hinks, 1943: 45–47
Cyprinus carpio, Dill, 1944: 151–153
Cyprinus carpio, Dimick and Merryfield, 1945: 36
Cyprinus carpio, Miller and Alcorn, 1946: 180–181
Cyprinus carpio, Eddy and Surber, 1947: 145–147
Cyprinus carpio, Carl and Clemens, 1948: 70, 72–73
Cyprinus carpio, Shapovalov and Dill, 1950: 386
Cyprinus carpio, Simon, 1951: 67–68
Cyprinus carpio, Wallis, 1951: 89
Cyprinus carpio, Beckman, 1952: 38–39
Cyprinus carpio, La Rivers, 1952: 97
Cyprinus carpio, La Rivers and Trelease, 1952: 116
Cyprinus carpio, Miller, 1952: 17, 29
Cyprinus carpio, Gabrielson and La Monte, 1954: 108–109
Cyprinus carpio, Scott, 1954: 44

Cyprinus carpio, Dymond, 1955: 543–544
Cyprinus carpio, Shomon, 1955: 49–50
Cyprinus carpio, Harlan and Speaker, 1956: 85, 87
Cyprinus carpio, Pitt, Garside and Hepburn, 1956
Cyprinus carpio, Eddy, 1957: 80
Cyprinus carpio, Koster, 1957: 54–55
Cyprinus carpio, Moore, 1957: 98
Cyprinus carpio, Mraz and Cooper, 1957
Cyprinus carpio, Trautman, 1957: 283–285
Cyprinus carpio, Slastenenko, 1958A: 182–183
Cyprinus carpio, Carl, Clemens and Lindsey, 1959: 100–101
Cyprinus carpio, Shapovalov, Dill and Cordone, 1959: 171

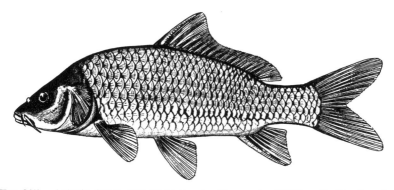

FIG. 205. Asiatic Carp, *Cyprinus carpio*. Drawn by William L. Brudon for Miller 1952. Courtesy California Department of Fish and Game and Leo Shapovalov.

ORIGINAL DESCRIPTION

"Carpio. 2. C. pinna ani radiis 9, cirris 4, pinnae dorsalis radio secundo postice serrato. D. 2/24. P. 16. V. 9. A. 9. C. 19.

Art. gen. 4. *Syn.* 3. *Spec.* 25. Cyprinus cirris 4. ossiculo tertio pinnarum dorsi anique unicinulis armato. D. 24. P. 16. V. 9. A. 9. C. 19.

Fn. Svec. 317. idem. *Gron. mus.* I. *n.* 19. idem. D. 23. P. 17. V. 8. A. 8. C.—*Habitat in* Europa. *Nobilis piscis saepius in piscinis educatus, circa a.* 1600 *primum in Angliam introductus"* (Linnaeus 1758: 320).

DIAGNOSIS

Head about 3.5 times in body length, body depth about 3.0 or less times into body length; head rather conical, with a pronounced slope or taper to snout tip, which is bluntly pointed; mouth oblique, somewhat small, terminal, its maxillary not extending posterior of nostril; two pairs of maxillary barbels, the lower pair the longer; pharyngeal teeth prominent, coarse, with knobbed, flattened molar surfaces—in 3 rows, 1/1–2/3–3/1–2/1; lateral line complete, often sinuous; scales large, 5–6/35–38/5–6,[1] about 11 before the dorsal fin; fins—dorsal very

---

[1]Except in the partially scaled "Mirror," and the scaleless "Leather" types which occur regularly as parts of almost any given population of carp.

long, beginning with a large spine toothed along its hind edge, 17–21 rayed, longest in anterior portion—anal similarly spined, 5–6 rayed—pectorals almost reaching to front of pelvics—caudal prominent, deeply forked; caudal peduncle stout, least depth 8–10 times into body length; intestine exceeding body in length; peritoneum gray, smooth or spotted; color olivaceous, darkest on back, grading to a yellowish cast on the belly; fins of breeding individuals, especially on lower part of body, often with a rosy tinge; attains a large size, Pyramid Lake specimens growing in excess of three feet in length. European accounts contain records of individuals exceeding 60 pounds in weight, but these must be exceptional monsters, for 20–25 pounds seem to be the usual upper weight limits.

TYPE LOCALITY

Europe.

RANGE

Originally an Asiatic species, the carp was introduced into Europe several centuries before it was brought to the United States.[2] The time of its first arrival in this country is usually stated as 1877 when Hessel brought in 345 specimens: "This method I followed with the carp which I imported from Europe for the purpose of breeding in the winter of 1876–77" (1878: 869). R. A. Poppe (1880) records carp as having been brought into California as early as 1872 when J. A. Poppe started out from Reinfeld (Holstein), Germany, with 83 carp of all sizes only 8 of which were left when he landed in New York. He finally got home to Sonoma Valley, California, with 5 fish, the smallest of the original lot; from this nucleus, the early-day carp culture in California seems to have started.

Dymond (1955: 543) states that carp "was taken to the United States under private auspices as early as 1831. By 1877 it was being widely distributed by the U. S. Fish Commission. The regard in which the fish was held at that time was indicated by the fact that in 1850 legislation was passed imposing a fine of $50.00 for destroying carp."

The Asiatic Carp is widespread in Nevada, as well as the United States, being found in nearly all available waters except the higher, colder streams and lakes, even having been planted in many of the smaller, isolated spring-pool systems, warm and cold.

Carp were first recorded as a Nevada introduction in 1881 (McDonald 1884: 1126), while 9 counties—west-central Nevada and those along the Humboldt River except Lander—received carp in 1885 and 1886 (Cary 1887: 8). What may have happened earlier, we do not know, but Poppe (1880: 665) pointed out that California carp—established there in 1872—had been distributed to other States.

According to Cal Allan (1958), carp are rare in Walker Lake and apparently the population there is sustained mainly from river recruitment.

TAXONOMY

Etymology—"carpio," carp.

---

[2]Hessel (1878) gives a date of 1227 for the European introduction.

LIFE HISTORY

The Asiatic Carp is a large, spring and summer spawning herbivorous species with omnivore tendencies. Its living habits are equally varied. It does well in shallow-to-moderately deep waters—fresh-to-brackish—with or without much vegetation. It is more at ease in the warmer, murky environments but can survive temperature extremes which include temporary freezing.[3]

Egg-laying is a matter of a single female being attended by several males and laying her enormous numbers of eggs in batches of several hundred at a time. The generative capacity of the species is prodigious, as shown by such figures as a half million eggs for a five-pound fish and up to better than 20 million eggs from a 20-pounder.

No nest of any kind is built, the eggs being broadcast loosely among vegetation, rocks or coarse gravel. This spawning or "rolling" takes place in shoal water, often with backs of the fish exposed and their activity marked by much splashing. The migrations of carp to such areas are often impressive sights.

Hatching of the eggs depends, of course, upon temperatures to which they are subjected, but in our latitude, 10 days to two weeks or slightly less seems a reliable figure, the time decreasing as temperature increases. Lengths of nine inches may be achieved in their first year if food supplies are unlimited and temperatures optimum, a growth potential considerably exceeding that of most of our native fishes. However, growth rates are usually just a little over half this. Individuals come of spawning age early in their third year.

Because of its deep body, the carp is a heavy fish for its length. In our latitude, a 10-pounder is a good-sized fish, although Pyramid Lake specimens get much larger as do those in the lower Carson drainage. Strapping individuals in this country run to 50 pounds, with 60 pounds and better recorded from European waters. Lampman (1946) and Carl and Clemens (1948) note 45 and 50 pounders for the Pacific Northwest where temperatures and food conditions are no more optimum than they would be in, say, Pyramid Lake.

---

[3]Robert Sumner has given me the following more specific notes on carp in our area:

"Hatching time of carp eggs is probably closer to one week under optimum temperature conditions. In our area spawning can and often does continue more-or-less progressively from late April through July or even August. Peak spawning in shallow waters of the Fallon region (i.e., Stillwater, etc.) is during May and June, whereas in deeper waters (as Lahontan) it is June and July.

"Growth the first year averages about 5", however varies from 3" to 8" or 9", depending upon conditions. Males of 7" and up (Class I & II fish) are often capable of fertilizing eggs. In general, I would say 2-year-old males and 3-year-old females spawn successfully.

"Carp are probably better adapted to waters of the lower Carson River system than elsewhere in the State. From some waters (as Lahontan, Stillwater, etc.) these fish often range in size from 10 to 15 pounds. However, individuals of 20 pounds are sometimes encountered (1958)."

Schoffman (1957) found the following age-group lengths and weights for carp in Reelfoot Lake, Tennessee:

Age group	Average length inches	Average weight ounces
4	14.7–15.3	28.9– 38.2
5	16.0–16.7	35.1– 39.3
6	17.5–17.7	41.3– 45.7
7	19.3–19.4	55.3– 60.7
8	21.9–22.0	71.9– 90.0
9	22.5–23.2	97.0–107.1
10	25.0–25.1	118.0–130.4
11	26.1–26.4	146.8–149.8
12	27.9–28.6	170.0–198.6

Where growth was considered fast for streams, Purkett (1958: 123) found the following for Salt River, Missouri:

Age group	Average length inches
1	4.9– 5.8
2	10.6–11.6
3	14.3–15.3
4	17.1–17.5
5	18.1–19.2
6	20.2–20.9
7	21.7–22.9
8	25.7

ECONOMICS

No other fish has been so cussed and discussed by those who cast their lines into the water. To most, friend carp is the Great American Tragedy and its becoming so tells a story that often repeats itself.

In the early days of the United States Fish Commissioner—tall pioneer of the 1870s—that worthy gentleman became enamoured of the "German" carp and its possibilities as a cheap source of meat. During this decade, carp were brought from Germany and planted at both ends of the country—California in 1872 and Washington, D. C. in 1877.

The eagerness with which State fish commissioners and private individuals sought the husky immigrant is shown well enough by Nevada examples. Said Nevada Fish Commissioner Parker in 1881: "Carp, as a food fish, have no superior; . . . One of my great aims has been to stock our waters with the best species of carp . . ." We may be pardoned our silent thoughts when we call to mind's eye that the 1880 angler in western Nevada could, figuratively, walk dry shod across rivers on the backs of a most magnificent strain of Cutthroat Trout.

In similar vein, a citizen of the Toquima Range in central Nevada, one Pasco, remarked that "I persuaded the man above me on my stream not to go to Reese River after trout, because I hoped sooner or later to get carp, and I did not want trout in the stream to eat the young" (1882). The pendulum was far to one side, the point of release. Its swing brought swift reaction.

In the Nevada of 1887, Fish Commissioner Cary had faint misgivings about this "wonder" fish, although George T. Mills, his successor of four years later, seemed pleased that many Nevada residents had become carp culturers. But by 1897 Mr. Mills was no longer a friend of the carp. Said he with some bitterness:

"Several years ago, during the carp furore, the General Government, while not entirely to blame, was *particeps criminis* in foistering upon this State and in polluting our waters with that undesirable fish, the carp." From all quarters of the country people were carping about the situation. Even Canada joined in: "It is generally conceded that the promiscuous introduction of carp on this continent has been attended with nothing but evil results" (Ontario Fisheries Report for 1899).

The gist of the whole matter was this: Once widely distributed over the country, carp were accused of ruining waters for the feeding and breeding of other fish (this by roiling it and uprooting vegetation), as well as for water birds; of eating game fish spawn; of annoying other fish so they refused to spawn. A gloom settled over the placid countenance of the American fisherman that has never quite lifted.

And as Time, Sewage, Water Diversion and Watershed Destruction teamed with Ye Increased Angler to destroy native fish stocks as carp could never do, the Wayfaring Stranger took the blame. Seldom has there been a more glorious scapegoat for the anglerfolks' woes (unless it be the hapless fishery technician), and always these stories were tailored to the local needs. In Nevada the carp has been accused of everything short of stopping up the drain whereby Pyramid Lake empties into the Pacific Ocean.

So completely did guilt of the carp take hold that few bothered to ask themselves what—among people with thinking filters—should be inevitable: Is there another side to the question? Among the most concise of such past studies is that of Forbes and Richardson, first published in 1908 and more fully elaborated in 1920, concerning the impact of carp on Illinois fisheries. They summed up the case succinctly:

"However, it is becoming more and more to be believed that its good qualities more than overbalance the other side of the account, the most serious of the charges against it appearing to rest on uncertain or gratuitously assumed premises" (p. 108).

Purkett's statement also lends aid to the credit side of the ledger.

"Carp were important hook-and-line fish in the Salt River. They ranked first in total weight in the creel. Although considered undesirable in many waters, they make up a high proportion of the creel in turbid streams of Missouri. A considerable amount of fishing is especially for carp" (1958: 122).

Robert Sumner (1958) says:

"You will perhaps be interested in knowing that carp in Stillwater do have some virtue. In the west marsh I have found no evidence to indicate that carp have been detrimental to game fish populations. In fact, they have provided the largest source of forage for adult bass I could find. In addition, I could find no evidence in these same waters that carp activity contributed materially to roily waters there."

And so it goes.

CRUCIAN CARPS

Genus *Carassius* Nilsson 1832

(Latinization of the vernacular names "Karass" or "Karausche," applied to the European Crucian carp, *Carassius carassius*)

THE GOLDFISH

Carassius auratus (Linnaeus)

Cyprinus auratus Linnaeus 1758: 527
Carassius auratus, Günther, 1868: 32–34 / 1880: 591–592
Carassius auratus, Jordan and Gilbert, 1882: 253
Carassius auratus, Jordan and Evermann, 1896B: 201
Carassius auratus, Regan, 1911: 177–178
Carassius auratus, Pratt, 1923: 64–65
Carassius auratus, Norman, 1931: 420 and other mentions
Carassius auratus, Hayes, 1935: 105
Carassius auratus, Jenkins, 1936: 285–286
Carassius auratus, Schultz, 1936: 145
Carassius auratus, Schrenkeisen, 1938: 98
Carassius auratus, Kuhne, 1939: 64
Carassius auratus, Miller and Alcorn, 1946: 181–182
Carassius auratus, Hubbs and Lagler, 1941: 44, 53 / 1947: 53, 62
Carassius auratus, Eddy and Surber, 1947: 147–148
Carassius auratus, Carl and Clemens, 1948: 70, 74
Carassius auratus, Shapovalov and Dill, 1950: 386
Carassius auratus, Simon, 1951: 68–69
Carassius auratus, Wallis, 1951: 88
Carassius auratus, Beckman, 1952: 40
Carassius auratus, La Rivers, 1952: 97
Carassius auratus, La Rivers and Trelease, 1952: 116
Carassius auratus, Miller, 1952: 17, 29–30
Carassius auratus, Scott, 1954: 45
Carassius auratus, Dymond, 1955: 544
Carassius auratus, Stokell, 1955: 143
Carassius auratus, Harlan and Speaker, 1956: 87–88
Carassius auratus, Eddy, 1957: 81
Carassius auratus, Koster, 1957: 55
Carassius auratus, Moore, 1957: 98
Carassius auratus, Trautman, 1957: 286–288
Carassius auratus, Slastenenko, 1958A: 184–185
Carassius auratus, Carl, Clemens and Lindsey, 1959: 102–103
Carassius auratus Shapovalov, Dill and Cordone, 1959: 171

ORIGINAL DESCRIPTION

"auratus. 8. C. pinna ani gemina, caudae transversa trifurca. *Fn. Suec.* p. 125. *t.* 2.
Act. Stockh. 1740. p. 403. *t.* I. *f.* I–8. idem. D. 2/18. P.16. V.9. A.8, 8. C.37

Gron. mus. I. *n.* 15. idem. D.8. P.11. V.7. A.—C.20 *mus.* 2. *n.* 150. C. pinna ani simplici, cauda trifurca. D.18. P.11. V.8. A.8. C.44
Edw. av. t. 209.
Pet. gaz. t. 78. *f.* 7.
Habitat in Chinae, Japoniae *fluviis.*
Piscis colitur in vasis murrhinis ob aureum fulgorem, tum multum pinnis varians. Radius secundus *ani postice serratus.*" (Linnaeus 1758: 322)

DIAGNOSIS

Head about 3.0 times, body depth about 2.5 times, into body length, respectively; head blunt, short, conical; mouth oblique, small, terminal, maxillary not reaching posterior to nostril; barbels lacking; pharyngeal teeth prominent, but not molar-like, in a single row, 4–4; lateral line complete, nearly straight; scales very large, 5–6/25–30/5–6, about 8

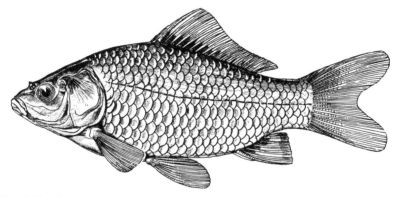

FIG. 206. Goldfish, *Carassius auratus.* Drawn by William L. Brudon for Miller 1952. Courtesy California Department of Fish and Game and Leo Shapovalov.

before the dorsal fin; fins—dorsal very long, with an anterior spine toothed along its hind margin and followed by about 18 soft rays, longest in front part of fin—anal single spined, 5–6 rayed—pectorals reaching behind origin of pelvics—caudal large, deeply forked; caudal peduncle very broad, least depth about 6 times into body length; intestine longer than body; color olivaceous, often with a brownish caste, becoming lighter ventrally (wild state)—domestic varieties run the gamut of colors from red to white to black with combinations of mottling of various shades; in size, smaller than carp, although Regan (1911: 178) speaks of English specimens of several pounds in weight, and Eddy and Surber (1947: 148) mention similarly heavy individuals in the northeastern United States. Maximum length seems to be about a foot and a half.

TYPE LOCALITY

China and Japan.

RANGE

Originally an eastern Asiatic species, the goldfish is now distributed pretty much over the north temperate world, chiefly as an aquarium fish. It has escaped from domestication in many places, where it has maintained itself in the wild state. When wild, it reverts to its olive-brown coloration. Günther (1880: 592) records the goldfish as being introduced to England in 1691, and it was known to parts of Europe much earlier.

Few records are now available of early goldfish plants in Nevada. I have been told by Diamond Valley residents (Eureka County) that the goldfish now common in Big Shipley Warm Springs on the western side of the valley, and those which existed—at least some years ago—at Bower's Mansion south of Reno, all came from an original population kept at the Eureka Consolidated Mine just outside the town of Eureka.

Mrs. James Sharp of Blue Eagle Ranch, east side of Railroad Valley, Nye County, has told me that the goldfish in their area were planted by a man named Johnson before 1900.

I also recall seeing some goldfish in a pond at Indian Springs, Clark County, many years ago (1940), and in a small pond at Warm Springs, Ely-Tonopah Highway, Nye County, in 1953.

Sumner (personal correspondence, 1958) reports goldfish in the Humboldt River (below Ryepatch Dam), Pershing County; Lakes Mead and Mohave, Clark County; and around Fallon (Churchill County).

Miller and Alcorn (1946: 181–182) list various other occurrences, such as McGill, the Campbell Ranch (18 miles north of Ely) and 1.3 miles south of Cherry Creek, White Pine County; and Steve Collins Ranch in Ash Meadows, Nye County.

Ted Frantz (personal correspondence, 1958) writes that he has seen goldfish in warm springs at the Goicoechea Ranch in Newark Valley and at the site of the old town of Schellbourne (White Pine County).

TAXONOMY

Etymology—"auratus," gilded, referring to the golden domesticated colors.

LIFE HISTORY

In the wild state quite similar to that of the Asiatic Carp (*Cyprinus carpio*); in fact, the goldfish may be considered a "little carp" in many respects. The goldfish reaches a "wild" length of about 18 inches and a weight of "several" pounds as maximal figures, is prolific in egg laying and omnivorous in appetite. In Nevada it has done best as an escapee in certain warm spring areas, such as those on the west side of Diamond Valley in Eureka County north of the town of Eureka.

BLACKFISH
Genus *Orthodon* Girard 1856
("Straight tooth")

THE SACRAMENTO BLACKFISH

Orthodon microlepidotus (Ayres)

(Blackfish Carp, Greaser Blackfish)

Gila microlepidota Ayres 1855C: 20–21
Orthodon microlepidota, Girard, 1856B: 182
Orthodon microlepidotus, Girard, 1859: 237–238
Orthodon microlepidotus, Cooper, in Cronise, 1868: 496
Orthodon microlepidotus, Günther, 1868: 275
Orthodon microlepidotus, Jordan and Copeland, 1878: 418
Orthodon microlepidotus, Jordan and Gilbert, 1882: 152 / 1894: 139
Orthodon microlepidotus, Jordan and Evermann, 1896B: 207
Orthodon microlepidotus, Rutter, 1908: 125
Orthodon microlepidotus, Pratt, 1923: 65
Orthodon microlepidotus, Jordan, Evermann and Clark, 1930: 112
Orthodon microlepidotus, Walford, 1931: 43
Orthodon microlepidotus, Schrenkeisen, 1938: 99
Orthodon microlepidotus, Murphy, 1941: 167 / 1949: 160 / 1950
Orthodon microlepidotus, Shapovalov and Dill, 1950: 386
Orthodon microlepidotus, La Rivers, 1952: 97–98
Orthodon microlepidotus, La Rivers and Trelease, 1952: 116
Orthodon microlepidotus, Kimsey, 1954: 395
Orthodon microlepidotus, Eddy, 1957: 85
Orthodon microlepidotus, Moore, 1957: 99
Orthodon microlepidotus, Shapovalov, Dill and Cordone, 1959: 171

ORIGINAL DESCRIPTION AND DIAGNOSIS

"*Gila microlepidota*.—Ayres. This species which is brought, not unfrequently, to our markets, appears seldom to exceed twelve inches in length. It is sold by many of the fishermen under the name of Fantail, from the peculiar form of the caudal fin; like the others of the *Cyprinidae*, it is not much esteemed.

"*Form* elongated, subcompressed, rather slender, tapering most posteriorly. Greatest depth contained about five and a half times in the total length; length of the head about four and a half times in the same; depth anterior to the caudal not quite one-third of the greatest depth. Head tapering regularly from the back, with a straight dorsal outline; nape not elevated.

"*Mouth* small, the tip of the maxillary by no means reaching the border of the orbit; lower jaw received beneath the upper.

"Border of the opercular apparatus forming a smooth and regular curve.

"*Lateral line* curving gently downward, passing nearer the ventrals than the dorsal fin, thence rising, and at length running straight to the caudal fin.

"*Scales* small, numbering about a hundred and ten along the lateral line, and twenty-four in an oblique line above it at the origin of the dorsal fin; they are strongly impressed with radiating striae.

"The *dorsal* fin arises a little nearer to the caudal rays than to the snout. Its length equals the distance from the snout to the border of the preoperculum, being contained seven times in the length to the tip of the central caudal rays; the height of the fifth ray, which is the longest, is greater than the length of the fin.

"The *anal* arises posterior to the termination of the dorsal; it resembles that fin in form, but is smaller, its length equalling only the distance from the snout to the middle of the eye.

"The origin of the *ventrals* is posterior to that of the dorsal, which fin they very nearly equal in height.

"The *pectorals* are rounded, and slightly exceed the ventrals in height.

"The *caudal* is large, deeply concave; the height of the external rays exceeding the greatest depth of the body; the height of the central rays half that of the external. The great number and prominence of the accessory rays causes the fin to spring out suddenly from the caudal portion of the body, thus giving occasion for the name by which the fish is designated, as already mentioned.

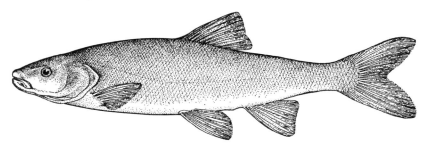

FIG. 207. Sacramento Blackfish, *Orthodon microlepidotus*. Drawn by Silvio Santina.

"C. 4–10; A. 3–8; V. 1–10; P. 17; C. 19, with twelve accessories.

"*Color* dark grayish brown above, lighter on the sides and beneath, a darker band passing from the base of one pectoral across the nape to the base of the other.

"*G. microlepidota* is taken in the lower waters of the Sacramento and San Joaquin, in company with the other species of this family described in the Proceedings of the Academy. The form of its head indicates the propriety of a different generic position, and it is also separated from *Gila* by the structure of its inferior pharyngeal bones. But from the same reason that has been given in previous instances (the absence of any means in California of comparison with established forms, and the lack even of works of reference containing the divisions of the Cyprinidae, as at present recognized), it has been deemed advisable not to propose at present a new generic name" (Ayres 1855C: 20–21).

TYPE LOCALITY

Sacramento and San Joaquin Rivers, California.

RANGE

Lower portions of the Sacramento and San Joaquin River systems, and Clear Lake, California. Introduced into the Truckee Meadows

(Reno) region of west-central Nevada on at least one known occasion (as noted in La Rivers and Trelease 1952: 116), where it produced hybrids with the Lahontan Tui Chub, *Siphateles bicolor obesus* in a small pond at the northwest edge of Reno.[1] All cyprinids in the pond winterkilled during 1948–1949, as did the Green Sunfish there (*Lepomis cyanellus*), only the Black Bullheads (*Ictalurus melas*) surviving. The pond was subsequently drained in subdivision developmental work, but there is a possibility that the species may exist in drain ditches east of Sparks. Robert R. Miller, of the University of Michigan, who identified the material, suggested that the Sacramento Blackfish may have been brought in with fish rescue material from the Sacramento Valley, and noted the *Orthodon-Siphateles* combination as a new hybridization record. See *Addendum* on page 592.

TAXONOMY

Etymology—"microlepidotus," small-scaled.

LIFE HISTORY

A comprehensive study of the Sacramento Blackfish has been published by Murphy (1950), based on the Clear Lake, California, populations. The following account is a resumé of Murphy's findings.

The Sacramento Blackfish is an animal of lowland, sluggish waters and is pronouncedly herbivorous in its feeding habits, a fact which would be suspected on anatomical evidence alone—108 inches of intestine were found in a 17-inch fish.

Although designed to subsist largely on vegetal material, blackfish have fine, numerous and close-set gillrakers of typical plankton feeders, and as such consume large quantities of both phyto-and zoö-plankton, obtaining concentrated protein foods from the latter.

A lake spawner, the species is prolific (Murphy estimated some 350,000 eggs from a 17-inch specimen), the female laying her adhesive eggs among vegetation in shallow water, where the male fertilizes them. Sexual color differences, black for males and olive for females, were noted. Spawning in Clear Lake extended over a two-and-a-half month period, from April to June; evidence indicates that most blackfish spawn at the end of their third year and die after their second spawning a year later. Maximum lengths achieved seem to be about 20 inches.

ECONOMICS

The Sacramento Blackfish appeared regularly in the San Francisco markets since pioneer days as a limited dietary staple until 1948 when it was placed on the protected list to rebuild its numbers. It appears to have other desirable features, and has been planted in California beyond the confines of its original area because of its forage food value and its adaptations to fluctuating reservoirs.

In the first matter, the species appears to constitute a good forage source for certain of the warm water fishes as blackbass which have been extensively and very successfully introduced to California since it converts vegetal material to proteins, does not compete with game

---

[1]This pasture pond is now the pond of the Lake Park heights subdivision.

fish for food and remains of "forage size" and habit (staying in shallow water) for about its first year.

Its favorableness for reservoirs lies largely in the fact that its early spawning habits makes it possible for the species to complete its life cycle before the annual drawdown of reservoirs generally takes place.

THE SPINEFINS

This small but very distinctive cyprinid group, formed of three unique genera, is restricted to portions of the southwestern United States where they may be only locally common. They were first discovered and described during the early government western surveys in the third quarter of the last century. Recently (1960), Miller and Hubbs have added more species and subspecies to the least differentiated of the three genera involved, *Lepidomeda*. The other two genera are monotypic—the Woundfin (*Plagopterus argentissimus* Cope 1874) and the Spinefin (*Meda fulgida* Girard 1856).

DIAGNOSIS

It is sufficient here to note the major ways in which this group (Tribe Plagopterini, Subfamily Leuciscinae) differs from the remainder of our native cyprinids: (1) the peculiar double anterior spine(s) of the dorsal fin; (2) the spinose pelvic fins; and (3) the adherence of the pelvic fins to the body along their inner edges. In addition two of the genera, *Meda* and *Plagopterus,* are scaleless, and all genera show marked and bright silvery coloration.

Cope, the describer of two of the three genera in the group observed that "the only other instance of this ossification of the ventral rays is to be seen in the extinct family of the *Saurodontidae* of the cretaceous period, the nearest approach among recent fishes being the internal spine in the ventral fin of *Amphacanthus*"[1] (1874A: 129).

KEY TO GENERA

1. Maxillary barbels (2) present (body scaleless)
    (*Plagopterus argentissimus*) .... WOUNDFIN
   .... Maxillary barbels absent, but some dermal papillation may
    be evident 2
2. Body scaled (*Lepidomeda*) .... SPINEDACE
   (..... Body scaleless (*Meda fulgida* Girard 1856) .... Spinefin)

SPINEDACE
Genus *Lepidomeda* Cope 1874
("Scale Meda," or a Meda-like fish with scales; "Meda" is a classical feminine name, and its application to fishes is obscure)

---

[1]Family Teuthidae, Suborder Amphacanthoidei, Order Perciformes.

KEY TO SPECIES[2]

1. Mouth more oblique (a line from uppermost tip of premaxillary to middle of caudal peduncle passes above middle of pupil). Snout sharper. Dorsal spines stronger. Dorsal fin higher (depressed length 1.8 to 2.1 in predorsal length), and more sharply pointed than in any other form. Head more compressed (width 1.9 to 2.0 in its length). Almost no pigment on shoulder girdle in advance of scapular bar. Additional characters—pigment on opercles about as in *L. mollispinis;* a band of coarse pigment crossing chin behind upper lip. Pelvic-fin length 1.4 to 1.5 in head. Upper-jaw length 1.15 to 1.3 in postorbital. (Formerly) cooled, swift outflow from Ash Spring and Upper Pahranagat Lake, Pahranagat Valley, Nevada (in course of Pluvial White River) (*Lepidomeda altivelis*)_____
_____ PAHRANAGAT SPINEDACE

   Mouth less oblique (a line from uppermost tip of premaxillary to middle of caudal peduncle passes below middle of pupil). Snout more rounded. Dorsal spines variably weaker. Dorsal fin low to moderate (depressed length 2.2 to 2.4 in predorsal length, except in *L. mollispinis pratensis*), and varyingly less pointed. Head broader (width 1.5 to 1.85 in its length). Pigment on shoulder girdle extending variably forward beyond scapular bar_____ 2

2 (1). Melanophores typically extending across opercle and subopercle, and to angle of preopercle; lower half of outer face of shoulder girdle in adults with considerable pigment in front of vertical from pectoral insertion. Size larger (commonly 80–90 mm., largest 103 mm. in standard length) and colors particularly bright. Dorsal spines stronger. Pharyngeal arch and teeth much more massive (in an average adult about 0.25 percent weight of fish); the whole arch thicker (less flattened dorsoventrally); anterior angle usually less conspicuous, more evenly rounded; edge of dorsal surface usually broadly rounded. Additional characters—pelvic-fin length 1.4 to 1.8 in head length. Upper-jaw length 1.1 to 1.5 in postorbital. Springs and spring-fed creeks in Recent White River Valley (upper headwaters of Pluvial White River system), Nevada (*Lepidomeda albivallis*)_____ ___
_____ WHITE RIVER SPINEDACE

   Melanophores confined to upper half of opercle and to upper part of upper limb of preopercle; lower half of outer face of shoulder girdle in adults lacking pigment in front of vertical from pectoral insertion. Size smaller (only rarely more than 80 mm. in standard length) and

---

[2]This is the just-published key of Miller and Hubbs (1960: 15–16), with slight rearrangement.

colors less bright. Dorsal spines weaker. Pharyngeal arch and teeth much smaller and more delicate (in an average adult about 0.10 percent weight of fish); the whole arch flatter (more compressed dorsoventrally); anterior angle usually sharp and conspicuous; edge of dorsal surface medially and on anterior limb usually rather sharply ridged (the pharyngeal arch and tooth distinctions are not sharp in the localized form *L. mollispinis pratensis*, of which no large adults are known). Virgin River system in Utah, Nevada, and Arizona, and (formerly) Big Spring in Meadow Valley, Nevada (*Lepidomeda mollispinis*)........
.... COLORADO RIVER SPINEDACE.... 3

3 (2). Dorsal fin lower and more rounded (its depressed length 1.45 to 1.65 in distance from dorsal origin to occiput); when the fin is erected at about 45° the outer edge in the adult usually slopes downward and backward. Pelvic fin shorter (length 1.5 to 1.85, usually 1.65 to 1.8, in head length). Mouth smaller (upper-jaw length 1.4 to 1.8, usually 1.45 to 1.6, in postorbital), and less oblique (a line from uppermost tip of premaxillary to middle of caudal peduncle passes below pupil). Virgin River system in Utah, Nevada, and Arizona (*L. m. mollispinis*)................
.... VIRGIN RIVER SPINEDACE

.... Dorsal fin higher and more pointed (its depressed length 1.2 to 1.45 in distance from dorsal origin to occiput); when the fin is erected at about 45° the outer edge in the adult is usually about vertical. Pelvic fin longer (length 1.35 to 1.6, usually about 1.5, in head length). Mouth larger (upper-jaw length 1.25 to 1.4 in postorbital), and more oblique (a line from uppermost tip of premaxillary to middle of caudal peduncle passes through lower part of pupil). (Formerly) outflow of Big Spring, in meadow adjacent to Meadow Valley Wash, on course of Pluvial Carpenter River, Nevada (*L. m. pratensis*)........................
.... BIG SPRING SPINEDACE

WHITE RIVER SPINEDACE
Lepidomeda albivallis Miller and Hubbs

Lepidomeda species, Sumner and Lanham, 1942: 319
Lepidomeda vittata, La Rivers, 1952: 97
Lepidomeda vittata, La Rivers and Trelease, 1952: 118
Lepidomeda species, Miller, 1952: 19, 35
Lepidomeda species, Moore, 1957: 138
Lepidomeda albivallis Miller and Hubbs 1960: 15–16, 24–25

ORIGINAL DESCRIPTION AND DIAGNOSIS

"DIAGNOSIS.—A species of *Lepidomeda* distinguished from others in having 5-4 teeth in the main row, lateral-line scales typically fewer than 90, mouth moderately oblique, dorsal fin of moderate height, melanophores extending well below level of lateral line, and in other details of pigmentation (see key . . . .).

"CHARACTERS.—. . . Principal dorsal rays 7 (4), 8 (97); anal rays 8 (24), 9 (75), 10 (2); pelvic rays 7–7 (82), 7–6 (1), 6–7 (5), 7–4 (1). The dental formula is typically 2, 5–4, 2, varying as follows: 2, 5–4, 2 (20), 2, 5–5, 2 (1). Lateral-line scales number about 79 to 92: 79 (1), 80 (1), 82 (5), 83 (3), 84 (2), 85 (5), 86 (3), 87 (2), 88 (1), 89 (2), 90 (3), 92 (2). The vertebrae range from 42 to 44 . . .

"This appears to be the most brightly colored of the four species of *Lepidomeda*. Life colors of postnuptial males were noted at the time of collection as follows: The body is bright brassy green to olive above, brassy over bright silvery on sides, and silvery white below, splashed with sooty on the sides. Dorsal and caudal fins pale olive-brown to

FIG. 208. Holotype of *Lepidomeda albivallis*, UMMZ 173781, 69.5 mm. S. L. Photograph by William L. Brudon for Miller and Hubbs 1960. Courtesy University of Michigan Press and Dr. Robert R. Miller.

pinkish brown, with the rays often deep-olive and with the rather clear interradial membranes faintly flushed with rosy color; pectorals yellowish with orange-red axils; anal and pelvic fins bright orange-red, in the young only toward the base anteriorly, but in adults over most of these fins, which otherwise are whitish. Lower edge of caudal peduncle with a speckled diffusion of orange-red in adults. Some coppery-red to red on side of face, at upper end of gill opening, on preorbital just behind mouth, and along upper arm of opercle. Cheeks and opercles with rather strong gilt reflections; the gular membranes watery yellow. Lateral line more strongly gilt than adjacent parts of body. In females the coloration is similar but less intense.

"Remnants of tubercles on postnuptial males collected on August 27 indicate that this pattern in the species is similar to that of *L. m. mollispinis*.

"This is the largest species of *Lepidomeda*, commonly attaining a total length over 4 inches; the largest specimen (UMMZ 124980) is nearly 5 inches long (103 mm. S.L.)" (Miller and Hubbs 1960: 24–25).

TYPE LOCALITY AND RANGE

"TYPES.—The holotype, UMMZ 173781, an adult 69.6 mm. long, was collected by C. L. Hubbs and family from White River, just below the mouth of Ellison Creek, about 5 miles NW of Preston, along the Tonopah-Ely highway (T. 13, N, R. 61 E), White Pine County,

Nevada, on September 10, 1934. The 428 paratopotypes, UMMZ 132180 (16–98 mm.) were taken with the holotype. An additional 579 specimens were secured as follows: White Pine Co.: UMMZ 124980 (61, 37–103 mm.), from Preston Big Spring, 2 mi. NW of Preston, Aug. 26, 1938; UMMZ 124984 (17, 17–88 mm.), from Lund Spring at Lund, Sept. 15, 1938; UMMZ 124973 (14, 39–90 mm.), from Nicholas Spring in Preston, Aug. 26, 1938; UMMZ 124977 (371, 24–83 mm.), outflow of Preston Big Spring and Nicholas Spring, Preston, Aug. 26, 1938; UMMZ 138331 (5, 73–93 mm.), from spring near Preston, Sept. 28, 1941 (U. N. Lanham, coll.). Nye Co.: UMMZ 124990 (111, 12–96 mm.), from springs at Hendrix Ranch (southern ranch of the Sunnyside group), Aug. 27, 1938" (Miller and Hubbs 1960: 24).

Specimens from White River below the outlet of the Adams-McGill (Sunnyside) Reservoir, northeastern Nye County, were transplanted by Ted Frantz to Potash Well No. 1 and the government artesian well just to the west of No. 1, Railroad Valley, in the same county, in August, 1957.

TAXONOMY

Lepidomeda albivallis has gone by the name of *L. vittata* since the first collections were made in eastern Nevada.

Etymology—"The name *albivallis* is from the Latin *albus*, white, and *vallis*, valley, in reference to the restriction of the new species to the White River Valley, in White Pine and Nye counties, Nevada" (Miller and Hubbs 1960: 25).

LIFE HISTORY

"HABITAT.—This species occurs in cool springs (65–71° F.), their outflows, and in White River, in the upper part of the ancient White River system of eastern Nevada. In all the spring habitats the water was clear in the source pools, which varied from 15 by 25 feet to 60 by 80 feet in major dimensions. The bottom was mostly gravel and sand, with some mud; the fish were generally captured in water not over 2 feet deep (up to 5 feet in Lund Spring); watercress, a fine-leafed *Potamogeton,* and rushes were the common aquatic plants, often dense. The current in the spring-fed ditches and in White River was swift to moderate.

"ASSOCIATES.—The following species were associated with *Lepidomeda albivallis: Pantosteus intermedius* (Tanner), *Rhinichthys osculus* (Girard), and *Crenichthys baileyi* (Gilbert). *Salmo gairdneri* (Richardson), *S. trutta* Linnaeus, *Salvelinus fontinalis* (Mitchill), and *Archoplites interruptus* (Girard) have been planted in the same waters, and centrarchids, likely bluegill, *Lepomis macrochirus* Rafinesque, and largemouth bass, *Micropterus salmoides,* have been stocked in a reservoir in White River Valley at Sunnyside, according to information received from Ted Frantz" (Miller and Hubbs 1960: 25).

ECONOMICS

In addition to the usual food chain importance, there has been some use of the species for bait in the Colorado River fishery. Miller (1952) has given a resumé of the traffic in spinedace, which has not been great. Of *L. albivallis* (then undescribed) he wrote:

"On February 3, 1951, Richard Beland obtained four adults from Murphy's Windmill Camp, California. This constitutes the only record to date (May, 1951) of the use of this species as a bait fish along the Colorado River. The specimens must have come from the upper White River.

"There is little likelihood that the White River spine-dace will become established in the Colorado River. However, because of its very restricted range, and hence its interest to science, the use of this species for bait should be discouraged" (Miller 1952: 35).

COLORADO RIVER SPINEDACE
Lepidomeda mollispinis Miller and Hubbs

"DIAGNOSIS.—A species of *Lepidomeda* distinguished by having 5-4 teeth in the main row, a relatively weak and soft-tipped (second) dorsal spine, 9 anal rays, typically fewer than 90 lateral-line scales, the depressed length of the dorsal fin less than the head length, the sides of the body mostly silvery, and with melanophores confined to upper half of opercle and to upper part of ascending limb of preopercle" (Miller and Hubbs 1960: 18).

VIRGIN RIVER SPINEDACE
Lepidomeda mollispinis mollispinis Miller and Hubbs

Lepidomeda vittata, Tanner, 1932: 135 / 1936: 171
Lepidomeda vittata, Wallis, 1951: 87
Lepidomeda vittata, La Rivers, 1952: 97
Lepidomeda vittata, La Rivers and Trelease, 1952: 118
Lepidomeda species, Miller, 1952: 18, 35
Lepidomeda vittata, Eddy, 1957: 104
Lepidomeda vittata, Moore, 1957: 138
Lepidomeda mollispinis mollispinis Miller and Hubbs 1960: 18-21

ORIGINAL DESCRIPTION AND DIAGNOSIS

"DIAGNOSIS.—This subspecies is most closely related to *L. m. pratensis*, from which it differs in having the dorsal fin less elevated and more rounded, the pelvic fin shorter, and the mouth smaller and less oblique.

"CHARACTERS.—. . . Principal dorsal rays 8 (in 46 specimens); anal rays, 8 (4), 9 (41), 10 (1); pelvics 7-7 (36), 6-7 (1). The dental formula is typically 2, 5-4, 2, varying as follows: 2, 5-4, 2 (19), 2, 4-5, 2 (1), 2, 4-4, 2 (3), and 3, 5-4, 2 (1). Lateral-line scales about 77 to 91, varying as follows: 77 (1), 81 (1), 82 (2), 83 (4), 84 (1), 85 (4), 86 (2), 87 (2), 88 (3), 91 (1). Vertebrae 42 to 44 . . .

"Life colors were noted in the field as follows: body silvery, with a more or less brassy sheen and with sooty specklings on sides; axils of paired fins and basal band on anal fin orange-red to translucent orange-pink; a little spot of golden red at upper end of gill-slit. Younger fish look whitish in the water and some adults have the sides blackish, especially conspicuous in the water.

"The nuptial tubercles, best developed on males in a collection of June 17, but evident also in July specimens, are distinctive. Those on the head are almost wholly confined to the dorsal surface, extending onto the sides only in a definite patch across the upper part of the opercle. They are irregularly scattered over the dorsal surface, from near the occipital edge forward to the upper part of the snout and outward to the orbital margins. They are of moderate size, and their spiny tips are weakly curved forward. Tubercles occur on the scales over the entire body, but become obsolescent on the midsides and belly. Those near the margin of the head, between the lateral lines, are considerably strengthened. Here the partly fused tubercles form a single straightish [sic] transverse (or vertical) series, with the points essentially erect. On the caudal peduncle the points are smaller and form a more curved series on each scale. On the breast the tubercles are somewhat strengthened, and usually single on each scale. The scales in a

FIG. 209. Holotype of *Lepidomeda mollispinis mollispinis*, UMMZ 141673, 88 mm. S. L. Photograph by William L. Brudon for Miller and Hubbs 1960. Courtesy University Michigan Press and Dr. Robert R. Miller.

band just behind the shoulder girdle, above the pectoral fin, have the soft tissue considerably swollen, and are weakly tuberculate. On the first pectoral ray there is a single file of tubercles, and on the outer part of several succeeding rays a file branched once near the base. Each tubercle has one to several weakly antrorse spiny points. Weak tubercles line pelvic and anal rays. Despite our large collections none of the other species seems to be represented by nuptial males.

"This subspecies often attains a total length of nearly four inches. The largest specimen we have seen is the holotype, about 4.25 inches long (88 mm. in standard length). It is the only specimen among 718 that is longer than 88 mm." (Miller and Hubbs 1960: 19–20).

TYPE LOCALITY AND RANGE

"TYPES.—The holotype, UMMZ 141673, an adult 88 mm. long, was seined by C. L., L. C., and E. L. Hubbs from Santa Clara River, 3 mi. SE of Shivwitz and 4.5 mi. NW of Santa Clara, Washington Co., Utah, on July 29, 1942. Secured with the holotype were 103 paratopotypes, 23–80 mm. long, UMMZ 141674. An additional 604 specimens were examined from localities in Arizona, Nevada and Utah as follows: Arizona: UMMZ 141662 (381, 22–77 mm.), Beaver Dam Cr at U. S. 91 crossing near Littlefield, Coconino Co., July 28, 1942. Nevada: UMMZ 105496

(68, 36–78 mm.), Virgin R., Clark Co. (near Utah border), Sept. 1938; UMMZ 125013 (12, 39–65 mm.), Virgin R. W of Bunkerville, Clark Co., Aug. 31, 1938. Utah (Washington Co.): UMMZ 85955 (2, 48–54 mm.), Santa Clara R., June, 1928; UMMZ 124764 (17, 20–62 mm.), trib. Virgin R. near La Verkin, July 3, 1938; UMMZ 124772 (38, 22–71 mm.), Virgin R. at mouth of La Verkin and Ash creeks, July 3, 1938; UMMZ 162849 (86, 22–74 mm.), Santa Clara R., 2.5 mi. below Gunlock, June 17, 1950 . . ." (Miller and Hubbs 1960: 19).

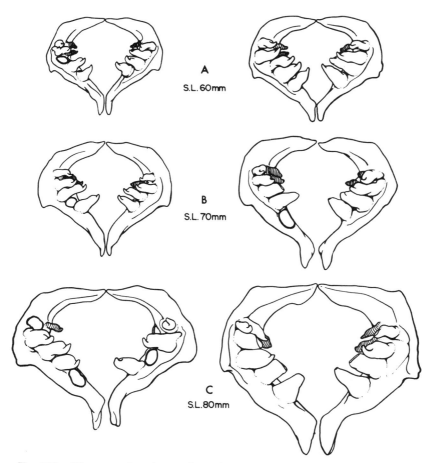

FIG. 210. Pharyngeal arches and teeth of *Lepidomeda m. mollispinis* (left column) and *Lepidomeda albivallis* (right column) at comparable sizes drawn by William L. Brudon for Miller and Hubbs 1960. Courtesy Dr. Robert R. Miller.

TAXONOMY

Like the other new species and subspecies described by Miller and Hubbs (1960), *Lepidomeda mollispinis mollispinis* has previously gone under the name *L. vittata*.

Etymology—"The name *mollispinis* is derived from the Latin *mollis,* soft, and *spina,* spine, in reference to the relatively weak and soft-tipped main (second) dorsal spine" (Miller and Hubbs 1960: 21).

LIFE HISTORY

"HABITAT.—This subspecies is common in the Virgin River and its tributaries in Utah, Arizona and Nevada, in moderate to swift current, chiefly in pools. Where the collections were made the bottom was usually sand and gravel, often with stones and occasionally with boulders and some mud. Green algae and sparse pondweed were often associated, and depth of capture varied from 1 to 3 feet. The water was either clear or, as in the Virgin River, very milky; with bottom visibility from about 3 inches to 3 feet.

"ASSOCIATES.—Fish species taken with *L. m. mollispinis* include *Pantosteus delphinus utahensis* (Tanner), *Catostomus latipinnis* Baird and Girard, *Rhinichthys osculus* (Girard), and *Plagopterus argentissimus* Cope. *Micropterus salmoides* (Lacepede) was caught at the type locality" (Miller and Hubbs 1960: 20).

ECONOMICS

Concerning its limited use as a Colorado River bait fish:

"The Virgin River spine-dace was noted as a bait fish by Carl L. Hubbs when he interviewed Clarence Alexander, bait dealer in Las Vegas, on August 31, 1938. Two specimens were secured by O. L. Wallis from the Lake Mead Boat Dock on December 31, 1948, and three more were picked up from a bait box on Lake Mead by Al Jonez, of the Nevada Fish and Game Commission, in February, 1951.

"The specimens for sale by Mr. Alexander, for use on Lake Mead, were seined by him in the Virgin River west of Bunkerville, Nevada. Those picked up by Mr. Jonez were reported to have come from the St. George, Utah, area, likely from Santa Clara River where this species abounds and from which Milt Holt seines his bait and delivers it to the Lake Mead area" (Miller 1952: 35).

BIG SPRING SPINEDACE

Lepidomeda mollispinis pratensis Miller and Hubbs

Lepidomeda mollispinis pratensis Miller and Hubbs 1960: 21–24
Lepidomeda mollispinis pratensis, Miller, 1961: 380–381

ORIGINAL DESCRIPTION AND DIAGNOSIS

"DIAGNOSIS.—Like *L. m. mollispinis,* but differing in the higher and more pointed dorsal fin, the longer pelvic fins, and the smaller and more oblique mouth" (Miller and Hubbs 1960: 21) (See key for more specific details).

"CHARACTERS.—. . . Principal dorsal rays 8, anal rays 9, pelvics 7–7 (6), 6–7 (1). The teeth number 2, 5–4, 2 in 3, and the lateral-line scales vary from 82 to 90 (about 89 in the holotype). Vertebrae number 42 or 43 (only 4 specimens countable).

"In life *L. m. pratensis* was bright silvery. Some specimens were lemon to orange in the axils of the paired fins, on the basal part of the anal fin, near the upper edge of the shoulder girdle, on the vertical arm of the preopercle, and above the mouth" (Miller and Hubbs 1960: 21).

TYPE LOCALITY AND RANGE

"TYPES.—The holotype, UMMZ 124799, an adult 55 mm. long, was seined by C. L. Hubbs and family, R. R. Miller and A. J. Calhoun from Big Spring, about 1 mile NE of Panaca, Lincoln County, Nevada, on July 10, 1938. Taken with this specimen were six paratopotypes, 48 to 56 mm. long, UMMZ 136097" (Miller and Hubbs 1960: 21).

"DISTRIBUTION, LOCALIZATION, AND EXTINCTION.—Our field explorations have indicated that this fish has very recently become extinct in the one spring-fed marsh in which the last few individuals persisted until 1938, or a few years later. This marsh is fed chiefly from a single source, Big Spring, which issues from the base of low hills about 1 mile northeast of Panaca, Lincoln County, Nevada. The

FIG. 211. Holotype of *Lepidomeda mollispinis pratensis*, UMMZ 124799, 55 mm. S. L. Photograph by William L. Brudon for Miller and Hubbs 1960. Courtesy University Michigan Press and Dr. Robert R. Miller.

meadow occupies a basin off the east side of Meadow Valley Wash, which here dissects the remnants of a lacustrine fill. Meadow Valley Wash follows the ancient course of Pluvial Carpenter River, which was the main, eastern affluent of Pluvial White River (Hubbs and Miller, 1948: 96–100).

"Prior to recent agricultural modifications Big Spring discharged onto and spread over the large wet meadow, and doubtless provided a more favorable fish habitat than presently exists. The natural outflow course can still be followed for about a mile across the meadow. Between 1938, when the seven types of *Lepidomeda mollispinis pratensis* were collected, and 1959, when the subspecies was found to have become extinct, the source was dammed and a contoured ditch was constructed to divert the spring effluent to the upper part of the meadow. In 1938 the outflow followed the natural channel until it spread out. The channel was generally 1 to 3 feet wide and as deep as 2 feet. It contained watercress (*Nasturtium*) above, pondweeds of the *Potamogeton pectinatus* and floating types, and rushes, below. The bottom was of firm to soft clay, with some gravel. The current was slight over most of the course, but occasionally swift. The water was clear, but easily roiled. On July 10, with the air at 94° F., the water temperature was 84°. Here, in 1938, in the stream within the meadow area, we seined,

Fig. 212. Radiographs of the seven forms of the Plagopterini:
A. *Lepidomeda vittata*, UMMZ 137082, 55 mm. S. L.
B. *Lepidomeda m. mollispinis*, UMMZ 162849, 56 mm. S. L.
C. *Lepidomeda m. pratensis*, UMMZ 136097, 56 mm. S. L.
D. *Lepidomeda altivelis*, UMMZ 125004 (holotype), 56 mm. S. L.
E. *Lepidomeda albivallis*, UMMZ 132180, 54 mm. S. L.
F. *Meda fulgida*, UMMZ 162738, 49 mm. S. L.
G. *Plagopterus argentissimus*, UMMZ 141669, 53 mm. S. L.
Radiographs by William L. Brudon for Miller and Hubbs 1960. Courtesy Dr. Robert R. Miller.

by prolonged effort, along with the seven types of *Lepidomeda mollispinis pratensis*, 31 half-grown mountain suckers, *Pantosteus* sp., and 312 young to adult speckled dace, *Rhinichthys osculus* (Girard), both representing local forms characteristic of the remnants of Pluvial Carpenter River.

"By 1959 the generally abandoned natural channel had become very largely clogged with silt and a variety of submerged and emergent plants, chiefly chara, *Nasturtium, Decodon,* a broad-leafed *Potamogeton*, green and blue-green algae, and *Scirpus*. Except for about a half-dozen stretches of 10 to 20 feet the stream was almost completely choked and partially dammed, with silt and weeds filling the stream bed from the original more or less gravelly bottom to the surface. The water was rather clear but very readily roiled. The flow, of seepage origin, was slight; in most places, barely perceptible. On July 4, 1959, when the air was 84.5° F., the water temperature varied from 75° to 81°. In the open-water stretches, where the width averaged 3 to 4 feet, intensive and thorough seining, following removal by hand of the excess of water weeds and muck, yielded only one of the three native species that held out here 21 years previously. This was the *Rhinichthys,* which had not only survived but had even become concentrated—to confirm the ubiquity and adaptability of this dace. In addition we found, in considerable density, a large population of the western mosquitofish, *Gambusia affinis affinis* (Baird and Girard), a scourge of native fishes. Another introduction, the bullfrog, *Rana catesbeiana* Shaw, now also abounded, greatly outnumbering the native *Rana pipiens* Schreber.

"The diversion of water, with, no doubt, the occasional stoppage of flow in the ditch (which, furthermore, originates above the former habitat of the native fishes), presumably led to the rapid and catostrophic deterioration of the original habitat and to the decline in the population of the two more susceptible native fishes, the *Pantosteus* and the *Lepidomeda*. The introduced species almost certainly contributed to the extirpation of the *Pantosteus* and to the extermination of the *Lepidomeda*.

"The flow of the Spring is almost surely less than it was formerly. According to Angel (1881: 491), writing of Panaca: 'The water supply is abundant, being taken from Warm Spring, which is about one and one-half miles east of the town. A large stream of water, about three feet deep and six feet wide, is thrown out from the spring, and the quantity is not affected by the seasons. This is the principal source of water supply for the whole valley.'

"Our field studies in 1938 and 1959 indicated, with an approach to certainty, that no populations of *Lepidomeda* have persisted in any other remnants of Pluvial Carpenter River. In July, 1938, about two miles below Panaca, the flow of Meadow Valley Wash was found to contain only a fishless trickle of slightly alkaline seepage from the irrigated fields above. Just above the box canyon, above the section of Meadow Valley in which Caliente lies (about 14 miles southwest of Panaca), enough water issued from springs to restore a stream that averaged eight feet wide, continuing as a surface flow through Rainbow Canyon below Caliente. We fished in this canyon on the same day as at Big Spring, but caught no specimens of *Lepidomeda;* in the lowermost, permanent section of Meadow Valley Wash (13 miles by road above Moapa) we took only *Pantosteus* and *Rhinichthys* in a thorough collection on July 13, 1938. On March 1, 1938, the wash had been subjected to a very severe flood, which may have had a marked effect on the fish fauna.

"Absence of *Lepidomeda* was also indicated by our 1938 collections along the former upper course of Pluvial Carpenter River and in Lake or Duck Valley, which may have discharged into this Pluvial stream. Collections from the old stream course were made (1) between Camp Valley and Eagle Valley in the headwaters of the creek of Camp Valley Wash, about 25 miles due northeast of Pioche, (2) in adjacent springs, and (3) in the canyon between Spring Valley and Eagle Valley, about 21 miles northeast of Pioche. Furthermore, thorough collections made in 1950 and 1959 in Clover Creek, the main eastern tributary of Meadow Valley Wash, at two points in Nevada, contained no *Lepidomeda*, and no fish of any kind were visible in the deeply entrenched stream in the northern outskirts of Caliente" (Miller and Hubbs 1960: 21–23).

TAXONOMY

Etymology—"The name *pratensis* is derived from the Latin, meaning pertaining to or growing in a meadow, in reference to the extensive meadowland about Big Spring" (Miller and Hubbs 1960: 23).

PAHRANAGAT SPINEDACE

Lepidomeda altivelis Miller and Hubbs

Lepidomeda vittata, Gilbert, 1893: 231
Lepidomeda vittata, Jordan and Evermann, 1896B: 328
Lepidomeda jarrovii, ibid
Lepidomeda vittata, Jordan, Evermann and Clark, 1930: 147
Lepidomeda jarrovii, ibid
Lepidomeda vittata, Tanner, 1936: 171
Lepidomeda vittata, La Rivers, 1952: 97
Lepidomeda vittata, La Rivers and Trelease, 1952: 118
Lepidomeda jarrovi, Eddy, 1957: 104
Lepidomeda species, Moore, 1957: 138
Lepidomeda altivelis Miller and Hubbs 1960: 26–28
Lepidomeda altivelis, Miller, 1961: 380

ORIGINAL DESCRIPTION AND DIAGNOSIS

"DIAGNOSIS.—A species of *Lepidomeda* distinguished by the combination of 5–4 teeth in the main row, 9 anal rays, an oblique mouth, high and sharp dorsal fin, and a compressed head . . . *L. altivelis* has the longer mandible of *L. albivallis*, the silvery coloration of *L. mollispinis*, and is distinctive in its very oblique mouth, high and expansive dorsal fin, more compressed head, and generally finer scales.

"CHARACTERS.—. . . Principal dorsal rays 8 (24); anal rays 8 (1), 9 (21), 10 (2); pelvics 7–7 (21), 5–7 (1), 8–7 (1). The pharyngeal teeth number 2, 5–4, 2 in ten specimens, and the lateral-line scales are 84 to 95 in ten (90 or more in six), approximately 92 in the holotype. Vertebrae 42 to 44. This species attains a maximum length of about 3.75 inches (66 mm. S.L.)" (Miller and Hubbs 1960: 26). See key for coloration characteristics.

TYPE LOCALITY AND RANGE

"TYPES.—The holotype, UMMZ 125004, an adult 56 mm. in standard length, was collected by Carl L. Hubbs and family from the outflow of Ash Spring, from .25 to 2.5 miles (in straight line) below the spring source, in Pahranagat Valley, Lincoln Co., Nevada, on August 29, 1938. Sixteen paratopotypes, UMMZ 125005 (33–66 mm.), were taken with the holotype. Seven other specimens, UMMZ 124813 (33–41 mm.), were seined in Upper Pahranagat Lake, Lincoln Co., Nevada, by Hubbs and party on July 12, 1938.

"EXTERMINATION OF THE SPECIES.—Field study on July 2–3, 1959, undertaken to test the point, indicated that this highly

FIG. 213. Holotype of *Lepidomeda altivelis*, UMMZ 125004, 56 mm. S. L. Photograph by William L. Brudon for Miller and Hubbs 1960. Courtesy University Michigan Press and Dr. Robert R. Miller.

localized species, like *Lepidomeda mollispinis pratensis* of Big Spring, near Panaca, had become exterminated during the previous 21 years. The thorough collecting in 1959 failed to reveal a single specimen of *Lepidomeda* in the outlet of Ash Spring or in or about the Pahranagat Lakes, where the species was collected in 1891 and in 1938. A spring-fed ditch in the bed of Maynard Lake, southernmost of the Pahranagat chain, yielded on July 2, 1959, only the local form of *Rhinichthys osculus*. Upper Pahranagat Lake and its outlet on the same day yielded only *Cyprinus carpio*. In the pools of Ash Spring proper, where the water temperature is 89° F., the local (typical) form of *Crenichthys baileyi* still occurred commonly, though for some reason in less abundance than in 1938. In an examination of the outlet stream this cyprinodont occurred down to and into the diversion ditches 5 to 7 miles north of Alamo. Seining at several points in the outlet, supplemented by a visual examination along nearly the whole length of the natural flow and by a thorough application of rotenone in the lower half-mile of the natural creek and the upper mile of the diversion ditch then in operation, on July 3, 1959, indicated the fish population of the outlet, at water temperatures of 86 to 88° F. (air temperature 88.5° F.), to consist of moderate numbers of the *Crenichthys* and of the *Gila*, hordes of carp, and a considerable number of another introduced pest,

Gambusia affinis affinis. Not a single sucker (*Pantosteus intermedius*) or a single *Lepidomeda* was secured. Since for about 2.5 miles the stream has retained essentially its original bed and condition, with only limited diversions, and since the water source has remained uniform, the probable local extirpation of the *Pantosteus* and the almost certain complete extermination of the *Lepidomeda* is most plausibly attributed largely to the increased abundance of carp and to the establishment of the mosquitofish (*Gambusia*). Both of these introduced species have had a deleterious effect on native fish life in many western waters

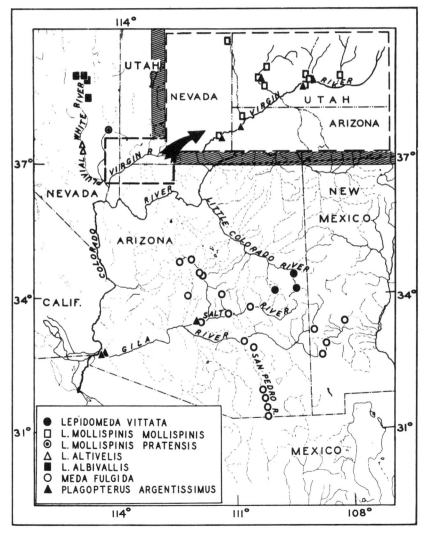

FIG. 214. Distribution of the plagopterine species. From Miller and Hubbs 1960. Courtesy University of Michigan Press and Dr. Robert R. Miller.

(Sigler, 1958; Miller, in press). The establishment of bullfrogs, *Rana catesbeiana*, may also have played a role in the modification of the habitat" (Miller and Hubbs 1960: 27–28).

TAXONOMY

Etymology—"The name *altivelis* is from the Latin *altus*, meaning high, and *velum*, a sail, in reference to the high, expanded dorsal fin" (Miller and Hubbs 1960: 28).

LIFE HISTORY

"In the outflow of Ash Spring, from .25 to 2.5 miles in a straight line below its head, the water was very clear (bottom clearly visible in six feet), the greatly meandering channel 5 to 20 feet wide, and the current generally swift (rarely quiet; in places very swift). The bottom comprised sand, gravel, mud and boulders; the depth of water was up to 6 feet. Vegetation consisted in general of abundant bulrushes and submerged weeds, but the stream was partly clear of weeds; the banks (about 5 feet high) were wooded, and beyond were grass and farmland. The collection was made between noon and evening on August 29, 1938, using a seine in the upper part, and derris root below (where *Lepidomeda* was taken). In the upper part the water was 88° F., but below was much cooler.

"Species taken with *Lepidomeda altivelis* in the outflow from Ash Spring were: *Cyprinus carpio* Linnaeus, *Gila robusta jordani* Tanner, *Rhinichthys osculus velifer* Gilbert, *Pantosteus intermedius* (Tanner), and *Crenichthys baileyi* Gilbert. The native fishes are all endemic as species or subspecies to the immediate region or to the whole Pluvial White River system.

"At Upper Pahranagat Lake on July 12, 1938, the water was dirty olive-green and was alkaline to the taste, with bottom visibility about 2 feet. Vegetation comprised a heavy growth of flooded cockleburs. The shore was an alkali flat. The bottom was firm to soft clay soil and we seined to 100 feet from shore, where the water depth did not exceed 3 feet. Associated with *L. altivelis* here were *Pantosteus intermedius* and many carp; the lake was worked between 8:30 and 9:00 A.M., when the water was 74° F., the air 89° F." (Miller and Hubbs 1960: 27).

WOUNDFINS
Genus *Plagopterus* Cope 1874
("Wound fin")

THE WOUNDFIN
Plagopterus argentissimus Cope
(Barbelled Desert Minnow)

Plagopterus argentissimus Cope 1874A: 130
Plagopterus argentissimus, Cope and Yarrow, 1875: 640–641
Plagopterus argentissimus, Jordan and Copeland, 1877A: 155
Plagopterus argentissimus, Jordan and Gilbert, 1882: 253

Meda argentissima, Jordan, 1885: 122 / 1889: 821
Plagopterus argentissimus, Evermann and Rutter, 1895
Plagopterus argentissimus, Jordan and Evermann, 1896B: 329
Plagopterus argentissimus, Gilbert and Schofield, 1898: 496–497
Plagopterus argentissimus, Snyder, 1915: 584
Plagopterus argentissimus, Jordan, Evermann and Clark, 1930: 147
Plagopterus argentissimus, Schrenkeisen, 1938: 157
Plagopterus argentissimus, Shapovalov and Dill, 1950: 386
Plagopterus argentissimus, La Rivers, 1952: 97
Plagopterus argentissimus, La Rivers and Trelease, 1952: 118
Plagopterus argentissimus, Miller, 1952: 18, 36
Plagopterus argentissimus, Böhlke, 1953: 36
Plagopterus argentissimus, Illick, 1956: 215–218
Plagopterus argentissimus, Eddy, 1957: 105
Plagopterus argentissimus, Moore, 1957: 138
Plagopterus argentissimus, Shapovalov, Dill and Cordone, 1959: 172
Plagopterus argentissimus, Miller and Hubbs, 1960: 17, 33–35

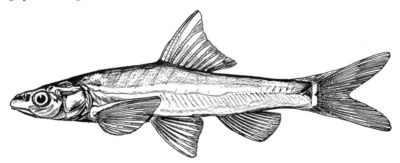

FIG. 215. Woundfin, *Plagopterus argentissimus*. Drawn by William L. Brudon for Miller 1952. Courtesy California Department of Fish and Game and Leo Shapovalov.

ORIGINAL DESCRIPTION

"This is a small fish of slender proportions, with a rather broad head, with slightly depressed muzzle overhanging by a little a horizontal mouth of moderate size. The caudal peduncle is of medium depth, and the caudal fin is deeply forked. The eye is somewhat oval, and enters the length of the side of the head 4.2 times, and the interorbital width 1.5 times. The greatest depth (near the ventral fin) enters the total length nearly six times, or five and three quarters, exclusive of the caudal fin. The latter measurement is four times the length of the head. The origin of the dorsal is entirely behind the proper basis of the ventral; its first spine is curved and longer than the second, and its basis is intermediate between the base of the caudal and the end of the muzzle. The dorsal rays behind the spine have the basal two-thirds to one-half thickened and completely ossified, the articulated portions issuing from the apices of the spines. Radial formula, D. II. 7; C. 19; A. I. 10–9; V. 2. V; P. 16. The first or osseous ray of the anal is rudimental; the fifth spinous ray of the ventral is bound by nearly its entire length to the abdomen by a membrane. The pectoral rays from the second to the sixth exhibit a basal osseous spinous portion, which is not nearly so marked as in the ventrals. The pectorals reach the basis of the latter.

"The lateral line is complete and is slightly deflexed opposite the dorsal fin. The lips are thin, and the end of the maxillary bone extends to the line of the front of the orbit. Total length M. 0.071; ditto to middle of basis of caudal fin .0565; ditto to anterior basis of anal fin .040; ditto to basis ventral .021; ditto of head .0145; of muzzle .004; width at posterior nares .006; at middle of pterotic .0078. Color, pure silver for a considerable width above the lateral line. Dorsal region somewhat dusky from minute chromatophorae" (Cope 1874A: 130).

FIG. 216. Three species of plagopterine fishes:
  A. *Lepidomeda vittata*, UMMZ 124754, 62 mm. S. L.
  B. *Meda fulgida*, UMMZ 125030, 51 mm. S. L.
  C. *Plagopterus argentissimus*, UMMZ 141669, 58 mm. S. L.
  Photographs by William L. Brudon for Miller and Hubbs 1960. Courtesy of Dr. Robert R. Miller.

DIAGNOSIS

Head about 4 times, body depth about 6 times, into body length; head somewhat broad, snout overhanging the small, horizontal mouth; maxillary attaining front edge of eye; lips thin, barbels present at end of maxillary; pharyngeal teeth in two rows, 2/5–4/2, hooked and lacking crushing surfaces; lateral line complete, slightly decurved; *scales entirely lacking;* fins—dorsal with two anterior spines, one grooved to receive the other, followed by 7 soft rays—anal 10-rayed—pelvics joined to abdomen along inner edges—caudal large, deeply forked; caudal peduncle rather stout, least depth about 10 times into body length; intestine moderate in length, slightly less than body length, with two main loops; peritoneum silvery; color silvery all over, darkening along back; small fishes, about 2½ inches in maximum length.

TYPE LOCALITY

"Numerous specimens from the San Luis Valley, Western Colorado" (Cope 1874A: 130). There is some reason now to believe this is an error (Miller 1958).

RANGE

"This streamlined minnow, which shines like burnished silver when first taken from the water, is now known to inhabit only the Virgin River and its tributaries in Arizona, Nevada, and Utah. It formerly was found in the Gila River basin, from which it has been recorded only three times, the last in 1894" (Miller 1952: 36).

TAXONOMY

Plagopterus argentissimus is the only species in its genus and is without synonyms. The peculiarities of the closely-knit group to which it belongs have been discussed above.

Etymology—"argentissimus," most silvery.

LIFE HISTORY

Essentially unknown.

ECONOMICS

Of little importance other than the minor part it undoubtedly plays in the food chain. Only one instance has been recorded of its use as a bait fish in the Colorado River fishery:

"On June 16, Miller and Winn obtained one specimen from the boat dock on Lake Mead. Otherwise it has not turned up in the bait tanks along the river.

"This is an interesting and gradually vanishing species which should be protected from further reduction" (Miller 1952: 36).

Division SILURI
Suborder SILUROIDEI
Superfamily SILUROIDEA

NORTH AMERICAN CATFISHES
Family *ICTALURIDAE*[1]

This family of strictly freshwater catfishes is widespread in the New World from Central America into Canada. It is a small group, not many more than 40 species being known, in contrast to some of the much larger tropical catfish families. No ictalurids are native to the United States west of the Rocky Mountains, but several species have been very successfully introduced to that area and form a valuable adjunct to the fishing in these waters.

As would be expected of a geologically recent group, few catfishes of any family occur in the fossil record and although they are largely scaleless animals, undoubtedly they had scaled ancestors. They are related to the Cyprinidae or Minnow family.

North American catfish, like so many others, are mainly nocturnal in habit and live essentially as scavengers over muddy bottoms. They are adapted to life in situations unfavorable to most fishes and some species can thrive in stagnant waters when other fish have died. When oxygen becomes scarce, the bullhead periodically breaks the surface and gulps air. Their hardiness in other ways is equally striking. They can be shipped alive for considerable distances in blocks of ice or wrapped in wet sacking, and some South American catfish are known to travel overland from drying ponds in search of water.

Ictalurids are probably not as important in the food chain as a casual knowledge of their habits might suggest. True, they convert much of what would otherwise be waste materials in their scavengerous feeding, but little of this seems in turn to be available to larger predatory game fishes because of the relative rarity with which they use the catfish for food. Several factors are undoubtedly involved here, but the foremost seem to be their nocturnal habits and their natural protection in the way of poisoned dorsal and pectoral spines.

DIAGNOSIS

Body elongate-to-short, scaleless; mouth terminal, large-to-medium, bearing 4 pairs of barbels in its immediate vicinity; jaws toothed; pharyngeal teeth or jaws present; branchiostegals 8 to 11; gill membranes not connected to skin of isthmus, free along their posterior margins; weberian apparatus present (a chain of small bones connecting the ear and the air bladder); fins—adipose present, pectorals latero-ventral, spined—dorsal spined—pelvics abdominal—caudal blunt or forked; lateral line present; air bladder large.

---

[1]This has been, and is being, called "Ameiuridae" by other authors. Recently (1956) Bailey added his acceptance of Taylor's work (1954) in which the latter recommended abandoning the genus *Ameiurus* entirely as an entity distinct from *Ictalurus*.

CATFISH AND BULLHEADS

Genus *Ictalurus* Rafinesque 1820
("Fish Cat")

KEY TO SPECIES

1. Tail or caudal fin distinctly forked, the 2 lobes well-defined and often sharply pointed; animal predominantly colored silver-blue through silver to white, breeding individuals showing more bluish; supra-occipital bone extending rearward and connecting with a forward-projecting process of the dorsal fin; lower jaw distinctly shorter than upper jaw_____ 2

   Tail or caudal fin square, convex or weakly concave, not forked; species predominantly colored in yellows, browns and blacks; supra-occipital bone extending rearward but distinctly falling short of union with the dorsal fin anterior process; jaws about equal in length_____ 3

2 (1). Anal fin rays from 24 to 30 in number, counting all rudiments; pelvic fins, when extended rearward, overlapping origin of long anal fin; black-spotting conspicuous on all except larger, older fishes—from 5 pounds and above, spotting may be lacking; young usually have dark edges along top of tail fin, leading edge of dorsal and distal edge of caudal fins____(*Ictalurus punctatus*)____CHANNEL CATFISH

   Anal fin rays from 19 to 23, counting rudiments (in some cases there are 24); pelvic fins, when extended rearward, not reaching origin of shorter anal fin (i.e., shorter than in the Channel Catfish); never any black-spotting on body, although breeding specimens may be blue-black as in the Channel Cat; young without the darker fin edges as described for the Channel Cat (*Ictalurus catus*)_____ WHITE CATFISH

3 (1). Anal rays generally 18–19 (rarely 16–22), counting rudiments; pectoral fin spine without strong, definite barbs or teeth on posterior edge, although occasionally some individuals will show 5–10 poorly developed teeth on this posterior edge, teeth which are little more than bumps (see test for teeth of Brown Bullhead in next couplet); inter-radial membranes of anal fin jet black—fin is never base-striped nor spotted nor uniformly pigmented on both rays and membranes as in the following species—outer two-thirds of these membranes always darker than the rays themselves; ventral color in adults contrasting with darker sides, giving the individual a bicolored look with a lightening of color vertically in front of the caudal fin base (*Ictalurus melas*)_____ BLACK BULLHEADS

   Anal rays generally 22–23 (rarely 19–24), counting rudiments; pectoral fin spine with strong barbs on posterior edge (in the young these number from 5–10 prominent teeth whose lengths average more than half the pectoral spine diameter—in adults and older fish, these teeth or barbs become smaller and more numerous, from 10–25; a practical test for this is to grasp the spine in the plane of the fin between thumb and forefinger, hold tightly, and pull outward—if your grasp holds, or the spines prick, it is this species); inter-radial membranes of anal fin not black, the spotting characteristically densest near free margins of fin, or producing a vague longitudinal stripe near fin base, or showing as faint mottlings scattered unevenly on both rays and membranes—if individuals are washed out in color, pigmentation is about equally distributed on both rays and membranes; ventral color in adults usually fusing gradually into the mottled sides with no lightening of color vertically in front of caudal fin base (*Ictalurus nebulosus*)_____ BROWN BULLHEAD

THE CHANNEL CATFISH

Ictalurus punctatus (Rafinesque)

Silurus punctatus Rafinesque 1818: 355
Pimelodus caudafurcatus Le Sueur 1819: 152 (Mem. Mus.)
Pimelodus maculatus Rafinesque 1820 (Quart. Jour. Sci., London)
Pimelodus gracilis Hough 1852: 26 (Ann. Rept. Regent's, Albany)
Pimelodus vulpes Girard 1858: 170 (Acad. Nat. Sci., Phila. Proc. 10)
Synechoglanis beadlei Gill 1859: 40 (Lyc. Nat. Hist., N. Y. Ann. 7)
Pimelodus houghi Girard 1859: 159 (Acad. Nat. Sci. Phila., Proc. 11)
Pimelodus megalops ibid
Pimelodus graciosus ibid
Pimelodus hammondi Abbott 1860: 568–569 (Acad. Nat. Sci., Phila. Proc. 12)
Pimelodus notatus ibid
Ictalurus simpsoni Gill 1862: 42 (Boston Soc. Nat. Hist., Proc. 8)
Pimelodus vulpes, Günther, 1864: 98
Pimelodus megalops, ibid
Amiurus cauda-furcatus, ibid
Ichthaelurus punctatus, Jordan and Copeland, 1878: 415
Ictalurus punctatus, Jordan and Gilbert, 1882: 108–109
Ictalurus punctatus, Jordan and Evermann, 1896B: 134–135 / 1902: 21–22
Ictalurus punctatus, Forbes and Richardson, 1920: 180–183
Ictalurus punctatus, Pratt, 1923: 92
Ictalurus punctatus, Hildebrand and Towers, 1928: 120
Ictalurus punctatus, Coker, 1930: 175–177
Ictalurus punctatus, Jordan, Evermann and Clark, 1930: 152
Ictalurus punctatus, Fish, 1932: 349
Ictalurus punctatus, Schultz, 1936: 150
Ictalurus punctatus, Schrenkeisen, 1938: 159–160
Ictalurus lacustris punctatus, Kuhne, 1939: 63–64

Ictalurus lacustris punctatus, Hubbs and Lagler, 1941: 63 / 1947: 72
Ictalurus punctatus, Moffett, 1942: 82 / 1943: 182
Ictalurus lacustris, Hinks, 1943: 58–59
Ictalurus lacustris punctatus, Dill, 1944: 155–159
Ictalurus lacustris punctatus, Miller and Alcorn, 1946: 182
Ictalurus lacustris punctatus, Eddy and Surber, 1947: 170–171
Ictalurus lacustris punctatus, Bailey and Harrison, 1948
Ictalurus lacustris punctatus, Appleget and Smith, 1951
Ictalurus lacustris punctatus, Simon, 1951: 92–93
Ictalurus punctatus, Beckman, 1952: 65–70
Ictalurus punctatus, La Rivers, 1952: 91
Ictalurus lacustris punctatus, La Rivers and Trelease, 1952: 118
Ictalurus lacustris punctatus, Marzolf, 1952 / 1955
Ictalurus punctatus, Miller, 1952: 7
Ictalurus punctatus, Speirs, 1952
Ictalurus punctatus, Scott, 1954: 67
Ictalurus lacustris punctatus, Shomon, 1955: 45–46
Ictalurus punctatus, Bailey, 1956: 335
Ictalurus punctatus, Harlan and Speaker, 1956: 106–108
Ictalurus punctatus, Clemens and Sneed, 1957
Ictalurus punctatus, Eddy, 1957: 148
Ictalurus punctatus, Koster, 1957: 74–75
Ictalurus punctatus, Marzolf, 1957
Ictalurus punctatus, Moore, 1957: 141
Ictalurus punctatus, Trautman, 1957: 415–417
Ictalurus punctatus, Slastenenko, 1958A: 239–241
Ictalurus punctatus, Davis, 1959
Ictalurus punctatus, Shapovalov, Dill and Cordone, 1959: 172
Ictalurus punctatus, McCammon and LaFaunce, 1961

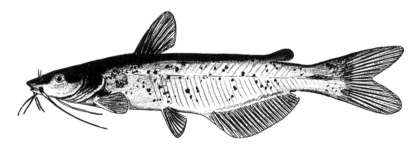

FIG. 217. Channel Catfish, *Ictalurus punctatus*. Courtesy California Department of Fish and Game and Leo Shapovalov.

ORIGINAL DESCRIPTION

"Sp. 8. *Silurus punctatus*, Raf. Body whitish with gilt shades and many brown unequal dots on the sides, 8 barbels, 4 underneath, 2 lateral long and black, dorsal fin 7 rays, 1 spiny, pectoral fins 6 rays, 1 spiny, anal 27 rays, lateral line a little curved beneath at the base, tail forked unequal, upper lobe longer" (Rafinesque 1818: 355).

DIAGNOSIS

Head 3.6 to 4.0 times, body depth 4.2 to 5.0 times, into body length, respectively; head small and compressed dorso-ventrally; mouth nearly terminal, the upper jaw only slightly longer than the lower; maxillary barbels long, usually reaching beyond gill openings; eye large; fins—dorsal high and narrow, its origin located slightly nearer to snout than to origin of adipose fin, its conspicuous spine followed by 6 soft rays; dorsal spine serrated in young fish, becoming smooth in adults by the time they are 12 to 14" long; serration posterior—anal long (even though it is comparatively short for the genus), 24–30 rayed (counting rudiments)—pectoral spine about equal in size to dorsal spine, more serrate or toothed posteriorly than is dorsal, becoming smoother as the fish becomes older but always with some remaining serration; spine followed by 9 rays—caudal strongly forked, the upper lobe longer and narrower and more pointed than the lower—humeral process about half as long as pectoral spine; color—above dark olive through yellow-gray to gray-blue, depending upon age and environment; sides gray through yellow-green; venter gray-white—occasionally a copper hue on sides, which are conspicuously and irregularly black-spotted in young fish, this spotting tending to disappear with age. A large catfish, reaching some 25 pounds in weight, although most individuals seldom exceed five pounds. The largest one on record appears to be a 55-pounder (Hutt 1956).

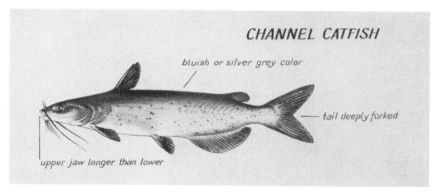

FIG. 218. Identifying characteristics of the Channel Catfish. Courtesy Wisconsin Conservation Department, Jens von Sivers and N. H. Hoveland.

TYPE LOCALITY

Ohio River.

RANGE

Native to the Mississippi drainage from the Great Lakes region south to the Gulf of Mexico; have been planted successfully into the West on numerous occasions. The species is well established in the Colorado River system of southern Nevada and environs—Arizona is known to have brought the Channel Catfish to the Colorado River as early as 1892 according to Worth (1895: 127). "We have been unable to determine what happened to these fish. According to Game Warden Frank

Wait of Las Vegas, this species was first introduced into the lower Colorado River about 1906 (personal interview)" (Miller and Alcorn 1946: 182). Moffett (1942: 82, 1943: 182), Wallis (1951: 90) and Jonez and Sumner (1954: 50–54) have all indicated the abundance of this fish in Lake Mead and vicinity.

TAXONOMY

The synonymicon, as can be seen, is quite complex, the species having acquired a multitude of synonyms in its 142 years of formal existence. Only recently it has been shown that the name *lacustris*, as used in *Ictalurus*, is properly a subspecies of the Burbot (*Lota lota* (Linnaeus)) and that *punctatus* should be returned again as the specific term for the Channel Catfish (Speirs 1952).

Etymology—"punctatus," spotted.

LIFE HISTORY

This fish is remarkably tolerant of varying environmental conditions. Dill (1944) gives a good account of its capacity to do well in the contrasting habitats offered by the lower Colorado River and its adjacent drainages as well as the unique Salton Sea in California. In this area it is commonly found in swift, cold and fairly clear water as well as in the warm, sluggish and silt-carrying sections—Salton Sea, with a greater concentration of materials in solution than is found in the ocean, is also part of its expanding range.

Its size attainment in the Colorado River is comparable to the best growths reported anywhere in its range—natural or planted—a 22-pounder being known from Lake Havasu behind Parker Dam, Arizona. In his work on the lower Colorado, Dill (1944) found some individuals only slightly over 7 inches in length to be in spawning condition. Spawning there occurs during spring and into summer, with, as would be supposed, the larger fish contributing the most eggs per individual. Dill's figures for these were: an 8.5 inch specimen had approximately 1,600 eggs while one 26 inches long contained something like 34,500 eggs.

In feeding habits, the Channel Catfish is omnivorous, almost anything suitable having been reported from its stomach in past surveys. Its game fish eating potential is small, however. Forbes and Richardson in Illinois, found insects to be the main food, with vegetation (principally algae) and mollusks following in that order. The Channel Catfish, when feeding on snails and bivalves, crushes and rejects the shells (1920: 182–183).

Dill's study showed similar habits within the framework of the environment, with insects constituting the bulk of food for the Colorado River specimens while plants and other arthropods were second and third in importance, respectively. In Lake Mead, Sumner found them actively feeding on shad, which were introduced in 1953, as bass food.

Harms (1959) found the following parasites on the Channel Catfish in northeastern Kansas:

TREMATODA
Cleidodiscus floridanus
Phyllodistomum lacustri
Crepidostomum ictaluri
Clinostomum marginatum

CESTODA
Corallobothrium fimbriatum
Corallobothrium giganteum
Proteocephalus ambloplitis

NEMATODA
Spinitectus gracilis
Camallanus oxycephalus
Dichelyne robusta

HIRUDINEA
Piscicolaria reducta

COPEPODA
Ergasilus versicolor

ECONOMICS

This is the most common game fish in the lower Colorado River drainage, below Nevada, and the one most often caught. Sumner ranks it second after bass in Lake Mead (1954). From the fishery biologist's point-of-view, the Channel Catfish has additional value in that it is a consumer of some food materials consistently avoided by other game fishes, such as organic debris and plant materials and so an efficient converter of vegetation to proteins upon which other animals can in turn feed. Dill and others have also speculated upon the very considerable probability that its predation upon eggs and possibly fry of the native cyprinids has contributed materially to the scarcity of these latter fishes in the lower Colorado River.

Similarly, this species is an important commercial fish in its home range—". . . it may be said to constitute the bulk of the catfishes (not including bullheads) of the annual Illinois River catch . . ." (Forbes and Richardson 1920: 183).

The Channel Catfish is also an important fishery to the lower Humboldt River.

THE WHITE CATFISH

Ictalurus catus (Linnaeus)

Silurus catus Linnaeus 1758: 305
Pimelodus albidus Le Sueur 1818: 148 (Mem. Mus.)
Pimelodus lynx Girard 1859: 160 (Acad. Nat. Sci. Phila. Proc. 11)
Amiurus catus, Günther, 1864: 99
Ictalurus macaskeyi Stauffer 1869: 578 (Hist. Lancaster Co., Penn.)
Ictalurus kevinskii ibid

Amiurus niveiventris Cope 1870: 486 (Amer. Phil. Soc. Proc. 11)
Amiurus catus, Jordan and Copeland, 1878: 415
Amiurus catus, Jordan and Gilbert, 1882: 104–105 (in part)
Ictalurus albidus, ibid: 107 (in part)
Ictalurus niveiventris, ibid (in part)
Ameiurus catus, Jordan and Gilbert, 1894: 140
Ameiurus catus, Jordan and Evermann, 1896B: 138 / 1902: 25
Ameiurus catus, Pratt, 1923: 94
Haustor catus, Jordan, Evermann and Clark, 1930: 153
Ameiurus catus, Schrenkeisen, 1938: 161
Ictalurus catus, Murphy, 1941: 166
Ictalurus catus, La Monte, 1945: 163
Ictalurus catus, Miller and Alcorn, 1946: 182–183
Ictalurus catus, Shapovalov and Dill, 1950: 386
Ictalurus catus, La Rivers, 1952: 91
Ictalurus catus, La Rivers and Trelease, 1952: 118
Ictalurus catus, Eddy, 1957: 148
Ictalurus catus, Moore, 1957: 141
Ictalurus catus, Trautman, 1957: 418–420
Ictalurus catus, Shapovalov, Dill and Cordone, 1959: 172
Ictalurus catus, McCammon and Seeley, 1961

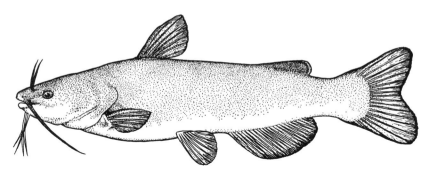

FIG. 219. White Catfish, *Ictalurus catus*. Drawn by Silvio Santina.

ORIGINAL DESCRIPTION

"Catus. 9. S. pinna dorsali postica adiposa, ani radiis 20, cirris 8.
B. 5. D. 1/6, o. P. 1/11. V. 8. A. 20. C. 17.
Catesb. car. 2. *p.* 23. *t.* 23. Bagre 2. Marcgr. affinis.
Marcgr. bras. 173. Bagre species 2.
Habitat in America, Asia.
Ex Asia vidi pinnis ventr, radiis 6."

(Linnaeus 1758: 305–306).

DIAGNOSIS

Head 3.0 to 3.5 times, body depth 4.5 to 5.0 times, into body length, respectively; head small, compressed dorso-ventrally in specimens under 2–3 pounds—large individuals have disproportionally large heads and wide mouths, surpassing the Channel Catfish in these respects; mouth subterminal, the upper jaw distinctly but only slightly longer than lower jaw; maxillary barbels long, generally extending behind the gill cover; eye medium in size; fins—dorsal narrow, high, origin a bit closer to snout tip than to origin of adipose fin, its anterior spine followed by 5-6 soft rays—anal fin long, but shorter than the Channel Catfish, having from 18 to 24 rays, counting rudiments (19-23 otherwise)—pectoral spine subequal to dorsal spine, generally toothed on inner or posterior margin and followed by about 9 rays—caudal fin less prominently forked than in the Channel Catfish, its upper lobe usually somewhat shorter and narrower than lower lobe, both lobes rounded—humeral process about one-half as long as pectoral spine; color—varied from washed out bluish or blue-gray to olive above and on sides except in large individuals which may become dark blue on back and upper sides, grading to shades of white-to-gray on venter—lacks the spotting of the Channel Catfish, but often shows an unevenness or mottling, particularly along the sides; this is a sizeable catfish, smaller than the Channel Cat, but reaching weights of up to 15 pounds, with 2-3 pounders being the usual run.

TYPE LOCALITY

"Northern America."

RANGE

Originally the Atlantic Coast of the United States, being particularly abundant in the lower portions of streams, from Delaware to Florida. It has been successfully introduced into many parts of the Middle West and West, particularly California.

Regarding the Nevada introductions:

"About 1938 Mr. William A. Powell, Jr., Secretary of the Nevada Fish and Game Commission, caught on hook and line 39 large white catfish from Clear Lake (near Kelseyville), Lake County, California, and transplanted them to the vicinity of Fallon (personal interview). In 1939, 500 'blue catfish' were transferred from the California State Central Valleys Hatchery to the State of Nevada (Anonymous, 1941b: 88-89) [36th Biennial Rept. Div. Fish & Game of Calif.]. Mr. Dill informs us that these fish were *Ictalurus catus*. According to Mr. Powell, he obtained 500 'rescued' fish of this species from the Elk Grove Hatchery, California, in 1939, and planted them in the vicinity of Fallon, both above and below the Lahontan Dam.

"We have seen only two specimens of this catfish from Nevada. One was taken 6 miles north of Stillwater, Churchill County, on April 5, 1943, by J. R. Alcorn and V. L. Mills; the other specimen was collected by Alcorn in a drainage ditch 1 mile north of Stillwater, on August 15, 1941. Both were caught on hook and line. Vernon L. Mills, State Game Warden, reports that fishermen, as well as himself, have taken a number of the small white catfish.

"Since this species is particularly adapted to large rivers, any further plants in Nevada should be made in large reservoirs. This species is called the 'fork-tail catfish' in California" (Miller and Alcorn 1946: 182-183).

The White Catfish is now common throughout the Lahontan Valley system—Lahontan Reservoir, Indian Lakes and Stillwater Marsh.

TAXONOMY

Etymology—"catus," cat.

LIFE HISTORY

Surprisingly little seems to have been published on the biology of the White Catfish. A recent, intensive study of the catfish industry in California provides the bulk of the information here given.[1]

The White Catfish is not limited in its habitats, but can successfully live in a variety of situations, seeming to prefer slower or standing waters—fresh or brackish—but doing well in creeks if these contain large pools. It spawns in both running and still water, requiring some

FIG. 220. White Catfish *Ictalurus catus*, from Stillwater Marsh. Photo by Robert Sumner.

shelter and producing eggs according to the size of the female. A fish eight inches long will have a little over a thousand eggs.[2] This length represents 3-year old fish (i.e., those in their 4th year) which are spawning for the first time. They begin to mature at a forklength of six inches.

Optimum spawning temperatures seem to be around 70° F. Latitude will determine the time of year any given water reaches this temperature; in the Delta area of the Sacramento-San Joaquin Rivers of central California, spawning takes place in June and July. Data on egg hatch-time and larval development are lacking, but it was found that ages are best calculated from vertebral sections, and the California

---

[1] California Dingle-Johnson Project F2R, "A study of the catfish fishery in California," unpublished but to appear shortly in the California Fish and Game periodical. We are indebted to George W. McCammon for these data.

[2] Number of eggs $= 10.96 \times$ fork length $^{2.20152}$.

investigations gave the following age-length relationships for the Delta (growths were more rapid in Clear Lake to the north):

Age group	Fork length in inches
Age 0	3.0
Age 1	5.2
Age 2	6.9
Age 3	8.6
Age 4	10.2
Age 5	11.6
Age 6	13.0
Age 7	14.0

California weights are recorded up to a 15-pound maximum, with 5–6 pounders being common. In Nevada, the largest fish I have any record for is one of 7 pounds, taken from the Fallon area, with weights of 1–2 pounds being usual.[3]

In food habits, the White Catfish seems "intermediate between the Channel Catfish and bullheads. In the Delta, their food consists of amphipods, snails and other invertebrates. There are practically no insects in the Delta and the forage fish situation is poor, so the white cats are limited in their choice of food. In Clear Lake, the small cats subsist primarily on gnat larvae (*Tendipes*) and other insects. The larger fish eat a lot of small fish, such as sculpins, hitch, blue-gills and viviparous perch. White Catfish are often taken on artificial lures in Clear Lake" (McCammon 1957).

ECONOMICS

Little published information seems to exist concerning this species and its importance as a sport or commercial fish. In the West, California, with its experience of nearly 90 years with the White Catfish (it was introduced there in 1874), is the only reliable source of long-term data.

For most of this period, the White Catfish supported both a commercial and a sport fishery in California. The former was at its height around 1900 and had declined very considerably by the early 1950s when commercial utilization was prohibited in favor of sport fishing for the species. Since no commercial fishery involving the White Catfish has ever existed in Nevada, it is not possible to compare relative values between the two areas. In our State, this species is an important sports fish in the heavily populated west-central section (as it is in California) where much of the fishable waters are muddy reservoirs suited to its tastes.

THE BROWN BULLHEAD
Ictalurus nebulosus (Le Sueur)
(Horned Pout, Bull Pout)

Pimelodus nebulosus Le Sueur 1819: 149 (Mem. Mus.)
Pimelodus atrarius DeKay 1842: 185 (N. Y. Fauna: Fishes)

---

[3]Length-weight relationship can be determined by the following formula: Log wt. = —5.65374 plus 3.33063 Log forklength.

Pimelodus felis Agassiz 1850: 281 (Lake Superior)
Pimelodus hoyi Girard 1859: 159 (Acad. Nat. Sci. Phila. Proc. 11)
Pimelodus vulpeculus ibid:. 160
Amiurus nebulosus, Günther, 1864: 101
Ameiurus mispilliensis Cope 1870: 486 (Amer. Phil. Soc. Proc. 11)
Amiurus catus, Jordan and Copeland, 1878: 415
Amiurus catus, Jordan and Gilbert, 1882: 104–105 (in part)
Ameiurus nebulosus, Gilbert, 1893: 229
Ameiurus nebulosus, Jordan and Gilbert, 1894: 140
Ameiurus nebulosus, Jordan and Evermann, 1896B: 140–141 / 1902: 26–30
Ameiurus nebulosus, Snyder, 1917A: 85
Ameiurus nebulosus, Forbes and Richardson, 1920: 187–190
Ameiurus nebulosus, Pratt, 1923: 94
Ameiurus nebulosus, Hildebrand and Towers, 1928: 121
Ameiurus nebulosus, Jordan, Evermann and Clark, 1930: 154
Ameiurus nebulosus, Fish, 1932: 350
Ameiurus nebulosus, Hatton, 1932: 37–38
Ameiurus nebulosus, Breder, 1935
Ameiurus nebulosus, Schultz, 1936: 151
Ameiurus nebulosus, Tanner, 1936: 172
Ameiurus nebulosus, Schrenkeisen, 1938: 161–162
Ameiurus nebulosus, Kuhne, 1939: 64–68
Ameiurus nebulosus, Hubbs and Lagler, 1941: 62, 64 / 1947: 71–72
Ameiurus nebulosus, Murphy, 1941: 166
Ameiurus nebulosus, Curtis, 1942: 6 / 1949: 256
Ameiurus nebulosus, Hinks, 1943: 59–61
Ameiurus nebulosus, Dill, 1944: 160–161
Ameiurus nebulosus nebulosus, Dimick and Merryfield, 1945: 40
Ameiurus nebulosus, La Monte, 1945: 164
Ameiurus nebulosus nebulosus, Miller and Alcorn, 1946: 183–184
Ameiurus nebulosus nebulosus, Eddy and Surber, 1947: 170, 177
Ameiurus nebulosus, Carl and Clemens, 1948: 85–87
Ameiurus nebulosus, Everhart, 1950: 39
Ameiurus nebulosus nebulosus, Shapovalov and Dill, 1950: 386
Ameiurus nebulosus nebulosus, Beckman, 1952: 67–68
Ameiurus nebulosus, La Rivers, 1952: 92
Ameiurus nebulosus, La Rivers and Trelease, 1952: 118
Ameiurus nebulosus, Gabrielson and La Monte, 1954: 107–108
Ameiurus nebulosus, Scott, 1954: 66
Ictalurus nebulosus, Bailey, 1956: 335, 363
Ictalurus nebulosus, Harlan and Speaker, 1956: 109–111
Ictalurus nebulosus, Eddy, 1957: 150
Ameiurus nebulosus, Koster, 1957: 77
Ictalurus nebulosus, Moore, 1957: 141
Ictalurus nebulosus, Trautman, 1957: 424–426
Amiurus nebulosus, Slastenenko, 1958A: 243–244
Ictalurus nebulosus, Carl, Clemens and Lindsey, 1959: 131–133
Ictalurus nebulosus, Shapovalov, Dill and Cordone, 1959: 172
Ictalurus nebulosus, McCammon and Seeley, 1961

ORIGINAL DESCRIPTION

"2. Pimelode Yellow Belly. (*P. nebulosus.*)

"*Caract. spéc.* Couleur jaune cuivrée avec une teinte brune disposée en nuage sur le dos et les côtés; iris blanc; abdomen blanchâtre.

"Le corps des poissons de cette espèce a quatre fois et demie ou environ la longueur de la tête. Sa forme se rapproche beaucoup de celle de la précédente espèce; mais ce pimelode est moins épais, moins large, et d'une plus petite taille.

"Première nageoire dorsale ronde et moyenne; deuxième nageoire adipeuse et arrondie; pectorales et ventrales petites et rondes; anale allongée, arrondie; caudale épaisse, peu échancrée, presque droite et arrondie aux extrémités des lobes; toutes ces nageoires épaisses; les premiers rayons de la dorsale et des pectorales fort osseux et cachés sous la peau.

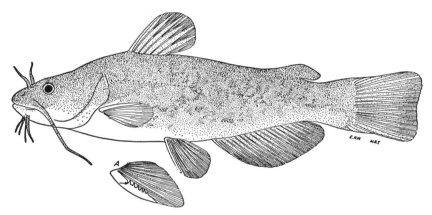

FIG. 221. Brown Bullhead, *Ictalurus nebulosus.* (A) Pectoral fin, showing serrations on posterior or inner edge of pectoral spine. Reproduced from *The Fishes of Ohio*, by Milton B. Trautman (Columbus, Ohio: Ohio State University Press, 1957), p. 424. Used by permission of the publisher.

"Mâchoires inégales, la supérieure plus longue, toutes deux armées de petites dents en forme de carde. Dans la gorge sont de petites dents forts pointues implantées dans des tubercules pisiformes.

"Narines antérieures tubulées; les postérieures linéaires, un peu élevées et surmontées par de longs cirrhes charnus dans leur partie antérieure. D'autres barbillons se trouvent à chaque angles des mâchoires et sous la mâchoire inférieure.

"Vessie natatoire en forme de coeur, avec une dépression dans sa partie supérieure, pour embrasser l'épine dorsale. De sa base part le canal qui conduit l'air de l'estomac à cette vessie, en se rattachant à son centre. L'estomac est tourné sur lui-même. Le canal intestinal présente plusieurs plis. Sa longueur étendue étoit de vingt-deux pouces dans un individu qui avoit neuf pouces de longueur. Cette espèce est trèsnonbreuse à Philadelphie. On lavoit depuis le commencement du mois de mai jusqu'aux premiers froids de l'hiver; on la pêche dans la Delaware. Sa chair est blanche et trèsestimée.

"Rayons: B. 8; P. 8; D. 6; V. 8; A. 21; C. 18;

"C'est un poisson qui a la vie extrêmement dure" (La Sueur 1819: 149–150).

DIAGNOSIS

Head 3.2 to 3.6, body depth 3.5 to 4.3, times into body length, respectively; head somewhat narrow and depressed; snout longer and more distinctly rounded than in *A. melas;* upper jaw generally noticeably longer than lower jaw; maxillary barbels generally distinctly reaching posteriorly to operculum, occasionally beyond the humeral process; eye medium in size; fins—dorsal moderate in size, its anterior base slightly nearer adipose fin than tip of snout, its front spine followed by about 6 soft rays—anal relatively moderate in size for the

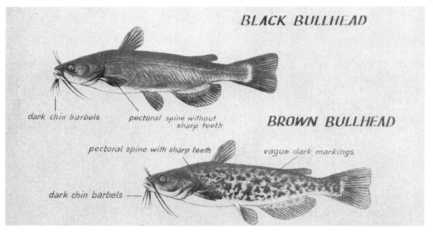

Fig. 222. Comparisons between the two species of bullheads. Courtesy Wisconsin Conservation Department, Jens von Sivers and N. H. Hoveland.

group, 21–24 rayed, the inter-radial membranes brownish, free fin edge essentially straight, with little curvature, from about the 6th to 16th rays, the rays somewhat slender and generally divided less than a third of their distance to bases—pectoral spine usually long, curved, retrorsely toothed posteriorly, these teeth decreasing in size and increasing in number with age (from 5–10 to 10–25, usually), the spine followed by about 9 rays—caudal weakly, broadly forked (i.e., with an incipient median, very shallow concavity along posterior or free edge)—humeral process relatively long and sharp, more than half the length of the pectoral spine; color somewhat variable, light to dark brownish-yellow on back, lightening on sides and varying from smooth white to pale gray or pinkish below—an overlaid mottling of darker color also varies in its development—inter-radial membranes of anal fin not black, the spotting characteristically densest near free fin margins, or producing a vague longitudinal stripe near fin base, or showing as faint mottlings scattered unevenly on both rays and membranes—if individuals are washed out in color, pigmentation is about equally distributed on both rays and membranes; size—grows to a somewhat larger size than the Black Bullhead, reaching lengths of about 18 to 20 inches with some monsters of 7 pounds having been recorded.

TYPE LOCALITY
Lake Ontario.

RANGE
Eastern United States, from the Great Lakes region to Maine, south to Texas and Florida. It has been widely and successfully introduced into the West, since the early days. It was first brought into Nevada by Parker (see p. 18) in 1877 (1879: 3).

TAXONOMY
Etymology—"nebulosus," clouded.

LIFE HISTORY
The now little-read Thoreau (a good man with a pen) felt that:
"The horned pout are dull and blundering fellows, fond of the mud and growing best in weedy ponds and rivers without current. They stay near the bottom, moving slowly about with their barbels widely spread, watching for anything eatable. They will take any kind of bait, from an angleworm to a piece of tomato can, without hesitation or coquetry, and they seldom fail to swallow the hook. They are very tenacious of life, opening and shutting their mouths for half an hour after their heads have been taken off. They spawn in spring and the old fishes lead the young in great schools near the shore, caring for them as a hen cares for her chickens. A bloodthirsty and bullying set of rangers with ever a lance at rest and ready to do battle with their nearest neighbor."

The Brown Bullhead is a spring-spawning species, building its nest and laying its eggs in shoal water often only a few inches deep, over mud or sand bottoms, or attached among vegetation. The eggs are often a striking creamcolor and resemble those of frogs in their mass structure. They are guarded and fin-fanned by both parents for a time, although the male often assumes this responsibility both over eggs and young. Often, he sucks both eggs and young in and out of his mouth, possibly for aeration in the case of the former. The eggs hatch in about a week.

Their food is variable, running the gamut of whatever meat and vegetable matter they happen upon; arthropods, mollusks, algae, etc.

The tenacity of the bullhead is pronounced, though no more remarkable than that of some other catfishes; they can live in waters too stagnant for other fishes by gulping surface air, and will survive in shallow ponds where other species winterkill, by the simple expedient of burying themselves in the bottom mud.

ECONOMICS
In addition to its importance to the angler, the Brown Bullhead is extensively taken by commercial fishermen, particularly in its native

area, for market sales—in Nevada, this latter aspect of its utilization is lacking.

Since there is no fish which is not accused by someone somewhere of eating eggs of another fish the accuser dearly loves, we are happy to report the bullhead is no exception. However, evidence for the crime is not convincing. In a now obscure publication Fuhrmann wrote many years ago that whitefish eggs (for instance) were not eaten except immediately after they were laid; i.e., the hardening action of the water, which takes place very rapidly, seems to make the eggs unpalatable to the bullhead (Soc. Acclim. Bull. 51: 351, 1904). No doubt they do eat some fish eggs, along with many other fish.[1]

THE BLACK BULLHEAD
Ictalurus melas (Rafinesque)

Silurus melas Rafinesque 1820: 51
Silurus xanthocephalus ibid
Pimelodus pullus DeKay 1842: 184
Pimelodus confinis Girard 1859: 159 (Acad. Nat. Sci. Phila. Proc. 11)
Amiurus obesus Gill 1862: 45 (Boston Soc. Nat. Hist. Proc. 8)
Pimelodus pullus, Günther, 1864: 98
Amiurus melas, Jordan and Copeland, 1878: 415
Amiurus melas, Jordan and Gilbert, 1882: 104
Amiurus xanthocephalus, ibid
Amiurus cragini Gilbert 1884: 10 (Washburn Lab. Nat. Hist. Bull. 1)
Ameiurus melas, Jordan and Evermann, 1896B: 141 / 1902: 30–31
Ameiurus melas, Forbes and Richardson, 1920: 190–192
Ameiurus melas, Pratt, 1923: 94–95
Ameiurus melas, Jordan, Evermann and Clark, 1930: 154
Ameiurus melas, Hatton, 1932: 39–41
Ameiurus melas, Schultz, 1936: 150
Ameiurus melas, Tanner, 1936: 172
Ameiurus melas, Schrenkeisen, 1938: 163
Ameiurus melas, Kuhne, 1939: 64–68
Ameiurus melas melas, Hubbs and Lagler, 1941: 62–63 / 1947: 71–72
Ameiurus melas, Hinks, 1943: 59–61
Ameiurus melas, Dill, 1944: 160
Ameiurus melas, La Monte, 1945: 163–164
Ameiurus melas melas, Miller and Alcorn, 1946: 183
Ameiurus melas melas, Eddy and Surber, 1947: 176–177
Ameiurus melas, Carl and Clemens, 1948: 85, 88–89
Ameiurus melas, Curtis, 1949: 256
Ameiurus melas melas, Shapovalov and Dill, 1950: 386
Ameiurus melas melas, Simon, 1951: 93–94
Ameiurus melas, Wallis, 1951: 90
Ameiurus melas, Beckman, 1952: 66–67
Ameiurus melas, La Rivers, 1952: 92

---

[1] "Only Sunday, May 5, 1958 I caught some bullheads (black) at Washoe Lake that contained many hundreds of what I believe were perch or bullhead eggs" (Tom Trelease).

Ameiurus melas, La Rivers and Trelease, 1952: 118
Ameiurus melas, Gabrielson and La Monte, 1954: 108
Ictalurus melas, Bailey, 1956: 335, 364
Ictalurus melas, Harlan and Speaker, 1956: 111–112
Ictalurus melas, Eddy, 1957: 151
Ameiurus melas, Koster, 1957: 76–77
Ictalurus melas, Moore, 1957: 141
Ictalurus melas, Trautman, 1957: 427–429
Amiurus melas, Slastenenko, 1958A: 244–246
Ictalurus melas, Carl, Clemens and Lindsey, 1959: 133–134
Ictalurus melas, Shapovalov, Dill and Cordone, 1959: 172

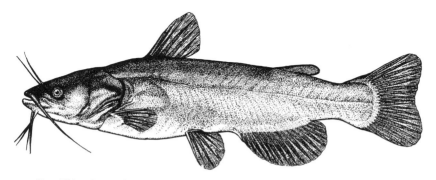

FIG. 223. Black Bullhead, *Ictalurus melas*. Drawn by Silvio Santina.

ORIGINAL DESCRIPTION

"8. *Silurus Melas*. Black cat-fish. Body blackish, jaws and barbels unequal, the lateral barbs shorter than the head, lateral line straight, eyes rounded, spinous rays short and smooth, anal fin with twenty rays, tail semi-truncate.

"A small species from three to ten inches long, throat and belly hardly pale, iris black slightly elliptical. D. 1 and 6. P. 1 and 7. Abd. 8. An. 20. C. 24" (Rafinesque 1820: 51).

DIAGNOSIS

Head and body depth each from 3.1 to 3.5 times into body length; head short and broad; snout short and broader than in *A. nebulosus*; jaws nearly equal in size; maxillary barbels extending usually only about to tip of opercle, rarely long enough to reach base of humeral process; eye medium in size; fins—dorsal moderately proportioned, its anterior base about midway between snout and base of adipose fin, its spine more variable than that of the Brown Bullhead, the spine followed by about 6 soft rays—anal rather small for the group, 17–20 rayed, the inter-radial membranes jet black, free fin edge rounded, with no flat portions to its outline, the rays stout and, especially those in the middle, usually divided about halfway to bases—pectoral spine stubbier than in the Brown Bullhead, less curved, without strong, definite barbs or teeth on posterior edge, although occasionally some adult specimens will show 5–10 poorly developed teeth on this edge, teeth which are little more than bumps, spine followed by about 9 rays—caudal more nearly truncate than is that of *A. nebulosus,* but still with a slight concavity in middle—humeral process relatively short, roughened, blunt, half or less than half length of the pectoral spine; color rather variable, generally black grading into deep brown or green on back, the flanks becoming cast with gold or green—below, the fish may be yellowish, greenish or gray, but does not show the smooth white of many of the Brown Bullheads—inter-radial membranes of anal fin jet black, not mottled or equally pigmented on both membranes and rays as in the Brown Bullhead; size—a smaller species than *A. nebulosus,* with 12 inches generally the maximum length with many individuals only attaining half this size—two pounds is about the maximum weight.

TYPE LOCALITY

Ohio River.

RANGE

Generally distributed in the United States east of the Rocky Mountains, the species has been introduced into many western localities with marked success, and is well established in western Canada also. There appear to be no reliable records indicating its first arrival in Nevada, but it has long been common in certain parts of west-central Nevada, particularly the Fallon region and the Walker and Humboldt Rivers (see following subspecific sections for more distributional information). Very probably both Black and Brown Bullheads were indiscriminately mixed in early shipments.

TAXONOMY

Etymology—"melas," black.

LIFE HISTORY

Most of what has been said of the Brown Bullhead will apply equally well to this species. The slightly smaller Black Bullhead seems to exhibit more preference for the shallower, smaller, better aerated streams while its larger relative is often more characteristically at home in markedly stagnant, more extensive waters.

A spring spawner, the Black Bullhead makes rounded, shallow nests, typically on muddy bottoms in shoal water. Both bullheads will use natural depressions such as caved-in and abandoned muskrat nests. The number of eggs laid in each nest varies with the size of the spawning female, the larger specimens depositing in excess of 8,000 eggs. Hatching is also variable, depending upon temperatures, but a week can be considered an average time. Typically, young bullheads remain in rather dense schools for a while, eventually scattering to fend for themselves. During the early schooling period they are generally guarded by one or both parents.

The omnivorous bullhead eats anything that comes along. Such studies as the following, while showing that food details may vary from locality-to-locality, indicate the lack of selectivity—"The food of 36 specimens, doubtless composed of these two species commingled [Brown and Black Bullheads], is distinguished by the fact that nearly a fourth of it consists of aquatic vegetation of various kinds, including distillery refuse eaten by one of the fishes. Two of the bullheads had filled themselves with other fish, a sunfish and a perch among them. Small bivalve mollusks made a fifth of the food, and river snails and aquatic insects—the latter somewhat more than a fourth of the entire quantity—together with crawfishes and other crustaceans, were the other more important elements" (Forbes and Richardson 1920: 192).

ECONOMICS

In its native area, the Black Bullhead has never been a prominent part of the commercial catch in spite of its commonness because of its preference for the smaller bodies of water. It is, however, more often taken by the angler than is the less abundant but larger Brown Bullhead. In spite of their feeding and living habits, the bullheads as a group are generally good eating fish unless they have been living excessively off the muddy bottom.

NORTHERN BLACK BULLHEAD

Ictalurus melas melas (Rafinesque)

(See *I. melas* synonymicon)

This is the subspecies to which our west-central Nevada Black Bullheads all seemed referable according to Miller and Alcorn (1946: 183). In addition to the range listed under *I. melas,* they note a 1942 introduction by Vernon L. Mills into Reese River, "about 10 miles west-southwest of Austin, from near Fallon" (loc. cit.).

SOUTHERN BLACK BULLHEAD

Ictalurus melas catulus (Girard)

Pimelodus catulus Girard 1858: 208
Pimelodus catulus, Günther, 1864: 99
Amiurus melas, Jordan and Gilbert, 1882: 104 (in part)
Ameiurus nebulosus catulus, Jordan and Evermann, 1896B: 141
Ameiurus nebulosus catulus, Pratt, 1923: 94
Ameiurus nebulosus catulus, Jordan, Evermann and Clark, 1930: 154
Ameiurus nebulosus catulus, Schrenkeisen, 1938: 163
Ameiurus melas catulus, Miller and Alcorn, 1946: 183
Ameiurus melas catulus, Wallis, 1951: 190
Ameiurus melas catulus, La Rivers and Trelease, 1952: 118

The only recorded instance of this subspecies' presence in Nevada seems to be the notation: "Two young of this subspecies were taken in Las Vegas Creek at Las Vegas, Clark County, on August 30, 1938. No information regarding their introduction is available" (Miller and Alcorn 1946: 183).

Etymology—"catulus," diminutive of *catus,* cat.

Order *CYPRINODONTIFORMES* (Microcyprini, Cyprinodontes)
Suborder CYPRINODONTOIDEI (Poecilioidei)
Superfamily CYPRINODONTOIDEA

KILLIFISHES

Family *CYPRINODONTIDAE*

This rather distinctive family of small fishes occurs over wide areas of the world, the tropics being their chief population center. They are lacking in Australia, northern North America and northern Asia. They run the gamut of food habits, from plant to animal feeders, and have a size range from about a foot down to species less than two inches in length. All members of the family are egglayers (= oviparous) in contrast to the closely related poeciliids such as the Mosquitofish (*Gambusia*). While generally inhabiting fresh water, many species live in brackish, estuarine situations or are landlocked in interior regions in warm, highly mineralized waters such as commonly occur in Nevada.

DIAGNOSIS

Body variable in shape, from rather elongate to short and chubby; head at least with dorsal scales; mouth terminal or supraterminal, usually small and with protruding lower jaw, markedly protractile since the upper jaw consists of only the premaxillaries; no barbels; jaws toothed, teeth often of considerable complexity; pharyngeal teeth present; branchiostegals 4–6; no pseudobranchiae; gill membranes free from isthmus; fins—adipose lacking, pectorals ventro-lateral, pelvics abdominal, lacking in some species, caudal not forked; lateral line obsolescent or lacking; stomach siphonal; air bladder undifferentiated or absent; pyloric caeca wanting.

Species of small size and variable feeding habits.

KEY TO GENERA OF NEVADA CYPRINODONTIDAE

1. Jaw teeth tricuspid; pelvic fins usually present (*Cyprinodon*) _____ PUPFISH

   Jaw teeth not tricuspid; pelvic fins always absent _____ 2

2. Jaw teeth bicuspid; jaws equal in length (i.e., the lower jaw not projecting forward beyond the upper jaw); lower pharyngeal teeth conical; intestine markedly coiled (*Crenichthys*) _____ SPRINGFISH

   Jaw teeth conical; jaws unequal in length, the lower jaw projecting forward beyond the upper jaw; lower pharyngeal teeth molar-like; intestine merely S-curved (*Empetrichthys*)_____ POOLFISH

PUPFISHES

Genus *Cyprinodon* Lacépède 1803
("Carp tooth")

KEY TO SPECIES

1. Pelvic fins absent; tail fin bilobed; body dwarfed; found only in Devil's Hole, Ash Meadows (*Cyprinodon diabolis*)... DEVIL PUPFISH

    Pelvic fins usually present, occasionally lacking on one side, rarely lacking on both sides; tail fin rounded or truncate; body generally normal-sized for the genus, rarely slightly dwarfed; lower Amargosa River system (*Cyprinodon nevadensis*)... AMARGOSA PUPFISH.... 2

2. Pectoral fin rays usually 17; scales average 9.79 between dorsal and pelvic fins; scales average 15.44 around peduncle; scales average 24.75 around the body; preopercular pores average 13.36 (*Cyprinodon nevadensis pectoralis*)... LOVELL SPRING AMARGOSA PUPFISH

    Pectoral fin rays usually 16; scales average 9.25 between dorsal and pelvic fins; scales average 14.17 around peduncle; scales average 22.67 around the body; preopercular pores average 12.48 (*Cyprinodon nevadensis mionectes*)... BIG SPRING AMARGOSA PUPFISH

THE AMARGOSA PUPFISH

Cyprinodon nevadensis Eigenmann and Eigenmann
(Nevada Pupfish)

Cyprinodon nevadensis Eigenmann and Eigenmann 1889: 270
Cyprinodon macularius, Gilbert, 1893: 232
Cyprinodon macularius, Jordan and Evermann, 1896B: 674 (in part)
Cyprinodon nevadensis, Jordan, 1924: 23
Cyprinodon nevadensis, Jordan, Evermann and Clark, 1930: 181
Cyprinodon nevadensis, Wales, 1930: 62, 69–70
Cyprinodon nevadensis, Evermann and Clark, 1931: 21, 56
Cyprinodon nevadensis, Schrenkeisen, 1938: 186
Cyprinodon sp., Sumner and Sargent, 1940: 47–50
Cyprinodon nevadensis, Shapovalov, 1941: 444
Cyprinodon nevadensis, Miller, 1943A: 18–20 and various mentions / 1943B: 72–73, 77 / 1948: 22–82 / 1949A: 475
Cyprinodon nevadensis, Miller and Alcorn, 1946: 190–191
Cyprinodon nevadensis, Hubbs and Miller, 1948B: 83
Cyprinodon nevadensis, Shapovalov and Dill, 1950: 387
Cyprinodon nevadensis, La Rivers, 1952: 90 / 1953B: 92
Cyprinodon nevadensis, La Rivers and Trelease, 1952: 118–119
Cyprinodon, La Rivers, 1956: 157
Cyprinodon nevadensis, Eddy, 1957: 164
Cyprinodon nevadensis, Moore, 1957: 156
Cyprinodon nevadensis, Shapovalov, Dill and Cordone, 1959: 172

ORIGINAL DESCRIPTION

"Types No. 580, 3 specimens, .04–.044 m. W. H. Shafer.

"Head $3\frac{1}{4}$–$3\frac{2}{7}$ ($3\frac{5}{6}$–4 in total); depth $2\frac{3}{4}$–$2\frac{2}{5}$ ($2\frac{7}{8}$–$3\frac{1}{3}$); D. 10–11; A. 11; Lat. 1. 26; tr. 9–10.

"Form of *C. variegatus*, the back in front of the dorsal conspicuously broader and less trenchant.

"Eye 3–$3\frac{3}{5}$ in the head, $1\frac{2}{5}$–$1\frac{1}{2}$ in interorbital.

"Exposed portion of the humeral scale not larger than some of the other scales. Intestinal canal little more than twice the entire length. *Origin of dorsal equidistant from base of caudal and middle of orbit or much nearer base of caudal.* Base of dorsal little if any longer than snout and orbit. Highest dorsal ray $1\frac{2}{3}$–2 in head. Caudal rather broadly rounded $1\frac{4}{5}$ in head. Highest anal ray $1\frac{3}{5}$–2 in the head. Ventral very short, inserted in front of the dorsal.

"Color in spirits, dark grey with an indistinct darker lateral band, and indistinct lighter vertical streaks in the female. Dorsal in the male almost uniform black; in the female lighter with a blackish spot on its posterior rays. Caudal margined with white, within which is a broader blackish crescentiform band, most conspicuous in the male. Pectorals, ventrals and posterior half of anal, blackish in the male, light gray in the female" (Eigenmann and Eigenmann 1889: 270).

DIAGNOSIS

Head 2.7 to 3.5, body depth 2 to 3, times in body length, respectively; head rather blunt, steeply sloping in front; mouth small, oblique; maxillary reaching only about halfway to front edge of orbit; teeth tricuspid, resembling incisors, only a single row in each jaw, the central cusp about twice as wide at the base as either of the lateral cusps are at their base (in contrast to *C. macularius* in which the center cusp is approximately three times as wide at the base as are either of the lateral cusps); scales 23–28 in the lateral line, usually 24–27—7–11 from dorsal fin to anal fin, usually 8–10—15–24 before the dorsal, usually 16–21—18–32 around the body, usually 19–30—circuli lacking spinelike projections, the interspaces between them densely reticulate (in contrast to *C. macularius* in which the projections are present and the interspaces lack noticeable reticulations)—prehumeral length (distance between snout and hind margin of humeral scale) averaging 1.8 divided into the predorsal length (1.6 to 2.0) (in contrast to *C. macularius* which averages 1.5 (1.3 to 1.7); fins—dorsal usually closer to base of caudal than to snout (in *C. macularius* it is about equidistant), 8–12 rayed, usually 9–11—anal 9–11 rayed, rarely 8—pectorals 14–18 rayed, rarely down to 11—pelvics often lacking on one or both sides—when present, they are usually 4–7 rayed, but may vary from 1–9 rays, usually averaging 6/6 and are small (in contrast to *C. macularius* in which the average is 7/7, size is large, and they are never lacking)—caudal small, short, blunt and slightly convex on hind surface; 14–22 rayed, usually 16–20, the 14 and 22 counts seemingly very rare; caudal peduncle very stout, least depth about 5 times into body length; intestine much exceeding body length, considerably convoluted, a little over two times as long as body; color—rather nondescript dark above, light below except in breeding males, which are sharp, deep blue overall (lacking the yellow or orange tints on the caudal fin and peduncle characteristic of breeding males of *C. macularius*)—some mottling or lateral barring along sides may or may not be present; a tiny species, rarely attaining 2 inches in length.

TYPE LOCALITY

"The new species of Cyprinodon described below was collected in a hot spring locally known as 'Saratoga Spring,' in the south arm of Death Valley, Inyo County, California.

"The types are in the collection of the California Academy of Sciences" (Eigenmann and Eigenmann 1889: 270).

The locality is now in San Bernardino County.

RANGE

The lower portion of the Amargosa drainage, from Ash Meadows, Nye County, Nevada to Saratoga Springs, Death Valley, California. Miller and Alcorn (1946: 190–191) list one instance of the species

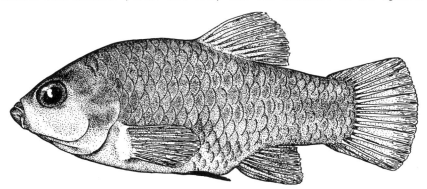

FIG. 224. Amargosa Pupfish, *Cyprinodon nevadensis*.
Drawn by Silvio Santina.

having been transferred to an adjacent basin: "Between December 31, 1940, and February 18, 1941, 20 specimens of a *Cyprinodon* collected near Las Vegas were sent for identification to the University of Michigan Museum of Zoology. At first they were believed to represent an undescribed form, but on thorough study their identity with *nevadensis* became certain. This species is confined to the Amargosa River basin of eastern California and southwestern Nevada (Miller, 1943). The specimens were all taken from a 'spring' on the Las Vegas golf course in Clark County. An intensive search for additional specimens was made at this locality in October, 1942, but none could be found. The so-called spring is a seepage area fed in part by artesian overflow and in part by overflow from ponds at the Las Vegas Station of the U. S. Fish and Wildlife Service, just to the west. Presumably, this small species was introduced from the Amargosa basin, California, as a forage fish for bass and sunfish propagated at the station."

TAXONOMY

Shortly after its initial description, *C. nevadensis* was submerged in the synonymy of *C. macularius* Baird and Girard 1853, and although mentioned by several interim writers as probably not distinct from *C. macularius,* it was not until many years later that it was placed on a substantial footing and suitable criteria established for distinguishing between the two species (Miller 1943A).

Etymology—the name *"nevadensis"* indicates a desire on the part of the describers to honor the region east of California, but the reasons for such a desire apparently are lost in obscurity, for the Eigenmanns were well aware that the type locality was in California.

LIFE HISTORY

As common as they are in their localities, very little specifically is known of their life cycles. In Nevada they can be considered typical of warm water habitats—isolated warm springs of high mineral content in alkaline desert areas—and in such seem to breed the year around. The anatomist would deduce an herbivorous feeding pattern from the long, twisted intestine (which the author has verified from stomach examinations), while the ecologist would also assume the same thing after a look at their environments which, in many cases, contain little but vegetation upon which to feed.

ECONOMICS

Undoubtedly of no importance in the usual sense; in many of their situations, the Amargosa Pupfish represent the peak of the food chain as adults, whereas the young certainly are used for food by the larger aquatic insects with which many are associated.

DISTINCTIONS BETWEEN SUBSPECIES

Cyprinodon nevadensis has been divided into six subspecies, four occurring in the California portion of the Amargosa drainage, two in Nevada's Ash Meadows. Since these distinctions are quite finely drawn, being based upon small shifts of the mean in many characteristics, it seems best to consider them in tabular form (Miller 1948).

MEASUREMENTS

	Standard length (mm.)	Body depth	Head length	Depth caudal peduncle	Opercle length	Preorbital margin to preorbital angle
			(these figures in thousandths of standard length)			
mionectes	26–40	364–460	302–359	165–221	90–147	153–185
pectoralis	27–37	388–473	324–349	160–222	106–121	166–199

SCALE COUNTS

	Lateral scales	Dorsal-anal	Pre-dorsal	Peduncle circumference	Body circumference	Pre-opercular pores
mionectes	23–27	7–11	15–21	11–18	18–30	8–16
pectoralis	24–27	8–10	16–21	13–16	21–28	11–16

FIN RAY COUNTS (means)

	Dorsal	Anal	Pectoral	Caudal
mionectes	$9.83 \pm .02$	$9.90 \pm .02$	$15.65 \pm .02$	$17.82 \pm .04$
pectoralis	$9.51 \pm .05$	$10.03 \pm .04$	$17.10 \pm .04$	$17.72 \pm .08$

BIG SPRING AMARGOSA PUPFISH

Cyprinodon nevadensis mionectes Miller

Cyprinodon nevadensis mionectes Miller 1948: 44–58
Cyprinodon nevadensis mionectes, La Rivers, 1952: 90
Cyprinodon nevadensis mionectes, La Rivers and Trelease, 1952: 118

ORIGINAL DESCRIPTION AND DIAGNOSIS

The original description is largely in the nature of diagnostic and comparative statements and figures:

"Diagnosis.—*C. n. mionectes* is characterized by having scale and fin-ray counts lower than average for the species; a reduced size; a

FIG. 225. *Cyprinodon nevadensis mionectes*. *Top*, adult male. *Bottom*, adult female. From Forest Spring Ash Meadows, Nye County, Nevada. Photographs by Clarence Flaten for Miller 1948. Courtesy Dr. Robert R. Miller.

short, deep, and slab-sided body with a greatly arched and rather compressed predorsal profile; and a very long head and opercle, and a greater than usual distance between the preorbital margin and the preopercular angles . . . Among the low counts, the count of the scales around the caudal peduncle particularly stands out. The dorsal is posterior, its origin varying from directly over the origin of the pelvics to well behind it.

"Comparisons.—*C. n. mionectes* differs from all other subspecies of *Cyprinodon nevadensis* in the characteristics outlined above. Since some of these features are difficult to measure and no single count or measurement will separate all individuals of this subspecies from any other subspecies, it is advisable to employ a character index for distinguishing *mionectes* from other forms. By adding together the head length, opercle length, and the distance between the preorbital margin to the preopercular angle (all expressed in hundredths of the standard length), and subtracting the number of scales around the body, indices result which enable one to separate *mionectes* readily from nearly all of the other subspecies of *C. nevadensis*.

"Variation.—The many Ash Meadows populations of *Cyprinodon nevadensis* are variable in meristic characters, but uniform in head and other measurements and in the characteristic chubby body form. Some of them constitute rather distinct races, and no 2 are identical. It seems best at this time, however, to refer all but 1 of these populations to *mionectes*, pending experimental analysis of some of the particular traits to determine if they have a genetic basis" (Miller 1948: 45, 47).

TYPE LOCALITY

"Types.—The holotype, a mature male 34 mm. long, was seined on September 27, 1942, by Robert R., Ralph G., and Francis H. Miller in Big Spring, north-central Ash Meadows, Nye County, Nevada; U.M.M.Z. No. 141775. Four hundred and twenty-two paratypes, U.M.M.Z. No. 140460, were collected with the holotype. An additional 17 paratypes were taken by R. R. and R. G. Miller on June 5, 1937, U.M.M.Z. No. 132900" (Miller 1948: 44–45).

RANGE

Confined to the warm springs of Ash Meadows, in the southeastern section of the Amargosa Desert, Nye County, Nevada, occurring in the following known springs: Big Spring, Deep Spring, Eagle Spring, Point-of-Rocks (Kings) Spring, Tubb Spring, Bradford Spring, Hidden Spring and Fairbanks Springs and their tributaries. During the writer's last visit to Fairbanks Springs (April 1960), the once teeming population of pupfish in the springs by the house was gone. Only great numbers of the introduced crawfish *Procambarus clarkii* (Girard) were evident, and bullfrogs made the welkin ring in nearby ponds and marshes—species which are both detrimental to these tiny fishes.

TAXONOMY

Etymology—"The name *mionectes* is derived from the Greek meaning 'one who has less,' in reference to the reduced size and lowered number of fin rays and scales" (Miller 1948: 48).

LOVELL SPRING AMARGOSA PUPFISH
Cyprinodon nevadensis pectoralis Miller

Cyprinodon nevadensis pectoralis Miller 1948: 58, 69–70
Cyprinodon nevadensis pectoralis, La Rivers, 1952: 90
Cyprinodon nevadensis pectoralis, La Rivers and Trelease, 1952: 118

ORIGINAL DESCRIPTION AND DIAGNOSIS

"Diagnosis and comparisons.—*C. n. pectoralis* is characterized principally by a higher number (usually 17) of pectoral rays. Otherwise, it is very similar to *C. n. mionectes*. It differs further from that subspecies on the average in having more scales between dorsal and pelvic fins, around peduncle, and around body, and more preopercular pores. In measurements, it has a deeper, broader body, a longer and deeper head, a greater distance between preorbital margin and preopercular angle, and a wider mouth. The pectoral and pelvic fins of the females appear to be shorter than they are in females from other populations in Ash Meadows. This may result from the fact that these fish were not in peak spawning condition, for longer pectoral and pelvic fins in both males and females characterize the specimens taken by Sumner and Sargent in May, 1939, when the fish were near the height of spawning. The shorter basal length of the dorsal and anal fins of the females may be due in part to the same situation or may represent a true reduction in these measurements.

"On the basis of the number of pectoral rays alone, 86 per cent of the individuals of this subspecies are identifiable (using a line of separation between 16 and 17 rays). Of all populations of *mionectes* combined 93 per cent are identifiable on this character, and the percentage does not fall below 85 for any single population of that subspecies. In 2 populations of *mionectes*, Bradford's Spring and Fairbanks Spring, 100 per cent of the specimens examined can be separated from *pectoralis* by this 1 character.

"The wider head of this subspecies accounts in part for the wider mouth. It will be recalled, however, that the races inhabiting Eagle Spring and its outlet also have a wide mouth.

"When compared with the nearest population of *mionectes* in Big Spring, *pectoralis* agrees rather closely in measurements, except that the body is deeper, and the head and mouth are wider than in the Big Spring sample. Of the individuals from Big Spring 95 per cent can be identified on the number of pectoral rays. The difference in temperature between these 2 springs is only about 2 degrees Centigrade" (Miller 1948: 58, 69).

TYPE LOCALITY

"Types.—The holotype, a mature male 36 mm. in standard length, was collected by Robert R. and Ralph G. Miller on January 9, 1939, in the concrete pool just below Lovell's Spring, Ash Meadows, Nye County, Nevada; U.M.M.Z. No. 141779. Two hundred and fifty-five paratypes were collected with the holotype; U.M.M.Z. No. 132908" (Miller 1948: 58).

RANGE

Restricted to the type locality.

TAXONOMY

Etymology—"This subspecies is named *pectoralis* in reference to the increased number of pectoral fin rays" (Miller 1948: 70).

THE DEVIL PUPFISH
Cyprinodon diabolis Wales
(Devil's Hole Pupfish)

Cyprinodon nevadensis, Gilbert, 1893: 232
Cyprinodon macularius, ibid: 233
Cyprinodon diabolis Wales 1930: 61–70
Cyprinodon diabolis, Sumner and Sargent, 1940: 47
Cyprinodon diabolis, Miller, 1943A: 2, 19 / 1943B: 70, 72–73, 77 / 1948: 83–86 / 1949A: 451, 475
Cyprinodon diabolis, Hubbs and Miller, 1948B: 83–84
Cyprinodon diabolis, Chandler, 1949: 136
Cyprinodon diabolis, La Rivers, 1950B: 105–106 / 1952: 90
Cyprinodon diabolis, La Rivers and Trelease, 1952: 119
Cyprinodon diabolis, Böhlke, 1953: 53
Cyprinodon diabolis, Eddy, 1957: 165
Cyprinodon diabolis, Moore, 1957: 156
Cyprinodon diabolis, Miller, 1961: 386–387

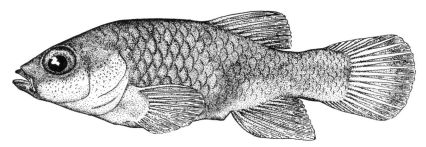

FIG. 226. Devil Pupfish, *Cyprinodon diabolis*. Drawn by Silvio Santina.

ORIGINAL DESCRIPTION AND DIAGNOSIS

"Standard length, 21 mm.; total length, 25 mm.; head length, 7 mm., 3 in body; greatest depth, 7 mm., 3 in body; depth of caudal peduncle, 3.5 mm., 2 in head; tip of snout to upper insertion of pectoral 7 mm., 3 in body; tip of snout to insertion of dorsal, 13 mm., 1.64 in body; tip of snout to insertion of anal, 14 mm., 1.5 in body; diameter of eye, 2 mm., 3.5 in head.

"Lateral median row of scales, 27; oblique row of scales from insertion of dorsal, 9. Scales ctenoid. The nuclei of the scales are left uncovered by the preceding ones.

"Rows of pores along entire border of preopercle, 7 in number. A pore on posterior median border of eye. Five pores in a row from superior posterior border of eye to snout. Another pore inside this row on interorbital space opposite posterior half of eye. Four pores in a Y formation anterior to eye.

"Pectoral rays 17; greatest length of pectoral, 5 mm., 4.2 in body. Dorsal rays 12; height of fin, 4 mm., 1.43 in head; first two dorsal rays simple, the first 0.75 of the second which is 4 mm. long; other rays branched. Caudal rays 28; the fin convex.

"Body subovate to posterior ends of dorsal and anal bases. The caudal peduncle is almost rectangular. Top of head flat. Mouth low, on level with ventral edge of caudal peduncle. Premaxillary entirely below middle of eye. Teeth on jaws in single series, 16 teeth in upper and 16 in lower. Width of tooth crown slightly less than one-half length of tooth, without root. The three long cusps of approximately equal length, the middle cusp but slightly wider than others. Teeth of upper approximately the same size as those on the lower jaw. Gill-rakers 16, longest about ⅓ mm. Upper pharyngeal bone with about 5 rows of teeth. The center row with largest teeth, about 10 in number. Each tooth much compressed and in shape resembling the toe of a cat, the point of the tooth being the cat's claw. Dentigerous surface of upper pharyngeal nearly round. Surface of lower pharyngeal triangular, the posterior and inner border with a marginal row of about 18 teeth. Those teeth on the posterior border the largest, the other teeth placed irregularly.

"The following color notes were taken from newly caught live specimens. Male: Sides of a metallic bluish iridescence which changes to metallic greenish or golden in different light. Back dark brownish. All fins edged with blackish. Dorsal with a golden iridescence. Anal whitish towards base. Opercles especially iridescent with a violet sheen on upper posterior portion. Iris iridescent blue. Female: General color more yellowish than male. Back yellowish brown. Caudal and pectoral yellowish on basal half. Dorsal edged with a black band. An indication of a lateral, dark bar through middle of caudal peduncle. Opercles metallic green. Eyes of a faint metallic blue. Young—color much like that of female but with faint vertical bars on sides. Indication of a lateral band" (Wales 1930: 68–69).

TYPE LOCALITY

"Holotype: Male from Devil's Hole, Ash Meadows, Nevada; No. 23928, Stanford University Fish Collection" (Wales 1930: 68). The locality is in Nye County. Eight paratypes were listed as being in the Stanford collection also, all collected by Joseph H. Wales and George S. Myers.

RANGE

Restricted to the type locality, which has recently (1952) been annexed to Death Valley National Monument.

TAXONOMY

Although the dwarfed fish of Devil's Hole had been known for many years before Wales' formal characterization of *C. diabolis*, it had always been considered either *C. macularius* or *C. nevadensis* (Gilbert 1893). As has been pointed out, the essentially dominant characteristics of *C. diabolis* are those attendant upon retention of juvenile traits and expressed, among other ways, in dwarfing (Miller 1948: 83). Apparently the species represents *C. nevadensis* stock which became isolated in a small, very warm pool and, through Sewell Wright effects, developed different mean particulars.

The pupfish entered Devil's Hole either at a time when the Amargosa Desert was the site of a Pleistocene lake, or possibly when the Hole was actively overflowing into Ash Meadows. Wales (1930: 61) mentioned local testimony that the water level of Devil's Hole had slowly dropped during the previous 20 years, indicating a considerable fluctuation over long periods of human time.

After his exhaustive study of the *C. nevadensis* populations of Ash Meadows, Miller (1948) concluded that the Devil Pupfish most closely resembled his *C. n. pectoralis* race of the Amargosa Pupfish, which latter was also the geographical component of *C. nevadensis* closest to Devil's Hole. The dwarfing which is so conspicuous in *C. diabolis* is also detectable in varying degrees among *C. nevadensis* populations in the Meadows.[1]

Etymology—named for the type locality.

LIFE HISTORY

As would be suspected, breeding is continuous in the warm environment, for pupfish of all sizes are seen during the year. The associated

---

[1] Whether or not *C. diabolis* represents a period of time greater than Late Pleistocene is a moot question. On the basis of geologic changes (tilting) and geomorphologic alterations (topographic dissection) in the lake beds of the area, Miller (1948: 85–86) felt that the period involved might be as far removed as Early Pleistocene. However, this could have occurred relatively recently, and other evidence is not necessarily corroborative.

Until the collecting season of 1951, the unique aquatic "Riffle" beetle *Stenelmis calida* Chandler 1949 (Dryopidae: Elminae) was known only from Devil's Hole, the type locality. At the time of its description it was the only *Stenelmis* known in southern Nevada, and denoted a segment of the genus singularly detached from the main stock to the east. This beetle, an algal feeder like the Devil Pupfish, with an abundance of food in the Hole; a warm, stable environment small enough to restrict its population to a point where the somewhat controversial Sewell Wright aftermaths would be operative; and an incapacity to voluntarily distribute itself (the beetle is flightless), it would, as a marked differentiate, indicate considerable antiquity for Devil's Hole, in line with expectations based on the fish data.

However, *Stenelmis calida* is now known to be a common part of the fauna of hardscrabble creeks wherever they have been investigated in Ash Meadows (such streams are not common there), occurring at several places in addition to Devil's Hole; and the subspecies *S. c. moapa* La Rivers 1949 extends the species' range east and north into the southern part of the White River drainage (type locality = Warm Springs, Clark County, Nevada). As a consequence, the species has lost its significance as a time indicator where Devil's Hole is concerned. We do not have to assume that *Stenelmis* and *Cyprinodon* came into the Amargosa system at even approximately the same time, but it can reasonably be held that both were present during the occupancy of the area by the latest Pleistocene lake, and have been subjected to the same general isolating influences since the last widespread desiccation of the basin.

The fact that *Cyprinodon* stock isolated in Devil's Hole has diverged sufficiently from Ash Meadows populations to be considered a species, while *Stenelmis* within the Meadows, including the Hole, constitutes a homogeneous group not taxonomically divisible, cannot safely be used as either (1) a criterion for advocating an older age for Devil's Hole or (2) an indication that *Cyprinodon* has been longer in the basin, since the metabolic rate can be reasonably held to be much greater in the fish than the beetle (the latter is a lethargic, slow-moving type even in warm waters) and the fish population (in Devil's Hole) is considerably smaller than that of the beetle, which would accelerate the Sewell Wright changes; both presumably breed during the entire year (the beetle most certainly does). The comparative slowness of *Stenelmis* to respond to isolating influences can also, of course, be presumed to favor the antiquity of Devil's Hole.

life is qualitatively very sparse, but quantitatively very important and includes, in addition to what has been mentioned; an amphipod crustacean, a copepod crustacean (*Cyclops*), a pelecypod mollusk, a turbellarian flatworm, diatoms and other algae. The few stomachs the author has examined contained principally algae with such remains of flatworms and beetle larvae as might be expected to be accidentally ingested while cropping algae from the rocks.

FIG. 227. Devil's Hole, Ash Meadows, Nye County. Looking north and down. To descend to the water level, permission must be obtained from the Superintendent, Death Valley National Monument, into which Devil's Hole has been incorporated for protection of the unique little fish which it contains. The structures shown were not present when the author first saw the Hole in 1947, and are evidently part of a device to measure water fluctuation or movement. Photograph by the author in 1960.

Fishes can occasionally be seen in the depths of the pool or well, in 25 or 30 feet of water, but most of them spend their time feeding along the shallow south shelf, where the water grades from a few inches to three feet in depth—at the north edge of this shelf the bottom abruptly drops to a slanting talus slope some 30 feet beneath the surface level, and disappears northerly into a steep crevice or cavern which is unpenetrated by light. When unmolested, a good part of the fish population will accumulate on this shelf to feed painstakingly over the surfaces of the algal covered shelf and the many loose rocks of varying sizes scattered about on the shelf, like many suckers or chub. When disturbed sufficiently, they disappear over the north edge of the shelf into deeper water—a relatively short wait on the part of the observer will see them appear on the shelf, swimming up the small narrow and sloping cañon-like crevices and keeping close to the rock surfaces. Very probably spawning takes place on the shelf.

The limnologic characteristics of this limestone water have only briefly been investigated by the present writer, showing the following:

Temperature .... 33.4° C. (92° F.)
pH 7.3
Dissolved oxygen .... 1.8–3.3 parts per million (ppm)
Carbonates .... lacking
Bicarbonates .... 256 ppm
Carbon dioxide .... 18 ppm
Chlorides and sulfates .... present
Taste .... good

Miller (1948: 85) gives figures-of-variation of 32.8° C. to 33.9° C. representing seasonal readings over a 15-year period, which indicates the thermal stability of the environment in human time. Fluctuations in the amounts of dissolved oxygen in the water are to be expected for two reasons: (1) there is obviously considerable circulation of the water of Devil's Hole from subterranean inlets and outlets as evidenced by the clarity and purity of the water (both Wales 1930, and Miller 1948 have commented on this), and (2) the large quantity of algae growing on the rocks tends to make some local differences, particularly over the shallow south shelf where the fish spend most of their time. The small amounts of oxygen present are to be expected in warm water, which holds less oxygen in solution than colder waters.

The author has had considerable success keeping *C. diabolis* in aquaria, when maintaining temperatures of 92° F. and keeping heavy algal growths going.

ECONOMICS

Since *Cyprinodon diabolis* is the largest animal living in Devil's Hole, it occupies the top of the limited food chain, and is of no practical importance from the standpoint of human economics.

SPRINGFISHES

Genus *Crenichthys* Hubbs 1932

("Spring fish")

KEY TO SPECIES

Lateral dark spots in two series (White River system)
  (*Crenichthys baileyi*) .... WHITE RIVER SPRINGFISH
Lateral dark spots in a single series (Railroad Valley)
  (*Crenichthys nevadae*) ...... RAILROAD VALLEY SPRINGFISH

WHITE RIVER SPRINGFISH

Crenichthys baileyi (Gilbert)

(Bailey Minnow, White River Killifish)

Cyprinodon macularius baileyi Gilbert 1893: 233
Cyprinodon baileyi, Jordan and Evermann, 1896B: 675
Cyprinodon baileyi, Jordan, Evermann and Clark, 1930: 182
Cyprinodon baileyi, Schrenkeisen, 1938: 186
Crenichthys baileyi, Sumner and Sargent, 1940: 46, 50
Crenichthys baileyi, Hubbs, 1941: 68
Crenichthys baileyi, Hubbs and Miller, 1941: 1-3 / 1948A: 8, 10
Cyprinodon macularius baileyi, ibid, 1941: 1-3
Crenichthys baileyi, Tanner, 1942: 31
Crenichthys baileyi, Sumner and Lanham, 1942: 319, 326
Crenichthys baileyi, Miller, 1943A: 9, 19
Cyprinodon macularius baileyi, ibid
Crenichthys baileyi, Kopec, 1949, 56-61
Crenichthys baileyi, La Rivers, 1949: 222 / 1950C: 372 / 1952: 91
Cyprinodon macularius baileyi, Tanner, 1950: 32, 35
Crenichthys baileyi, La Rivers and Trelease, 1952: 118
Cyprinodon macularius baileyi, Böhlke, 1953: 53
Crenichthys baileyi, Eddy, 1957: 162
Crenichthys baileyi, Moore, 1957: 151

ORIGINAL DESCRIPTION

"Eleven immature specimens from Pahranagat Valley, Nevada, show no trace of ventral fins. They are olivaceous above, bright silvery on the lower half of sides and below, and have two lengthwise series of coarse black spots, one along middle line of body, the other on a level with the lower edge of caudal peduncle. The anal fin is larger than in typical *macularius*, the eleven specimens having each 13 rays instead of 10 or 11, as constantly in the latter. The material is insufficient to fully decide the status of this form. Except in the characters noted it agrees in proportions and formulae with *macularius*" (Gilbert 1893: 233).

DIAGNOSIS

Head about 3.2, body depth about 3.2, times in body length, respectively; head large, blunt, flat in profile in young specimens, concave in older ones, steeply sloping, very broad in older individuals when viewed from above; mouth, in lateral view, small and straight; teeth bicuspid; scales about 27 in lateral line, 18 before the dorsal, 10 from dorsal to anal; fins—dorsal and anal set far back, less than their own lengths removed from caudal base and about 11 and 14-rayed, respectively— dorsal about exactly over anal—pectorals about 16-rayed—pelvics lacking—caudal about 28-rayed, straight in terminal outline; color pattern —the two lateral rows of dark spots against a lighter background are unusually distinctive, although some individuals show but one line or none; size small, reaching a maximum of about 2½ inches.

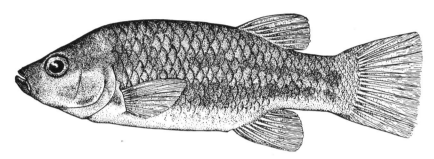

FIG. 228. White River Springfish, *Crenichthys baileyi*.
Drawn by Silvio Santina.

TYPE LOCALITY

"Pahranagat Valley, Nevada, collected by C. Hart Merriam and Vernon Bailey, May 25, 1891" (Gilbert 1893: 233). The locality is in Lincoln County, with Alamo the principal town in the valley.

RANGE

Found in suitable warm springs and their effluents at intervals along the now disrupted Pluvial White River drainage system of extreme eastern and southern Nevada, from the vicinity of Preston and Lund in the north (White Pine County), south to Warm Springs, Clark County (headwaters of the Moapa or Muddy River), a lineal distance of some 150 miles.

TAXONOMY

The species has had some nomenclatorial instability of minor importance in the 60-odd years since its initial description, but has accumulated no synonyms. Gilbert originally assigned it to subspecific status under the then widespread *Cyprinodon macularius* Baird and Girard 1853; Jordan and Evermann (1896B) considered it a species, as is its current status. In 1940, Sumner and Sargent, on the advice of Hubbs, associated *baileyi* with its present genus, *Crenichthys*. There has been only one interim lapse in the terminology, that of Tanner's return to *Cyprinodon macularius baileyi* in 1950.

As Hubbs noted when describing *Crenichthys* in 1932, the genus is closely related to Gilbert's *Empetrichthys* of the Amargosa system. Both of these fundulines agree in their loss of the pelvic fins, the extreme posterior position of both dorsal and anal fins, the lack of vomerine teeth and pseudobranchiae, and in oviduct characteristics. They differ in intestinal lengths (the intestine of the herbivorous *Crenichthys* being long and coiled, while that of the omnivorous *Empetrichthys* is much shorter). Other minor differences are detectable in the jaw and tooth structure.

Etymology—Named in honor of the famous U. S. Bureau of Biological Survey field naturalist, Vernon L. Bailey, who was one of the collectors of the type series.

FIG. 229. Adams-McGill Reservoir (Sunnyside Reservoir) in the Sunnyside Wildlife Management Area, looking east to the Egan Range. Cave Valley and Ely Range lie just beyond. The reservoir is in White River Valley and is an impoundment of White River which is fed by numerous springs as well as by the small river. This photograph, taken by Tom Trelease on March 9, 1961, also shows the parallel arrangement of north-south striking valleys and ranges. Note the marked stratification of the limestone mountains.

LIFE HISTORY

More varied and detailed work has been done with this species than any other of our cyprinodonts. Kopec (1949) has described the breeding activities and immature stages, while Sumner et al have published data on certain phases of their metabolism (1940 and 1942).

In general, *C. baileyi* is common wherever found. As Kopec observed them, the male exhibited characteristic courting behavior, and, after sufficient excitation to induce the female to extrude eggs, fertilized them singly as they appeared. Only one egg was laid and fertilized at a time. From 10 to 17 eggs seemed to constitute a spawning. Each egg measured 1.9 mm. in diameter, and the incubation period was from 5 to 7 days. The species apparently lacks cannibalistic tendencies, for he reared the young safely in the presence of adults. His paper contained detailed descriptions of the *prolarva* (newly hatched), *postlarva*

(87 hours old), *juvenile* (15 days old) and *young* (17.7 mm. long), the first two stages being illustrated. Infusoria and *Daphnia* were used as food for these immatures, and Kopec's conclusions were that the species was essentially carnivorous. His material came from Warm Springs, Clark County.

Sumner and Sargent (1940) made field studies on *Crenichthys baileyi* populations in the vicinity of Preston and Lund, White Pine County, to determine the relation of temperature to respiratory metabolism. Their experiments consisted of immersing specimens in thousandth-molar potassium cyanide (KCN) solutions and noting survival times at different temperatures.

FIG. 230. Crystal Springs, Pahranagat Valley, Lincoln County.
Photo by the author in July 1937.

"At different temperatures, the reciprocals of the times of survival of fishes in thousandth-molar KCN solution are in very nearly the same ratio as the measured rates of oxygen consumption at the same temperatures . . ." Briefly, they found that populations of *C. baileyi* living in a warmer spring (35.5° C. to 37° C. = 96° F. to 98.5° F.) had less than a fourth the survival time of the same species in a cooler spring (21° C. = 70° F.), each series being tested at its own temperatures of water, the indication being that fish from the warmer spring had metabolic rates some four times those of fish from cooler water.

They also found that specimens could be transferred from warm to cooler water and returned to the warm water without ill-effects; however, when the reverse transferal was attempted, that of placing individuals native to cooler water in warm water, they rapidly succumbed.

"The death of the cold-spring individuals in warm water seems to have been due neither to the greatly reduced oxygen content of the latter nor to the heightened oxygen requirements of the fishes themselves. This is rendered probable by two circumstances: (1) the oxygen content, at the points where the experiments were actually performed, was nearly or quite as great in Mormon Spring [warm water] as in Preston Spring [cooler water]; (2) fishes of the warm-spring population swim freely into the immediate vicinity of the point of outflow, and thus into water having an oxygen content not more than a sixth that of the stream a short distance below. Many fishes are known to be able to adjust themselves to widely different oxygen concentrations in their medium" (p. 53).

In 1942, Sumner and Lanham conducted further field tests on *C. baileyi* in the neighborhood of Preston. In these experiments, the metabolic constants were based upon titration for oxygen content of water, before and after occupancy by test specimens of *Crenichthys*.

"At the close of each test of oxygen consumption, the fishes involved were dried upon paper towels and weighed. The aggregate weight of each lot, together with the estimated original oxygen titre of the water, the oxygen titre at the close of the test, the volume of water in the tube (allowing for volume occupied by the fishes), and the exact duration of the test, are the data upon which the metabolic rates (ml./gm./hr.) are based" (p. 316).

Their conclusions were that the populations in warmer springs had a much higher demand for oxygen than those in cooler springs (the same two springs were used as in the 1940 studies), those of the former consuming about twice as much oxygen in a given period as the latter. Transferring warm spring fishes to the cooler spring resulted in a drop of oxygen demand to the approximate level of the native cool spring population in about 24 hours. Their data also indicated that the oxygen consumption of smaller specimens was much more dependent upon temperature changes than was that of the larger individuals.

ECONOMICS

Although the White River Springfish occurs with other species of fishes,[1] most of these are no larger than it is, or are not predaceous.[2] The White River Colorado Gila (*Gila robusta jordani*) of Pahranagat Valley, and the Colorado Gila in the upper Moapa River are the only native carnivores of any ichthyological importance as possible predators of the springfish, and they are not common. The springfish has no value from the standpoint of human economics.

---

[1]White River Speckle Dace (*Rhinichthys osculus velifer*), Moapa Dace (*Moapa coriacea*), Spinedace (*Lepidomeda* species), and Mosquitofish (*Gambusia affinis*).

[2]White River Mountainsucker (*Pantosteus intermedius*).

RAILROAD VALLEY SPRINGFISH

Crenichthys nevadae Hubbs

(Railroad Valley Killifish)

Crenichthys nevadae Hubbs 1932: 3–5 / 1941: 66–67
Crenichthys nevadae, Brues, 1932: 280–281
Crenichthys nevadae, Sumner and Sargent, 1940: 46
Crenichthys nevadae, Hubbs and Miller, 1941: 1–2 / 1948B: 90–91, 93
Crenichthys nevadae, Kopec, 1949: 56
Crenichthys nevadae, La Rivers, 1952: 91
Crenichthys nevadae, La Rivers and Trelease, 1952: 118
Crenichthys nevadae, Eddy, 1957: 162
Crenichthys nevadae, Moore, 1957: 151
Crenichthys nevadae, Frantz, 1958: 7
Crenichthys nevadae, Miller, 1958: 206

ORIGINAL DESCRIPTION AND DIAGNOSIS

"The body is massive, two-thirds as wide as deep, especially heavy and turgid forward. The greatest depth (in female) enters the standard length 3.3 times (3.2 times in paratype). The caudal peduncle is rather slender, though only one-fifth (one-third) longer than deep; the least depth enters the head 2.6 (2.7) times.

"The head is very heavy, almost as wide as it is deep below the occiput. Its upper profile carries forward the nuchal hump, but flattens

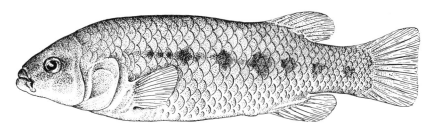

FIG. 231. Railroad Valley Springfish, *Crenichthys nevadae*.
Drawn by Silvio Santina.

to a nearly straight line above the eye, to become moderately convex again on the snout. The length of the head is contained 3.1 (3.2) times in the standard length. The bony width of the somewhat convex interorbital enters the head 2.5 (2.4) times. The preorbital is 0.6 (0.7) as wide as the orbit, which enters the head length 5.3 (4.0) times. The combined length of snout and eye is equal to (or a little longer) than the postorbital length.

"The mouth in anterior view is twice as wide as in the lateral projection. The upper lip is heavier than in *Empetrichthys merriami*, about one-fourth as long (on midline) as wide. The general structure of the lips, and also of the nostrils, is alike in the two species.

"Scales in 30 rows from the moderately constricted upper end of gill-slit to caudal base, and in 13 (11) rows between origins of dorsal and anal fins.

"Dorsal rays 12 and anal 13 (counting small anterior rays, and the last ray as doubled).

"The color pattern of the holotype is well shown on the figure. In the paratype the scale centers, in pale bars between the dark blotches, are pearly.

"The eggs are minute, and numerous, indicating an oviparous habit" (Hubbs 1932: 4–5).

TYPE LOCALITY

"Holotype.—A maturing female 44 mm. long to caudal, collected by Dr. and Mrs. C. T. Brues in an isolated warm spring at Duckwater, Nye County, Nevada, on July 21, 1930; field number 58. The location of the spring is given by Dr. Brues as being 16 miles east and 46 miles south of Ely; in Township 12 N., Range 56 E.; near the north end of Warm Spring Valley.[1] The species abounds in this spring, according to the collectors. The only other specimen taken, also a maturing female, is 29 mm. long. The holotype is retained as Cat. No. 32, 948 in the Museum of Comparative Zoology of Harvard University, while the paratype is deposited (as Cat. No. 95, 024) in the Museum of Zoology of the University of Michigan, through the kind permission of Dr. Thomas Barbour" (Hubbs 1932: 4).

FIG. 232. Teeth of the Railroad Valley Springfish (*Crenichthys nevadae*). Courtesy University of Utah Press.

RANGE

Railroad Valley drainage immediately adjacent to the more easterly White River system, and consisting of Railroad Valley and its tributary, Duckwater Valley, in Nye County, Nevada. Successfully transplanted by Tom Trelease into artificial ponds at Sodaville, southeastern Mineral County on September 4, 1947. This natural seep of warm water on the west side of U. S. Highway 95 was later enlarged by bulldozer action into a small pond. The transplant was motivated because of the possibility that blackbass would be planted in the springfish's home locality. The 6 original fish have now become many.

TAXONOMY

This species was described in 1932 as the genotype of the new genus *Crenichthys,* and its only species, since its describer did not then recognize the relationship of *Cyprinodon baileyi* to the new genus. The two species of *Crenichthys* are very close, occupying adjoining drainage systems with no interconnections.

---

[1]The spring location as cited above is wrong—it should be 16 miles *south* and 46 miles *west*.

LIFE HISTORY

Presumably much like that of its congener the White River Springfish (*C. baileyi*), but it has not been specifically worked out. Both are thermal endemics, and Brues, the collector of the type specimens, has published the following limnologic characteristics of the type locality, Duckwater Springs (1932: 280):

Temperature—31.8° C. (88° F.)
pH—8.0
Specific gravity—1.0032

The springs typically have a substantial fauna, of which Brues makes the following partial listing:

INSECTA
 Ambrysus sp. (Hemiptera: Naucoridae).
 Paracymus subcupreus Say (Coleoptera: Hydrophilidae)
 Enochrus nebulosus Say (Coleoptera: Hydrophilidae)

FIG. 233. Railroad Valley Springfish, *Crenichthys nevadae*.
Photograph by Thomas Trelease.

 Heterocerus collaris Kiesenwetter (Coleoptera: Heteroceridae)
 H. brunneus Melsheimer

MOLLUSCA
 Lymnaea sp. (Gastropoda: Lymnaeidae)
 Amnicola sp. (Gastropoda: Amnicolidae)

The temperature at Locke's Springs, farther south in Railroad Valley, where specimens have been secured, is somewhat warmer, 35.5° C. (96° F.).

Ted Frantz tells me that he has found newly hatched young at

Locke's, Railroad Valley, during the latter part of December (1956) and the following January.

ECONOMICS

Beyond its position in the food pyramid, this species seems to have no economic importance.

POOLFISHES

Genus *Empetrichthys* Gilbert 1893

("Within rock fish," referring to the hard pharyngeals)

KEY TO SPECIES

1. Lateral line scales usually 30 or less in number; downward slope of snout quite marked so that mouth lies below the plane of longitudinal section; found only in Ash Meadows (*Empetrichthys merriami*)............ ASH MEADOWS POOLFISH
   ...... Lateral line scales usually 31 or more; downward slope of snout distinctly less abrupt so that mouth lies approximately on the plane of longitudinal section; found only in Pahrump Valley (*Empetrichthys latos*)......... PAHRUMP POOLFISH.... 2
2. Distance between anal origin and caudal base shorter, averaging 346[1] in males and 328 in females; head depth less, averaging 267 in males and 281 in females; caudal peduncle length less, averaging 199 in males and 206 in females (*Empetrichthys latos latos*)............ MANSE RANCH PAHRUMP POOLFISH
   ...... Not as above.. 3
3. Length of middle caudal rays shorter, averaging 221 in males and 214 in females (*Empetrichthys latos pahrump*)................
   .... PAHRUMP RANCH PAHRUMP POOLFISH
   ...... Length of middle caudal rays longer, averaging 232 in males and 228 in females (*Empetrichthys latos concavus*).................
   .... RAYCRAFT RANCH PAHRUMP POOLFISH

ASH MEADOWS POOLFISH

Empetrichthys merriami Gilbert

(Ash Meadows Killifish)

Empetrichthys merriami Gilbert 1893: 234
Empetrichthys merriami, Jordan and Evermann, 1896B: 667
Empetrichthys merriami, Jordan, 1925: 419 (erroneously referred to "Death Valley," Calif.)
Empetrichthys merriami, Jordan, Evermann and Clark, 1930: 182
Empetrichthys merriami, Hubbs, 1932: 1, 3, 5
Empetrichthys merriami, Schrenkeisen, 1938: 186–187
Empetrichthys merriami, Shapovalov, 1941: 444

---

[1]Measured in thousandths of the standard length. Data from Miller 1948.

Empetrichthys merriami, Miller, 1948: 99-111
Empetrichthys merriami, La Rivers, 1952: 91
Empetrichthys merriami, La Rivers and Trelease, 1952: 118
Empetrichthys merriami, Böhlke, 1953: 53
Empetrichthys merriami, Eddy, 1957: 165
Empetrichthys merriami, Moore, 1957: 150
Empetrichthys merriami, Miller, 1961: 386

ORIGINAL DESCRIPTION

"In form and general appearance much resembling the mud minnow (*Umbra limi*), though somewhat deeper and more compressed.

"Head compressed, its upper surface slightly convex. Mouth very oblique, with a distinct lateral cleft, the maxillary free at tip only, reaching slightly behind front of eye. Length of gape (measured from tip of snout to end of maxillary), $3\frac{1}{8}$ in head; interorbital width, $2\frac{1}{2}$; length of snout (from front of orbit to middle of upper jaw), $3\frac{3}{4}$. Eye small, its greater oblique diameter 5 to $5\frac{1}{2}$ in head.

"Distance from front of dorsal to middle of base of tail equals one-half its distance from tip of snout. The dorsal begins slightly in

FIG. 234. Ash Meadows Poolfish, *Empetrichthys merriami*.
Drawn by Silvio Santina.

advance of anal, and ends above its posterior third. Its greatest height equals length of snout and eye.

"Caudal truncate when spread. Pectorals broadly rounded, reaching halfway to vent. D., 11 or 12 (13 in one specimen); A., 14 (from 13 to 15). Lat. l., 30 or 31, counted to base of caudal rays; 33 or 34 in all.

"In spirits the color is dark brown above, sides and below lighter, often irregularly blotched with brown and white. The belly often appears checkered, having centers of scales brown and margins white, or the reverse. Fins all dusky, the basal portions of dorsal and caudal with elongated brown spots on the interradial membranes" (Gilbert 1893: 234).

DIAGNOSIS

Head 2.8 to 3.4, body depth 2.8 to 3.7, times into body length, respectively; head chunky, heavy—viewed dorsally, somewhat narrow, constricting to snout; *downward slope of snout abrupt so that mouth lies below the plane of longitudinal section;* mouth narrower, jaws dissimilar, the lower protruding; jaw teeth conical, biserial to weakly triserial, a few of the outer ones somewhat enlarged; pharyngeal bones enlarged, the molars tuberculate; scales—29–30 in the lateral series, 13–15 from dorsal to anal fins, 24–30 before the dorsal, 29–35 around the body; fins—dorsal set back near caudal, its front edge about over the vent and front edge of anal fin, usually 11-rayed, but may vary from 9–12—anal usually 14-rayed, varying from 12–15—pectorals 16–18 rayed, rarely 15—pelvics lacking—caudal rather small, 18–20 rayed, rarely 17; caudal peduncle stout, least depth about 7 times into body length; intestine moderate in length, about one and a half times the body length, with a sigmoid curve; color quite nondescript, either characteristically darker above and lighter below, or rather irregularly mottled and spotted, some specimens even with faint traces of lateral barring; a tiny species, seldom exceeding 2½ inches in length.

TYPE LOCALITY

"Ash Meadows, Amargosa Desert, on boundary between California and Nevada" (Gilbert 1893: 234). The locality is in Nye County, Nevada.

RANGE

Known only from the type locality, where it is not common.

TAXONOMY

The Ash Meadows Poolfish has been considered a component of three different families during the 67 years it has been known. Gilbert, its describer, assigned it to what is currently considered the correct family, *Cyprinodontidae;* Jordan and Evermann (1896) substituted Poeciliidae in place of Cyprinodontidae for the group name. Eigenmann (1920) felt the animal was a member of the family Orestiidae (associating it with the South American High Andean genus *Orestias*), and Jordan, Evermann and Clark, in 1930, reconstituted it as the sole member of its own family, the Empetrichthyidae. Myers (1931) placed it in its present family and subfamily.

The closest generic relative of *Empetrichthys* is the eastern Nevada *Crenichthys.* Garman (1895) felt *Empetrichthys* to be derived from *Fundulus,* a point on which Miller (1948: 100) concurred: "The fossil evidence also supports this view for the Death Valley species, *Fundulus curryi* (Miller, 1945: 316–19, Fig. 1), appears to have been strikingly like *Empetrichthys.*" In his original diagnosis of *F. curryi,* Miller had mentioned only its similarity to *Crenichthys.*

Etymology—named in honor of C. Hart Merriam, one of the collectors, and a famous elder naturalist associated with the development of the United States Bureau of Biological Survey which later became the United States Fish and Wildlife Service. He was largely responsible for formulation of the Life Zone concept.

LIFE HISTORY

Beyond the fact that the Ash Meadows Poolfish is a thermal isolate, little is known of its life cycle. The notation ". . . frequenting the deeper holes; usually uncommon in shallow spring-fed ditches or marshy areas" of Miller (1948: 100) agrees with the author's subsequent observations. The examination of contents of a single stomach indicates an omnivorous species, to which the teeth and intestinal length lend strong support.

This poolfish is associated with the Amargosa Pupfish (*Cyprinodon nevadensis*) and the Amargosa Speckle Dace (*Rhinichthys osculus nevadensis*). The pupfish is the overwhelmingly dominant fish in Ash Meadows and it is possibly the effects of their strong competition that has made the poolfish the rarity it now is in its type locality—where this competition is lacking, as in nearby Pahrump Valley, poolfish are very abundant.

ECONOMICS
Nothing beyond its obviously very minor part in the food chain.

THE PAHRUMP POOLFISH
Empetrichthys latos Miller

Empetrichthys merriami Gilbert 1893: 234 (in part)
Empetrichthys latos Miller 1948: 101–103
Empetrichthys latos, La Rivers, 1952: 91
Empetrichthys latos, La Rivers and Trelease, 1952: 118
Empetrichthys latos, Eddy, 1957: 165
Empetrichthys latos, Moore, 1957: 150
Empetrichthys latos, Miller, 1961: 386

ORIGINAL DESCRIPTION

"Diagnosis.—A rather slender species of *Empetrichthys* with a gently sloping to convex predorsal profile, a relatively short and slender head, a comparatively broad mouth, a weak mandible, and usually 31 or 32 scales in the lateral series. The sides are marked by a narrow axial streak (Pls. X and XI), which is faint to obsolescent in *E. l. concavus*. The anal rays usually number 12 or 13.

"Comparison.—*Empetrichthys latos* differs from *E. merriami*, the only other known species of the genus, principally in mouth structure, body shape, and color pattern (Pls. X and XI). In *merriami* the head constricts abruptly in the preorbital region so that the 2 sides of the snout slope markedly toward the tip of the mandible. As a result of this constriction the mouth is narrower. In *latos* the sides of the head are almost parallel all the way to the tip of the mandible, and the mouth is consequently broader. When the mouth is forced open, as with a pair of forceps, the horizontal gape is definitely evident in *merriami*, but is almost eliminated in *latos*. The bones of the premaxillaries and mandible are much weaker and less firmly connected in *latos*.

"In *merriami* the predorsal region is broadly convex. The change in slope of the predorsal profile takes place farther forward so that the head is deeper than it is in *latos*. The body is also thicker in *merriami*. The differences in color pattern between the 2 species are well shown in Plate X. In particular the narrow axial streak of *latos* contrasts with the disrupted lateral band of *merriami*.

"The large difference in head depth and opercle length between *merriami* and *latos* can be expressed by a character index in which these measurements, expressed in thousandths of the standard length, are added together" (Miller 1948: 102).

DIAGNOSIS

Head 2.9 to 3.4, body depth 3.0 to 3.6, times into body length, respectively; head comparatively slender, viewed dorsally, rather wide, more-or-less parallel to snout, making snout and mouth relatively broad; *downward slope of snout not so abrupt, mouth lying about on plane of longitudinal section;* mouth wider, jaws dissimilar, the lower protruding; jaw teeth conical, biserial to weakly triserial, a few of the outer ones somewhat enlarged; pharyngeal bones enlarged, the molars tuberculate; scales—29-33 in the lateral series (most commonly 30-32), 12-15 from dorsal to anal fins (rarely 16), 23-31 before the dorsal (most commonly 24-28), 27-38 around the body (usually 29-36); fins—dorsal set back near caudal, its front edge about over the

FIG. 235. Pahrump Poolfish, *Empetrichthys latos*. Drawn by Silvio Santina.

vent and front edge of anal fin, usually 10–12 rayed, occasionally 13—anal commonly 11–13 rayed, rarely 10 or 14—pectorals generally 16–18 rayed, varying from 15 to 20—pelvics lacking—caudal rather small, 16–23 rayed, usually 18–20; caudal peduncle stout, least depth about 7 times in body length; intestine and color about as in *E. merriami*, size a trifle smaller, 2 inches apparently being about the maximum length.

TYPE LOCALITY

"The holotype of the typical form, *E. latos latos*, is an adult female, 43 mm. long, seined by Robert R. and Frances H. Miller on October 5, 1942, from the main spring pool on Manse Ranch, Pahrump Valley, Nye County, Nevada (U. S. Geological Survey, Las Vegas Quadrangle); U.M.M.Z. No. 141855. The type specimens of each subspecies are designated in the subspecific descriptions" (Miller 1948: 102).

RANGE

Confined to Pahrump Valley, where it occurs in three separate spring areas, each the type locality of a different subspecies of *E. latos*.

TAXONOMY

The Pahrump Poolfish constitutes the second known species in the genus, occupying a valley immediately adjacent to the south and east to that portion of the Amargosa drainage (Ash Meadows) inhabited by the Ash Meadows Poolfish. It apparently breaks down into three subspecies which, while scarcely distinguishable from each other, can be provisionally accepted on the basis of their distribution.

Etymology—the specific name is derived from the Latin *latus*, "wide" and *os*, "mouth," in reference to the wideness of the mouth as compared to *E. merriami*.

LIFE HISTORY

Knowledge of the life history of this species is practically unknown, and while it presumably follows the same pattern, insofar as it is known, of the Ash Meadows Poolfish, unlike the latter, the Pahrump species lives without interspecific competition, being the only fish native to Pahrump Valley. Perhaps as a consequence of this lack of competition, as Miller suggested, it is much commoner in its environment than is *E. merriami* in Ash Meadows. The full effects upon *E. latos* of the introduced Asiatic Carp, *Cyprinus carpio*, and the bullfrog, have yet to be evaluated, but in keeping with their habits and reputation, such a relationship can only operate to the disadvantage of the poolfish. (See *E. l. concavus*.)

Unfortunately, the environment in which the species lives is being altered to an extent which has already restricted its populations and could eventually result in its extinction. Some of the source pools of the type localities have been completely drained by continued pumping and the populations affected are precariously living farther down in the fields, some in concrete-lined ditches.

Like its congener, *E. latos* is a thermal species, living in warm springs varying in temperature from 23.3° C. (74° F.) to 25.3° C. (77° F.), each with little annual fluctuation. For information on specific springs, see the subspecies following.

ECONOMICS

Originally, before the introduction of carp at least the Pahrump Poolfish represented, ichthyologically, the top of the food pyramid, competing, no doubt, with some of the larger aquatic insects. Otherwise of no practical economic importance.

KEY TO SUBSPECIES

As in the case of Miller's *Cyprinodon nevadensis* units (which see), it will be more intelligible to tabulate some of the chief distinctions between his *Empetrichthys latos latos*, *E. l. pahrump* and *E. l. concavus* than attempt to key them out. His data on *E. merriami* are included for comparison.

MEASUREMENTS

(In thousandths of the standard length; parenthetical figures are averages)

	E. l. latos	E. l. pahrump	E. l. concavus	E. merriami
Standard length (mm.)	26– 47 (35)	28– 42 (33)	25– 39 (32)	20– 59 (35)
Predorsal	669–693 (681)	658–687 (673)	645–688 (663)	650–694 (675)
Body depth	274–332 (306)	281–333 (309)	271–328 (293)	288–348 (314)
Head length	306–337 (316)	294–326 (312)	298–320 (310)	307–353 (329)
Caudal peduncle, depth	143–164 (155)	147–171 (157)	143–160 (153)	138–162 (149)
Preörbital, least width	24– 33 (29)	20– 28 (24)	19– 28 (24)	24– 36 (28)
Preörbital margin—preöpercular angle	147–167 (155)	144–156 (147)	143–158 (151)	130–159 (147)
Opercle length	107–128 (117)	108–124 (118)	115–129 (122)	123–140 (132)
Snout length	78– 89 (83)	71– 87 (79)	66– 79 (73)	73– 88 (80)
Mouth width	104–125 (113)	102–113 (105)	102–114 (106)	82–107 (97)
Anal origin to caudal base	♂ 332–356 (346)	355–382 (367)	359–377 (366)	351–376 (362)
	♀ 310–344 (328)	338–358 (348)	335–359 (350)	334–359 (342)
Head depth	♂ 252–277 (267)	269–290 (280)	273–293 (284)	286–308 (298)
	♀ 271–302 (281)	272–287 (277)	265–296 (276)	285–309 (299)
Anal fin, basal length	♂ 130–153 (139)	151–167 (158)	142–165 (155)	151–179 (160)
	♀ 112–131 (123)	123–143 (133)	127–142 (134)	124–147 (134)
Anal fin, length	♂ 214–239 (223)	225–251 (239)	236–263 (247)	216–240 (230)
	♀ 181–210 (199)	195–218 (207)	213–226 (221)	191–213 (204)
Middle caudal rays, length	♂ 199–221 (211)	215–228 (221)	221–246 (232)	181–215 (197)
	♀ 194–227 (210)	206–224 (214)	220–241 (228)	179–191 (184)
SCALE COUNTS				
Lateral series	30– 33 (31–32)	29– 33 (30–32)	30– 32 (31)	29– 30
Dorsal to anal	12– 16 (12–15)	12– 16 (13–15)	12– 15 (13–14)	13– 15 (13–14)
Predorsal	24– 29 (25–27)	23– 31 (24–27)	24– 30 (26–27)	24– 30 (26–27)
Peduncle circumference	16– 22 (17–20)	16– 22 (18–20)	18– 24 (20)	16– 21 (18–20)
Body circumference	27– 38 (30–33)	28– 38 (32–35)	29– 36 (30–33)	29– 35 (30–34)
HEAD PORE COUNTS				
Preöpercular	13– 16 (14)	13– 16 (14)	13– 14 (14)	13– 15 (14)
Preörbital	7– 9 (8)	7– 8 (8)	8	8
Mandibular	7– 8 (8)	6– 8 (8)	7– 8 (8)	7– 8 (8)
FIN RAY COUNTS				
Dorsal	10– 13 (11–12)	10– 13 (11–12)	10– 12 (11)	9– 12 (11)
Anal	10– 14 (12–13)	11– 14 (12–13)	12– 14 (13)	12– 14 (14)
Pectoral	16– 20 (17–18)	15– 18 (16–17)	16– 18 (16–17)	15– 18 (16–18)
Caudal	17– 22 (18–20)	16– 23 (18–20)	17– 20 (18–20)	17– 20 (18–20)

Miller's previously mentioned character indices for these subspecies, as well as for *E. merriami,* give the following results (see Original Description):

	Range (and average)
E. l. latos	372–426 (391)
E. l. pahrump	382–413 (396)
E. l. concavus	383–421 (402)
E. merriami	414–449 (431)

MANSE RANCH PAHRUMP POOLFISH
Empetrichthys latos latos Miller

Empetrichthys latos latos Miller 1948: 103–104
Empetrichthys latos latos, La Rivers, 1952: 91
Empetrichthys latos latos, La Rivers and Trelease, 1952: 118

ORIGINAL DESCRIPTION AND DIAGNOSIS

"Diagnosis and comparisons.—A subspecies of *E. latos* characterized by a relatively short distance between anal origin and caudal base. In this character the males of *E. l. latos* show little or no overlap with the males of either *E. l. pahrump* or *E. l. concavus;* in the females the values show only slight overlap. The comparatively short distance between the anal origin and the caudal base appears to be influenced by the more posterior position and the short basal length of the anal fin. The head depth is less in the males, and the width of the preorbital is broader in both sexes than in those 2 subspecies, and the snout is longer than it is in *concavus.* The basal length of the anal fin is shorter, but overlaps the figures for the other 2 subspecies. The caudal fin of *latos* is much shorter than it is in *concavus.* On the average the mouth is broader and the dorsal fin more posterior in position than in either *pahrump* or *concavus.* The darker pigmentation of *latos* may be due, in part at least, to the darker habitat" (Miller 1948: 103–104).

TYPE LOCALITY

"The holotype has been designated in the description of the species. There are 143 paratypes, U.M.M.Z. No. 140489, 15 to 48 mm. long, taken with the holotype at Manse Ranch, and 34 paratypes, U.M.M.Z. No. 132915, 10 to 50 mm. long, collected by R. R. Miller and Alex J. Calhoun on July 16, 1938, from the outlet of the main spring pool" (Miller 1948: 103).

RANGE

Confined to the type locality.

PAHRUMP RANCH PAHRUMP POOLFISH
Empetrichthys latos pahrump Miller

Empetrichthys latos pahrump Miller 1948: 104–105
Empetrichthys latos pahrump, La Rivers, 1952: 91
Empetrichthys latos pahrump, La Rivers and Trelease, 1952: 118

ORIGINAL DESCRIPTION AND DIAGNOSIS

"Diagnosis and comparisons.—A subspecies of *Empetrichthys latos* closely resembling *E. l. latos,* differing principally in the longer distance between anal origin and caudal base. From *concavus* it is readily separated by the much shorter and more nearly truncate caudal fin. In the length of the anal and caudal fins *pahrump* is somewhat intermediate between *latos* and *concavus.* It also appears to be intermediate in color pattern, but the paleness is very probably influenced by the clay and silt bottom over which the sample was collected. The head is only very slightly concave" (Miller 1948: 104).

TYPE LOCALITY

"The holotype, an adult female, 35 mm. long, was seined by R. R. and F. H. Miller from the marshy overflow of a spring-fed ditch on Pahrump Ranch, 6 miles northwest of Manse Ranch, in Pahrump Valley, Nye County, Nevada; U.M.M.Z. No. 141856. One hundred and forty-two paratypes, U.M.M.Z. No. 140490, 14 to 36 mm. long, were taken with the holotype" (Miller 1948: 104).

TAXONOMY

Etymology—named for the type locality.

RANGE

Known only from the type locality.

LIFE HISTORY AND ECONOMICS

See *Empetrichthys latos.*

RAYCRAFT RANCH PAHRUMP POOLFISH
Empetrichthys latos concavus Miller

Empetrichthys latos concavus Miller 1948: 105
Empetrichthys latos concavus, La Rivers, 1952: 91
Empetrichthys latos concavus, La Rivers and Trelease, 1952: 118

ORIGINAL DESCRIPTION AND DIAGNOSIS

"Diagnosis and comparisons.—A subspecies of *Empetrichthys latos* most closely resembling *E. l. pahrump,* from which it differs chiefly in the much longer and more nearly rounded caudal fin. The rays along the upper and lower borders of this fin are definitely shorter than they are in either *pahrump* or *latos.* The profile between snout and occiput is most strongly concave in this subspecies. The anal fin of *concavus* is longer than it is in *pahrump,* especially in the female, and much longer than the anal fin of *latos.* In the males the snout is shorter and the body is deeper than in the 2 other subspecies. The cheek is deeper than it is in either *pahrump* or *latos.* The axial streak is generally finer and much less conspicuous than it is in the other 2 subspecies, and in some specimens of *concavus* it is obsolescent" (Miller 1948: 105).

TYPE LOCALITY

"The holotype is an adult female, U.M.M.Z. No. 141857, 39 mm. long. It was collected by R. R. and F. H. Miller on October 5, 1942, in a spring on the Raycraft Ranch, about one-half mile north of Pahrump

FIG. 236. Four forms of Empetrichthys, top to bottom: *E. merriami*, adult male, Eagle Spring, Ash Meadows, Nye County, Nevada. *E. latos concavus*, adult male, Raycraft Ranch, Pahrump Valley, Nye County. *E. latos pahrump*, adult male, Pahrump Ranch, Pahrump Valley. *E. latos latos*, adult male, Manse Ranch, Pahrump Valley. Photographs by F. W. Ouradnik for Miller 1948. Courtesy Dr. Robert R. Miller.

Ranch, Pahrump Valley, Nye County, Nevada. This spring is named on a map in Waring's report (1920). Twenty-six paratypes, U.M.M.Z. No. 140491, 17 to 40 mm. long, were collected with the holotype" (Miller 1948: 105).

RANGE

Only the type locality.

TAXONOMY

Etymology—the comparatively pronounced concavity of the top of the head of this subspecies gives it its name.

LIFE HISTORY

The habitat of *E. l. concavus* is described as a: "spring-fed pond, 5 to 25 feet wide and about 40 feet long, and its outlet ditch, 1 to 4 feet wide. The temperature of this spring on October 5, 1942, was 25.3° C., slightly warmer than were the springs on either Manse or Pahrump ranches. The water in the spring pond and outlet was clear but easily roiled because of a bottom of silt and trash. Vegetation noted in 1942 was water cress, *Typha*, and grass. The current in the spring was slight, but rather swift in the outlet. The depth of water was not over 1½ feet. The shore consisted of low banks, willows, and meadowland. According to Waring (1920: 76) this spring has a flow of about 10 gallons a minute. *Empetrichthys* was not common, perhaps because introduced carp were also present" (Miller 1948: 105).

Superfamily POECILIOIDEA
THE TOPMINNOWS
Family *POECILIIDAE*

Genus *Gambusia* Poey 1854

(From "Gambusino," a Cuban word; fishing for Gambusinos is to fish for nothing)

This rather sizeable and largely tropical family of live-bearers is entirely American in original distribution, their range including an area from the mid-Mississippi valley to the Argentine of South America. Because of the popularity of many species with aquarists, they have been carried with the trade to many parts of the world, particularly Europe, while many strictly tropical species are well known now in northern North America.

Other types, such as the Mosquitofish (*Gambusia affinis*) and the Guppy (*Lebistes reticulatus*) have been planted far and wide because of their mosquito-feeding propensities.

Size range in the family is not great, most species being on the order of two or three inches in length; the largest is reported to be about eight inches long. Males are often considerably smaller than females, and the tiny males of the American Topminnow (*Heterandria formosa* Agassiz 1855) once had the distinction of being the smallest living vertebrate. Certain oviparous Philippine gobies now hold that title, so that the topminnow's minimal ⅝ of an inch makes it only the smallest of the viviparous fishes.

According to Regan, there were some 26 genera and 82 species in the family in 1913. While most are fresh water species, some are habituées of salt water, living in coastal ocean situations. As the term "live-bearer" indicates, these fishes bring forth living young and have appropriate reproductive modifications for so doing. Several rays in the anal fin (rays 3–5) are elongated and form a groove or tube for the conduction of spermatozoa from male to female. These form the so-called "intromittent organ" which, because of structural differences, has been very useful to the systematist intent on distinguishing between the various species.

Unlike the preceding family, poeciliids are virtually unknown in the fossil record, and would also appear on other grounds (i.e., anatomical[1]) to be recent differentiates of cyprinodontid stock.

Some of the most popular topminnows for home aquaria are the: Guppy (*Lebistes reticulatus* (Peters) 1859), Molly (*Mollienesia latipinna* Le Sueur 1821), Swordtail (*Xiphophorus helleri* Heckel 1848) and Moonfish or Platy (*Xiphophorus maculatus* (Günther) 1866).

DIAGNOSIS
See Cyprinodontidae; the two families are grossly similar enough that the one diagnosis will suffice for both.

---
[1]Specialized male intromittent organ, cephalad movement of pelvic fins, etc.

MOSQUITOFISH

Gambusia affinis (Baird and Girard)

(Gambusia, Viviparous Topminnow)

Heterandria affinis Baird and Girard 1853B: 390
Heterandria patruelis ibid
Gambusia holbrooki Girard 1858: 61
Gambusia speciosa ibid: 121
Gambusia gracilis ibid
Gambusia patruelis, ibid: 72
Gambusia affinis, ibid
Gambusia holbrooki, Günther, 1866: 334
Gambusia humilis ibid
Gambusia affinis, ibid: 336
Haplochilus melanops Cope 1870: 457. (Proc. Amer. Philos. Soc.)
Zygonectes atrilatus, Jordan and Brayton, 1878: 84
Gambusia holbrooki, Jordan and Copeland, 1878: 433
Gambusia affinis, ibid: 434
Gambusia patruelis, ibid
Zygonectes melanops, ibid: 1878: 52 (Bull. Illinois Nat. Hist.)
Zygonectes brachypterus, ibid
Zygonectes brachypterus Cope 1880: 34 (U.S.N.M. Bull. 20)
Zygonectes atrilatus, Jordan and Gilbert, 1882: 340
Zygonectes brachypterus, ibid: 341
Gambusia humilis, ibid: 345
Gambusia affinis, ibid: 346
Gambusia patruelis, ibid
Zygonectes inurus, ibid: 892
Gambusia affinis, Evermann and Kendall, 1894: 107 (U. S. Fish Comm. Bull. 12)
Gambusia affinis, Jordan and Evermann, 1896B: 680–682
Gambusia affinis, Kuntz, 1913
Gambusia affinis, Hildebrand, 1917, 1921
Gambusia affinis, Forbes and Richardson, 1920: 215–217
Gambusia modesta Ahl 1923: 220 (Blätter für Aquarium—und Terrariakunde 36)
Gambusia affinis, Pratt, 1923: 105–106
Gambusia myersi Ahl 1925: 36 (Blätter für Aquarium—und Terrariakunde 38)
Gambusia patruelis, Hildebrand and Towers, 1928: 124–125
Gambusia patruelis, Jordan, Evermann and Clark, 1930: 185
Gambusia holbrooki, ibid
Gambusia patruelis holbrooki, Brues, 1932: 279
Gambusia affinis, Lindberg, 1934
Gambusia patruelis, Schrenkeisen, 1938: 187–188
Gambusia holbrooki, ibid: 188
Gambusia affinis affinis, Self, 1940
Gambusia affinis, Barnickol, 1941
Gambusia affinis affinis, Hubbs and Miller, 1941: 3 / 1948A; 10
Gambusia affinis, Rice, 1941
Gambusia affinis affinis, Dill, 1944: 162–163

Gambusia affinis affinis, Miller and Alcorn, 1946: 184
Gambusia affinis affinis, Krumholz, 1948
Gambusia affinis affinis, Kopec, 1949: 56
Gambusia affinis affinis, Shapovalov and Dill, 1950: 387
Gambusia affinis affinis, Wallis, 1951: 90
Gambusia affinis affinis, Beckman, 1952: 74–75
Gambusia affinis, La Rivers, 1952: 89
Gambusia affinis affinis, La Rivers and Trelease, 1952: 119
Gambusia affinis affinis, Miller, 1952: 12, 37
Gambusia affinis, Eddy, 1957: 176
Gambusia affinis, Koster, 1957: 85–86
Gambusia affinis, Moore, 1957: 158
Gambusia affinis affinis, Trautman, 1957: 458–460
Gambusia affinis affinis, Shapovalov, Dill and Cordone, 1959: 173

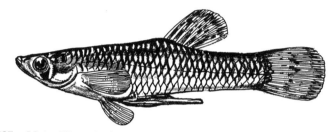

FIG. 237. Male Mosquitofish, *Gambusia affinis*. Drawn by William L. Brudon for Miller 1952. Courtesy California Department of Fish and Game and Leo Shapovalov.

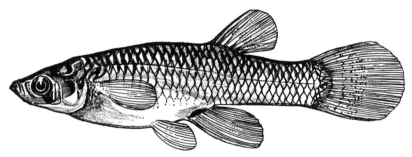

FIG. 238. Female Mosquitofish, *Gambusia affinis*. Drawn by William L. Brudon for Miller 1952. Courtesy California Department of Fish and Game and Leo Shapovalov.

ORIGINAL DESCRIPTION

(*Heterandria affinis*)—
"Body elongated, subfusiform and compressed. Head forming about one-fifth of the entire length. Body yellowish brown above, orange beneath. Fins unicolor, except the caudal which has two narrow bands of black.

"D 6. A 8. C 3. I. 7. 6. I. 2. V 5. P 12.

"Rio Medina and Rio Salado" (Baird and Girard 1853B: 390).
(*Heterandria patruelis*)—

"Body rather elongated, compressed. Head stouter than in *H. affinis,* though forming the fifth of the entire length. Reddish brown above, yellowish beneath.

"D 5. A 8. C 3. I. 7. 6. I. 2. V 6. P 11.

"Inhabits the Hydrographic basin of the Rio Nueces; specimens were collected in the Rio Sabinal, Rio Leona and Rio Nueces, and Elm Creek" (Baird and Girard 1853B: 390).

DIAGNOSIS

Head 3 to 4, body depth 3.5 to 4.0, times in body length; head moderate in size, very wide dorsally, depressed, almost spatulate in side view; mouth dorsal, the lower jaw projecting beyond the upper and up-turned; each jaw with a villiform row of immovable teeth; eye moderate in size; scales 29–33 in the lateral series, 6–10 from front of dorsal fin to front of anal fin; fins—dorsal small, considerably closer to caudal fin base than to snout, 7–9 rayed, rarely 6—anal large in female, in male developed into a narrow, long copulatory process, 8–10 rayed in the former—caudal moderate in size, about 24-rayed; caudal peduncle quite stout, between 7 and 8 times into body length; intestine moderate in length, about as long as the body; color—olivaceous tan dorsally, lighter ventrally, often with a thin dark lateral line—blackish spot or bar below the eye; in size quite small, the females seldom exceeding 2 inches in length, the males much smaller, usually with a maximum length of about an inch.

TYPE LOCALITY

Rio Medina and Rio Salado, Texas.

RANGE

Originally restricted to the eastern United States and New Mexico from New Jersey and Illinois south to the Gulf of Mexico. In the interest of mosquito control, it has been planted in many parts of the world outside its native range. While it has undoubtedly been brought into Nevada on more than one unrecorded occasion, the known implantations were summarized by Miller and Alcorn (1946: 184): "This little viviparous fish, now one of the most widely distributed species in the world, was introduced from the vicinity of Los Angeles into the lower Carson River about 1934, by J. H. Kispert (personal interview). Vernon L. Mills told us that he also brought in a stock from the Sacramento Valley in 1934 or 1935. From Fallon they have been transplanted to Lyon and Douglas counties. The mosquito fish is also common in the Moapa River and along the shore of Lake Mead near the mouth of the Moapa River (observations of 1938–1942—in the last year by Carl L. Hubbs)." Moffett (1943: 182), Wallis (1951: 90) and Jonez and Sumner (1954) have all reported this species from Lake Mead and vicinity. In 1949 the author found schools numbering into the thousands in the waters of the Virgin River above the lake. Since the original western introductions into California in 1922 (Dill 1944: 162) for mosquito control, the Mosquitofish has found its way to various parts of Nevada other than those mentioned above. It was brought into the Lovelock area from Stillwater and Wabuska Hot Springs in June 1958 by Glen Griffin.

TAXONOMY

Examination of the description upon which this species was based, plus its wide distribution and consequent variability, makes the long and confusing synonymy rather understandable. Both *affinis* and *patruelis* were inadequately characterized on the same page, the former having line priority. From the original description in 1853 to the 1890's, the synonymy was quite confused, and the use of names for this species rather indiscriminate. With the appearance of Jordan and Evermann's authoritative work in 1896, *Gambusia affinis*, as a name, enjoyed nomenclatural stability for some 30 years, being replaced by *patruelis*, again with the Jordan and Evermann blessing (Jordan, Evermann and Clark, 1930). Some decade afterwards, the names were again reversed.

Etymology—"affinis," related (to *Gambusia holbrooki*).

LIFE HISTORY

This now widespread and abundant little fish has some peculiarities which make it one of the biologically more interesting species in the State. Members of the family to which it belongs are distinctly segregated from the preceding spring- and pup-fish family (Cyprinodontidae) in being "live-bearers," i.e., they give birth to living young rather than laying eggs. The sexes are dimorphic, the males and females differing in external physical appearance, and fertilization is internal rather than the typical external broadcasting of eggs and sperm of most fishes.

The anal fin of the male is elongated and modified into a copulatory organ for transfer of sperm to the female. The female produces her several broods from a single fertilization by the male.

Size and growth ratios differ between the sexes, the female being the larger and continuing to grow throughout life whereas the male's growth is imperceptible after maturity is reached. A single fertilization may suffice for all the broods a female will produce during a short life-span not exceeding a year-and-a-half and usually much shorter.

According to Krumholz (1948: 41), the females are physiologically hardier than the males and outnumber and outlive the latter although the sexes are about equal at birth. He found the largest single brood to total 315 embryos, the smallest but 1, showing the great numerical diversity here. The average gestation period is about 24 days. About five broods seem the maximum number in a lifetime.

As is well known in the case of egg-layers, the larger females produce the more offspring. Contrary to earlier beliefs, Krumholz's observations indicate the young are born singly, rather than in batches, and are considered prey immediately by the female parent.

The much smaller numbers of young produced by live-bearers in contrast to egg-laying species still results in fish who can become numerically dominant in their environment—by nurturing the young to the point where they can immediately fend for themselves after birth, the Mosquitofish greatly cuts down on early mortality normal to egg-layers.

There are marked differences between northern and southern populations in their ability to withstand cold—many unsuccessful attempts to stock Mosquitofish into new areas have been the result of utilizing southern specimens who were not acclimated to cold weather. In Nevada, the species has done well in both warm and cold waters, but winter kills rather easily.

As its common name implies, this fish is partial to the larvae of aquatic insects and will feed exclusively on mosquito larvae and pupae under certain circumstances. They also consume vegetable matter such as algae.

They are somewhat social and commonly school together, often in great numbers. Their hardiness has allowed them to prosper in warm springs in Nevada even in competition with the equally small native species there; ponds, streams, sloughs and even lakes (Mead) in the State now have their prolific populations of Mosquitofish.

ECONOMICS

The rapid artificial spread of the Mosquitofish from its native southeastern United States to a range including Europe, India, Japan, southwest Pacific, New Zealand, South America and Africa has been almost entirely motivated by its reputation as a mosquito destroyer. Indeed, at one time, as so often happens, this fish was regarded as a panacea for the world's mosquito ills—and, like similar biological ventures, it soon proved that it is no exception to the natural law which holds that living things are always in balance of one kind or another with their associates.

Today, the Mosquitofish has settled down to the role of an auxiliary to Man's still relentless (in some parts of the world) attempts to find answers to the mosquito problem. Under some circumstances, the fish can do much damage to developing mosquito populations—but as everyone knows who lives in both mosquito- and mosquitofish-infested areas, the former continue to thrive unless further substantial checks are added to the control measures.

In Nevada, although mosquito control was the prime mover for at least some of the original introductions (i.e., in the Fallon area), not much result was obtained. In warm water regions of the State where the mosquitofish presumably could offer severe competition to the native fishes, such does not seem to have occurred.

The importance of the species to the aquarium trade has been mentioned.

* * * * *

Limited establishment of three other species of popular aquarium poeciliids is known in southern Nevada:

(1) Ash Meadows, Nye County—Black Mollies (*Mollienesia latipinna* Le Sueur 1821). These can be found in small numbers in warm water roadside drains near the Ash Meadows Ranch in the southeast part of the Meadows.

(2) Rogers Springs, Clark County (Overton Arm of Lake Mead)— Black Mollies (*Mollienesia latipinna* Le Sueur 1821), Swordtail (*Xiphophorus helleri* Heckel 1848) and Platys (*Xiphophorus maculatus* (Günther) 1866) have all been successfully released in here, according to Robert Sumner.

Order *PERCIFORMES*
Suborder PERCOIDEI
Superfamily PERCOIDEA

THE TRUE PERCHES
Family *PERCIDAE*

The true perches, of which we have one introduced species in Nevada at this writing, constitute a goodly-sized Holarctic group most of whose members show marked preference for at least moderately cold water. The nearly two dozen genera contain about five times that many species; in American waters, the small perches known as "Johnny Darters" (Etheostominae) are by far the most numerous from the standpoint of species. These are certainly of some importance in the food chains of larger sporting fishes but in addition to these, several of the country's important game species belong here: the Yellow Perch, Sauger and Walleyed Pike.

The perches are not much given to "hovering" in the water after the manner of so many fishes; indeed, the smaller types such as the darters have many of the habit preferences of such secretive bottom-loving forms as the sculpins, resting on their ventral fins and sculling with the large pectoral fins when disturbed (the tail is used very little by these species for swimming).

Some even bury themselves in sand and others, because of a habit of slightly turning the head in relation to the body, present the rather droll appearance of "peering" at objects. As we might expect in such bottom-dwellers, the air bladder is nonfunctional and they have consequently lost their buoyancy.

European types largely parallel those in America (or vice versa) and the family has always been a problem to the ichthyological taxonomist. The smaller darters are as brilliantly colored as any fish, but the larger game species are comparatively colorless.

DIAGNOSIS

Rather attenuate in shape, the scales closely fixed, small and ctenoid; lateral line usually prominent; mouth terminal or subterminal, variable in size; commonly the jaws as well as the palatine and vomerine bones bear teeth; head scaly-to-bare; opercles usually posteriorly spined; branchiostegals 6–7; gills 4, with a slit posterior to the 4th gill, membranes not connected to isthmus; gill rakers thin, hooked; pseudobranchiae reduced or wanting; pharyngeals armed with prominent, sharp teeth; anal papilla generally conspicuous; fins prominent— first dorsal, anal and pelvics with spines; air bladder reduced and nonfunctional, or absent; pyloric caeca greatly reduced in number; vertebrae 30–48.

Freshwater species of the northern hemispheres, preferring at least moderately cold waters; carnivorous feeders, mainly from the bottom.

PERCHES
Genus *Perca* Linnaeus 1758
(The ancient Greek name for Perch)

YELLOW PERCH
Perca flavescens (Mitchill)
(Striped Perch, Ringed Perch, Redfin)

Perca americana Schranck 1792: 100 (Abh. Privatges. Oberdeutschland 1)
Morone flavescens Mitchill 1814: 18 (Rept. Fishes N.Y.)
Centropomus luteus Rafinesque 1814: 19 (Precis des Decouvertes Somiologiques)
Bodianus flavescens, Mitchill, 1815: 421 (Trans. Lit. & Philos. Soc. N. Y.)
Perca notata Rafinesque 1818: 205 (Amer. Monthly Mag. 2)
Perca serrato-granulata Cuvier and Valenciennes 1828: 34 (Hist. Nat. Poiss. 2)
Perca granulata ibid: 35
Perca acuta ibid: 36
Perca gracilis ibid
Perca flavescens, ibid: 46
Perca flavescens, Richardson, 1836: 1 (colored plate p. 74)
Perca flavescens, Günther, 1859: 59
Perca americana, Jordan and Copeland, 1878: 437
Perca fluviatilis flavescens, Steindachner, 1878: 24 (Sitzgber Ak. Wien 78)
Perca americana, Jordan and Gilbert, 1882: 524
Perca flavescens, Jordan and Evermann, 1896B: 1023–1024
Perca flavescens, Forbes and Richardson, 1920: 276–278
Perca flavescens, Pearse and Achtenberg, 1920: 295–366
Perca flavescens, Pratt, 1923: 122–123
Perca flavescens, Coker, 1930: 204–205
Perca flavescens, Jordan, Evermann and Clark, 1930: 281
Perca flavescens, Neale, 1931: 5, 11
Perca flavescens, Fish, 1932: 362–367
Perca flavescens, Hatton, 1932: 42–44
Perca flavescens, Schultz, 1936: 161
Perca flavescens, Tanner, 1936: 173
Perca flavescens, Schrenkeisen, 1938: 208–209
Perca flavescens, Hubbs and Lagler, 1941: 69, 72–73 / 1947: 82, 86
Perca flavescens, Murphy, 1941: 168
Perca flavescens, Hinks, 1943: 70–71
Perca flavescens, Dimick and Merryfield, 1945: 42
Perca flavescens, La Monte, 1945: 130
Perca flavescens, Miller and Alcorn, 1946: 184
Perca flavescens, Eddy and Surber, 1947: 204, 206–208
Perca flavescens, Harrington, 1947:
Perca flavescens, Carl and Clemens, 1948: 95–97

Perca flavescens, Curtis, 1949: 258, 269
Perca flavescens, Everhart, 1950: 45
Perca flavescens, Shapovalov and Dill, 1950: 387
Perca flavescens, Simon, 1951: 97–99
Perca flavescens, Beckman, 1952: 90–91
Perca flavescens, La Rivers, 1952: 87
Perca flavescens, La Rivers and Trelease, 1952: 119
Perca flavescens, Miller, 1952: 10, 38
Perca flavescens, Gabrielson and La Monte, 1954: 102
Perca flavescens, Scott, 1954: 83
Perca flavescens, Shomon, 1955: 43–44
Perca flavescens, Harlan and Speaker, 1956: 145–147
Perca flavescens, Eddy, 1957: 200
Perca flavescens, Koster, 1957: 100–101

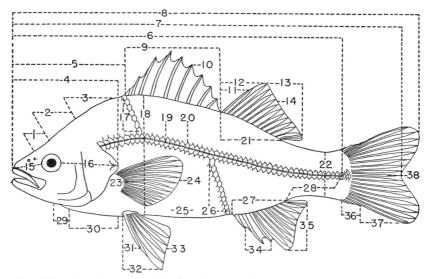

FIG. 239. A spiny-rayed fish showing morphometry or counting-and-measurement details and techniques. (1) interorbital. (2) occipital. (3) nape. (4) head length. (5) predorsal length. (6) standard length. (7) fork length. (8) total length. (9) length of base of spinous or first dorsal fin. (10) one of the spines of dorsal fin. (11) one of the spines of second or soft dorsal fin. (12) height of second dorsal fin. (13) length of distal, outer or free edge of second dorsal fin. (14) one of the soft rays of second dorsal fin. (15) snout length. (16) postorbital head length. (17) scales above lateral line. (18) body depth. (19) one of the lateral line pores. (20) one of the lateral scales which, with the remainder, form the lateral series. (21) length of base of second or soft dorsal fin. (22) least depth of caudal peduncle. (23) pectoral fin. (24) one of the soft rays of pectoral fin. (25) abdominal region. (26) scales below the lateral line. (27) length of base of anal fin. (28) length of caudal peduncle. (29) isthmus. (30) breast. (31) pelvic spine. (32) height of pelvic fin. (33) one of the soft rays of pelvic fin. (34) spines of anal fin. (35) soft rays of anal fin. (36) rudimentary rays. (37) one of the principal rays of caudal fin. (38) caudal fin. Reproduced from *The Fishes of Ohio*, by Milton B. Trautman (Columbus, Ohio: Ohio State University Press, 1957), p. 58. Used by permission of the publisher.

Perca flavescens, Moore, 1957: 177–178
Perca flavescens, Trautman, 1957: 531–533
Perca flavescens, Slastenenko, 1958A: 305–307
Perca flavescens, Carl, Clemens and Lindsey, 1959: 145–147
Perca flavescens, Shapovalov, Dill and Cordone, 1959: 173

ORIGINAL DESCRIPTION

"3. *Morone Flavescens**—Yellow Perch, or Yellow Basse of New-York.

"Lives in fresh water, both stagnant and running. Head rather small, and tapering toward the snout; body deep and thick, but becoming slender towards the tail; colors brown or olive on the back, turning to yellow on the sides, and white on the belly; faint brown zones, to the number of four or more, diversifying the sides from back

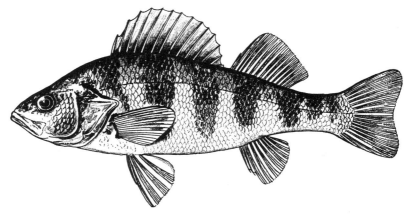

FIG. 240. Yellow Perch, *Perca flavescens*. Drawn by William L. Brudon for Miller 1952. Courtesy California Department of Fish and Game and Leo Shapovalov.

to belly; dorsal and pectoral fins brown; ventral and anal scarlet; gill-covers tripartite; lower and after edges of the foremost acutely serrated; middle one serrated at the lower edge, and striated radially on the broad side; there is one serrated bone on the thorax, immediately above the pectoral fin, at the posterior margin of the branchial opening, and two other bones with serrated edges above the first, near the upper part of the branchial opening; vent near the tail; lateral line almost straight; tail rather concave; scales rather hard and rough; eyes large; irides yellowish; both jaws roughened with very small teeth. The first ventral ray is spinous; so are the two first anal rays, all the rays of the foremost dorsal fin, and the foremost of the second dorsal. Is a beautiful fish. Length 10 inches, depth 2½.

"Rays, P. 14: V. 5: D. 12–14: A. 10: C. 19" (Mitchill 1814: 18–19).

DIAGNOSIS

Body elongate, somewhat flattened; head about 3.25, body depth about 3.0 times into body length, respectively; cheeks scaly, opercle nearly bare, operculum single-spined; mouth terminal; vomerine, palatine and jaw teeth in villiform rows, canines lacking; branchiostegals 7; pseudobranchiae present, but small; anal papillae lacking; scales medium 6–7/57–62/15–18; fins—dorsal split, the anterior half 13-spined, the posterior half 15-rayed with the first two rays spinous—anal 10-rayed, first two spinous—dorsal spine I shorter than anal spine I—caudal slightly emarginate; air bladder present; pyloric caeca 3; color—back dark greenish, sides lighter yellow with dark greenish vertical bars, 6–8 in number, belly white—ventral fins with an orange caste, dorsal fins greenish; in size, seldom exceeding a foot in length and 2 pounds in weight, although weights of 4 pounds or over are on record.

TYPE LOCALITY

Vicinity of New York City, N. Y.

RANGE

Originally generally distributed in the northern half of the eastern United States. Widely introduced throughout the West, and it now occurs abundantly in certain Nevada localities. The Nevada Fish and Game Commission planted 27 specimens in the West Carson River in the vicinity of Genoa in 1930 (1931: 23, 27), and also listed 100 additional individuals, 5 to 6 inches long, at Verdi for distribution, but no subsequent details were published regarding their disposition. Leach (1932: 683) mentioned the sending of 150 Yellow Perch to Nevada applicants during the 1930–1931 biennium. It is now common in the lower Humboldt River system and in the Fallon area, from which it has been transplanted into the vicinity of Reno and Washoe Lake.

Like most of our introduced species, it is not possible to determine most of the plantings. Ted Frantz has also told me of their presence in Basset Lake in Steptoe Valley (White Pine County).

TAXONOMY

Etymology—*Perca*, a classical name, literally meaning "dusky." *flavescens* means yellowish.

LIFE HISTORY

A middle-to-late spring spawner, the Yellow Perch seems to prefer spawning temperatures of about 45° F. to 50° F. (7° C. to 10° C.). Spawning is either by pairs or with one female attended by several males and is usually a nighttime activity. Sand or pebble bottoms near shore are utilized, and the eggs are laid in flat, hollow chains which may be several inches in width. The number of eggs produced depends upon the size of the fish, and varies from 10,000 to nearly 50,000 for a single individual. No redd or nest is made, the egg bands merely settling to the bottom.

The following admirable summary of Yellow Perch biology is that of Pearse and Achtenberg, 1920:

(1) The habits of perch in a small, shallow, and muddy lake were compared with those of perch in a neighboring large, deep, and clean lake, Lakes Wingra and Mendota, Wisconsin. Perch were the most abundant fishes in both, but, in proportion to the size of the lake, there were more in the larger lake.

(2) The perch is a versatile feeder but usually gets its food on or near the bottom. The percentage by volume of the foods eaten by 1,147 adults was as follows: Chironomid larvae, 25.2; cladocerans, 22.1; *Corethra* larvae, 6.4; silt and bottom debris, 6; chironomid pupae, 5.9; fish, 5.2; amphipods, 3.6; *Sialis* larvae, 3.4; caddisfly larvae, 2.1; oligochaetes, 1.5; crawfishes, 1.5; odonate nymphs, 1.4; clams, 1.2; algae, 1.2; snails, 1.1; ephemerid nymphs, 0.9; calcium carbonate crystals, 0.5; leeches, 0.4; hemipterous adults, 0.3; mites, 0.3; chironomid adults, 0.2; *Corethra* adults, 0.2; *Corethra* pupae, 0.2; copepods, 0.1; ostracods, 0.09.

(3) There are more or less marked seasonal variations in all constituents of the perch's food. In general, foods are eaten in proportion to their abundance and availability; but this is not always the case.

(4) An adult perch eats about 7 percent of its own weight each day. Digestion is three times more rapid in summer than in winter.

(5) Perch do not take any abundant food but select certain things. There are daily and seasonal variations. Individuals feeding in shallow water eat a greater variety than those from greater depths. Perch contain food which is available at the depths where they are caught, which indicates that extensive vertical migrations are infrequent.

(6) Food varies with age. During youth there is a change from *Cyclops* and other entomostracans to *Hyalella* and insect larvae. At the end of the first summer the food of young perch is much like that of adults.

(7) As judged by the rate of increase in young perch when fed on a single food the following varieties rank in the order given, the best being first: Earthworms, entomostracans, chironomid larvae, amphipods, fish, small amounts of various foods, liver and flour, adult insects.

(8) In the small lake investigated insects were the most important constituent of the food. In the larger lakes insects and entomostracans were equally important.

(9) Compared with crappie, the perch eats a greater variety and shows other specificities of behavior.

(10) Though perch are able to recognize the proportions of oxygen and carbon dioxide in water, they enter regions where conditions are unfavorable for respiration and may remain in oxygen-free water for as much as two hours without dying. When in water without oxygen perch use part of the oxygen in the swim bladder.

(11) Perch may become sexually mature in two years. In the smaller lake investigated they generally become mature when of much smaller size than do those in the larger lake.

(12) During the spawning season the males come into shallow water and remain for some time. The females remain on the spawning grounds only long enough to breed.

(13) Except during the spawning season and when the deeper water is stagnant, most of the perch in the large lake remain in deep water through the year. In the smaller lake similar migrations take place.

(14) There appears to be an upward migration at night.

(15) Perch swim more or less in schools throughout the year and apparently do not remain in one locality but move along the shore.

(16) Perch have many predaceous enemies.

(17) Perch are very generally infected with parasites. Those in the two lakes investigated contained cestodes and cestode larvae (one or more species), trematodes (5), acanthocephalans (2), nematodes (1). Leeches and an insect larva were found on the outside of the body.

(18) Perch are more abundant in inland lakes than other species because they are more versatile.

(19) Large inland lakes will generally contain more fishes per unit of volume than those of smaller size.

(20) Judging by the data presented in this paper the reason why the fishes in certain inland lakes attain a rather small maximum size is because there are various adverse conditions which prevent growth. In the present instance food does not appear to be as important as other factors such as shallowness, exposure to wind, etc.

ECONOMICS

While the Yellow Perch is one of the finest pan fishes in its native habitat, and can be caught at almost any time of the year on a great variety of baits, it is often regarded as a nuisance, particularly in areas where larger and more desirable game and food fishes occur and where the Yellow Perch's readiness to take nearly every bait offered makes it difficult to catch anything else. These very characteristics, however, make the species an excellent fish for the youthful angler.

SUNFISHES, BLACKBASSES AND CRAPPIES
Family *CENTRARCHIDAE*

This is a small North American family of less than 30 species, probably derived from Asiatic stocks. The three groups into which the family can be divided are indicated by the common names "sunfish, crappie and blackbass." The term "warm water" is often applied to members of the family in allusion to the well-known fact that, without exception, its species prefer waters distinctly warmer than those which are optimum for, say, salmonoids.

Nest-building follows a habit pattern characteristic of the family and shows its least development in the most primitive species, the Sacramento Perch (*Archoplites interruptus*). The nest is made by the male who sweeps out a depression with vigorous tail movements. After

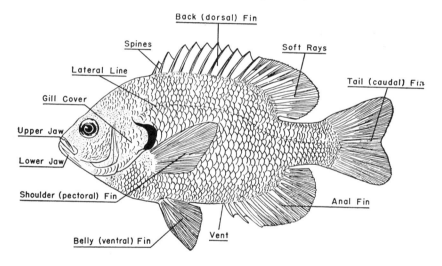

Blue Gill, typical warm water game fish with descriptive terms.

FIG. 241. Courtesy Oregon State Game Commission and Harold P. Smith.

spawning, males drive away the females and guard the nest pugnaciously from intruders. All species are spring spawners and some of the smaller sunfishes apparently have two spawning periods during the year, particularly in the southern United States.

In their habitat preferences, centrarchids show some interesting diversity. The classic work of Forbes and Richardson on the fishes of Illinois (1908 and 1920) contains some general information, from the "heart of the sunfish country," on this aspect—closely allied species of the large sunfish genus *Lepomis* were seldom found together in their collections over the State, but rather with more distantly related family members, a selective device which reduces interspecific competition between species with highly similar environmental demands.

They also point out some of the correlations between structure and function in the sunfishes: "Those with large mouths have a large ratio of fishes and crawfishes in the food, those with long gill-rakers take more *Entomostraca*, and those with broad and heavy pharyngeal bones, bearing stout blunt teeth, live more largely on mollusks" (1920: 235).

The family is an eastern and midwestern one, only a single species being native in the vast expanse west of the Rocky Mountains. This, the Sacramento Perch (*Archoplites interruptus*), is a remnant member of an earlier fish fauna which has been replaced by the modern species which now inhabit the United States, and originally occurred only in central California.

DIAGNOSIS

Body oval in lateral view, blunted anteriorly-posteriorly, laterally compressed and dorso-ventrally expanded; scales moderately ctenoid as a general rule; mouth essentially terminal, with pronounced dorsal upswing, variable in size; barbels lacking; jaws toothed; branchiostegals 6–7, usually the former; pseudobranchiae present, small, rather imperfect in that they may be skin-covered; gills 4 in number, the fourth followed by a slit; gill membranes unconnected, unattached to the isthmus; gillrakers variable, equipped with minute teeth; pharyngeal teeth numerous, sharp-to-flattened tuberculate, set on widened pharyngeal bones; fins—adipose lacking, pectorals lateral, pelvics thoracic, dorsal entire-to-bilobed, nearly bifurcated in *Micropterus;* dorsal, pelvics and anal with spinous rays preceding the soft rays; caudal usually weakly forked; lateral line present; intestine short; air bladder prominent, lacking a pneumatic duct in adults; pyloric caeca 5–10.

KEY TO GENERA OF NEVADA CENTRARCHIDAE

1. Anal fin spines 5 or more in number 2
   Anal fin spines 3 in number 3
2 (1). Dorsal fin spines 12 to 13 in number; dorsal fin base much longer than anal fin base; gillrakers short (one species) (*Archoplites interruptus*) .... SACRAMENTO PERCH
   Dorsal fin spines 6 to 8 in number; dorsal fin base subequal to anal fin base (i.e., both about the same length); gillrakers long and slender (two species) (*Pomoxis*) .... CRAPPIES
3 (1). Scales small, 58 or more in the lateral line; body depth about one-third of the standard length (two species) (*Micropterus*) .... BLACKBASSES
   Scales large, 53 or less in the lateral line; body depth usually about one-half of standard length (two species) (*Lepomis*) .... SUNFISHES

SACRAMENTO PERCH
Genus *Archoplites* Gill 1861
("Anal armature")

SACRAMENTO PERCH
Archoplites interruptus (Girard)
(Sacramento Bass, Striped Bass, Striped Perch)

Centrarchus interruptus Girard 1854: 129
Centrarchus maculosus Ayres 1854: 7 / 1855A
Ambloplites interruptus, Girard, 1859: 10–11
Centrarchus interruptus, Günther, 1859: 257
Archoplites interruptus, Jordan and Copeland, 1878: 435
Ambloplites interruptus, Jordan and Henshaw, 1878: 199
Archoplites interruptus, Jordan and Gilbert, 1882: 466 / 1884: 140
Archoplites interruptus, McKay, 1882: 87
Archoplites interruptus, Bollman, 1892: 560
Archoplites interruptus, Jordan and Evermann, 1896B: 991

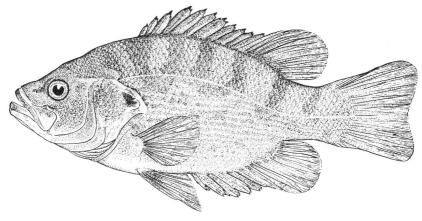

FIG. 242. Sacramento Perch, *Archoplites interruptus*.
Drawn by Silvio Santina.

Archoplites interruptus, Rutter, 1908: 119, 143
Archoplites interruptus, Snyder, 1917A: 86
Archoplites interruptus, Pratt, 1923: 116
Archoplites interruptus, Jordan, Evermann and Clark, 1930: 303
Archoplites interruptus, Neale, 1931A: 11–12 / 1931B: 409–411
Archoplites interruptus, Walford, 1931: 84
Archoplites interruptus, Hutchinson, 1937: 92
Archoplites interruptus, Schrenkeisen, 1938: 255
Archoplites interruptus, Sumner, 1940: 220
Archoplites interruptus, Murphy, 1941: 169 / 1948
Archoplites interruptus, Curtis, 1942: 5 / 1949: 256, 258, 265

Archoplites interruptus, Miller and Alcorn, 1946: 186–187
Archoplites interruptus, Shapovalov and Dill, 1950: 387
Archoplites interruptus, Simon, 1951: 103
Archoplites interruptus, La Rivers, 1952: 88
Archoplites interruptus, La Rivers and Trelease, 1952: 120
Archoplites interruptus, Dineen and Stokely, 1956
Archoplites interruptus, Eddy, 1957: 190
Archoplites interruptus, Moore, 1957: 174
Archoplites interruptus, Miller, 1958: 200
Archoplites interruptus, Shapovalov, Dill and Cordone, 1959: 173
Archoplites interruptus, Miller, 1961: 388

ORIGINAL DESCRIPTION

"General form rather elongated, very much compressed. Nuchal region swollen; oculo-cephalic region subconcave. Snout tapering; lower jaw longest. Posterior extremity of upper maxillary reaching a vertical line drawn back of the pupil. Head forming a little less than the third of total length. Eyes large and circular; their diameter being comprised four times in the length of side of the head. Scales on cheeks rather small; a little larger on the opercle than on the cheek.

"D XIII, 11. A VII. 10. C 5. 1. 8. 7. 1. 4. V I. 5. P 13.

"The origin of the spiny dorsal is situated opposite the base of the pectorals, and the origin of the anal, opposite the space between the eleventh and twelfth dorsal spines. The tip of rays, as well as the base of anal, extends a little farther back than the dorsal. The base of ventrals falls upon the same vertical line as that which would intersect the base of pectorals. Scales of medium size; minutely serrated.

"Greyish brown above, silver grey beneath. Irregular transverse bands of dark brown or black, interrupted along the lateral line, the portion of the band above it is somewhat alternating with the portion beneath it. A large black spot may be seen at the upper angle of opercle.

"Specimens from Sacramento River, Cal." (Girard 1854: 129).

DIAGNOSIS

Head and body depth each about $2\frac{1}{2}$ times into body length; mouth large and oblique; maxillary broad and long, reaching posteriorly to about hind level of orbit; operculum blunt angulate posteriorly; teeth present on jaws, vomer, tongue palatines, ectopterygoids and entopterygoids—two groups of lingual teeth, pharyngeal teeth pointed; branchiostegals usually 7; scales 7–9/38–42/13–16, about 8 series on cheek; fins—dorsal with 12–13 anterior spines, about 10 posterior soft rays, the fin very long and undivided, the spines low and strong—anal prominent, with 6–7 anterior spines and about 10 posterior rays—caudal weakly emarginate; caudal peduncle stout, 7–8 times into body length; color—blackish over back, mottled black-brown and white on sides, silvery below—often with brassy reflections—usually with about 7 vertical lateral dark bars—opercle with blackish spot; a rather large fish, attaining lengths of about 2 feet and weights of nearly ten pounds.

TYPE LOCALITY

The Sacramento River, California.

RANGE

Originally restricted to the Sacramento and San Joaquin Rivers and their tributary streams and lakes in central California, the Sacramento Perch has been widely and successfully introduced into west-central Nevada since the days of the State's first Fish Commissioner; as far as I can determine, all these early plants were made in Nevada and later some were taken to western Utah—how much later, no one seems to know. The species has long been common in Pyramid and Walker Lakes and the intervening Fallon area. Ted Frantz writes me that Sacramento Perch are quite common in Pruess Reservoir south of Garrison in Millard County, Utah. There is a substantial population of these perch in Bassett Lake, Steptoe Valley, White Pine County, which he says are numerous but small. His information also indicates that Sacramento Perch were present in Spring Valley, White Pine County, but died during a severe drought, according to several ranchers. They may still exist there, however.

Mr. Frantz also noted small perch in the Wabuska Drain, Lyon County, in 1952.

Miller and Alcorn (1946: 187) state that the Sacramento Perch "was . . . common at that time [August of 1938] in a lowland slough at Lusetti Ranch, about 16 miles north of Ely, White Pine County."

TAXONOMY

There have been no nomenclatural problems involving this very distinct and primitive crappie-like centrarchid, the only perciform native to the West Coast. An apparent relict of the family, the Sacramento Perch seems to have been left as a remnant of a once-flourishing but somewhat more primitive fish fauna which dominated North American fresh waters before such modern types as the blackbass and sunfish appeared.

In the east and midwest, this archaic fauna gave way to the present species so dominant there, but, presumably because of its isolation, the Sacramento Perch persisted behind the Sierran barrier to the present day.

Etymology—Archoplites = archos, "anus," plus oplites, "armature." "Interruptus," interrupted.

LIFE HISTORY

In addition to briefer accounts in some of the more comprehensive publications on the California fauna (Neale 1931A, Walford 1931, Curtis 1949), Neale (1931B) and Murphy (1948) have written short reports on aspects of the species' life cycle.

Murphy's account—the most detailed—was based on observations made in Clear Lake, Lake County, California. He found the species spawning in one-to-two feet of water where temperature was about 75° F. (23.7° C.). Spawning beds were in an area of boulders which were heavily encrusted with algal growths. Most spawners were paired, with little apparent exchanging of mates; preliminary activity consisted of increased restlessness graduating into energetic "nudging and butting," and terminating in egg deposition in which both fish were often vertically positioned with the vents close together. Each pair alternately rested and spawned in an area of about one-and-a-half feet in diameter, the male remaining on guard if the female momentarily strayed off. No nest building of any kind was noted.

On the particular day of observation, evidence indicated that spawning had begun about 9 o'clock in the morning—by late afternoon (4 p. m.) all activity had ceased and the fish departed. The long strings of eggs were laid conspicuously over the boulders, to which they

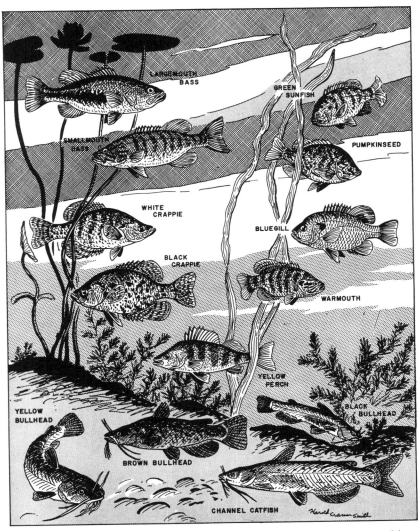

FIG. 243. Warm water game fishes. With the exception of the Pumpkinseed, Warmouth and Yellow Bullhead, these species have all been successfully established in Nevada. Courtesy Oregon State Game Commission and Harold P. Smith.

adhered; their exposure, plus the fact that the Sacramento Perch shows little or no nest-guarding proclivities, seems to account for the decline of this species in competition with introduced blackbass, sunfish and catfish.

Murphy noted that on the day of spawning, eggs were very much in evidence; the following day few were to be found, and he assumed the many small ½-inch Bluegills (*Lepomis macrochirus*) had consumed most of them. He specifically listed the case of Thurston Lake, near Clear Lake: Here Sacramento Perch were living successfully with two introduced eastern species—a centrarchid, the Largemouth Blackbass (*Micropterus salmoides*) and a catfish, the Brown Bullhead (*Ictalurus nebulosus*). Then, shortly following the 1933 implantation of White Catfish (*Ictalurus catus*), Sacramento Perch disappeared from the lake.

In Clear Lake, Sacramento Perch have the following growths:

	First year	Second year	Third year	Fourth year
Total lengths (inches)	3½	7	8	9
Total weights (ounces)	½	3½	6	8

Spawning takes place at the end of the second year (Murphy 1948: 93–100).

In Washoe Lake, Washoe County, Nevada, the author has seined fish in April with the following characteristics: Length 6¼ inches, weight 4½ ounces, apparent age 3 years, in nearly full spawning condition. Unlike Clear Lake, Washoe Lake is a more-or-less permanent pan whose waters are continually roiled and muddy.

Published lengths for the Sacramento Perch record an upper limit of two feet, which is exceptional. The heaviest specimen taken from Walker Lake, Mineral County, Nevada, was supposed, according to verbal but reasonably reliable testimony, to have weighed 8 pounds. A rather emaciated 12-inch female from Pyramid Lake (10 April 1951) weighed 2 pounds, and was seemingly 9 years old.

The species is carnivorous, after the manner of centrarchids, being insectivorous while small and piscivorous in later life. The above-mentioned Washoe Lake specimens were found to be feeding on aquatic insects of the families Dytiscidae (Coleoptera) and Corixidae (Hemiptera), and many of them were gorged. Of three 12-inch specimens taken at 70 feet in a gill net in Walker Lake in midwinter (31 December 1950) by Tom Trelease and the writer, two had empty stomachs, while the third stomach contained only a fragment of a fish fin. The water was cold (50.1° F. = 10.2° C.) and the fish were evidently doing little feeding.

Mr. Trelease (1951) has found that small (1–2-inch) specimens kept in aquaria consistently refuse the usual type of prepared fish food, but greedily eat smaller living fishes as soon as the latter are introduced into the tank.

Like most centrarchids, the Sacramento Perch at times is rather heavily infested with external and internal worms. Pyramid Lake specimens are commonly taken with several fish leeches, while an unidentified nematode is often present in large numbers in the mesenteries of a single individual. The leeches are occasionally found congregated in small colonies on the under surface of small tufa slabs in shoal water at Pyramid Lake—this last fact seems to corroborate conclusions that the Sacramento Perch is essentially a "lurker," spending most of its time under overhanging rocks, or between them, and in partial contact, a habit conducive to the acquiring of leeches. The more free-moving forms in the same environment such as the chub *Siphateles,* only rarely are found with leeches.

ECONOMICS

While the importance of the Sacramento Perch as a game and food fish has decreased in California to the point where it is negligible as a fishery species, it has consistently been an adjunct to the limited fisheries in Nevada. This decrease of the species in its native habitat has been due, as mentioned above, to the competition of introduced types such as catfish which devour the unprotected eggs and more aggressive, modern centrarchids such as blackbass and sunfish who directly compete for the same food. In Nevada, where blackbass, sunfish and catfish have also been introduced, they have not always proliferated in the same habitats as have the Sacramento Perch, and so the interspecific effects of competition here have been unimportant. Pyramid and Walker Lakes, foci for large Sacramento Perch populations, have no other centrarchids in them, apparently being unsuitable for blackbass (the only member of the family which could have early gained access to them) and catfish. These all do well together in Lahontan, Stillwater and Indian Lakes.

In its native habitat, the Sacramento Perch has been recorded as taking worms, live bait, spinners, and wet flies apparently at all times of the year. In Walker Lake (Mineral County), where it has been of most importance in Nevada, it is caught with almost anything which will sink and can be dragged slowly through their spawning areas—fish taken in this manner have empty stomachs and seemingly take a hook only to remove it from the spawning bed, and not because it represents food. As a consequence, the Walker Lake perch fishery only lasts for periods varying from six weeks to two months while spawning is in progress, and fish are seldom caught during the remainder of the year.

In other parts of west-central Nevada, such as Pyramid and Washoe Lakes, Sacramento Perch are very common but provided little fishing until recently. In the former lake they rival if not exceed the Walker Lake populations, but due to a variety of circumstances, were seldom fished for until the last few years. Washoe Lake, as mentioned, is a roiled pan periodically famous for its catfishing, but seldom fished for perch. Sacramento Perch are occasionally caught in heavily fished Lahontan Reservoir, near Fallon.

Robert Sumner has told me that the Sacramento Perch is on the verge of being regarded as a pest fish in parts of the lower Carson River area, partly because their abundance seems to be a potential problem in attempts to establish more desirable fisheries. In some of these waters—as in Washoe Lake—Sacramento Perch get to be numerous, but not sizeable, resembling the Green Sunfish in some respects.

It may be unlikely that a year-around fishery can be established on this species, in which case it can provide fishing only during its spawning season, and then only if the spawning areas or beds are approximately known. Several initial cooperative attempts by the Nevada Fish and Game Commission and the University of Nevada Biology Department to locate spawning beds in extensive Pyramid Lake were not intensive enough for practical results. However, with the great increase in trout angling on Pyramid Lake in the last several years, the Sacramento Perch take has also been stepped up. Many areas are now known in the lake where it is lucrative to fish for the once elusive perch. In addition to their apparent wariness, the species in Nevada occupies habitats where food is so abundant that it need not be sought but can be obtained readily by "lurking" tactics—as a consequence it is correspondingly difficult to fish for such a species, since the bait must be placed within feeding distance (i.e., it cannot be depended upon to attract "cruising" individuals as in the case of the Lahontan Cutthroat Trout), an entirely haphazard procedure in Nevada's opaque waters, and must be quite life-like to overcome natural suspicion; and even if these conditions are satisfied, the general abundance of its normal food adds the further difficulty that fish would be seldom encountered hungry enough to want such bait.

BLACKBASSES

Genus *Micropterus* Lacépède 1802

("Small fin")

KEY TO SPECIES

Dorsal fin deeply notched, almost divided into two fins, the shortest spine of the notch less than half as long as the longest spine of the fin; upper jaw (maxillary) extending behind hind margin of eye in adults (measured with mouth closed); from 58 to 69 scales in the lateral line; cheek scales in 9 to 12 rows (*Micropterus salmoides*)_____ LARGEMOUTH BLACKBASS

Dorsal fin less deeply notched, the shortest spine at the emargination being more than half the length of the longest fin spine; upper jaw extending beyond middle of eye pupil but not to hind margin of eye; from 68 to 81 scales in the lateral line; cheek scales in 14 to 18 rows (*Micropterus dolomieui*)_____
_____ SMALLMOUTH BLACKBASS

LARGEMOUTH BLACKBASS

Micropterus salmoides (Lacépède)

(The Blackbasses have an extraordinary number of common names, depending upon the part of the country one is in.)

Labrus salmoides Lacépède 1802: 716
Cichla floridana Rafinesque 1820: 39
Cichla floridana, Le Sueur, 1822: 219 (Jour. Acad. Nat. Sci. Philadelphia 2)
Huro nigricans Cuvier, in Cuvier and Valenciennes, 1828: 93(124) (Hist. Nat. Poiss. 2)
Perca (Huro) nigricans, Richardson, 1836: 4–8
Grystes nobilis Agassiz 1854: 298 (Amer. Journ. Sci. & Arts 17)
Grystes nuecensis Baird and Girard 1854: 25 (Proc. Acad. Nat. Sci. Philadelphia 7)
Grystes megastoma Garlick 1857: 108 (Treat. Art. Prop. Fish.)
Huro nigricans, Günther, 1859: 255
Micropterus salmoides, Jordan and Copeland, 1878: 435
Micropterus salmoides, Jordan and Gilbert, 1882: 484–485
Micropterus salmoides, Bollman, 1892: 578
Micropterus salmoides, Jordan and Evermann, 1896B: 1012 / 1902: 355, 357–359
Micropterus salmoides, Forbes and Richardson, 1920: 267–269
Micropterus salmoides, Pratt, 1923: 121
Micropterus salmoides, Hildebrand and Towers, 1928: 134
Micropterus salmoides, Coker, 1930: 202
Huro floridana, Jordan, Evermann and Clark, 1930: 297
Micropterus salmoides, Neale, 1931A: 8–9
Micropterus salmoides, Walford, 1931: 87
Aplites salmoides, Fish, 1932: 380–382
Huro floridana, Hatton, 1932: 45–47
Aplites salmoides, Schultz, 1936: 162
Huro floridana, Tanner, 1936: 173
Micropterus floridanus, Schrenkeisen, 1938: 237–240
Huro salmoides, Kuhne, 1939: 94
Huro salmoides, Hubbs and Bailey, 1940: 37–39
Huro salmoides, Sumner, 1940: 220
Huro salmoides, Hubbs and Lagler, 1941: 77–78
Huro salmoides, Murphy, 1941: 169
Huro salmoides, Curtis, 1942: 4–5 / 1949: 258, 260–263
Huro salmoides, Moffett, 1942: 82 / 1943: 182
Huro salmoides, Hinks, 1943: 78–79
Huro salmoides, Dill, 1944: 174–177
Huro salmoides, Dimick and Merryfield, 1945: 43, 46
Huro salmoides, La Monte, 1945: 135
Huro salmoides, Miller and Alcorn, 1946: 185–186
Huro salmoides, Eddy and Surber, 1947: 228, 231–234
Micropterus salmoides, Hubbs and Lagler, 1947: 91, 93
Micropterus salmoides, Carl and Clemens, 1948: 98, 101–102
Micropterus salmoides, Miller, 1948: 94 / 1952: 7
Micropterus salmoides, Everhart, 1950: 48–49

Micropterus salmoides, Shapovalov and Dill, 1950: 387
Huro salmoides, Simon, 1951: 104–105
Micropterus salmoides salmoides, Wallis, 1951: 91
Micropterus salmoides salmoides, Beckman, 1952: 78–79
Micropterus salmoides, La Rivers, 1952: 89
Micropterus salmoides, La Rivers and Trelease, 1952: 119
Micropterus salmoides, Scott, 1954: 98
Micropterus salmoides, Gabrielson and La Monte, 1954: 89
Micropterus salmoides, Shomon, 1955: 14–15
Micropterus salmoides salmoides, Harlan and Speaker, 1956: 127–130
Micropterus salmoides, Koster, 1957: 89–90
Micropterus salmoides, Eddy, 1957: 184
Micropterus salmoides, Moore, 1957: 168
Micropterus salmoides salmoides, Trautman, 1957: 493–495
Micropterus salmoides, Slastenenko, 1958A: 284–287
Microptereus salmoides, Carl, Clemens and Lindsey, 1959: 151–152
Micropterus salmoides, Shapovalov, Dill and Cordone, 1959: 173

ORIGINAL DESCRIPTION

"SUPPLÉMENT
"Au Tableau et a la Synonymie
"Du Genre des Labres

"PREMIER SOUS-GENRE

La nageoire de la queue, fourchue, ou échancrée en croissant

Espèces.	Caractères.
49. LE LABRE SALMOIDES. (*Labrus salmoïdes.*)	Neuf rayons aiguillonnés et treize rayons articulés à la nageoire du dos; treize rayons à la nageoire de l'anus; l'opercule composé de quatre lames, et terminé par une prolongation anguleuse; deux orifices à chaque narine; la couleur générale d'un brun noirâtre.

LE LABRE SALMOÏDE;
LE LABRE IRIS;
et
SUPPLÉMENT A LA SYNONYMIE
DU LABRE SPAROIDE

"On devra au citoyen Bosc la connoissance du labre salmoïde et du labre iris, qui tous les deux habitent dans les eaux de la Caroline.

"Le salmoïde a une petite élévation sur le nez; l'ouverture de la bouche fort large; la mâchoire inférieure un peu plus longue que la supérieure; l'une et l'autre garnies d'une grande quantité de dents très-menues; la langue charnue; le palais hérissé de petites dents que l'on voit disposées sur deux rangées et sur une plaque triangulare; le gosier situé au-dessus et au-dessous de deux autres plaques également hérissées; l'oeil grand; les côtés de la tête, revêtus de petites écailles; la ligne latérale parallèle au dos; une fossette propré à recevoir la partie antérieure de la dorsale; les deux thoracines réunies par une membrane; l'iris jaune, et le ventre blanc.

"On trouve un très-grand nombre d'individus de cette espèce dans toutes les rivières de la Caroline; on leur donne le nom de *trant* ou *truite*. On les prend à l'hameçon; on les attire par le maoyen de morceaux de *cyprin*. Ils parviennent à la longueur de six ou sept décimètres; leur chair est ferme, et d'un goût très-agréable" (Lacépède 1802: 716–718).

DIAGNOSIS

Head and body depth each 3.0 to 3.5 into body length; mouth large, the maxillary reaching beyond the eye orbit in adult specimens; lower jaw very prominent, protruding; teeth entirely lacking on tongue; cheek scales large, in from 9–12 rows; branchiostegals usually 6; lateral line complete; scales 7–9/58–70/14–18, 58–67 with pores—scales

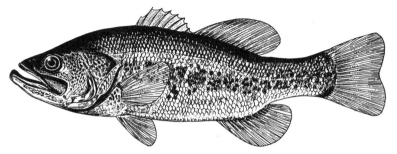

Fig. 244. Largemouth Blackbass, *Micropterus salmoides*. Courtesy California Department of Fish and Game and Leo Shapovalov.

around caudal peduncle 24–30, usually 26–28; fins—dorsal almost split between spinous and soft portions (shortest dorsal spine 2.4–3.9 times into longest spine), anterior half 9–10 spined, posterior half 12–13 rayed (rarely 11–14)—anal 3-spined, 11-rayed (rarely 10–12)—anal and soft dorsal lacking scales near bases of inter-radial membranes (excluding basal sheath of scales)—caudal broadly emarginate; pyloric caeca usually bifid near base; color—back dark green, becoming lighter along sides and white on belly—sides traversed by a broken dark green stripe—basal caudal spot particularly conspicuous in young, but still noticeable in adults; in size, a rather large species under optimum conditions, attaining extreme weights of nearly 25 pounds and lengths in excess of 3 feet, with average maxima much less than these—5 pounds is a more usual weight.

TYPE LOCALITY

South Carolina, probably Charleston, according to Hubbs and Bailey (1940: 37), since Bosc, the original collector, centered his activities there.

RANGE

Originally confined to the eastern United States from the Great Lakes region to northeastern Mexico and Florida, thence north along the coast to Virginia—in streams and ponds. It has been widely planted in the West, as well as into Europe, and very successfully in both areas. It is the warm water counterpart of the Smallmouth Blackbass (*Micropterus dolomieui*).

The first reliable records for Nevada postdate 1900, although there is good reason to believe that the Largemouth—like the Smallmouth—was brought into the State well before the turn of the century. By 1909, the Nevada Commission could report that largemouth were well established in ponds in the Reno area, from whence they were being distributed to other counties. Prior to 1900, Fish Commission and Commissioner reports did not differentiate between the two species of blackbass.

At the present writing (1962), Largemouth Blackbass constitute the most prominent fishery in Lakes Mead and Mohave (Clark County), Stillwater Marsh and vicinity (Churchill County) and Ruby Marsh (Elko County) as well as numerous other parts of the State. The major Lahontan rivers all support largemouth populations in their lower reaches, at least, particularly where reservoirs are present such as Lahontan Reservoir (lower Carson and Truckee Rivers), Rye Patch Reservoir (lower Humboldt River) and Weber Reservoir (lower Walker River). Fish Lake Valley (Esmeralda County), Fish Creek Springs (Eureka County) and Railroad Valley (Nye County) are among minor areas containing largemouth. They are the commonest fish in Manzanita Pond, on the University of Nevada campus, where they are overcrowded, stunted and provide considerable fishing and experience in fishing for many small fry of the town.

TAXONOMY

The nomenclature of this popular game and food fish has remained relatively stable since Jordan's publications of the late 1870s; by that time the specific synonymy had been established, and the changes since then have been those of deciding which genus the animal belonged in. Those who feel that the differences between the Largemouth Blackbass and the smallmouth are important enough to warrant generic separation have available the 1828 genus *Huro* of Cuvier for the former, a view which Hubbs and Bailey preferred in their comprehensive 1940 revision of the blackbasses. However, current trends are conservative, and most ichthyologists seem satisfied that variation within generic bounds can logically include both blackbasses in the single genus *Micropterus*. Both species have wide enough ranges to show some geographical segregation, particularly the smallmouth.

Etymology—"salmoides," trout-like, in reference to the use of the name "trout" for this fish in the southern United States.

LIFE HISTORY

A carnivorous fish, eating Crustacea, insects and aquatic vertebrates, particularly fishes, in that order as it gets larger. It spends much time lurking under overhanging banks, logs, etc., in wait for prey. It is a spring spawner, the male building a nest in shoal water, either on sand, gravel, rock or silt bottoms or among aquatic vegetation. Nest-building consists of the male rooting into the bottom and hollowing out a depression with the aid of rapid tail vibrations; such activities begin around

60° F. (15.5° C.) and egg laying is usually initiated when the temperatures are between 60° F. and 65° F. (18.5° C.).[1]

Sumner[2] (1954) reports spawning spring temperature extremes of 55° F. to 75° F. in Lake Mead, with the greatest activity centered about 57° F.–65° F. Nest building may go on at depths exceeding 15 feet and may take as long as two days. When the nest is finished, the male herds the female into it and spawning takes place with the vents of both fishes closely approximated. After egg-deposition, the male generally drives the female away, and guards the nest until the eggs have hatched (a week or ten days) and the young have left the nest (another ten days).

Yearly growth rates vary with the localities involved, being higher in warmer, southern United States waters than they are about the Great Lakes, for instance. The following tabular summary demonstrates this, showing the lessened growth rate for the colder, more northern Ruby Marsh as compared with more southern waters:

	1 year (in.)	2 years (in.)	3 years (in.)	4 years (in.)	5 years (in.)	6 years (in.)	7 years (in.)
Norris Reservoir, Tennessee (Stroud 1948)	6.9	12.2	14.7	16.1	17.5	19.3	20.8
Lake Mead, Nevada (Moffett 1943)	10.3	12.6	13.4	14.6	15.2	16.4	20.1
Lake Mead, Nevada (Sumner 1954)	5.3	10.3	13.7	16.4	18.5	20.5	
Stillwater Marsh, Nevada (Sumner 1948)	2.0	10.0	15.0	17.0	19.0	19.5	20.0
Stillwater Marsh, Nevada (Sumner 1951)	2.7	7.9	11.1	13.5	15.4		
Ruby Marsh, Nevada (Trelease 1948)	....	....	7.0	8.8	12.8	15.8	

Norris, Stillwater and Mead approximate each other as largemouth environments, but the colder Ruby Marsh is a more precarious habitat for this warm water fish, for Ruby is subject to more extreme winter conditions and, in occasional years, may freeze vertically solid except for the few deeper holes. Stillwater Marsh also experiences freezing temperatures nearly every year, but its deeper waters plus the general lack of snow over the ice eliminates the winter killing factor which occasionally decimates Ruby Marsh. Such growth conditions not only produce a long winter lapse, but may, as happened in Ruby Marsh during the winter of 1947–1948, eliminate most of the blackbass population.

Similarly, in Minnesota and Wisconsin waters, growths may be only half the Norris rates for the first two years, with more equality at later ages. Stroud's estimate of longevity for Norris largemouth was that

---

[1] Sumner (Jonez and Sumner 1954: 131) notes considerable versatility in nest placement in Lake Mead, from the usual sites to the tops of submerged hills and even on large boulders; only mud and sand bottoms seemed generally avoided. Two to ten feet seemed to be the preferred depths.

[2] Robert Sumner has evidence that other environmental factors produce different temperature spawning responses. "Most optimum temperatures for spawning are varied, I believe, with the area. In other words, they seem to have preferred temperatures at various locales which are not necessarily the same. At Mead it was between 57° F. and 65° F. At Stillwater it is between 63° F. and 68° F. And at Ruby I believe it is slightly different" (personal correspondence 1958).

"it is probable that a large proportion . . . die of 'old age' before reaching the age of 5 years, although a few live as long as 7 years or perhaps longer" (1948).

The maximum recorded length and weight for this species is that of a Florida specimen 37½ inches long and weighing 23½ pounds. The species seems to thrive best in waters which reach a temperature of 80° F. (26.6° C.) for several months of the year.

ECONOMICS

The Largemouth Blackbass is of prime importance wherever it is found, native or introduced, as the source of a warm water sport fishery. In gaminess, it is usually considered slightly inferior to the smallmouth, which it exceeds in size, but it is unexcelled as a pan fish.

FIG. 245. Ruby Marshes, Elko County, looking northerly.
Photo by the author, 1948.

In Nevada, this fish provides a major part of the small warm water fishery which exists here. Lakes Mead and Mohave, Stillwater Marsh and Ruby Marsh, and the major reservoirs in the Lahontan system all support sizeable populations of this blackbass.

The taking of blackbass in Nevada apparently was sporadic until the early 1940s, when previous plants in the Ruby Marshes began to produce good local fishing. About this same time another good blackbass fishery developed at Lake Mead in southern Nevada and later at Stillwater in the west-central part of the State near Fallon. Ruby Marsh fishing has been cyclic due to lack of suitable food for larger fishes and to periodic stiff winterkills.

Robert Sumner points out that "the Lake Mead fishery first received acclaim as a bass fishery in 1939 and 1940. It was well established by 1942 and was mentioned nationally at that time.

"The bass population in Lahontan Reservoir seems to be building up, and there are high hopes for the creation of an excellent bass fishery here as well. There are also potentialities in Weber (Schurz) Reservoir" (1959).

Lake Mead is currently the best blackbass fishing waters in the State, although its productivity problems are limiting fish numbers as is the case with reservoirs in general. Sumner's data (Jonez and Sumner 1954) show a slight decrease in average length of fish taken over the several-year period of his study in the early 1950's, which is to be expected where angling pressure is rapidly on the upswing.

The last year of the study, 1954, produced an average length of 11.1 inches, about an inch under the 1951–1952 averages. In the area below Lake Mead, trout have decreased markedly in numbers because of the development of another reservoir—Lake Mohave—and blackbass fishing has superseded the former very good trout fishery as a result.

SMALLMOUTH BLACKBASS
Micropterus dolomieui Lacépède

(numerous common names)

Micropterus dolomieu Lacépède 1802: 325
Bodianus achigan Rafinesque 1817: 120
?Lepomis pallida Rafinesque 1820: 30
Lepomis trifasciata ibid: 31
Lepomis flexuolaris ibid
Lepomis salmonea ibid: 32
Lepomis notata ibid
?Etheostoma calliura ibid: 36
Cichla fasciata Le Sueur 1822: 216 (Acad. Nat. Sci. Phila. Jour. 2)
Cichla ohiensis ibid: 218
Cichla minima ibid: 220
Centrarchus obscurus DeKay 1842: 30 (N. Y. Fauna, Fishes)
Centrarchus fasciatus, Günther, 1859: 258
Centrarchus obscurus, ibid
Dioplites variabilis Vaillant and Bocourt 1874: 152 (Mission Scientifique, Mexique, Poissons)
Micropterus pallidus, Jordan and Copeland, 1878: 435
Micropterus dolomieu, Jordan and Gilbert, 1882: 485 / 1894: 140
Micropterus dolomiei, ibid, 1882B: 916
Micropterus dolomieu, Bollman, 1892: 577
Micropterus dolomieu, Jordan and Evermann, 1896B: 1011–1012 / 1902: 355–357
Micropterus dolomieu, Forbes and Richardson, 1920: 263–266
Micropterus dolomieu, Pratt, 1923: 121
Micropterus dolomieu, Coker, 1930: 203
Micropterus dolomieu, Jordan, Evermann and Clark, 1930: 298

Micropterus dolomieu, Neale, 1931A: 9–10
Micropterus dolomieu, Walford, 1931: 86
Micropterus dolomieu, Fish, 1932: 378–380
Micropterus dolomieu, Schultz, 1936: 162
Micropterus dolomieu, Tanner, 1936: 173
Micropterus dolomieu, Schrenkeisen, 1938: 240–242
Micropterus dolomieu, Kuhne, 1939: 94–98
Micropterus dolomieu, Hubbs and Bailey, 1940: 29–39
Micropterus dolomieu dolomieu, Murphy, 1941: 169
Micropterus dolomieu, Curtis, 1942: 4 / 1949: 258, 260–263
Micropterus dolomieu, Hinks, 1943: 77–78
Micropterus dolomieu dolomieu, Miller and Alcorn, 1946: 184–185
Micropterus dolomieu, Carl and Clemens, 1948: 98–100
Micropterus dolomieu, Everhart, 1950: 48
Micropterus dolomieu dolomieu, Shapovalov and Dill, 1950: 387
Micropterus dolomieu dolomieu, Simon, 1951: 103
Micropterus dolomieui dolomieui, Beckman, 1952: 77–78
Micropterus dolomieui, La Rivers, 1952: 89
Micropterus dolomieu, La Rivers and Trelease, 1952: 119
Micropterus dolomieu, Gabrielson and La Monte, 1954: 89–90
Micropterus dolomieu, Scott, 1954: 99–100
Micropterus dolomieu, Shomon, 1955: 16–17
Micropterus dolomieui, Harlan and Speaker, 1956: 124–127
Micropterus dolomieui, Eddy, 1957: 185
Micropterus dolomieui, Koster, 1957: 90–91
Micropterus dolomieui, Moore, 1957: 167
Micropterus dolomieui dolomieui, Trautman, 1957: 486–489
Micropterus dolomieui, Slastenenko, 1958A: 282–284
Micropterus dolomieui, Carl, Clemens and Lindsey, 1959: 149–150
Micropterus dolomieui, Shapovalov, Dill and Cordone, 1959: 173

ORIGINAL DESCRIPTION

"CENT DIX–HUITIÈME GENRE.
LES MICROPTÈRES.

"*Un ou plusieurs aiguillons, et point de dentelure, aux opercules; un barbillon, ou point de barbillon, aux mâchoires; deux nageoires dorsales; la seconde très-basse, très-courte, et comprenant au plus cinq rayons.*

Espèce.	Caractères.
I. LE MICROPTÈRE DOLOMIEU. (*Micropterus dolomieu.*)	Dix rayons aiguillonés et sept rayons articulés a la première nageoire du dos; quatre rayons à la seconde; deux rayons aiguillonés et onze rayons articulés à la nageoire de l'anus; la caudale en croissant; un ou deux aiguillons à la seconde pièce de chaque opercule.

"LE MICROPTÈRE DOLOMIEU.[1]

"Je desire que le nome de ce poisson, qu'aucun naturaliste n'a encore décrit, rappelle ma tendre amitié et ma profonde estime pour l'illustre Dolomieu, dont la victoire vient de briser les fers. En écrivant mon Discours sur la durée des espèces, j'ai exprimé la vive douleur que m'inspiroit son affreuse captivité, et l'admiration pour sa constance héroiquè, que l'Europe mêloit a ses voeux pour lui. Qu'il m'est doux de ne pas terminer l'immense tableau que je tâche d'esquisser, sans avoir senti le bonheur de le serrer de nouveau dans mes bras!

"Les microptères ressemblent beaucoup aux sciènes: mais la petitesse très-remarquable de leur seconde nageoire dorsale les en sépare; et c'est cette petitesse que désigne le nom générique que je leur ai donné.[2]

"La collection due Muséum national d'histoire naturelle renferme un bel individu de l'espèce que nous décrivons dans set article. Cette espèce, qui est encore le seule inscrite dans le nouveaux genre des microptères, que nous avons cru devoir établir, a les deux mâchoires, le palais et la langue, garnis d'un très-grande nombre de rangées de dents petites, crochues et serrées; la langue est d'ailleurs très-libre dans ses mouvemens; et la machoire inférieure plus avancée que celle d'enhaut. La membrane branchiale disparoit entièrement sous l'opercule, qui présente deux pièces, dont la première est arrondie dans son contour, et la seconde anguleuse. Cet opercule est couvert de plusieurs écailles; celles du dos sont assez grandes et arrondies. La hauteur du corps proprement dit excède de beaucoup celle de l'origine de la queue. La ligne latérale se plie d'abord vers le bas, et se relève ensuite pour suivre le courbure du dos. Les nageoires pectorales et cell de l'anus sont très-arrondies; la première du dos ne commence qu'à une assez grande distance de la queue.

"Elle cesse d'être attachée au dos de l'animal, à l'endroit ou elle parvient au-dessus de l'anale; mais elle se prolonge en bande pointue et flottante jusqu'au-dessus de la seconde nageoire dorsale, qui est trèsbasse et très-petite, ainsi que nous venons de le dire, et que l'on croiroit au premier coup d'oeil entièrement adipeuse.*"

DIAGNOSIS

Head and body depth each 2.5 to 3.5 times into body length; mouth moderately large, the maxillary not reaching hind margin of eye orbit in adult specimens; lower jaw very prominent, protruding; teeth present on posterior portion of tongue; cheek scales small, in more than 12 rows, usually 17 rows; branchiostegals usually 6; lateral line complete; scales 10–14/67–85/17–25, 67–78 with pores—scales around caudal peduncle 28–32, usually 29–31; fins—dorsal nearly split, but not as much as in *M. salmoides* (shortest dorsal spine 1.1–2.5 times into longest spine), anterior half 9–10 spined, posterior half 13–15

"[1]Micropterus dolomieu.
"[2]Micros en grec, signifie *petit*.
"*5 rayons à la membrane branchiale.
"16 rayons a chaque pectorale.
"1 rayon aiguillonné et 5 rayons articulés a chaque thoracine.
"17 rayons à la nageoire de la queue" (Lacépède 1802: 324–326).

rayed (rarely 12)—anal 3-spined, 10–12 rayed (rarely 9)—anal and soft dorsal with scales near bases of inter-radial membranes (excluding basal sheath of scales)—caudal rather broadly emarginate; pyloric caeca usually unbranched; color—dull greenish gold, with bronze luster, weak lateral barring of a slightly darker color along the sides, most prominent in young—belly white, caudal fin usually darker than rest of body; a moderate sized species, attaining extreme weights approaching 15 pounds in the southeastern United States, but averaging less than 5 pounds.

TYPE LOCALITY

The United States. No locality was specified in the original description, and older authors have suggested South Carolina as the probable type locality (Jordan, Evermann and Clark 1930: 298). However, Hubbs and Bailey (1940: 36) pointed out that the smallmouth is not

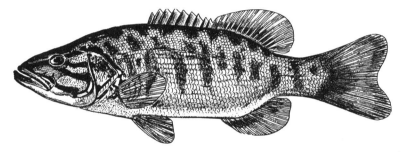

FIG. 246. Smallmouth Blackbass, *Micropterus dolomieui*. Courtesy California Department of Fish and Game and Leo Shapovalov.

native to South Carolina and thought that Lake Champlain would be a more logical type locality since it "was the only readily accessible part of the range of the northern smallmouthed bass in the time of Dolomieu."

RANGE

Native to the eastern United States region from Quebec to the Dakotas, south to Arkansas and Alabama—did not originally occur east of the Allegheny Mountains. The species has been introduced into other parts of the southern United States, such as Florida, and also into the West, as well as Europe.

In Nevada, smallmouth have been planted in several areas, beginning with the 1888 introductions in the Carson City region (Carson River and Washoe Lake). In 1911, Nevada Fish Commission records indicate some 400 were sent to Stone Cabin in Nye County. The Hubbs sight record of a specimen in Lake Mead in 1942 is the most recent one, but has not been confirmed by later observations or specimens.

Apparently, with the exception of a September 1956 stocking of smallmouth in the Humboldt River from Battle Mountain to Elko by the Nevada Commission, this species is extinct in the State. Currently (1962), these southern California immigrants are at least holding their own, as attested by fingerling production.

TAXONOMY

The long and complex synonymy was put in shape by Hubbs and Bailey in 1940, at which time they distinguished several previously unnamed segregates of the *dolomieui* complex. The specific synonymy had been fairly well stabilized since Henshall's contribution in 1881. As presently constituted, the *dolomieui* complex consists of three species, *dolomieui*, *punctulatus* and *coosae*, and several subspecies: *M. d. velox*, *M. p. wichitae* and *M. p. henshalli*. Since some of these may be desirable forms to introduce to certain Nevada waters, it will be appropriate to elaborate somewhat on them.

M. punctulatus, the Spotted Blackbass, was described by Hubbs in 1927 as the new species *M. pseudaplites*, and synonymized with Rafinesque's poorly described *Calliurus punctulatus* (1819) by Hubbs and Bailey in 1940. It occurs native from Ohio and West Virginia to Oklahoma (*M. p. wichitae* Hubbs and Bailey 1940) and Georgia (*M. p.*

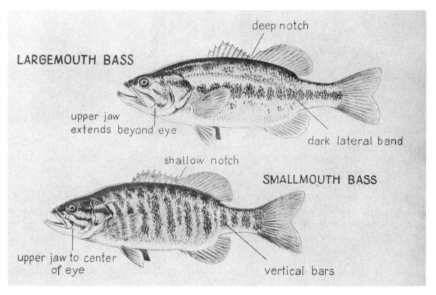

Fig. 247. Our two blackbasses compared. Courtesy Wisconsin Conservation Department, Jens von Sivers and N. H. Hoveland.

henshalli Hubbs and Bailey 1940), and prefers quieter waters and tolerates man-made impoundments and turbid water better than *M. dolomieui*. In spite of the frequent stocking of *M. salmoides* into Ohio waters, for example, *M. punctulatus* is the dominant blackbass in the area, the former maintaining itself successfully only in impoundments in the southern part of the State.

M. coosae Hubbs and Bailey 1940, the Redeyed Blackbass, is a marked geographical upland stream differentiate with many of the characteristics of both *dolomieui* and *punctulatus*, restricted to parts of Alabama and Georgia. It is a respected game fish. The maximum sizes of *punctulatus* and *coosae* are not matters of record, since lengths and weights in the literature are quite obviously below the species' potentials: Hubbs and Bailey (1940) list an Eschmeyer specimen of *punctulatus* from Norris Lake, Tennessee, 17½ inches long and weighing 3 pounds, 15 ounces (p. 16) and quote 2 pounds as the maximum recorded weight of *coosae* (p. 26). The largest specimen they examined was only a little over ten inches in total length.

While all introductions of *M. dolomieui* stock into Nevada seem referable to *M. d. dolomieui* (Miller and Alcorn 1946: 184–185), the southwestern variant of the species, *M. d. velox* Hubbs and Bailey 1940, from Arkansas, Oklahoma and Missouri, might be a desirable stream type for some parts of Nevada.

Etymology—named for the French mineralogist M. Dolomieu, for whom the rock dolomite was also named.

LIFE HISTORY

Much like that of the largemouth, except that it is usually credited with being a more gamey fish, probably as a result of its colder water habitat, since the smallmouth prefers colder, swifter and clearer waters, generally, than those occupied by the largemouth, although the two often occur together, particularly in larger lakes where deep waters are cool and shoal waters provide a warm water habitat for the largemouth. While both are carnivorous, the smallmouth does not seem to be as strongly piscivorous as its relative, feeding more upon the available invertebrates, which are usually insects.

ECONOMICS

While the smallmouth is one of the most important of the eastern game fish, and a substantial part of the warm water fishery in California, where it has been introduced, its doubtful establishment in Nevada makes its value here a moot question. Perhaps with more judicious stocking, it might, in future, provide a limited fishery in certain sections of Nevada.

THE SUNFISHES

Genus *Lepomis* Rafinesque 1819

("Scale operculum")

KEY TO SPECIES

Pectoral fins short and rounded, their length about one-fourth that of the standard length; mouth large, the upper jaw or maxillary reaching behind front edge of eye (in adults, the upper jaw is two or more times the width of the eye); usually lacking any definite isolated black spot on dorsal fin rays, except for a darkening at ray bases (*Lepomis cyanellus*)
GREEN SUNFISH

Pectoral fins long and pointed, their length about one-third that of the standard length; mouth small, the maxillary not reaching the front edge of eye (in adults, the maxillary is but slightly longer than the width of the eye); possesses a prominent black isolated spot on dorsal fin rays (*Lepomis macrochirus*) .... BLUEGILL SUNFISH

BLUEGILL SUNFISH

Lepomis macrochirus Rafinesque

(numerous common names)

Labrus palladus Mitchill 1815: 407
Lepomis macrochirus Rafinesque 1819: 420
Pomotis incisor Cuvier and Valenciennes 1831: 350 (466)
Pomotis gibbosus ibid: 351 (467)
Pomotis obscurus Agassiz 1854: 302 (Amer. Jour. Sci. Arts 17)
Pomotis speciosus Baird and Girard 1854: 24 (Acad. Nat. Sci. Phila. Proc. 7)
Pomotis luna Girard 1857: 201 (ibid 9)
Pomotis machrochir Günther, 1859: 263
Lepomis longispinis Cope 1865: 83 (ibid 17)
Lepomis ardesiacus Cope 1868: 222 (Acad. Nat. Sci. Phila. Jour. 6)
Lepomis nephelus ibid
Lepomis purpurascens Cope 1870: 454 (Amer. Philos. Soc. Proc. 11)
Lepiopomus pallidus, Jordan and Copeland, 1878: 397–398
Lepomis macrochirus, Jordan and Gilbert, 1882: 475
Lepomis pallidus, ibid: 479–480
Lepomis macrochirus, Bollman, 1892: 572
Lepomis macrochirus, Jordan and Evermann, 1896B: 1005 / 1902: 345, 348
Lepomis pallidus, ibid: 1005–1006 / 1902: 349–350
Lepomis pallidus, Snyder, 1917A: 86
Lepomis pallidus, Forbes and Richardson, 1920: 257–259
Lepomis macrochirus, Pratt, 1923: 119
Lepomis pallidus, ibid
Lepomis pallidus, Coggeshall, 1924
Lepomis pallidus, Potter, 1925
Lepomis incisor, Hildebrand and Towers, 1928: 132
Lepomis incisor, Coker, 1930: 203
Lepomis pallidus, Jordan, Evermann and Clark, 1930: 299
Helioperca macrochira, ibid: 301
Heliperca incisor, Neale, 1931A: 10
Helioperca incisor, Walford, 1931: 85
Helioperca incisor, Schultz, 1936: 162
?Lepomis auritus?, Tanner, 1936: 173
Helioperca incisor, Schrenkeisen, 1938: 249–251
Helioperca macrochira, ibid: 251
Helioperca nephela, ibid
Lepomis macrochirus, Kuhne, 1939: 108
Lepomis macrochirus, Leonard, 1939
Lepomis macrochirus, Curtis, 1941: 259, 263–264
Lepomis macrochirus macrochirus, Hubbs and Lagler, 1941: 78–79 / 1947: 92–94

Lepomis macrochirus, Murphy, 1941: 169
Lepomis macrochirus, Moffett, 1943: 182
Lepomis macrochirus, Ricker, 1943
Lepomis macrochirus, Dill, 1944: 169–171
Lepomis macrochirus macrochirus, Dimick and Merryfield, 1945: 46
Lepomis macrochirus, La Monte, 1945: 138–139
Lepomis macrochirus macrochirus, Miller and Alcorn, 1946: 186
Lepomis macrochirus macrochirus, Eddy and Surber, 1947: 238–240
Lepomis macrochirus macrochirus, Shapovalov and Dill, 1950: 387
Lepomis macrochirus macrochirus, Simon, 1951: 106–107
Lepomis macrochirus macrochirus, Wallis, 1951: 91
Lepomis macrochirus macrochirus, Beckman, 1952: 84–85
Lepomis macrochirus, La Rivers, 1952: 89
Lepomis macrochirus, La Rivers and Trelease, 1952: 120
Lepomis macrochirus, Miller, 1952: 11, 37
Lepomis macrochirus macrochirus, Stokely, 1952
Lepomis macrochirus, Gabrielson and La Monte, 1954: 91–92
Lepomis macrochirus, Scott, 1954: 102
Lepomis macrochirus macrochirus, Shomon, 1955: 25–29
Lepomis macrochirus, Harlan and Speaker, 1956: 132–136
Lepomis macrochirus, Eddy, 1957: 194
Lepomis macrochirus, Koster, 1957: 95–96
Lepomis macrochirus, Moore, 1957: 171
Lepomis macrochirus macrochirus, Trautman, 1957: 502–504
Lepomis macrochirus, Slastenenko, 1958A: 294–296
Lepomis macrochirus macrochirus, Shapovalov, Dill and Cordone, 1959: 173

ORIGINAL DESCRIPTION

". . . . . 2. *L. macrochirus*. Corps ovale, points bruns, point d'auricule, tache oblongue toute noire, pectorales très-longues atteignant l'anale, queue fourchue" (Rafinesque 1819: 420).

This is improved somewhat by an immediately preceding generic description:

"13. Lepomis. (Thoracique.) Corps arrondi, ovale ou oblong, très-comprimé. Tête et opercules écailleux, ceux-ci mutiques, le postérieur flexeux, membraneux, quelquefois auriculé. Bouche petite, mâchoires à petites dents, lèvre supérieure à peine extensible. Une nageoire dorsale, nageoire thoracique à 6 rayons dont l'épineux, sans appendices. Anus au milieu" (Rafinesque 1819: 420).

DIAGNOSIS

Head about 3.0, body depth about 2.0, times into body length, respectively; mouth small, oblique; jaws subequal; maxillary not quite, to just, reaching vertical line from anterior edge of eye; palatine teeth lacking; opercle with a dorso-caudally projecting blackish flap about on the median line, its center line slightly below center line of eye; cheek scales large, in about 5 rows; lateral line complete; scales 6/38–50/13–14, 40–50 with pores; fins—dorsal with 10 anterior spines, 11–12 posterior rays—anal with 3 anterior spines, 10–12 posterior rays —pectorals relatively long and narrow, pointed, about subequal in length to the head, reaching to or beyond base of anal spine I—caudal moderately and roundly emarginate; color—a medium olive green dorsally, becoming lighter to whitish on the belly—sides with greenish vertical bars—older specimens may have flanks and belly reddish; in size, reaching lengths of about a foot, and weights of about 1½ pounds.

TYPE LOCALITY
Ohio, Wabash, Green and Licking Rivers.

RANGE
Originally confined to the eastern United States from the Great Lakes region to New Jersey, south to Texas and Florida. The species has been introduced into many parts of the western United States, with great success.

The first known Nevada introduction was in 1909–1910, when 150 specimens of varying ages were stocked into Olsen's Lake at Ely, in

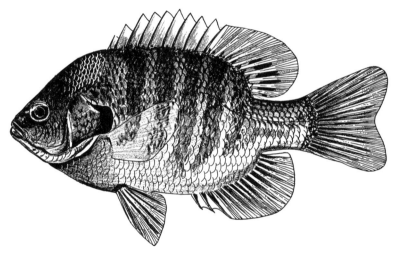

FIG. 248. Adult Bluegill Sunfish, *Lepomis macrochirus*. Courtesy California Department of Fish and Game and Leo Shapovalov.

White Pine County (U. S. Comm. Fish. 1911: 102). A 1918–1919 planting of another 150 fish, supposedly bluegills, was made in the Fallon area (Cottonwood Cañon Creek), where the species is now common (U. S. Comm. Fish. 1920: 69). A larger plant of 640 fish, 4 to 7 inches long, was made near Genoa, Douglas County, in the West Carson River during 1929–1930 (Nev. Fish & Game Comm. 1931: 23–37).

In addition to these, several other implantations of the species have been made into west-central Nevada during the intervening years. As Miller and Alcorn have summed it up (1946: 186): "According to a letter from O. Lloyd Meehean, the 280,000 sunfish sent to Nevada in 1940 and 1941 (Leach, James, and Douglass, 1941: 569, 603; 1942: 10, 24; and 1943: 14, 25) were of this species, and most if not all of those shipped between 1924 and 1938 were also bluegills."

At the time of their writing, however, bluegills were not as abundant in the Fallon vicinity as they are now, as attested by their statement:

"One specimen was taken on January 28, 1943, from the stomach of an American Merganser collected from the lower Carson River, where bluegills apparently are not abundant. We have not seen specimens from elsewhere in the state, although this species is common in Lake Mead (Moffett, 1943: 182)."

At the present time, Bluegill Sunfish are found in the lower portions of all the major Lahontan-system rivers as well as their sumps; Reno and vicinity; Lakes Mead and Mohave; Fish Lake Valley and Silver Peak (Esmeralda County); Adams-McGill Reservoir, Nye County (Ted Frantz), etc.

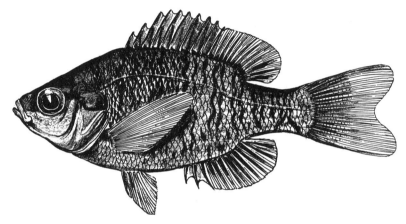

FIG. 249. Young Bluegill Sunfish, *Lepomis macrohirus*. Drawn by William L. Brudon for Miller 1952. Courtesy California Department of Fish and Game and Leo Shapovalov.

TAXONOMY

As indicated in the extensive synonymicon, the long and involved terminology of the Bluegill Sunfish was not generally resolved until about 1940, 125 years after its formal description.

Etymology—"macrochirus," large hand, in reference to the long, pointed pectoral fin.

LIFE HISTORY

A pond and large river species, the bluegill is quite gregarious, generally being found in schools. Its late-spring to early-fall spawning is colonial, with typical centrarchid redd-building (see the Largemouth Blackbass), and it spawns somewhat later in the year than the blackbasses, seeming to prefer water around 75° F. (23.8° C.) in temperature over most of its native range. Bluegill and Largemouth Blackbass are often found together, having apparently about the same habitat requirements.

Carbine's interesting observations, reported in 1939, of centrarchid nesting patterns in Michigan, showed considerable use of each others' nests by the four centrarchids studied[1] as well as different individuals of the same species.

He reported bluegill to use all bottom types available for nests, including mud, in about 2½ feet of water, and to be distinctly colonial. They were more timid in guarding the nest than associated species, but spawned longer and produced more fry. His figures gave an average number of fry amounting to 17,914 per nest with extremes of 4,670 and 61,815 noted for individual nests. Such high figures, and particularly those of Coggeshall (1924) who found one bluegill nest with 224,000 fry, indicate the multiple use of a single nest by several females.

For his particular study, Carbine estimated that bluegill produced a total of 6,610,000 fry in the 15-acre lake, while the blackbass totalled only 164,000 fry. Little wonder the bluegill is so common wherever found. Such productivity can be expected of fishes producing some 67,000 eggs for a single female (Eddy and Surber 1947: 240) and the use of the same guarded nest by several females.

As a general habit, bluegills seem to prefer the shallower, marginal water (not exceeding 15 to 20 feet in depth) where aquatic plants are plentiful. The bluegill diet consists largely of invertebrates, occasionally small fishes and some vegetation. It is the largest sunfish, about one foot being maximum length.[2]

In Nevada, bluegill have been most intensively studied in the Lake Mead area where the Nevada Fish and Game Commission recently completed a several years' study. Robert Sumner (Jonez and Sumner 1954) reported bluegills as making up some 2.4 percent of the gillnet sampling catch, but estimated them to be more abundant on other grounds than that figure indicated.

"Bluegills were observed spawning in June in water whose temperature was 68° F. and warmer. In 1954, bluegills were found spawning during the first week of June. They were digging nests in about six feet of water in locations similar to those described under the life history of bass. The nests, which usually were round, ranged from about six inches to about 12 inches in diameter,. These warm-water game fish were much more gregarious than the bass; they allowed others of their kind to spawn within several inches of their nests. As many as 60 nests were observed in an area of about 60 square feet" (p. 137).

The largest bluegill noted for Lake Mead was a 9-inch specimen. Much the same parasites were found in the bluegill as in the Largemouth Blackbass. Studies in the midwest indicate that parasitization of the bluegill (as is also the case with most centrarchids) is extensive. This involves, as the commoner parasites, such organisms as the flat flukeworm genera *Crepidostomum, Gyrodactylus, Neascus, Clinostomum* and *Bunodera;* tapeworms belonging to the genus *Proteocephalus;* roundworms of the genera *Spinitectus, Contracaecum* and

[1]Bluegill; Pumpkinseed or Common Sunfish (*Lepomis gibbosus* (Linnaeus) 1758) ; Rock Bass (*Ambloplites rupestris* (Rafinesque) 1817) ; and Largemouth Blackbass).

[2]A 24-inch extreme length given by Forbes and Richardson (1920) is certainly an error.

Capillaria; the copepod crustacea *Lernaea, Cyclochaeta* and *Ergasilus; Saprolegnia* fungi and numerous Protozoa.

ECONOMICS

In its native habitat, the bluegill is usually the commonest species of sunfish, and the one most often taken by commercial river fishermen. Although small, it is a gamey fish, providing considerable light-tackle sport; as a food fish, it is one of the best.

It has done very well in California, where specimens as heavy as 2½ pounds are known, with one-half pounders being average. In Nevada, it is becoming an increasingly important element in the State's limited warm-water fishery. Robert Sumner pointed out the utilization of bluegill in Lake Mead by both blackbass and Rainbow Trout in considerable numbers. This is an instance in which a game fish—not highly regarded in southern Nevada by the angling fraternity—is currently of more importance as food for the more eagerly sought blackbass and trout.

GREEN SUNFISH
Lepomis cyanellus Rafinesque
(numerous common names)

Lepomis cyanellus Rafinesque 1819: 420
Icthelis melanops ibid 1820: 28
Pomotis longulus Baird and Girard 1853: 391 (Proc. Acad. Nat. Sci. Philadelphia 6)
Calliurus diaphanus Girard 1857: 200 (ibid 9)
Calliurus formosus ibid
Calliurus microps ibid
Calliurus murinus ibid
Bryttus signifer ibid: 201
Calliurus longulus, Girard, 1859: 5
Bryttus longulus, Günther, 1859: 259
Bryttus melanops, ibid: 260
Bryttus murinus, ibid
Bryttus mineopas Cope 1865: 84 (Acad. Nat. Sci. Phila. Proc. 17)
Apomotis cyanellus, Jordan, 1878C: 435
Apomotis cyanellus, Jordan and Copeland, 1878: 398–399
Lepomis cyanellus, Jordan and Gilbert, 1882: 473
Lepomis murinus, ibid: 475–476
Lepomis cyanellus, Bollman, 1892: 569–570
Apomotis cyanellus, Jordan and Evermann, 1896B: 996 / 1902: 343–344
Lepomis cyanellus, Forbes and Richardson, 1920: 248–250
Lepomis cyanellus, Pratt, 1923: 118
Apomotis cyanellus, Hildebrand and Towers, 1928: 131–132
Lepomis cyanellus, Coker, 1930: 203
Apomotis cyanellus, Jordan, Evermann and Clark, 1930: 298–299
Apomotis cyanellus, Neale, 1931A: 10
Apomotis cyanellus, Walford, 1931: 85

Apomotis cyanellus, Hatton, 1932: 48–49
Apomotis cyanellus, Schultz, 1936: 161
Apomotis cyanellus, Tanner, 1936: 173
Apomotis cyanellus, Schrenkeisen, 1938: 242–243
Lepomis cyanellus, Kuhne, 1939: 112
Lepomis cyanellus, Hubbs and Lagler, 1941: 77, 79 / 1947: 91, 93
Lepomis cyanellus, Murphy, 1941: 169
Lepomis cyanellus, Curtis, 1942: 5 / 1949: 259, 263–264
Lepomis cyanellus, Dill, 1944: 172–173
Lepomis cyanellus, La Monte, 1945: 138
Lepomis cyanellus, Miller and Alcorn, 1946: 186
Lepomis cyanellus, Eddy and Surber, 1947: 235–236
Lepomis cyanellus, Shapovalov and Dill, 1950: 387
Lepomis cyanellus, Simon, 1951: 105–106
Lepomis cyanellus, Wallis, 1951: 91
Lepomis cyanellus, Beckman, 1952: 81–82
Lepomis cyanellus, La Rivers, 1952: 89
Lepomis cyanellus, La Rivers and Trelease, 1952: 119
Lepomis cyanellus, Miller, 1952: 11, 37–38
Lepomis cyanellus, Gabrielson and La Monte, 1954: 93
Lepomis cyanellus, Harlan and Speaker, 1956: 131
Lepomis cyanellus, Eddy, 1957: 190
Lepomis cyanellus, Koster, 1957: 93–94
Lepomis cyanellus, Moore, 1957: 170
Lepomis cyanellus, Trautman, 1957: 499–501
Lepomis cyanellus, Slastenenko, 1958A: 288–289
Lepomis cyanellus, Shapovalov, Dill and Cordone, 1959: 173

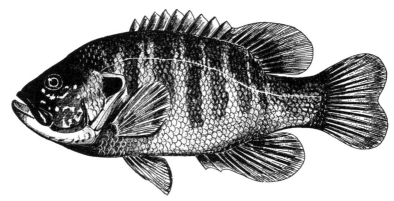

Fig. 250. Adult Green Sunfish, *Lepomis cyanellus*. Courtesy California Department of Fish and Game and Leo Shapovalov.

ORIGINAL DESCRIPTION

". . . . . 1. *L. cyanellus*. Corps oblong, tout couvert de points bleus, joues à lignes flexueuses bleues, opercule sans auricule, tache oblongue, queue bilobée" (Rafinesque 1819: 420). See the original description of the Bluegill Sunfish (*L. macrochirus*) for the additional generic description.

DIAGNOSIS

Head about 3.0, body depth about 2.5, times into body length, respectively; mouth large, oblique; lower jaw distinctly forward-jutting beyond upper jaw (prognathous); maxillary prominent and long, reaching about to a vertical line from the middle of the eye; palatine teeth present; opercle with a dorso-caudally projecting black-spotted flap which lies above the median line (about directly behind the eye) and is sharper than in the bluegill; cheek scales small, in from 7 to 10 rows; lateral line complete; scales 6–7/45–49/15–16, 40–48 with pores; fins—dorsal with 9–10 anterior spines, 10–12 posterior rays—anal with 3 anterior spines, 9–10 posterior rays—pectorals relatively short and broad, rounded, shorter than head (about 1.5–1.7 times into head length), not reaching to base of anal spine I—caudal moderately and roundly emarginate; color—a brassy green above, grading into yellow below—some darker vertical barring on sides; a small species, seldom exceeding 8 inches in length and ¼ pound in weight.

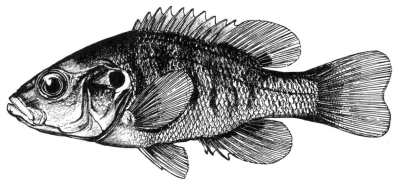

FIG. 251. Young Green Sunfish, *Lepomis cyanellus*. Drawn by William L. Brudon for Miller 1952. Courtesy California Department of Fish and Game and Leo Shapovalov.

TYPE LOCALITY

The Ohio River.

RANGE

Native to the eastern United States from the Great Lakes to Mexico, and the Alleghenies to the Mississippi. The species has been introduced into many western states, with marked success. Although the Green Sunfish is well established in parts of Nevada, little information is available regarding specific introductions. A 1941 California Division of Fish and Game biennial report lists the 1939 transference of 300 Green Sunfish from the California State Central Valleys Hatchery to Nevada (pp. 88–89), but there seems to have been no record made of their subsequent disposition. Since the species is regularly confused with the bluegill, it may have been, as Miller and Alcorn suggested (1946: 186), brought in with, or as, bluegills on more than one occasion. It now occurs in the Reno, Fallon and Boulder Lake areas.

TAXONOMY

In addition to being described many times as supposedly different fish from over its extensive range, the Green Sunfish has occupied several genera in the nearly 140 years it has been known as a named entity.

Etymology—"cyanellus," from the Greek meaning "blue." Diminutive form.

LIFE HISTORY

This small and rather lean sunfish is commonly mistaken for bluegill by many people because of a superficial resemblance including the prominent spotted opercular flap behind the eye and the fin-spotting. Similarly, the life cycles of the two species are much alike in that the

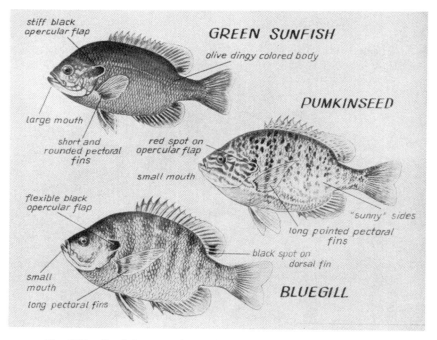

FIG. 252. Sunfish comparisons. Courtesy Wisconsin Conservation Department, Jens von Sivers and N. H. Hoveland.

Green Sunfish is a colonial breeder, constructing its rounded, shallow nests (averaging ten to twelve inches in diameter) in shallow water and utilizing mud as well as sanded bottoms.

The males fight among themselves during the early nest-building activities, but soon settle down in a restricted area to the sunfish habit of guarding the nests from intruders. Maturity is reached at two years of age, with the fish being less than two inches in maximum length at the end of the first season. Extreme lengths of 10 inches have been cited (Harlan and Speaker 1956: 131) but half or less of this figure is the common rule.

Sumner's studies in Lake Mead (Jonez and Sumner 1954) showed the Green Sunfish to be quite common, although disregarded as a game fish by anglers—gillnet sampling gave an incidence figure of 4.3 percent of all catches. Selective sampling of fish taken by anglers showed an average size of 6.4 inches in 1954, the last year of the study. Observations on their food habits here corroborated the well-known fact that this sunfish, despite its small size, is more of a fish-eater than its close relatives—indeed, its large mouth gives it the voracious appearance substantiated by its activities.

Purkett (1958: 126) gives the following growth rates in the Salt River of Missouri:

	YEAR				
	1	2	3	4	5
Length in inches	1.7	3.2	4.7	5.9	7.6

ECONOMICS

In the eastern United States, this species is the commonest sunfish in the smaller streams and bodies of water. Like most of its group, it is an excellent pan fish when sizeable enough, and the larger specimens provide considerable sport of sorts on light tackle, taking a variety of lures. Along with Yellow Perch and many other sunfishes, it is considered ideal for the "small fry" fisherman.

However, because of its strongly developed fish-eating tendencies and its abundance, it is often considered a nuisance, even in its native range, where larger and more desirable game fishes may not be too plentiful and any predation on their young would be resented by angler and biologist alike. Many implantations of Green Sunfish have been made by people who thought they were handling bluegills, particularly in the western United States.

"The further distribution of the Green Sunfish, *Lepomis cyanellus*, is not recommended as this species is a serious competitor and does not reach a size suitable for game fishing" (Miller and Alcorn 1946: 173, 186). It appears that the substantial populations that now exist in certain parts of Nevada began from individuals brought in with bluegills, or as bluegills, from California. In some instances, however, this is offset for all practical purposes, by the use of young Green Sunfish (along with bluegills) as food for their associated blackbass, etc., as in Lake Mead.

THE CRAPPIES

Genus *Pomoxis* Rafinesque 1818

("Opercle, sharp")

KEY TO SPECIES

Dorsal fin larger, its base length about equal to distance from origin of dorsal fin (front end) to eye; dorsal fin spines usually 7 to 8 in number; mandible shorter than pectoral fin; maxillary (including supramaxillary), shorter, reaching posteriorly to about a vertical line from posterior edge of eye pupil; body speckled (*Pomoxis nigromaculatus*) .... BLACK CRAPPIE

Dorsal fin smaller, its base length much less than distance from origin of dorsal fin to eye; dorsal fin spines usually 6 in number; mandible about equal in length to pectoral fin; maxillary (including supramaxillary) longer, extending posteriorly to about a vertical line which is as near posterior edge of eye as it is to posterior edge of eye pupil; body banded (*Pomoxis annularis*) _____ WHITE CRAPPIE

THE BLACK CRAPPIE
Pomoxis nigromaculatus (Le Sueur)
(numerous common names)

Labrus sparoides Lacépède 1802: 517–518
Cantharus nigro-maculatus Le Sueur 1829: 65 (88), in Cuvier and Valenciennes.
Centrarchus hexacanthus Cuvier and Valenciennes 1831: 344 (459) (Histoire Naturelle des Poissons 7)
Pomoxis sparoides, Girard, 1858: 5–6
Pomoxis nigromaculatus, ibid: 6
Centrarchus hexacanthus, Günther, 1859: 257
Centrarchus irideus, ibid
Hyperistius carolinensis Gill 1864: 93 (Amer. Jour. Sci. & Arts 36)
Pomoxys sparoides, Jordan and Gilbert, 1882: 465
Pomoxis sparoides, Bollman, 1892: 559
Pomoxis sparoides, Jordan and Evermann, 1896B: 987–988 / 1902: 335–336
Pomoxis sparoides, Forbes and Richardson, 1920: 238, 240–241
Pomoxis sparoides, Pratt, 1923: 115
Pomoxis sparoides, Jordan, Evermann and Clark, 1930: 305
Pomoxis sparoides, Schultz, 1936: 162
Pomoxis sparoides, Schrenkeisen, 1938: 259–260
Pomoxis sparoides, Kuhne, 1939: 103
Pomoxis nigro-maculatus, Hubbs and Lagler, 1941: 78, 80 / 1947: 93–94
Pomoxis nigro-maculatus, Curtis, 1942: 5–6 / 1949: 258, 264–265
Pomoxis nigro-maculatus, Hinks, 1943: 74
Pomoxis nigro-maculatus, Moffett, 1943: 182
Pomoxis nigro-maculatus, Dill, 1944: 167–168
Pomoxis nigro-maculatus, Dimick and Merryfield, 1945: 48
Pomoxis nigro-maculatus, La Monte, 1945: 143
Pomoxis nigro-maculatus, Eddy and Surber, 1947: 242–244
Pomoxis nigro-maculatus, Shapovalov and Dill, 1950: 387
Pomoxis nigro-maculatus, Simon, 1951: 109
Pomoxis nigro-maculatus, Wallis, 1951: 91
Pomoxis nigromaculatus, Beckman, 1952: 87–88 (sic)
Pomoxis nigromaculatus, La Rivers, 1952: 88
Pomoxis nigromaculatus, La Rivers and Trelease, 1952: 120
Pomoxis nigro-maculatus, Gabrielson and La Monte, 1954: 91
Pomoxis nigromaculatus, Scott, 1954: 106
Pomoxis nigro-maculatus, Shomon, 1955: 22–24

Pomoxis nigromaculatus, Eddy, 1957: 186
Pomoxis nigromaculatus, Koster, 1957: 97–98
Pomoxis nigromaculatus, Moore, 1957: 173
Pomoxis nigromaculatus, Trautman, 1957: 480–482
Pomoxis nigromaculatus, Slastenenko, 1958A: 301–302
Pomoxis nigromaculatus, Shapovalov, Dill and Cordone, 1959: 173

ORIGINAL DESCRIPTION

"Le *CENTRARCHUS SPAROÏDE*.
"(*Centrarchus sparoides*. nob.; *Labrus sparoides*, Lacép.)

"Notre troisième *centrarchus* est celui qui a été gravé dans M. de Lacépède (tom. III, pl. 24, fig. 2), d'après un dessin de M. Bosc, sous le nom de *labre sparoïde*. Il nous a été envoyé récemment par M. Lesueur, qui l'avait pris dans la rivière d'Ouabache, et qui le nommait *cantharus nigro-maculatus*. Il a bien, en effet, comme tous les *centrarchus*, quelque apparence de canthère; mais ses dents au palais et sur la langue, et les deux pointes de so opercule, ne permettent pas de le rapporter à la famille des spares.

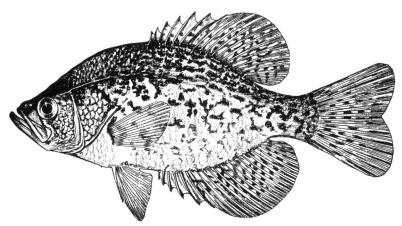

FIG. 253. Black Crappie, *Pomoxis nigromaculatus*. Courtesy California Department of Fish and Game and Leo Shapovalov.

"Il se distingue du centrarchus bronzé par la forme de sa dorsale, qui est plus basse en avant, plus élevée en arrière, et qui n'a que huit rayons épineux. Son anale est aussi bien plus haute et plus longue. L'angle et le bord inférieur de son préopercule ont quelques dentelures irrégulières.

"D. 8/16; A. 6/18; C. 17; P. 12; V. 1/5.

"Tout son corps est irrégulièrement marbré et tacheté de noirâtre sur un fond qui paraît avoir été argenté. Les points noirâtres d'entre les rayons sont irréguliers comme les taches du corps.

"Nos individus sont longs de près d'un pied.

"M. Bosc dit que ces poissons abondent dans les eaux douces de la Caroline, et qu'on les recherche principalement au printemps" (Le Sueur, in Cuvier & Valenciennes, 1829: 88–89).

DIAGNOSIS

Head about 3, body depth about 2.75, times into body length, respectively; mouth mediumly large, oblique; maxillary slightly shorter than in *P. annularis*, extending caudad about to a vertical line from posterior edge of eye pupil; mandible shorter than pectoral fin; palatine, vomerine, lingual and entopterygoidal teeth present; branchiostegals 7; cheek scales smaller, in about 6 rows; lateral line complete; scales 6/38–45/12; fins—dorsal relatively larger than in *P. annularis*, length of fin base about equal to distance from front end or origin of fin to eye, with 7–8 anterior spines, 15 posterior rays—anal with 6 anterior spines, 16–18 posterior rays—pelvics with the first ray a strong spine—caudal angulately and moderately emarginate; color—dark green dorsally, uniformly green anteriorly, mottled with lighter yellow posteriorly and along sides, grading into the whitish belly—dorsal, anal and caudal fins heavily black-spotted; a moderate-sized species, reaching lengths of about a foot and maximum weights of about 1½ pounds.

TYPE LOCALITY

The Wabash River.

RANGE

Native to the eastern United States from the upper Mississippi Valley east to New Jersey and south to Texas and Florida. The species has been introduced widely into the western states, as well as into Europe. "Crappies," a name which applies to both *Pomoxis nigromaculatus* and *P. annularis*, have been brought into Nevada at different times since 1924, but since only the Black Crappie was definitely known to be established in the State until recently, it must be assumed that such "crappies" were this species.

TAXONOMY

The rather confused early terminology for this species did not become reasonably stabilized until about 20 years ago.

Etymology—"nigromaculatus," black-spotted.

LIFE HISTORY

A typically carnivorous centrarchid, mildly piscivorous as an adult, more of a plankton feeder than the blackbasses, and inhabiting a great variety of waters in its native habitat. It exhibits considerable local variation in color, which has resulted in many common names. The species is a spring spawner, often cited as preferring water slightly above 65° F. (18.5° C.) in temperature for egg-deposition, although Sumner (1954) found a temperature range of 55° to 65° F. during spawning in Lake Mead. Its nests are much like those of the blackbass, often in deeper waters.

Although growth rates vary with the area involved, Norris Reservoir in Tennessee provides figures which seem to represent growths under near optimum conditions:

	1 year (in.)	2 years (in.)	3 years (in.)	4 years (in.)	5 years (in.)	6 years (in.)
Norris Reservoir, Tennessee (Stroud 1948)	2.5	9.2	11.5	12.7	14.2	
Lake Mead, Nevada (Sumner 1954)	2.9	6.5	8.8	10.8	12.2	13.4

Of 698 Black Crappie sampled at Norris Reservoir, only one was found to exceed five years of age. As in the cited case of blackbass, this probably represents the average death point from old age. Of the 540 specimens studied by Robert Sumner, 13 were considered to be 6 years old. He reported the same three parasites from these fishes as he found on the Largemouth Blackbass in Lake Mead.[1]

From the observations of many investigators, this—the commonest of the two crappies—prefers clear and not too expansive waters, often being absent from the larger lakes but not occupying many creek environments as extensively as does the White Crappie. While spawning nests of the Black Crappie are often placed on muddier bottoms than is usual for the family, it seems a little fonder of hard bottoms than does the White Crappie.

Spawning can take place at a very early age in these fishes—maturity has been reported for 1- and 2-year-old specimens—and they are prolific egg layers. Two-pounders will have more than 150,000 eggs. In size range, 4-pound specimens are known, but the average seems to be slightly less than that for the White Crappie so that 1½ pounds would be the most generally found maximum. Few specimens grow to more than 12 inches in length. The two species of crappie grow at about equal rates for the first two years, with the White Crappie surpassing the other in subsequent years (Ricker and Lagler 1943: 96).

ECONOMICS

In its original range, Black Crappie usually ranks with blackbass, White Crappie, and Bluegill Sunfish as the most prized panfish. Having a softer mouth than the blackbasses, it is not as gamey. In the northern midwestern states, crappies are exceeded only by bluegills in the anglers' catch, and some three-quarters of fish taken through winter ice are Black Crappies. This is probably due, in part, to the fact that this species—contrary to usual family habits—is an active feeder during winter. The winter catch is greater than that of summer (Eddy and Surber 1947: 244).

In California it is an important addition to the introduced warmwater fishery, surpassing blackbass in numbers caught, and sunfish in poundages taken. Robert Sumner (Jonez and Sumner 1954) at Lake Mead found that Black Crappie was third in importance in the fishery (from which White Crappie is absent), being led by Largemouth Blackbass and Channel Catfish, in that order.

However, where blackbass is the most highly prized game fish, crappie is often looked upon with disfavor, since all members of this family can be markedly piscivorous and so strong competitors of the blackbass where the latter may be critically balanced in its environment and overfished.

---

[1] The trematode flatworm *Contracaecum*, the cestode tapeworm *Proteocephalus* and the copepod crustacean *Lernaea*.

WHITE CRAPPIE

Pomoxis annularis Rafinesque

(numerous common names)

Pomoxis annularis Rafinesque 1818: 41
Cichla storeria Kirtland 1838: 191 (Rpt. Zool. Ohio)
Pomoxis nitidus Girard 1857: 200 (Acad. Nat. Sci. Phila. Proc. 9)
Pomoxis annularis, Girard, 1859: 6
Pomoxis nitidus, ibid
Centrarchus nitidus, Günther, 1859: 257–258
Pomoxys brevicauda Gill 1865: 64
Pomoxys intermedius ibid: 65
Pomoxys protacanthus ibid: 66
Pomoxys annularis, Jordan and Copeland, 1878: 437
Pomoxys annularis, Jordan and Gilbert, 1882: 464–465
Pomoxis annularis, Bollman, 1892: 560
Pomoxis annularis, Jordan and Evermann, 1896B: 987 / 1902: 334–335
Pomoxis annularis, Forbes and Richardson, 1920: 238–240
Pomoxis annularis, Pratt, 1923: 115
Pomoxis annularis, Hildebrand and Towers, 1928: 126–127
Pomoxis annularis, Coker, 1930: 202
Pomoxis annularis, Jordan, Evermann and Clark, 1930: 304
Pomoxis annularis, Neale, 1931A: 10
Pomoxis annularis, Walford, 1931: 83
Pomoxis annularis, Hatton, 1932: 50–51
Pomoxis annularis, Schultz, 1936: 162
Pomoxis annularis, Tanner, 1936: 173
Pomoxis annularis, Schrenkeisen, 1938: 259
Pomoxis annularis, Kuhne, 1939: 103
Pomoxis annularis, Hubbs and Lagler, 1941: 78, 80 / 1947: 93–94
Pomoxis annularis, Murphy, 1941: 168
Pomoxis annularis, Curtis, 1942: 5–6 / 1949: 258, 264–265
Pomoxis annularis, Hansen, 1943: 259–260 / 1951
Pomoxis annularis, Dill, 1944: 167–168
Pomoxis annularis, Dimick and Merryfield, 1945: 46, 48
Pomoxis annularis, La Monte, 1945: 143
Pomoxis annularis, Miller and Alcorn, 1946: 187
Pomoxis annularis, Eddy and Surber, 1947: 241–242
Pomoxis annularis, Shapovalov and Dill, 1950: 387
Pomoxis annularis, Beckman, 1952: 86–87
Pomoxis annularis, La Rivers, 1952: 88
Pomoxis annularis, La Rivers and Trelease, 1952: 120
Pomoxis annularis, Gabrielson and La Monte, 1954: 91
Pomoxis annularis, Scott, 1954: 107
Pomoxis annularis, Shomon, 1955: 21–22
Pomoxis annularis, Eddy, 1957: 186–187
Pomoxis annularis, Koster, 1957: 97
Pomoxis annularis, Moore, 1957: 173
Pomoxis annularis, Trautman, 1957: 477–479
Pomoxis annularis, Slastenenko, 1958A: 299–300
Pomoxis annularis, Shapovalov, Dill and Cordone, 1959: 173

ORIGINAL DESCRIPTION

"... *Pomoxis annularis*. (Ring-tail Pomoxis.) Body silvery, scales ciliated caduc, back and fins olivaceous, a gilt ring at the base of the tail, lateral line straight, lower jaw longer, tail forked; anal, dorsal and caudal fins tipped with blackish, pectoral fins extended beyond the vent. A curious small fish of the Ohio, rather scarce, length 2 or 3 inches, vulgar name Silver Perch. The number of rays in the fins is as follows: dorsal fin 20 rays, whereof 6 are spinescent; anal fin 22, whereof 6 are spinescent; thoracic fin 6, whereof one is spinescent; pectoral fin 15; caudal 28" (Rafinesque 1818: 41).

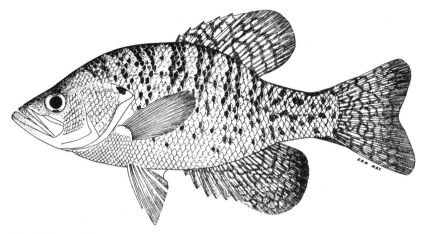

FIG. 254. White Crappie, *Pomoxis annularis*. Reproduced from *The Fishes of Ohio*, by Milton B. Trautman (Columbus, Ohio: Ohio State University Press, 1957), p. 477. Used by permission of the publisher.

A preceding generic description adds a bit more:

"2. N. G. Pomoxis. (A fish. Natural family of *Leiopomes*.) Body oblong, compressed, one dorsal fin opposed to the anal, vent nearer to the head than to the tail, no appendage to the thoracic fins, mouth toothless, gills without scales and mutic" (Rafinesque 1818: 41).

DIAGNOSIS

Head and body depth each about 2.75 times into body length; mouth rather large, oblique; maxillary slightly longer than in *P. nigromaculatus*, extending caudad to a vertical line about midway between posterior edge of eye and posterior edge of eye pupil; mandible about as long as pectoral fin; palatine, vomerine, lingual and entopterygoidal teeth present; branchiostegals 7; cheek scales larger, in 4–5 rows; lateral line usually complete; scales 6/36–48/12; fins—dorsal relatively smaller than in *P. nigromaculatus*, length of fin base much less than distance from front end or origin of the fin to eye, with usually 6 anterior spines, 15 posterior rays—anal with 5–6 anterior spines, 17–19 posterior rays—pelvics and caudal as in *P. nigromaculatus*; color— back dark greenish, sides light yellow with faint vertical stripes, grading into the white venter—dorsal, caudal and anal fins black-speckled

—in general, much lighter in color than the Black Crappie; in size about as the latter species, but apparently reaching a slightly greater weight, some specimens in excess of 2½ pounds being reported from Illinois, with 4 pounds the maximum; average weight is about a pound.

TYPE LOCALITY

Falls of the Ohio River.

RANGE

About the same as the Black Crappie. It has been introduced into many localities in the west, and like its relative, has prospered in California. The Nevada implantation is a very recent one and the species seems to be established only in the lower Humboldt and Carson River areas—Ryepatch Reservoir (Pershing County), Lahontan Reservoir (Churchill-Lyon Counties) and Indian Lakes (Churchill County). These introductions were made early in 1956.

TAXONOMY

Etymology—"annularis," having rings.

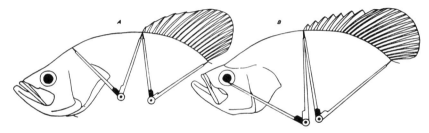

FIG. 255. The crappies, showing the dorsal fin shorter in the White Crappie (A) than in the Black Crappie (B). Reproduced from *The Fishes of Ohio*, by Milton B. Trautman (Columbus, Ohio: Ohio State University Press, 1957), p. 477. Used by permission of the publisher.

LIFE HISTORY

Essentially the same as that of the Black Crappie as concerns food, spawning times and habits, etc. There are some differences in habitat preferences which, while not marked, have been commented on by most investigators of these species. The White Crappie—slightly the larger of the two—is more often found in clouded or silty waters than is the Black Crappie, and the former, as would be expected, occurs commonly in waters over mud. On the other hand, the White Crappie is the one most frequently living in creeks, being outnumbered or replaced entirely in many lake environments by its companion species, which latter is more at home over a hard bottom.

Hansen (1943) has made some interesting observations on the nesting of this species. At Lake Springfield, Illinois, he located a spawning site in a shallow, protected arm of the lake where nests were being protected in just a few inches of water. In this instance, no excavated nests were found, and such eggs as Hansen noted were strung or clumped in water vegetation. The fish present exhibited unmistakable territorial guarding behavior. He pointed out that blackbass and sunfish spawning habits and procedures were common knowledge in the region, but few observations of crappie spawning seemed to have been available because: "(1) they nest more or less apart from one another, (2) they sometimes fail to excavate a nest, (3) they nest among water weeds and (4) they nest under objects" in addition to probably nesting in deeper waters (p. 260).

Both species of crappies are hardy, and both, unlike most of their

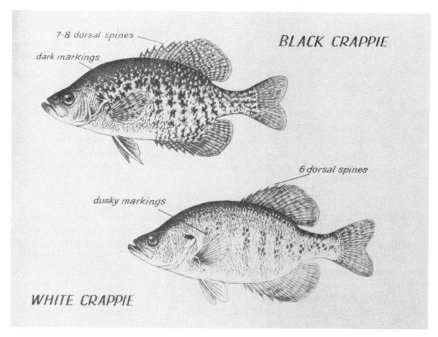

FIG. 256. Comparisons between our two species of crappie. Courtesy Wisconsin Conservation Department, Jens von Sivers and N. H. Hoveland.

family, are active wintertime feeders. White Crappies have been recorded up to a rare maximum of 4 pounds, but half that is the usual upper limit. Lengths seldom top 12 inches.

ECONOMICS

Fishing statistics only occasionally differentiate between the two species of crappies, but lump them together so that separate fishery figures are generally not available for either crappie. They are among the most popular pan fishes in their native areas, and afford a good deal of ice fishing in the northern midwestern states (see Black Crappie).

THE COTTOID FISHES
Suborder COTTOIDEI
SCULPINS, Family *COTTIDAE*

This is a large[1] family of small, distinctive-looking bottom and shore inhabiting species; "weird" is the word that best describes their appearance to most people. The most unusual-looking species are marine, although our common freshwater sculpin looks odd enough to the average person.

As a group, cottids are typical of northern waters, and a few are deep-sea forms. A primitive sculpin—*Jordania*—is a colorful little Puget Sound species which is one of many fishes named for the country's one-time premier ichthyologist, David Starr Jordan.

Scales are well-developed only in primitive genera, and are lost to varying degrees in most sculpins. Some species make sounds by rapidly vibrating their gill covers against their heads, and others have a poisonous principle in their eggs although their flesh appears to be harmless (Schultz and Stern 1948: 68).

The group has appeared sparingly in the fossil record, being definitely known as far back as the Oligocene (approximately 30 million years ago). The controversial *Eocottus*, which may be either cottoid or gobioid, is Lower Eocene (50 million years) in age.

DIAGNOSIS

Body rounded or dorso-ventrally flattened in cross-section, about mediumly elongate, rapidly decreasing in size behind the head, which is large and very pronounced; body lacking any condition approaching uniform scalation, usually scaleless or with prickles; mouth terminal; jaws equally toothed, as may be the palatines and vomer; protractile premaxillaries; eyes high on head, large and prominent; no barbels on jaw angles; dorsal angle of preopercle generally bearing one to several spines of various sizes; pseudobranchiae present; gills $3\frac{1}{2}$ to 4, the terminating slit often obsolete; gill membranes widely connected, often to the isthmus; fins—adipose lacking; pectorals lateral; pelvics thoracic or lacking; dorsal quasi-entire or separate, anteriorly spinous (these may be buried in the skin), posteriorly soft-rayed; caudal rounded; anal long, soft-rayed; lateral line present; intestine short; air bladder generally lacking; pyloric caeca 4–8.

SCULPINS, MILLER'S THUMBS, BULLHEADS, RIFFLEFISHES
Genus *Cottus* Linnaeus 1758
(The old name for a European species, from "Kótta," head, referring to the large head)

KEY TO SPECIES

No spines (at most, only a tubercle or two) below single spine at

---

[1] The old figures, 60 genera and 250 species, of Jordan and Evermann (1898: 1879), are still quoted but it is certain that modern revision of these genera would considerably alter these statistics.

preopercular angle; Lahontan and Columbia River systems
(*Cottus beldingi*) _____ BELDING SCULPIN
One or two spines below the single spine at preopercular angle;
Bonneville system (*Cottus bairdi semiscaber*)_____
_____ BONNEVILLE BAIRD SCULPIN

BAIRD SCULPIN
Cottus bairdi Girard

Cottus bairdii Girard 1850: 410
Cottus gracilis, Günther, 1860: 157–158
Uranidea richardsoni bairdi, Jordan and Gilbert, 1882: 697
Cottus ictalops, Jordan and Evermann, 1898A: 1950–1952
Cottus ictalops, Gill, 1908: 115
Cottus ictalops, Forbes and Richardson, 1920: 326–327
Cottus ictalops, Pratt, 1923: 141
Cottus ictalops, Jordan, Evermann and Clark, 1930: 385
Cottus ictalops, Schrenkeisen, 1938: 268–269
Cottus bairdii bairdii, Hubbs and Lagler, 1941: 81–82 / 1947: 97
Cottus bairdii bairdii, Eddy and Surber, 1947: 251–252
Cottus bairdii, Bailey and Dimick, 1949: 11
Cottus bairdii, Shapovalov and Dill, 1950: 387
Cottus bairdi, La Rivers, 1952: 87
Cottus bairdi, La Rivers and Trelease, 1952: 120
Cottus bairdi, Eddy, 1957: 235
Cottus bairdi, Koster, 1957: 105
Cottus bairdi, Moore, 1957: 204
Cottus bairdi bairdi, Trautman, 1957: 611–613
Cottus bairdi, Slastenenko, 1958A: 340–341

ORIGINAL DESCRIPTION
"Cottus Bairdii Grd.—*Cottus gobio* Kirtl. Bost. Jour. Nat. Hist. v. 1847, p. 342—Body subcylindrical, short; mouth comparatively large. —Pennsylvania, tributaries of the Ohio; Prof. Baird,—Mohoning River; T. P. Kirtland" (Girard 1850: 410).

DIAGNOSIS
See the following subspecies.

TYPE LOCALITY
Mahoning River at Poland, Ohio.

RANGE
Apparently widespread in the northern United States and adjacent Canada from the intermontane west to the east coast.

TAXONOMY
As far as the published literature is concerned, most species of *Cottus* have led checkered existences, and the Baird Sculpin is no exception. For many years, *Cottus bairdi* was submerged in the synonymy of

Cottus ictalops (Rafinesque) 1820 which was considered to be the most common, widespread and variable sculpin, and was extended westward after its general separation to include the western Great Basin sculpin. *Cottus bairdi* appears to be more restricted than its former usage indicates, and more work is necessary to unravel the classification of these difficult little fishes.

Etymology—Named for the famous biologist Spencer Fullerton Baird, one-time Secretary of the Smithsonian Institution and the first United States Fish Commissioner, who described many western United States cold-blooded vertebrates with Charles Girard.

LIFE HISTORY AND ECONOMICS

See the following subspecies.

BONNEVILLE BAIRD SCULPIN
Cottus bairdi semiscaber (Cope)

Cottopsis semiscaber Cope 1872C: 476
Uranidea wheeleri Cope 1874A: 138
Uranidea vheeleri, Cope and Yarrow, 1875: 696–697
Potamocottus wheeleri, Jordan and Copeland, 1878: 441
Cottopsis semiscaber, ibid
Cottopsis semiscaber, Jordan and Gilbert, 1881B: 459
Uranidea semiscabra, ibid, 1882: 695
Uranidea wheeleri, ibid: 697–698
Cottus bairdi punctulatus, Jordan, 1890C: 53 / 1891: 29, 36
Cottus semiscaber, Jordan and Evermann, 1898A: 1949–1950
Cottus semiscaber, Pratt, 1923: 140
Cottus semiscaber, Jordan, Evermann and Clark, 1930: 384
Cottus semiscaber, Hatton, 1932: 52
Cottus punctulatus, Schultz, 1936: 179
Cottus semiscaber, Tanner, 1936: 172
Cottus semiscaber, Schrenkeisen, 1938: 269
Cottus semiscaber, Simon and Brown, 1943: 41
Cottus bairdii semiscaber, Bailey and Dimick, 1949: 11
Cottus bairdi semiscaber, Miller and Sigler, 1950
Cottus semiscaber, Simon, 1951: 111–112
Cottus bairdi semiscaber, Zarbock, 1951: 42 / 1952: 249
Cottus bairdi semiscaber, La Rivers, 1952: 87
Cottus bairdi semiscaber, Miller, 1952: 9, 38–39
Cottus bairdi semiscaber, Eddy, 1957: 235

ORIGINAL DESCRIPTION

"Radii, D. VII–18; A. 13; V. 1–4; first ray of anal below third of second dorsal. Skin prickly above the lateral line, smooth below it posteriorly. Body compressed, profile rising rather steeply to the basis of first dorsal fin. Eye 4.5 times in head, .75 time in interorbital space. Muzzle contracted, maxillary bone reaching to below middle of pupil. Two spines on preoperculum. On an inferior anterior angle of operculum [Sic!]. Lateral line discontinued on last fourth of caudal peduncle. Head one-third length without caudal fin.

"Below yellow; dorsal line with a series of black spots; sides with large, dark clouds" (Cope 1872C: 476).

DIAGNOSIS

Head about 3.0, body depth about 4.5, in body length, respectively; head large, surmounting a body regularly tapering caudally; eye large, about 5.0 times into head length; maxillary 2.0 to 2.5 into head length, extending caudally about to hind eye margin; jaws broadly toothed, as are the palatines and vomer; interorbital space restricted, one-half or less the eye-width; spines of preoperculum a bit blunt and short, two in number; lateral line incomplete; variable development of body prickling from smooth (few) to rough (many); fins—dorsals usually

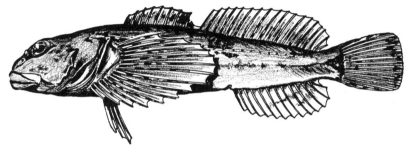

FIG. 257. Bonneville Baird Sculpin, *Cottus bairdi semiscaber*. Drawn by William L. Brudon for Miller 1952. Courtesy California Department of Fish and Game and Leo Shapovalov.

separate, with just a suggestion of union in occasional individuals, VIII to 17–18 (i.e., about 8 anterior spines and 17–18 following softer rays)—pectorals extending caudally to anterior end of anal—pelvics not reaching vent, which latter is closer to end of tail than to snout tip—anal 12–13 rayed—caudal peduncle stout, about equal in width to snout length; color—variable, but laterally broadly mottled, typically darker above and grading to lighter or even yellow ventrally; size—about 5 inches is maximum.

TYPE LOCALITY

Fort Hall, Idaho.

RANGE

Bonneville drainage of western Utah and eastern Nevada. During his population inventories of the Snake Range area of White Pine County, Ted Frantz found sculpins only in Lake Creek (Big Spring) and Spring Creek.

TAXONOMY

Etymology—"semiscaber," half rough.

LIFE HISTORY

Zarbock (1952) has reported extensive observations on the most generally interesting aspects of the life history of this fish, placing particular emphasis on its economic inter-relationships with trout.

Aquatic insects make up most of the sculpin's food. Zarbock's study showed these insects to be important in the following order (frequencies of occurrence): Flies (Diptera), 54; Stoneflies (Plecoptera), 37; Mayflies (Ephemeroptera), 34; Caseflies (Trichoptera), 30. By comparison, vertebrates (which were fishes, almost entirely sculpins) amounted to only 2 on the scale.

Zarbock's age distribution data gave the following average results: yearlings were 55–60 millimeters in length (about 2¼ inches); two-year-olds were 60–80 mm. (about 2¾ inches); three-year-olds were 70–100 mm. (3½ inches); four-year-olds were 95–105 mm. (3⅞ inches); five-year-olds were 95–130 mm. (4⅜ inches).

His data from dissected females indicated a spawning period in the Logan River of Utah from February to May, although he was unsuccessful in finding nests. The general spawning sites included rock and/or gravel bottoms in either swift or slow waters with depths varying from 6 inches to 3 feet. An altitude of 7,000 feet represented their highest occurrence in the river.

ECONOMICS

Although the idea persists among fishermen that sculpins are death on trout eggs and young trout, the habitually bottom-feeding sculpins have been exonerated of predation on game fishes in all food habit studies made on them. Zarbock's observations indicated that this sculpin gets its own eggs and young more often than that of any other fish. In fact, the two trout eggs he found came from a single sculpin stomach out of a total of 275 stomachs examined, and these probably were eggs washed from a redd and which would have failed to develop anyway. These conclusions are in agreement with the results of other investigators such as Surber (1920) and Koster (1937).

Probably the large head, cavernous mouth and generally forbidding appearance of the sculpin encourage superstitions about it.

THE BELDING SCULPIN

Cottus beldingi Eigenmann and Eigenmann

(Lahontan Sculpin, Pahute Sculpin)[1]

Uranidea gulosa, Jordan and Henshaw, 1878: 199–200 (Fallen Leaf Lake, California)
Cottus beldingii Eigenmann and Eigenmann 1891: 1132
Cottus beldingii, Jordan and Evermann, 1898A: 1958–1959
Cottus beldingii, Rutter, 1903: 148 / 1908: 147
Cottus beldingi, Snyder, 1917A: 81–83
Cottus beldingi, Pratt, 1923: 140–141
Cottus beldingii, Hay, 1927: 152
Cottus beldingii, Jordan, Evermann and Clark, 1930: 384

---

[1]Although this has been more commonly known as the "Lahontan Sculpin," and "Pahute Sculpin" also has its advocates, the term "Belding Sculpin" is distinctive and avoids the pitfalls inherent in the implication that the distribution of this species coincides with that of the Lahontan Basin or the original Pahute tribes—neither of which is true.

Cottus beldingii, Hubbs and Schultz, 1932: 5–6
Cottus beldingii, Schultz, 1936: 178
Cottus beldingii, Schrenkeisen, 1938: 269
Cottus beldingi, Shapovalov, 1941: 444
Cottus beldingii, Miller and Miller, 1948: 184
Cottus beldingii, Bailey and Dimick, 1949: 15
Cottus bairdii beldingii, Shapovalov and Dill, 1950: 387
Cottus bairdi beldingi, La Rivers, 1952: 87
Cottus bairdi beldingi, La Rivers and Trelease, 1952: 120
Cottus beldingi, Eddy, 1957: 234
Cottus beldingi, Moore, 1957: 204
Cottus beldingii, Shapovalov, Dill and Cordone, 1959: 173

ORIGINAL DESCRIPTION

"Head 2¾–4; depth 4–5; D. VI–VIII. 15½–18; A. 11–13; V. I. 4.

"Head rather short and broad, the profile convex, more steep from eye forward; eye large, orbit 4–5 in head; interorbital concave, 2 in orbit; mouth large; maxillary reaching at least to below the pupil, about 2 in the head. Preopercle with a simple, backward-directed spine,

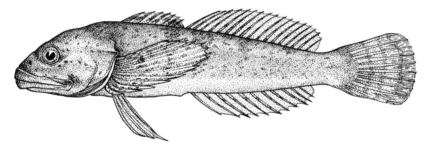

FIG. 258. Belding Sculpin, *Cottus beldingi*. Drawn by Silvio Santina.

very slightly curved upwards. Teeth on jaws and vomer, none on palatines. Skin smooth. Pectorals reaching vent, or further in young; ventrals 1⅓–2 in head. Distance of anal from caudal 1⅓ in its distance from snout. Mottled with black and white. About six blackish crossbars on back; the first across head just behind eyes, next at origin of dorsal. First dorsal tinged with rust, the second less so. All the fins except the ventrals spotted with dark. The ground color varies greatly with the bottom over which these fishes live.

"In October, 1889, Mr. L. Belding obtained three specimens of a species of Cottus in Lake Tahoe, California. During June, 1890, we obtained a much larger number at the same place. A series of these was sent to the British Museum. The rest are in the collections of the California Academy of Sciences, No. 504. Mr. Belding's specimens are also in the collections of the Academy, No. 702. We also obtained a number from Donner Lake, California, No. 505, California Academy of Sciences.

"These specimens represent a variety of species distinct from the Alaskan *Cottus minutus*, with which it is most closely related.

"The other species found at Lake Tahoe were *Phoxinus montanus* Cope [*Richardsonius egregius*], *Agosia oscula* Girard [*Rhinichthys osculus robustus*], *Algansea obesa* Girard [*Siphateles bicolor obesus*], *Coregonus williamsoni* Girard, *Catostomus tahoensis* Gill & Jordan, *Salmo mykiss henshawi* Gill & Jordan [*Salmo clarki henshawi*]. Besides these we obtained *Algansea olivacea* Cope [*Siphateles bicolor obesus*], from Donner Lake.—C. H. and R. S. Eigenmann" (1891: 1132).

DIAGNOSIS

Head 2.0 to 3.0, body depth 2.0 to 4.0, in body length, respectively; head rather small for the type; eye variable, 3 to 5 times into head length; maxillary 1.5 to 2.5 into head length, reaching posteriorly as far as below eye center in some cases; jaws and vomer always toothed, the palatines only occasionally, when a short row may be present which is about as long as half the eye width; interorbital width from one-half to three-quarters eye width; preopercular spines variable, usually only one is present, and this is large, upturned, sharp and flat, but occasionally a smaller secondary spine shows with point directed downward; lateral line incomplete, generally terminating beneath caudal one-third of the rayed portion of the dorsal fin; body smooth, with no development of prickles; fins—dorsal variable from distinctly separated into two parts (an anterior spinous and a posterior soft-rayed) to weakly united, VII–VIII to 15–17—pectorals extending caudally to anterior end of anal—pelvics I to 3–4—anal 11–13 rayed—caudal abruptly rounded; caudal peduncle very broad, always distinctly exceeding snout length; color—variable spotting, reticulations or broad light and dark suffusions dorsally and laterally—lateral banding often seen, which generally takes the pattern of two below anterior spinous dorsal and three beneath posterior soft rayed dorsal; caudal base usually with a distinct, but thin, band; fins variously dotted, mottled or barred; general background color varies from grayish to dark olive; length up to about 5 inches.

TYPE LOCALITY

Lake Tahoe, California.

RANGE

Lahontan and Bonneville and Columbia River drainage systems of Nevada, California, Utah, Idaho, Oregon and Washington. In the eastern part of the State, Nevada Fishery Biologist Donald B. Thurston found sculpin in the following streams: Currant Creek (Snake Mountains), Gance Creek (Independence Mountains), Goose Creek, Martin Creek (Tennessee Mountain), North Fork of the Humboldt River (Independence Mountains), Tabor Creek (Snake Mountains), Tea Creek (Mary's River Peak) and Wilson Creek. These localities are all in Elko County.

TAXONOMY

With no synonyms, the nomenclature of this animal has been quite simple, the only apparent problem being whether it should be considered a species or a subspecies of *Cottus bairdi*. Robert R. Miller, the latest student of this problem, believes *C. beldingi* should have undisputed specific standing outside the *C. bairdi* complex.

Etymology—Named for the collector, L. Belding.

LIFE HISTORY

Snyder's few remarks on this fish are—

"On account of the roily condition of the streams when most of the observations in the Lahontan system were made, specimens of this species were seldom seen until they became entangled in the net. Several were secured with hook and line. None was caught in the lower course of the Truckee River. Examples were most often taken while dragging the net over the rough bottom in the more rapid parts of the stream. In the Truckee River near Floriston an individual was occasionally seen either under a boulder or close by its sheltering edge. The partly digested body of a young sucker was found in the stomach of one specimen" (1917A: 82).

The writer's data from Truckee River and Lake Tahoe specimens indicate a spawning period in late spring and early summer, and a diet consisting largely of insects for the river populations particularly. The lake is depauperate in insects and presumably here the fish must depend more on its own young, the young of other fishes and upon fish eggs. This last item—fish eggs—seems as often to consist of its own eggs as of those from other species.

The Belding Sculpin, as in the case of its relatives, is a bottom-dwelling fish and is well adapted to such a life in its broad, rather flattened shape; its high density due to lack of an air bladder; its wide, cavernous mouth; and upturned vision. Its hunting movements consist of short, swift dashes at prey, and there is little of the usual grace of the swimming fish in its motion.

Its bottom-haunting preferences are evident whether the fish is living in shallow stream or deep lake, and in both situations it prefers the colder waters. It is not found in the lower, warmer waters of the rivers. It is a very common species in Lake Tahoe, and is also found in the higher tributaries coming into the lake. A situation such as that presented by Pyramid Lake is totally unsuited to the species, for here the waters are too warm and spawning sites almost entirely lacking— it demands crevices under or between rocks for egg laying.

ECONOMICS

The Belding Sculpin seems similar to other species of the genus in its economic food chain position. The widely held ideas that sculpins make inroads on eggs and young of trout have never been upheld in any of the several studies made on the subject (see discussion under *Cottus bairdi semiscaber*).

On the other hand, observations at Lake Tahoe by several investigators (Corlett; La Rivers; Miller 1951) show well enough that the sculpin is instead a most valuable link in the food chain of the Lake Trout (*Salvelinus namaycush*) which, under usual conditions, is the most important game fish in the lake. The sculpin is the most readily obtainable food for the Lake Trout, in one investigation comprising nearly half the trout's food for at least part of the year (Miller 1951: 22).

In streams, such as the Truckee River and others like it, sculpin may be considered an indirect competitor with trout for food, but here the former feeds largely on immature stages of the same insects that the more surface-frequenting trout utilizes after these insects emerge as winged adults. However, this need not concern even the most rabid trout fisherman since there are insects for everybody. In actuality, the bottom-feeding sculpin is converting insect protein—which is normally unavailable to the surface-feeding rainbow—into a concentrated protein product, namely itself, which is used to some extent by the trout and to a much greater degree by the benthic or deep-living Lake Trout.

Addendum:

THE SACRAMENTO BLACKFISH

Orthodon microlepidotus (Ayres)

The following information on this fish was received too late to be included in the text concerned with the species:

The Sacramento Blackfish has not been detected in Nevada waters since the 1948 occurrence until it was recently picked up in some numbers at Lahontan Dam, Churchill County. Robert C. Sumner and Cal Allan of the Nevada State Fish and Game Commission obtained several specimens, ranging from medium to large in size, while seining in the pool on the downstream side of Lahontan Dam, March 27, 1962. Several age groups were present. Unlike the Reno occurrence, this would indicate that a population of some size is established and maintaining itself.

All specimens were heavily infested with *Lernaea*.

CHAPTER VII—APPENDIX

CONTENTS

	PAGE
A—Lake Survey Report Outline and Field Forms	595
B—Stream Survey Procedures and Forms	603
C—Alkalinity and Acidity	611
D—Determining the Age of Fishes by Scale Readings	631
E—Trout Hybridization	635
F—Inventory of Game Fishes by Specific Waters	639
G—List of Stream and Lake Reports	653
H—Dingell-Johnson Fisheries Projects	667
I—Invertebrate Biota of Pyramid Lake and Lake Tahoe	673
J—Biology of *Crenichthys baileyi*	681
K—Biology of *Richardsonius balteatus*	687
L—Biology of *Siphateles bicolor obesus*	697
M—Glossary	701
N—Map of Pluvial Lakes and Rivers	774

A.
LAKE SURVEY REPORT OUTLINE AND FIELD FORMS

The following lake survey report outline, developed by University of Michigan fishery biologists,[1] has been used for several years with some modification by the author's students in Limnology at the University of Nevada.

Introduction
    Location and accessibility
    Town, Range and Section
    Township
    County
    Near what other geographical features
        Mountains
        Lakes
    Drainage affinities
    Accessibility to the general public
    Roads, Trails and Railroads
Map and Survey
    Map
    Source of map
    Biological Survey
        Personnel
        Time of survey
        Purpose or instigation
History and Recreational Development
    Amount of recreational development
    Amount and nature of industrial development
    History of fishing
        Past
        Present, including creel census records and angling pressure
    Boat liveries
    Hotels and resorts
    Importance as a public fishing water
        Summer
        Winter
    Importance other than as a fishing water
    Desirability of maintaining or altering present state of
        development
Physical Characters
    General physical characters
        Shape of basin
        Surface area
        Volume
        Elevation

---

[1]Lagler, Karl F. 1952. Freshwater Fishery Biology. Wm. C. Brown Company, Dubuque, Iowa.

LAKE SURVEY REPORT—*Continued*
>   Length and width (greatest)
>   Length of shoreline
>   Compass direction of main axis
> Depth
>   Maximum
>   Average
>   Minimum
> Shoreline development
> Dropoff (presence, absence, location)
> Shoal areas
>   Width around the lake
>   Percentage of surface area of lake made up by such
> Bottom types and their extents
> Water color
> Transparency
>   Secchi disc readings
>   Sources of turbidity
> Geologic origin and present stage of the lake's evolution
> Drainage
> Extent of drainage basin in square miles
> Nature of the drainage basin
>   Soil
>   Ground cover
>   Woods
>   Marginal vegetation
> Fluctuations of water level
> Inlets
> Size, flow, dimensions, characteristics
> Points of entrance
> Differences among inlets
>   Use by lake fishes for spawning, etc.
> Outlets (same data as for inlets)
> Dams in the drainage system
> Temperature and chemical characteristics
> Significance of these
> Water temperatures
>   Surface
>   Middle
>   Bottom
>   Thermocline
> Oxygen
>   Epilimnion
>   Thermocline
>   Hypolimnion
>   Top
>   Bottom

LAKE SURVEY REPORT—*Continued*
- Other Chemical Characteristics
  - Carbon dioxide
  - Carbonates
  - Bicarbonates
  - Hydroxides
  - (Or phenolphthalein—methyl orange alkalinity)
  - pH
  - Pollution
- Biological Characters
  - Vegetation (including species lists, relative abundance and bathymetric distribution)
  - Fish foods
    - Plankton (species lists, abundance and distribution)
    - Dominant forms
    - Relative abundance
      - Quantitatively
      - Qualitatively
    - Bottom foods (same data as for plankton)
    - Larger species utilized by fish as food (same data as for plankton)
  - Fishes present
    - Game species (species, abundance, sizes and weights)
  - Non-game fishes (same data as above)
    - Forage fishes (same data as above)
    - Obnoxious fishes (same data as above)
  - Growth rate and condition for game fishes
    - Compare with regional averages
    - Explain differences
  - Adequacy of spawning facilities and natural propagation
  - Previous management history
    - Stocking records
    - Fate of introduced species
  - Management Suggestions
    - What is wrong with fish or fishing?
    - What is wrong with the environment?
    - How may this environment be improved?
    - Stocking recommendations
    - What have been standard procedures by local fish and game organizations?
    - What further studies are needed?
    - How and by whom should these be conducted?

SOME OF THE COMMONEST FIELD CHEMICAL PROCEDURES USED IN LIMNOLOGY

BASIC WINKLER DISSOLVED OXYGEN PROCEDURE

(1) 250–300 ml sample taken with usual precautions; add
(2) 1 ml $MnSO_4$, with pipette tip below surface; at once, add
(3) 1 ml alkaline iodine soln. in similar manner. Close and
(4) Shake well for 1 min.; when ppt is half settled, add
(5) 1–2 ml H_2SO_4 (or HCl), conc.; close bottle immediately and
(6) Shake well several times (inversion). O_2 is now fixed and titration may be delayed for several hours.
(7) Titrate 200 ml of soln. over white surface with $Na_2S_2O_3$ until only pale straw color remains; add
(8) 5 drops starch soln. and
(9) Titrate until blue color disappears for first time. This is the true endpoint.
(10) Mls of $Na_2S_2O_3$ used = ppms of O_2 if $Na_2S_2O_3$ is exactly N/40. If not, the titration factor of $Na_2S_2O_3$ must be used in the calculations.

RIDEAL–STEWARD MODIFICATION OF BASIC WINKLER

(1) 250–300 ml sample taken with usual precautions; add
(2) 0.7 ml conc. H_2SO_4 below HOH surface; at once, similarly
(3) Add 1 ml K_2MnO_4: if violet color not permanent for 20', add another ml; etc. Then add
(4) 0.5 ml potassium oxalate; invert several times; allow to stand 5'; add another 0.5 ml to eliminate K_2MnO_4 color, if necessary, allowing to stand; then add
(5) 1 ml $MnSO_4$ and 3 ml alkaline iodide; mix by inversion, then allow ppt to settle at least halfway down; then
(6) Proceed as by BASIC WINKLER method, using steps 5 and on.
(7) Calculations are the same as for the BASIC WINKLER.

FREE CARBON DIOXIDE (CO_2)

(1) 100 ml Nessler tube sample, collected with usual precautions and run immediately; add
(2) 5–10 drops of phenolphthalein (phph) indicator; then
(3) Titrate with N/44 NaOH until a faint permanent pink appears.
(4) No. of ml of N/44 NaOH times 10 = ppm of free CO_2.

The LINGERING PINK which flashes all through the liquid for but 1–2 seconds is the more accurate endpoint (than the PERMANENT PINK), but is difficult to get with ordinary quality glassware delivery.

IT IS IMPORTANT TO REMEMBER THAT ALL FREE CO_2 TESTS MUST BE RUN IMMEDIATELY THE SAMPLE IS COLLECTED FOR SATISFACTORY RESULTS.

PHENOLPHTHALEIN ALKALINITY

(1) 100 ml sample in flask against white background; add
(2) 5 drops phenolphthalein (phph) indicator. IF COLOR

APPEARS IN SOLN., HYDROXIDE (OH) OR NORMAL CARBONATE (CO_3) IS PRESENT. IF sample colors (pink), titrate with
(3) N/50 H_2SO_4 until pink color just disappears;
(4) No. of ml of N/50 H_2SO_4 times 10 = ppm of $CaCO_3$ phph alkalinity. THEN, add, to find the

METHYL ORANGE ALKALINITY,

(5) 5 drops of methyl orange (MO) indicator to sample; if YELLOW COLOR IS PRODUCED, HYDROXIDE, NORMAL CARBONATE OR BICARBONATE IS PRESENT. Add
(6) N/50 H_2SO_4 until color shows first change from pure yellow; endpoint is a faint orange tinge (difficult to detect);
(7) No. ml N/50 H_2SO_4 times 10 = ppm of $CaCO_3$ MO alkalinity.

EXPLORATORY TESTS

Chlorides: to a 5 ml sample add 2 drops 2% $AgNO_3$ in 1 to 3 HNO_3. White ppt indicates chlorides, which will clear upon addition of 2 drops conc. NH_4OH. Thiocyanates and cyanides will invalidate this test.

Sulfates: to a 5 ml sample add 0.5 ml benzidine soln. and allow to stand for few minutes. A ppt is positive.

Phosphates: to a 5 ml sample add 0.5 ml NH_4 molybdate and 1 drop of 1% $SnCl_2$ in 1–5 HCl. A blue color is positive; silicates and/or arsenates may interfere.

Sulfides: to a 5 ml sample add 1 drop conc. NH_4OH and 5 drops of 10% Na nitroferricyanide; violet color positive.

Chlorine: to 5 ml sample add 2 drops KI soln. and 2 drops starch soln. Blue color positive. Manganics, ferries and nitrites may interfere.

Iron: to 5 ml sample add 2 drops 0.6% $KMnO_4$, 2 drops conc. HCl and 0.5 ml 60% K thiocyanate; pink or red color is positive.

ROUTINE DATA SUMMARY

LOCATION:

HOH Temp.	Depth	Air Temp.	Therm. No.	Date	Sta. No.	DO PPM	CO_3 PPM	HCO_3 PPM	pH

No._____ **BOTTOM FOOD SUMMARY** First Part
LOCATION:
Station No.:
Date:
Depth:
Sample area:
Bottom type:
Vegetation:
Vol. of sample:
No. organisms:
 Turbellaria:
 Nematoda:
 Oligochaeta:
 Hirudinea:
 Gastropoda:

No._____ **bottom food summary** Second Part
LOCATION:
 Pelecypoda:
 Amphipoda:
 Cladocera:
 Copepoda:
 Ostracoda:
 Hydracarina:
 Ephemeroptera:
 Bioptera:
 Trichoptera:
 Coleoptera:
 Tendipedidae:
 Other Orders:

B.
STREAM SURVEY PROCEDURES AND FORMS

By THEODORE C. FRANTZ
Wheeler District Fisheries Manager[1]
Nevada Fish and Game Commission

POPULATION INVENTORY

Population inventories are conducted on streams most frequently during the summer months. The first initial phase of a stream survey constitutes a population inventory of that creek. A 31.5 pound portable generator and a pair of electrodes are used for this purpose. The generator is a two-cycle, 400-watt plant which operates at 3,500 RPM producing 110 volts of alternating current. Thirty weight oil and low grade or "regular" gasoline are mixed together at a ratio of one-half pint oil to one gallon of gas for operation and lubrication of the engine. The tank has a five-pint capacity which is more than adequate during any one-day operation.

Each electrode is 6.5 feet long with a bamboo pole securely attached by clamps at the terminal end with a screen disc. The disc is 11 inches in diameter with 4 pieces of copper screen soldered to the one-fourth inch copper frame—two screens to each side. One hundred feet of number 14 gauge, rubber insulated cord are attached to the electrodes.

Shocking efficiency of a stream is dependent on the width of the creek, mineral content of the water, and obstacles present. Creeks 6 feet or less in width are shocked more efficiently with better results when debris in the stream and other obstacles are at a minimum, and carbonates in solution exceed 75 parts per million. Low volume flows are beneficial but often necessitate picking up the fish by hand or with a dip net as they are shocked.

Two men are required to transport and operate the equipment. When roads no longer follow the stream where a vehicle can be utilized, one person carries the generator on a light-weight metal back-pack, and a several-foot seine; the other transports the electrodes, cord, small equipment bag, and clip board.

At each shocking section, which may vary in length from 30 to 200 feet, the seine is securely fastened to the lower end of the section. The electrodes are then worked from the upper end of the creek slowly down toward the seine—one man attending the electrodes and the other keeping the cord from becoming entangled and picking up those stunned fish that do not readily float or move down toward the seine. Fish in the seine are then removed and placed in a net, dip net, or seine for identifying the species, obtaining the lengths (total length), weight, and for procurement of fish scales. These data are recorded on a special sheet attached to the clip board with other information such as the number, length, and location of the section; shocking efficiency;

---

[1] Now Sierra District Fisheries Manager.

the condition of the fish; and such stomach contents as may be analyzed. Air and water temperatures are also recorded at this time along with the elevation (altimeter reading). The trout are then released preferably in one or more pools.

The number of shocking sections set up along a stream depends primarily on the length of the creek. These may vary from two to eleven sections or more and occasionally one section. Many of the streams are short and require only four shocking sections. Sections may be placed at intervals of one-half mile or less on short streams and one mile or more on longer streams. Waters of wide width are not often inventoried with this portable generator as its efficiency drops considerably with waters over seven feet.

Field Investigations—

Streams previously inventoried for trout population are followed up during the late summer and fall with intensive investigations of the physical, biological, and chemical conditions of the creek. At this time the entire length of stream is often covered from its point of dissemination to the source. Distances along a stream are recorded by a truck's odometer or when traversing the distance by foot, one or two pedometers are used. Such information is mileaged as diversion points, total and fishable lengths, dry sections, tributaries, springs, seeps, types of vegetation, beaver cuttings and impoundments, types of stream gradient, outcrops, erosion, piscivorous predators, forest boundaries, fence lines and gates, camps and picnic areas, bridges, trespass signs, and possible reservoir sites.

Photographs are taken of diversions, dams, and portions of the watershed. At the source or initial flow of the creek, air and water temperatures, altimeter reading, and a pH reading are taken. Sample areas are set up along the creek. The number of areas used depends on the length of the stream. Three are most frequently used but may vary from one to eleven. Stream Survey forms are used for obtaining all the essential data at each sample area. In addition, specific comments are made in reference to the cover, pools, spawning areas, erosion, recommendations for stream improvement, etc., and to include lengths of stream other than the sample areas.

The accumulation of other information is necessary before a report can be written up on a stream. Individuals closely connected with a creek such as ranchers and others are contacted to obtain additional data. Such questions are used as: 1. Your name, occupation and name of ranch. 2. How long have you resided at the present address? 3. How much property do you own along the creek? 4. How many diversions, ponds, or dams do you have and where located? 5. What has been the volume flow of water during past years—low flow, high flow, cloudbursts, etc.? 6. Do you have water rights on the creek, how much, other water users, amount? 7. Have you found any fish in fields—what kind? 8. Do you object to fishermen fishing on property and if so why? 9. Other names for creek. 10. Past stocking of trout.

Property ownership along a creek is obtained from the plat books at the county assessor's office; available information on water rights, water use and volume flows is gathered from the State Engineer's Office, the

Ground and Surface Water Branch of the U. S. Geological Survey, Water Master's Office or/and Irrigation District Office; local U. S. Forest Service offices advance such information upon request as numbers and species of trout planted in years gone by. Also available are such data as grazing information on the forest. The Soil Conservation Service records may be available for a watershed in reference to snow surveys; past planting records are obtained from the Nevada Fish and Game Commission office, county fish and game boards, and/or sportsmen's organizations.

The Report—

Since each creek is an entity in itself, the material gathered during the surveys and incorporated into a report identifies the creek now as something more tangible whereby an understanding of the stream's attributes and problems are now at hand. From the report proper management of the stream can be determined. Collectively, the reports afford a basis upon which to formulate future stocking policies for the state.

Each report is broken down into twelve major headings with survey sheets, stream profile, planimetric map and photographs included at the end of the dissertation. These headings are summarized below as to what constitutes a Stream Survey Report for Nevada.

(1). *Introduction—*

Under this caption the upper and lower altitudes of the creek are given with the name of the range or mountain and the drainage, the type of watershed and general features of the stream, whether it is on public or private land, listing the governmental agencies having jurisdiction if the land is public and/or the names of property owners having land along the creek. Mileages for the total and fishable lengths are indicated and what portions of the water are included under the fishable length with the average width and depth. Of importance is the distance of the stream from one or more populated centers with a description of the types of roads, if any, along the creek and what type of conveyance can travel the road. Other general information covers fishing pressure, picnic grounds and camps, trespassing, accessibility and grazing.

(2). *Water Type—*

Under Water Type there are several subheadings—Flow (Velocity and Volume), Temperatures, Turbidity, Pools and Riffles, and Chemical Analyses. Here the average velocity and approximate volume of flow are indicated. These are taken from the Sample Area sheets which were recorded at each sample area by the float method at the time of the field survey. Velocities are listed in feet per second and also correlated in terms of *sluggish* through *torrential* as follows: 0.2 to 0.9, sluggish; 1.0 to 1.4, slow; 1.5 to 2.4, fast; 2.5 to 3.4, rapid; and 3.5 and above, torrential. The approximate volume of flow is given in cubic feet per second and also in gallons per minute. Comparisons in flows are frequently shown for different elevations such as the upper, central and lower elevations or specifically for each sample area. Estimated flows are given for springs and incoming tributaries if measurements have not previously been taken. Other aspects are taken into consideration as extensiveness of dry sections of stream, dry and wet years, volume flows obtained from other agencies, amount of water diverted, and amount of snowfall as obtained from the Snow Survey bulletins.

Recorded air and water temperatures at the source of flow, springs, shocking sections, and sample areas are entered with the time shown and possible weather conditions.

Turbidity is noted in the report and the source causing the discoloration when such is of unnatural origin.

The number and the quality of the pools are listed for each elevation with a description of the type of pools. Stream bottom coverage by riffle areas is included.

Water analyses in part are taken at one or more sample areas and are normally shown in parts per million. The pH values are shown for the source and sample areas. Carbonates and bicarbonates appear as part of the data important to the contents and dissolved oxygen and carbon dioxide in some instances are taken and also entered. Additional tests are made on some waters for more complete analyses to determine the amount of chlorides, sulphates, calcium, magnesium, iron and aluminum oxides, boron, silica and arsenic present. Such water samples collected have been sent to the State Food and Drug Laboratory or the Nevada Soil and Water Testing Laboratory.

(3). *Bottom Types—*

Bottom compositions along the stream are made up in percentages to include rocks, rubble, gravel, sand, mud or silt. The amount of adhesion of bottom material is inserted under the subheading and the factors responsible.

(4). *Vegetation Types and Abundance—*

Vegetation is broken down into aquatic and bank vegetation. The types of aquatic plants are listed and their density. Those plants appearing along the banks are classified and those species that predominate are indicated wherever predominance occurs. The scientific names do not usually appear in the text but are listed on the survey sheets. Description of the watershed is sometimes included under this heading.

(5). *Cover—*

This section deals with the type of cover over and in a stream, how extensive and the amount of cover it affords. Types of cover include woody, herbaceous and aquatic plants, rocks, debris, undercut banks, pools, water depth, artificial structures, etc.

(6). *Spawning Conditions—*

To what extent spawning beds are present and the suitability of such spawning areas are shown here.

(7). *Animal Types and Abundance*—

Under this category the animals which inhabit the water or have a close relationship to the stream are subjects of great importance. The aquatic and surface food supplies are listed for the various elevations and their abundance. The scientific names are usually indicated on the survey sheets but not in the text—only common or vernacular names appear here. The abundance is arrived at by estimating what can be observed along various bottom sections of the stream and along the banks.

Beaver dams and impoundments and beaver cuttings are included in reference to their beneficial or detrimental qualities. Fish-eating birds, garter snakes and other predators are entered.

A table is made up showing the game fish(es) present in the stream (and the number of fish per mile, each species, and species combined) according to size group (2" to 6", 6" to 12", 2" to 12") for the entire length of stream and for each elevation (upper, central, and/or lower). The number of shocking sections set up, their length and contribution toward shocking efficiency are also included with the last planting of trout shown prior to the inventory if such a plant was made during the year. The number of fish less than two inches are noted but not included in the table. The condition of the trout shocked is commented upon and stomachs may be opened to determine the type of food consumed. Age and growth studies are made on some streams with the data appearing herein. The species of nongame fishes are listed and their density.

All available information from past stocking records is shown in a table with the following captions: Date (stocking made), Species, Average Size, Number Planted, Source of Information. Ranchers frequently contribute information on the estimated number of fish found in the fields during irrigation.

(8). *Barriers*—

Waterfalls, beaver dams, debris and artificial structures are noted particularly when they interfere with the upward migration of spawning fish from a lake or reservoir.

(9). *Diversions and Water Uses*—

The number of diversions and the type of diverting structures are listed with the amount of flow diverted and to what use the water is put. The water users and the amount of water they are entitled to for any given time are shown here.

(10). *Pollution*—

When pollution is found to occur in a stream, the pollutant is described, its effect upon the stream, its point of entry and distribution, and the agency responsible for polluting the water.

(11). *Rating to Key Stream*—

Each stream is given a numerical rating of 1 through 5 in relation to its fishable qualities. A rating of one is considered a very good stream and a rating of five is classified as a nonfishable body of water.

Ratings are actually temporary as conditions change throughout the years. When more than one stream is surveyed on the same side of the watershed and such streams drain into a common valley, that stream with the highest numerical rating is selected as the Key Stream, or that stream which is the most popular and accessible.

(12). *Recommendations—*

Such recommendations are made as species and numbers of fish to be stocked; stream improvement; preventative measures to stop or curtail pollution; suggested reservoir sites; land acquisition; construction of roads; replanting watersheds where areas have been over-logged and fires have caused considerable damage; suggestive legislation to prohibit the destruction of watersheds by logging and other concerns on private lands; the purchase of private land by the U. S. Forest Service within the boundary of the National Forest when such land is available. Before any public domain land is appropriated for private use, provisions should be made to guarantee a public right-of-way along the banks of fishable streams where such streams pass over public domain land; acquisition of stream rights-of-way on private land.

In addition to the text, the stream system map, survey sheets, stream or planimetric map, stream profile, and photographs are attached. When the waters from two or more streams come together or an important drainage system is present, a stream system map is drawn up to show part or all of the drainage. A planimetric map is made up of the stream, utilizing Soil Conservation Service, U. S. Geological Survey, U. S. Forest Service and U. S. Army Map Service aerial photographs; U. S. Geological Survey sheets, Bureau of Land Management and Forest Service maps, U. S. Reclamation Service maps and State of Nevada highway maps, and others. Such location data are plotted on the map as shocking sections, sample areas, upper, central and lower elevations, springs, and seepage areas, types of prominent vegetation, picnic and camping areas, fence land and cattle guards, forest boundary, county lines and state line, meadows and fields, ranch houses and other land marks, diversion points, bridges and culverts, erosion, beaver dams and cuttings, reservoirs, lakes and ponds, mineral workings and mines, intermittent flow, mouth of canyon, narrows, corrals, waterfalls, incoming flows and outcrops. The stream profile shows the profile of a fishable stream with elevations taken from the U. S. G. S. quadrangles or altimeter readings. The fishable length, average gradient, and ownership of land are shown and the total drop of the stream is given in feet from the source to the end of its total length. Photographs $3\frac{1}{2}''$ x $5''$ are attached to sheets. Photographs taken are of sample areas, watersheds, beaver impoundments, ponds and reservoirs, diversions, erosion, waterfalls, etc.

STREAM SURVEY SHEET   Fisheries Form #16

Sample Area (a, b, etc.)   Report Number

Date   Name(s)   Stream System

Tributary to   County or Counties   District Number

Total Length   Fishable Length   From: To:
                                                Altitude Range

Sample Area Location and Length   Elevation

Stream Type: Torrential__Rapid__Fast__Slow__Sluggish
Bottom Type: Rock__Rubble__Gravel__Sand__Mud__   Temperature:
Turbidity : Clear__Cloudy__Murky__Muddy__   Air   °F
                                                    Water   °F
Food Types   Abundance   Overall Rating   Time   .M.
                                                 Flow Average:
                                                    Width   " or
                                                    Depth   " or
                                                    Velocity   f/s
                                                    Volume   cfs
                                                                   gpm
                                                 Measurements:
                                                    Good
                                                    Fair
Vegetation Types   Abundance   Overall Rating   Poor
                                                 Note: Correction factor
                                                       = 0.8
                                                 Water Analyses:
                                                    D.O.   ppm
                                                    CO3(Phph)   ppm
                                                    HCO3(M.O.)   ppm
                                                    CO2   ppm
                                                    pH
Pool Abundance   : Excellent__Good__Fair__Poor__
Spawning Areas   : Excellent__Good__Fair__Poor__
Riffle Areas   : Excellent__Good__Fair__Poor__

POPULATION DATA

S.S.	Lgth. of Sec.	Species	No. Per Mile	Size - Weight Class Data					
				2"-4"	4"-6"	6"-8"	8"-10"	10"-12"	12" +
				# :Wt.	# :Wt.	# :Wt.	# :Wt.	# :Wt.	# :Wt.
				:	:	:	:	:	:
				:	:	:	:	:	:
				:	:	:	:	:	:
				:	:	:	:	:	:

Stream Rating (1), (2), (3), (4), (5): _____   Key Stream:
Remarks: (Pollution, Barriers, Burned Areas, Habitat Improvements, Predators, Photos, Terminal Reservoirs, Availability of Land, Rights-of-Way, etc., Water Rights)

Use Reverse Side

FIG. 259. (Stream survey sheet, Fisheries Form No. 16.)

Fisheries Form #24 POPULATION INVENTORY SHEET SAMPLE

Date	Name (s)	County

Sec.	Ft.	Species	No. Per Mile	Size - Weight Class Data -2"	2"-4"		4"-6"		6"-8"		8"-10"		10"-12"		12" +	
				#	#	Wt.	#	Wt.	#	Wt.	#	Wt.	#	Wt.	#	Wt.
		TOTAL														

Sec.	A-°F	W-°F	Time	Elev.	Location

Sec.	S. Efficiency	Cond. of Fish	Scales-# Fish	Stomach Analyses

FIG. 260

C.
ALKALINITY AND ACIDITY[1]

CONTENTS

	PAGE
Collection and Preparation of Samples	612
Analytical Procedures for Dissolved Oxygen	614
Alkalinity and Acidity	617
Methods	619
Alkalinity	619
Acidity	619
Calculations of Alkalinity	620
Rawson's Nomogram	623
Special Determinations	624
Reagents Used in Special Determinations	625
Evaluation of Dissolved Oxygen Determinations	626
Evaluation of Hydrogen-ion Determinations	627

The accurate determination of dissolved oxygen is highly important in water-quality studies not only because oxygen is indispensable to fishes and other aquatic organisms, but also because knowledge of the exact amount of dissolved oxygen (DO) is essential in solving many problems which confront the aquatic biologist, the sanitary engineer, and the industrial chemist. The measurement of the effect of specific oxyphilic pollutants, the charting of stream pollution, the computation of a maximal load of organic pollution which any given mass of water can tolerate and still remain habitable for aquatic life, and the biochemical exploration of deep lakes and impoundments are among the many field investigations requiring precise analyses for DO.

Although the degree of accuracy requisite in the various types of field and laboratory studies varies, any determination of DO requires the careful observance of several precautions if a dependable value is to be obtained. The amount of oxygen gas in solution in the water can be altered easily by either improper preparation or handling of the sample; and the analytical methods for DO have certain chemical limitations which cannot be disregarded. In addition, oxygen determinations must be made as soon as possible after the sample is taken, which in field studies necessitates portable apparatus for streamside use. However, accurate determinations of DO can be made if reasonably simple procedures of sampling and analysis are followed. A system of dissolved oxygen studies that will meet the conditions encountered in

---

[1] Extracted from: Ellis, M. M., B. A. Westfall and Marion D. Ellis. 1948. Determination of water quality. United States Department Interior Fish and Wildlife Service Research Report 9: 1–122.

both good and polluted water is presented. It is recognized that many other procedures could be followed, but the methods offered here have proved dependable and practical in a variety of situations.

In this system there are choices of sampling equipment depending upon the degree of accuracy required and the conditions under which the samples are to be taken. The selection of an analytical sequence is not optional, however, as these procedures have been developed for specific water types. Consequently orientation tests (q.v.) should be made before any quantitative DO sample analyses are attempted.

COLLECTION AND PREPARATION OF SAMPLES

Careless manipulation of the sampling devices or improper handling of the sample once it is obtained can change DO content of water so that analysis is useless. It must be borne in mind while sampling for DO that slight changes in pressure or temperature to which the sample may be subjected, agitation of the water in the presence of air, and entrainment of bubbles of air or other gases with the sample, readily alter the amount of DO in any given sample. Therefore, it is of great importance that the sample be taken without undue agitation, bubbling, or reduction of pressure, and that it be held as nearly as possible at the temperature of the water at the time of sampling until the reagents for oxygen fixation have been added. Ideally, the sample should be fixed immediately after collecting and for this purpose the writers use a small box containing the required chemicals which can be carried to the actual site of the sampling operations.

It is not always possible in the field to meet these specifications absolutely but both the samplers and the sample bottles have been designed to reduce errors from pressure, bubbling, and agitation, to the minimum, and should it be necessary to hold unfixed samples for a short time, they can usually be kept near the proper temperature by placing them in a bucket of water taken from the same source as the sample or by covering the outside of the bottles with wet cloths if the air temperature is high.

Samplers

Either the Foerst modification of the Kemmerer, or the bottle-train sampler will provide accurate and uncontaminated samples for DO under most field conditions. In laboratory tests, recourse must be had to either the siphon or some other device which will meet the specific difficulties of sampling under experimental conditions.

Clarification

If the water to be sampled carries a large amount of mud or other suspended matter which does not settle out readily or rise to the top on standing, so that a suitable sample can be obtained from a large sample by siphoning, it is usually desirable to remove the suspended matter by flocculation before DO analyses are attempted. A large sample, 1–3 litres, is procured in a glass-stoppered sample bottle, with the usual precautions. The glass-stopper is carefully removed and 10 ml of 10% alum solution (potassium aluminum sulfate) added per litre of sample from a glass pipette, the tip of which has been lowered at least a cm below the surface of the sample. The alum solution is followed immediately by 1–2 ml of approximately 35% NaOH solution. The bottle is then carefully stoppered without entrainment of air and gently rotated until the solutions are thoroughly mixed. A heavy precipitate of white aluminum hydroxide will be formed and will appear in curds throughout the bottle. Place the bottle in an upright position and allow the precipitate to settle. As the aluminum hydroxide moves to the bottom of the bottle, the sample will be clarified by flocculation (Ruchhoft and Moore, 1940).

When the top two-thirds of the sample bottle has cleared, a clear sample can be drawn through the siphon into a smaller sample bottle. If very accurate results are required, the dilution of the original sample by the addition of reagents of flocculation must be computed. For general work, this correction may be disregarded. It is very important that the quantities of alum solution and NaOH be added exactly as directed for an excess of alkali could raise the alkalinity of the sample to pH 11 or above, at which levels of alkalinity many organic substances will take up DO. If too much alkali is used, therefore, in the flocculation process, the final DO reading will be too low owing to the loss of DO when organic substances are present. Ruchhoft (1941) states that this procedure for clarification by flocculation is very effective for raw sewage and river-mud suspensions and that using the supernatant fluid from this clarification, the DO may be determined by the rapid Winkler method.

A second procedure, using a copper-sulfamic acid mixture (Ruchhoft, 1941), for clarification of samples containing activated sludge not only clarifies the sample, but dissolves the bottle as well:

Glass-stoppered Sample Bottles

Bottles, with stoppers ground to fit, preferably of pyrex glass, of 200–300 ml capacity make excellent sample bottles for DO if the end of the ground portion of the glass stopper is held against a revolving emery wheel and rotated until the edge is definitely beveled so that air bubbles will not be caught under the stopper when it is inserted into a completely filled bottle. The standard glass-stoppered, biochemical oxygen-demand (B.O.D.) bottles used by sanitary engineers, although more costly, are excellent for this purpose as these bottles are provided with a tapered-end glass stopper.

Occasionally it is necessary to collect a large sample, 1–3 litres, for special treatment, as mentioned above. For this purpose, large, glass-stoppered acid bottles are quite satisfactory if the ends of the stoppers are tapered or pointed.

The glass-stoppered sample bottle of whatever kind must have a round shoulder so that no bubbles will be caught in the bottle when it is being filled, must be provided with a ground glass stopper which fits accurately and securely into the neck of the bottle, and which is tapered sufficiently at the end to prevent the trapping of air bubbles below the stopper when it is lowered into the neck of the completely filled bottle.

Glass-stoppered sample bottles should be used in both field and laboratory when an accuracy of at least 0.01 parts per million (ppm) of DO is expected.

In filling these bottles care must be taken to flush the bottle with at least three times its volume of water and to withdraw the entry tube slowly from the bottle while the water is still flowing.

ANALYTICAL PROCEDURES FOR DISSOLVED O_2

Winkler Method

The method originally described by Winkler (1888), based on the formation of manganous compounds that are oxidized to manganic compounds by the DO in the water sample, and which subsequently release iodine quantitatively from KI when the sample is acidified, still remains in principle the most useful method of both field and laboratory determinations of DO in water. Numerous modifications have been proposed to meet specific needs and to offset the action of substances in the water affecting the Winkler reagents. These interfering substances are discussed under *Orientation Tests*.

Basic Winkler Method

This method assures accuracy only with waters of high purity, free from interfering substances and organic pollutants and containing less than 0.05 ppm of nitrate and 1.0 ppm of ferrous iron.

Procedure

(1) A 250–300 ml sample is desirable, and the sample bottle should be placed in a lead-lined box of appropriate size so that the corrosive chemicals and overflow will not foul other equipment.

(2) Open sample bottle with great care, avoiding aeration of sample surface by bubbling or agitation, and add EXACTLY 1 ml OF $MnSO_4$ (or $MnCl_2$) SOLUTION FROM A GLASS PIPETTE. The point of the pipette must be lowered an inch or more into the bottle and the heavy $MnSO_4$ soln. introduced rather slowly so that it will sink to the bottom.

(3) ADD AT ONCE 1 ml OF ALKALINE IODINE SOLN. in a similar manner from another glass pipette.

(4) CLOSE BOTTLE IMMEDIATELY AND MIX THOROUGHLY BY INVERSION FROM 40–60 SECONDS. During this mixing, a milk-white precipitate (ppt) of manganous hydroxide will form and as this compound combines with the DO in the water, yellowish-brown manganic compounds will appear. If the milk-white color persists, very little DO is present. After shaking the bottle for approx. 1 minute, return bottle to lead-lined box and allow ppt to settle at least halfway down the bottle; i.e., *there should be no ppt in the top half of bottle when the next step is started.*

(5) Open bottle carefully as before and ADD 2 mls OF CONCENTRATED H_2SO_4 (or HCl) BY BRINGING PIPETTE POINT AGAINST NECK OF BOTTLE A FEW MLS BELOW SURFACE OF LIQUID SO ACID WILL FLOW DOWN INSIDE WALL OF BOTTLE. Close bottle immediately after acid is added and mix thoroughly by inversion several times. During this mixing, the ppt, either white or brown, should dissolve completely and the sample will become clear, developing a yellowish-brown color due to the liberated iodine.

The depth of this color is proportional to amount of DO present so that the analyst may estimate approx. the amount of DO present in the sample in advance of titration. Sometimes part of the ppt will not dissolve owing to the formation of compounds other than the manganic compounds involved in the liberation of iodine, by the reaction of the reagents with impurities in the water. If such an insoluble ppt forms, it may be disregarded *but care must be taken not to include any of the undissolved ppt in the aliquot portion of the sample taken subsequently for titration.*

(6) When the ppt has dissolved and the sample has become uniformly colored, fill a 100 ml graduated cylinder and pour this portion into a 250 ml Ehrlenmeyer flask for titration. *200 mls may be used if greater accuracy in titration is desired.* To the aliquot taken from the first sample add, from a pipette, the exact amount of treated sample necessary to correct for the volume of sample lost by displacement when the various test solutions were added during the process of oxygen fixation. The volume of this correction can be computed by substituting actual values for the following equation:

$$\frac{ab}{b-c} = d$$

where a = mls of treated sample taken, b = ml capacity of sample bottle, c = combined mls of $MnSO_4$ and alkaline iodide solutions and d = total volume of treated sample required.

Pour the corrected aliquot sample into a 250 ml container and wash out the original flask with a few mls of distilled HOH, adding the washings thus obtained to the sample for final titration.

(7) TITRATE OVER A WHITE SURFACE WITH STANDARDIZED SODIUM THIOSULFATE SOLUTION ($Na_2S_2O_3$) until only a pale straw-yellow color remains. Add 0.5 ml (about 10 drops) of starch soln. and continue titration until the blue color completely disappears for the first time. This must be taken as the endpoint as various impurities in HOH may cause a return of blue color a few seconds to a few minutes after it is first discharged and if the titration is continued, an erroneously high reading for DO will result.

Computation of Results

If properly standardized before use, the $Na_2S_2O_3$ used in titration will be a direct reading of the ppm of DO in the HOH. I.E., if 10 mls of thiosulfate are used to reach the endpoint, then the HOH contains 10 ppms of DO. This is usually all the accuracy required of field work, but an exact rendering of all the above-mentioned factors would have to be handled with the following formula:

$$\frac{1000}{A} \times B \times F = X$$

A = mls of aliquot (uncorrected 100–200 mls) of treated sample
B = mls of $Na_2S_2O_3$ used in titration
F = factor value of 1 ml of thiosulfate soln. in milligrams of DO (if the thiosulfate is properly adjusted, this is "1")
X = parts per million (ppms) of DO.

Limitations

Approx. 1 minute of mixing time must be allowed after adding $MnSO_4$ and the alkaline iodide soln. so that consumption of DO present in the HOH by the manganous hydroxide formed from the reagents can be completed. For uniformity, therefore, the mixing should be timed rather accurately to be not less than 40 seconds nor more than 60 seconds. After the acid is added, titration should be made as soon as possible for the sample is not dependable without special treatment for more than 1 hour in this phase if even very small quantities of ferric iron or various organic substances are present. Although several procedures for "fixing" the sample after the iodine has been liberated have been proposed, they should be avoided if possible as this procedure not only adds an additional step to the routine but may introduce errors of various sorts.

Rapid, Unmodified Winkler Method

This method must be used if the sample contains considerable organic pollution but is otherwise free from disturbing material. However, it does not correct for nitrites, iron, sulfites, and similar interfering substances. This procedure will give valid, dissolved oxygen determinations in the presence of organic matter up to the approximate equivalent of 5,000 ppms of dextrose.

Procedure

Follow the same routine as given for the basic Winkler method *except:*

(1) Use 2 mls of both $MnSO_4$ and alkaline iodide reagents.

(2) Upend the bottle and agitate for not more than 20 seconds.

(3) Add 2 mls of acid as soon as ppt has settled a half inch from top of bottle, or not later than 15 seconds after mixing is completed.

(4) Titrate as soon as possible after acid has dissolved the brown ppt.

Rideal-Stewart Modification of Winkler Method

The Winkler methods described above will yield satisfactory results, as mentioned, except in the presence of certain impurities. If nitrites, iron salts or certain organic compounds are present, the Rideal-Stewart modification should be used. Since in work on unknown waters it may be uncertain whether or not this modification is necessary, it is common practice to use it as a regular procedure on all waters, the only disadvantage being the use of the additional steps required in the analysis.

Procedure

(1) Remove stopper from the 250 ml sample bottle and by means of a volumetric pipette, add exactly 0.7 ml of concentrated H_2SO_4 just below water surface.

(2) Then with a similar pipette add at once and well below the surface, enough of the potassium permanganate soln. (usually 1 ml) to yield a typical permanganate violet color which will remain permanent for at least 20 minutes after the bottle has been mixed throughly by

repeated inversions. If the permanganate color does not persist for 20 minutes, add a small additional amount of the permanganate soln.; mix by inversion and allow to stand; keep adding permanganate soln. until this procedure produces the required color for 20 minutes.

(3) Add 0.5 ml of potassium oxalate soln.; replace stopper and invert several times; allow to stand for 5 minutes; if, after this interval, the color of permanganate persists, add another 0.5 ml of the oxalate soln. and mix as before.

(4) After the permanganate color has disappeared completely, add 1 ml of manganous sulfate soln. and 3 mls of the alkaline iodide soln.; insert stopper and repeatedly invert bottle. Allow the ppt to settle about halfway in bottle, then mix again and allow to settle; allow sample to stand then for at least 5 minutes.

(5) Then proceed as in the regular Winkler Method first described above, beginning with step (5).

General Considerations

In field work, the analyses of the DO should be made immediately after the sample is secured and the reagents added up to the point where storage and transportation to the laboratory is safe (after the adding of the acid).

For dependable work, it is necessary that the temperature of the water at each sampling position be measured.

If appreciable amounts of iron salts are present, the reduction of the permanganate color, even with the proper amounts of oxalate soln., may be very slow. To remedy this condition, add 2 mls of potassium fluoride (40% soln.) along with the permanganate and decolorize in the dark.

ALKALINITY AND ACIDITY

Alkalinity of waters, as usually interpreted, refers to the quantity and kinds of compounds present which collectively shift the pH to the alkaline side of neutrality. As the alkalinity of natural waters is generally the result of carbonates, although other compounds may be involved in polluted waters and waters from various mineralized areas, alkalinity is usually expressed in terms of carbonates and of calcium carbonate in particular. Hence, 3 kinds of alkalinity are usually designated; viz., hydroxide (OH), normal carbonate (CO_3), and bicarbonate (HCO_3) and the entire group summed as total alkalinity.

The determination of the particular alkalinities present and the quantity of each in any sample of HOH is made by titrating with a standard acid (.02N H_2SO_4), using two indicators, phenolphthalein and methyl orange. As the pink color of the phenolphthalein (phph) is due to the presence of OH or CO_3 but not HCO_3 (for convenience referred to as phph alkalinity) and as methyl orange is yellow in the presence of any of the 3 (OH, CO_3 or HCO_3), but is red in an acid medium at approx. pH 4.4, the desired alkalinities can be separated, and specific values calculated.

Various combinations of the different kinds of alkalinity occur. CO_3 (sometimes referred to as *normal carbonate*) and either OH or HCO_3 may be present in the same sample, but OH and HCO_3 are not found together due to the reaction between them. Hence, there are 5 conditions of alkalinity possible in the sample, viz., CO_3, HCO_3 or OH alone or mixtures of CO_3 and OH, or CO_3 and HCO_3.

As the CO_3s have a definite influence on the fauna of freshwaters and as these compounds exist in 3 forms, viz., free CO_2, which in HOH forms the weak acid hydrogen carbonate or carbonic acid, bicarbonate and carbonate, it is advisable to consider each form separately.

Carbonate and Bicarbonate

CO_3 and HCO_3 in natural unpolluted HOHs supporting good fish faunae, range from 0 to 350 ppm (expressed as $CaCO_3$), with the usually expected values lying between 45 and 200 ppm. In these quantities the CO_3 and HCO_3 have little direct effect on the fishes. There are, however, 3 indirect effects of these compounds that do influence the life of fishes materially.

First, the free CO_2 and the HCO_3 are necessary to plant life in a stream, upon which the fauna depends for complete food chains, but, even though the variation in CO_3 content is large in streams with excellent fauna, the small quantity of CO_3 necessary for phytoplankton and other aquatic plants is practically always present. This is evident from the excellent fish fauna in many mountain streams and lakes, the waters of which have extremely low CO_3 content. The buffering and neutralizing effects, i.e., the chemical protection afforded by the CO_3 against acids is often a major factor in determining the severity of acid pollution. The CO_3 also serve as precipitants of many elements and compounds thereby tending to hold the heavy metals as iron, manganese, and zinc to nontoxic low levels, and precipitating out many harmful substances which enter the stream as pollutants, e.g., copper sulfate ($CuSO_4$) in hard HOH is much less toxic to fishes than the same concentration of $CuSO_4$ in soft HOH (Ellis, 1937), because the Cu is largely precipitated out of the hard HOH as a CO_3. As the free CO_2 in HOH acts as a weak acid, it is discussed under *acidity*.

Hydroxides (OH) are rarely found in natural HOHs unless added through pollutants or HOHs from mineral springs and mineral fields. In some streams, however, receiving certain types of chemical wastes, great damage to the fauna may result from the OH alkalinity. Although the OH is important as a pollutant, it is of negligible value in most fresh HOHs.

Acidity

Acidity in natural unpolluted HOHs is usually due to the presence of CO_2 and several organic acids as tannic and humic. However, in polluted HOHs, both those receiving runoff from mineralized areas and those polluted by industrial effluents, almost any kind of acid or hydrolizable salt can be expected. The total acidity is of importance, not only because of the toxic effects of the various acids (discussed under hydrogen-ion concentration), but as a protection against strongly alkaline pollutants, and because the quantities of soluble salts of heavy metals are raised in acid HOHs which dissolve these elements out of rocks, silts and muds.

The free CO_2 in most HOHs is seldom present in large quantities due to the reaction with other compounds in soln. and to constant aeration. It has been observed that 90 percent of the HOHs containing good fish fauna have less than 2 ml of free CO_2 per litre. If 3 ml or more per litre are found, organic pollution is suspected, although some lakes, ponds, and bogs may approach, or even exceed, this figure under normal conditions.

In HOH-quality studies, the determination of free CO_2 is of importance because, since this gas is used more readily by the phytoplankton than either HCO_3 or CO_3, it may indicate organic pollution, and if present, in sufficient amounts it is specifically toxic to most aquatic organisms.

Mineral acids and many hydrolizable salts are often harmful in very small quantities due to the changes in pH which they may produce and to the specific toxicity of the individual compounds. Consequently, these substances must be considered individually although total acidity determinations will give valuable data concerning the amounts present.

METHODS

Alkalinity

(1) Titrate a 100 ml sample in a 250 ml Erlenmeyer flask, with 0.02 N sulfuric acid soln. (N/50) using phenolphthalein (phph) (5 drops) as an indicator until the pink color just disappears and record the number of mls of acid used.

(2) Add methyl orange (MO) indicator (5 drops), and continue the titration until the yellow color characteristic of the alkaline phase of this indicator turns to a definite orange, and record the number of mls of acid required. It is often advisable to carry a blank of distilled HOH and MO for comparison, as the endpoint of MO is very difficult for many persons to detect, especially anyone who borders on color blindness. This group of persons can use to great advantage the indicator known as xylene cyanole-methyl orange which has a definite blue component disappearing at the endpoint.

If, on the addition of the phph to the sample, it remains clear and does not turn pink, immediately proceed with the MO titration.

Acidity

Free Carbon Dioxide (CO_2)

Fill a 100 ml Nessler tube with the sample and using a glass rod with a foot almost as large as the tube, stir the soln. gently to avoid loss of CO_2, and titrate to a faint permanent pink color, using phph, with N/44 NaOH soln.

The number of mls required for this titration multiplied by 10 gives the CO_2 content in ppms. This value in ppms of CO_2 can also be obtained somewhat more accurately from the pH and total alkalinity using Moore's (1939) nomograms. Parts-per-million (ppms) of free CO_2 multiplied by 2.272 gives ppm acidity as $CaCO_3$.

Total Acidity

(1) Titrate a 100 ml sample in a flask to faint permanent pink, using phph with N/50 NaOH soln. The amount of OH used is equivalent in $CaCO_3$. To convert from mls of acid to ppms of acidity expressed as $CaCO_3$, multiply the mls of acid by 10.

In using this method for total acidity in the presence of aluminum and iron salts (these salts do not enter into the reaction with any degree of speed in ordinary temperatures) the reaction may be quite slow and the first extremely faint persistent pink color should be taken as the endpoint, especially if aluminum is present, as a partial resolution of $Al(OH)_2$ may occur.

If acidity resulting from free mineral acids is to be determined, use MO indicator and titrate with N/50 NaOH until the pink color characteristic of the acid phase of MO just disappears. If ferric and aluminum salts are present in appreciable quantities, the result for mineral acids will be too high. To correct for this error, determine the total acidity caused by mineral acids, and the salts of iron and aluminum by titrating at boiling temperature using phph as an indicator. From this value subtract the acidity due to the presence of salts of iron and aluminum from calculations based on the previously determined amounts of these compounds. This gives a more accurate value for the acidity resulting from mineral acids. This procedure is especially necessary in the study of HOHs from coal mines, slag piles and metal pickling works.

If no acidity resulting from mineral acids is present, the total acidity is usually taken as a measure of free CO_2. However, in pollution studies other weak acids or compounds which will neutralize NaOH may be present even though the pH is above the MO endpoint. In these situations, a spuriously high value for free CO_2 will be obtained. Caution, therefore, must be exercised in assessing a value for free CO_2 from any unknown water.

CALCULATIONS OF ALKALINITY

Several sets of relations between hydroxides, carbonates and bicarbonates in soln. exist and have been carefully worked out by physical chemists. However, 5 of these relations are of interest to the aquatic biologist in HOH-quality studies, to enable him to calculate the various units of alkalinity in a given HOH. These relations and the calculations based on them are set forth below.

IF THERE IS PHPH ALKALINITY PRESENT AND NO MO ALKALINITY, THEN THE TOTAL ALKALINITY IS DUE SOLELY to OH. This follows because HCO_3 and OH cannot exist together in any appreciable amount due to the reaction between them. Were CO_3 present, there would be an MO titration to satisfy the bicarbonate formed on titrating the CO_3 with acid to the phph endpoint. THE OH ALKALINITY CAN BE EXPRESSED AS PPMS OF $CaCO_3$, BY MULTIPLYING THE NUMBER OF MLS OF N/50 H_2SO_4 used in the phph titration by 10.

IF THE NUMBER OF MLS OF STANDARD ACID USED TO DISCHARGE THE PINK COLOR OF PHPH IS EQUAL TO THAT

NECESSARY TO ATTAIN THE MO ENDPOINT AFTER THE PHPH ENDPOINT HAS BEEN REACHED, ALL OF THE ALKALINITY IS NORMAL CARBONATE. This is true because the amount of acid used in titrating the CO_3 to the HCO_3 form is equal to that necessary to change the HCO_3 so produced over to carbonic acid in the MO titration. Therefore, THE TOTAL NUMBER OF MLS OF ACID USED FOR BOTH PHPH AND MO TITRATIONS IS MULTIPLIED BY 10 TO OBTAIN THE NORMAL CARBONATE ALKALINITY IN PPMS OF $CaCO_3$.

IF THERE IS NO PHPH ALKALINITY BUT SOME MO ALKALINITY, ONLY HCO_3 IS PRESENT FOR NORMAL CARBONATE AND OH DO NOT EXIST IN ANY APPRECIABLE QUANTITIES IN A SOLN. MORE ACID THAN ABOUT pH 8.3, the approximate endpoint for phph in titrations of many natural HOHs.

If the phph alkalinity is greater than one-half the total alkalinity, both OH and CO_3 are present. As HCO_3, which cannot exist in appreciable quantities with OH, is absent, this phph titration represents the amount of acid necessary to neutralize the OH plus that required to change the CO_3 to HCO_3. The MO titration, in this case, gives the amount of acid necessary to change the HCO_3 thus formed to carbonic acid. Therefore, THE TOTAL NUMBER OF MLS OF ACID, MINUS THE MLS USED IN THE phph TITRATION GIVES THE NUMBER OF MLS OF ACID REQUIRED FOR THE MO TITRATION, WHICH REPRESENTS THE HCO_3 FORMED ON TITRATION OF THE CO_3 TO THE phph ENDPOINT. As the reduction of this HCO_3 to CO_2 requires the same number of mls of acid as were required to reduce the CO_3 to HCO_3, the normal CO_3 can be calculated by multiplying the difference between the total alkalinity and the phph titration by two and this product by 10 to give ppms of $CaCO_3$. Therefore, twice the phph titration gives the number of mls of acid necessary to neutralize the OH.

This figure multiplied by 10 expresses the OH alkalinity in ppms $CaCO_3$.

If the phph titration is less than one-half the total, then CO_3 is present plus some HCO_3. As previously shown, the CO_3 requires the same amount of acid to form HCO_3 as the HCO_3 needs to produce carbonic acid. Therefore, if the MO titration is greater than the phph titration (as in this case where twice the phph is less than the total), then the extra amount of acid needed in the MO titration was required by HCO_3 present in the sample. Therefore, normal CO_3 as ppms $CaCO_3$ is equal to twice the phph titration in mls of acid times 10. Likewise, HCO_3 in ppms of $CaCO_3$ is equal to the total alkalinity titration minus twice the phph titration plus 10.

If more accurate determinations are desired, other procedures are available. During recent years many investigations have been made to check the method of determining free CO_2 and normal CO_3, HCO_3 and OH alkalinity in which phph is used as an indicator. Also, certain assumptions currently used have been found to be inaccurate. The NaOH titration method for estimating free CO_2 content of HOH tends to give higher values than actually exist when only a few ppms are present (De Martini, 1938). For waters high in CO_2, the reverse is true. As the phph endpoint varies with different concentrations of alkali, the assumed pH at the endpoint does not always agree with the hydroxyl-ion concentration as computed from the hydrogen-ion concentration determined by thymol blue or by glass electrode pH meter (De Martini, 1938). Besides, the assumptions used in the phph-MO titrations that neither CO_2 and normal CO_3, nor HCO_3 and OH can exist together are not strictly quantitative (Moore, 1939).

Consequently, Moore (1939) using the formula of De Martini (1938) devised a method in which it is possible to calculate from the hydrogen-ion concentration and total alkalinity of a HOH the amounts of CO_2, HCO_3, normal CO_3 and OH present. Graphs were made based on the formulae as influenced by temperature and the concentration of dissolved salts. By reading the pH with a glass electrode, making a total alkalinity titration potentiometrically, and then reading the nomograph, more accurate results can be obtained.

Calculations of Results With Winkler Method

If a 200 ml HOH sample is titrated with N/40 $Na_2S_2O_3$ soln., the number of mls of thiosulfate used is numerically equal to the dissolved oxygen content in PARTS PER MILLION and no additional calculation is necessary. This value, multiplied by 0.698 will yield the value in terms of MILLILITRES PER LITRE. Results in terms of PERCENTAGE OF SATURATION are determined by dividing the titration value in mls by the solubility value as determined by the temperature of the sample.

For altitudes other than sea level, correction for barometric pressure must be made if precise results are desired. This is done by determining the oxygen solubility value at the pressure involved by the use of the following formula:

$$s_1 = s \frac{p}{760} = s \frac{p_1}{29.92}$$

in which

s_1 = solubility at p or p_1
s = solubility at 760 mm. or 29.92 in.
p = barometric pressure in mm.
p_1 = barometric pressure in inches.

For quick reference, Rawson's nomogram (which see below) for obtaining oxygen saturation values at different pressures is useful.

FISHES AND FISHERIES OF NEVADA

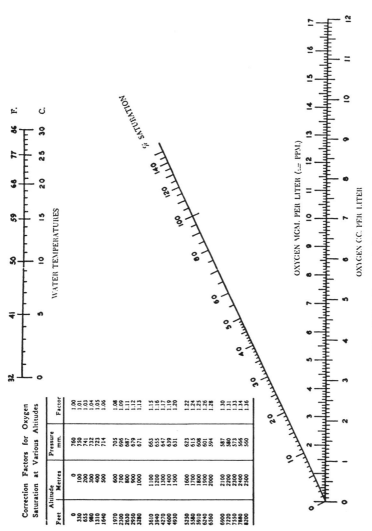

FIG. 261. Rawson's Nomogram.

SPECIAL DETERMINATIONS

Chlorides

To a 5 ml sample in a test tube add
(1) 2 drops of a 2% soln. of $AgNO_3$ in 1–3 nitric acid soln. A WHITE PPT INDICATES CHLORIDES.
(2) Add 2 drops of concentrated NH_4OH. If chlorides are present, the soln. will clear. If thiocyanate or cyanide is present, a white ppt forms, thus invalidating the test for chlorides in the presence of these compounds. However, thiocyanates and cyanides are seldom found in waters except in wastes from a few industries, such as electroplating, and the gold cyanide extraction process. This test is positive for chlorides in a concentration of THREE OR MORE PPMS. The treated sample should be checked for a ppt immediately on the addition of $AgNO_3$, especially if in bright sunlight, as the light will free some colloidal silver resulting in a purplish-blue color.

Sulfates

To a 5 ml sample in a test tube add 0.5 ml benzidine soln. and allow to stand for a few minutes. A ppt indicates sulfates. This test is positive and practical in concentrations of two parts or more of sulfates per million.

Phosphates

To a 5 ml sample in a test tube add 0.5 ml ammonium molybdate soln. and one drop of 1% stannous chloride in 1–5 HCl soln. A blue color develops in the presence of phosphates in a concentration of 0.05 ppm or more. Silicates in large quantities or arsenates of 0.1 ppm will interfere by forming a blue color.

Sulphides

To a 5 ml sample in a test tube add 1 drop of concentrated NH_4OH and 5 drops of 10% sodium nitroferricyanide. A violet color will develop in the presence of sulfides.

Chlorine

To a 5 ml sample in a test tube add 2 drops of potassium iodide soln. and 2 drops of starch soln. A blue color indicates chlorine. Manganic manganese, ferric iron and nitrites interfere in solutions more acid than pH 4.2. If the temperature of the sample during the analysis is kept below 20° C., 0.1 ppm chlorine can be detected.

Iron

To a 5 ml sample in a test tube add 2 drops of 0.6% potassium permanganate, 2 drops of conc. HCl and 0.5 ml of 60% potassium thiocyanate. A pink or red color indicates Fe. This method is sensitive to 0.1 ppm total iron.

REAGENTS USED IN SPECIAL DETERMINATIONS

Chlorides

(1) SILVER NITRATE: dissolve 2.5 gms of reagent-quality $AgNO_3$ in 0.5% HNO_3 to make 1 litre. Standardize against NaCl and store in a brown glass-stoppered bottle.

(2) SODIUM CHLORIDE STANDARD: dissolve exactly 1.6486 gms pure NaCl in distilled HOH to make 1 litre. 1 ml of this carries 1 milligram of chloride chlorine.

Sulfates

(1) BENZIDINE HYDROCHLORIDE SOLN.: dissolve 11.2 gms benzidine hydrochloride in 1% HCl to make 500 mls.

Phosphates

(1) AMMONIUM MOLYBDATE SOLN.: dissolve 10 gms C.P. ammonium molybdate in 80 mls distilled HOH and make up to 100 ml volume with distilled HOH. This soln. may be kept in a dark bottle for several months with safety.

(2) STANNOUS CHLORIDE SOLN.: dissolve 2.15 gms good stannous chloride ($SnCl_2.2HOH$) in 20 mls of concentrated C.P. HCl. When the soln. is complete, add sufficient distilled HOH to make 100 mls and place a small piece of mossy metallic tin in the bottle. Under these conditions, the soln. will last for weeks.

Sulfides

(1) 10 gms sodium nitroferricyanide dissolved in 100 mls distilled HOH.

Chlorine

(1) POTASSIUM IODIDE SOLN.: 7.5% in 10% ethyl alcohol.
(2) STARCH SOLN.: same as for DO reagents.

Iron

(1) POTASSIUM PERMANGANATE SOLN.: dissolve 0.32 gms $KMnO_4$ in distilled HOH to make 100 mls. This gives a 0.1 N soln.

ORIENTATION TESTS FOR DISSOLVED OXYGEN

Tests for Nitrites, Ferric Iron and Other Substances Causing the DO Reading To Be Too High

(1) Obtain a sample from the HOH to be tested in a clean 250 ml beaker. No precautions against aeration are necessary, i.e., the sample may be dipped up directly in the beaker. If the HOH is clear and free from suspended matter, it can be used without further preparation. However, as even clear HOH often contains plankton, it is almost always desirable to filter it.

(2) Filter some of the sample through clean paper or absorbent cotton or glass wool using a glass funnel. It is necessary only to remove the larger particles so rapid filtration can be used.

(3) Into a small Ehrlenmeyer flask measure 10 mls of filtered sample and add
    a) 1 ml of 7% KI soln., then
    b) 1 ml of conc. HCl.

Rotate flask to insure mixing. If a yellow-to-brown color develops, substances which would cause the determination with Winkler reagents to be too high are present.

Tests for Sulfites, Ferrous Iron and Other Substances Causing the DO Reading To Be Too Low

(1) Measure 10 ml of filtered sample into a small flask and 10 ml distilled HOH into a second similar flask.

(2) To each flask add exactly 1 ml of dilute iodine soln.

(3) Titrate each flask with standard N/40 sodium thiosulphate soln., to the first starch endpoint. If the titration of the sample is less than that of the distilled HOH blank, reducing substances are present which would cause the DO determination to be too low.

EVALUATION OF DISSOLVED O_2 DETERMINATIONS

The aquatic biologist must evaluate DO determinations by both biological and chemical standards, i.e., both in terms of the requirements of living fishes and of the demands which various pollutants can make upon the available oxygen supply.

The lethal limit of DO for freshwater fishes is subject to some individual and species variation and to the modifying action of numerous physical and chemical factors. It is not surprising, therefore, that the voluminous literature on this subject contains many apparently conflicting statements due to differences in the conditions under which the fishes were exposed to low O_2, and to the methods by which the determinations were made. However, if the usual stream and lake complex of aquatic conditions is specified and the water is free from acutely dangerous pollution, certain limits of DO for freshwater fishes can be defined with considerable dependability.

From observations on fishes under field conditions made at several thousand stations in the rivers and streams of the U. S. over a period of 5 years, during the months of June to September, inclusive, Ellis (1937) states that in general *3 ppm of DO in water at 25° C. (77° F.) is the upper limit of DO at which asphyxia from low oxygen can be expected, and that to maintain in good condition a varied fauna of warm-water fishes in freshwater streams, 5 ppm are required.* These limits for river and stream conditions have been confirmed by Brinley (1942) and by the U.S.P.H.S. (1944). They found that fishes, excepting an occasional buffalo, carp or sunfish, were absent from heavily polluted waters carrying not more than 3 ppm of DO even during the daytime; that fishes, in polluted HOH having DO between 3 and 5 ppm, showed tendencies to sickness, deformity and parasitism; and that the fishes were varied, plentiful and healthy in the less polluted waters with DO not below 5 ppm.

Moore (1942) studying the O_2 requirements of freshwater fishes in Minnesota lakes, found that at temperatures between *15° C. (59° F.)* and *26° C. (78.5° F.)*, fishes of most species in those lakes died in 24 hours or less if the DO was less than 3.5 ppm, and that 5 ppm was not lethal to any fishes examined. Also, during the winter season that 3 ppm was not lethal but that fishes of most species died in 48 hours or less if the DO was reduced to 2 ppm.

Excepting waters carrying strong pollution, which in itself would kill or would act as an adjuvant to the detrimental action of low DO, it may be stated in general that DO at levels of 3 ppm or lower should be regarded as hazardous-to-lethal under average stream and lake conditions; and that 5 ppm or more should be present in waters, if conditions are to be favorable for freshwater fishes.

It is not proper, however, to accept waters of 5 ppm or more as suitable for fishes on that basis alone, for DO is only one of several vital requirements which must be maintained within limits if fishes are to thrive. Often the simultaneous alteration of 2 or more factors toward the unfavorable levels for those factors will produce conditions which, although not definitely unfavorable alone, will produce conditions of lethality. An increase in acidity and a synchronous reduction in DO is an example of such a synergy. (See pH discussion.)

The amount of DO which it is possible for any given water to carry varies directly with the amount of O_2 available and the pressure it is standing under, and inversely with the salinity and temperature of the water.

In most fieldwork, the amount of O_2 can be regarded as constant as pure air carries 20.96% O_2. The partial pressure of the O_2 in the air varies, however, with local barometric fluctuations and with the altitude of the stream or lake. As water supporting fish life will have in general a range between *0° C.* and *35° C.* (32° and 95° F.), temperature must also be considered. Unless the water is definitely brackish, the effect of a lesser salt content of fresh water on O_2-carrying power can be disregarded.

From the DO determinations, the expected saturations, the stream flow and the effluent volume available, computations can be made easily showing how serious the drain on the available DO of the water by the effluent is at the time or may become during other seasons and under different water conditions.

EVALUATION OF HYDROGEN-ION DETERMINATIONS (pH)

In water quality studies, it must be borne in mind that the hydrogen-ion concentration is the resultant of several, often many, factors, some synergistic, some antagonistic. It is not unusual, therefore, to find that two waters, although carrying different loads of dissolved materials, have identical pH concentrations. As CO_2 and the carbonates and sulfates of calcium and magnesium, all normal components of most stream and lake waters, collectively cause marked changes in the pH of the HOH in which they are dissolved, if the amounts of any or all of these compounds in solution vary even slightly, it is not surprising that unpolluted HOHs supporting good fish fauna may vary in pH concentration from *6.0* to *8.7*. Variations within this range or a slightly smaller one, pH *6.3–8.5* are readily tolerated by most freshwater fishes if these variations result from the relatively small amounts of CO_2 and carbonates and sulfates of calcium and magnesium that are usually found in fresh waters.

However, Ellis and Jones (1945) have found that the susceptibility of trout to changes in temperature and DO is markedly increased by shifting the pH from 8 to 6 with either HCl or H_2SO_4. The actual amounts of acid used were small so that this change in sensitivity to temperature and reduced DO is not due to the specific toxicity of the acid. Several factors are involved, but these findings were interpreted to point out that it is neither accurate to state that any given pH is or is not harmful to fish life, nor sound to dismiss the pH determination as of no consequence merely because it falls in the range of pH commonly accepted as harmless to fishes.

There are certain gross limits to the pH concentration beyond which rapidly lethal results are almost inevitable if fishes are exposed to such waters for even a short time; viz., *acidity in excess of pH 4.0 and alkalinity greater than pH 9.5*. Many acids and their salts, however, kill at pHs much less acidic than 4.0, and various alkalis and their salts are lethal at pHs less than 9.0 (Ellis, 1937).

In general, strong acids and alkalis kill by damaging the gills as well as by the systemic action of these compounds on the fish. Sulfuric, nitric and tannic acids react with the mucous, coating gill filaments so that O_2 absorption and branchial circulation are hampered and death may result before specific toxic systemic effects are observed (Ellis, 1937; Westfall, 1945). Acetic acid causes the filaments to swell which interferes with O_2 absorption and gill circulation, producing petechial, and later, general hemorrhages. Some acids as lactic, and alkalis as NH_4OH have pronounced systemic effects which are usually lethal before gross changes in the gills are noted.

Generally, pH determinations lying outside the ranges 6.0–8.7 indicate the usual balance of CO_2 and the carbonates-sulfates of Ca and Mg has been disturbed either by the presence of too large a quantity of 1 or more of these compounds or by the addition of other substances to the physico-chemical complex of the HOH. Generally such deviations point to conditions at least unfavorable to fish life. It must be remembered that many animals, including man, can exist under conditions which are definitely harmful for varying periods of time, so the presence of fishes in HOHs having pH concentrations beyond this rather narrow range of pH does not prove these waters are favorable to fish life.

The most conspicuous exceptions to HOHs having pHs outside these limits and supporting certain species of fishes will be found in HOHs of low conductivity from some mountain streams and various bogs and swamps, as well as certain alkaline lakes. Small amounts of CO_2 can alter the pH appreciably if the water is of low conductivity and poorly buffered. It is not surprising therefore that mountain streams carrying melted snow water are often slightly acidic and that bog and swamp waters which receive large amounts of CO_2 from disintegrating plant material frequently have pHs between 6.0 and 4.5. However, if the total acidity of these waters of low conductivity is determined, it will be found to be very low.

In reporting pH concentration, the method by which it was obtained and the temperature at which the determination was made should always be stated.

ALKALINITY AND ACIDITY

Natural waters exhibit wide variations in relative acidity and alkalinity, not only in actual pH values, but also in the total amount of dissolved material producing the acidity or alkalinity. The concentration of these compounds and the ratio of 1 to another determine the actual pH and the efficiency of buffering in a given water. As the lethal effects of most acids begin to appear near pH 4.5, and of most alkalis near pH 9.6, it is evident that buffering in the stream water capable of preventing certain sudden shifts to these extremes on the addition of strong acidic or alkaline effluents can be of paramount importance in the maintenance of life. Therefore, the determination of total acidity and alkalinity of the water is necessary among other things to establish the capacity of the stream to neutralize many types of wastes.

D.

DETERMINING THE AGE OF FISHES BY SCALE READINGS[1]

The determination of the age of fishes by means of a microscopic examination of their scales has been used very extensively. It depends upon the fact that the rate of growth of the fish varies materially at different times of the year. During the spring and summer, in general, growth is more rapid, and during the fall and winter it is slower. The scales of many fishes, including the salmon and trout, bear series of concentrically arranged ridges on their outer surfaces. On account of their concentric arrangement these are known as "RINGS" or "CIRCULI" (singular, "circulus"). The scales increase in size with the growth of the fish, and circuli are added at the margins. The scales are never normally shed, but increase in diameter by these accretions to their margins, and in thickness by additions to the inner surfaces. The markings formed by the circuli are therefore persistent throughout the life of the fish. The arrangement of these circuli is characteristically modified by variations in the rate-of-growth of the fish. During the more rapid growth of the spring and summer the rings are spaced relatively widely but during the period of slow or of perhaps no growth, which obtains during the winter and fall, the rings are crowded together closely and are frequently more-or-less broken and imperfect. The complete year's growth, therefore, consists in a *summer* band of relatively wide rings followed by a *winter* band of narrow rings. By counting the number of summer or winter bands, the age can readily be determined.

Weymouth (1923) has given a brief though comprehensive discussion of the various methods of determining age. In answering certain criticisms that have been made of the method of age determination of fishes by means of scales, he says:

"These objections are not valid, however, in the case of a number of species of fish, where the annual nature of the marks rests on no assumption of any kind, but of direct observation. In some a study of scales throughout the year has clearly shown that the ring is formed during the winter and only once each year. In others it has been shown that the number of rings agrees with the known age of fish kept in captivity or of marked fish recaptured after known periods. The soundness of these conclusions is not affected by the fact that in certain species the rings are less distinct and hence in these cases may be an unreliable guide to age, nor that incompetent or hasty workers may have reached incorrect conclusions in any species."

Although this method of determining the age of fishes appears to be simple and of easy application, in practice serious difficulties are frequently encountered. The relative approximation of the rings is merely a reflection of the rate of growth prevailing at the time the rings were

---

[1] From: Rich, 1925. Growth and degree of maturity of Chinook salmon in the ocean. Bull. U. S. Bur. Fish. 41: 15–90.

formed. Not infrequently factors cause variations in the rate of growth which are unassociated with the seasonal variations responsible for the formation of summer and winter bands. As a rule, the cause of these incidental variations is unknown, although Rich (1920) has suggested causes for some of the minor variations observed in the scale growth

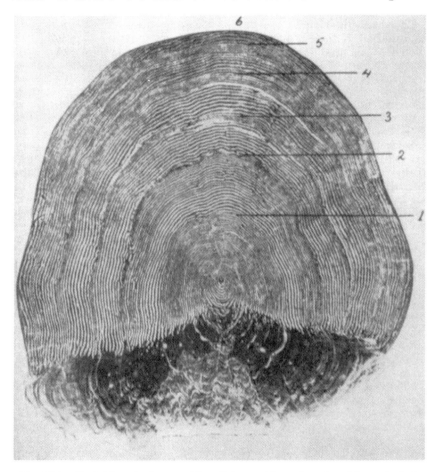

FIG. 262. Cycloid scale from a King Salmon (*Oncorhynchus tshawytscha*) in its sixth year (from Rich 1925, fig. 33). Numerals along side represent the end of each year of life. Note changing widths between individual lines or circuli to produce different appearing fields of circuli denoting the separate ages.

of Chinook Salmon previous to or during their seaward migration. Snyder (1922, 1923) has also discussed the causes responsible for unusual checks on the scales of Chinook Salmon in the Klamath River. In the case of the Chinook Salmon, the determination of age is further complicated by the fact that the young fishes, at least in the Columbia River, may migrate seaward at any time after they emerge from the gravel of the spawning beds up to an age of 18 months or more. There are, therefore, many difficult and puzzling characters in the scale growth which must be worked out before a completely satisfactory analysis can be made, but a full consideration of the problems connected with age determination lies beyond the scope of this discussion.

With the exception of age, the only other feature of the life history which has been used as a basis of segregation has been the general type of the early growth. Gilbert has shown (1913) that the chinooks of the Columbia River exhibit two distinct types of "NUCLEAR GROWTH" (the growth of the first year), forming what he terms the "OCEAN TYPE" and the "STREAM TYPE" of nuclei. The ocean type is large, with rings that are relatively widely separated, and indicates that the fish migrated to the ocean early in its first year. The stream type, on the other hand, is small, of closely crowded and more delicate rings, and characterizes the scales of those fishes that have spent the entire first year in fresh water.

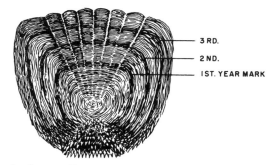

FIG. 263. Ctenoid scale from a Largemouth Blackbass (*Micropterus salmoides*), showing the "clumping" of circuli to form the year or annulus marks. Lines cutting across the circuli and ending on the upper, fluted edge are the radii. Courtesy California Department of Fish and Game.

An additional, more recent summary of value is the following from a U. S. Fish and Wildlife circular:

The foundation of the scale method rests upon the following fundamental facts:

(1) The number of scales on the body of a fish, after all the scales have been formed, remains constant throughout life;

(2) The same scales are retained throughout life, unless lost accidentally whereupon a new scale forms (the new or regenerated scales do not exhibit the record of growth up to that time, but instead have a mottled opaque area in the center—the "REGENERATION CENTER." Regenerated scales cannot be used, obviously, for growth studies.).

(3) The growth of the scale and the growth of the body of the fish are roughly proportional.

(4) The rate of growth of a fish, except in tropical and subtropical latitudes, varies throughout the year. Growth is most rapid during the warm summer months and practically ceases during the winter. This seasonal variation in rate of growth is recorded in the spacing of the

circuli on the scale, giving rise to the bands of growth ("ANNULI"), made up of groups of circuli, which represent each year of life.

On the basis of these fundamental postulates, rather precise methods of calculating the length of the fish at the end of each year of life have been developed. Details of the method are not, however, applicable to this brief resumé. Certain difficulties due to the presence of accessory checks, spawning marks, or false year marks, as well as occasional abnormalities in the configuration of the scale make precise age determinations difficult at times. An inexperienced person working with fish scales should have his results checked and his conclusions verified by a trained investigator.

Some fishes do not have scales or have epidermal (ganoid or placoid) scales that do not possess the structural features of dermal (cycloid or ctenoid) scales which are employed in age and growth determinations. For such fishes the age may be determined by examining the "OTOLITHS" or "ear stones" which correspond to a part of the inner ear of mammals. The otolith has ridges arranged in a somewhat concentric fashion which may be interpreted in the same manner as the circuli and annuli of fish scales. The bones of fishes also exhibit characteristics which may be used for age determinations. Cross-sections of vertebrae and bony finrays exhibit concentric markings which have been found to correspond to the annular markings of scales and otoliths. Sometimes the age as determined from the scales is verified by examining also the otoliths and vertebrae. However, because of the distinctness of the scale structure in comparison with those of the bony parts of the body, scales generally give the most satisfactory and reliable results if they can be employed.

An experimental method of checking the reliability of the foregoing methods of age determination is by tagging and releasing the fishes of known age after they have been measured. The length of the recaptured fish and the time that elapsed between tagging and recapture afford not only information on growth but on migration as well.

E.
TROUT HYBRIDIZATION

For a good many years, the fact that various trout would hybridize has been common knowledge among hatcherymen and experimenters. Long before any sound studies had been made of the extent to which such hybridization was factual—as well as of the predominant peculiarities of the hybrid offspring—trout literature became the repository for offhand statements about hybridization, which, as far as data were concerned, were based largely on rumor. Most authors passed on the information that certain species cross and produce sterile offspring without being aware that data existed (where it did) to prove or disprove such statements.

In the last few years, however, a considerable amount of work has been done with the subject of trout hybrids—initiated by the probability that some of these hybrids might produce better game fishes than either of their parents. This is a powerful stimulus, especially to fish and game groups, and discoveries have come rapidly.

Trout crosses appear in the literature well back into the last century, and had even attracted the attention of professionals by their possibilities. Scott (1956) has summarized the highlights of certain of these efforts. Bean (1887: 216) noted that the Pennsylvania Fish Commission was producing hybrids between Lake and Brook Trout and was pleased with the results. Just after the turn of the present century, Ottawa fishery officials experimented briefly with the same cross, but abandoned their work. Canadian fishery biologists and ichthyologists continued to pioneer in this field, with highly productive results in the 1950s.

Probably the best recent resumé of results of hybridizing several species of trout (except cutthroat) and salmon can be found in Buss and Wright (1956), where excellent illustrations (here reproduced) are available on various hybrids. The subject of trout hybridization is timely here since most of the species which have successfully produced hybrids of some promise have been planted in Nevada and, in a few cases, have provided the best fishing in certain areas (such as the Lake Trout in Lake Tahoe, Brook Trout over much of the State, etc.). From about 1905 to 1912, the Nevada State Fish Commission enthusiastically experimented with the hybrid of Lahontan Cutthroat and Rainbow Trout at the Verdi Hatchery, feeling, at the time, that they might have the answer to the problem of increasing productivity in two favored fishes to compensate for the even then deteriorating trout streams of the State.

When experimentation became intensive, it was apparent that contrary to the older idea that all hybrids were sterile, one in particular was found to be sufficiently fertile to show some promise of commercial production. This was the

SPLAKE or WENDIGO TROUT

Named from SPeckled and LAKE, the splake's name was suggested by Ontario fishery biologists. Wendigo is also a Canadian name. This is the hybrid between male Brook and female Lake Trout (the reciprocal cross is virtually unproductive) and has been developed to the point where it is beginning to be rather widely planted in both Canada and the United States. In addition to eastern provinces and states, the splake has been brought to California, but is as yet untried in Nevada. Scott's account (1956) is perhaps the best popular one.

FIG. 264. Splake Trout, hybrid between male Brook and female Lake Trout. Courtesy Pennsylvania Fish Commission and Keen Buss.

The splake appears to have suitable nesting and mating patterns to hold some hope of its partial maintenance by natural reproduction, but since the second and third generations produce throwbacks to the original parents, any planted stock would fairly quickly lose its distinctive "splakeness" by reversion to brook and lake patterns. Scott points out that four types would result—brook, lake, splake and a combination of these three.

Kmiotek and Oehmcke (1959) have reported on splake planted in Wisconsin from Michigan stock in 1957. Their habitat preferences are intermediate between those of their parents in that they will successfully live in upper regions of lakes shunned by the Lake Trout and, because of their rapid growth and aggressiveness, are popular with fishermen.

Among others, Fry and Gibson (1953) have investigated lethal temperatures in the splake, while Slastenenko (1954) has published on its relative growth. These are both Canadian studies.

Fig. 265. Tiger Trout, hybrid between male Brook and female Brown Trout. Courtesy Pennsylvania Fish Commission and Keen Buss.

Fig. 266. Brownbow Trout, hybrid between male Brown and female Rainbow Trout. Courtesy Pennsylvania Fish Commission and Keen Buss.

TIGER TROUT

Crosses between male brooks and female browns produce the most strikingly-marked of all the hybrid trout. Beyond determining that the reciprocal cross is unpromising, little work has been done with the Tiger Trout. Because of their outstanding appearance and relative ferocity, they could be a very popular game fish but their development has not reached the stage where it is practical to think of large-scale plants.

BROWNBOW TROUT

This is the name given the cross between male Brown and female Rainbow Trout. The result is a distinctive and heavily spotted fish and, like the Tiger Trout, little can yet be said about the possibilities of developing a fishable population from this type.

CUTBOW TROUT

This is probably as suitable a name as any for crosses between the closely-related Cutthroat and Rainbow Trout, although "Rainthroat" might be a bit more euphonious. The literature has many notations on cutthroat-rainbow hybridization, with particular reference to the fact that such mixing eliminates the native cutthroat from streams where they supposedly have an adaptive edge over the rainbow. This elimination is effected, apparently, because of sterility of their hybrids. With constant rainbow stocking, this would work cutthroat blood out of the stream.

It might be supposed that two such closely related strains of trout would produce comparatively fertile offspring and probably with close genetic culling this would be true. Little work has been done on this hybrid in the past and little interest seems to attach to it at present. Practically all knowledge of which I am aware on the sterility of Cutbow Trout stems from the field observations of fishery biologists where often the individual is reporting an impression and not statistics. And while the impression may be quite true in its trending, we still lack much concrete and objective data on the sterility problem. Needham and Gard (1959) have a recent discussion on cutthroat-rainbow hybridization (pp. 71–73). Unfortunately, as they noted, none of these experiments (many of which have been regularly done at hatcheries where no attempt was made to organize an experiment and obtain data) have been systematically prosecuted so as to derive sound genetic conclusions therefrom.

SAM BROWN TROUT

Crosses between Brown Trout and Atlantic Salmon have been successfully accomplished at the St. John, New Brunswick, hatchery, and it is claimed that such hybrids can still be found in New Brunswick waters (Wiggins 1950: 21–22).

F.

INVENTORY OF GAME FISHES BY SPECIFIC WATERS IN NEVADA AS DETERMINED BY RECENTLY COMPLETED SURVEYS OF THE NEVADA FISH AND GAME COMMISSION

(A) SALMON AND TROUT FAMILY
Family *SALMONIDAE*

(1) Brook Trout, *Salvelinus fontinalis*

CHURCHILL COUNTY
Edwards Creek, Edwards Valley, Desatoya Mountains.
Horse Creek, Dixie Valley, Clan Alpine Mountains.

DOUGLAS COUNTY
Cary Creek, West Fork of the Carson River, Carson Range.
Daggett Creek (Haines or Kingsbury Creek), West Fork of the Carson River, Carson Range.
Genoa Creek, Carson River, Carson Range.
Jacks Valley Creek, Carson River, Carson Range (recently eliminated by logging operations).
Job's Creek, West Fork of the Carson River, Carson Range.
Miller Creek (Scossa Creek), West Fork of the Carson River, Carson Range.
Mott Cañon Creek, West Fork of the Carson River, Carson Range.
Sheridan Creek, West Fork of the Carson River, Carson Range.
Sierra Creek, Carson River, Carson Range.

ELKO COUNTY
Currant Creek, Tea Creek, Snake Mountains.
Gilbert Creek, Corral Creek, Ruby Mountains.
North Fork of the Humboldt River, Humboldt River, Independence Mountains.
Lamoille Creek, South Fork of the Humboldt River, Ruby Mountains.
Loomis Creek, Thousand Springs Creek, Snake Mountains.
Jarbidge River, Bruneau River, Jarbidge Mountains.
Martin Creek (Penrod Creek), East Fork of the Owyhee River, Tennessee and Pine Mountains.
Merritt Creek, Bruneau River, Merritt Mountains.
East Fork of the Owyhee River, Owyhee River, Table Mountain.
Pearl Creek, Huntington Creek, Ruby Mountains.
Tabor Creek, Humboldt River, Snake Mountains.
Talbot Creek, Lamoille Creek, Ruby Mountains.
Taylor Cañon Creek, South Fork of the Owyhee River, Independence Mountains.
Tea Creek, Mary's River, Jarbidge Mountains.
Thorpe Creek, Lamoille Creek, Ruby Mountains.

Toyn Creek, Corral Creek, Ruby Mountains.
Trail Creek, East Fork of the Owyhee River, Northern Independence Mountains.
Waterpipe Cañon Creek, Taylor Cañon Creek, Independence Mountains.
Boulder Lake, Boulder Creek, East Humboldt Range, at 9,500 feet.
Cold Lakes, Cold Creek, Ruby Mountains, at 9,900 feet elevation.
Echo Lake, North Fork of the South Fork of the Humboldt River, Ruby Mountains, at 9,800 feet elevation.
Favre Lake, Kleckner Creek, Ruby Mountains.
Hidden Lake, Soldier Creek, Ruby Mountains, at 9,500 feet elevation.
Island Lake, Lamoille Creek, Ruby Mountains, at 9,800 feet elevation.
Lamoille Lake, Lamoille Creek, Ruby Mountains, at 9,700 feet elevation.
Liberty Lake, Kleckner Creek, Ruby Mountains, at 9,700 feet elevation.
Robinson Lake, Robinson Creek, Ruby Mountains, at 9,000 feet elevation.
Steel Lake, Steel Creek, East Humboldt Range, at 9,500 feet elevation.

ESMERALDA COUNTY

Chiatovich Creek, Fish Lake Valley, White Mountains.
Davis Creek, Chiatovich Creek, White Mountains.
Indian Creek (McNett Creek), Fish Lake Valley, White Mountains.
Leidy Creek (Cord Creek), Fish Lake Valley, White Mountains.
Middle Creek, Chiatovich Creek, White Mountains.
Perry Aiken Creek, Fish Lake Valley, White Mountains.
Trail Cañon Creek, Fish Lake Valley, White Mountains.

EUREKA COUNTY

Denay Creek, Denay Valley.
Roberts Creek, Diamond Valley, Roberts Mountains.
Trout Creek, Pine Creek, Piñon Range.

HUMBOLDT COUNTY

Alder Creek, Onion Valley, Pine Forest Range.
Alta Creek, Quinn River, Pine Forest Range.
Bartlett Creek, Black Rock Desert, Black Rock Range.
Buffalo Creek, Quinn River, Santa Rosa Mountains.
Cañon Creek, Quinn River, Santa Rosa Mountains.
Cottonwood Creek, Martin Creek, Santa Rosa Mountains.
Dutch John Creek, Martin Creek, Santa Rosa Mountains.
Flat Creek, Quinn River, Santa Rosa Mountains.
Granite Creek, Quinn River, Pine Forest Mountains.
Leonard Creek, Black Rock Desert, Pine Forest Mountains.
Little Cottonwood Creek, Cottonwood Creek, Santa Rosa Mountains.
Little Humboldt River (North Fork), Humboldt River, Santa Rosa Mountains.
Lye Creek, Dutch John Creek, Santa Rosa Mountains.
Martin Creek, Little Humboldt River, Santa Rosa Mountains.
McConnel Creek, Quinn River, Santa Rosa Mountains.
McDermitt Creek, Quinn River, Disaster Peak.

Pole Creek, Humboldt River, Sonoma Range.
Poplar Creek (Cabin Creek), Kings River, Trident Peak.
Quinn River, Black Rock Desert, Santa Rosa Mountains.
Raster Creek, Poplar Creek, Trident Peak.
Rebel Creek, Quinn River, Santa Rosa Mountains.
Siard Creek, Cottonwood Creek, Santa Rosa Mountains.
Singas Creek, Cottonwood Creek, Santa Rosa Mountains.
Stone House Creek, Martin Creek, Santa Rosa Mountains.
Wash O'Neil Creek (Riley Creek), Martin Creek, Santa Rosa Mountains.
Wilder Creek, Trout Creek, Trident Peak.
Willow Creek, Quinn River, Santa Rosa Mountains.

LANDER COUNTY

Bade Creek, Smoky Valley, Toiyabe Range.
Big Creek, Reese River, Toiyabe Range.
Birch Creek, Smoky Valley, Toiyabe Range.
Boone Creek, Reese River, Toiyabe Range.
Bowman Creek (Clear Creek), Smoky Valley, Toiyabe Range.
Callahan Creek, Grass Valley, Toiyabe Range.
Carseley Creek, Bowman Creek, Toiyabe Range.
Cowboy Rest Creek, Grass Valley, Toiyabe Range.
Indian Creek, Crescent Valley, Shoshone Range.
Kingston Creek, Smoky Valley, Toiyabe Range.
Lewis Creek, Reese River, Shoshone Range.
Mill Creek, Reese River, Shoshone Range.
Peterson Creek, Smith Creek Valley, Shoshone Range.
Silver Creek, Reese River, Toiyabe Range.
Skull Creek, Grass Valley, Toiyabe Range.
Smith Creek, Smith Creek Valley, Desatoya Range.
Steiner Creek, Grass Valley, Simpson Park Range.
Trout Creek, Reese River, Shoshone Range.

LINCOLN COUNTY

Eagle Valley Creek, Meadow Valley Wash.

NYE COUNTY

Barker Creek (Shipley Creek), Smoky Valley, Toquima Range.
Barley Creek, Monitor Valley, Monitor Range.
Broad Creek, Smoky Valley, Toiyabe Range.
Clear Creek, Little Fish Lake Valley, Monitor Range.
Clear Creek, Reese River Valley, Toiyabe Range.
Cottonwood Creek, Barley Creek, Monitor Range.
Cottonwood Creek, Reese River, Toiyabe Range.
Danville Creek, Little Fish Lake Valley, Monitor Range.
George's Creek, Stone Cabin Valley, Monitor Range.
Hooper Creek, Railroad Valley, Quinn Cañon Range.
Hunts Creek, Ralston Valley, Monitor Range.
Illinois Creek, Reese River Valley, Toiyabe Range.
Jefferson Creek, Smoky Valley, Toquima Range.

Jett Creek, Smoky Valley, Toiyabe Range.
Marysville Creek, Reese River Valley, Toiyabe Range.
Meadow Creek, Monitor Valley, Toquima Range.
Moore's Creek, Smoky Valley, Toquima Range.
Mosquito Creek, Monitor Valley, Monitor Range.
Ophir Creek, Smoky Valley, Toiyabe Range.
Peavine Creek, Smoky Valley, Toiyabe Range.
Pine Creek, Monitor Valley, Toquima Range.
Upper Reese River, Reese River, Toiyabe Range.
San Juan Creek, Reese River, Toiyabe Range.
Sawmill Creek, Little Fish Lake Valley, Monitor Range.
Six Mile Creek, Hot Creek Valley, Hot Creek Range.
Stewart Creek, Reese River, Toiyabe Range.
Stoneberger Creek, Monitor Valley, Toquima Range.
Summit Creek, Smoky Valley, Toiyabe Range.
Tierney Creek (Gross Creek), Reese River, Toiyabe Range.
Troy Creek, Railroad Valley, Grant Range.
Twin Rivers, Smoky Valley, Toiyabe Range.
Wisconsin Creek, Smoky Valley, Toiyabe Range.

ORMSBY COUNTY

Ash Creek (Gregory Creek), Carson River, Carson Range.
Clear Creek, Carson River, Carson Range.
Kings Cañon Creek, Carson River, Carson Range.

PERSHING COUNTY

Clear Creek, Grass Valley, Sonoma Range.
Cottonwood Creek, Buena Vista Valley, Humboldt Range.
Coyote Creek, Buena Vista Valley, Humboldt Range.
Hoffman Creek, Buffalo Valley, Tobin Range.
Indian Creek, Buena Vista Valley, Humboldt Range.
Rocky Creek, Humboldt River, Humboldt Range.
Sonoma Creek, Grass Valley, Sonoma Range.
Star Creek, Buena Vista Valley, Humboldt Range.
Unionville Creek, Buena Vista Valley, Humboldt Range.

WASHOE COUNTY

Bronco Creek, Truckee River, Carson Range.
Browns Creek, Steamboat Creek, Carson Range.
Evans Creek, Steamboat Creek, Carson Range.
Franktown Creek, Washoe Valley, Carson Range.
Galena Creek, Steamboat Creek, Carson Range.
Gray Creek, Truckee River, Carson Range.
Hobart Creek, Franktown Creek, Carson Range.
Hunter Creek, Truckee River, Carson Range.
Ophir Creek, Washoe Valley, Carson Range.
Thomas Creek, Steamboat Creek, Carson Range.
White Creek, Steamboat Creek, Carson Range.
Hobart Reservoir, Hobart Creek, Carson Range, at 7,650 feet elevation.
Hunter Lake, Hunter Creek, Carson Range, at 8,240 feet elevation.

Joy Lake, Browns Creek, Carson Range, at 5,850 feet elevation.
Marlette Lake (partly in Ormsby County), Lake Tahoe, Carson Range, at about 7,900 feet elevation.
Price Lake, Ophir Creek, Carson Range, at 7,240 feet elevation.

WHITE PINE COUNTY

Big Indian Creek, Steptoe Valley, Schell Creek Range.
Big Wash Creek, Pruess Reservoir, Snake Range.
Bird Creek, Duck Creek, Schell Creek Range.
Cave Creek, Cave Lake, Schell Creek Range.
Cleve Creek, Spring Valley, Schell Creek Range.
Currant Creek, Railroad Valley, White Pine Range.
Duck Creek, Steptoe Valley, Schell Creek Range.
Huntington Creek, South Fork of the Humboldt River, Huntington Valley.
Illipah Creek, Jakes Valley, White Pine Range.
Kalamazoo Creek, Spring Valley, Schell Creek Range.
Lehman Creek, Snake Valley, Snake Range.
Mattier Creek, Steptoe Valley, Schell Creek Range.
Meadow Creek, Spring Valley, Schell Creek Range.
Muncy Creek, Spring Valley, Schell Creek Range.
North Creek, Lake Valley, Schell Creek Range.
Paris Creek, Butte Valley, Cherry Creek Mountains.
Piermont Creek, Spring Valley, Schell Creek Range.
Snake Creek, Snake Valley, Snake Range.
Steptoe Creek, Steptoe Valley, Schell Creek Range.
Strawberry Creek, Weaver Creek, Snake Range.
Sunkist Creek, Spring Valley, Schell Creek Range.
Taft Creek, Spring Valley, Schell Creek Range.
Tailings Creek, Bassett Lake, Steptoe Valley.
Timber Creek, Duck Creek, Schell Creek Range.
Vipont Creek (Stephens Creek), Spring Valley, Schell Creek Range.
White River, White River Valley, White Pine Range.
Baker Lake, Baker Creek, Snake Range, 10,600 feet elevation.
Johnson Lake (Emerald Lake, Snake Lake, Treasure Lake), Snake Creek, Snake Range, at 10,800 feet elevation.

(2) Brown Trout, *Salmo trutta*

CHURCHILL COUNTY

Horse Creek, Dixie Valley, Clan Alpine Mountains.
Lahontan Reservoir, Carson and Truckee Rivers, Sierra Nevada Mountains.

CLARK COUNTY

Lakes Mead and Mohave, Colorado River.

DOUGLAS COUNTY

East Fork of the Carson River, Sierra Nevada Mountains.
West Fork of the Carson River, Sierra Nevada Mountains.
Topaz Lake, West Walker River, Sierra Nevada Mountains.

ELKO COUNTY

North Fork of the Humboldt River, Humboldt River, Independence Mountains.

HUMBOLDT COUNTY

Little Humboldt River, Humboldt River, Santa Rosa Mountains.
Martin Creek, Little Humboldt River, Santa Rosa Mountains.
Quinn River, Black Rock Desert, Santa Rosa Mountains.

LANDER COUNTY

Big Creek, Reese River, Toiyabe Range.
Birch Creek, Smoky Valley, Toiyabe Range.
Campbell Creek, Smith Creek Valley, Desatoya Range.
Kingston Creek, Smoky Valley, Toiyabe Range.
Silver Creek, Reese River, Toiyabe Range.
Skull Creek, Grass Valley, Toiyabe Range.
Smith Creek, Smith Creek Valley, Desatoya Range.

LYON COUNTY

East Drain, Walker River, Mason Valley.
East Walker River, Bridgeport Reservoir, Sierra Nevada Mountains.
Lahontan Reservoir, Carson and Truckee Rivers, Sierra Nevada Mountains.
Wabuska Drain, Walker River, Mason Valley.

NYE COUNTY

Barley Creek, Monitor Valley, Monitor Range.
Corcoran Creek, Monitor Valley, Toquima Range.
Hunts Creek, Ralston Valley, Monitor Range.
Jefferson Creek, Smoky Valley, Toquima Range.
Jett Creek, Smoky Valley, Toiyabe Range.
Moore's Creek, Smoky Valley, Toquima Range.
Ophir Creek, Smoky Valley, Toiyabe Range.
Peavine Creek, Smoky Valley, Toiyabe Range.
Pine Creek, Monitor Valley, Toquima Range.
Lower Reese River, Humboldt River.
Stewart Creek, Reese River, Toiyabe Range.

PERSHING COUNTY

Coyote Creek, Buena Vista Valley, Humboldt Range.
Ryepatch Reservoir, Humboldt River.
Star Creek, Buena Vista Valley, Humboldt Range.

WASHOE COUNTY

Steamboat Creek, Truckee River, Carson Range.
Truckee River, Lake Tahoe-Pyramid Lake, Sierra Nevada Mountains.
Washoe Lake, Washoe Valley, Carson Range.

WHITE PINE COUNTY

Baker Creek, Snake Valley, Snake Range.
Cave Creek, Cave Lake, Schell Creek Range.

Cleve Creek, Spring Valley, Schell Creek Range.
Duck Creek, Steptoe Valley, Schell Creek Range.
Huntington Creek, South Fork of the Humboldt River, Diamond Mountains.
Illipah Creek, Jakes Valley, White Pine Range.
Kalamazoo Creek, Spring Valley, Schell Creek Range.
Lake Creek (Big Spring Creek), Pruess Reservoir, Snake Valley.
Lehman Creek, Snake Valley, Snake Range.
Big Nigger Creek, Spring Valley, Snake Range.
Piermont Creek, Spring Valley, Schell Creek Range.
Shingle Creek, Spring Valley, Snake Range.
Silver Creek, Snake Valley, Snake Range.
Snake Creek, Snake Valley, Snake Range.
Tailings Creek, Bassett Lake, Steptoe Valley.
White River, White River Valley, White Pine Range.
Willow Creek, Steptoe Valley, Egan Range.

(3) Cutthroat Trout, *Salmo clarki*

CHURCHILL COUNTY
Edwards Creek, Edwards Creek Valley, Desatoya Range.

ELKO COUNTY
Coon Creek, Big Goose Creek, Goose Creek Mountains.
Currant Creek, Tea Creek, Snake Mountains.
Foreman Creek, North Fork of the Humboldt River, Independence Mountains.
Gance Creek, North Fork of the Humboldt River, Independence Mountains.
Goose Creek, Snake River, Goose Creek Mountains.
Jarbidge River, Bruneau River, Jarbidge Mountains.
Lewis Creek, Willow Creek, Tuscarora Mountains.
Nelson Creek, Willow Creek, Tuscarora Mountains.
North Fork of the Humboldt River, Humboldt River, Independence Mountains.
Pearl Creek, Huntington Creek, Ruby Mountains.
Pine Creek, Goose Creek.
Rock Creek, Humboldt River, Tuscarora Mountains.
Talbot Creek, Lamoille Creek, Ruby Mountains.
Tea Creek, Mary's River, Jarbidge Mountains.
Thorpe Creek, Lamoille Creek, Ruby Mountains.
Toe Jam Creek, Rock Creek, Tuscarora Mountains.
Toyn Creek, Corral Creek, Ruby Mountains.

ESMERALDA COUNTY
Leidy Creek (Cord Creek), Fish Lake Valley, White Mountains.
Middle Creek, Chiatovich Creek, White Mountains.

EUREKA COUNTY
Birch Creek, Denay Valley, Roberts Mountains.

HUMBOLDT COUNTY

Andorna Creek, Quinn River, Santa Rosa Mountains.
Bilk Creek, Quinn River, Trident Peak.
Dutch John Creek, Martin Creek, Santa Rosa Mountains.
Little Humboldt River, Humboldt River, Santa Rosa Mountains.
Lye Creek, Dutch John Creek, Santa Rosa Mountains.
Martin Creek, Little Humboldt River, Santa Rosa Mountains.
Mullinix Creek, Cottonwood Creek, Santa Rosa Mountains.
Quinn River, Black Rock Desert, Santa Rosa Mountains.
Solid Silver Creek, Martin Creek, Santa Rosa Mountains.
Trout Creek, Desert Valley, Jackson Mountains.

LANDER COUNTY

Italian Creek, Reese River, Toiyabe Range.
Santa Fé Creek, Smoky Valley, Toiyabe Range.
Shoshone Creek, Smoky Valley, Toiyabe Range.

NYE COUNTY

Crane Creek, Reese River, Toiyabe Range.
Moore's Creek, Smoky Valley, Toquima Range.
Ophir Creek, Smoky Valley, Toiyabe Range.
Peavine Creek, Smoky Valley, Toiyabe Range.
Pine Creek, Monitor Valley, Toquima Range.
Upper Reese River, Reese River, Toiyabe Range.
San Juan Creek, Reese River, Toiyabe Range.
Tierney Creek (Gross Creek), Reese River, Toiyabe Range.

PERSHING COUNTY

Hoffman Creek, Buffalo Valley, Tobin Range.

WHITE PINE COUNTY

Baker Creek, Snake Valley, Snake Range.
Bassett Creek, Spring Valley, Schell Creek Range.
Deep Cañon Creek, Deadman Creek, Snake Range.
Garden Creek, Spring Valley, Schell Creek Range.
Hendrys Creek, Snake Valley, Snake Range.
Illipah Creek, Jakes Valley, White Pine Range.
Lehman Creek, Snake Valley, Snake Range.
McCoy Creek, Spring Valley, Schell Creek Range.
Meadow Creek, Spring Valley, Schell Creek Range.
Muncy Creek, Spring Valley, Schell Creek Range.
Big Nigger Creek, Spring Valley, Snake Range.
Odgers Creek, Spring Valley, Schell Creek Range.
Piermont Creek, Spring Valley, Schell Creek Range.
Pine Creek, Spring Valley, Snake Range.
Ridge Creek, Pine Creek, Snake Range.
Silver Creek, Snake Valley, Snake Range.
Steptoe Creek, Steptoe Valley, Schell Creek Range.
Strawberry Creek, Weaver Creek, Snake Range.
Taft Creek, Spring Valley, Schell Creek Range.
Willard Creek, Spring Valley, Snake Range.

(4) Dolly Varden Trout, *Salvelinus malma*
ELKO COUNTY
Jarbidge River, Bruneau River, Jarbidge Mountains.

(5) Rainbow Trout, *Salmo gairdneri*

Just about every fishable stream in the State has, or has had, Rainbow Trout in it at one time. For these, see the listings of streams and lakes. At the present time, of all the major waters, Rainbow Trout are conspicuously absent only from Walker Lake in contrast to their success in Pyramid Lake.

(B) WHITEFISH FAMILY
Family *COREGONIDAE*

Mountain Whitefish, *Prosopium williamsoni*
ELKO COUNTY
Bruneau River, Snake River, Stag Mountain.
Jarbidge River, Bruneau River, Jarbidge Mountains.
Meadow Creek, Bruneau River, Tennessee Mountain.
Owyhee River, Snake River.
Salmon Falls Creek, Snake River, O'Neil Basin.
Shoshone Creek, Salmon Falls Creek.

The Mountain Whitefish is also common to the Lahontan system rivers—Carson, Humboldt, Truckee and Walker, all of which cut across, or border upon, more than one county.

(C) SUCKER FAMILY
Family *CATOSTOMIDAE*

Cui-ui Lakesucker, *Chasmistes cujus*

Confined to Pyramid Lake, Washoe County, and originally getting a short distance up the Truckee River during its spawning runs. It is now restricted to the lake because of insufficient river inflow.

(D) CATFISH FAMILY
Family *ICTALURIDAE*

(1) Channel Catfish, *Ictalurus punctatus*
CHURCHILL COUNTY
Lahontan Reservoir, Carson and Truckee Rivers, Sierra Nevada Mountains.
CLARK COUNTY
Lakes Mead and Mohave, Colorado River.
LYON COUNTY
Lahontan Reservoir, Carson and Truckee Rivers, Sierra Nevada Mountains.

PERSHING COUNTY
Ryepatch Reservoir, Humboldt River.

WHITE PINE COUNTY
Lake Creek (Big Springs Creek), Pruess Reservoir, Snake Valley. Ted Frantz writes that 152 Channel Catfish (6" to 16" long) were brought in from Lake Mead in 1954, but nothing is currently known of their success or failure here.

(2) White Catfish, *Ictalurus catus*

CHURCHILL COUNTY
Indian Lakes, Carson and Truckee River drains, Lahontan Reservoir.
Lahontan Reservoir, Carson and Truckee Rivers, Sierra Nevada Mountains.
Stillwater Marsh, Carson and Truckee River drains, Lahontan Reservoir.

LYON COUNTY
Lahontan Reservoir, Carson and Truckee Rivers, Sierra Nevada Mountains.

PERSHING COUNTY
Ryepatch Reservoir, Humboldt River.

(3) Black Bullhead, *Ictalurus melas*

CHURCHILL COUNTY
Indian Lakes, Carson and Truckee River drains, Lahontan Reservoir.
Stillwater Marsh, Carson and Truckee River drains, Lahontan Reservoir.

CLARK COUNTY
Lakes Mead and Mohave, Colorado River.

WASHOE COUNTY
Various ponds in the Truckee Meadows (Reno), part of the irrigation system draining into the Truckee River.
Washoe Lake, Washoe Valley, Carson Range.

(4) Brown Bullhead, *Ictalurus nebulosus*

CHURCHILL COUNTY
Lahontan Reservoir, Carson and Truckee Rivers, Sierra Nevada Mountains.
Stillwater Marsh, Carson and Truckee River drains, Lahontan Reservoir.

EUREKA COUNTY
Flynn Pond, Diamond Valley.

LYON COUNTY
Lahontan Reservoir, Carson and Truckee Rivers, Sierra Nevada Mountains.

PERSHING COUNTY
Ryepatch Reservoir, Humboldt River.

WASHOE COUNTY
Washoe Lake, Washoe Valley, Carson Range.

WHITE PINE COUNTY
Bassett Lake, Steptoe Valley.
Warm Springs Pond, Newark Valley.

(E) PERCH FAMILY
Family *PERCIDAE*

Yellow Perch, *Perca flavescens*

CHURCHILL COUNTY
Indian Lakes, Carson and Truckee River drains, Lahontan Reservoir.
Lahontan Reservoir, Carson and Truckee Rivers, Sierra Nevada Mountains.
Stillwater Marsh, Carson and Truckee River drains, Lahontan Reservoir.

PERSHING COUNTY
Ryepatch Reservoir, Humboldt River.

WASHOE COUNTY
Various ponds in the Truckee Meadows (Reno), part of the irrigation system draining into the Truckee River.
Washoe Lake, Washoe Valley, Carson Range.

WHITE PINE COUNTY
Bassett Lake, Steptoe Valley.

(F) SUNFISH FAMILY
Family *CENTRARCHIDAE*

(1) Smallmouth Blackbass, *Micropterus dolomieui*

LANDER, EUREKA AND ELKO COUNTIES
In September of 1956, the State Commission planted Smallmouth Blackbass in this stretch of the Humboldt River between Battle Mountain and Elko. The success of this venture is still unknown, and it is certain that all other plants have long since failed, such as those in the last part of the Nineteenth Century in Washoe Lake and vicinity.

(2) Largemouth Blackbass, *Micropterus salmoides*

CHURCHILL COUNTY
Indian Lakes, Carson and Truckee River drains, Lahontan Reservoir.
Lahontan Reservoir, Carson and Truckee Rivers, Sierra Nevada Mountains.
Stillwater Marsh, Carson and Truckee River drains, Lahontan Reservoir.

CLARK COUNTY
Lakes Mead and Mohave, Colorado River.

ESMERALDA COUNTY
Fish Lake, Fish Lake Valley, White Mountains.

EUREKA COUNTY
Fish Creek Springs, Little Smoky Valley. A few blackbass from Manzanita Pond on the University of Nevada campus in Reno were planted here by Ted Frantz in September of 1952, but the success of this is unknown.

LINCOLN COUNTY
Geyser Spring, Cove Valley.

LYON COUNTY
Lahontan Reservoir, Carson and Truckee Rivers, Sierra Nevada Mountains.

MINERAL COUNTY
Weber Reservoir, Walker River, Sierra Nevada Mountains.

NYE COUNTY
Adams-McGill Reservoir (Sunnyside Reservoir), White River, White Pine Range. Ted Frantz doubts that they are still present here.
Locke's Pond No. 2, Railroad Valley.

PERSHING COUNTY
Ryepatch Reservoir, Humboldt River.

WASHOE COUNTY
Various ponds in the Truckee Meadows (Reno), part of the irrigation system draining into the Truckee River.
Wadsworth Slough (between Wadsworth and Nixon), Truckee River, Sierra Nevada Mountains.
Washoe Lake, Washoe Valley, Carson Range.

WHITE PINE COUNTY
Bassett Lake, Steptoe Valley.

(3) Bluegill Sunfish, *Lepomis macrochirus*

CHURCHILL COUNTY
Indian Lakes, Carson and Truckee River drains, Lahontan Reservoir.
Stillwater Marsh, Carson and Truckee River drains, Lahontan Reservoir.

CLARK COUNTY
Lakes Mead and Mohave, Colorado River.

ESMERALDA COUNTY
Silver Peak Pond, Clayton Valley, Silver Peak Range.

NYE COUNTY

Adams-McGill Reservoir (Sunnyside Reservoir), White River, White Pine Mountains.
Railroad Valley Ponds.

PERSHING COUNTY

Ryepatch Reservoir, Humboldt River.

WASHOE COUNTY

Various ponds in the Truckee Meadows (Reno), part of the irrigation system draining into the Truckee River.

WHITE PINE COUNTY

Bassett Lake, Steptoe Valley.

(4) Sacramento Perch, *Archoplites interruptus*

CHURCHILL COUNTY

Indian Lakes, Carson and Truckee River drains, Lahontan Reservoir.
Lahontan Reservoir, Carson and Truckee Rivers, Sierra Nevada Mountains.
Stillwater Marsh, Carson and Truckee River drains, Lahontan Reservoir.

LYON COUNTY

Lahontan Reservoir, Carson and Truckee Rivers, Sierra Nevada Mountains.

MINERAL COUNTY

Walker Lake, Walker River, Sierra Nevada Mountains.

PERSHING COUNTY

Ryepatch Reservoir, Humboldt River.

WASHOE COUNTY

Pyramid Lake, Truckee River, Lake Tahoe.
Various ponds in the Truckee Meadows (Reno), part of the irrigation system draining into the Truckee River.
Washoe Lake, Washoe Valley, Carson Range.

WHITE PINE COUNTY

Bassett Lake, Steptoe Valley.
Little Meadow Lake, Spring Valley. According to Ted Frantz, Sacramento Perch were present at one time in this Sacramento Ranch locale. In 1934, a dry year, many were transplanted to Pruess Reservoir across the line into Utah's Millard County, where they are still common.

(5) White Crappie, *Pomoxis annularis*

CHURCHILL COUNTY

Indian Lakes, Carson and Truckee River drains, Lahontan Reservoir.
Lahontan Reservoir, Carson and Truckee Rivers, Sierra Nevada Mountains.

Stillwater Marsh, Carson and Truckee River drains, Lahontan Reservoir.

LYON COUNTY
Lahontan Reservoir, Carson and Truckee Rivers, Sierra Nevada Mountains.

PERSHING COUNTY
Ryepatch Reservoir, Humboldt River.

(6) Black Crappie, *Pomoxis nigromaculatus*

CHURCHILL COUNTY
Lahontan Reservoir, Carson and Truckee Rivers, Sierra Nevada Mountains.

CLARK COUNTY
Lakes Mead and Mohave, Colorado River.

EUREKA COUNTY
Fish Creek Springs, Little Smoky Valley.

LYON COUNTY
Lahontan Reservoir, Carson and Truckee Rivers, Sierra Nevada Mountains.

G.
LIST OF STREAM AND LAKE REPORTS ISSUED BY THE NEVADA FISH AND GAME COMMISSION

INDEX
By Counties

STREAM AND LAKE SURVEY REPORTS
State of Nevada

CHURCHILL COUNTY

Bench Creek, Clan Alpine Mountains, Dixie Valley drainage.
Cherry Creek, Clan Alpine Mountains, Edwards Creek Valley drainage.
Edwards Creek, Desatoya Mountains, Edwards Creek Valley drainage.
Horse Creek, Clan Alpine Mountains, Dixie Valley drainage.

DOUGLAS COUNTY

Bryant Creek (Barney Riley Creek), Sierra Nevada Mountains, East Fork Carson River drainage.
Carson River, Carson Valley to Lahontan Reservoir.
Carson River, East Fork, Sierra Nevada Mountains, Carson River drainage.
Carson River, West Fork, Sierra Nevada Mountains, Carson River drainage.
Cary Creek, Carson Range, West Fork Carson River drainage.
Daggett Creek (Haines Creek, Kingsbury Creek), Carson Range, West Fork, Carson River drainage.
Genoa Creek, Carson Range, Carson River drainage.
Jacks Valley Creek, Carson Range, Carson River drainage.
James Creek, Carson Range, Carson River drainage.
Job's Creek, Carson Range, West Fork Carson River drainage.
Miller Creek (Scossa Creek), Carson Range, West Fork Carson River drainage.
Mott Cañon Creek, Carson Range, West Fork Carson River drainage.
Sheridan Creek, Carson Range, West Fork Carson River drainage.
Sierra Creek, Carson Range, Carson River drainage.

ELKO COUNTY

Ackler Creek, East Humboldt Range, Boulder (Starr) Creek drainage.
Allegheny Creek, Merritt Mountain, East Fork Owyhee River drainage.
Angel Creek, South Fork, East Humboldt Range, Clover Valley drainage.
Badger Creek, Northern Independence Mountains, East Fork Owyhee River drainage.
Beaver Creek, Northern Independence Mountains, East Fork Owyhee River drainage.
Beaver Creek, Tuscarora Mountains, Maggie Creek drainage.

Boulder Creek (Starr Creek), East Humboldt Range, Humboldt River drainage.
Boyd Creek, Independence Mountains, Jack Creek drainage.
Brown Creek, Ruby Mountains, Huntington Creek drainage.
Bruneau River, Stag Mountain, Snake River drainage.
Bull Camp Creek, Snake Mountains, Dry Creek drainage.
Bull Run Creek (Blue Jacket Creek), Bull Run Mountains, South Fork Owyhee River drainage.
Bull Run Creek, East Fork (Columbia Creek), Bull Run Mountains, South Fork Owyhee River drainage.
Bull Run Creek, South Fork (Cap Winn Creek), Northern Independence Mountains, South Fork Owyhee River drainage.
Burns Creek, Independence Mountains, Jerritt Creek drainage.
California Creek, Merritt Mountains, East Fork Owyhee River drainage.
Camp Creek, O'Neil Basin, Salmon Falls River drainage.
Cañon Creek, O'Neil Basin, Salmon Falls River drainage.
Carville Creek, Ruby Mountains, Corral Creek drainage.
Chicken Creek, Independence Mountains, Jack Creek drainage.
Cold Creek, Ruby Mountains, Lamoille Creek drainage.
Coon Creek, Big Goose Creek drainage.
Copper Creek, Jarbidge Mountains, Bruneau River drainage.
Corral Creek, Ruby Mountains, Smith Creek drainage.
Cottonwood Creek, O'Neil Basin, Salmon Falls River drainage.
Cottonwood Creek, Ruby Mountains, Smith Creek drainage.
Coyote Creek, Tuscarora Mountains, Maggie Creek drainage.
Currant Creek, Snake Mountains, Tea Creek drainage.
Dave's Creek, Jarbidge Mountains, East Fork Jarbidge River drainage.
Deep Creek, Independence Mountains, South Fork Owyhee River drainage.
Deering Creek, East Humboldt Range, Boulder (Starr) Creek drainage.
Dixie Creek, Piñon Range, South Fork Humboldt drainage.
Dry Creek, Snake Mountains, Salmon Falls River drainage.
Evans (Spring) Creek, Midas Area, Humboldt River drainage.
Foreman Creek, Independence Mountains, North Fork Humboldt River drainage.
Furlong Creek, North, Ruby Mountains, Long Cañon Creek drainage.
Gance Creek, Independence Mountains, North Fork Humboldt River drainage.
Gennette Creek, Ruby Mountains, Smith Creek drainage.
Gilbert Creek, Ruby Mountains, Corral Creek drainage.
Goose Creek, Big, Snake River drainage.
Goose Creek, Little, Golliher Mountain, Big Goose Creek drainage.
Green Mountain Creek, Ruby Mountains, Toyn Creek drainage.
Herder Creek, East Humboldt Range, Boulder (Starr) Creek drainage.
Horse Creek, East Humboldt Range, Pole Creek drainage.
Humboldt River
Humboldt River, North Fork, Independence Mountains, Humboldt River drainage.
Humboldt River, South Fork, Ruby Mountains, Humboldt River drainage.

Humboldt River, North Fork of South Fork, Ruby Mountains, South Fork Humboldt River drainage.
Humboldt River, South Fork of South Fork, Ruby Mountains, South Fork Humboldt River drainage.
Jack Creek, Independence Mountains, South Fork Owyhee River drainage.
Jack Creek, Little, Tuscarora Mountains, Maggie Creek drainage.
Jackstone Creek, Elko Mountains, Humboldt River drainage.
Jakes Creek, Snake Mountains, Salmon Falls River drainage.
Jarbidge River, West Fork, Jarbidge Mountains, Bruneau River drainage.
Jerritt Creek, Independence Mountains, South Fork Owyhee River drainage.
Kelly Creek, Snowstorm Mountain, Humboldt River drainage.
Kleckner Creek, Ruby Mountains, South Fork Humboldt River drainage.
Lamoille Creek, Ruby Mountains, South Fork Humboldt River drainage.
Leach Creek, East Humboldt Range, Clover Valley drainage.
Lewis Creek, Tuscarora Mountains, Willow Creek drainage.
Lime Creek, O'Neil Basin, Wilson Creek drainage.
Long Cañon Creek, Ruby Mountains, South Fork Humboldt River drainage.
McCann Creek, Tuscarora Mountains, South Fork Owyhee River drainage.
McCutcheon Creek, Ruby Mountains, Smith Creek drainage.
McDonald Creek, Merritt Mountain, Bruneau River drainage.
Maggie Creek, Independence-Tuscarora Mountains, Humboldt River drainage.
Mahala Creek, Independence Mountains, North Fork Humboldt River drainage.
Marsh Creek, Independence Mountains, Jack Creek drainage.
Martin Creek (Penrod Creek, Big Bend Creek, Gold Creek, Hay Meadow Creek, Mill Creek, Sweet Creek, Thompson Creek, combined), Tennessee and Pine Mountain, East Fork Owyhee River drainage.
Mary's River, Jarbidge Mountains, Humboldt River drainage.
Meadow Creek (Telephone Creek, Tennessee Creek, combined), Tennessee Mountain, Bruneau River drainage.
Merritt Creek, Sheep Creek (combined), Merritt Mountains, Bruneau River drainage.
Nelson Creek, Tuscarora Mountains, Willow Creek drainage.
Owyhee River, East Fork, Table Mountain, Owyhee River drainage.
Owyhee River, South Fork, Tuscarora-Independence Mountains, Owyhee River drainage.
Pearl Creek, Ruby Mountains, Huntington Creek drainage.
Pie Creek, Independence Mountains, North Fork Humboldt River drainage.
Pine Creek, Goose Creek drainage.
Pole Creek, East Humboldt River, Ruby Valley drainage.
Pole Creek, Snake Mountains, Tabor Creek drainage.

Pratt Creek, Independence Mountains, North Fork Humboldt River drainage.
Rattlesnake Creek, Ruby Mountains, South Fork Humboldt River drainage.
Rock Creek, Northern Tuscarora Mountains, Humboldt River drainage.
Salmon Falls River, O'Neil Basin, Snake River drainage.
Salmon and Little Salmon Creeks, Merritt Mountain, Sheep Creek drainage.
Schoonover Creek, Independence Mountains, Jack Creek drainage.
Secret Creek, Ruby Mountains, Humboldt River drainage.
Seitz (Rabbit) Creek, Ruby Mountains, Lamoille Creek drainage.
76 Creek, Jarbidge Mountains, Bruneau River drainage.
Sherman Creek, Elko Mountains, Humboldt River drainage.
Shoer Creek, East Humboldt Range, Clover Valley drainage.
Shoshone Creek, Salmon Falls River drainage.
Smith Creek, Ruby Mountains, South Fork Humboldt River drainage.
Snow Creek, Independence Mountains, South Fork Owyhee River drainage.
Soldier Creek, Ruby Mountains, Secret Creek drainage.
Spring Creek, Tuscarora Mountains, Maggie Creek drainage.
Sun Creek, O'Neil Basin, Salmon Falls River drainage.
Tabor Creek, Snake Mountains, Humboldt River drainage.
Talbot Creek, Ruby Mountains, Lamoille Creek drainage.
Taylor Cañon Creek, Independence Mountains, South Fork Owyhee River drainage.
Tea Creek, Jarbidge Mountains, Mary's River drainage.
Thorpe Creek, Ruby Mountains, Lamoille Creek drainage.
Thousand Spring Creek (Crittenden Creek, Loomis Creek, Dakes Creek, Montello Reservoir, Tecoma Reservoir, combined), Snake Mountains, Bonneville drainage.
Toe Jam Creek, Tuscarora Mountains, Rock Creek drainage.
Toyn Creek, Ruby Mountains, Corral Creek drainage.
Trail Creek, Van Duzer Creek (combined), Northern Independence Mountains, East Fork Owyhee River.
Trout Creek, Big Goose Creek drainage.
Trout Creek, Granite Mountains, Salmon Falls River drainage.
Waterpipe Cañon Creek, Independence Mountains, Taylor Cañon Creek drainage.
Weeks Creek, East Humboldt Range, Clover Valley drainage.
Willow Creek, Ruby Mountains, Huntington Creek drainage.
Wilson Creek, O'Neil Basin, Salmon Falls River system.

LAKES, RESERVOIRS AND PONDS

Angel Lake, East Humboldt Range, Willow Creek drainage.
Boulder Lake, East Humboldt Range, Boulder Creek drainage.
Cold Lakes, Ruby Mountains, Cold Creek drainage.
Echo Lake, Ruby Mountains, North Fork of South Fork Humboldt River drainage.
Emerald Lake, Jarbidge Mountains, East Fork Jarbidge River drainage.
Favre Lake, Ruby Mountains, Kleckner Creek drainage.

Greys Lake, East Humboldt Range, Greys Creek (Clover Valley) drainage.
Hidden Lake, Ruby Mountains, Soldier Creek drainage.
Island Lake, Ruby Mountains, Lamoille Creek drainage.
Lamoille Lake, Ruby Mountains, Lamoille Creek drainage.
Liberty Lake, Ruby Mountains, Kleckner Creek drainage.
Robinson Lake, Ruby Mountains, Robinson Creek (Ruby Valley) drainage.
Ruby Marshes, Ruby Valley.
Soldier Lake, Ruby Mountains, Soldier Creek drainage.
Steel Lake, East Humboldt Range, Steel Creek (Clover Valley) drainage.
Verdi Lake, Ruby Mountains, Talbot Creek drainage.

ESMERALDA COUNTY

Chiatovich Creek, White Mountains, Fish Lake Valley drainage.
Davis Creek, White Mountains, Tributary to Chiatovich Creek.
Indian Creek (McNett Creek), White Mountains, Fish Lake Valley drainage.
Leidy Creek (Cord Creek), White Mountains, Fish Lake Valley drainage.
Middle Creek, White Mountains, Tributary to Chiatovich Creek.
Perry Aiken Creek (Patterson Creek key stream), White Mountains, Fish Lake Valley drainage.
Spring Creek, White Mountains, Tributary to Chiatovich Creek.
Trail Cañon Creek, White Mountains, Fish Lake Valley drainage.

LAKES, RESERVOIRS AND PONDS
Fish Lake, Fish Lake Valley.

EUREKA COUNTY

Allison Creek, Monitor Range, Antelope Valley drainage.
Ardens Creek (Torre Creek), Diamond Mountains, Diamond Valley drainage.
Birch Creek, Roberts Mountains, Denay Valley drainage.
Cedar Creek, Monitor Range, Antelope Valley drainage.
Chimney Creek, Sulphur Spring Range, Henderson Creek drainage.
Cottonwood Creek, Diamond Mountains, Diamond Valley drainage.
Denay Creek (Tonkin Creek), Denay Valley drainage.
Fish Creek Springs, Little Smoky Valley.
Hildebrand Creek, Diamond Mountains, Diamond Valley drainage.
Hunter Creek (Simpson Creek, key stream), Diamond Mountains, Diamond Valley drainage.
McKlusky Creek, Simpson Park Mountains, Grass Valley drainage.
Minnow Creek, Diamond Mountains, Diamond Valley drainage.
Pete Hanson Creek, Roberts Mountains, Denay Valley drainage.
Pine Creek, Cortez Mountains, Humboldt River drainage.
Roberts Creek, Roberts Mountains, Diamond Valley drainage.
Schaefer Creek (Hoosac Canyon), Hoosac Mountain, Little Smoky Valley drainage.

Spring Creek, Diamond Valley drainage.
Trout Creek, Piñon Range, Pine Creek drainage.
Vinini Creek, Roberts Mountains, Henderson Creek drainage.
Williams Creek, Diamond Mountains, Diamond Valley drainage.
Willow Creek, Piñon Range, Pine Creek drainage.
Willow Creek, Roberts Mountains, Denay Valley drainage.

LAKES, RESERVOIRS AND PONDS

Frazer Pond, Crescent Valley.

HUMBOLDT COUNTY

Alder Creek, Pine Forest Range, Onion Valley drainage.
Alta Creek, Pine Forest Range, Quinn River drainage.
Andorna Creek, Santa Rosa Mountains, Quinn River drainage.
Bartlett Creek, Black Rock Range, Black Rock Desert drainage.
Big or Alexander Creek, Jackson Mountains, Desert Valley Desert drainage.
Big Creek, Pine Forest Range, Quinn River drainage.
Bilk Creek, Trident Peak, Quinn River drainage.
Bottle Creek, Jackson Mountains, Desert Valley Desert drainage.
Buffalo Creek, Santa Rosa Mountains, Quinn River drainage.
Cabin Creek, Santa Rosa Mountains, Martin Creek drainage.
Cañon Creek, Santa Rosa Mountains, Quinn River drainage.
Cottonwood Creek, Santa Rosa Mountains, Martin Creek drainage.
Cottonwood Creek, Little, Santa Rosa Mountains, Cottonwood Creek drainage.
Dutch John Creek, Santa Rosa Mountains, Martin Creek drainage.
Eightmile Creek, Santa Rosa Mountains, Quinn River drainage.
Elder Creek, Battle Mountain Range, Humboldt River drainage.
Flat Creek, Santa Rosa Mountains, Quinn River drainage.
Granite Creek, Pine Forest Mountains, Quinn River drainage.
Happy Creek, Jackson Mountains, Quinn River drainage.
Horse Creek, Santa Rosa Mountains, Quinn River drainage.
Humboldt River, Little (North Fork), Santa Rosa Mountains, Little Humboldt River drainage.
Jackson Creek, Jackson Mountains, Black Rock Desert drainage.
Leonard Creek (Chicken Creek, New York Cañon Creek, Sagehen Creek, Snow Creek, combined), Pine Forest Mountains, Black Rock Desert drainage.
Lye Creek, Santa Rosa Mountains, Dutch John Creek drainage.
McConnel Creek, Santa Rosa Mountains, Quinn River drainage.
McDermitt Creek, Disaster Peak, Quinn River drainage.
Martin Creek, Santa Rosa Mountains, Little Humboldt River drainage.
Mary Sloan Creek, Jackson Mountains, Black Rock Desert drainage.
Mullinix Creek, Santa Rosa Mountains, Cottonwood Creek drainage.
Pole Creek, Sonoma Range, Humboldt River drainage.
Poplar Creek (Cabin Creek), Trident Peak, Kings River drainage.
Quinn River, Santa Rosa Mountains, Black Rock Desert drainage.
Raster Creek, Trident Peak, Poplar Creek drainage.
Rebel Creek, Santa Rosa Mountains, Quinn River drainage.
Rock Creek, Sonoma Range, Humboldt River drainage.

Siard Creek, Santa Rosa Mountains, Cottonwood Creek drainage.
Singas Creek, Santa Rosa Mountains, Cottonwood Creek drainage.
Solid Silver Creek, Santa Rosa Mountains, Martin Creek drainage.
Stone House Creek, Santa Rosa Mountains, Martin Creek drainage.
Threemile Creek, Santa Rosa Mountains, Quinn River drainage.
Trout Creek, Jackson Mountains, Desert Valley Desert drainage.
Trout Creek (Crawley Creek), Disaster Peak, Quinn River drainage.
Wash O'Neil Creek (Riley Creek), Santa Rosa Mountains, Martin Creek drainage.
Wilder Creek, Trident Peak, Trout Creek drainage.
Willow Creek, Jackson Mountains, Desert Valley Desert drainage.
Willow Creek, Santa Rosa Mountains, Quinn River drainage.

LANDER COUNTY

Bade Creek, Toiyabe Range, Smoky Valley drainage.
Big Creek, Toiyabe Range, Reese River drainage.
Birch Creek, Toiyabe Range, Smoky Valley drainage.
Blakely Creek, Toiyabe Range, Smoky Valley drainage.
Boone Creek, Toiyabe Range, Reese River drainage.
Bowman Creek (Clear Creek), Toiyabe Range, Smoky Valley drainage.
Callahan Creek (Woodward and Hunt Creeks), Toiyabe Range, Grass Valley drainage.
Campbell Creek, Desatoya Mountains, Smith Creek Valley drainage.
Carseley Creek (South Fork Bowman Creek), Toiyabe Range, Tributary to Bowman Creek.
Cowboy Rest Creek, South Fork, Toiyabe Range, Grass Valley drainage.
Crum Creek (Crum Cañon Creek), Shoshone Mountains, Humboldt River drainage.
Fish Creek, Fish Creek Mountains, Reese River drainage.
Frenchman Creek, Toiyabe Range, Smoky Valley drainage.
Galena Creek, Battle Mountain Range.
Hall Creek, Toiyabe Range, Carico Lake Valley drainage.
Indian Creek, Shoshone Mountains, Crescent Valley drainage.
Iowa Creek and Reservoir, Toiyabe Range, Carico Lake Valley drainage.
Italian Creek, Toiyabe Range, Reese River drainage.
Kingston Creek (key stream), Toiyabe Range, Smoky Valley drainage.
Lewis Creek, Shoshone Mountains, Reese River drainage.
Mill Creek, Shoshone Mountains, Reese River drainage.
Peterson Creek (Sawmill Creek), Shoshone Mountains, Smith Creek Valley drainage.
Reeds Cañon Creek, Toiyabe Range, Reese River drainage.
Santa Fe Creek, Toiyabe Range, Smoky Valley drainage.
Shoshone Creek, Toiyabe Range, Smoky Valley drainage.
Silver Creek, Toiyabe Range, Reese River drainage.
Skull Creek (key stream), Toiyabe Range, Grass Valley drainage.
Smith Creek, Desatoya Mountains, Smith Creek Valley drainage.
Steiner Creek, Simpson Park Mountains, Grass Valley drainage.
Trout Creek, Shoshone Mountains, Reese River drainage.
Willow Creek, Battle Mountain Range, Reese River drainage.

LINCOLN COUNTY

Beaver Dam Wash, Virgin River drainage in Arizona.
Caselton Creek, Pumped flow from Combined Reduction Metal Plant.
Clover Creek (Big Springs Creek), Meadow Valley Wash drainage.
Crystal Springs Creek, Frenchie(y) Lake and Nesbitt Lake, Pahranagat Valley drainage.
Eagle Valley Creek, Meadow Valley Wash drainage.
Gyser Creek, Schell Creek Range, Lake Valley drainage.

LYON COUNTY

Desert Creek, Sweetwater Mountains, West Walker River drainage.
East Drain, Mason Valley, Walker River drainage.
Fernley Drains, Truckee Canal, Truckee River drainage.
Sweetwater Creek, Sweetwater Mountains, East Walker River drainage.
Wabuska Drain, Mason Valley, Walker River drainage.
Walker River, Walker Lake drainage.
Walker River, East Fork, Walker River drainage.
Walker River, West Fork, Walker River drainage.
West Drain, Mason Valley, Walker River drainage.

MINERAL COUNTY

Rough Creek, Sierra Nevada Mountain Range, East Walker River drainage.

NYE COUNTY

Andrews Creek, Toquima Range, Monitor Valley drainage.
Barker Creek (Shipley Creek), Toquima Range, Smoky Valley drainage.
Barley Creek, Monitor Range, Monitor Valley drainage.
Belcher Creek, Toiyabe Range, Smoky Valley drainage.
Big Creek, Quinn Cañon Range, Railroad Valley drainage.
Blairs Creek (Segura Creek), Antelope Valley drainage.
Broad Creek, Toiyabe Range, Smoky Valley drainage.
Cherry Creek, Quinn Cañon Range, Garden Valley drainage.
Clear Creek, Monitor Range, Little Fish Lake Valley drainage.
Clear Creek, Toiyabe Range, Reese River Valley drainage.
Clover Creek, Monitor Range, Little Fish Lake Valley drainage.
Copenhagen Creek (Martin Creek), Monitor Range, Antelope Valley drainage.
Corcoran Creek, Toquima Range, Monitor Valley drainage.
Cottonwood Creek, Monitor Range, Barley Creek drainage.
Cottonwood Creek, Quinn Cañon Range, Garden Valley drainage.
Cottonwood Creek, Toiyabe Range, Reese River Valley drainage.
Crane Creek, Toiyabe Range, Reese River drainage.
Danville Creek, Monitor Range, Little Fish Lake Valley drainage.
Deep Creek, Quinn Cañon Range, Railroad Valley drainage.
Georges Cañon Creek, Monitor Range, Stone Cabin Valley drainage.
Green Monster Creek, Monitor Range, Little Fish Lake Valley.
Hooper Cañon Creek, Quinn Cañon Range, Railroad Valley drainage.

Hunts Cañon Creek, Monitor Range, Ralston Valley drainage.
Illinois Creek, Toiyabe Range, Reese River Valley drainage.
Jefferson Creek, Toquima Range, Smoky Valley drainage.
Jett Creek, Toiyabe Range, Smoky Valley drainage.
Last Chance Creek, Toiyabe Range, Smoky Valley drainage.
Marysville Creek, Toiyabe Range, Reese River Valley drainage.
Meadow Creek, Toquima Range, Monitor Valley drainage.
Mohawk Creek, Toiyabe Range, Reese River Valley drainage.
Moore's Creek, Toquima Range, Smoky Valley drainage.
Morgan Creek, Monitor Range, Monitor Valley drainage.
Mosquito Creek, Monitor Range, Monitor Valley drainage.
Ophir Creek, Toiyabe Range, Smoky Valley drainage.
Pablo Creek, Toiyabe Range, Smoky Valley drainage.
Peavine Creek, Toiyabe Range, Smoky Valley drainage.
Pine Creek, Quinn Cañon Range, Garden Valley drainage.
Pine Creek, Toquima Range, Monitor Valley drainage.
Reese River, Lower, Reese River Valley, Humboldt River drainage.
Reese River, Upper (key stream), Toiyabe Range, Lower Reese River drainage.
San Juan Creek, Toiyabe Range, Reese River drainage.
Sawmill Creek, Monitor Range, Little Fish Lake Valley drainage.
Six Mile Creek, Hot Creek Range, Hot Creek Valley drainage.
Stewart Creek, Toiyabe Range, Reese River drainage.
Stoneberger Creek (Wilson Creek), Toquima Range, Monitor Valley drainage.
Summit Creek, Toiyabe Range, Smoky Valley drainage.
Tierney Creek (Gross Creek), Toiyabe Range, Reese River drainage.
Troy Cañon Creek, Grant Range, Railroad Valley drainage.
Tulle Cañon Creek, Monitor Range, Little Fish Lake Valley drainage.
Twin River, North, Toiyabe Range, Smoky Valley drainage.
Twin River, South, Toiyabe Range, Smoky Valley drainage.
Washington Creek, Toiyabe Range, Reese River drainage.
Willow Creek, Monitor Range, Stone Cabin Valley drainage.
Willow Creek, Quinn Cañon Range, Railroad Valley drainage.
Willow Creek, Toquima Range, Smoky Valley drainage.
Wisconsin Creek, Toiyabe Range, Smoky Valley drainage.

LAKES, RESERVOIRS AND PONDS

Adams-McGill Reservoir, White River Valley.
Clear Lake, Monitor Range, drains into Clear Creek.
Little Fish Lake (Lakefield Lake, Williams Lake), Little Fish Lake Valley.
Manhattan Dredge Pond, Toquima Range.
Pine Lake, Toquima Range, drains into Pine Creek.
Railroad Valley Ponds, Series "A," Railroad Valley.

ORMSBY COUNTY

Ash Cañon Creek (Gregory Creek), Carson Range, Carson River drainage.
Clear Creek (key stream), Carson Range, Carson River drainage.

Kings Cañon Creek, North Fork, Carson Range, Carson River drainage.
Sawmill Creek, Carson Range, tributary to Ash Cañon.

PERSHING COUNTY

Clear Creek, Sonoma Range, Grass Valley drainage.
Cottonwood Creek, Humboldt Range, Buena Vista Valley drainage.
Coyote Creek, Humboldt Range, Buena Vista Valley drainage.
Hoffman Cañon Creek (Hoffman Creek), Tobin Range, Buffalo Valley drainage.
Indian Creek, Humboldt Range, Buena Vista Valley drainage.
Rocky Creek, Humboldt Range, Humboldt River drainage.
Sonoma Creek, Sonoma Range, Grass Valley drainage.
Star Creek, Humboldt Range, Buena Vista Valley drainage.
Unionville Creek, Humboldt Range, Buena Vista Valley drainage.

WASHOE COUNTY

Bronco Creek, Carson Range, Truckee River drainage.
Brown Creek, Carson Range, Steamboat Creek drainage.
Cottonwood Creek, Granite Range, Black Rock Desert drainage.
Deep Cañon Creek, Carson Range, Truckee River drainage.
Evans Creek, Carson Range, Steamboat Creek drainage.
Franktown Creek, Carson Range, Washoe Valley drainage.
Galena Creek, Carson Range, Steamboat Creek drainage.
Granite Creek, Granite Range, Black Rock Desert drainage.
Gray Creek, Carson Range, Truckee River drainage.
Gray Creek, West Fork, Carson Range, Gray Creek drainage.
Hunter Creek (key stream), Carson Range, Truckee River drainage.
Nigger Creek, Granite Range, Black Rock Desert drainage.
Ophir Creek, Carson Range, Washoe Valley drainage.
Red Mountain Creek, Granite Range, Black Rock Desert drainage.
Thomas Creek, Carson Range, Steamboat Creek drainage.
Truckee River, Sierra Nevada Mountain Range, Pyramid Lake and Stillwater drainage.
White Creek, Carson Range, Steamboat Creek drainage.

LAKES, RESERVOIRS AND PONDS

Gray Lake, Carson Range.
Hobart Reservoir and Creek, Carson Range, Franktown Creek drainage.
Hunter Lake, Carson Range.
Joy Lake, Carson Range.
Price Lake, Lower, Carson Range.
Price Lake, Upper, Carson Range.
Red Mountain Reservoir, Gerlach Area.
Stony Lake (Rock Lake), Carson Range.

WHITE PINE COUNTY

Baker Creek (key stream), Snake Range, Snake Valley drainage.
Baker Creek, South Fork, Snake Range, Baker Creek drainage.

Bassett Creek (key stream), Schell Creek Range, Spring Valley drainage.
Bastian Creek, Schell Creek Range, Spring Valley drainage.
Berry Creek, Lower, Schell Creek Range, Duck Creek Valley drainage.
Berry Creek, North Fork or Upper Berry, Schell Creek Range, Berry Creek drainage.
Big Wash and South Fork, Snake Range, Pruess Reservoir drainage.
Bird Creek, Schell Creek Range, Duck Creek Valley drainage.
Cave Creek, Schell Creek Range, Steptoe Creek drainage.
Cherry Creek, Cherry Creek Mountains, Steptoe Valley drainage.
Chin Creek, Antelope Range, Antelope Valley drainage.
Cleve Creek, Schell Creek Range, Spring Valley drainage.
Cleve Creek, North Fork, Schell Creek Range, Cleve Creek drainage.
Cleve Creek, South Fork, Schell Creek Range, Cleve Creek drainage.
Cold Spring (Hunt Ranch), Diamond Mountains, Newark Valley drainage.
Currant Creek, White Pine Range, Railroad Valley drainage.
Deadman Creek, Snake Range, Smith Creek drainage.
Deep Cañon Creek, Snake Range, Deadman Creek drainage.
Duck Creek (key stream), Schell Creek Range, Steptoe Valley drainage.
East Creek, Schell Creek Range, Duck Creek Valley drainage.
Egan Creek, Cherry Creek Mountains and Egan Range, Steptoe Valley drainage.
Eightmile Creek, Snake Range, Spring Valley drainage.
Ellison Creek, White Pine Range, White River drainage.
Freehill Creek, Schell Creek Range, Spring Valley drainage.
Frenchman Creek, Schell Creek Range, Spring Valley drainage.
Garden Creek, Schell Creek Range, Spring Valley drainage.
Gold Cañon Creek, Egan Range, Steptoe Valley drainage.
Gordon Creek, Schell Creek Range, Spring Valley drainage.
Goshute Creek, Cherry Creek Mountains, Steptoe Valley drainage.
Hampton Creek, Snake Range, Snake Valley drainage.
Hendrys Creek, Snake Range, Snake Valley drainage.
Horse Cañon Creek, Snake Range, Snake Valley drainage.
Huntington Creek, Huntington Valley, South Fork Humboldt drainage.
Illipah Creek, White Pine Range, Jakes Valley drainage.
Indian Creek, Big, Schell Creek Range, Steptoe Valley drainage.
Indian Creek, Little, Schell Creek Range, Spring Valley drainage.
Kalamazoo Creek, Schell Creek Range, Spring Valley drainage.
Lake Creek (Big Spring Creek), Snake Valley, Pruess Reservoir drainage.
Lehman Creek, Snake Range, Snake Valley drainage.
Lexington Creek, Snake Range, Snake Valley drainage.
McCoy Creek, Schell Creek Range, Spring Valley drainage.
Mattier Creek, Schell Creek Range, Steptoe Valley drainage.
Meadow Creek, Schell Creek Range, Spring Valley drainage.
Muncy Creek, Schell Creek Range, Spring Valley drainage.
Nigger Creek, Big, Snake Range, Spring Valley drainage.

Nigger Creek, Little, Schell Creek Range, Spring Valley drainage.
North Creek, Schell Creek Range, Lake Valley drainage.
North Creek, Schell Creek Range, Duck Creek Valley drainage.
North Creek, Schell Creek Range, Spring Valley drainage.
Odgers Creek, Schell Creek Range, Spring Valley drainage.
Paris Creek, Cherry Creek Mountains, Butte Valley drainage.
Piermont Creek, Schell Creek Range, Spring Valley drainage.
Pine Creek, Snake Range, Spring Valley drainage.
Pinto Creek, Diamond Mountains, Newark Valley drainage.
Ridge Creek, Snake Range, Tributary to Pine Creek.
Sage Creek, Snake Range, Tributary to Weaver Creek.
Schell Creek (Schellbourne Creek), Schell Creek Range, Spring Valley drainage.
Schist Creek (S. W. Tributary), Snake Range, Tributary to Hendrys Creek.
Seigle Creek, Schell Creek Range, Spring Valley drainage.
Shingle Creek, Snake Range, Spring Valley drainage.
Short Creek (Cold Creek), Schell Creek Range, Tributary to Steptoe Creek.
Silver Creek, Snake Range, Snake Valley drainage.
Silver Creek, Main Fork, Snake Range, Tributary to Silver Creek.
Silver Creek, Second Fork, Snake Range, Tributary to Silver Creek.
Smith Creek, Snake Range, Snake Valley drainage.
Snake Creek, Snake Range, Snake Valley drainage.
Snake Creek, North Fork (Johnson Creek), Snake Range, Tributary to Snake Creek.
Snake Creek, South Fork, Snake Range, Tributary to Snake Creek.
Snow Creek, Cherry Creek Mountains, Butte Valley drainage.
Spring Creek, Snake Range, Tributary to Silver Creek.
Spring Creek, Snake Range, Tributary to Snake Creek.
Stephens Creek (Vipont Creek), Schell Creek Range, Spring Valley drainage.
Steptoe Creek, Schell Creek Range, Steptoe Valley drainage.
Strawberry Creek, Snake Range, Tributary to Weaver Creek.
Sunkist Creek (North Creek), Schell Creek Range, Spring Valley drainage.
Swallow Creek, Snake Range, Spring Valley drainage.
Taft Creek, Schell Creek Range, Spring Valley drainage.
Taft Creek, Middle Fork (Ranger Creek), Schell Creek Range, Spring Valley drainage.
Taft Creek, South Fork (South Creek), Schell Creek Range, Spring Valley drainage.
Tailings Creek, Bassett Lake drainage.
Timber Creek, Schell Creek Range, Duck Creek Valley drainage.
Timber Creek, North Fork, Schell Creek Range, Tributary to Timber Creek.
Vipont Creek (Stephens Creek), Schell Creek Range, Spring Valley drainage.
Water Cañon Creek, Diamond Mountains, Newark Valley drainage.
Water Cañon Creek, Egan Range, White River drainage.
Weaver Creek, Snake Range, Snake Valley drainage.

Whiteman Creek (Lambert Creek), Schell Creek Range, Steptoe Valley drainage.
White River, White Pine Range, White River Valley drainage.
Willard Creek, Snake Range, Spring Valley drainage.
Williams Creek, Snake Range, Spring Valley drainage.
Willow Creek and Reservoir, Egan Range, Steptoe Valley drainage.

LAKES, RESERVOIRS AND PONDS

Baker Lake, Snake Range.
Bassett Lake, Steptoe Valley.
Blanch Lake, Lehman Lake II (Teresa Lake), Snake Range.
Johnson Lake (Treasure Lake, Snake Lake), Snake Range.
Me-Me-Lake, Lehman Lake III, Snake Range.
Stella Lake, Lehman Lake I, Snake Range.

Total streams surveyed	430
Total lakes, reservoirs and ponds surveyed	38
Total	468

H.

DINGELL–JOHNSON FISHERIES PROJECTS IN NEVADA

FISHERIES PROJECTS CONDUCTED IN NEVADA UNDER THE FEDERAL AID IN FISH RESTORATION ACT
1951 to 1961

(Courtesy Nils Nilsson)

INTRODUCTION

Behind the Dingell-Johnson Act is the recognition that fish restoration problems are closely tied in with the land and waters.

This Act was passed and approved by Congress on August 9, 1950.

Every nickel that is spent on a D–J project comes from the sportsmen. Every time a fisherman buys sport fishing tackle, he chips in to improve his sport by paying an excise tax of 10 percent. These moneys are then apportioned to our 50 states each fiscal year.

Nevada's first D–J project was activated on July 1, 1951.

DINGELL–JOHNSON PROJECTS

The following are D–J Projects that have been set up and approved for Nevada since 1951:

1. Type of project ..... Investigations.
   Name ..... F–1–R, Lakes Mead and Mohave.
   Location ..... Clark County.
   Species ..... Bass and Trout.
   Purpose ..... To gather data upon which to base an efficient fisheries program for both lakes and to improve and prolong the productivity of these lakes.
   Beginning date ..... July 1, 1951.
   Termination date ..... June 30, 1954.
   Personnel ..... Al Jonez, Fisheries Technician; Ed Dwyer, Fisheries Technician; Bob Sumner, Fisheries Technician (Replaced Dwyer).

2. Type of project ..... Investigations.
   Name ..... F–2–R, Stream and Lake Survey.
   Location ..... Statewide.
   Species ..... All game fish.
   Purpose ..... To investigate and study all fishable waters of the State for the purpose of developing a sound management and stocking program.
   Beginning date ..... November 16, 1951.
   Termination date ..... June 30, 1958.
   Personnel ..... Ted Frantz, Fisheries Technician; Don Thurston, Fisheries Technician; Don King, Fisheries Technician (Replaced Thurston).

3. Type of project......Investigations.
   Name......................F-3-R, Lahontan Project.
   Location...............Lahontan Reservoir, Stillwater Marsh and Indian Lakes, Churchill County; Ryepatch Reservoir, Pershing County.
   Species................Namely, warm water fishes.
   Purpose................To make limnological and related studies of all above-mentioned waters.
   Beginning date......July 15, 1954.
   Termination date...June 30, 1957.
   Personnel.............Bob Sumner, Fisheries Technician; Don King, Fisheries Technician.

4. Type of project......Investigations.
   Name....................F-4-R, Lakes Pyramid, Walker, Tahoe and Topaz; reservoir investigations.
   Location...............Washoe, Mineral, Douglas and Ormsby Counties.
   Species................Trout, Cui-ui and Sacramento Perch.
   Purpose...............A study of the natural and artificial fish production in the above waters.
   Beginning date......July 1, 1954.
   Termination date...June 30, 1958.
   Personnel.............Ray Corlett, Fisheries Technician; Cal Allen, Fisheries Technician; Virgil Johnson, Fisheries Technician; Norman Wood, Fisheries Technician (Replaced Ray Corlett).

5. Type of project......Development.
   Name....................F-5-D, Colorado River Shad.
   Location...............Lakes Mead and Mohave, Clark County.
   Species................Threadfin shad.
   Purpose...............To provide a suitable forage fish to the food chain in both of the above lakes.
   Beginning date......November 15, 1954.
   Termination date...June 30, 1955.
   Personnel.............Tom Trelease, Chief of Fisheries; Wallace Rabenstine, Senior Engineer; Al Jonez, Fisheries Technician; Bob Sumner, Fisheries Technician.

6. Type of project......Development.
   Name....................F-6-D, Likes Lake Restoration.
   Location...............Churchill County.
   Species................Warm water game fish species.
   Purpose...............To replace washed out control structure with a new one so water levels can be better manipulated.
   Beginning date......May 15, 1955.
   Termination date...August 31, 1955.
   Personnel.............Robert Sumner, Fisheries Technician; Howard Garratt, Assistant Engineer.

7. Type of project......Investigation.
   Name...................FW-1-R, National Hunting and Fishing Survey.
   Location...............Statewide—(in cooperation with other states).
   Purpose................To obtain statistical data on the number of resident fishermen in the State of Nevada and the number of days and amounts of money they spent in the pursuit of these recreational activities.
   Beginning date.........February 1, 1956.
   Termination date.......March 31, 1956.
   Personnel..............Crossley S-D Surveys, Inc., New York, N. Y.

8. Type of project......Development.
   Name...................F-8-D-1, Tahoe Fisherman Access.
   Location...............Cave Rock, Lake Tahoe, Douglas County.
   Purpose................To provide a boat launching and loading facility.
   Beginning date.........June 1, 1958.
   Termination date.......October 31, 1958.
   Personnel..............Tom Trelease, Chief of Fisheries; Delmer Davis, Senior Engineer.

9. Type of project......Development.
   Name...................F-8-D-2, Wildhorse Fisherman Access.
   Location...............Wildhorse Reservoir, Elko County.
   Purpose................To provide a boat launching and loading facility.
   Beginning date.........January 1, 1959.
   Termination date.......June 30, 1959.
   Personnel..............Tom Trelease, Chief of Fisheries; Delmer Davis, Senior Engineer.

10. Type of project.....Investigations and Management.
    Name..................F-7-R, Sierra District Fisheries Management Investigations.
    Location..............Sierra District.
    Species...............All game fish.
    Purpose...............To carry on a program of inventory and investigations on all phases of fisheries management.
    Beginning date........July 15, 1957 and currently active.
    Personnel.............Robert Sumner, Fisheries District Manager; Cal Allen, Fisheries District Manager.

11. Type of project.....Investigations and Management.
    Name..................F-10-R, Wheeler District Fisheries Management Investigation.
    Location..............Wheeler District.
    Species...............All game fish.
    Purpose...............To carry on a program of inventory and investigations on all phases of fisheries management.
    Beginning date........July 1, 1958 and currently active.

Personnel.................Ted Frantz, District Fisheries Manager (Replaced by Dale Lockard)

12. Type of project .....Investigations and Management.
    Name......................F–11–R, Owyhee District Fisheries Management Investigations.
    Location................Owyhee District.
    Species ....All game fish.
    Purpose.................To carry on a program of inventory and investigations on all phases of fisheries management.
    Beginning date ....July 1, 1958 and currently active.
    Personnel...............Donald King, District Fisheries Manager; William Nisbet, Fieldman II.

13. Type of project .....Investigations and Management.
    Name......................F–12–R, Black Rock District Fisheries Management Investigations.
    Location ....Black Rock District.
    Species .... All game fish.
    Purpose.................To carry on a program of inventory and investigations on all phases of fisheries management.
    Beginning date ....July 1, 1958 and currently active.
    Personnel...............Virgil Johnson, District Fisheries Manager; Darrel Harold, Fieldman II.

14. Type of project .....Investigations and Management.
    Name ....F–13–R, Charleston District Fisheries Management Investigations.
    Location ....Charleston District.
    Species .... All game fish.
    Purpose.................To carry on a program of inventory and investigations on all phases of fisheries management.
    Beginning date ....July 1, 1958 and currently active.
    Personnel .... Norman Wood, District Fisheries Manager.

15. Type of project.....Development.
    Name ....W–14–D, Schroeder Dam.
    Location ....Beaver Dam State Park, Lincoln County.
    Species .... Rainbow or Brook Trout.
    Purpose ....To create a 20-acre reservoir for recreational purposes.
    Beginning date ....May 23, 1960.
    Termination date .. November 30, 1960.
    Personnel...............Tom Trelease, Chief of Fisheries; Delmer Davis, Senior Engineer.

16. Type of project .....Investigations and Management.
    Name......................F–15–R, Lake Tahoe Fisheries Investigations.
    Location ....Lake Tahoe, Nevada and California.
    Species .... Mainly Rainbow, Mackinaw and Kokanee.
    Purpose.................A cooperative project between California and Nevada, which will attempt to determine the best and most efficient method for improving

the quality of fishing by evaluation of various stocking techniques.
Beginning date......July 1, 1960.
Personnel...............Ted Frantz, Wildlife Manager II.

17. Type of project...... Development.
Name...................F–16–D–1, Fishery Rehabilitation.
Location..............Washoe Lakes, Washoe County.
Species................Carp, Catfish, Perch, Bullhead and Chubs.
Purpose...............To eradicate present rough fish populations by chemical treatment and restock waters next spring.
Beginning date......November 9, 1960.
Termination date...November 30, 1960.
Personnel..............Tom Trelease, Chief of Fisheries; Jack Dieringer, Assistant Chief of Fisheries.

18. Type of project...... Development.
Name ....F–8–D–3, Washoe Lake Fisherman Access.
Location ....Washoe Lake, Washoe County.
Purpose ....To provide a boat launching and loading facility.
Beginning date......January 1, 1961.
Termination date..March 1, 1961.
Personnel..............Tom Trelease, Chief of Fisheries; Delmer Davis, Senior Engineer.

19. Type of project ..... Development.
Name...................F–8–D–4, Wildhorse Fisherman Access.
Location...............Wildhorse Reservoir, Elko County.
Purpose................To provide an additional area for boat launching and parking area.
Beginning date......July 1, 1961.
Termination date ..December 31, 1961.
Personnel..............Tom Trelease, Chief of Fisheries; Harry Brown, Conservation Engineer I.

20. Type of project ..... Development.
Name ....F–16–D–2, Fishery Rehabilitation.
Location...............Crittenden Reservoir, Elko County.
Species..................Rainbow Trout.
Purpose................To eradicate present rough fish population by chemical treatment and restock with trout.
Beginning date .....October 1, 1961.
Termination date .December 31, 1961.
Personnel...............Donald King, Fisheries Manager.

21. Type of project...... Development.
Name....................F–16–D–3, Fishery Rehabilitation.
Location................Dutch Bill and Ward Lakes, Churchill County.
Species..................Carp, catfish, bullhead, bass and chub.
Purpose ....To eradicate present rough fish populations by chemical treatment.

Beginning date........October 10, 1961.
Termination date....October 30, 1961.
Personnel................Bob Sumner, Wildlife Manager III.

22. Type of project......Development.
    Name......................F–16–D–4, Fishery Rehabilitation.
    Location..................Rye Patch Reservoir, Pershing County.
    Species....................Carp, catfish, bullhead, channel catfish, crappie, Sacramento perch and bass.
    Purpose..................To eradicate populations of rough fish by chemical treatment.
    Beginning date........October 19, 1961.
    Termination date....October 30, 1961.
    Personnel................Darrel Harold, Fisheries Manager.

23. Type of project......Maintenance.
    Name......................F–17–D–1, Tahoe Fisherman Access.
    Location..................Cave Rock, Lake Tahoe, Douglas County.
    Purpose..................To make repairs and maintain boat launching facility.
    Beginning date........March 15, 1961.
    Termination date....May 15, 1961.
    Personnel................Tom Trelease, Chief of Fisheries; Delmer Davis, Senior Engineer.

24. Type of project......Investigations and Management.
    Name......................Fisheries Research.
    Location..................Statewide.
    Species....................Warm water fish.
    Purpose..................To restore and improve the warm water fisheries that have been lost to the State through continued drought.
    Beginning date........August 1, 1961.
    Personnel................Bob Sumner, Wildlife Manager III.

25. Type of project......Coordination.
    Name......................FW–9–C, Coordination.
    Purpose..................To coordinate D–J Projects with over-all state fisheries management program.
    Beginning date........July 1, 1952 and currently active.
    Personnel................Tom Trelease, Chief of Fisheries; Nils Nilsson, Coordinator, Nevada Federal Aid.

I.
INVERTEBRATE BIOTA OF PYRAMID LAKE AND LAKE TAHOE

(1) PYRAMID LAKE

The following listing is offered as a contribution to our knowledge of the biology of this important and unique body of water and can also be used as an example for many other types of lakes in Nevada whose biota cannot be listed in a work of this kind because of lack of space.

Animals

JOINT-LEGGED ANIMALS, Phylum ARTHROPODA
    INSECTS, Class INSECTA
        FLIES, Order DIPTERA[1, 3]
            MIDGES, Family *Tendipedidae*

1. *Hydrobaenus* sp. (*Smittia*)

            SHORE FLIES, Family *Ephydridae*
2. *Gymnopa bidentata*
3. *Lamproscatella nivosa*
4. *Parydra tibialis*
5. *Parydra nitida*
6. *Parydra halteralis*
7. *Paracoenia bisetosa*

        BEETLES, Order COLEOPTERA
           PREDACEOUS WATER BEETLES, Family *Dytiscidae*
8. *Hygrotus fastidiosus* (Brues 1932, La Rivers 1951[2])

           HERBIVOROUS WATER BEETLES, Family *Hydrophilidae*
9. *Laccobius ellipticus* (Brues 1932, La Rivers 1954)
10. *Enochrus carinatus* (same)
11. *Tropisternus lateralis*[3]

           CRAWLING WATER BEETLES, Family *Dryopidae*
12. *Microcylloepus similis* (La Rivers 1950)[3]

           MUD BEETLES, Family *Heteroceridae*
13. *Heterocerus gemmatus*[3]

---
[1]Determined by Dr. Willis W. Wirth, U. S. National Museum.
[2]These are the published records.
[3]These are first records.

CASEFLIES, Order TRICHOPTERA
Family *Psychomyiidae*
14. *Polycentropus halidus*[3]

BUGS, Order HEMIPTERA
CRAWLING WATER BUGS, Family *Naucoridae*
15. *Ambrysus mormon* (La Rivers 1951)

DRAGONFLIES–DAMSELFLIES, Order ODONATA
CLUBTAILS, Family *Gomphidae*
16. *Ophiogomphus morrisoni* (La Rivers, 1940, 1946)
17. *Herpetogomphus compositus**

DARNERS, Family *Aeshnidae*
18. *Anax junius**
19. *Aeshna multicolor**
20. *Aeshna constricta**

SKIMMERS, Family *Libellulidae*
21. *Macromia magnifica**
22. *Libellula saturata**
23. *Libellula forensis**
24. *Libellula pulchella**
25. *Libellula composita**
26. *Plathemis lydia subornata**
27. *Sympetrum corruptum* (La Rivers 1940, 1946)
28. *Pachydiplax longipennis**
29. *Erythemis simplicicollis**
30. *Pantala flavescens**
31. *Tramea onusta**

DAMSELFLIES, Family *Coenagriidae*
32. *Lestes congener**
33. *Argia emma**
34. *Amphiagrion saucium**
35. *Enallagma clausum* (La Rivers 1940, 1946)
36. *Enallagma carunculatum**
37. *Ischnura denticollis**
38. *Ischnura cervula**

CRUSTACEANS, Class CRUSTACEA
CRAWFISH, CRABS, etc., Order DECAPODA
WESTERN CRAWFISH, Family *Astacidae*
39. *Pacifastacus leniusculus*

---

*Starred species are not known to breed in the lake, but adults are common flying over its waters.

[3]First records.

SCUDS, Order AMPHIPODA
FRESHWATER SHRIMPS, Family *Gammaridae*
40. *Gammarus* sp.

WATERFLEAS, Order CLADOCERA
WATERFLEAS, Family *Daphniidae*
41. *Daphnia pulex* (Hutchinson 1937)
42. *Moina hutchinsoni* (Winnemucca Lake, Hutchinson 1937)

COPEPODS, Order COPEPODA
Family *Diaptomidae*
43. *Diaptomus sicilis* (very common)

Family *Cyclopidae*
44. *Cyclops vernalis* (not common)

Family *Canthocamptidae*
45. *Marshia albuquerquensis* (Hutchinson 1937, as *Cletocamptus*)

CLAM SHRIMPS, Order OSTRACODA
These minute crustacea are very common in Pyramid Lake, but none have as yet been identified.

SEGMENTED WORMS, Phylum ANNELIDA
EARTHWORMS, Class OLIGOCHAETA
A small oligochaete is not uncommon in the mud bottoms of the lake, but is so far unidentified.

LEECHES, Class HIRUDINEA
A moderate-sized leech is fairly common on some of the sedentary fishes in the lake, such as the Sacramento Perch, but has not been studied.

ROUNDWORMS, Phylum NEMATODA
There is undoubtedly a large and substantial population of this group in the lake—free-living and parasitic—but are largely unknown. The parasitic forms are common in many of the fishes.

ROTIFERS, Phylum ROTIFERA[4]
Class MONOROTIFERA
Order FLOSCULARIFORMES
Family *Filiniidae*
46. *Pedalia jenkiniae*

---

[4] Listed for Winnemucca Lake by Hutchinson (1937), and these are undoubtedly a part of the Pyramid fauna as well.

Order PLOIFORMES
Family *Brachionidae*

47. *Brachionus plicatilis*

GASTROTRICHS, Phylum GASTROTRICHA

These minute, interesting little animals, not much larger than many Protozoa, have been detected in the sands and algae of the lake.

NEMERTEANS, Phylum NEMERTEA

48. *Prostoma rubrum* (the only species in the United States)[3]

FLATWORMS, Phylum PLATYHELMINTHES
PLANARIA, Class TURBELLARIA

One of these free-living flatworms has been found in Pyramid.

SPONGES, Phylum PORIFERA
Class DEMOSPONGEA
Order HAPLOSCLERIFORMES
FRESH-WATER SPONGES, Family *Spongillidae*

49. *Meyenia mülleri* (the common alkaline water species, widespread on tufa in the lake).[3]

PROTOZOA, Phylum PROTOZOA
CILIATES, Class CILIATA
Order PERITRICHA
Family *Vorticellidae*

50. *Zoöthamnium* sp.

Other, noncolonial, ciliates have been frequently observed, and doubtless the total protozoan fauna of the lake will be found to be sizeable when adequately investigated.

Plants

ALGAE, FUNGI and BACTERIA, Subkingdom THALLOPHYTA
GREEN ALGAE, Division CHLOROPHYTA[5]
Class CHLOROPHYCEAE
Order VOLVOCALES
Family *Polyblepharidaceae*

1. *Dunaliella viridis*

Order ULOTRICHALES
Family *Ulotrichaceae*

2. *Stichococcus subtilis* (a very common species, abundant on fish-trap ropes down to about 6 feet below the surface; see No. 5).

---

[3]First records.
[5]Determined by Dr. Francis Drouet unless otherwise credited. These are all first records.

Order ULVALES
Family *Ulvaceae*
3. *Enteromorpha prolifera*

Order OEDOGONIALES
Family *Oedogoniaceae*
4. *Oedogonium* sp.

Order CLADOPHORALES
Family *Cladophoraceae*
5. *Cladophora glomerata* (the common species, with No. 2, in the lake, growing everywhere).
6. *Rhizoclonium hieroglyphicum*

Order ZYGNEMATALES
Family *Zygnemataceae*
7. *Spirogyra varians*

Family *Desmidiaceae*
8. *Closterium acerosum*

Class CHAROPHYCEAE
Order CHARALES
Family *Characeae*
9. *Chara* sp. (a large, branching, nonalgal-looking alga)

Division CHRYSOPHYTA[6]
DIATOMS, Class BACILLARIOPHYCEAE
Order CENTRALES
Family *Coscinodiscaceae*
10. *Cyclotella quillensis*

Order PENNALES
Family *Fragilariaceae*
11. *Synedra pulchella*

Family *Naviculaceae*
12. *Navicula gastrum*[7] (associated with *Oscillatoria*, *Phormidium* and sulfur bacteria in warm spring water up to 150° F.).
13. *Amphiprora alata*
14. *Tropidoneis lepidoptera* (*Amphiprora*)
15. *Caloneis liber*
16. *Pleurosigma strigilis*
17. *Anomoeoneis sculpta*

---
[6] From Hanna and Grant 1931.
[7] Warm-water species.

Family *Gomphonemataceae*
18. *Gomphonema* sp.

Family *Cymbellaceae*
19. *Cymbella mexicana*

Family *Surirellaceae*
20. *Surirella nevadensis* (type locality is Pyramid Lake)
21. *Surirella testudo*
22. *Surirella utahensis* (type locality is Great Salt Lake)
23. *Campylodicus clypeus*

BLUE–GREEN ALGAE, Division CYANOPHYTA[8]
Class MYXOPHYCEAE
Order OSCILLATORIALES
Family *Oscillatoriaceae*
24. *Oscillatoria terebriformis*[7]
25. *Oscillatoria simplicissima*
26. *Oscillatoria chalybea*
27. *Oscillatoria brevis*
28. *Phormidium tenue*[7]
29. *Phormidium treleasei*[7]
30. *Phormidium valderianum*[7]
31. *Spirulina subsalsa*[7]

Family *Nostocaceae*
32. *Nodularia spumigena* (this species produced a September bloom in 1951 that covered the entire lake; it was several inches thick on the windward west side, thinning to the east, where it became patchy).
33. *Nostoc* sp. (Jones 1925)

Family *Rivulariaceae*
34. *Calothrix* (*thermalis* or *parietina*[7]—Jones 1925, identified by C. L. Brown).

BACTERIA, Division SCHIZOPHYTA
BACTERIA, Class SCHIZOMYCETES
SULFUR BACTERIA, Order THIOBACTERIALES
Family *Nitrobacteriaceae*
35. *Thiothrix nivea* (common in 54° F. spring water coming into the lake at Camp Foster).
36. *Lamprocystis ?roseo-persicina*[7]

(2) LAKE TAHOE

Undoubtedly the microscopic plants and animals which form the basic substance of the fish food chain in Lake Tahoe are more numerous

---
[7]Warm-water species.
[8]Determined by Dr. Francis Drouet unless otherwise credited.

than these few species would indicate, but a fuller exposition of these important animals will have to wait on a more thorough investigation of this facet of the lake's biota.

Animals
JOINTED-LEGGED ANIMALS, Phylum ARTHROPODA
CRUSTACEANS, Class CRUSTACEA
WATERFLEAS, Order CLADOCERA
WATERFLEAS, Family *Daphniidae*
1. *Daphnia pulex* (*pulicaria*) (Juday 1907, Kemmerer et al 1923).
2. *Daphnia longispina hyalina* (Juday 1907—as *D. hyalina richardi* —Kemmerer et al 1923).

Family *Macrothricidae*
3. *Ilyocryptus acutifrons* (Juday 1907).

Family *Chydoridae*
4. *Eurycerus lamellatus* (Juday 1907).
5. *Acroperus harpae* (Juday 1907).
6. *Alona affinis* (Juday 1907).
7. *Chydorus sphaericus* (Juday 1907).

COPEPODS, Order COPEPODA
Family *Diaptomidae*
8. *Diaptomus* sp. (Juday 1907, Kemmerer et al 1923; probably same as No. 9).
9. *Diaptomus sicilis.*
10. *Epischura nevadensis* (Juday 1907, Kemmerer et al 1923).

ROTIFERS, Phylum ROTIFERA
Class MONOROTIFERA
Order PLOIFORMES
Family *Brachionidae*
11. *Notholca longispina* (Juday 1907, Kemmerer et al 1923).

Plants
ALGAE, FUNGI and BACTERIA, Subkingdom THALLOPHYTA
GREEN ALGAE, Division CHLOROPHYTA
Class CHLOROPHYCEAE
Order ZYGNEMATALES
Family *Zygnemataceae*
1. *Mougeotia* sp. (Kemmerer et al 1923).
2. *Zygnema* sp. (ibid).

Family *Desmidiaceae*
3. *Staurastrum* sp. (ibid).

J.
BIOLOGY OF *CRENICHTHYS BAILEYI* (reprint)

Ecology, Breeding Habits and Young Stages of *Crenichthys baileyi*, a Cyprinodont Fish of Nevada

By JOHN A. KOPEC

(Reprinted from COPEIA, 1949, No. 1, April 15)

Among the most interesting fishes of the isolated drainages of the arid American west are those that constitute certain highly localized endemic genera (Hubbs and Miller, 1948 *a–b*). Since very little is known about their ecology and life history, a study was made of one of the species, *Crenichthys baileyi* (Gilbert).

This fish was originally described as *Cyprinodon macularius baileyi* by Gilbert (1893: 233), who emphasized the lack of pelvic fins but failed to note that the teeth are bifid, rather than trifid as in *Cyprinodon*. Jordan and Evermann (1896: 675) recognized the species as distinct, but still retained it in *Cyprinodon*, not appreciating its relationship with the genus *Empetrichthys* (Gilbert, 1893: 233–234, pl. 5; Miller, 1948: 99–111, pls. 10–11), which is confined to the springs of Ash Meadows and Pahrump Valley, Nevada. The species *baileyi* was shown by Hubbs (1941: 68) and Hubbs and Miller (1941: 1–2) to be wide-spread in warm springs throughout the remnants of the Pluvial White River system, Nevada (Hubbs and Miller, 1948*b*: 7–8, pls. 3 and map 2), and to be referable to the genus *Crenichthys*, which Hubbs (1932) had established for a newly discovered species, *C. nevadae*, of Railroad Valley, Nevada. On the basis of identifications by Hubbs, *Crenichthys baileyi* was also reported by Sumner and Sargent (1940) and by Sumner and Lanham (1942), in their studies on the adaptation of fishes to warm-spring waters.

Previous attempts to culture *Crenichthys* in aquaria have proved unsuccessful, although desert forms of *Cyprinodon* have been reared, as by Cowles (1934), Miller and Miller (1942) and Miller (1948: 122–126).

DISTRIBUTION AND ECOLOGY

Crenichthys baileyi occurs not only in a number of isolated warm springs and their creek outflows in the White River and Pahranagat valleys, but also in the warm-spring headwaters of the Moapa, or Muddy River, which flows into the Virgin River arm of Lake Mead. In the Moapa springs it is associated with *Moapa coriacea,* which has just been described by Hubbs and Miller (1948*b*: 1–14, Map 2, pl. 1, fig. 1) as a general relict confined to these restricted waters. Other associates are the introduced *Gambusia affinis affinis* Baird and Girard and, in the spring-fed streams, a local form of *Gila robusta* Baird and Girard.

The ecological characteristics of the Moapa River spring headwaters, as well as the past and present hydrographic and faunal relationships of this river, have been discussed by Hubbs and Miller (1948: a–b). Their account of the ecology includes some results from my field studies of December, 1947.

The many warm springs of the Moapa Valley make up the headwaters of the Moapa, or Muddy River. Some of these springs have a rather slow current with large pools measuring up to 7 feet in depth and 25 feet in diameter. Others form streams that run swiftly and cool rapidly. The water maintains a nearly uniform temperature at the spring sources. In December, 1947, the readings were 90° F. at each spring tested. The pH. ranged from 7.4 to 7.5. The bottoms of the streams are generally muddy with some rocks protruding.

A grass-like plant, *Eleocharis acicularis*, abounds in the faster currents. Some of these plants live entirely out of the water, others are partly submerged and still others are wholly submerged with at least 3 feet of water over them.

Two species of mollusks also inhabit these warm waters, a small elongate spiral form, *Tryonia clathrata* (Stimpson), and a round form, *Amnicola micrococcus* (Pilsbry).

Many schools of *C. baileyi* were observed in one large pool where most of the collecting was done. The schools consisted of not more than 25 individuals; however, in the smaller and swifter streams the schools comprised as many as 250 fish. This greater abundance in the swifter water may be due to a more ample food supply.

In the larger pools the fish were observed spawning and one egg was noted on the aquatic fibrous roots of a smoke tree, *Parosela spinosa*.

Live specimens were transported to Los Angeles in 5 gallon water jugs, with a mortality of only 3 fish out of about 150. Specimens of *Moapa coriacea* were brought at the same time and proved to live well in aquaria.

SPAWNING ACTIVITIES

In an effort to induce spawning, an adult male spring fish measuring 70 mm. and 2 adult females, slightly smaller, were placed in a standard 15-gallon aquarium, with a 75-watt heater regulated by a thermostat set at 90° F. A small aquarium air pump supplied the necessary aeration. The aquarium was heavily planted with *Anacharis, Myriophyllum, Vallisneria* and other standard aquarium plants. The fish were fed *Tubifex* worms, *Daphnia,* and prepared fish food throughout the experiment.

The male began his courting behavior the same evening, approaching a female distended with eggs at a 45° angle, head down, at a distance of 1 to 3 inches. Usually his courtings took place in front of the female where he would presumably be seen. He displayed intense colors, the mid-dorsal markings becoming very dark gray—almost black—in striking contrast to the almost white sides above the black lateral streaks or fused rows of spots. The fins were all trimmed with black margins. The female's colors were less intense but with the same general markings. Her dorsal line was merely brownish.

The male approached along the side of the female and tried to corner her in some thick vegetation. Soon they went into an S-shaped clasp, both fish vibrating very fast as they lay on their sides. The anal fin of the male was folded under the female's now enlarged ovipositor, supposedly to insure a pathway for the sperm directly onto the egg as it was being deposited. The adhesive egg fell onto the nearest vegetation and adhered tightly. After the sex act the female lay for a moment as if in a coma after which she swam quickly out across the tank. The act of copulation took about one second. Mating and egg-laying were repeatedly observed to follow the same general pattern.

At times the female, ready to deposit an egg when the male was not nearby, went into the floating vegetation, lay on her side and vibrated. This activity appeared to stimulate the male to take part in the spawning process.

Only one egg is layed and fertilized at a time. The act is repeated until from 10 to 17 eggs are deposited at each spawning. The eggs measure 1.9 mm. in diameter. By isolating freshly laid eggs singly in floating vials in the aquarium, the incubation period was found to vary from 5 to 7 days.

The male does not seem to be very pugnacious except when his intentions are interrupted by another fish. He will then turn and chase the intruder away, nudging it ferociously. In general, the fish are peaceful and show no sign of cannibalism. All the young may be raised safely in the same tank with the adults.

YOUNG STAGES

Newly hatched young and later stages (fed on Infusoria and *Daphnia*) were sampled for description. The specimens were mounted on depression slides after being preserved in alcohol. Measurements were made by means of a mechanical stage from the tip of the snout to the vertical passing through each given point, according to the method devised by Hubbs and Cannon (1935: 8–9) and applied by Koster (1948) to a similar study of the young stages of another cyprinodont, *Plancterus kansae* (Garman).

PROLARVA

The yolk sac of the newly hatched larva is very prominent. Ten caudal rays are developed. The pectoral fin could not be found although one is suggested mid-ventro-laterally on the yolk sac. There are 6 myotomes anterior and 20 posterior to the anus. Neither dorsal nor anal fins are evident. The dorsal and anal finfolds are continuous with the caudal fin. One patch of melanophores is present between the eyes. Immediately over the brain is a second larger patch, the cells of which diminish progressively in number laterally and ventrally. These melanophores over the brain are the largest of the entire larva, measuring up to 0.1 mm. in spread. The lateral surface of the yolk sac has interrupted linear chromatophores with streamers running vertically and connected with one another dorsally by a horizontal line. The pigment of the streamers is interrupted more and more toward the ventral side and becomes less intense toward the posterior end of the yolk sac near the anus. The mid-dorsal and mid-ventral surfaces are rather heavily pigmented. The dorsolateral surface of the abdomen contains approximately 2 to 4 melanophores per myotome and the ventrolateral surface has a few scattered melanophores ranging from 0 to 2 per myotome.

The pattern of the melanophores seems to follow irregularly that of the myotomes. The vertebral line is seen as a black interrupted line running anteriorly to the top of the eye and posteriorly to almost the end of the caudal peduncle. On the base of the caudal fin there is a single vertical row of melanophores, each on an inter-radial membrane. Small melanophores outline the rays on the caudal fin. Pigment cells extend onto the finfolds in two places, near the anus on the anal finfold and above the caudal peduncle on the dorsal finfold.

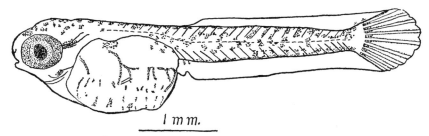

FIG. 267.   Newly hatched prolarva *Crenichthys baileyi*, 4.3 mm. in standard length.

POSTLARVA

The mouth and head regions of the 87-hour postlarva have developed remarkably. The operculum is evident. There are now 17 rays in the caudal fin. The pectoral fin is visible, with 8 rays, each outlined with melanophores. There are 9 myotomes before and 17 after the anus. The anal fin bud is streaked with chromatophores which are penetrating the anal finfold. Three melanophores are located near the point where the dorsal fin should develop. Otherwise the pigmentation is essentially the same as that of the prolarva, except that the melanophores have grown more intense.

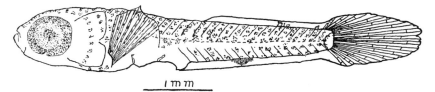

FIG. 268.   Postlarva of *Crenichthys baileyi*, 87 hours old, 4.6 mm. in standard length.

JUVENILE

At the age of 15 days the caudal fin contains 29 rays, the anal 15 rays, the dorsal 11, the pectoral 13. Remnants of the finfolds are still present anteriorly to the anal fin, anteriorly, both dorsally and ventrally on the caudal fin and posteriorly on the dorsal fin. There are 8 myotomes anterior and 16 posterior to the anus. The iris contains blue chromatophores and the pupil is dark blue. Teeth are present. The lips are covered dorsally with small melanophores. A small patch above and posterior to the snout covers the eye region. The brain is covered densely with melanophores. Pigment cells are evenly distributed on the rest of the dorsal surface, but become less intense and fewer in number toward the ventral surface. The melanophores themselves are smaller than in the prolarva but are more numerous.

YOUNG OF 17.7 MM. (STANDARD LENGTH)

The caudal fin contains 30 rays, the anal 16, the dorsal 12, and the pectoral 17. This seems to be the full quota of rays although they are not fully developed. The caudal fin rays are branched twice, instead of 4 times as they are in the adult. Ten heavily pigmented spots extend dorsolaterally along the body behind the head. A slightly smaller number of equivalent spots are evident on the ventrolateral surface. In the specimen described, 2 of these spots have fused with a vertical pigmented bridge. In an older specimen all of the spots are bridged vertically. The joining pigment bridges have considerably less pigmentation than the rudimentary spots. The head region has evenly distributed melanophores about the dorsal part of the snout around the eye. The dorsolateral surface of the 17.7 mm. specimen has more pigmentation than the ventrolateral surface. The dorsal melanophores are grouped evenly under the now well-developed scales. The ventral surface has no evident melanophores. Four prominent unpigmented spots appear in a semicircular pattern anteriorly to the eye on each side of the snout.

Measurements of the several young stages described are recorded in Table I.

TABLE I

MEASUREMENTS OF YOUNG STAGES OF *Crenichthys baileyi* IN MILLIMETERS AND, IN PARENTHESES, IN HUNDREDTHS OF THE STANDARD LENGTH

	STAGES			
	Prolarva (newly hatched)	Postlarva (87 hours)	Juvenile (15 days)	Young (17.7 mm.)
To base of caudal fin	4.3 (100)	4.6 (100)	7.3 (100)	17.7 (100)
To end of caudal fin	5.3 (123)	5.9 (128)	7.8 (107)	21.2 (119)
To front of yolk sac	0.7 (16)			
To rear of yolk sac	2.0 (47)			
To origin of dorsal finfold	2.3 (53)	2.8 (61)		
To origin of anal finfold		2.4 (50)		
To front of dorsal fin base			4.8 (66)	12.0 (67)
To rear of dorsal fin base			5.7 (78)	14.2 (80)
To front of anal fin base			5.0 (68)	11.8 (66)
To rear of anal fin base			6.0 (82)	14.6 (82)
To origin of pectoral fin		1.4 (30)	2.7 (37)	6.4 (35)
To anus	2.1 (48)	2.4 (50)	4.9 (67)	11.8 (66)
Head length	0.8 (19)	1.3 (28)	2.3 (31)	5.2 (29)
Snout length	0.1 (2)	0.2 (4)	0.4 (5)	0.9 (3)
Eye length	0.5 (12)	0.6 (13)	0.9 (12)	1.7 (9)
Body depth thru origin of dorsal finfold	0.4 (9)	0.6 (13)		
Greatest body depth	1.2 (27)*	0.9 (20)	1.6 (22)	5.0 (28)
Caudal peduncle depth	0.2 (5)	0.3 (6)	0.9 (12)	2.5 (19)
Dorsal fin height	0.2 (5)**	0.1 (2)**	0.4 (5)	1.8 (10)
Longest dorsal ray			1.0 (13)	2.3 (13)
Anal fin height	0.2 (5)**	0.1 (2)**	0.3 (4)	1.9 (10)
Longest anal ray				2.1 (12)
Pectoral fin length		0.7 (15)	0.8 (11)	2.6 (14)

*Depth through center of yolk sac.
**Height of finfold.

ACKNOWLEDGMENT

The author wishes to express his thanks to Dr. Carl L. Hubbs, of Scripps Institution of Oceanography, and to Dr. A. Weir Bell and Dr. Sherwin F. Wood, both of Los Angeles City College, for their helpful suggestions and criticisms in preparing this manuscript.

LITERATURE CITED

COWLES, RAYMOND B.
    1934 Notes on the ecology and breeding habits of the desert minnow, *Cyprinodon macularius* Baird and Girard. COPEIA, 1934: 40–42.

GILBERT, CHARLES H.
    1893 Report on the fishes of the Death Valley Expedition collected in southern California and Nevada in 1891, with descriptions of new species. *N. Am. Fauna*, 7: 229–234, 400–402, pls. 5–6.

GREGG, WENDELL O.
    1941 Fluminicola avernalis and Fluminicola avernalis carinifera from Nevada. *The Nautilus*, 54(4), April: 117–118.

HUBBS, CARL L.
    1932 Studies of the fishes of the order Cyprinodontes. XII. A new genus related to *Empetrichthys*. *Occ. Pap. Mus. Zool., Univ. Mich.*, 252: 1–5, pl. 1.
    1941 Fishes of the desert. *The Biologist*, 22, "1940": 61–69, figs. 1–7.

HUBBS, CARL L., and MOTT DWIGHT CANNON.
    1935 The darters of the genera *Hololepis* and *Villora*. *Misc. Publ. Mus. Zool., Univ. Mich.*, 30: 1–93, pls. 1–3.

HUBBS, CARL L., and ROBERT R. MILLER.
    1941 Studies of fishes of the order Cyprinodontes. XVII. Genera and species of the Colorado River system. *Occ. Pap. Mus. Zool., Univ. Mich.*, 433: 1–9.
    1948a *II*. The zoological evidence; correlation between fish distribution and hydrographic history in the desert basins of western North America. *In*: The Great Basin, with emphasis on Glacial and Postglacial times. *Bull. Univ. Utah*, (Biol. Serv.) 10(7): 17–165, figs. 10–29, Map 1.
    1948b Two new, relict genera of cyprinid fishes from Nevada. *Occ. Pap. Mus. Zool., Univ. Mich.*, 507: 1–30, maps 1–2, pls. 1–3.

JORDAN, DAVID STARR, and BARTON WARREN EVERMANN.
    1896 Fishes of North and Middle America. Part I. *Bull. U.S. Nat. Mus.*, 47(1): i-lx, 1–1240.

KOSTER, WILLIAM J.
    1948 Notes on the spawning activities and the young stages of *Plancterus kansae* (Garman). COPEIA, 1948(1): 25–33.

MILLER, ROBERT R.
    1948 The cyprinodont fishes of the Death Valley system of eastern California and southwestern Nevada. *Misc. Publ. Mus. Zool., Univ. Mich.*, 68: 1–155, figs. 1–5, maps 1–3, pls. 1–15.

MILLER, ROBERT R., and RALPH MILLER.
    1942 Rearing desert fish in garden pools. *Aqu. Journ.*, 15: 96–97.

SUMNER, F. B., and URLESS N. LANHAM.
    1942 Studies of the respiratory metabolism of warm and cool spring fishes. *Biol. Bull.*, 82: 313–327, figs. 1–4.

SUMNER, F. B., and M. C. SARGENT.
    1940 Some observations on the physiology of warm spring fishes. *Ecology*, 21: 45–54, figs. 1–2.

ALLAN HANCOCK FOUNDATION OF SOUTHERN CALIFORNIA, LOS ANGELES 7, CALIFORNIA.

K.

BIOLOGY OF *RICHARDSONIUS BALTEATUS* (reprint)

Breeding Habits, Development and Early Life History of *Richardsonius balteatus*, a Northwestern Minnow[1]

George F. Weisel and H. William Newman

(Reprinted from Copeia, 1951, No. 3, August 31)

One of the most abundant, widespread, and brightly colored fishes found in the Columbia River drainage is the redside shiner, *Richardsonius balteatus balteatus* (Richardson). Aside from the interest paid to the high variability of its anal ray counts by such early American ichthyologists as Eigenmann (1895: 10–25) and Gilbert and Evermann (1894: 196–197), practically nothing has been recorded concerning this minnow. Few investigations have been made on the life histories of the freshwater fishes of the American Northwest. Since the redside shiner is one of the commoner species and undoubtedly plays an important role as a bait and forage fish, the following observations are considered to be of value.

SPAWNING PLACE AND TIME

Richardsonius has been taken in all the lakes, sloughs and quieter parts of large streams which we have seined in western Montana. On April 2, 1949, the junior author noticed a crowded school of these shiners splashing in a small spring which empties into a slough near Bearmouth, Montana. These fish were evidently spawning. The slough is about ¼ of a mile long and 20 to 40 feet wide. Water cress, *Chara*, *Myriophyllum* and *Lemna* grow in profusion. It is fed by a number of warm springs that well up along its west bank and flow over a few feet of riffles before entering the slough, which discharges directly into the Clark Fork of the Columbia River. *Richardsonius* spawns in the riffles leading to the springs and in the welling water of the springs. The water here is from 1 to 3½ inches deep over a bottom of gravel and rocks about the size of the fist. In the particular spring hole where observations were made, there is no plant growth. The temperature of the water in both the spring and slough remained constantly 17° to 18° C. throughout the period of spawning and early development.

No breeding activities were noticed in parts of the slough other than the springs and their outlets. No eggs or prolarvae were found along the mud banks and short stretches of gravel beach or among the aquatic plants and the deeper parts of the slough.

In the spring of 1950 a close watch was kept on this site. None of the minnows taken on March 5 had ripe gonads. Milt was stripped from a single male on March 20, but not from other adults collected. On the afternoon of April 8, shiners were congregated in the springs and riffles along the slough's shore, and eggs were found adhering to rocks. Breeding activities continued until June 17. By this date the shiners had left the breeding grounds and eggs were no longer present.

---

[1] Aided by a grant from the Research Committee, Montana State University.

Richardsonius spawns earlier in this warm spring slough than it does in colder waters. Samples taken from Post Creek, in the Flathead River Valley, Montana, were ripe from May 20 to June 30, and both sexes could be stripped. Post Creek is typical of the cold streams in this region, with spring temperatures of 9° to 15° C. *Richardsonius* was still ripe in Flathead Lake, Montana, the last of June. Consequently shiners in these waters must have a later or more extended spawning period than those in the slough.

In Glacier Park, Schultz (1941: 32) found ripe redside shiners in June and stated that their breeding season was in the spring and early summer. The subspecies *R. b. hydrophlox* (Cope) in Jackson Hole and Green River, Wyoming, is said by Simon (1946: 82) to spawn during late June and early July. However, the localities where the observations by these authors were made are at a considerably higher elevation and have a later spring than the areas considered here.

Since our observations were made only from 12:00 to 5:00 p. m., it is not known whether the shiners spawn at night as well as during the day.

DESCRIPTION OF SPAWNING FISH

Females with ripe gonads ranged from 7.0 to 11.1 cm. in total length and males from 6.0 to 10.4 cm.[2] Length frequencies of two large collections indicate that the 1-year group includes specimens from 3.0 to 5.9 cm., and the 2-year individuals from 6.0 to 8.5 cm. Not enough of those in the older age-groups were taken to more than guess at their relative sizes. It is evident that *Richardsonius balteatus* does not spawn until it is two or more years old.

The bright coloration of the redside shiner is intensified in the spring, especially in the males. Ripe males are brassy on all fins, along the sides ventral to the dark lateral band, and in a narrow strip above this dark area. They also have a brassy half-moon below the eye. There is a dark, rosy wash just behind the operculum and at the base of the pectorals, which continues posteriorly in a narrower band to just above the origin of the anal fin. The top of the head, the back, and the lateral band are dark olive to black. The small but numerous tubercles are distributed most profusely over the top of the head and the back, from the nostrils to the origin of the dorsal fin. The pectoral, pelvic and dorsal fins also are frequently tuberculate, and a roughness can be felt on the scales. The gravid females show the same general coloration as the males but it is not so brilliant. Instead of having the brassy hue below the eye and on the sides, they are pale gold. Tubercles are generally lacking; if present, they are only weakly developed. No difficulty was experienced in distinguishing mature males from females by these characters. However, the nuptial colors of the fish from the warm spring slough were more intense than those of specimens from neighboring silty streams.

---

[2]The largest specimen of *R. balteatus* taken by us came from the Clark Fork River at Missoula. It is 13.3 cm. long, with a standard length of 11.3 cm. Eigenmann's (1895) record of a specimen 14.0 cm. from Mission, British Columbia, is the only larger individual that we found recorded in the literature.

SPAWNING BEHAVIOR

In the early part of the spawning period, males outnumbered females about ten to one, but the proportion became nearly even toward the last of the season. No activity was observed that properly could be described as fighting or courtship, and no attempt was made by the fish to defend territories or to build any type of a nest.

The shiners were present in schools of thirty to fifty individuals in a pool just below the spring's riffles. They swarmed about in the pool and then, in small groups of two to fifteen, entered the riffles and the spring hole to spawn. There was no "pairing off." Minnows in the riffles remained quietly upstream, but from time to time they would be seen to arch their backs and dip under the edges of rocks, either in the act of eating or spawning. A water glass was placed over the riffles. In a few minutes the fish swam beneath it with no apparent fright. Through this viewer, groups of two to five minnows were seen to drift a few inches downstream with their tails down and heads elevated, and then to move upstream again in the same position, as if trying to rub their genital papillae on the rocks. Some fish, apparently males, tried to crowd others against rocks. The fish in a single group kept changing position, leaving the group and being replaced by others. After remaining but a few seconds in a spawning huddle, they would leave to remain quietly on the riffles or to return to the school in the pool below. Apparently they take repeated turns in spawning and as many as six or eight crowd together for the act.

Few eggs are deposited at one time. From ten to twenty eggs were found in clusters on rocks and since egg counts for six females averaged 1,852, it probably requires several days for a female to become spent. The number of eggs per fish varied from 829 in a specimen 8.0 cm. total length to 3,602 in one 10.4 cm. long.

The eggs are demersal and adhesive. Since most of them are found on the undersurface of rocks and in narrow crevices, where the fish could not squeeze, they must be swept into that position by the current. Direct observations, as well as stomach analyses, showed that exposed eggs were promptly eaten.

FEEDING HABITS DURING SPAWNING

Richardsonius balteatus feeds during the spawning period. Of eighteen mature males and females taken in the riffles on May 8, three had empty stomachs, one contained molluscs (*Physella* and *Gyraulus*) and algae, four had small water beetles and gammarids, one had an adult dipteran and algae, and nine contained eyed eggs and prolarvae of their own species along with some other partially digested plant and animal residue. A total of 79 fish eggs and larvae were eaten. Sixteen stomachs from minnows collected May 19 revealed no cannibalism. Three were empty and the others contained water beetles, adult diptera, dragonfly nymphs, gammarids, algae and sand. There was no particular preference for any of the foods taken. Four adult shiners taken on June 3 from Post Creek had digested insect remains in their digestive tracts. All the fish from the slough were parasitized with small roundworms and spiny-headed worms, but they appeared to be in good condition.

The slough is extremely rich in many types of fish food—molluscs, diptera, water beetles, gammarids, dragonfly nymphs, small forage fish, algae, etc.—but the food taken by *Richardsonius* was not selective. They are principally insectivorous, at least at this time of year and in this locality. The small amount of algae found in their digestive tracts may have been ingested while consuming other food. Notes on the feeding habits of *Richardsonius balteatus* by Carl and Clemens (1948: 85) and of *R. egregius* by Snyder (1917: 57) support the evidence, derived from our few analyses, that fish of the genus are insectivorous.

Redside shiners are definitely cannibalistic and are probably their own worst egg predators. Squawfish (*Ptychocheilus oregonensis*), suckers (*Catostomus catostomus*) and large brown trout (*Salmo trutta*) were abundant in the same slough, but none of them was seen on the spawning ground of the shiners. Stomachs of five 8- to 10-inch squawfish and two brown trout, which weighed over 2 pounds apiece, contained no eggs, young or adults of *Richardsonius*.

In addition to our record of the redside shiner consuming its own eggs, other members of the genus have been noticed to follow suckers and feed on their eggs (Snyder, 1917: 56) and to prey on newly released grayling fry (Simpson, 1941, Master's thesis, quoted by Simon, 1946: 82).

DEVELOPMENT AND LARVAL STAGES

For a close study of the development and larval stages of *R. balteatus*, milt was stripped over eggs of two females on April 19. All but a few of the eggs became fertile. They were placed in aerated jars covered with wrapping paper to exclude most of the light. The water temperature was 21° to 23° C. Eggs were removed at intervals and examined under a dissecting microscope. After hatching, the larvae were placed in 10-gallon aquaria. The descriptions and illustrations that follow were made from living material, which is much more satisfactory than preserved specimens.

Eggs—The freshly fertilized eggs are from 1.9 to 2.2 mm. in diameter, have a pale yellow yolk, and are more adhesive than the unfertilized eggs. They adhere to the sides and bottoms of glassware and tend to clump together.

Early Embryos (26-hour).—The diameter of the eggs, although variable, are the same as freshly fertilized eggs. The embryo develops rapidly and encircles ¾ of the yolk's circumference by this time. The eyes are barely discernible and 8 to 10 somites are formed (Fig. 269a).

Embryos (60-hour).—The eggs are slightly elliptical. One measured 1.9 mm. in its long axis and 1.4 mm. in the short axis. The embryos are well formed 60 hours after fertilization (Fig. 269b). The tail reaches to the back of the head. Eye lenses and ear vesicles, as well as many somites, are present. The heart is tubular and beats with strong regularity. In this stage the embryos are all actively turning within their enveloping membranes.

Late Embryos (78-hour).—This is the period of development just prior to hatching (Fig. 269c). Only ten of about 300 eggs hatched at approximately 74 hours. The embryo is larger than that of the previous stage and the yolk sac is considerably smaller. There is no pigment in the eye or elsewhere on the embryo. The size of the egg remains constant.

Hatching.—The embryos generally break from their restraining membranes instantaneously. A few take a minute or so to become entirely free, and from these it can be noticed that the tail comes out first, leaving the head and yolk sac last to come out of the enveloping vitelline membrane. Hatching occurs between 74 and 168 hours after

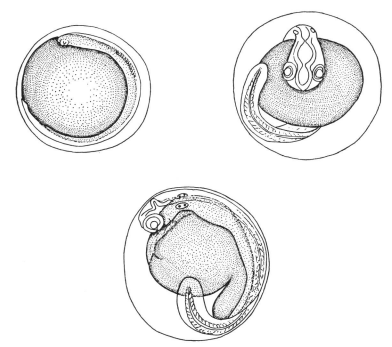

FIG. 269. A, B, C.

fertilization, which is a rather prolonged hatching period. (This was observed in spawn taken from a single female, fertilized by one male, and kept in one container, so that the eggs were subjected to similar conditions.) It takes place during the night as well as in the day. Agitations of the jar containing the eggs brings on a flurry of hatching.

All of the embryos are in the same state of development at any one time. Those which hatch early are not so well developed as those which break free of the egg membrane later. The yolk sac of the late hatching fish is more absorbed and, unlike the early hatchers, the eyes are pigmented. Since these eggs were reared at temperatures higher than those which prevail under natural conditions, the development was undoubtedly faster than normal. In nature it probably takes from 5 to 10 days for the eggs to hatch, rather than the 3 to 7 days noted in these experiments.

Early Prolarvae (3.9 to 5.5 mm.).—Newly hatched prolarvae have a slightly pigmented eye, but no chromatophores on the body. There is a continuous dorsal and ventral fin fold. The pelvic fins are lacking and no rays are developed in the caudal, dorsal or anal fins (Fig. 269d). The yolk sac measures 1.4 mm. in length, the eye 0.4 mm. in diameter, and the postanal tail 1.9 mm. long in prolarvae 5.5 mm. total length. Although varying in size, the prolarvae are in about the identical state of development.

In this stage the larvae remain quietly on the bottom of their aquaria unless disturbed. When startled, they dart erratically for short distances and then settle to the bottom. They avoid light, preferring to remain under rocks and vegetation. In the field, the early prolarvae are found resting on the rocky bottom where spawning has occurred. Their transparency and habit of remaining sheltered under rocks undoubtedly protects them from most predators.

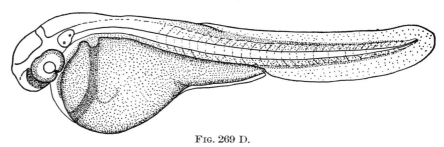

FIG. 269 D.

Prolarvae (6.0 to 8.0 mm.).—Four days after hatching most of the young fish measure 7.0 mm. in total length. The pectorals are well developed, but the continuous dorsal and ventral fin folds persist. Rays are apparent only in the pectorals. The eyes are darkly pigmented and large melanophores have developed on the top of the head, along the margins of the fin folds, and are especially abundant around the air bladder and dorsal surface of the digestive tract. The yolk sac is largely absorbed. The relatively small air bladder is not constricted in the middle as it is in the adult.

These prolarvae swim rapidly but jerkily for short distances, 2 to 8 inches, with short pauses between movements. When they cease swimming, they sink to the bottom head foremost. They tend to remain motionless on the bottom or to adhere to the algae along the sides of the aquaria. Due to their small size and transparency, they are difficult to see unless they are in motion. They do not feed at this time and they show no particular avoidance of light.

Late Prolarvae (8.0 to 8.5 mm.).—Eight days after hatching the yolk is practically absorbed. The original transparency is lost because of a general orange tinge about the head, green color along the back, and the addition of more melanophores. The extreme tip of the caudal vertebrae is barely upturned and caudal fin rays are just evident. The gill arches can be seen through the operculum.

Late prolarvae remain in the open water of the aquarium and swim less jerkily. However, constant swimming motions are necessary or they slowly sink to the bottom. There is no attempt to hide among the rocks and plants.

Early Postlarvae (8.7 to 9.0 mm.).—In ten days there is very little yolk left (Fig. 269e). Except for the more numerous melanophores and further development of the caudal fin rays, this stage is very similar to the previous one. The tiny fish feed along the sides of the aquarium and void feces, and are now sufficiently buoyant so that they do not sink when swimming movements cease. This results from absorption of the yolk and enlargement of the air bladder.

Postlarvae (10.4 to 10.6 mm.).—Seventeen days after hatching the yolk is completely absorbed and the air bladder is larger, but still

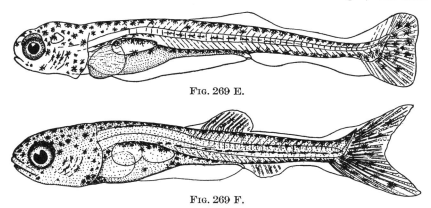

FIG. 269 E.

FIG. 269 F.

undivided. Although dorsal and ventral fin folds are continuous, there are 6 rays in the dorsal fin. The anal fin has no rays. The caudal rays and the hypural plates are well developed. The eyes are 0.8 mm. in diameter, the head 2.2 mm. long, the air bladder 1.1 mm., and the postanal tail 3.8 mm.

Postlarvae (11.0 to 11.8 mm.).—The air bladder of 24-day old fish is constricted into an anterior and a posterior chamber (Fig. 269f). The dorsal and ventral fin folds are present but reduced. There are 9 rays in the dorsal fin and 7 to 10 forming in the anal. The pelvics still are not evident. The chromatophores in this stage are distributed as in the preceding one.

Late Postlarvae (12.8 to 13.4 mm.).—Forty-six days after hatching, postlarvae have 10 dorsal rays and 11 to 15 anal rays. The caudal and pectoral fins also have fully developed rays; the pelvic fins and rays are just commencing to form (Fig. 269g). All but a tip of the dorsal fin fold near the caudal is lost, but a larger portion of the ventral fin fold, from the anus to below the middle of the gut, remains. Melanophores are more numerous and are particularly abundant along the base of the anal fin. One of the critical periods in the life of redside shiners reared in aquaria is between the third and fourth week after hatching. During this time many become emaciated and swim weakly.

Juveniles (17.1 to 21.5 mm.).—The specific characters of *R. balteatus* that are three months old may be easily recognized (Fig. 269h). The lower part of the head, belly and sides are silvery. A narrow band of melanophores extends along the middle of the back. There is a wider dark band in the posterior half of the lateral line and another above the base of the anal fin. Such characteristics as the long, black-bordered anal base, the lack of a definite black spot at the caudal base, and the relatively deep body readily separate it from other minnows of the Clark Fork drainage.

SUMMARY

The redside shiner, *Richardsonius balteatus balteatus* (Richardson), is one of the commonest fishes found in the upper Columbia River. In

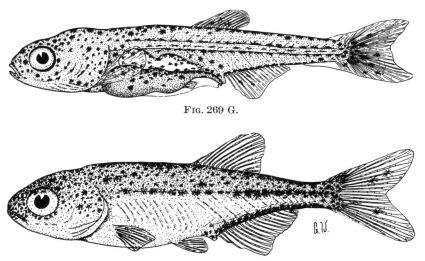

FIG. 269 G.

FIG. 269 H.

western Montana they spawn in their second year, from the first of April to July. The spawning grounds are in the shallow water of rocky riffles or in welling springs. They exhibit no obvious courtship behavior and spawn in groups. No nest is constructed; the demersal eggs are broadcast over the bottom and adhere to rocks or detritus. Although the minnow is essentially insectivorous, it will eat its own eggs.

At laboratory temperatures of 21° to 23° C., the eggs hatch in 3 to 7 days. The young remain as prolarvae for 8 days. During this period they are quietly hidden in the rocks of the spawning area. The postlarval stage lasts until they are about 46 days old. They take food when 10 days old. Although the yolk sac is absorbed in these young, the pelvic fins are not developed and a portion of the dorsal and ventral fin folds persists. The postlarvae are active swimmers and apparently make no effort to remain hidden. Juveniles are essentially similar to the adults.

LITERATURE CITED

EIGENMANN, CARL H.
- 1895 *Leuciscus balteatus* (Richardson), a study in variation. *Amer. Nat.*, 29: 10–25, figs. 20–22, pls. 1–3.

GILBERT, CHARLES H., and BARTON W. EVERMANN.
- 1894 A report upon investigations in the Columbia River basin, with descriptions of four new species of fishes. *Bull. U. S. Fish Comm.*, 14: 169–208, pls. 16–25.

CARL, G. CLIFFORD, and W. A. CLEMENS.
- 1948 The fresh-water fishes of British Columbia. *Brit. Col. Prov. Mus., Dept. Ed., Handbook* No. 5: 1–132, figs. 1–6, 63 illustrations.

SCHULTZ, LEONARD P.
- 1941 Fishes of Glacier National Park, Montana. *U. S. Dept. Interior, Conserv. Bull.* No. 22: i-v, 1–42, figs. 1–26.

SIMON, JAMES R.
- 1946 Wyoming fishes. *Bull. Wyo. Game and Fish Dept.*, No. 4: 1–129, frontis., figs. 1–92.

SNYDER, JOHN OTTERBEIN.
- 1917 The fishes of the Lahontan system of Nevada and northeastern California. *Bull. U. S. Bur. Fish.*, 35 (1915–16): 31–86; figs. 1–9, pls. 3–4, 1 map.

UNIVERSITY OF MONTANA, MISSOULA, MONTANA.

L.

BIOLOGY OF *SIPHATELES BICOLOR OBESUS*
(reprint)

The Embryonic and Early Larval Stages of the Tui Chub, *Siphateles Bicolor* (Girard), From Eagle Lake, California[1]

By ROBERT R. HARRY, Stanford University

(Reprint from CALIFORNIA FISH AND GAME, Vol. 37, No. 2, April 1951)

INTRODUCTION

The complete early development has not been described for any native western North American cyprinid. The postembryonic development of one species, the venus roach, *Hesperoleucus venustus* Snyder, has been discussed in detail by Fry (1936), but its embryonic development is completely unknown.

The present study on the tui chub, *Siphateles bicolor* (Girard), was made by the writer while in the employ of the California Division of Fish and Game. It was carried out at the suggestion of Mr. Harry A. Hanson of the Division and forms an integral part of an investigation into the entire life cycle of the species in Eagle Lake, Lassen County, being conducted by Mr. J. Bruce Kimsey, also of the Division of Fish and Game.

Fertilized eggs of the tui chub were obtained at Eagle Lake by Mr. Kimsey and the writer and their development studied by the latter from 70 hours after fertilization to 12 hours after hatching. It is unfortunate that circumstances caused the development of the eggs under conditions that did not lead to unquestionably normal results and that there was no chance to repeat the work under more normal conditions. However, since there is so little recorded on the early development of fishes, the publication of the present study, even if incomplete, seemed worthwhile.

METHODS

Ripe adults were caught in a gill net on the western side of Eagle Lake near Webb's resort (formerly Spaulding's) over their spawning grounds, which were approximately six feet in depth. The females were stripped into a bowl containing plants and the eggs were artificially inseminated at 8 a. m. on June 24, 1949. Approximately 20,000 eggs were obtained and immediately transported to the Lake Almanor State Fish Hatchery on Clear Creek, Lassen County. They arrived at noon the same day and were placed in quart jars partly submerged in a trout hatching trough. The eggs were kept in the highly mineralized Eagle Lake water at all times. The water was partially changed at frequent intervals and most of the eggs were kept at an even temperature of 45 degrees F., approximately 20 degrees below that normal for the development of *Siphateles* eggs in Eagle Lake.

---

[1] Submitted for publication October, 1950.

All eggs developed normally for the first 100 hours, forming the blastodermal cap and the segmentation cavity. After the latter stage the eggs at 45 degrees either remained dormant or continued growth with irregular cell development. Some of the eggs clumped together in large masses and these eggs formed the blastodermal cap, but developed abnormally thereafter. The yolk became granular and shrunken, and after about 100 hours these eggs also seemed to become dormant, without further organized development. However, they did not die or appear to be more susceptible to fungus than the properly developing eggs.

Three days after stripping 100 eggs were removed and allowed to stay at room temperature, which varied from 34 degrees to 84 degrees F. These eggs developed rapidly and appeared to be almost fully formed within six days. At this time most of the eggs died from fungus. The remaining embryos continued to develop, and formed the pigment pattern characteristic of this species. It is possible that the embryos were so weakened by abnormal conditions for development that they were unable to break out of the eggs at the proper time.

In the following description of embryonic development the eggs at 45 degrees are used for the first 100 hours. The remainder of the description is compiled from embryos kept at room temperature. Only one larva hatched and the description of the early larval stages is from this specimen.

DESCRIPTION OF EGGS

The freshly stripped eggs of *Siphateles* are translucent, yellowish, spherical, adhesive, and rather large for cyprinid eggs, measuring 1.8–2.0 mm. in diameter. Their specific gravity is considerably more than that of fresh water and the free eggs sink quickly to the bottom. The membrane is thick and tough but fairly smooth. The surface is covered with minute crenulations, which at first are hardly visible under 160x power, and which become covered with the floating matter in the water as development progresses. The yolk is filled with numerous oil globules and appears to be separated into small cells, but is so opaque that it is impossible to determine whether or not the divisions are limited to the surface.

The eggs are emitted as a sticky, fluid mass and adhere to each other or other objects soon after coming in contact with fresh water. This adhesive character is soon lost. The membrane becomes flattened over a wide surface wherever the eggs come in contact with each other.

EARLY EMBRYONIC DEVELOPMENT

By 70 hours the fully developed blastodermal cap (Figure 270, A) has become a symmetrical dome. It is very large and its outer surface forms a gentle curve almost continuous with the outline of the yolk. Its irregular inner surface pushes into the yoke. At this time the cells are so small as to be indistinguishable. During all stages of development the perivitelline space is small.

The segmentation cavity (Figure 270, B), partly developed beneath the central area of the blastodermal cap, is formed by the thinning of the central blastodermal wall. This cavity could be distinguished by looking down on the upper surface of the blastoderm in the early stages (Figure 270, C).

At about 100 hours the segmentation cavity is narrower and slightly eccentric. This is the first indication of the main axis of the future embryo; the thicker area marks the posterior pole.

Continued growth and development of the segmentation cavity causes a thinning of the central portion of the blastodermal cap and a thickening of the peripheral germ ring, which is very difficult to distinguish and appears as a broad, slightly darker ring surrounding a more translucent area. From the time of its origin until the closure of

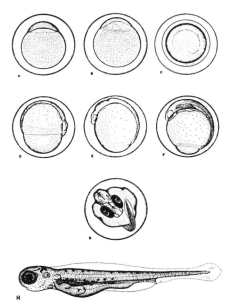

FIG. 270. Development of *Siphateles bicolor* (Girard). A. 70 hours after fertilization. B. 85 hours. C. 100 hours. D. 120 hours. E. 136 hours. F. 152 hours. G. 170 hours. H. larva about 6 hours after hatching. From Harry 1951, courtesy California Department of Fish and Game.

the blastopore it marks the advance of the blastoderm over the yolk sphere. The embryonic shield is fairly narrow and somewhat wedge-shaped. During the formation of the embryonic axis the blastoderm covers half of the yolk sphere at about 120 hours (Figure 270, D).

At about the same time that the germ ring forms an equatorial belt, Kupffer's vesicle appears imbedded in the yolk at the posterior end of the embryo and metameric segmentation is detected at the middle of the body. The notochord also becomes apparent.

LATE EMBRYONIC DEVELOPMENT

Near the time of the closure of the blastopore at 136–140 hours (Figure 270, E and F) the eyes become visible. Metameric segmentation is well developed caudally and the heart begins to pulsate below the middle of the left eye (viewed from the dorsal aspect). Immediately behind the eyes the auditory capsules are barely visible as small oval vesicles. Kupffer's vesicle now appears to be obsolete.

At 170 hours (Figure 270, G) the blastopore is closed and the head as a whole has increased markedly in size. The pupils and lenses in the eyes are visible and the otoliths are formed. The tail twitches from side to side at frequent intervals and passes over the snout and eyes. The continuous fin fold is clearly visible.

About the ninth day pigment begins to appear in the form of several rows of conspicuous melanophores on the dorsal side, particularly on the head. The eyes are heavily pigmented. The extent of pigmentation increases during the remainder of embryonic development. The yolk becomes reduced to less than half its original size.

EARLY LARVAL DEVELOPMENT

After hatching, which took place during the night of the twelfth day, the larva was exceptionally active and was able to swim rapidly and effectively (Figure 270, H). The yolk was almost completely absorbed. The auditory capsules were enlarged and the cells of the notochord could be seen along its entire length. The tail had begun to develop rays (ceratotrichia) and to change shape. Chromatophores were concentrated along the ventral midline of the body, exclusive of the gut. Two irregular longitudinal rows of large melanophores extended along each side of the dorsal midline. A large triangular blotch of giant melanophores was present on the nape. The eye and lens were solid black. The pectoral fins were very small and the rays only slightly developed. The anus was far posterior in position.

The only larva that hatched died approximately 12 hours after it emerged. So far as could be seen, it had no special structures on the head for breaking the egg membrane.

REFERENCE

FRY, DONALD H., JR.
    1936   Life history of *Hesperoleucus venustus* Snyder. California Fish and Game, vol. 22, no. 2, p. 65–98, 8 figs.

M.
GLOSSARY

ABDOMEN, belly, containing the viscera or digestive organs and the reproductive organs.

ABDOMINAL, relating to the abdomen. Some ventral fins are abdominal or pelvic in position; other ventral fins are moved forward near the position usually occupied by the pectoral or shoulder fins, in which case the latter are above them.

ABORTIVE, decreased in size or scope.

ACCESSORY CAUDAL RAYS, shortened rays on dorsal and ventral areas of the caudal peduncle.

ACCESSORY PELVIC SCALES, noticeable scaly or fleshy appendages on the upper base of the pelvic fin. Synonym of SCALY APPENDAGES.

ACUMINATE, tapering to a point gradually.

ACUTE, tapering to a point suddenly.

ADIPOSE FIN, a fin-like fleshy extension of the back between dorsal and caudal fins.

ADNATE, grown together.

AFFLUENT STREAMS are those going into a lake.

AGAPE, with jaws open.

AGGRADING, geological, is the building up of its channel by a stream which is flowing too slowly to carry its load.

AIR BLADDER, a membranous sac containing gas and lying in the body cavity against the backbone; not homologous (same as) the lungs of higher vertebrates. Synonymous with SWIM BLADDER.

ALGAE, simple plants with green chlorophyll, commonly called "moss" or "pond scum" by most people.

AMPHICOELIAN, vertebrae which are concave on both ends.

ANADROMOUS, fishes which leave ocean waters and run up into fresh waters to spawn.

ANAL FIN, ventral median fin just behind the anus or vent.

ANAL FIN II-8, meaning that this fin has two anterior hard spines and eight posterior, following, soft rays.

ANAL PAPILLA (plural papillae), fleshy bump anterior to the genital pore and posterior to the vent; this functions as the penis of more advanced vertebrates.

ANCHYLOSED, growing solidly together.

ANNULI (singular annulus), are yearly growth lines formed by groups of circuli which lie closer together than usual.

ANGULAR, small bone on the back end of the mandible.

ANTECEDENT STREAMS are those in place before the rising of mountain chains. As the mountains rise, the streams cut through at the same rate and so maintain their positions. The Truckee River is one such example.

ANTRORSE, turned forward.

ANUS or vent, the posterior opening of the digestive tract.

ARTICULAR, mandibular bone holding the DENTARY and attached to the QUADRATE.

ARTICULATE, jointed, as in certain soft fin rays.

ATROPHY, decrease in size.

ATTENUATE, long and drawn-out.

AUDITORY CAPSULE, a swelling on the skull which holds part of the inner ear. Ventro-lateral in position.

AXIL, the area below or behind the base of the pectoral or pelvic (ventral) fin.

BARBEL, thin, fleshy extension or tentacle about the mouth.

BICOLOR, two-colored.

BICUSPID, two-pointed, as of teeth.

BIOTA are all living forms, plants and animals in a given area or habitat.

BRANCHIA (plural branchiae), gill, the fish's breathing organ.

BRANCHIOSTEGAL, bony, enlarged ray holding the branchiostegal membrane on the fish head, below the opercle. Attached to the hyoid arch.

BUCCAL, mouth.

CADUCOUS, falling off readily or early in life.

CAECUM (plural caeca), elongated blind sac originating from the digestive tract between stomach and intestine.

CANINES, eye teeth behind the incisors. Used for any distinctly enlarged conical teeth.

CARINATE, keeled, showing as a sharpened ridge along the middle line.

CATADROMOUS, fish running from fresh water into salt water to spawn.

CAUDAL FIN, tail fin.

CAUDAL PEDUNCLE, that part of the body behind the anal fin and terminating in the tail.

CAUDAL PEDUNCLE SCALE COUNT is made by setting up a peduncle circumference line which will result in the lowest count of scale lines crossing it.

CENOZOIC ERA began about 70 million years ago and is continuing at the present time. Mammals, birds and modern fishes made their great developmental strides during this era.

CENTRUM, vertebra main body.

CHEEK SCALES, are the scales counted from the eye obliquely downward and backward to the preopercular angle.

CHIN, area between the rami of the lower jaws.

CILIATED, bordered with hairs.

CIRCULI (singular circulus), are numerous, fine, closely spaced lines on the fish scale which encircle the focus point in increasingly large circles from anterior-to-posterior.

CIRCUMFERENCE SCALE COUNT is used frequently for cyprinids and is made by setting up a body circumference line around the specimen just in front of the origin of the dorsal fin and counting the scale lines that cross it.

COMPRESSED, flattened from side-to-side.

CRETACEOUS PERIOD is the terminal period of the Mesozoic Era. Ending some 70 million years ago and having a duration of 60 million years, it has yielded one species of fossil fish for the Nevada area.

CTENOID, rough-edged. Cycloid scales that are toothed along their hind margins.

CYCLOID, smooth-edged, without teeth, referring to scales.

DATA (singular datum); this is the information which accompanies each museum or research specimen. In its simplest form it must include LOCALITY, DATE and COLLECTOR. For much limnologic and fisheries work a great deal of additional information might be required such as physical, chemical, and biological data.

DECIDUOUS, temporary.

DECURVED, curving downward.

DEGRADING, geologically, is the downcutting of its channel by a stream whose waters are still flowing swiftly enough to carry all its detritus.

DEMERSAL, sinking in water.

DENTARY, the foremost body of the mandible (lower jaw), usually having teeth.

DENTICLE, a small tooth.

DEPRESSED, flattened dorso-ventrally.

DIAPHANOUS, translucent.

DISTAL, removed from the point of attachment.

DORSAL FIN, principal fin of the back.

DORSAL FIN 2 + 8, referring to two simple or rudimentary rays anteriorly and eight major or branched rays following them. In salmonids, catostomids and cyprinids the enumeration would simply be eight, since the two simple rays are usually omitted from the count. These simple rays are one-half the length or less of the major, branched rays.

DORSAL FIN VI–16, refers to a fin with six anterior hard spines and sixteen posterior soft rays in which the spined and ray sections are separate from each other. If they are connected, the designation would be VI 16.

EFFLUENT STREAMS are those leaving a lake.

EKMAN DREDGE, a brass instrument consisting of a box to which are attached spring-loaded jaws which can be set open so that a messenger sent down the line will close them. In this way samples of lake bottoms can be obtained for physical, chemical and biological analyses.

EMARGINATE, notched, forked.

EPIBRANCHIAL, bone above the branchial arch angle.

EPILIMNION is the uppermost zone of summer circulating water in a lake. It is separated from the Hypolimnion by the Thermocline.

ERECTILE, capable of being erected.

ETHMOID, a bone above the Vomer on the median anterior line of the skull.

EUTROPHIC LAKES are those high in nutrient materials and, other things being satisfactory, furnishing a rich environment for the development of aquatic life.

FALCATE, scythe-shaped.

FAUNA, all the animals living in a given region or under given conditions.

FILIFORM, thread-like in structure.

FIN HEIGHT is measured as the length of the longest ray.

FIN LENGTH is measured along the fin base.

FLORA, all the plants living in a given region or under given conditions.

FONTANELLE, open (unossified) spot on top of the head, bordered by the parietals and covered by a membrane.

FORAMEN, an opening.

FORFICATE, strongly forked. Synonymous with FURCATE.

FORK LENGTH, that length measured from snout tip to middle indentation of tail fin when the latter is in a relaxed condition.

FRENUM, a small fleshy binding between jaw edge and lip along the median line.

FRONTAL, a median anterior bone on the top of the head.

FRY are fish up to the time their yolk sac is absorbed.

FUSIFORM, spindle-shaped. The taper involves both ends, but is usually more pronounced anteriorly.

GAPE, mouth opening.

GIBBOUS, abruptly rounded and convex.

GILL MEMBRANES are thin skin walls bordering the gill chamber ventrally and supported by the Branchiostegal rays.

GILL RAKER, a bony appendage or tooth-like structure lying on the inner or anterior face of the gill arch which bears the gill posteriorly or outwardly. Gill raker count is made only on the first arch.

GILL RAKERS 2 + 5 refers to the method of counting, which includes those rakers above the arch angle (the 2) and those below the angle (the 5). These are called upper and lower limbs of the arch, respectively.

GILL SLIT is the opening between the gill arches.

GLABROUS, smooth.

GLOSSOHYAL, the bone of the tongue.

GRAVIMETRIC METHOD, of fish food analysis, involves sorting the different types of organisms found in the fish stomach, weighing them and expressing their percentages in terms of the total weight mass of food in the stomach.

GULAR, the area between chin and Isthmus.

HEAD LENGTH is measured from snout tip to the hind margin of the bony opercle, including spines.

HUMERAL SPINE, above the Pectoral Fin base.

HYDROGEN–ION CONCENTRATION, simply interpreted, is the terminology for the acidity-alkalinity scale. The essentials of the scale can be read in terms of 1 to 14, with 7 being the mid or neutral point. From 7 toward 1 is increasingly acid, and from 7 toward 14 is increasingly alkaline. Synonymous with pH.

HYOID APPARATUS, a group of bones supporting the tongue.

HYOMANDIBULAR, a bone which supports the mandible, hyoid and opercular elements.

HYPOLIMNION is the summer stagnating lower levels of lake water lying below Thermocline.

ICHTHYOLOGY, the study of fishes.

IMBRICATE, shingle-like overlapping.

IMPERFORATE, not pierced.

INARTICULATE, not jointed.

INCISORS, front teeth flattened into a cutting edge.

INSERTION, point of origin, as of a fin.

INTEROPERCLE, the bone connecting the Preopercle and the Branchiostegals.

INTERORBITAL SPACE, area between the eyes measured across the top of the head.

ISTHMUS, the area upon which the gill membranes converge at the front end of the breast, or the soft parts between the gill openings.

JUGULAR, lower throat region. When ventral fins have moved forward beyond the pectorals, as in centrarchids, they are said to be jugular in position.

KEELED, sharply ridged along the median line.

KEMMERER WATER BOTTLE, a brass cylinder with movable rubber corks at both ends adjusted in such a way that when the bottle is lowered into the water a messenger can be sent down the line to close the corks and obtain a sample of water at that particular depth for such determinations as oxygen, carbonates, bicarbonates, temperature, etc.

LACUSTRINE, pertaining to lakes.

LATERAL LINE, a row of tiny holes opening to the outside from a deeper sensory canal along the side of a fish.

LENTIC WATERS, ponds or lakes (standing water).

LIMNOLOGY is a study of the aquatic environment and its life.

LINGUAL, referring to the tongue.

LOTIC WATERS are streams (flowing waters).

LUNATE, new moon-shaped.

MARBLED, variegated.

MAXILLA, the upper jaw.

MAXILLARY, an outermost or hindmost bone of the upper jaw which joins the premaxillaries in front or below and usually exceeds them in backward length. It may lie above the Premaxillaries.

MESOZOIC ERA began some 185,000,000 years ago and terminated about 70,000,000 years ago and encompassed a span of 115,000,000 years. The greatest development of reptiles took place during this time, as well as significant advances among the fish.

MIOCENE EPOCH is a subdivision of the Tertiary Period (Cenozoic Era) covering 16 million years somewhat closer to modern times than the period's half-way point. Several fossil fishes have been described from the Nevada area for this epoch.

MOLAR, grinding tooth. Also refers to posterior jaw tooth or any flat-topped tooth.

MOUTH INFERIOR refers to a mouth which projects downward due to overhanging upper lips or skull structure.

MOUTH OBLIQUE refers to a mouth situated so that the jaws are at about a 40 degree or more angle with respect to the long axis of the body.

MOUTH VENTRAL indicates an extreme development from the inferior mouth condition.

MYOTOME, muscle segment.

NAPE, the upper neck region, adjacent to the Occiput.

NARES, nostrils.

NUCHAL, referring to the nape.

OBTUSE, blunt.

OCCIPUT, the back of the head.

OLIGOTROPHIC LAKES are those low in nutrient materials and consequently poor areas for the development of extensive aquatic floras and faunas.

OPERCULUM, the gill cover.

OPISTHOCOELIAN, referring to vertebrae with ball and socket joints, or a posterior concavity to the Centrum.

ORBICULAR, circular.

ORBIT, eye socket.

FIN ORIGIN, that line along which fins articulate with the body.

OSSEOUS, bony.

OTOLITH, a bone of the inner ear.

OVIPAROUS, egg-producing fishes.

OVOVIVIPAROUS, those fish which bear living young.

OVUM, the egg.

PALATE, the mouth roof.

PALATINE, a membrane bone in the roof of the mouth.

PALEONTOLOGY is the study of fossil or extinct life.

PALEOZOIC ERA is a period of some 315 million years beginning with the CAMBRIAN PERIOD and closing with the PERMIAN PERIOD. The Paleozoic began 500 million years ago and terminated 185 million years ago. The fossil record begins essentially with the Paleozoic.

PALUSTRINE, pertaining to swamps.

PAPILLA (plural papillae), a small, fleshy bump.

PAPILLOSE, covered with papillae.

PARIETAL, a prominent bone forming part of the skull case on top and back.

PARR MARKS, broad, vertical color bands in the young of salmonids.

PECTINATE, with comb- or teeth-like projections.

PECTORAL FIN, one of the anterior-most pair of ventral fins; in centrarchids and others these are high up on the sides.

PELAGIC, living in the open sea or water.

PERITONEUM, the membranous lining of the Abdominal cavity.

PERMIAN PERIOD is the terminal subdivision of the Paleozoic Era; lasting for some 25 millions of years, it contains the only Paleozoic fishes (sharks) known to have occurred in the area now called Nevada.

pH. See Hydrogen-ion Concentration.

PHARYNGEAL BONES lie behind the gills and in front of the esophagus.

PHARYNGEAL TEETH 2–4/4–2 refers to teeth borne on the pharyngeal bones and counted according to numbers on the main or larger row (the 4's) and numbers on the inner or lesser row (the 2's) of both sides. Also written 2,4–4,2, 2/4–4/2 and 2–4–4–2.

PHYSIOGRAPHY is the systematic study of natural landforms.

PHYSOCLISTIC, lacking connections between Air Bladder and esophagus.

PHYSOSTOMOUS, with connection between the Air Bladder and the esophagus.

PHYTOPLANKTON are microscopic floating plants, mainly Algae.

PLANKTON are the minute floating forms of microscopic plants and animals in water which cannot get about to any extent under their own power. They form the important beginnings of food chains for larger animals.

PLICATE, folded, such as having transverse wrinkling.

PLIOCENE EPOCH, the last subdivision of the Tertiary Period, began 12 million years ago and ended 11 million years later. Several fossil fishes from Nevada are known for this epoch.

PLUMBEOUS, lead colored.

POST-, a prefix meaning "behind."

PRE-, a prefix meaning "in front of."

PREFRONTALS, the bones forming lateral projections on the anterior edges of the orbits.

PREMAXILLARIES, bones forming the upper jaw front, one on each side.

PREOPERCLE, a membrane bone situated before the Opercle and essentially parallel with the latter.

PRESERVATIVE. In preserving fish, make a small incision on the right side near the mid-belly line and put the specimen in 10 percent formalin (this is really about 4 percent). For museum work the specimen should subsequently be transferred to alcohol.

PROCOELIAN, vertebrae with anterior concavity of the Centrum.

PROXIMAL, basal; that part which is nearest.

PSEUDOBRANCHIAE, the small gills on the Opercle inner side near its union with the Preopercle.

PTERYGOID, one of several roof mouth bones behind the palatines.

PUNCTATE, dotted with small pits.

PYLORUS, the passage between stomach and intestine.

RADII (singular radius), are the major lines radiating backward from the central or focus point of the scale across the ANNULI and CIRCULI.

RAY, a cartilaginous rod which supports the fin membrane. It may be either simple or branched, soft or spiny.

RECURVED, bending upward.

RETICULATE, with a network of lines.

RETRORSE, bending backward.

RICTUS, the hind mouth corner.

ROSTRAL PLATE, a small structure on the snout tip.

RUGOSE, wrinkled.

SCALE SAMPLE AREA is that from which scales are scraped for Annuli age counts. In the Spiny-Rayed fishes, this is below the middle area of the anterior spinous portion of the dorsal fin and below the lateral line. In others, it is below the origin of the dorsal fin and above the lateral line.

SCALES ABOVE THE LATERAL LINE. These are counted obliquely downward and backward from the anterior origin of the dorsal fin to the lateral line.

SCALES BELOW THE LATERAL LINE. These are counted from the anterior origin of the pelvic or ventral fin obliquely upward and forward to the lateral line.

SCALE FORMULA 12–16/80–120/14–12 is interpreted as follows: 12–16 refers to scales above the lateral line, 80–120 to scales in the lateral line and 14–12 to scales below the lateral line.

SCALES IN THE LATERAL LINE refers to the count made along the lateral line from the Opercle to the posterior part of the Caudal Peduncle. This latter point is determined (without dissection) by moving the tail from side-to-side and noting where the crease so produced occurs. If the crease involves a majority of a particular scale, that scale is counted, but none posterior to it on the peduncle even though the lateral line pores are well developed there. If the fish has no lateral pores, scales are counted along the mid-lateral line which would normally be occupied by such pores.

SCUTE, external plate, often spiny or keeled.

SECCHI DISK, a round, white-enameled piece of metal, usually 8 inches in diameter, lowered into the water to determine the extent of visibility and so the water turbidity.

SECOND DORSAL FIN is the hindermost part of the long dorsal— the soft-rayed part—of the Spiny-Rayed fishes. Synonymous with SOFT DORSAL.

SEICHES ("say-sh-es") are oscillations or "tides" caused in lakes by wind action, barometric pressure changes, etc.

SEPTUM, a thin partitioning membrane.

SERRATE, saw-toothed.

SESSILE, meaning attached to a substrate, or lacking a stalk.

SETACEOUS, bristly.

SHOAL means shallow.

STANDARD LENGTH, the length from snout tip to the base of the tail fin rays.

SUBULATE, awl-shaped.

SUPRA, a prefix meaning "above."

SUPRAORBITAL SPINE, a prominent spine above the eye.

SWIM BLADDER. Synonym of AIR BLADDER.

SYNONYMICON is a listing of the synonyms for a given species.

TAXONOMY is the science of naming fishes (or all other animals and plants).

TERETE, tapering and cylindrical.

TERTIARY PERIOD constitutes most of the Cenozoic Era and spanned all but 1 million of that era's 70 million years. Several types of fishes have been described from these Nevada rocks.

TESSELLATED, showing squares or checks.

THERMOCLINE is the intermediate summer zone in lakes between overlying Epilimnion and underlying Hypolimnion, defined as that region in which the temperature decrease reaches 1° C. or more for each meter of descent ($=0.55°$ F. per foot).

TOTAL LENGTH is measured from snout tip to end of longest lobe of tail fin when the latter is compressed between the fingers so that it extends backward its maximum distance.

TRIASSIC PERIOD was the first of three periods which make up the Mesozoic Era and spanned some 30 million years, ending 155 million years ago. Several sharks and primitive fishes were described from Nevada rocks of this period.

TUFA, a deposit of calcium carbonate laid down in water.

TUFF, fine volcanic dust and ash blown into the air and deposited conspicuously as lake sediments.

TYPE LOCALITY, the locality from which the originally described specimens of a species came.

VENT, the anus.

VENTRAL FINS, the posterior Abdominal paired fins. The same as PELVIC FINS.

VILLIFORM, referring to compact or velvety patches of teeth.

VIVIPAROUS, bringing forth living young in the sense of mammalian embryogeny.

VOLUMETRIC METHOD of fish food analysis is done by using water displacement volumes for each type of food item in the stomach.

VOMER, the anterior part of the roof mouth. A bone usually with teeth and positioned behind the Premaxillaries.

WEBERIAN OSSICLES are small bones forming a chain connecting the ear with the Air Bladder in certain fishes.

ZOOPLANKTON are microscopic floating animals, chiefly protozoa and microcrustacea for limnological purposes.

BIBLIOGRAPHY

ABBOTT, CHARLES C. 1860 (1861). Descriptions of four new species of North American Cyprinidae. Academy Natural Sciences Philadelphia Proceedings 12: 437–474 (*Xyrauchen texanus*, etc.).

ADAMS, ANDREW LEITH. 1874. The lake trouts. United States Fish Commission Report 1872–1873, 2: 357–362.

ADELMAN, HARVEY M. and JOSEPH L. BINGHAM. 1956. Size discrimination in the brook, brown and rainbow trout. Progressive Fish-Culturist 18(1): 26–29.

AGASSIZ, LOUIS. 1855. Art. XII.—Synopsis of the ichthyological fauna of the Pacific slope of North America, chiefly from the collections made by the U. S. Expl. Exped. under the command of Capt. C. Wilkes, with recent additions and comparisons with eastern types. American Journal Science Arts, 2nd Series, 19(55): 71–99; 19(56): 215–231 (new genera *Acrocheilus* and *Ptychocheilus*, etc.).

ALCORN, J. RAY. 1943. Observations on the white pelican in western Nevada. Condor 45(1): 34–36.

ALDO, L. 1918. Mixing trout in western waters. American Fisheries Society Transactions 47(3): 101–102.

ALLEN, GEORGE H. 1956. Age and growth of brook trout in a Wyoming beaver pond. Copeia 1956(1): 1–9.

ALLEN, K. RADWAY. 1956. The geography of New Zealand's freshwater fish. New Zealand Science Review 14(3): 3–9 (some discussion of American and other fish introductions).

ALM, GUNNAR. 1955. Artificial hybridization between different species of the salmon family. Fishery Board Sweden, Institute Freshwater Resources, Annual Report 1954, No. 36: 13–56.

ANDERSON, R. A. 1909. Geology and oil prospects of the Reno region, Nevada. United States Geological Survey Bulletin 381: 475–489.

ANONYMOUS. 1875. Truckee River Trout. Forest and Stream 5: 308.

ANONYMOUS. 1878A. Trout fishing at Lake Bigler, California. Forest and Stream 10: 28.

ANONYMOUS. 1878B. Good news from California. Chicago Field 10: 84 (Truckee River fish ladders).

ANONYMOUS. 1879A. Lake Tahoe. Chicago Field 11: 260.

ANONYMOUS. 1879B. Catfish in the Susan River. Chicago Field 12: 85.

ANONYMOUS. 1879C. Trout in the Truckee. Chicago Field 12: 117.

ANONYMOUS. 1879D. Trout culture in Nevada. Chicago Field 12: 180 (from Virginia City Territorial Enterprise).

ANONYMOUS. 1952. Nevada pool has unique kind of fish. Nevada State Journal (Reno), Feb. 10 (Devil's Hole, Ash Meadows, Nye County).

ANONYMOUS. 1955. Nevada's Vast Desert Lake. Sunset Magazine, November, 1955: 38, 43–44.

ANONYMOUS. 1957. Squawfish predation study, progress report. Fishery development program of the Columbia River. United States Fish Wildlife Service, Regional director, Portland, Oregon. pp. 1–23, mimeographed.

ANONYMOUS. 1959. Today's rainbow trout replacing yesterday's cui-ui, cutthroat. Sierra Magazine 5(1): 6–9 (written by someone in the Nevada Fish and Game Comm.).

ANTEVS, ERNST. 1925. On the Pleistocene history of the Great Basin *IN* Quaternary Climates. Carnegie Institution Washington Publication 352: 51–114.

  1936. Pluvial and Postpluvial fluctuations of climate in the Southwest *IN* Quaternary Climates. Carnegie Institution Washington Publication Yearbook 35: 322–323.

1938. Climatic variations during the last glaciation in North America. American Meteorological Society Bulletin 19: 172–176.

1939. Precipitation and water supply in the Sierra Nevada, California. American Meteorological Society Bulletin 20: 89–91.

1945. Correlation of Wisconsin glacial maxima. American Journal Science 243–A: 1–39.

1948. Climatic changes and Pre-White Man, pp. 168–191, IN The Great Basin, with emphasis on the Glacial and Postglacial Times. University Utah Bulletin 38(20): 1–191 (Biological Series, Vol. 10, No. 7).

APPLEGATE, VERNON C. 1947. Growth of some Lake trout, *Cristivomer n. namaycush*, of known age in inland Michigan Lakes. Copeia 1947 (4): 237–241.

APPLEGET, JOHN and LLOYD L. SMITH, JR. 1951. The determination of age and rate of growth from vertebrae of the Channel catfish, *Ictalurus lacustris punctatus*. American Fisheries Society Transactions 80(1950): 119–139.

ATKINSON, N. J. 1931. The destruction of grey trout eggs by suckers and bullheads. American Fishery Society Transactions 61 (1931): 183–188.

AXELROD, DANIEL I. 1940A. The Pliocene Esmeralda flora of west-central Nevada. Washington Academy Science Journal 30: 163–174.

1940B. Late Tertiary floras of the Great Basin and border areas. Torrey Botanical Club Bulletin 67: 477–487.

1948. Climate and evolution in western North America during middle Pliocene time. Evolution 2: 127–144.

1949. Eocene and Oligocene formations in the western Great Basin. Geological Society America Bulletin 60: 1935–1936.

1950. Evolution of desert vegetation in western North America. Carnegie Institution Washington Publication 590: 215–306.

1956. Mio-Pliocene floras from west-central Nevada. University of California Publications Geological Sciences 33: 1–322.

1958. The Pliocene Verdi flora of Western Nevada. University California Publications Geological Sciences 34(2): 91–160.

AYRES, WILLIAM O. 1854. Descriptions of new species of fish. California Academy Science Proceedings 1: 7–8 (2d ed., pp. 7–8, 1873) (*Centrarchus maculosus = Archoplites interruptus*).

1855A. New species of California fishes. Boston Society Natural History Proceedings 5: 94–103 (a second mention of *Centrarchus maculosus*).

1855B. Descriptions of two new species of cyprinoids. California Academy Science Proceedings 1: 18–19 (2d ed. pp. 17–18, 1873) (*Catostomus occidentalis* and *Gila grandis = Ptychocheilus grandis*).

1855C. Description of a new cyprinoid fish. California Academy Science Proceedings 1: 21–22 (2d ed., pp. 20–21, 1873) (*Gila microlepidota = Orthodon*).

AYSON, L. F. 1910. Introduction of American Fishes into New Zealand. United States Bureau Fisheries Bulletin 28: 967–975.

BACON, EDWARD H. 1954. Field characters of prolarvae and alevins of Brook, Brown and Rainbow trout in Michigan. Copeia (1954) (3): 232.

BAILEY, REEVE M. 1951. A check-list of the fishes of Iowa, with keys for identification, pp. 185–237, IN Iowa Fish and Fishing by Harlan and Speaker (which see).

1956. A revised list of the fishes of Iowa, with keys for identification, pp. 326–377, IN Iowa Fish and Fishing by Harlan and Speaker (which see).

BAILEY, REEVE M. and MARY FITZGIBBON DIMICK. 1949. *Cottus hubbsi*, a new cottid fish from the Columbia River system in Washington and Idaho. University of Michigan Museum Zoölogy Occasional Paper 513: 1–18.

BAILEY, REEVE M. and WILLIAM A. GOSLINE. 1955. Variation and systematic significance of vertebral counts in the American fishes of the Family Percidae. University Michigan Museum Zoölogy Miscellaneous Publication 93: 1–44.

BAILEY, REEVE M. and HARRY M. HARRISON, JR. 1948. Food Habits of the southern channel catfish (*Ictalurus lacustris punctatus*) in the Des Moines River, Iowa. American Fishery Society Transactions 75 (1945) : 110–138.

BAILEY, REEVE M., HOWARD ELLIOTT WINN and C. LAVETT SMITH. 1954. Fishes from the Escambia River, Alabama and Florida, with ecologic and taxonomic notes. Academy Natural Sciences Philadelphia Proceedings 106: 109–164.

BAILEY, R. W. 1941. Land erosion—normal and accelerated—in the semiarid West. American Geophysical Union Transactions 1941(2) : 240–250.

BAIRD, SPENCER FULLERTON and CHARLES GIRARD. 1853A (1854). Descriptions of some new fishes from the River Zuñi. Academy Natural Science Philadelphia Proceedings 6: 368–369 (new genus and species *Gila robusta*, *G. elegans* and *G. gracilis*) (also issued the same year as a part of "32d Congress, 2d Session, Senate Executive Document No. 59" by Capt. L. Sitgreaves, Corps Topographical Engineers, on his expedition down the Zuñi and Colorado Rivers).

1853B (1854). Descriptions of new species of fishes collected by Mr. John H. Clark, on the U. S. and Mexican Boundary Survey, under Lt. Col. Jas. D. Graham. Academy Natural Science Philadelphia Proceedings 6: 387–390 (*Catostomus lapitinnis*, *Gila emoryi*, *G. grahami*, *Heterandria affinis* = *Gambusia*, etc.).

BALL, SIDNEY H. 1907. A geological reconnaissance in southwestern Nevada and eastern California. United States Geological Survey Bulletin 308: 1–218.

BANCROFT, HUBERT HOWE. 1890. History of Nevada, Colorado and Wyoming, 1540–1888. Volume 25: xxxii–828 (Nevada history = pp. 1–322.).

BARNICKOL, PAUL G. 1941. Food habits of *Gambusia affinis* from Reelfoot Lake, Tennessee, with special reference to malaria control. Reelfoot Lake Biological Station Report 5: 5–13.

BASU, SATYENDRA PRASANNA. 1959. Active respiration of fish in relation to ambient concentrations of oxygen and carbon dioxide. Fisheries Resources Board Canada Journal 16(2) : 175–212.

BEAN, TARLETON H. 1882 (1881). A partial bibliography of the fishes of the Pacific Coast of the United States and of Alaska, for the year 1880. United States National Museum Proceedings 1881, Vol. 4 : 312–317.

1888. Distribution and some characters of the Salmonidae. American Naturalist 22(257) : 306–314.

BECKMAN, WILLIAM C. 1943 (1942). Annulus formation on the scales of certain Michigan game fishes. Michigan Academy Science Arts Letters Papers 28: 281–312.

1948. The length-weight relationship, factors for conversions between standard and total lengths, and coefficients of condition for seven Michigan fishes. American Fisheries Society Transactions 75: 237–256 (1945).

1952. Guide to the Fishes of Colorado, pp. 1–110. Sponsored by: University of Colorado Museum, Colorado Agricultural and Mechanical College, Colorado Game and Fish Department, and United States Fish and Wildlife Service.

BECKWITH, LT. E. G. 1855. Report of explorations for a route for the Pacific Railroad of the line of the Forty-First Parallel of north latitude, by Lt. E. G. Beckwith, 1854: 1–132 IN Reports of explorations and surveys to ascertain the most practicable and economical route for a railroad from the Mississippi River to the Pacific Ocean, made under the direction of the Secretary of War, in 1853–4, according to Acts of Congress of March 3, 1853, May 31, 1854 and August 5, 1854. *Volume 2.* 33d Congress, 2d Session, Senate Executive Document No. 78. Washington, D. C., Beverley Tucker, printer.

BEHLE, WILLIAM H. 1942. Distribution and variation of the Horned larks (*Otocoris alpestris*) of western North America. University California Zoölogy Publication 46(3) : 205–316.

BELAND, RICHARD D. 1954. Report on the Fishery of the Lower Colorado River—the Lake Havasu Fishery. California Department Fish Game, Inland Fisheries Administrative Report No. 54–17, 42 pp. (mimeo).

BELDING, DAVID L. 1928. Water temperature and fish life. American Fisheries Society Transactions 58 : 98–105.

BELDING, D. V. 1929. The respiratory movements of a fish as an indicator of a toxic environment. American Fisheries Society Transactions 59 : 238–245.

BENDIRE, CHARLES. 1882 (1881). Notes on Salmonidae of the upper Columbia. United States National Museum Proceedings 1881, Vol. 4: 81–87.

BENSON, NORMAN G. 1953. Seasonal fluctuation in the feeding of brook trout in Pigeon River, Michigan. American Fisheries Society Transactions 83 : 76–83.

BENSON, SETH B. and ROBERT J. BEHNKE. 1961. *Salmo evermanni* a synonym of *Salmo clarkii henshawi*. California Fish and Game 47(3) : 257–259.

BERG, LEO S. 1947. Classification of fishes, both recent and fossil. J. W. Edwards, Ann Arbor, Michigan. Pages 87–517. (This is an offset reproduction of the 1940 Russian original and contains the Russian text on pages 87–345, the remainder being an English translation. Foreword by Karl F. Lagler.)

BERGMAN, R. 1942. Trout. William Penn Publishing Corporation, New York.

BERRY, E. W. 1927. Flora of the Esmeralda formation in western Nevada. United States National Museum Proceedings 72(23) : 1–15.

BERRY, FREDERICK H., MELVIN T. HUISH, and HAROLD MOODY. 1956. Spawning mortality of the Threadfin Shad, *Dorosoma petenense* (Günther), in Florida. Copeia 1956(3) : 192.

BESANA, GIUSEPPE. 1910. American Fishes in Italy. United States Bureau Fisheries Bulletin 28 : 947–954.

BILLINGS, W. DWIGHT. 1945. The plant associations of the Carson Desert region, western Nevada. Butler University Botanical Studies 7 : 89–123.

BISSONETTE, T. H. and J. WENDELL BURGER. 1940. Experimental modification of the sexual cycle of fish. Northeastern Fish Culturist's Meeting Abstracts 1940: 12.

BLACK, EDGAR C. 1940. The transport of oxygen by the blood of freshwater fish. Biological Bulletin 79 : 215–229.

1951. Respiration in fishes, *IN* Some aspects of the physiology of fish. Part III (by W. S. Hoar, V. S. Black and E. C. Black). University Toronto Studies, Biological Series No. 59. Ontario Fisheries Research Laboratory Publication No. 71: 91–111.

1953. Upper lethal temperatures of some British Columbia freshwater fishes. Fisheries Research Board Canada Journal 10(4) : 196–210.

BLACK, EDGAR C., F. E. J. FRY and VIRGINIA S. BLACK. 1954. The influence of carbon dioxide on the utilization of oxygen by some freshwater fish. Canadian Journal Zoology 32 : 408–420.

BLACK, EDGAR C., F. E. J. FRY and W. J. SCOTT. 1939. Maximum rates of oxygen transport for certain freshwater fishes. Anatomical Record 75, Supplement 80.

BLACKWELDER, ELIOT. 1931. Pleistocene glaciation in the Sierra Nevada and Basin Ranges. Geological Society America Bulletin 42: 865–922.

1934A. Origin of the Colorado River. Geological Society America Bulletin 45: 551–556.

1934B. Supplementary notes on Pleistocene glaciation in the Great Basin. Washington Academy Science Journal 24: 217–222.

1948. The Geological Background, pp. 3–16, IN The Great Basin, with emphasis on Glacial and Postglacial times. University Utah Bulletin 38(20) : 1–191 (Biological Series, Vol. 10, No. 7).

BLAKE, JAMES. 1872A. On the absence of a rim to the Great Basin to the north of the Pueblo Butte. California Academy Science Proceedings 4: 223–224.

1872B. Remarks on the topography of the Great Basin. California Academy Science Proceedings 4: 276–278.

BLAKE, WILLIAM P. 1857. Geological Report IN Routes to California, to connect with the routes near the Thirty-Fifth and Thirty-Second Parallels, explored by Lieut. R. S. Williamson, Corps Topographical Engineers, in 1853. Explorations and surveys to ascertain the most practicable and economical route for a railroad from the Mississippi River to the Pacific Ocean, 5(2) : xvi–310 (1856).

BÖHLKE, J. 1953. A catalogue of the type specimens of recent fishes in the Natural History Museum of Stanford University. Stanford Ichthyological Bulletin 5: 1–168.

BOLLMAN, CHARLES HARVEY. 1892. A review of the Centrarchidae, or freshwater sunfishes, of North America. Appendix 8 of United States Commissioner Fish Fisheries Report 1888: 557–579.

BOND, RICHARD M. 1940. Birds of Anaho Island, Pyramid Lake, Nevada. Condor 42(5) : 246–250.

BOULENGER, G. A. 1895A. Catalogue of the fishes in the British Museum. 2nd Edition. Vol. 1. Taylor and Francis, London.

1895B. Remarks on some cranial characters of the salmonoids. Zoölogical Society London Proceedings 1895 : 299–302.

BOUTWELL, JOHN M. 1933. The Salt Lake Region. 16th International Geological Congress, Guidebook 17—Excursion C–1 : v–149 (1933).

BRASCH, JOHN, JAMES McFADDEN and STANLEY KMIOTEK. 1958. The Eastern Brook Trout. Its life history, ecology and management. Wisconsin Conservation Department Publication 226: 1–11.

BREDER, C. M., JR. 1935. The reproductive habits of the common catfish, *Ameiurus nebulosus* (Le Sueur), with a discussion of their significance in ontogeny and phylogeny. Zoölogica 19: 143–185.

1936. The reproductive habits of the North American Sunfishes (Family Centrarchidae). New York Zoölogical Society Zoölogica 21(1) : 1–48.

1939. Variations in the nesting habits of *Ameiurus nebulosus* (Le Sueur). New York Zoölogical Society Zoölogica 24(3) : 367–378.

1947. A note on protective behavior in *Gambusia*. Copeia 1947(4) : 223–227.

1957. A note on preliminary stages in the fossilization of fishes. Copeia 1957 (2) : 132–135.

BREDER, C. M., JR. and F. HALPERN. 1946. Innate and acquired behavior affecting the aggregation of fishes. Physiological Zoölogy 19(2) : 154–190.

BREDER, C. M., JR. and PRISCILLA RASQUIN. 1950. A preliminary report on the role of the pineal organ in the control of pigment cells and light reactions in recent teleost fishes. Science 111(2871) : 10–12.

BREDER, C. M., JR. and JANET ROEMHILD. 1947. Comparative behavior of various fishes under differing conditions of aggregation. Copeia 1947(1) : 29–40.

BREHM, V. 1937. Zwei neue Moina-Formen aus Nevada, U. S. A. Zoologische Anzeiger 117: 91–96. Leipzig.

BRETT, J. R. 1944. Some lethal temperature relations of Algonquin Park fishes. University Toronto Studies, Biological Series No. 52, Ontario Fisheries Research Laboratory 63: 1–49.

1956. Some principles in the thermal requirements of fishes. Quarterly Review Biology 31(2) : 75–87.

BRIDGE, T. W. and G. A. BOULENGER. 1904. Fishes. Cambridge Natural History 7: 139–727.

BRIDGES, COLTON H. 1958. A compendium of the life history and ecology of the eastern brook trout. (*Salvelinus fontinalis* (Mitchill)). Massachusetts Division Fisheries Game, Fisheries Section, Fisheries Bulletin 23: 1–36.

BRIGGS, JOHN C. 1953. The behavior and reproduction of salmonid fishes in a small coastal stream. California Department Fish Game Fish Bulletin 94: 1–62.

BROECKER, WALLACE S. 1957. Evidence for a major climatic change close to 11,000 years B. P. Geological Society America (and others) Annual Meeting Program 1957: 33–34.

BROECKER, W. S., M. EWING and B. C. HEEZEN. 1960. Evidence for an abrupt change in climate close to 11,000 years ago. American Journal Science 258: 429–448.

BROECKER, W. S. and PHIL C. ORR. 1956. Late Wisconsin history of Lake Lahontan. Geological Society, etc. Program 1956 Annual Meetings Abstracts: 29–30.

1958. Radiocarbon chronology of Lake Lahontan and Lake Bonneville. Geological Society America Bulletin 69: 1009–1032.

BROECKER, W. S. and A. F. WALTON. 1959. Reevaluation of the salt chronology of several Great Basin lakes. Geological Society America Bulletin 70: 601–618.

BROWN, C. J. D. 1952. Spawning habits and early development of the mountain whitefish, *Prosopium williamsoni*, in Montana. Copeia 1952 (2) : 109–113.

BROWN, C. J. D. and CHARLES BUCK, JR. 1939. When do trout and grayling fry begin to take food. Wildlife Management Journal 3(2) : 134–140.

BROWN, C. J. D. and G. C. KAMP. 1942. Gonad measurements and egg counts of brown trout (*Salmo trutta*) from the Madison River, Montana. American Fisheries Society Transactions 72: 195–200.

BROWN, MARGARET E. 1946. The growth of brown trout (*Salmo trutta* L.). I. The factors influencing the growth of trout fry. Experimental Biology Journal 22: 118–129.

1957A. The physiology of fishes. Vol. I. Metabolism. Academic Press, Inc., New York. pp. xiii + 447.

1957B. The physiology of fishes. Vol. II. Behavior. Academic Press, Inc., New York. pp. xi + 526.

BRUES, CHARLES T. 1928. Studies on the fauna of hot springs in the Western United States and the biology of thermophilous animals. American Academy Arts Science Proceedings 63(4) : 139–228.

1932. Further studies on the fauna of North American hot springs. American Academy Arts Science Proceedings 67(7) : 185–303.

BRUNSON, ROYAL BRUCE. 1952. Egg counts of *Salvelinus malma* from the Clark's Fork River, Montana. Copeia 1952(3) : 196–197.

BRUNSON, ROYAL BRUCE, GORDON B. CASTLE and RALPH PIRTLE. 1952. Studies of Sockeye Salmon. Montana Academy Science Proceedings. 12: 35–43.

BUNGENBERG DEJONG, C. M. 1955. Cytological studies on *Salmo irideus*. Genetica 27(5–6) : 472–483.

BURDICK, MILTON E. and EDWIN L. COOPER. 1956. Growth rate, survival and harvest of fingerling Rainbow Trout planted in Weber Lake, Wisconsin. Wildlife Management Journal 20(3) : 233–239.

BURNETT, A. M. R. 1959. Electric fishing with pulsatory direct current. New Zealand Journal Science 2(1) : 46–56.
BUSS, KEEN and JAMES E. WRIGHT, JR. 1956. Results of species hybridization within the family Salmonidae. Progressive Fish-Culturist 18 (4) : 149–158.
CABLE, LOUELLA E. 1929. Food of bullheads. United States Commission Fish Report 1928, Doc. No. 1037 : 27–41.
CAINE, LOU S. 1949. North American freshwater sport fish. A. S. Barnes and Company, N. Y. pp. xii–212.
CALHOUN, ALEX J. 1940. Note on a hybrid minnow, *Apocope* X *Richardsonius*. Copeia 1940 : 142–143.
——— 1944. The food of the Black-spotted trout (*Salmo clarkii henshawi*) in two Sierra Nevada lakes. California Fish and Game 30(2) : 80–94.
CALL, RICHARD ELLSWORTH. 1884. On the Quaternary and Recent Mollusca of the Great Basin with descriptions of new forms. United States Geological Survey Bulletin 11 : 1–56.
——— 1899. Ichthyologia Ohiensis or natural history of the fishes inhabiting the River Ohio and its tributary streams by C. S. Rafinesque. A verbatim et literatim reprint of the original, with a sketch of the life, the ichthyologic work, and the ichthyologic bibliography of Rafinesque. Cleveland, the Burrows Brothers Co., pp. 1–175.
CARBINE, WILLIAM F. 1939. Observations on the spawning habits of centrarchid fishes in Deep Lake, Oakland County, Michigan. Fourth North American Wildlife Conference Transactions 1939 : 275–287.
CARL, G. CLIFFORD. 1936. Food of the coarse-scaled sucker (*Catostomus macrocheilus* Girard). Biology Board Canada Journal 3 : 20–25.
——— 1938. A spawning run of brown trout in the Cowichan River system. Fisheries Research Board Canada, Progress Report Pacific Biological Station No. 36 : 12–13.
——— 1949. The distribution of freshwater fishes in British Columbia. British Columbia Provincial Museum Report 1949 : B21–B23.
CARL, G. CLIFFORD and W. A. CLEMENS. 1948. The freshwater fishes of British Columbia. British Columbia Provincial Museum Handbook 5 : 1–132.
CARL, G. CLIFFORD, W. A. CLEMENS and C. C. LINDSEY. 1959. Idem. Third Edition, pp. 1–192.
CARLANDER, KENNETH D. 1950. Handbook of Freshwater Fishery Biology. Wm. C. Brown Company, Dubuque, Iowa. Pages v–281.
——— 1953. First supplement to handbook of freshwater fishery biology. Wm. C. Brown Co., Inc., Dubuque, Iowa.
CARPENTER, EVERETT. 1915. Ground water in southeastern Nevada. United States Geological Survey Water Supply Paper 365 : 1–86.
CARPENTER, EVERETT and F. O. YOUNGS. 1928. Soil Survey of the Las Vegas area, Nevada. United States Department Agriculture, Field Operations, Bureau Chemistry Soils, 25 : 201–245 (1923).
CARPENTER, RALPH G. and HILBERT R. SIEGLER. 1947. A sportsman's guide to the freshwater fishes of New Hampshire. New Hampshire Fish Game Commission, Concord. pp. 1–87.
CARY, W. M. See NEVADA, State of: BIENNIAL REPORTS FOR FISH AND GAME for the years 1887 and 1889.
CHANDLER, HARRY P. 1949. A new species of Stenelmis from Nevada (Coleoptera, Elmidae). Pan-Pacific Entomologist 25(3) : 133–136.
CHAPMAN, ROBERT H. 1906. The deserts of Nevada and Death Valley. National Geographic Magazine 17 : 483–497.
CHAPMAN, W. M. and E. QUISTORFF. 1938. The food of certain fishes of north central Columbia River drainage, in particular, young chinook salmon and steelhead trout. Washington Department Fisheries, Biological Report 37A : 1–14.

CHU, YUANTING T. 1935. Comparative studies on the scales and on the pharyngeals and their teeth in Chinese cyprinids, with particular reference to taxonomy and evolution. St. John's University Biological Bulletin 2: x–225.
CHURCHILL, E. P. and W. H. OVER. 1933. Fishes of South Dakota. South Dakota Department Fish and Game.
CHUTE, W. H. et al. 1948. A list of common and scientific names of the better known fishes of the United States and Canada. American Fisheries Society Special Publication 1: 1–45.
CLARK, W. O. and C. W. RIDDELL. 1920. Exploratory drilling for water and use of ground water for irrigation in Steptoe Valley, Nevada. United States Geological Survey Water Supply Paper 467: 1–70.
CLELAND, ROBERT GLASS. 1950. This Reckless Breed of Men. The Trappers and Fur Traders of the Southwest. Alfred A. Knopf, New York. Pages xv–361–xx.
CLEMENS, H. P. and K. E. SNEED. 1957. The spawning behavior of the channel catfish, *Ictalurus punctatus*. United States Fish Wildlife Special Science Report—Fisheries No. 219: 1–11.
CLEMENS, WILBERT A. 1928. The food of trout from the streams of Oneida County, New York State. American Fisheries Society Transactions, 58: 183–197.
1935–1949. Contributions to the life history of the Sockeye Salmon. Reports British Columbia Fish Department Papers 20–33.
CLEMENS, WILBERT A., JOHN R. DYMOND and N. K. BIGELOW. 1924. Food studies of Lake Nipigon fishes. Biological Board Canada, Progress Reports Pacific Station, No. 25: 103–165.
CLEMENS, WILBERT A., JOHN R. DYMOND, N. K. BIGELOW, F. B. ADAMSTONE and W. J. K. HARKNESS. 1923. The food of Lake Nipigon fishes. University Toronto Studies, Biological Series, No. 22; Ontario Fisheries Research Laboratory Publications, No. 16: 173–188.
CLEMENS, WILBERT A. and J. A. MUNRO. 1934. The food of the squawfish. Biological Board Canada, Progress Reports Pacific Station, No. 19: 3–4.
CLEMENS, WILBERT A. and G. V. WILBY. 1949. Fishes of the Pacific Coast of Canada. Fisheries Research Board Canada Bulletin 68: 1–368.
COBB, E. W. 1933. Results of trout tagging to determine migrations and results from plants made. American Fisheries Society Transactions 63: 308–318.
COCKERELL, THEODORE D. A. 1908. The fishes of the Rocky Mountain Region. University Colorado Studies 5: 159–179.
1913. Observations on Fish Scales. United States Bureau Fisheries Bulletin 32(779): 119–174.
COCKERELL, THEODORE D. A. and EDITH M. ALLISON. 1909. The scales of some American Cyprinidae. Biological Society Washington Proceedings 22: 157–163.
COGGESHALL, LOWELL T. 1924. A study of the productivity and breeding habits of the Bluegill, *Lepomis pallidus* (Mitchill). Indiana Academy Sciences Proceedings 33: 315–320.
COKER, ROBERT E. 1920. Progress in biological inquiries. Report Division Scientific Inquiry fiscal year 1920. United States Commission Fisheries Report 1920: 1–32.
1930. Studies of common fishes of the Mississippi River at Keokuk. United States Bureau Fisheries Bulletin 45: 141–225.
COLE, L. J. 1905. The German carp in the United States. United States Fish Commission Report 1904: 525–641.
COLE, R. B. 1913. Explorations for salines in Silver Peak Marsh. United States Geological Survey Bulletin 530: 330–346.

COLEMAN, GEORGE A. 1926. Conditions of existence of fish in Lake Tahoe and tributary streams. California Fish and Game 12(1) : 23–27.

1929. A biological survey of Salton Sea. California Fish and Game 15(3) : 218–227.

1930. A biological survey of Clear Lake, Lake County, California Fish and Game 16(3) : 221–227.

COOK, W. A. 1929. A brief summary of the work of the Bureau of Fisheries in the Lake Superior region. American Fisheries Society Transactions 59 : 56–62.

COOPER, EDWIN L. 1951. Validation of the use of scales of brook trout, *Salvelinus fontinalis*, for age determination. Copeia 1951(2) : 141–148.

1952. Body-scale relationship of brook trout, *Salvelinus fontinalis*, in Michigan. Copeia 1952(1) : 1–4.

1953A. Mortality rates of brook trout and brown trout in the Pigeon River, Otsego County, Michigan. Progressive Fish-Culturist 15(4) : 163–169.

1953B. Growth of brook trout (*Salvelinus fontinalis*) and brown trout (*Salmo trutta*) in the Pigeon River, Otsego County, Michigan. Michigan Academy Science, Arts, Letters Paper 38 : 151–163.

1953C. Periodicity of growth and change of condition of brook trout (*Salvelinus fontinalis*) in the three Michigan trout streams. Copeia 1953(2) : 107–114.

COOPER, EDWIN L. and NORMAN G. BENSON. 1951. The coefficient of condition of brook trout, brown and rainbow trout in the Pigeon River, Otsego County, Michigan. Progressive Fish-Culturist 14(4) : 181–192.

COOPER, J. G. 1868. Chapter on Zoology, pp. 434–501, of which pp. 487–498 deal with fishes, *IN* T. F. Cronise's "The Natural Wealth of California," pp. xvi–696. H. H. Bancroft and Co., San Francisco, California.

COOTS, MILLARD. 1956. The Yellow perch, *Perca flavescens* (Mitchill), in the Klamath River. California Fish and Game 42(3) : 219–228.

COPE, EDWARD DRINKER. 1871A. On the fishes of a freshwater Tertiary in Idaho, discovered by Capt. Clarence King. American Philosophical Society Proceedings 11 : 538–547 (1870).

1871B. Observations on the systematic relations of the fishes. American Association Advancement Science Proceedings 20 : 317–343.

1872A. On the Tertiary coal and fossils of Osino, Nevada. American Philosophical Society Proceedings 12 : 478–481. (Descriptions of *Amyzon mentalis* and *Trichophanes hians*)

1872B. Part VIII. Recent reptiles and fishes. Report on the reptiles and fishes obtained by the naturalists of the expedition, pp. 432–443. *IN* Part IV : Special Reports *of* Preliminary report of the United States Geological Survey of Wyoming and portions of contiguous territories (being a second [actually the fourth!] annual report of progress) conducted under the authority of the Secretary of the Interior, by F. V. Hayden, United States Geologist, 511 pages.

1872C. Part VI. Report on the recent reptiles and fishes of the survey, collected by Campbell Carrington and C. M. Dawes, pp. 467–476 *IN* Part IV : Zoology and Botany *of* Preliminary report of the United States Geological Survey of Montana, and portions of adjacent territories; being a fifth annual report of progress, by F. V. Hayden, United States Geologist, vi–538 pages (new species *Apocope carringtonii*, *Clinostomus hydrophlox* and *Salmo pleuriticus*).

1873. On the extinct vertebrata of the Eocene of Wyoming, pp. 545–649, *IN* United States Geological Survey Territories embracing portions of Montana, Idaho, Wyoming and Utah; being a report of progress of the explorations for the year 1872 (Hayden) Sixth Annual Report: xi + 844.

1874A. On the Plagopterinae and the Ichthyology of Utah. American Philosophical Society Proceedings 14 : 129–139 (reprinted in the Wheeler Surveys West of the 100th Meridian the same year) (new genus and species *Plagopterus argentissimus*, *Lepidomeda vittata*, etc.).

1874B. Report on the vertebrate paleontology of Colorado, pp. 427–533; of Part 2.—Special reports on Paleontology IN Annual Report of the United States Geological and Geographical Survey of the Territories, embracing Colorado, being a report of progress of the exploration for the year 1873, by F. V. Hayden, United States Geologist. Conducted under the authority of the Secretary of the Interior. Washington: Government Printing Office. 1874: xii–718 (mention of *Amyzon mentalis*).

1878. Article II.—Descriptions of fishes from the Cretaceous and Tertiary Deposits west of the Mississippi River. Pages 67–77 of Bulletin 1: 1–311 of United States Geological and Geographical Survey of the Territories (Hayden survey) Bulletin, Volume 4: 1–908 (mention of *Trichophanes hians* from Osino, Nevada on p. 73).

1879A. The Amyzon Tertiary beds. American Naturalist 13: 332.

1879B. The fishes of Klamath Lake, Oregon. American Naturalist 13: 784–785.

1883. On the fishes of the recent and Pliocene lakes of the western part of the Great Basin, and of the Idaho Pliocene lake. Academy Natural Science Philadelphia Proceedings 35: 134–166 (*Leucus olivaceous, L. dimidiatus, Siphateles vittatus* [all = *Siphateles bicolor obesus*], *Squalius galtiae* [= *Richardsonius egregius*] and *Chasmistes cujus*).

1884. The Vertebrata of the Tertiary formations of the West. Book I. United States Geological Survey of the Territories, Report Volume 3: 1–1009.

1893. Review of "The Report of the Death Valley Expedition." American Naturalist 27: 990–995.

COPE, E. D. and H. C. YARROW. 1875. Report upon the collections of fishes made in portions of Nevada, Utah, California, Colorado, New Mexico, and Arizona during the years 1871, 1872, 1873, and 1874. Chapter 6: 635–700 IN United States Army Engineer Department Report on the Geography and Geology of the Explorations and Surveys West of the 100th Meridian, in charge of George M. Wheeler. Volume 5 (Zoölogy): 1–1021 (new species *Catostomus fecundus*, new genus *Pantosteus*).

COWLES, RAYMOND B. 1934. Notes on the ecology and breeding habits of the desert minnow, *Cyprinodon macularius* Baird and Girard. Copeia 1934: 40–42.

CRAWFORD, DONALD R. 1925. Field characters identifying young salmonoid fishes in fresh waters of Washington. University Washington Publications Fisheries 1(2): 64–76.

CROSS, FRANK B. 1953. Nomenclature in the Pimephalinae, with special reference to the Bullhead minnow, *Pimephales vigilax perspecuus*. (Girard). Kansas Academy Science Transactions 56: 92–96.

CURTIS, BRIAN. 1935A. The Golden Trout of Cottonwood Lakes (*Salmo aquabonita* Jordan). American Fisheries Society Transactions 34: 259–265.

1935B. The Golden Trout of Cottonwood Lakes. California Fish and Game 21: 101–109.

1942. The general situation and the biological effects of the introduction of alien fishes into California waters. California Fish and Game 28(1): 1–8.

1949A. The warm-water game fishes of California. California Fish and Game 35(4): 255–273.

1949B. The life story of the fish. Harcourt, Brace and Co., New York. Pages xii–284.

CURTIS, BRIAN and JOHN C. FRASER. 1948. Kokanee in California. California Fish and Game 34(3): 111–114.

DARLINGTON, PHILIP J., JR. 1957. Freshwater fishes, Chapter 2: 39–127, IN Zoögeography: The Geographical Distribution of Animals (same author), pp. xi–675, 1957, John Wiley and Sons. (Good Summary, best at present (1958) of fish groups).

DAVID, LORE R. 1941. Leptolepis nevadensis, a new Cretaceous fish. Journal Paleontology 15(3) : 318–321.

1943. Miocene fishes from Southern California. Geological Society America Special Paper 43: xiii–193.

DAVIDSON, F. A. and S. J. HUTCHINSON. 1938. The geographic distribution and environmental limitations of the Pacific salmon (genus *Oncorhynchus*). United States Bureau Fisheries Bulletin 48(26) : 667–692.

DAVIDSON, PIRIE. 1919. A cestraciont spine from the Middle Triassic of Nevada. University California Geology Publications 11(4) : 433–435.

DAVIS, JACKSON. 1959. Management of Channel catfish in Kansas. University Kansas Museum Natural History Miscellaneous Publication 21: 1–56.

DAVIS, R. E. 1955. Heat transfer in the goldfish, *Carassius auratus*. Copeia 1955(2) : 207–209.

DAVIS, WILLIAM M. 1903. The mountain ranges of the Great Basin. Harvard University Museum Comparative Zoölogy Bulletin 42: 129–177 (also reprinted in Geographical Essays).

DEAN, BASHFORD. 1895. Fishes, living and fossil. Columbia University Biological Series 3: 1–300. Macmillan, New York.

1916. Bibliography of fishes. Authors: A–K. Enlarged and edited by C. R. Eastman. American Museum Natural History, New York, vol. I: xii–718.

1917. Bibliography of fishes. Authors: L–Z. Enlarged and edited by C. R. Eastman. Ibid II: 1–702.

1923. Bibliography of fishes. Indices, bibliographies, etc. Extended and edited by E. W. Gudger with the cooperation of A. W. Henn. Ibid III: xiv–707.

DeKAY, J. E. 1842. Natural History of New York. Part I, Zoölogy; Reptiles and Fishes. Albany, 1842.

DeLACY, ALLAN C. and W. MARKHAM MORTON. 1942. Taxonomy and habits of the charrs, *Salvelinus malma* and *Salvelinus alpinus*, of the Karluck drainage system. American Fisheries Society Transactions 72: 79–91.

DENCE, WILFORD A. 1928. A preliminary report on the trout streams of southwestern Cattaraugus Co., New York. Roosevelt Wild Life Bulletin (5) 1, Syracuse University.

DENYES, H. ARLISS and JEANNE M. JOSEPH. 1956. Relationships between temperature and blood oxygen in the Largemouth Bass. Wildlife Management Journal 20(1) : 56–64.

DILL, WILLIAM A. 1944. The fishery of the Lower Colorado River. California Fish and Game, 30: 109–211.

DILL, WILLIAM A. and LEO SHAPOVALOV. 1939A. An unappreciated California game fish, the Rocky Mountain Whitefish, *Prosopium williamsoni* (Girard). California Fish and Game 25(3) : 226–227.

1939B. California freshwater fishes and their possible use for aquarium purposes. California Fish and Game 25(4) : 313–324.

DIMICK, R. E. and FRED MERRYFIELD. 1945. The fishes of the Williamette River system in relation to pollution. Oregon State Engineering Experiment Station Bulletin Series No. 20: 1–58.

DIMICK, R. E. and D. C. MOTE. 1934. A preliminary survey of the food of Oregon trout. Oregon State Agricultural Experiment Station Bulletin 323: 1–23.

DINEEN, CLARENCE F. and PAUL S. STOKELY. 1957 (1956). The osteology of the Sacramento perch, *Archoplites interruptus* (Girard). Copeia 1956(4) : 217–230.

DINSMORE, A. H. 1934. Effect of heredity on the growth of brook trout. American Fisheries Society Transactions 64: 203–205.

DOBBIN, CATHERINE N. 1941. A comparative study of the gross anatomy of the air bladders of ten families of fishes of New York and other eastern states. Morphology Journal 68(1) : 1–30.

DORF, ERLING. 1959. Climatic changes of the past and present. University Michigan Museum Paleontology Contributions 13(8) : 181–210.
DOUGLAS, PHILIP A. 1952. Notes on the spawning of the Humpback sucker, *Xyrauchen texanus* (Abbott). California Fish and Game 38(2) : 149–155.
DREWRY, GEORGE E., EXALTON A. DELCO, JR. and CLARK HUBBS. 1958. Occurrence of the Amazon Molly, *Mollienesia formosa*, at San Marcos, Texas. Texas Journal Science 10(4) : 489–490.
DURRANT, STEPHEN D. 1935. A survey of the waters of the Humboldt National Forest. United States Bureau Fisheries Mimeographed Report, 33 pages.
DUWE, ARTHUR EDWARD. 1952. The embryonic origin of the gas bladder in the centrarchid fish *Lepomis macrochirus macrochirus*. Copeia 1952 (2) : 92.
    1955. The development of the gas bladder in the Green Sunfish, *Lepomis cyanellus*. Copeia 1955(2) : 92–95.
DYMOND, JOHN R. 1926. The fishes of Lake Nipigon. University Toronto Studies, Biological Series, No. 27; Ontario Fisheries Research Laboratory Publication 27 : 3–107.
    1928. Some factors affecting the production of lake trout (*Cristivomer namaycush*) in Lake Ontario. University Toronto Studies, Biological Series, No. 31; Ontario Fisheries Research Laboratory Publication 33 : 29–41.
    1936. Some freshwater fishes of British Columbia. Royal Ontario Museum Zoölogy Contribution 9 : 60–73.
    1955. The introduction of foreign fishes in Canada. International Association Theoretical Applied Limnology Proceedings 12 : 543–553.
DYMOND, JOHN R. and V. D. VLADYKOV. 1934. The distribution and relationship of the salmonoid fishes of North America and North Asia. Pacific Science Congress Proceedings 5 : 3741–3750.
EASTMAN, CHARLES R. 1917. Fossil fishes in the collection of the United States National Museum. United States National Museum Proceedings 52 : 235–304.
EATON, THEODORE H., JR. 1935. Evolution of the upper jaw mechanism in teleost fishes. Morphology Journal 58(1) : 157–172.
    1956. Notes on the olfactory organs in Centrarchidae. Copeia 1956(3) : 196–199.
EDDY, SAMUEL. 1957. How to know the freshwater fishes. William C. Brown Company, Dubuque, Iowa. pp. 1–253.
EDDY, SAMUEL and THADDEUS SURBER. 1947. Northern fishes, with special reference to the upper Mississippi Valley. Second edition. University Minnesota Press, Minneapolis, xii–276 pages. First edition, 1943.
EGE, R. and A. KROUGH. 1915. On the relation between the temperature and the respiratory exchange in fishes. International Review Hydrobiology Hydrography 7 : 48–55.
EICHER, GEORGE J., JR. 1946. Lethal alkalinity for trout in waters of low salt content. Wildlife Management Journal 10(2) : 82–85.
    1947. Trout and the Colorado River. Arizona Wildlife-Sportsman 4(4) : 7–11, 18–19, April.
EIGENMANN, CARL H. 1895. *Leuciscus balteatus* (Richardson), a study in variation. American Naturalist 29 : 10–25.
    1920. On the genera *Orestias* and *Empetrichthys*. Copeia 1920 : 103–106.
EIGENMANN, CARL H. and ROSA SMITH EIGENMANN. 1889. Description of a new species of *Cyprinodon*. California Academy Science Proceedings (2d series) 1 : 270 (*Cyprinodon nevadensis*).
    1891. *Cottus beldingi*, spec. nov. American Naturalist 25 : 1132.
    1893. Preliminary descriptions of new fishes from the Northwest. American Naturalist 27 : 151–154 (*Catostomus columbianus*, etc.).

ELLIOTT, BOB. 1950. The eastern brook trout. W. W. Norton and Company, Inc., N. Y. 242 pages.
ELLIS, MAX M. 1914. Fishes of Colorado. University Colorado Studies 11 (1) : 1–136.
ELLIS, MAX M., B. A. WESTFALL and MARION D. ELLIS. 1948. Determination of water quality. United States Department Interior Fish Wildlife Service Research Report 9: 1–122.
EL-ZARKA, SALAH EL-DIN. 1959. Fluctuations in the population of Yellow Perch, *Perca flavescens* (Mitchill), in Saginaw Bay, Lake Huron. United States Fish Wildlife Service Fishery Bulletin 151: 365–415 (from Fishery Bulletin of the Fish and Wildlife Service Volume 59).
EMBODY, G. C. 1921. Concerning high water temperatures and trout. American Fisheries Society Transactions 51: 58–64.
  1934. Relation of temperature to the incubation periods of eggs of four species of trout. American Fisheries Society Transactions 64: 281–292.
EVERHART, W. HARRY. 1950. Fishes of Maine. Department Inland Fisheries Game, State of Maine, Augusta. pp. 1–53.
EVERMANN, BARTON WARREN. 1892. A reconnaissance of the streams and lakes of western Montana and northwestern Wyoming. United States Fishery Commission Bulletin 11: 3–60 (1891).
  1906. The Golden Trout of the southern High Sierras. United States Bureau Fisheries Bulletin 25: 1–51.
  1916. Fishes of the Salton Sea. Copeia 1916: 61–63.
EVERMANN, BARTON WARREN and HAROLD C. BRYANT. 1919. California Trout. California Fish and Game 5(3) : 105–135.
EVERMANN, BARTON WARREN and H. W. CLARK. 1931. A distributional list of the species of freshwater fishes known to occur in California. California Division Fish Game Fish Bulletin 35: 1–67.
EVERMANN, BARTON WARREN and THEODORE D. A. COCKERELL. 1909. Descriptions of three new species of cyprinoid fishes. Biological Society Washington Proceedings 22: 185–187 (*Richardsonius* substituted for *Leuciscus* on basis of examination of European specimens of the latter genus).
EVERMANN, BARTON WARREN and WILLIAM C. KENDALL. 1898 (1897). Descriptions of new or little-known genera and species of fishes from the United States. United States Fish Commission Bulletin 1897, Vol. 17: 125–133.
EVERMANN, BARTON WARREN and CLOUDSLEY RUTTER. 1895. The fishes of the Colorado Basin. United States Fish Commission Bulletin 14: 473–486 (1894).
EVERMANN, BARTON WARREN and HUGH M. SMITH. 1896. The whitefishes of North America. United States Commissioner Fish Fisheries Report 1894: Appendix 4: 283–324.
EYCLESHYMER, A. C. 1901. Observations on the breeding habits of *Ameiurus nebulosus*. American Naturalist 35: 911–918.
F., G. A. 1874. Fish in the hot springs of Nevada. Scientific American 30(10) : 148–149.
FENNEMAN, NEVIN M. 1931. Physiography of western United States. McGraw-Hill Co., New York. 534 pages.
  1932. Physiographic history of the Great Basin. Pan-American Geology 57: 131–142.
FERGUSON, J. K. W. and EDGAR C. BLACK. 1941. The transport of CO_2 in the blood of certain freshwater fishes. Biological Bulletin 80: 139–152.
FETH, J. H. 1959. Reevaluation of the salt chronology of several Great Basin lakes: a discussion. Geological Society America Bulletin 70: 637–640.
FIELD, JOHN B., C. A. ELVEHJEM, and CHANCEY JUDAY. 1943. A study of the blood constituents of carp and trout. Biological Chemistry Journal 148(2) : 261–269.

FISH, F. F. 1937. Furunculosis in wild trout. Copeia 1937(1) : 37–40.
FISH, M. P. 1932. Contributions to the early life histories of sixty-two species of fishes from Lake Erie and its tributary waters. United States Bureau Fisheries Bulletin 47(10) : 293–398.
FISHER, W. F. 1949. What's Killing the Truckee River? Nevada Hunting and Fishing 1(4) : 7–8, 19–20.
FLEENER, GEORGE G. 1952. Life history of the Cutthroat Trout, *Salmo clarki* Richardson, in Logan River, Utah. American Fisheries Society Transactions 81 : 235–248 (1951).
FLINT, R. F. 1947. Glacial geology and the Pleistocene epoch. John Wiley and Sons, Inc., New York.
FOERSTER, R. EARLE. 1925. Studies in the ecology of the sockeye salmon (*Oncorhynchus nerka*). Canadian Biologist 2(16) : 335–422.
　　1947. Experiment to develop sea-run from land-locked sockeye salmon (*Oncorhynchus nerka kennerlyi*). Fisheries Research Board Canada Journal 7(2) : 88–93.
FOERSTER, R. E. and A. L. PRITCHARD. 1934. The identification of the young of the five species of Pacific Salmon, with notes on the freshwater phase of their life history. Commissioner Fisheries Province British Columbia Report, pp. 103–105.
FOERSTER, R. E. and W. E. RICKER. 1941. The effect of reduction of predaceous fish on survival of young sockeye salmon at Cultus Lake. Fisheries Research Board Canada Journal 5(4) : 315–336.
FORBES, STEPHEN ALFRED. 1878. The food of Illinois fishes. Illinois State Laboratory Natural History Bulletin 1(2) : 71–89.
　　1880A. The food of fishes. Illinois State Laboratory Natural History Bulletin 1(3) : 18–65.
　　1880B. On the food of young fishes. Illinois State Laboratory Natural History Bulletin 1(3) : 66–79.
　　1883. The food of the smaller freshwater fishes. Illinois State Laboratory Natural History Bulletin 1(6) : 65–94.
　　1888A. Studies of the food of freshwater fishes. Illinois State Laboratory Natural History Bulletin 2 : 443–473.
　　1888B. On the food relations of freshwater fishes; a summary and discussion. Illinois State Laboratory Natural History Bulletin 2 : 475–538.
　　1888C. The food of the fishes of the Mississippi Valley. American Fisheries Society Transactions 17 : 1–17.
FORBES, STEPHEN A. and R. E. RICHARDSON. 1920. The fishes of Illinois. Illinois Natural History Division, cxxxvi–357 pages. Second Edition. (1st ed., 1908).
FOWLER, HENRY W. 1913A. Notes on catostomid fishes. Academy Natural Science Philadelphia Proceedings 65: 45–66 (new subgenus *Pithecomyzon* proposed for *Chasmistes cujus*).
　　1913B. Some type-specimens of the American cyprinoid fishes of the genus *Rutilus*. Academy Natural Sciences Philadelphia Proceedings 65 : 66–71.
FRANKE, JOHANN. 1910. Causes of Degeneration of American Trouts in Austria. United States Bureau Fisheries Bulletin 28 : 983–989.
FRANTZ, THEODORE C. 1958. Railroad Valley Ponds—Series A : 1–9, maps and illustrations. Nevada Fish and Game Commission Stream Lake Surveys.
FRANTZ, THEODORE C. and DONALD J. KING. 1958. Completion Report. Nevada Fish and Game Commission Stream Lake Survey, xiii–188.
FRASER, JACK C. and A. F. POLLITT. 1951. The introduction of kokanee red salmon (*Oncorhynchus nerka kennerlyi*) into Lake Tahoe, California and Nevada. California Fish and Game 37(2) : 125–127.
FREE, E. E. 1913. Progress in potash prospecting in Railroad Valley, Nevada. Mining Scientific Press 107 : 176–178.

1914. The topographic features of the desert basins of the United States with reference to the possible occurrence of potash. United States Department Agriculture Bulletin 54: 1–65.

FRÉMONT, JOHN CHARLES (Brevét Captain). 1845. Report of the exploring expeditions to the Rocky Mountains in the year 1842, and to Oregon and Northern California in the years 1843–1844. Printed by order of the Senate of the United States. Washington: Gale and Seaton, printers, 583 pages.

1849. Geographical Memoir upon upper California, in illustration of his map of Oregon and California. 30th Congress, 2nd Session, House of Representatives Miscellaneous No. 5: 1–40. Washington: Tippin and Streeper.

FREY, DAVID G. 1940. Growth and ecology of the carp, *Cyprinus carpio* Linnaeus, in four lakes of the Madison region, Wisconsin. Ph.D. Thesis, University of Wisconsin.

FRY, F. E. J. 1939. The position of fish and other higher animals in the economy of lakes. American Advancement Science Publication 10: 132–142.

1947. Effects of the environment on animal activity. University Toronto Studies, Biological Series No. 55. Ontario Fisheries Research Laboratory Publication No. 68: 1–62.

1949. Lake trout in our inland waters. Sylva, Lands and Forests Review, 5(3): 3–13 (Canada).

1951. Some environmental relations of the speckled trout (*Salvelinus fontinalis*). Report Proceedings Northeastern Atlantic Fisheries Conference.

1958. Temperature Compensation. Annual Review Physiology 20: 207–224.

FRY, F. E. J. and EDGAR C. BLACK. 1938. The influence of carbon dioxide on the utilization of oxygen by certain species of fish in Algonquin Park, Ontario. Anatomical Record 72: Supplement 47.

FRY, F. E. J., VIRGINIA S. BLACK and EDGAR C. BLACK. 1947. Influence of temperature on asphyxiation of young goldfish (*Carassius auratus* L.) under various tensions of oxygen and carbon dioxide. Biological Bulletin 92: 217–224.

FRY, F. E. J. and V. B. CHAPMAN. 1948. The lake trout fishery in Algonquin Park from 1936 to 1945. American Fisheries Society Transactions 75(1945): 21–35.

FRY, F. E. J. and M. B. GIBSON. 1953. Lethal temperature experiments with Speckled trout x Lake trout hybrids. Journal Heredity 44(2): 56–57.

FRY, F. E. J., J. S. HART, and K. F. WALKER. 1946. Lethal temperature relations for a sample of young speckled trout, *Salvelinus fontinalis*. University Toronto Studies, Biological Series, No. 54. Ontario Fisheries Research Laboratory Publication No. 66: 9–35.

FUNK, JOHN L. 1957. Movement of stream fishes in Missouri. American Fisheries Society Transactions 85(1955): 39–57.

GABRIELSON, IRA N. and FRANCESCA LaMONTE. 1954. The Fisherman's Encyclopedia. The Stackpole Company, Harrisburg, Pennsylvania. Pages xxiv–730.

GALE, HOYT S. 1913A. Notes on the Quaternary Lakes of the Great Basin, with special reference to the deposition of potash and other salines. United States Geological Survey Bulletin 540–N: 399–406.

1913B. Potash tests at Columbus Marsh, Nevada. United States Geological Survey Bulletin 540–N: 422–427.

1915. Geological history of Lake Lahontan. Science (new series) 41: 209–211.

GARDNER, J. A. and G. KING. 1922. Respiratory exchange in freshwater fish, Part IV. Further comparison of goldfish and trout. Biochemical Journal 16: 729–735.

GARDNER, J. A., G. KING and E. B. POWERS. 1922. The respiratory exchange in freshwater fish. III. Goldfish. Biochemical Journal 16: 523–529.

GARDNER, J. A. and C. LEETHAM. 1914A. On the respiratory exchange in freshwater fish. I. On brown trout. Biochemical Journal 8: 374–390.

1914B. On the respiratory exchange in freshwater fish. II. On brown trout. Biochemical Journal 8: 591–597.

GARLICK, LEWIS R. 1948. A discussion of fisheries problems on the Truckee River and suggestions for future management. Typewritten report, 9 pages.

GARMAN, SAMUEL. 1895. The Cyprinodonts. Harvard College Museum Comparative Zoölogy Memoir 19: 1–170.

GARSIDE, E. T. and J. S. TAIT. 1958. Preferred temperature of rainbow trout (*Salmo gairdneri* Richardson) and its unusual relationship to acclimation temperature. Canadian Journal Zoölogy 36: 563–567.

GERKING, SHELBY D. 1945. The distribution of the fishes of Indiana. Investigations Indiana Lakes Streams 3: 1–137.

GIBBONS, W. P. 1855. Description of a new trout. California Academy Science Proceedings 1: 36–37 (2d ed., pp. 35–36, 1873) (*Salmo iridea* = *Salmo gairdneri irideus*).

GIBBS, EARL D. 1956. A bisexual steelhead. California Fish and Game 42(3): 229–231.

GIBSON, E. S. and F. E. J. FRY. 1954. The performance of the lake trout, *Salvelinus namaycush*, at various levels of temperature and oxygen pressure. Canadian Journal Zoölogy 32: 252–260.

GILBERT, CHARLES H. 1893. Report on the fishes of the Death Valley expedition collected in southern California and Nevada in 1891, with descriptions of new species. United States Department Agriculture, Bureau Biological Survey, North American Fauna No. 7: 229–234 (*Empetrichthys merriami*, *Cyprinodon macularius baileyi* = [*Crenichthys baileyi*], *Rhinichthys nevadensis* and *R. velifer*).

1898. The fishes of the Klamath Basin. United States Fish Commission Bulletin 17: 1–13 (discusses the relationships between the Lahontan, Bonneville and Klamath basins) (1897).

1913. Age at maturity of the Pacific Coast salmon of the genus Oncorhynchus. United States Bureau Fisheries Bulletin 32: 1–22 (1912 (1914)).

1914–1921. Contributions to the life history of the sockeye salmon. British Columbia Commission Fisheries Reports, Papers 1 to 7.

GILBERT, CHARLES H. and BARTON WARREN EVERMANN. 1895. A report upon investigations in the Columbia River basin, with descriptions of four new species of fish. United States Fish Commission Bulletin 14: 169–202 (1894).

GILBERT, CHARLES H. and WILLIS H. RICH. 1927. Investigations concerning the Red-Salmon runs to the Karluk River, Alaska. United States Bureau Fisheries Bulletin 43(11), Document 1021: 1–69.

GILBERT, CHARLES H. and N. B. SCHOFIELD. 1898. Notes on a collection of fishes from the Colorado Basin in Arizona. United States National Museum Proceedings 20: 487–499.

GILBERT, GROVE KARL. 1875. Report on the geology of portions of Nevada, Utah, California and Arizona, examined in the years 1871 and 1872. United States Army Engineer Department Report on the Geography and Geology of the Explorations and surveys West of the 100th Meridian, in charge of George M. Wheeler. Volume 3: 17–187.

1890. Lake Bonneville. United States Geological Survey Monograph I: xx–438.

1928. Studies of Basin-Range structure. United States Geological Survey Professional Paper 153: vii–92.

GILL, THEODORE. 1861. Description of a new species of the genus *Tigoma* of Girard (abridged from the forthcoming report of Capt. J. H. Simpson). Boston Society Natural History Proceedings 8: 42 (*Tigoma squamata = Gila atraria*).

1865. Synopsis of the genus *Pomoxys*, Raf. Academy Natural Science Philadelphia Proceedings 17: 64.

1876. Report on Ichthyology, Appendix L: 383–431, of Explorations across the Great Basin of the Territory of Utah for a direct wagon route from Camp Floyd to Genoa, in Carson Valley, in 1859, by Capt. J. H. Simpson, Corps of Topographical Engineers, United States Army.

1878A. Credited, with Jordan, for the description of *Catostomus tahoensis* IN Jordan, United States National Museum Bulletin No. 12, which see.

1878G. Credited, with Jordan, for the description of *Salmo henshawi* IN Jordan's Manual of the Verbetrates, which see.

1882. Bibliography of the fishes of the Pacific Coast of the United States to the end of 1879. United States National Museum Bulletin 11: 1–73.

1894. The nomenclature of the family Poeciliidae or Cyprinodontidae. United States National Museum Proceedings 17: 115–116.

1895. The differential characters of the Salmonidae and Thymallidae. United States National Museum Proceedings 17: 117–122.

1905. The family of cyprinids and the carp as its type. Smithsonian Miscellaneous Collections 48(2): 195–217 (No. 1591).

1908. The Millers-thumb and its habits. Smithsonian Miscellaneous Collections 52: 101–116.

GIRARD, CHARLES. 1850. A monograph of the freshwater *Cottus* of North America. American Association Advancement Science Proceedings 2: 409–411.

1854 (1856). Description of new fishes, collected by Dr. A. L. Heermann, naturalist attached to the survey of the Pacific Railroad Route, under Lt. R. S. Williamson, U. S. Army. Academy Natural Science Philadelphia Proceedings 7: 129–140 (*Centrarchus interruptus = Archoplites*).

1856A (1857). Contributions to the Ichthyology of the western coast of the United States, from specimens in the museum of the Smithsonian Institution. Academy Natural Science Philadelphia Proceedings 8: 131–137 (1855) (*Coregonus williamsoni*, etc., and a general discussion).

1856B (1857). Researches upon the cyprinoid fishes inhabiting the fresh waters of the United States of America, west of the Mississippi Valley, from specimens in the museum of the Smithsonian Institution. Academy Natural Science Philadelphia Proceedings 8: 165–213 (1856) (with an instructive discussion, enumerating the major western surveys and collectors and describing numerous new species, those of importance to Nevada being: *Catostomus macrocheilus, Algansea bicolor* and *A. obesa* [*Siphateles*], *Argyreus nubilus* and *A. osculus* [*Rhinichthys*], *Tigoma humboldti* (undecipherable except for a misplaced *Richardsonius balteatus*), *Siboma atraria* [*Gila*] and *Ptychocheilus lucius*).

1856C (1857). Notice upon the species of the genus *Salmo* of authors, observed chiefly in Oregon and California. Academy Natural Science Philadelphia Proceedings 8: 217–220 (1856) (*Salar lewisi = Salmo*).

1857. Report upon fishes collected on the survey, No. 1: 9–34, in Part IV, Zoölogical Report, of Routes in California and Oregon explored by Lt. R. S. Williamson, Corps of Topographical Engineers, and Lt. Henry L. Abbott, Corps of Topographical Engineers, in 1855 IN Reports of explorations and surveys to ascertain the most practicable and economical route for a railroad from the Mississippi River to the Pacific Ocean, made under the direction of the Secretary of War, in 1854–1855, according to Acts of Congress of March 3, 1853, May 31, 1854, and August 5, 1854. *Volume 6.* 33rd Congress 2d Session, Senate Executive Document No. 78. Washington, D. C. Beverley Tucker, Printer.

1859. (a) Fishes, of the general report upon the zoölogy of the several Pacific Railroad routes—Part IV: xiv–400 (*1858*) (*Tigoma egregia = Richardsonius*).

(b) Report on fishes collected on the survey, No. 4: 21–27; Route near the 38th and 39th parallels, explored by Captain J. W. Gunnison, and near the 41st parallel, explored by Lt. E. G. Beckwith (*1857*).

(c) Report upon fishes collected on the survey, No. 5: 47–59; Route near the thirty-fifth parallel, explored by Lt. A. W. Whipple, Topographical Engineers, in 1853 and 1854 (*1859*).

(d) Report on fishes collected on the survey, No. 4: 83–91; Routes in California, to connect with the routes near the thirty-fifth and thirty-second parallels, explored by Lt. R. S. Williamson, Corps of Topographical Engineers in 1853 (*1859*).

ALL IN

Reports of explorations and surveys to ascertain the most practicable and economical route for a railroad from the Mississippi River to the Pacific Ocean, made under the direction of the Secretary of War, in 1853–1856, according to acts of Congress of March 3, 1853, May 31, 1854, and August 5, 1854. Volume 10. 33d Congress, 2d Session, Senate Executive Document No. 78. Washington, D. C. Beverley Tucker, printer. (The reports A, B, C, and D are listed in order as they appear in Volume 10, although they are not so lettered. All have their own pagination, and all were typeset in the years indicated at the ends, the entire volume appearing in 1859. The major report is (a), the others being minor and Girard referred the reader back to the major report in each instance.)

GOODE, G. BROWN. 1888. American Fishes; a popular treatise upon the game and food fishes of North America. Standard Book Co., New York. 496 pp. (New ed. Dana Estes and Co., pp. 562, 1903).

1891. The published writings of Dr. Charles Girard, Part V, of the Bibliographies of American Naturalists. United States National Museum Bulletin 41: vi–141.

GOODE, G. BROWN and THEODORE GILL. 1903. American Fishes. L. C. Page and Co., Boston. pp. lxviii–562.

GRAHAM, J. M. 1949. Some effects of temperature and oxygen pressure on the metabolism and activity of the Speckled trout, *Salvelinus fontinalis*. Canadian Research Journal 27: 270–288.

GRAINGER, E. H. 1953. On the age, growth, migration, reproduction potential and feeding habits of the arctic char (*Salvelinus alpinus*) of Frobisher Bay, Baffin Island. Fisheries Research Board Canada Journal 10: 326–370.

GREELEY, J. R. 1930. Fishes of the Lake Champlain watershed *IN* A biological survey of the Champlain watershed. New York Conservation Department Supplement to 19th Annual Report, 1929: 44–87.

1932. The spawning habits of brook, brown and rainbow trout and the problem of egg predators. American Fisheries Society Transactions 62: 239–248.

GREENE, C. WILLARD. 1935. The distribution of Wisconsin fishes. Wisconsin Conservation Commission, pp. 1–235.

GREGORY, WILLIAM K. 1933. Fish Skulls: a study of the evolution of natural mechanisms. American Philosophical Society Transactions New Series, 23, Article II: vii–, 75–481.

GRINNELL, JOSEPH. 1914. The Colorado River as a highway of dispersal and center of differentiation of species. University California Zoölogy Publications 12: 97–100.

GROVES, KENNETH ELLISON. 1951. Fishes of Moses Lake, Washington. Walla Walla College Publications Department Biological Sciences and Biological Station 1: 1–22.

GUERNSEY, J. E., JAMES KOEBER, C. J. ZINN and E. C. ECKMAN. 1919. Soil survey of the Honey Lake area, California. United States Department Agriculture, Field Operations, Bureau Soils 17: 2255–2314 (1915).

GÜNTHER, ALBERT. 1859. Catalogue of the fishes in the British Museum. Volume 1 : xxxix–524.

1860. Catalogue of the fishes in the British Museum. Volume 2 : xxi–548.

1861. Catalogue of the fishes in the British Museum. Volume 3 : xxv–586.

1862. Catalogue of the fishes in the British Museum. Volume 4 : xxi–534.

1864. Catalogue of the fishes in the British Museum. Volume 5 : xxii–455.

1866. Catalogue of the fishes in the British Museum. Volume 6 : xv–368.

1868. Catalogue of the fishes in the British Museum. Volume 7 : xx–512.

1870. Catalogue of the fishes in the British Museum. Volume 8 : xxv–549.

GURLEY, R. R. 1902. The habits of fishes. American Psychology Journal 13 : 408–425.

HADERLIE, EUGENE CLINTON. 1953. Parasites of the freshwater fishes of northern California. University California Publications Zoölogy 57(5) : 303–440.

HAGERMAN, F. B. 1951. An easy method of separating King and Silver Salmon. California Fish and Game 37(1) : 53–54.

HAGUE, ARNOLD. 1877. Truckee River region *IN* Descriptive Geology, United States Geological Exploration of the 40th Parallel, Volume 2 : 817–853. Washington, D. C.

1892. Geology of the Eureka District, Nevada. With an atlas. United States Geological Survey Monograph 20 : xvii–419.

HALL, E. RAYMOND. 1925. Pelicans versus fishes in Pyramid Lake. Condor 27 : 147–160.

1946. Mammals of Nevada. University California Press, 710 pages, Berkeley, California.

HALL, G. E. and R. M. JENKINS. 1952. The rate of growth of channel catfish in Oklahoma waters. Oklahoma Fisheries Research Laboratory Report 27 : 1–15.

HANKINSON, THOMAS L. 1927. A preliminary survey of the fish life of Allegany State Park in 1921. Roosevelt Wild Life Bulletin 4(3).

HANSEN, DONALD F. 1943. On nesting of the White Crappie, *Pomoxis annularis*. Copeia 1943(4) : 259–260.

1951. Biology of the White crappie in Illinois. Illinois Natural History Survey Bulletin 25(4) : 211–265.

HARDING, S. T. 1935. Changes in lake levels in the Great Basin area. Civil Engineer 5(2) : 87–90.

HARDMAN, GEORGE. 1948. The precipitation of Nevada. University Nevada Agricultural Experiment Station Department Irrigation Mimeograph Report, pp. 1–8.

HARDMAN, GEORGE and HOWARD G. MASON. 1949. Irrigated lands of Nevada. University Nevada Agricultural Experiment Station Bulletin 183 : 1–57.

HARDMAN, GEORGE and MEREDITH R. MILLER. 1934. The quality of the waters of southeastern Nevada, drainage basins and resources. University Nevada Agricultural Experiment Station Bulletin 136 : 1–62.

HARDMAN, GEORGE and ORVIL E. REIL. 1936. Relationship between tree growth and stream runoff in the Truckee River Basin, California-Nevada. University Nevada Agricultural Experiment Station Bulletin 141 : 1–38.

HARDMAN, GEORGE and CRUZ VENSTROM. 1941. A 100-year record of Truckee River runoff estimated from changes in levels and volumes of Pyramid and Winnemucca Lakes. American Geophysical Union Transactions 1941(1) : 71–90.

HARLAN, JAMES R. and EVERETT B. SPEAKER. 1951. Iowa Fish and Fishing. Iowa Conservation Commission, pp. 1–237.

1956. Iowa Fish and Fishing. Iowa Conservation Commission, pp. x plus 377 (3rd Edition).

HARMS, CLARENCE E. 1959. Checklist of parasites from catfishes of northeastern Kansas. Kansas Academy Science Transactions 62(4) : 262.

HARRINGTON, M. R. 1930. Archaeological explorations in southern Nevada. Southwest Museum Paper 4 : 1–25.

HARRINGTON, ROBERT W., JR. 1947. Observations on the breeding habits of the Yellow Perch, *Perca flavescens* (Mitchill). Copeia 1947(3) : 199–200.

— 1955. The osteocranium of the American cyprinid fish, *Notropis bifrenatus*, with an annotated synonymy of teleost skull bones. Copeia 1955(4) : 267–290.

HARRY, ROBERT R. 1951. The embryonic and early larval stages of the tui chub, *Siphateles bicolor* (Girard), from Eagle Lake, California. California Fish and Game 37(2) : 129–132.

HART, J. S. 1943. The cardiac output of four freshwater fish. Canadian Journal Research 21 : 77–84.

— 1952. Geographic variations of some physiological and morphological characters in certain freshwater fish. University Toronto Biology Series No. 60. Ontario Fisheries Research Laboratory Publication No. 72: 1–79.

HASKELL, DAVID C., LOUIS E. WOLF and LOYAL BOUCHARD. 1956. The effect of temperature on growth of brook trout. New York Fish and Game Journal 3(1) : 108–113.

HASKELL, WILLIAM L. 1959. Diet of the Mississippi Threadfin Shad, *Dorosoma petenense atchafalayae*, in Arizona. Copeia 1959(4) : 298–302.

HASLER, ARTHUR D. 1938. Fish biology and limnology of Crater Lake, Oregon. Journal Wildlife Management 2 : 92–103.

— 1956. Perception of pathways by fishes in migration. Quarterly Review Biology 31(3) : 200–209.

HASSELMAN, RONALD and ROBERT GARRISON. 1957. Studies on the Squawfish *Ptychocheilus oregonense* in Lookout Point and Dexter Reservoirs, 1957. Mimeographed Report prepared under supervision of the United States Fish and Wildlife Service and Oregon State College and issued by the U. S. Fish and Wildlife Service Regional Office at Portland, Oregon. pp. 1–41.

HATTON, S. ROSS. 1932. The fish fauna of Utah Lake. Master's thesis, Brigham Young University, Provo, Utah. pp. iii–64.

HAY, OLIVER P. 1907. A new fossil stickleback fish from Nevada. United States National Museum Proceedings 32: 271–273 (*Gasterosteus williamsoni leptosomus* = Jordan's *Merriamella doryssa*).

— 1927. The Pleistocene of the western region of North America and its vertebrated animals. Carnegie Institution Washington Publication 322B: 1–346.

HAYES, SHELDON PHIPPS. 1935. A taxonomical, morphological, and distributional study of the Utah Cyprinidae. Master's thesis, Brigham Young University, pp. iv–115.

HAYS, ALICE N. 1952. David Starr Jordan. A bibliography of his writings, 1871–1931. Stanford University Publications, University Series, Library Studies, Vol. 1 : xiii + 195.

HAZZARD, A. S. 1932. Some phases of the life history of the eastern brook trout, *Salvelinus fontinalis* Mitchill. American Fisheries Society Transactions 62(1932) : 344–350.

— 1933. Low water temperature, a limiting factor in the successful production of trout in natural waters. American Fisheries Society Transactions 63 : 204–207.

— 1936. A preliminary study of an exceptionally productive trout water, Fish Lake, Utah. American Fisheries Society Transactions 65(1935) : 122–128.

HAZZARD, A. S. and M. J. MADSEN. 1933. Studies of the food of the cutthroat trout. American Fisheries Society Transactions 63 : 198–203.

HEACOX, C. 1937. Introduction of the brown trout. Progressive Fish-Culturist 30: 39–40.
HENSHALL, JAMES A. 1881. Book of the Black Bass. Comprising its complete scientific and life history together with a practical treatise on angling and fly fishing and a full description of tools, tackle and implements, pp. vii–463. Clarke, Cincinnati (2d ed., 1904).
1889. More about the Black Bass. Being a supplement to the Book of the Black Bass, pp. 1–204. Cincinnati.
HENSHALL, J. A. 1903. Bass, pike, perch and others, pp. 410. Macmillan, New York.
HESS, ARCHIE D. and J. H. RAINWATER. 1939. A method for measuring the food preference of trout. Copeia 1939(3) : 154–157.
HESSEL, RUDOLPH. 1878. Carp and its culture in rivers and lakes; and its introduction in America. United States Fish Commission Report 1875–1876: 865–900.
HICKS, W. B. 1915. The composition of muds from Columbus Marsh, Nevada. United States Geological Survey Professional Paper 95A: 1–11.
HILDEBRAND, SAMUEL F. 1917. Notes on the life history of the minnows *Gambusia affinis* and *Cyprinodon variegatus*. United States Commission Fisheries Annual Report, Appendix 6, Document No. 857: 3–15.
1921. Top minnows in relation to malaria control, with notes on their habits and distribution. United States Public Health Bulletin 114: 3–34.
1955. The trouts of North America, general remarks on classification. United States Fish Wildlife Service, Fishery Leaflet 355 (reissued).
HILDEBRAND, SAMUEL F. and I. L. TOWERS. 1928. Annotated list of fishes collected in the vicinity of Greenwood, Mississippi, with descriptions of three new species. United States Bureau Fisheries Bulletin 43(1027) : 105–136.
HILE, RALPH. 1950. A nomograph for the computation of the growth of fish from scale measurements. American Fisheries Society Transactions 78: 156–162 (1948).
HINDS, NORMAN E. A. 1943. Geomorphology. The Evolution of Landscape. Prentice Hall, New York, pp. xi–894.
HINKS, DAVID. 1943. The fishes of Manitoba. Department Mines Natural Resources, Province Manitoba, Winnipeg. pp. x plus 117 (this is pagination for 1947 reprint with a supplement by J. J. Keleher and B. Kooyman).
HOBBS, DERISLEY F. 1937. Natural reproduction of Quinnat salmon, brown and rainbow trout in certain New Zealand waters. New Zealand Marine Department Fisheries Bulletin 6: 7–104.
HOFSTEETER, A. M., PAUL S. STOKELY and C. F. DINEEN. 1957. The auditory organ and its relation to the skull bones of a freshwater teleost. Anatomical Record 129(1) : 79–82.
1958. Skeletal differences between the black and white crappies. American Microsopical Society Transactions 77(1) : 19–21.
HOLTON, GEORGE D. 1953. A trout population study on a small creek in Gallatin County, Montana. Wildlife Management Journal 17(1) : 62–82.
HOOPER, E. J. 1875. Lake Tahoe, California, its scenery and trout fishing. Forest and Stream 5: 151.
HOOVER, EARL E. 1939. Age and growth of brook trout in northern breeder streams. Journal Wildlife Management 3: 81–91.
HOOVER, EARL E. and H. E. HUBBARD. 1937. Modification of the sexual cycle in trout by control of light. Copeia 1937(4) : 205–210.
HUBBS, CARL L. 1924. Studies of the fishes of the Order Cyprinodontes. II. An analysis of the genera of the Poeciliidae. University Michigan Museum Zoölogy Miscellaneous Publication 13.
1926A. The structural consequences of modifications of the developmental rate in fishes, considered in reference to certain problems of evolution. American Naturalist 60: 57–81.

1926B. Studies of the fishes of the Order Cyprinodontes. VI. Materials for a revision of the American genera and species. University Michigan Museum Zoölogy Miscellaneous Publication 16: 1–87.

1930. The specific name of the European trout, Salmo trutta L. Copeia 1930(172): 86–89.

1932. Studies of the fishes of the Order Cyprinodontes. XII. A new genus related to *Empetrichthys*. University Michigan Museum Zoölogy Occasional Paper 252: 1–5 (*Crenichthys nevadae*).

1940. Speciation of fishes. American Naturalist 74: 198–211.

1941A. Fishes of the Desert. Biologist 22: 61–69 (1940).

1941B. The relation of hydrological conditions to speciation in fishes IN A symposium on Hydrobiology. University Wisconsin Press, pp. 182–195. Madison, Wisconsin.

1948. Changes in the fish fauna of western North America correlated with changes in ocean temperature. Marine Research Journal 7(3): 459–482.

1955. Hybridization between fish species in nature. Systematic Zoölogy 4(1): 1–20.

1961. Isolating mechanisms in the speciation of fishes. Reprint, pages 5–23, from Vertebrate Speciation, a University of Texas Symposium.

HUBBS, CARL L. and REEVE M. BAILEY. 1940. A revision of the black basses (*Micropterus* and *Huro*) with descriptions of four new forms. University Michigan Museum Zoölogy Miscellaneous Publication 48: 1–51.

HUBBS, CARL L. and LAURA C. HUBBS. 1947. Natural hybrids between two species of catostomid fishes. Michigan Academy Science, Arts Letters Paper 31: 147–167.

HUBBS, CARL L., LAURA C. HUBBS and RAYMOND E. JOHNSON. 1943. Hybridization in nature between species of catostomid fishes. University Michigan Laboratory Vertebrate Biology Contribution No. 22: 1–76.

HUBBS, CARL L. and EUGENE R. KUHNE. 1937. A new fish of the genus *Apocope* from a Wyoming warm spring. University Michigan Museum Zoölogy Occasional Paper 343: 1–21.

HUBBS, CARL L. and KARL F. LAGLER. 1941. Guide to the fishes of the Great Lakes and tributary waters. Cranbrook Institute Science Bulletin 18: 1–100.

1947. Fishes of the Great Lakes Region. Cranbrook Institute Science Bulletin 26: xi–186.

HUBBS, CARL L. and ROBERT R. MILLER. 1941. Studies of the fishes of the Order Cyprinodontes. XVII. Genera and species of the Colorado River system. University Michigan Museum Zoölogy Occasional Paper 433: 1–9.

1942. Fish of the Catlow Valley IN L. S. Cressman, Archaeological researches in the northern Great Basin. Carnegie Institution Washington Publication 538: 152.

1943. Mass hybridization between two genera of cyprinid fishes in the Mohave Desert, California. Michigan Academy Science, Arts Letters Paper 28: 343–378.

1948A. Two new, relict genera of cyprinid fishes from Nevada. University Michigan Museum Zoölogy Occasional Paper 507: 1–30 (new genera and species *Moapa coriacea* and *Eremichthys acros*).

1948B. Correlation between fish distribution and hydrographic history in the desert basins of western United States, pp. 17–166, IN The Great Basin, with emphasis on Glacial and Postglacial times. University Utah Bulletin 38(20): 1–191 (Biological Series, Vol. 10, No. 7).

1951. *Catostomus arenarius*, a Great Basin fish, synonymized with *C. tahoensis*. Copeia. 1951(4): 299–300.

1953 (1952). Hybridization in nature between the fish genera *Catostomus* and *Xyrauchen*. Michigan Academy Science, Arts Letters Paper 38: 207–233.

HUBBS, CARL L. and LEONARD P. SCHULTZ. 1931. The scientific name of the Columbia River chub. University Michigan Museum Zoölogy Occasional Paper 232: 1–6.

1932A. *Cottus tubulatus*, a new sculpin from Idaho. University Michigan Museum Zoölogy Occasional Paper 242: 1–9.

1932B. A new catostomid fish from the Columbia River. University Washington Biology Publication 2(1): 1–14.

HUBBS, CARL L. and CHARLES L. TURNER. 1939. Studies of the fishes of the Order Cyprinodontes. XVI. A revision of the Goodeidae. University Michigan Museum Zoölogy Miscellaneous Publication 42: 1–80.

HUBBS, CLARK, GEORGE E. DREWRY and BARBARA NARBURTON. 1958. Occurrence and morphology of a phenotypic male of a gynogenetic fish. Science 129(3357): 1227.

HUNTINGTON, ELLSWORTH. 1914. The climatic factor as illustrated in arid America. Carnegie Institution Washington Publication 192: 1–262.

1915. The curtailment of rivers by desiccation. Carnegie Institution Washington Yearbook 14: 96.

1925. Tree growth and climatic interpretations IN Quaternary climates. Carnegie Institution Washington Publication 352: 155–204.

1945. Mainsprings of civilization. John Wiley and Sons, pp. xii–660. New York.

HUTCHESON, AUSTIN E. 1944. Overland in 1852: The McGuirk Diary. Pacific Historical Review 13(4): 426–432.

HUTCHINSON, G. EVELYN. 1931. The hydrobiology of arid regions. Yale Scientific Magazine 5: 13. New Haven.

1937. A contribution to the limnology of arid regions. Connecticut Academy Arts Science Transactions 33: 47–132.

1939. Ecological observations on the fishes of Kashmir and Indian Tibet. Ecological Monographs 9: 145–182.

HUTCHINSON, SAMUEL J. 1957. Personal correspondence on the Northern Squawfish *Ptychocheilus oregonensis*. United States Fish and Wildlife Service, Portland, Oregon.

HUTT, A. P. (Editor). 1956. Fisherman's Handbook, 3d edition. Fisherman's Press, Oxford, Ohio, pp. 512.

IDYLL, C. P. 1942. Food of rainbow, cutthroat and brown trout in the Cowichan River system, British Columbia. Fishery Research Board Canada Journal 5(5): 448–458.

ILLICK, HELEN J. 1956. A comparative study of the cephalic lateral-line system of North American Cyprinidae. American Midland Naturalist 56(1): 204–223.

INNES, WILLIAM T. 1951. Black Mollienisias. Aquarium 20: 219–220.

IRVING, L., EDGAR C. BLACK and V. SAFFORD. 1941. The influence of temperature upon the combination of oxygen with the blood of trout. Biological Bulletin 80: 1–17.

IRVING, ROBERT B. 1955. Ecology of the cutthroat trout in Henrys Lake, Idaho. American Fisheries Society Transactions 84(1954): 275–296.

JAHODA, W. J. 1947. Survival of Brook trout in water of low oxygen content. Wildlife Management Journal 11: 96–97.

JENDKINS, ROBERT M. (Editor). 1959. Bibliography of theses on Fishery Biology. Sport Fishing Institute, pages 1–80 (Washington, D. C.).

JENKINS, J. TRAVIS. 1936. The fishes of the British Isles. Frederick Warne and Company, Ltd., London and New York. Pages viii–408, Sel. Edit. First Edition, 1925.

JEWELL, MINNA E., EDWARD SCHNEBERGER and JEAN A. ROSS. 1933. The vitamin requirements of Goldfish and Channel Cat. American Fisheries Society Transactions 63: 338–347.

JOB, S. V. 1955. The oxygen consumption of *Salvelinus fontinalis*. University Toronto Biological Series No. 61. Ontario Fisheries Research Laboratory Publication No. 73: iv–39.

JONES, ALDEN M. 1941. The length of the growing season of largemouth and smallmouth black bass in Norris Reservoir, Tennessee. American Fisheries Society Transactions 70(1940): 183–187.

JONES, J. R. E. 1952. The reactions of fish to water of low oxygen concentration. Experimental Biology Journal 29: 403–415.

JONES, J. CLAUDE. 1914A. The geologic history of Lake Lahontan. Science (new series) 40: 827–830.

1914B. The tufa deposits of the Salton Sink IN the Salton Sea, a study of the geography, the geology, the floristics, and the ecology of a desert basin. Carnegie Institution Washington Publication 193: 79–83.

1925. The geologic history of Lake Lahontan IN Quaternary Climates. Carnegie Institution Washington Publication 352: 1–49.

1929. Age of Lake Lahontan. Geological Society America Bulletin 40: 533–540.

JONEZ, AL and ROBERT C. SUMNER. 1954. Lakes Mead and Mohave Investigations. A comparative study of an established reservoir as related to a newly created impoundment. Nevada Fish and Game Commission, Wildlife Restoration Division, pp. xiii–186 (unpublished).

JORDAN, DAVID STARR. 1875. Fishes of Indiana. Indiana Geological Survey 1874, 42 pp. (a reprint of two separate papers originally printed in the Annual Report of the Survey for 1874, published in 1875).

1876A. Concerning the fishes of the Ichthyologia Ohiensis. Buffalo Society Natural Sciences Bulletin 3: 91–97 (Vol. 3 extends from 1875 to 1897).

1876B. Manual of the vertebrates of the northern United States, including district east of the Mississippi River, and north of North Carolina and Tennessee, exclusive of marine species. Jansen, McClurg and Co., Chicago, 342 pages (first edition).

1877A. No. 9. Contributions to North American Ichthyology, No. 1. Review of Rafinesque's Memoirs on North American fishes. United States National Museum Bulletin 9: 1–53.

1877B. No. 10. Contributions to North American Ichthyology, based primarily on the collections of the United States National Museum. No. 2A. Notes on Cottidae, Etheostomatidae, Percidae, Centrarchidae, Aphoderidae, Dorysomatidae and Cyprinidae, with revisions of the genera and descriptions of new or little known species. United States National Museum Bulletin 10: 1–68.

1877C. No. 10. Contributions to North American Ichthyology, based primarily on the collections of the United States National Museum. No. 2B. Synopsis of the freshwater Siluridae of the United States. United States National Museum Bulletin 10: 69–116.

1877D. On the distribution of freshwater fishes. American Naturalist 11: 607–613.

1878A. No. 12. Contributions to North American Ichthyology, No. 3B. A synopsis of the Family Catostomidae, pp. 97–230. United States National Museum Bulletin 12: 1–237.

1878B. A catalogue of the fishes of Illinois. Illinois State Laboratory Natural History Bulletin 1(2): 37–70.

1878C. Notes on a collection of fishes from Clackamas River, Oregon. United States National Museum Proceedings. Separate No. 20, 17 pages (appeared in regular volume series in 1879, United States National Museum Proceedings 1(1878): 69–85).

1878D. Notes on a collection of fishes from the Rio Grande, at Brownsville, Texas. United States Geological Geographical Survey Territories Bulletin 4: 397–406, 663–667.

1878E. On the characteristics of trout (letter to the editor). Forest and Stream 10: 196.
1878F. Report on the collection of fishes made by Dr. Elliott Coues, U. S. A., in Dakota and Montana, during the seasons of 1873 and 1874. United States Geological Geographical Survey Territory Bulletin 4: 777–799 (also appearing as an author's separate, same year, Washington, D. C.).
1878G. Manual of the vertebrates of the northern United States, including the District east of the Mississippi River, and north of North Carolina and Tennessee, exclusive of marine species. Jansen, McClurg and Company, Chicago. 342 pp. 2nd Ed.
1879. On the distribution of freshwater fishes of the United States. New York Academy Sciences Annals 1(1877–79) : 92–120.
1880. Description of new species of North American fishes. United States National Museum Proceedings 2(1879) : 235–241.
1881. Notes on a forgotten paper of Dr. Ayres, and its bearing on the nomenclature of the cyprinoid fishes of the San Francisco markets. United States National Museum Proceedings 3(1880) : 325–327.
1883. McCloud River trout. Forest and Stream 20: 72.
1885A. A catalogue of the fishes known to inhabit the waters of North America, north of the Tropic of Cancer, with notes on species discovered in 1883 and 1884. United States Fish Commission Document 94, 185 pages (also appeared in (1) United States Commissioner Fish Fisheries Report. 1885, pt. 11, Appendix E, 1887 ; and (2) United States 48th Congress, 2d Session, Senate Miscellaneous Document 70, Serial 2270, pp. 789–973).
1885B. Identification of the species of Cyprinidae and Catostomidae, described by Dr. Charles Girard, in the Proceedings of the Academy of Natural Sciences of Philadelphia for 1856. United States National Museum Proceedings Separate No. 500, 10 pages (appeared in regular volume series in 1886, United States National Museum Proceedings 8(1885) : 118–127).
1885C. Note on Mr. Garman's paper on "The American salmon and trout." United States National Museum Proceedings 8(6) : 81–83.
1886. Rafinesque. Popular Science Monthly 29: 212–221.
1887. The geographical distribution of freshwater food fishes in the several hydrographic basins of the United States, pp. 133–141 IN G. B. Goode's (Editor) The fisheries and fishing industries of the United States, Section 3, Washington, 1887 (also appeared in: United States 47th Congress, 1st Session, Senate Miscellaneous Document 124, Pt. 4, Serial 2000).
1888A. The distribution of freshwater fishes. Forest and Stream 30: 516–517, 31: 9–10 (also appeared in American Fisheries Society Proceedings 17: 4–24 and in Michigan Fish and Game Commission Annual Report 8: 108–120, 1888).
1888B. On the occurrence of the Great Lake trout (*Salvelinus namaycush*) in the waters of British Columbia. United States National Museum Proceedings Separate No. 682, 1 page (appeared in regular volume series in 1889, United States National Museum Proceedings 11(1888) : 58).
1890A. The fishes of Yellowstone Park. Zoe 1: 38–40.
1890B. Notes on fishes of the genera Agosia, Algansea, and Zophendum. United States National Museum Proceedings Separate No. 822, 2 pages (appeared in regular volume series in 1891, United States National Museum Proceedings 13(1890) : 287–288).
1890C. A reconnaissance of the streams and lakes of the Yellowstone National Park, Wyoming, in the interest of the United States Fish Commission, United States Fish Commission Bulletin Separate, 23 pages (appeared in regular volume series in the following 3 sources (1) United States Fish Commission Bulletin 9(1889) : 41–63, 1891;

(2) United States 51st Congress, 2d Session, House Miscellaneous Document 131, Serial 2881; (3) United States Fish Commission Document, No. 149, 1891).

1891. Report of explorations in Colorado and Utah during the summer of 1889, with an account of the fishes found in each of the river basins examined. United States Fish Commission Bulletin separate, 40 pages (appeared in regular volume series in the following three sources (1) United States Fish Commission Bulletin 9(1889): 1–40, 1891; (2) United States 51st Congress, 2d Session, House Miscellaneous Document 131, Serial 2881; (3) United States Fish Commission Document, No. 148, 1891).

1892A. Forest and Stream, 39(19): 405, 10 November 1892. (*Salmo gairdneri kamloops* first described in this publication as "*Oncorhynchus kamloops*." An identical description appeared a month later as "*Salmo kamloops*" in Biennial Report for 1891–1892 of the California State Board of Fish Commissioners).

1892B. Description of a new species of trout (*Salmo kamloops*) from the lakes of British Columbia. California State Board Fish Commissioners Biennial Report for 1891–1892: 60–61.

1892C. Salmon and trout of the Pacific Coast. California State Board Fish Commissioners Bulletin 4, 15 pages (also appeared in the Board's biennial report for 1891–1892, pp. 44–58, 1892 and in California Legislature journals. This report was revised in 1894 and published in the above-listed sources).

1892D. Description of the Golden Trout of Kern County, California (*Salmo mykiss agua-bonita*). California State Board Fish Commissioners Biennial Report 1891–1892: 62–65 (also appeared in United States National Museum Proceedings 15: 481–483, 1893).

1901. Transplanting of California trout. American Naturalist 35: 225–226.

1902A. The colors of fishes. American Naturalist 36: 803–808.

1902B. Evolution of fishes. Popular Science Monthly 60: 556–564.

1903. The classification of fishes. Popular Science Monthly 63: 5–13.

1904A. The fishes of California. California Louisiana Purchase Exposition Commission. Sacramento, California, 1904, pp. 118–124.

1904B. The transplanting of trout in the streams of the Sierra Nevada. American Naturalist 38: 885–887.

1905A. A guide to the study of fishes. Henry Holt and Company, New York. Vol. I: xxvi–624. Vol. II: xxi–599.

1905B. The Loch Leven trout in California. Science, new series, 22: 714–715.

1906. Salmon hybrids (letter to editor). Science, new series 23: 434.

1907A. The fossil fishes of California with supplementary notes on other species of extinct fishes. University California Department Geology Bulletin 5(5): 95–144 (describes *Merriamella doryssa* [= *Gasterosteus doryssus*], from the "Lahontan" beds of western Nevada, placing it in the family Atherinidae).

1907B. Fishes. Henry Holt and Co., New York (a reprint of the nontechnical portion of the 1905, 2-volume "Guide to the Study of Fishes").

1908. Note on a fossil stickleback fish from Nevada. Smithsonian Miscellaneous Collections 52: 117.

1914. Third printing (or edition) of American Food and Game Fishes (1902).

1919A. The genera of fishes, Part II: From Agassiz to Bleeker, 1833–1858. Stanford University Publications, University Series, pp. ix, 163–284, xiii.

1919B. The genera of fishes. Part III. From Guenther to Gill, 1859–1880. Stanford University Publications, University Series, pp. 285–410, xv.

1919C. The trout of the Great West. American Angler 4: 363–372.

1920. The genera of Fishes. Part IV: From 1881 to 1920. Stanford University Publications, University Series, pp. 415–576, xviii.

1921A. The color of trout flesh. What causes the variation. American Angler 6: 419.

1921B. The fish fauna of the California Tertiary. Stanford University Publications, University Series, Biological Sciences, 1(4) : 237–300.

1923A. Fourth printing (or edition) of American Food and Game Fishes (1902).

1923B. A classification of fishes, including families and genera as far as known. Stanford University Publications, University Series, Biological Sciences 3(2) : 79–243, x.

1923C. The name of the Dolly Varden trout, *Salvelinus spectabilis* (Girard). Copeia 1923(121) : 85–86.

1923D. Name of the Steelhead. Ibid: 86.

1923E. Spencer Fullerton Baird and the United States Fish Commission. Scientific Monthly 17: 97–107.

1924A. Concerning the genus *Hybopsis* of Agassiz. Copeia (130) : 51–52.

1924B. A top minnow (*Cyprinodon browni*) from an artesian well in California. Academy Natural Science Philadelphia Proceedings 76: 23–24.

1924C. Concerning the American Dace allied to the genus *Leuciscus*. Copeia 1924: 70–72 (mentions *Siphateles* of Cope and *Leucidius* of Snyder as valid genera, distinct from *Rutilus*).

1924D. Description of a recently discovered fossil sculpin from Nevada regarded as *Cottus beldingi*. United States National Museum Proceedings 65: 1–2.

1924E. Description of Miocene fishes from southern California. Southern California Academy Sciences Bulletin 23: 42–50.

1925A. Fishes. D. Appleton and Company, New York, 773 pages (revised edition of the 1907 work published by Henry Holt and Company).

1925B. The fossil fishes of the Miocene of southern California. Stanford University Publications, University series, Biological Sciences 4(1), 51 pages.

1927A. The fossil fishes of the Miocene of southern California. Contribution No. 9. Stanford University Publications, University Series, Biological Sciences, 5: 89–99.

1927B. Carl H. Eigenmann. Science, new series, 65: 516.

1929. Manual of the vertebrate animals of northeastern United States inclusive of marine species. World Book Company, Yonkers, New York pp. xxxi plus 446 (13th edition).

1934. Reprint of 1923 Classification of fishes including families and genera, etc. Stanford University Publications, University Series, Biological Sciences, 3(2).

JORDAN, DAVID STARR and ALEMBERT WINTHROP BRAYTON. 1878. No. 12. Contributions to North American Ichthyology. No. 3A. On the distribution of the fishes of the Alleghany region of South Carolina, Georgia and Tennessee. United States National Museum Bulletin No. 12: 1–237.

JORDAN, DAVID STARR and HERBERT EDSON COPELAND. 1877A. Check list of the fishes of the fresh waters of North America. Buffalo Society Natural Sciences Bulletin 3: 133–164.

1877B. The genus *Pomoxys* Rafinesque. Academy Natural Sciences Philadelphia Proceedings 28: 68–71.

1878. A catalogue of the fishes of the fresh waters of North America. United States Geological Geographical Survey Territories (Hayden) Bulletin 4: 407–442 (replaced 1877 checklist).

JORDAN, DAVID STARR and BARTON WARREN EVERMANN. 1896A. A checklist of the fishes and fish-like vertebrates of North and Middle America. United States Commission Fish Fisheries. Report Commissioner 1894–1895, 21: 207–584 (also appeared in the following forms: (1) United States 54th Congress, 2d Session, House Miscellaneous Document 102, Vol. 42, Serial 3518 and (2) United States Fish Commission Document No. 336, 1896.

1896B. The fishes of North and Middle America. A descriptive catalogue of the species of fish-like vertebrates found in the waters of North America, north of the Isthmus of Panama. United States National Museum Bulletin 47, Part 1 : lx + 1–1240.

1898A. Ibid, Part 2 : xxx + 1241–2183.

1898B. Ibid, Part 3 : xxiv + 2183a–3136.

1900. Ibid, Part 4 : ci + 3137–3313, Plates i–cccxcii (1–392).

1902. American Food and game fishes. A popular account of all the species found in America north of the equator, with keys for ready identification, life histories and methods of capture. Doubleday, Page and Company, New York, 573 pp. (with minor changes, this was republished in 1905, 1914 and 1923).

1904. American Food and game fishes. Reprinted as Volume V of the Nature Library. 572 pages.

1911. A review of the salmonoid fishes of the Great Lakes, with notes on the whitefishes of other regions. United States Fish Commission Document 737, 41 pages (appeared serially in: (1) United States Bureau Fisheries Bulletin 29(1909) : 1–41. 1911 and (2) United States 61st Congress, 2d Session, House Document 122 Serial 5784).

1917. The genera of fishes. Part I : From Linnaeus to Cuvier, 1758–1833. Stanford University Publications, University Series, pp. 1–161.

1919. Fossil fishes of southern California. I. Fossil fishes of the Soledad deposits. II. Fossil fishes of the Miocene (Monterey) formations (with J. Z. Gilbert). III. Fossil fishes of the Pliocene formations (with J. Z. Gilbert). Stanford University Publications, University series, No. 38, 98 pages (= Jas. Zaccheus Gilbert).

1937. American food and game fishes, revised edition.

JORDAN, DAVID STARR, BARTON WARREN EVERMANN and HOWARD WALTON CLARK. 1930. Checklist of the fishes and fish-like vertebrates of North and Middle America north of the northern boundary of Venezuela and Columbia. United States Commissioner Fisheries report fiscal year 1928. Part II, Appendix 10 : 1–670 (also issued as United States Fish Commission Document No. 1055, 1930).

JORDAN, DAVID STARR and CHARLES H. GILBERT. 1878. On the genera of North American freshwater fishes. Academy Natural Sciences Philadelphia Proceedings 29(1877) : 83–104.

1881A. List of the fishes of the Pacific Coast of the United States, with a table showing the distribution of the species. United States National Museum Proceedings 3 : 452–458 (1880).

1881B. Notes on a collection of fishes from Utah Lake. United States National Museum Proceedings 3(175) : 459–464 (1880).

1881C. Description of a new species of *Ptychochilus* (*Ptychochilus harfordi*) from the Sacramento River. United States National Museum Proceedings 4 : 72–73 (1881).

1881D. Notes on the fishes of the Pacific Coast of the United States. United States National Museum Proceedings Separate 191, 20 pages (appeared in regular volume series in 1882, United States National Museum Proceedings 4 : 29–48 (1881)).

1882. Synopsis of the fishes of North America. United States National Museum Bulletin 16 : lvi–1018 (although many bibliographers and taxonomists use the 1883 date for this, which does appear on the outermost leaf, this was issued in 1882 according to Baird's introductory statement and Jordan's preface to his 1930 catalogue with Evermann and Clark. The Hays' Jordan bibliography (1952) states "according to Tarleton H. Bean, this paper appeared early in April 1883" (p. 107)).

1894. List of the fishes inhabiting Clear Lake, California. United States Fish Commission Document 288, 2 pages (appeared in regular volume series in 1895 in United States Fish Bulletin 14 : 139–140 (1894) and United States 53rd Congress, 3rd Session, House Miscellaneous Document 86, Vol. 12, Serial 3338).

JORDAN, DAVID STARR and JAMES Z. GILBERT. 1920. Fossil fishes of the diatom beds of Lompoc, California. Stanford University Publications, University series, No. 42, 44 pages.

JORDAN, DAVID STARR and HENRY W. HENSHAW. 1878. Report upon the fishes collected during the years 1875, 1876, and 1877, in California and Nevada. United States Geological Geographical Survey Territories of United States W. of 100th Meridian Report 1878: 187–200 (also appeared in Annual Report of the United States Chief of Engineers for 1878, part 3: 1609–1622; and in United States 45th Congress, 3d Session House Executive Document 1, Part 2, Volume 2, Serial 1846).

JORDAN, DAVID STARR and ROBERT E. RICHARDSON. 1907. Description of a new species of killifish, *Lucania browni*, from a hot spring in Lower California. United States National Museum Proceedings 33: 319–321.

JORDAN, DAVID STARR and JOHN O. SNYDER. 1909. Description of a new whitefish (*Coregonus oregonius*) from McKenzie River, Oregon. United States National Museum Proceedings 36(1677): 425–430.

JORDAN, DAVID STARR and EDWIN CHAPIN STARKS. 1895. The fishes of Puget Sound. Stanford University Publications. Contributions to Biology of Hopkins Laboratory of Biology No. 3: 785–855 (also published the next year in the California Academy Sciences Proceedings 1895, Series 2, Vol. 5: 785–855, 1896).

JORDAN, DAVID STARR and BALFOUR H. VAN VLECK. 1874. A popular key to the birds, reptiles, batrachians and fishes of the northern United States east of the Mississippi River. Appleton, Wisconsin, Reid and Miller, 85 pages.

JUDAY, CHANCEY. 1907. Notes on Lake Tahoe, its trout and trout-fishing. United States Bureau Fisheries Bulletin 26: 133–146 (1906).

KEMMERER, GEORGE, J. F. BOVARD, and W. R. BOORMAN. 1923. Northwestern lakes of the United States: Biological and chemical studies with reference to possibilities in production of fish. United States Bureau Fisheries Bulletin 39: 51–140 (1923–1924).

KENDALL, WILLIAM CONVERSE. 1910. American catfishes; habits, culture, and commercial importance. United States Bureau Fisheries Document No. 733: 1–39 (bound in with Report Commissioner Fisheries 1908).

1914. The fishes of New England. The Salmon family. Part I—the Trout or Charrs. Boston Society Natural History Memoirs 8(1): 1–103.

1919A. Concerning the generic name, *Cristivomer* vs *Salvelinus*, for the Great Lakes trout or Namaycush. Copeia 1919: 78–81.

1919B. Scotch sea trout in Maine. Copeia No. 75: 85–87.

1921. Peritoneal membranes, ovaries, and oviducts of salmonoid fishes and their significance in fish cultural practices. United States Bureau Fisheries Bulletin 37(1919–1920): 185–208.

1935. The fishes of New England. The salmon family. Part 2—the Salmons. Boston Society Natural History Memoirs 9(1): 1–166.

KENDALL, WILLIAM CONVERSE and WILFORD A. DENCE. 1927. A trout survey of the Allegheny State Park in 1922. Roosevelt Wild Life Bulletin 4(3).

1929. The fishes of the Cranberry Lake region. Roosevelt Wild Life Bulletin 5(2): 219–309.

KENNEDY, CLARENCE H. 1916. A possible enemy of the mosquito. California Fish and Game, 2: 179–182.

KEYES, CHARLES R. 1918. Lacustral records of past climates. Monthly Weather Review 46: 277–280.

KIMSEY, J. BRUCE. 1950. Some Lahontan fishes in the Sacramento River drainage, California. California Fish and Game 36(4): 438.

1951. Notes on Kokanee spawning in Donner Lake, California, 1949. California Fish and Game 37(3): 273–279.

1954. The life history of the Tui Chub, *Siphateles bicolor* (Girard) from Eagle Lake, California. California Fish and Game 40(4) : 395–410.

1955. Post-spawning behavior of the Kokanee, *Oncorhynchus nerka kennerlyi*, in Donner Lake, California. Copeia 1955(1) : 51–52.

1960. Note on spring food habits of the Lake Trout, *Salvelinus namaycush*. California Fish and Game 46(2) : 229–230.

KIMSEY, J. BRUCE and LEONARD O. FISK. 1960. Keys to the freshwater and anadromous fishes of California. California Fish and Game 46(4) : 453–479.

KIMSEY, J. BRUCE, ROBERT H. HAGY and GEORGE W. McCAMMON. 1957. Progress report on the Mississippi Threadfin shad, *Dorosoma petenensis atchafaylae* [sic], in the Colorado River for 1956. California Fish and Game, Inland Fisheries Administrative Report No. 57–23, 48 pp. (mimeographed).

KING, CLARENCE. 1877. United States Geological Exploration of the 40th Parallel. Vol. II (Descriptive Geology) : x–890. United States Army Engineers Department Professional Paper No. 18. Washington, D. C.

1878. Idem. Vol. I (Systematic Geology) : xii–803. Ibid.

KLAK, GEORGE E. 1940. The condition of brook trout and rainbow trout from four eastern streams. American Fisheries Society Transactions 70 : 282–289.

KMIOTEK, STANLEY and ARTHUR A. OEHMCKE. 1959. Will the Splake make good? Wisconsin Conservation Bulletin 24(6) : 23–24.

KNOWLTON, F. H. 1900. Fossil Plants of the Esmeralda Formation. pp. 209–220, with plates. United States Geological Survey 21st Annual Report, Part II : 1–522.

KOELZ, W. 1931. The coregonid fishes of Northeastern North America. Michigan Academy Science, Arts Letters, Paper 13(1930) : 303–432.

KOPEC, JOHN A. 1949. Ecology, breeding habits and young stages of *Crenichthys baileyi*, a cyprinodontid fish of Nevada. Copeia 1949(1) : 56–61.

KOSTER, WILLIAM J. 1937. The food of sculpins, Cottidae, in central New York. American Fisheries Society Transactions 66(1936) : 374–382.

1957. Guide to the fishes of New Mexico. University New Mexico Press, Albuquerque, New Mexico, pp. vii–116.

KROGH, A. and I. LEITCH. 1919. The respiratory function of the blood of fishes. Physiology Journal 52 : 288–300.

KRUMHOLZ, L. A. 1943. A comparative study of the Weberian Ossicles in North American Ostariophysine fishes. Copeia 1943(1) : 33–40.

1948. Reproduction in the Western mosquitofish, *Gambusia affinis affinis* (Baird and Girard), and its use in mosquito control. Ecological Monographs 18(1) : 1–43.

KUHNE, EUGENE R. 1939. A guide to the fishes of Tennessee and the Midsouth. Division Game and Fish, Tennessee Department Conservation, Nashville, Tennessee. pp. 1–124.

KUNTZ, ALBERT. 1913. Notes on the habits, morphology of the reproductive organs, and embryology of the viviparous fish (*Gambusia affinis*). United States Bureau Fisheries Bulletin 33 : 177–190.

KYLE, HARRY M. 1926. The biology of fishes. Sidgwick and Jackson, Ltd., London. pp. xvi–396.

LACHNER, E. A. 1952. Studies of the biology of the Cyprinid fishes of the chub genus *Nocomis* of northeastern United States. American Midland Naturalist 48(2) : 433–466.

LAGLER, KARL F. 1949. Studies in freshwater fishery biology. J. W. Edwards Bros., Ann Arbor, Michigan. pp. v–231.

1952. Freshwater fishery biology. Wm. C. Brown Co., Dubuque, Iowa. pp. x–360 (an expanded, hardback version of the preceding).

1956. Second edition, idem. pp. xii–421.

LA GORCE, JOHN O., et al. 1924. The Book of Fishes. National Geographic Society, Washington, D. C. pp. 1–243.
LA MOTTE, R. S. 1936. The upper Cedarville flora of northwestern Nevada and adjacent California. Carnegie Institution Washington Publication 455: 57–142.
LANE, CHARLES E., JR. 1953. Age and growth of the bluegill (*Lepomis machrochirus* Rafinesque) in a new Missouri impoundment. Master's thesis, University Missouri.
LA RIVERS, IRA. 1946A. An annotated list of Carabinae known to occur in Nevada (Coleoptera: Carabidae). Southern California Academy Science Bulletin 45(3): 133–140.
- 1946B. Some dragonfly observations in alkaline areas in Nevada. Entomological News 57(9): 209–217.
- 1948. A synopsis of Nevada Orthoptera. American Midland Naturalist 39(3): 652–720.
- 1949. A new subspecies of Stenelmis from Nevada. Entomological Society Washington Proceedings 51(5): 218–224.
- 1950A. The meeting point of Ambrysus and Pelocoris in Nevada (Hemiptera: Naucoridae). Pan-Pacific Entomologist 26(1): 19–21.
- 1950B. The Dryopoidea known or expected to occur in the Nevada area (Coleoptera). Wasmann Journal Biology 8(1): 97–111.
- 1950C. A new naucorid genus and species from Nevada (Hemiptera). Entomological Society America Annals 43(3): 368–373.
- 1951A. Nevada Dytiscidae (Coleoptera). American Midland Naturalist 45(2): 392–406.
- 1951B. A revision of the genus Ambrysus in the United States (Hemiptera: Naucoridae). University California Entomology Publications 8(7): 277–338.
- 1952. A key to Nevada fishes. Southern California Academy Science Bulletin 51(3): 86–102.
- 1953A. A lower Pliocene frog from western Nevada. Paleontology Journal 27(1): 77–81.
- 1953B. New Gelastocorid and Naucorid records and miscellaneous notes, with a description of the new species, *Ambrysus amargosus* (Hemiptera: Naucoridae). Wasmann Journal Biology 11(1): 83–96.
- 1954. Nevada Hydrophilidae (Coleoptera). American Midland Naturalist 52(1): 164–174.
- 1956. A new subspecies of *Pelocoris shoshone* from the Death Valley drainage (Naucoridae: Hemiptera). Wasmann Journal Biology 14(1): 155–158.

LA RIVERS, IRA and THOMAS J. TRELEASE. 1952. An annotated checklist of the fishes of Nevada. California Fish and Game 38(1): 113–123.
LARKIN, P. A. and S. B. SMITH. 1954. Some effects of introduction of the Redside Shiner on the Kamloops Trout in Paul Lake, British Columbia. American Fisheries Society Transactions 83: 161–175.
LARSON, E. RICHARD and JAMES B. SCOTT. 1955. *Helicoprion* from Elko County, Nevada. Paleontology Journal 29(5): 918–919.
LA RUE, E. C. 1916. Colorado River and its utilization. United States Geological Survey Water Supply Paper 395: 1–231.
LAYTHE, LEO L. 1957. Personal correspondence on the Northern Squawfish, *Ptychocheilus oregonensis*. United States Fish and Wildlife Service, Portland, Oregon.
LEACH, GLEN S. See United States Reports for Fish and Game for the years 1920, 1927, 1931 and 1932.
LE CONTE, JOHN. 1883. Physical studies of Lake Tahoe. Overland Monthly 2: 506–516, 595–612, November and December, respectively.

1884. Idem. 3: 41–46 (these are reprints of an earlier article published in the Free Press and the Mining and Scientific Press of San Francisco, during 1880 and 1881).

LEE, WILLIS T., RALPH W. STONE, HOYT S. GALE and others. 1915. Guidebook of the western United States. Part B. The overland route, with a side trip to Yellowstone Park. United States Geological Survey Bulletin 612: 1–224.

LENNON, ROBERT E. 1957. Observations on wild, long-jaw Rainbow Trout. Progressive Fish-Culturist 19(4): 179–181.

LEONARD, JUSTIN W. 1939. Further observations on the feeding habits of the Montana grayling (*Thymallus montanus*) and the bluegill (*Lepomis macrochirus*) in Ford Lake, Michigan. American Fisheries Society Transactions 69: 243–256.

1941. Some observations on the winter feeding habits of brook trout fingerlings in relation to natural food organisms present. Idem. 71: 219–227.

LIEBLING, A. J. 1955. A reporter at large. The Lake of the Cui-ui Eaters. The New Yorker magazine. Part I: 25–41, January 1 / Part II: 33–61, January 8 / Part III: 32–69, January 15 / Part IV: 37–73, January 22. (Liebling's theme was the Pyramid Lake Indians versus the late Senator Patrick McCarran, and very well done. To the biologist (for whom it was not written) some errors are evident—perpetuation of such myths as a 1,200-foot depth for Pyramid Lake, and an India relative for the Cui-ui Lakesucker. Needless to say, the United States Bureau of Indian Affairs being what it is, the plight of the Pyramid Lake Pahute with respect to water rights is as disgraceful now as ever.)

LINDBERG, G. U. 1934. On the systematics of Gambusia. Mag. Parasit. L'Institute Zoölogy L'Academy Sciences U. S. S. R. 4: 351–367.

LINDSEY, C. C. 1953. Variations in anal fin ray count of the Redside Shiner, *Richardsonius balteatus* (Richards). Canadian Zoölogy Journal 31: 211–225.

LINSDALE, JEAN M. 1938. Bird life in Nevada with reference to modification in structure and behavior. Condor 40: 173–180.

LOCKE, S. B. 1929. Whitefish, grayling, trout and salmon of the intermountain region. United States Commission Fisheries Report, Appendix 5: 173–190 (United States Bureau Fisheries Document 1062).

LOCKINGTON, W. N. 1881. Description of a new species of Catostomus (*Catostomus cypho*) from the Colorado River. Academy Natural Science Philadelphia Proceedings 32: 237–240 (= *Xyrauchen texanus*).

LORD, JOHN KEAST. 1866. The naturalist in Vancouver Island and British Columbia. Vol. I: xiv–358. Vol. II: vii–375. Richard Bentley, London.

LORD, RUSSELL F. 1933. Type of food taken throughout the year by brook trout in a single Vermont stream with special reference to winter feeding. American Fisheries Society Transactions 63: 182–197.

LOUD, LLEWELLYN L. and M. R. HARRINGTON. 1929. Lovelock Cave. University California American Archaeological and Ethnological Publication 25: viii–183.

LOUDERBACK, GEORGE D. 1923. Basin Range structure of the Great Basin. University California Geological Science Publication 14: 329–376.

LOVEJOY, EARL M. P. 1960. Structural implications of the shore lines of the Mio-Pliocene Lake Nevada. Program 1960 Annual Meetings, Geological Society America, Denver, Colorado. Pages 151–152.

1961. Basin Range origin. Program 1961 Annual Meetings, Geological Society America, Cincinnati, Ohio. Page 95A.

LOWDER, LYLE JUNIOR. 1951. A taxonomic study of the Catostomidae of Utah Lake with notes on the fish population. Master's thesis, Brigham Young University, pp. v–45.

LUCAS, FREDERIC A. 1900. Description of a new species of fossil fish from the Esmeralda formation, pp. 223–226, IN the Esmeralda Formation, a freshwater lake deposit, by H. W. Turner. United States Geological Survey, 21st Annual Report, pp. 191–226 (*Leuciscus turneri*).

    1901. A new fossil cyprinoid *Leuciscus turneri*, from the Miocene of Nevada. United States National Museum Proceedings 23(1212) : 333–334 (a reprint of the preceding item).

LYDELL, DWIGHT. 1902. The habits and culture of the black bass. United States Fish Commission Bulletin 22 : 39–44.

M., F. 1878A. Trout fishing in Truckee River. Chicago Field 10 : 20.

    1878B. Trouting in Nevada. Catching them in the water-works at Gold Hill and Virginia City. Ibid. 10.

    1879. The Nevada fish-hatchery. Ibid. 10 : 332 (from Carson City Appeal).

MACDONALD, MARSHALL. 1884. Report of distribution of carp, during the season of 1881–1882, by the United States Fish Commission. United States Commission Fish Fisheries Report 1881 : 1121–1126.

MACDOUGAL, D. T. 1912. North American deserts. Geographical Journal 39 : 105–123.

MACFARLANE, JOHN MUIRHEAD. 1923. Fishes the source of Petroleum. Macmillan Company, pp. 1–451.

MACLULICH, D. A. 1943A. *Proteocephalus parallacticus*, a new species of tapeworm from lake trout, *Cristivomer namaycush*. Canadian Journal Research 21 : 145–149.

    1943B. Parasites of trout in Algonquin Provincial Park, Ontario. Canadian Journal Research 21 : 405–412.

MACPHEE, CRAIG. 1960. Postlarval development and diet of the Largescale Sucker, *Catostomus macrocheilus*, in Idaho. Copeia 1960(2) : 119–125.

MADSEN, M. J. 1935A. A biological survey of streams and lakes of Coconino National Forest, Arizona. United States Department Commerce Bureau Fisheries Mimeograph Report, pp. 1–23.

    1935B. Report of a preliminary survey of Pyramid Lake, Nevada. United States Fish Wildlife Service, mimeographed, 6 pages, 2 tables, 1 map.

MAITLAND, J. R. G. 1884. Exchange of land-locked salmon eggs from Maine for Loch Leven trout from Scotland. United States Fish Commission Bulletin 4 : 114–115.

MALLOCH, P. D. 1910. Life history and habits of the salmon, sea trout, trout and other freshwater fish. Adam and Charles Black, London. Pages xvi–264 (Recommended, common-sense reading).

MARKUS, HENRY C. 1933. The extent to which temperature changes influence food consumption in largemouth bass (*Huro floridana*). American Fisheries Society Transactions 62(1932) : 202–210.

MARSHALL, W. S. and N. C. GILBERT. 1905. Notes on the food and parasites of some freshwater fishes from the lakes at Madison, Wisconsin. United States Fish Commission Bulletin 1904 : 513–522.

MARTIN, W. R. 1949. The mechanics of environmental control of body forms in fishes. University Toronto Studies, Biological Series, No. 58. Ontario Fisheries Research Laboratory Publication No. 70 : 1–72.

MARZOLF, RICHARD C. 1955. Use of pectoral spines and vertebrae for determining age and rate of growth of the channel catfish. Wildlife Management Journal 19(2) : 243–249.

    1957. The reproduction of channel catfish in Missouri ponds. Wildlife Management Journal 21(1) : 22–28.

MASON, JAMES E. 1953. The distribution of four anadromous members of the genus *Salmo* in the northern hemisphere. Progressive Fish-Culturist 15(2) : 51–56.

MATHER, F. 1887. Brown trout in America. United States Fish Commission Bulletin 7 : 21–22.

MATTHES, F. E. 1933. Geography and Geology of the Sierra Nevada, pp. 26–40 IN Guidebook 16, Excursion C–1, International Geological Congress XVI: 1–116.

1941. Rebirth of the glaciers of the Sierra Nevada during late post-Pleistocene times. Geological Society America Bulletin 52: 2030.

MAXFIELD, GALEN H., KENNETH L. LISCOM and ROBERT H. LANDER. 1959. Leading adult squawfish (*Ptychocheilus oregonensis*) within an electric field. United States Fish Wildlife Service Special Scientific Report—Fisheries No. 298: 1–15.

McCAMMON, GEORGE W. 1957. Personnal communication. Carmichael, California.

McCAMMON, GEORGE W. and DON A. LAFAUNCE. 1961. Mortality rates and movement in the Channel Catfish population of the Sacramento Valley. California Fish and Game 47(1): 5–26.

McCAMMON, GEORGE W. and CHARLES M. SEELEY. 1961. Survival, mortality, and movements of White Catfish and Brown Bullheads in Clear Creek, California. California Fish and Game 47(3): 237–255.

McCAY, C. M., L. A. MAYNARD, J. W. TITCOMB and M. F. CROWELL. 1930. Influence of water temperature upon growth and reproduction of brook trout. Ecology 11: 30–34.

McGRAW, B. M. 1950. Furunculosis in fish. Master's thesis, University of Toronto.

McCULLOUCH, FRANK. 1949. Ex-King cutthroat of Pyramid. Outdoor West 1(2): 16–17, 41.

McHUGH, J. LAURENCE. 1939. The whitefishes *Coregonus clupeaformis* (Mitchill) and *Prosopium williamsoni* (Girard) of the lakes of the Okanagan Valley, British Columbia. Canadian Fisheries Research Board Bulletin 56: 39–50.

1940. Food of the Rocky Mountain whitefish *Prosopium williamsoni* (Girard) Canadian Fisheries Research Board Journal 5(2): 131–137.

1941. Growth of the Rocky Mountain whitefish. Canadian Fisheries Research Board Journal 5(4): 337–343.

McKAY, CHARLES L. 1882 (1881). A review of the genera and species of the family Centrarchidae, with a description of one new species. United States National Museum Proceedings 1881, Vol. 4: 87–93.

MEEK, SETH EUGENE. 1904. The freshwater fishes of Mexico north of the Isthmus of Tehuantepec. Field Columbian Museum Publication 93, Zoölogical Series, Vol. 5: lxiii + 252.

MEEK, SETH EUGENE and SAMUEL F. HILDEBRAND. 1916. The fishes of the fresh waters of Panama. Field Museum Natural History Publication 191, Zoölogical Series, Vol. 10(15): 217–374.

MEINZER, OSCAR E. 1916. Ground water in Big Smoky Valley, Nevada. United States Geological Survey Water Supply Paper 375(d): 85–116.

1917. Geology and water resources of Big Smoky, Clayton, and Alkali Springs Valleys, Nevada. United States Geological Survey Water Supply Paper 423: 1–167.

1922. Map of the Pleistocene Lakes of the Basin-and-Range Province and its significance. Geological Society America Bulletin 33: 541–552.

1927. Plants as indicators of ground water. United States Geological Survey Water Supply Paper 577: 1–95.

MENDENHALL, WALTER C. 1909. Some desert watering places in southeastern California and southwestern Nevada. United States Geological Survey Water Supply Paper 224: 1–98.

MERRIAM, JOHN C. 1918. Evidence of mammalian paleontology relating to the age of Lake Lahontan. University California Geology Publications 10: 517–521.

MERRIMAN, DANIEL. 1935. Squam Lake trout. Boston Society Natural History Bulletin 75: 3–10.

1938. Peter Artedi—Systematist and Ichthyologist. Copeia 1938(1): 33–39.

METZELAAR, J. 1929. The food of the trout in Michigan. American Fisheries Society Transactions 59: 146–152.
MILLER, RICHARD B. 1954. Comparative survival of wild and hatchery-reared Cutthroat trout in a stream. American Fisheries Society Transactions 83(1953) : 120–130.
MILLER, RICHARD B. and W. A. KENNEDY. 1948. Observations on the lake trout of Great Bear Lake. Fisheries Research Board Canada Journal 7(4) : 176–189.
MILLER, RICHARD G. 1951. The natural history of Lake Tahoe fishes. Unpublished dissertation, Stanford University Library, pp. ix–160.
MILLER, ROBERT R. 1936. Classification of a Mohave River fish. Pomona College Entomology Zoölogy Journal 28 : 58–59.
1938A. Record of the freshwater minnow *Apocope nevadensis* from southeastern California. Copeia 1938 : 147.
1938B. Description of an isolated population of the freshwater minnow *Siphateles mohavensis* from the Mohave River Basin, California. Pomona College Entomology Zoölogy Journal 30: 65–67.
1943A. The status of *Cyprinodon macularius* and *Cyprinodon nevadensis*, two desert fishes of western North America. University Michigan Museum Zoölogy Occasional Paper 473 : 1–25.
1943B. *Cyprinodon salinus*, a new species of fish from Death Valley, California. Copeia 1943(2) : 69–78.
1945A. *Snyderichthys*, a new generic name for the Leatherside chub of the Bonneville and Upper Snake drainages in western United States. Washington Academy Science Journal 35: 28.
1945B. A new cyprinid fish from southern Arizona, and Sonora, Mexico, with the description of a new subgenus of *Gila* and a review of related species. Copeia 1945 : 104–110.
1945C. Paleontology.—Four new species of fossil cyprinodont fishes from eastern California. Washington Academy Science Journal 35(10) : 315–321.
1946A. Correlation between fish distribution and Pleistocene hydrography in eastern California and southwestern Nevada, with a map of the Pleistocene waters. Journal Geology 54(1) : 43–53.
1946B. The need for ichthyological survey of the major rivers of western North America. Science 104(2710) : 1–3.
1946C. Distributional records of North American fishes, with nomenclatorial notes on the genus *Psenes*. Washington Academy Science Journal 36(6) : 206–212.
1946D. *Gila cypha*, a remarkable new species of cyprinid fish from the Colorado River in Grand Canyon, Arizona. Washington Academy Science Journal 36(12) : 409–415.
1948. The cyprinodont fishes of the Death Valley system of eastern California and southwestern Nevada. University Michigan Museum Zoölogy Miscellaneous Publication 68 : 1–155 (new subspecies *Empetrichthys latos latos*, *E. l. pahrump*, *E. l. concavus*, *Cyprinodon nevadensis mionectes* and *C. n. pectoralis*).
1949A. Desert fishes—clues to vanished lakes and streams. American Museum Natural History Magazine 58(10) : 447–451.
1949B. Hot springs and fish life. Aquarium Journal 20(11) : 286–288.
1950A. Speciation in fishes of the genera *Cyprinodon* and *Empetrichthys*, inhabiting the Death Valley region. Evolution 4(2) : 155–163.
1950B. Notes on the cutthroat and rainbow trouts with the description of a new species from the Gila River, New Mexico. University Michigan Museum Zoölogy Occasional Paper 529 : 1–42.
1950C. A review of the American clupeid fishes of the genus Dorosoma. United States National Museum Proceedings 100(3267) : 387–410.
1952. Bait fishes of the lower Colorado River from Lake Mead, Nevada, to Yuma, Arizona, with a key for their identification. California Fish and Game 38(1) : 7–42.

1955. An annotated list of the American cyprinodontid fishes of the genus *Fundulus*, with the description of *Fundulus persimilis* from Yucatan. University Michigan Museum Zoölogy Occasional Paper 568: 1–25.

1958. Origin and affinities of the freshwater fish fauna of Western North America, pages 187–222, IN Zoögeography, American Association Advancement Science Publication 51: x + 509.

1960. Systematics and biology of the Gizzard Shad (*Dorosoma cepedianum*) and related fishes. United States Department Interior Fish and Wildlife Service Fishery Bulletin 173: 371–392.

1961. Man and the changing fish fauna of the American southwest. Michigan Academy Sciences, Arts, Letters Papers Vol. 46: 365–404.

MILLER, ROBERT R. and J. RAY ALCORN. 1946 (1945). The introduced fishes of Nevada, with a history of their introduction. American Fisheries Society Transactions 73(1943): 173–193.

MILLER, ROBERT R. and CARL L. HUBBS. 1960. The spiny-rayed cyprinid fishes (Plagopterini) of the Colorado River system. University Michigan Museum Zoölogy Miscellaneous Publication 115: 1–39.

MILLER, ROBERT R. and RALPH G. MILLER. 1948. The contribution of the Columbia River system to the fish fauna of Nevada: Five species unrecorded from the State. Copeia 1948(3): 174–187.

MILLER, ROBERT R. and WILLIAM M. MORTON. 1952. First record of the Dolly Varden, *Salvelinus malma*, from Nevada. Copeia 1952(3): 207–208.

MILLER, ROBERT R. and WILLIAM F. SIGLER. 1950. A list of the fishes of Utah with keys to identification. Utah State Agriculture College Wildlife Management Department, 13 pages (mimeographed).

MILLS, GEORGE T. See Nevada, State of: Biennial Reports for Fish and Game for the years 1891 and 1897.

MILNER, JAMES W. 1874. Report on the fisheries of the Great Lakes, the result of inquiries prosecuted in 1871 and 1872. United States Fish Commission Report 1872–1873(2): 1–78.

MITCHILL, SAMUEL L. 1815. The fisheries of New York, described and arranged. Literary Philosophical Society New York Transactions 1: 355–492.

MOFFETT, JAMES W. 1942. A fishery survey of the Colorado River below Boulder Dam. California Fish and Game 28(2): 76–86.

1943. A preliminary report on the fishery of Lake Mead. Eighth National American Wildlife Conference Transactions 1943: 179–186.

MOFFETT, JAMES W. and BURTON P. HUNT. 1943. Winter feeding habits of bluegills, *Lepomis macrochirus* Rafinesque, and yellow perch, *Perca flavescens* (Mitchill), in Cedar Lake, Washtenaw County, Michigan. American Fisheries Society Transactions 73: 231–242.

MOORE, E. 1922. *Octomitus salmonis*, a new species of intestinal parasite in trout. American Fisheries Society Transactions 52: 74–97.

1923. A report of progress in the study of trout diseases. American Fisheries Society Transactions 53: 74–81.

1924. The transmission of *Octomitus salmonis* in the egg of trout. American Fisheries Society Transactions 54: 54–56.

MOORE, GEORGE A. 1950. The cutaneous sense organs of barbeled minnows adapted to life in the muddy waters of the Great Plains region. American Microscopical Society Transactions 69(1): 69–95.

1957. Fishes, Part II, pp. 31–210 of Vertebrates of the United States, pp. ix–819, by W. Frank Blair, Introduction and Mammals; Albert P. Blair, Amphibians; Pierce Brodkorb, Birds; Fred R. Cagle, Reptiles; and George A. Moore, Fishes.

MOORE, WALTER C. 1942. Field Studies on the oxygen requirements of certain freshwater fishes. Ecology 23: 319–329.

MORGAN, DALE L. 1943. The Humboldt: Highroad of the West. Farrar and Rinehart, Inc., N. Y. (Rivers of America series). Pages x–374.

MOROFSKY, W. F. 1940. A comparative study of the insect food of trout. Journal Economic Entomology 33(3) : 544–546.
MORTON, WILLIAM MARKHAM and ROBERT R. MILLER. 1954. Systematic position of the Lake trout, *Salvelinus namaycush*. Copeia 1954(2) : 116–124.
MOTTLEY, C. McC. 1941. The effect of increasing the stock in a lake on the size and condition of rainbow trout. American Fisheries Society Transactions 70(1940) : 414–420.
MRAZ, DONALD and EDWIN L. COOPER. 1957. Natural reproduction and survival of carp in small ponds. Wildlife Management Journal 21(1) : 66–69.
MUELLER, J. F. and HARLEY A. VAN CLEAVE. 1932. Parasites of Oneida Lake Fishes. Roosevelt Wildlife Annals 5(2) : 79–137.
MULCH, ERNEST E. and WILLIAM C. GAMBLE. 1954. Game fishes of Arizona. Arizona Game and Fish Department, Phoenix, Arizona, pp. 1–19.
MULLAN, JAMES W. 1958. The sea-run or "Salter" brook trout (*Salvelinus fontinalis*) fishery of the coastal streams of Cape Cod, Massachusetts. Massachusetts Division Fisheries Game Bulletin 17 : 1–25.
MURPHY, GARTH I. 1941. A key to the fishes of the Sacramento-San Joaquin Basin. California Fish and Game 27(3) : 165–171.
 1942. Relationship of the freshwater mussel to trout in the Truckee River. California Fish and Game 28(2) : 89–102.
 1948. A contribution to the life history of the Sacramento Perch (*Archoplites interruptus*) in Clear Lake, Lake County, California. California Fish and Game 34(3) : 93–100.
 1949. The food of young largemouth black bass (*Micropterus salmoides*) in Clear Lake, California. California Fish and Game 35(3) : 159–163.
 1950. The life history of the greaser blackfish (*Orthodon microlepidotus*) of Clear Lake, Lake County, California. California Fish and Game 36(2) : 119–133.
MUTTKOWSKI, RICHARD A. 1925. The food of trout in Yellowstone National Park. Roosevelt Wild Life Bulletin 2(4) : 470–497.
 1929. The ecology of trout streams in Yellowstone Park. Roosevelt Wild Life Annals 2(3) (Syracuse University).
MUTTKOWSKI, RICHARD A. and GILBERT M. SMITH. 1929. The food of trout stream insects in Yellowstone Park. Roosevelt Wild Life Annals 2(2). (Syracuse University).
MYERS, GEORGE SPRAGUE. 1931. The primary groups of oviparous Cyprinodont fishes. Stanford University Publications, University series, Biological Sciences 6(3) : 1–14.
NEALE, GEORGE. 1931A. Spiny-rayed freshwater game fishes of California inland waters. California Fish and Game 17(1) : 1–16.
 1931B. Sacramento Perch. California Fish and Game 17(4) : 409–411.
NEEDHAM, PAUL R. 1930. Studies on the seasonal food of brook trout. American Fisheries Society Transactions 60 : 73–88.
 1938. Trout streams. Comstock Publishing Co., Ithaca, New York. pp. 1–233.
 1947. Survival of trout in streams. American Fisheries Society Transactions 77 : 26–31.
 1961. Observations on the natural spawning of Eastern Brook Trout. California Fish and Game 47(1) : 27–40.
NEEDHAM, PAUL R. and F. K. CRAMER. 1943. Movement of trout in Convict Creek, California. Wildlife Management Journal 7(2) : 142–145.
NEEDHAM, PAUL R. and RICHARD GARD. 1959. Rainbow trout in Mexico and California. University California Publications Zoölogy 67(1) : 1–124.

NEEDHAM, PAUL R., J. W. MOFFETT and D. W. SLATER. 1945. Fluctuations in wild Brown trout populations in Convict Creek, California. Wildlife Management Journal 9(1) : 9–25.

NEEDHAM, PAUL R. and A. C. TAFT. 1934. Observations on the spawning of steelhead trout. American Fisheries Society Transactions 64(1934) : 332–338.

NEEDHAM, PAUL R. and TIM M. VAUGHAN. 1952. Spawning of the Dolly Varden, *Salvelinus malma*, in Twin Creek, Idaho. Copeia 1952(3) : 197–199.

NEEDHAM, PAUL R. and ELDON H. VESTAL. 1938. Notes on growth of golden trout (*Salmo aqua-bonita*) in two high Sierra lakes. California Fish and Game 24(3) : 273–279.

NELSON, EDWARD M. 1948. The comparative morphology of the Weberian Apparatus of the Catostomidae and its significance in Systematics. Morphology Journal 83(2) : 225–245.

1955. The 2–3 invertebral joint in the genus *Catostomus*. Copeia 1955(2) : 151–152.

NELSON, PHILIP R. 1959. Effects of fertilizing Bare Lake, Alaska, on growth and production of Red Salmon (*O. nerka*). United States Fish Wildlife Service Fishery Bulletin 195 : iv–59–86 (from volume 60).

NEUHOLD, JOHN M. 1957. Age and growth of the Utah chub, *Gila atraria* (Girard) in Panguitch Lake and Navajo Lake, Utah, from scales and opercular bones. American Fisheries Society Transactions 85(1955) : 217–233.

NEVADA, STATE OF: Biennial Reports of the Surveyor General and State Land Register. (Some of these reports contain accounts purporting to be accurate on the rivers and lakes of Nevada. It is unfortunate that these have official sanction and as such have probably been consulted and quoted by other persons and agencies, both within and without the State; the statistics they contain on Nevada rivers and lakes are so erroneous in many cases as to be worthless, particularly when consulted by anyone who knows nothing about the State and consequently would have to depend *in toto* upon these figures).

NEVADA, STATE OF: Biennial Reports for Fish and Game for the years—

1879. First biennial report of the Fish Commissioner of the State of Nevada for 1877–1878, by H. G. Parker, Carson City. 8 pages.

1881 (second). Idem. 1879–1880. 13 pages.

1883 (third). Idem. 1881–1882. 15 pages.

1885 (fourth). Idem. 1883–1884. 9 pages.

1887 (fifth). Idem. 1885–1886, by W. M. Cary, Carson City. 10 pages.

1889 (sixth). Idem. 1887–1888. 6 pages.

1891 (seventh). Idem. 1889–1890, by George T. Mills, Carson City. 23 pages.

1893 (eighth). Idem. 1891–1892. 14 pages.

1895 (ninth). Idem. 1893–1894. 14 pages.

1897 (tenth). Idem. 1895–1896. 23 pages.

1899–1907 reports, covering the years 1897 to 1906, were never issued.

1909 (eleventh). Biennial Report of the State Fish Commission of Nevada for 1907–1908, by George T. Mills, E. B. Yerington and H. H. Coryell, Carson City. 20 pages.

1911 (twelfth). Idem. 1909–1910, by Mills, Yerington and James Clarke, Carson City. 46 pages.

1913 (thirteenth). Idem. 1911–1912. 32 pages.

1915 (fourteenth). Idem. 1913–1914, by Mills, Yerington and B. E. Nixon, Carson City. 35 pages.

1917 (fifteenth). Idem. 1915–1916, by Charles J. Miller, James P. O'Brien and George T. Mills, Carson City. 13 pages.

1919 (sixteenth). Idem. 1917–1918, by J. P. O'Brien, M. P. Macmillan and O. W. Tennant, Carson City. 16 pages.

1921 (seventeenth). Idem. 1919–1920, by O'Brien, J. E. Johnson and Tennant. 19 pages.
1923 (eighteenth). Idem. 1921–1922, by A. G. Meyers, Johnson and G. F. Dangberg. 16 pages.
1925 (nineteenth). Idem. 1923–1924, by Meyers, Johnson and J. E. Robbins. 29 pages.
1927 (twentieth). Idem. 1925–1926. 24 pages.
1929 (twenty-first). Idem. 1927–1928, by R. L. Douglass, E. M. Steniger, G. K. Elder, W. Inwood and E. J. Phillips, Carson City. 21 pages.
1931 (twenty-second). Ibid. 1929–1930, by R. L. Douglass, M. G. Bradshaw, G. K. Elder, E. J. Phillips and Hon. Noble H. Getchell, Carson City. 39 pages.
1932 (twenty-third). Report of the Fish and Game Commission for the period January 1, 1931, to June 30, 1932, inclusive, by R. L. Douglass, M. G. Bradshaw, E. J. Phillips, Noble H. Getchell and Howard S. Doyle, Carson City. 14 pages.
1935 (twenty-fourth). Biennial Report of the Fish and Game Commission for the period July 1, 1932, to June 30, 1934, inclusive, idem., Carson City. 10 pages.
1936 (twenty-fifth). Report of the Fish and Game Commission for the period July 1, 1934, to June 30, 1936, inclusive, by R. L. Douglass, E. J. Phillips, Noble H. Getchell, Howard S. Doyle and Andy C. Barr, Carson City. 14 pages.
1938 (twenty-sixth). Idem. July 1, 1936, to June 30, 1938, inclusive, Carson City. 17 pages.
1941 (twenty-seventh). Idem. July 1, 1938, to June 30, 1940, inclusive, by E. J. Phillips, Noble H. Getchell, Howard S. Doyle, Andy C. Barr and William A. Powell, Carson City. 16 pages.
1942 (twenty-eighth). Idem. July 1, 1940, to June 30, 1942, inclusive, by E. J. Phillips, Andy C. Barr, William A. Powell, Jr., F. L. Baker and Leroy Casady, Carson City. 17 pages.
1944 (twenty-ninth). Idem. July 1, 1942, to June 30, 1944, inclusive, with James Olin replacing Casady for the duration of the war, Carson City. 16 pages.
1946 (thirtieth). Idem. July 1, 1944, to June 30, 1946, inclusive, with Casady replacing Olin, Carson City. 17 pages.
1948 (thirty-first). Idem. July 1, 1946, to June 30, 1948, inclusive, W. O. Bay, Dan Evans, Jr., Leroy Casady, Cal Liles and Warren Monroe (= 5-man executive board of the now 17-man commission), Carson City. 23 pages.
1950 (thirty-second). Idem. July 1, 1948, to June 30, 1950, inclusive, by idem. Carson City. 26 pages.
1953 (thirty-third). Idem. July 1, 1950, to June 30, 1952, inclusive, by Frank W. Groves, Director, Carson City. 47 pages.
1954 (thirty-fourth). Idem. July 1, 1952, to June 30, 1954, inclusive, by Frank W. Groves, Director, Carson City. 63 pages.
1956 (thirty-fifth). Idem. July 1, 1954, to June 30, 1956, inclusive, by Frank W. Groves, Director, Carson City. 71 pages.
1958 (thirty-sixth). Idem. July 1, 1956, to June 30, 1958, inclusive, by Frank W. Groves, Director, Carson City. 103 pages.
1960 (thirty-seventh). Idem. July 1, 1958 to June 30, 1960, inclusive, by Frank W. Groves, Director, Carson City. 61 pages.

NEWELL, ARTHUR E. 1957. Two-year study of movements of stocked brook trout and rainbow trout in a mountain trout stream. Progressive Fish-Culturist 19(2) : 76–80.

NICHOLS, HUDSON M. 1953. The age and growth of the largemouth bass (*Micropterus salmoides* Lacépède), in a new flood control reservoir in Missouri. Master's thesis, University of Missouri.

NOBLE, L. F. 1926. Note on a colemanite deposit near Shoshone, California, with a sketch of the geology of a part of Amargosa Valley. Part (D): 63–73, OF Contributions to Economic Geology. Part I.—Metals and nonmetals except fuels. United States Geological Survey Bulletin 785: iii + 73.

NOLAN, THOMAS B. 1943. The Basin and Range Province in Utah, Nevada and California. United States Geological Survey Professional Paper 197–D: 141–196.

NOLAN, THOMAS B., C. W. MERRIAM and J. S. WILLIAMS. 1956. The stratigraphic section in the vicinity of Eureka, Nevada. United States Geological Survey Professional Paper 276: 1–77.

NORMAN, J. R. 1931. A history of fishes. A. A. Wyn, Inc., New York (1947 reprinting). pages xv–463.

ODUM, HOWARD T. and DAVID K. CALDWELL. 1955. Fish respiration in the natural oxygen gradient of an anaerobic spring in Florida. Copeia 1955(2): 104–106.

O'MALLEY, HENRY. See United States Reports for Fish and Game for the year 1919.

PARKER, H. G. See Nevada, State of: Biennial Reports for Fish and Game for the years 1879, 1881, 1883 and 1885.

1878. Land-locking the Quinnat Salmon. Chicago Field 10: 165 (experiments in Pyramid and Walker Lakes).

PARSONS, JOHN W. 1950. Life history of the yellow perch, *Perca flavescens* (Mitchill), of Clear Lake, Iowa. Iowa State College Science Journal 25: 85–97.

PARSONS, JOHN W. and J. BRUCE KIMSEY. 1954. A report on the Mississippi threadfin shad. Progressive Fish-Culturist 16(4): 179–181.

PASCO, I. D. 1882. A call for carp from Nevada. United States Fish Commission Bulletin 1: 29–30 (1881).

PEARSE, A. S. 1915. On the food of the small shore fishes in the waters near Madison, Wisconsin. Wisconsin Natural History Society Bulletin 13: 7–22.

1918. The food of the shore fishes of certain Wisconsin lakes. United States Bureau Fisheries Bulletin 35: 247–292.

1921. The distribution and food of the fishes of three Wisconsin lakes in summer. University Wisconsin Studies Science No. 3: 1–61.

PEARSE, A. S. and H. ACHTENBERG. 1920. Habits of yellow perch in Wisconsin lakes. United States Bureau Fisheries Bulletin 36(885): 295–366.

PHILIP, CORNELIUS B. 1947. A catalog of the blood-sucking fly family Tabanidae (Horseflies and Deerflies) of the Nearctic Region north of Mexico. American Midland Naturalist 37(2): 257–324.

PICKFORD, GRACE E. and JAMES W. ATZ. 1957. The physiology of the pituitary gland of fishes. New York Zoölogical Society, New York pp. xxiii–613.

PINCHER, CHAPMAN. 1948. A study of fish. Duell, Sloan and Pearce, New York. Pages 1–343.

PISTER, E. P. 1958. Personal communication. Bishop, California.

PITT, T. K., E. T. GARSIDE and R. L. HEPBURN. 1956. Temperature selection of the carp (*Cyprinus carpio* Linn.) Canadian Zoölogy Journal 34: 555–557.

POPPE, ROBERT A. 1880. The introduction and culture of the carp in California. United States Commission Fish Fisheries Report 1878: 661–666.

POTTER, GEORGE E. 1925. Scales of the bluegill *Lepomis pallidus* (Mitchill). American Microscopical Society Transactions 40(1).

POWERS, E. B. et al. 1922. The physiology of the respiration of fishes in relation to the hydrogen-ion concentration of the medium. General Physiology Journal 4: 305–317.

1932. The relation of respiration of fishes to environment. Ecological Monographs 2: 385–473.

PRATT, H. S. 1923. A manual of land and freshwater vertebrate animals of the United States (exclusive of birds). P. Blakiston's Sons and Company, Philadelphia. Pages xv + 422.

1935. 2nd edition, ibid, pp. xvii + 416.

1937. Population studies of the trout of the Gunnison River. University Colorado studies 24(2): 107–116.

PRICE, J. W. 1931. Growth and gill development in the small-mouthed black bass, *Micropterus dolomieu*, Lacépède. Stone Laboratory Contributions No. 4: 1–46 (Put-In-Bay, Ohio).

PURKETT, CHARLES A., JR. 1950. Growth rate of trout in relation to elevation and temperature. American Fisheries Society Transactions 80: 251–259.

1958. Growth rates of Missouri stream fishes. Missouri Conservation Commission, Fish and Game Division, Dingell-Johnson Series No. 1: 1–46.

RADBRUCH, DOROTHY H. 1957. Hypothesis regarding the origin of thinolite tufa at Pyramid Lake, Nevada. Geological Society America Bulletin 68: 1683–1688.

RAFINESQUE, CONSTANTINE SAMUEL. 1817. First decade of new North-American fishes. American Monthly Magazine Critical Review 2(2): 120–121.

1818. Discoveries in natural history, made during a journey through the western regions of the United States. American Monthly Magazine Critical Review 3(5): 354–356.

1819. Prodrome de 70 nouveaux Genres D'Animaux decouverts dans l'interieur des Etats-Unis D'Amerique, durant l'annee 1818. Journal Physique, Chemie D'Histoire Naturelle, Arts, etc. 88: 417–429.

1820. Ichthyologia Ohiensis, or natural history of the fishes inhabiting the River Ohio and its tributary streams, etc. Printed for the author by W. G. Hunt, Lexington, Kentucky, pp. 5–89.[1]

RAMASWAMI, L. S. 1955A. Skeleton of cyprinoid fishes in relation to phylogenetic studies. VI. The skull and Weberian apparatus in the subfamily Gobioninae (Cyprinidae). Acta Zoölogica 36: 127–158.

1955B. Idem. VII. The skull and Weberian apparatus of Cyprininae (Cyprinidae). Ibid: 199–242.

1957. Idem. VIII. The skull and Weberian ossicles of Catostomidae. Zoölogical Society (Calcutta) Proceedings, Mookerjee Memorial Vol.: 293–303.

RANEY, E. C. and E. A. LACHNER. 1941. Autumn food of recently planted young brown trout in small streams of central New York. American Fisheries Society Transactions 71: 106–111.

RAWSON, DONALD S. 1940. The eastern brook trout in the Maligne River system, Jasper National Park. American Fisheries Society Transactions 70: 221–235.

RAWSON, DONALD S. and C. A. ELSEY. 1950. Reduction in the longnose sucker population of Pyramid Lake, Alberta, in an attempt to improve angling. American Fisheries Society Transactions 78: 13–31.

REID, JOHN A. 1911. The geomorphology of the Sierra Nevada northeast of Lake Tahoe. University California Department Geology Bulletin 6: 89–161.

REIGHARD, JACOB. 1903. Function of the pearl organs of the *Cyprinidae*. Science, new series, 17: 531.

RICE, LUCILLE A. 1941. *Gambusia affinis* in relation to food habits from Reelfoot Lake, 1940, with special emphasis on malaria control. Reelfoot Lake Biological Station Report 5: 77–87.

---

[1] By 1899, according to Call, only eight copies of Rafinesque's *Ichthyologia Ohiensis* were known to exist. In that year, Call reprinted the work in an edition of 250 numbered copies, No. 98 being in the University of Nevada library at Reno.

RICH, WILLIS H. 1925. Growth and degree of maturity of Chinook Salmon in the ocean. United States Bureau Fisheries Bulletin 41, Document 974: 15–90.

RICHARDSON, C. H. 1915. Reptiles of northwestern Nevada and adjacent territory. United States National Museum Proceedings 48(2078): 403–435.

RICHARDSON, ROBERT EARL. 1904. A review of the sunfishes of the current genera *Apomotis*, *Lepomis*, and *Eupomotis*, with particular reference to the species found in Illinois. Illinois State Laboratory Natural History Bulletin 7: 27–35.

RICHARDSON, SIR JOHN. 1836. Fauna Boreali-Americana; or the Zoölogy of the northern parts of British America, containing descriptions of the objects of natural history collected on the late northern land expeditions under command of Captain Sir John Franklin, R.N. Part Third. The Fish: xv–327, 24 colored plates. 8vo. Richard Bentley, New Burlington-Street, London (descriptions of *Salmo clarkii*, *S. gairdnerii*, *Cyprinus oregonensis* (= *Ptychocheilus*), *Cyprinus balteatus* (= *Richardsonius*), etc.)

RICKER, WILLIAM E. 1930. Feeding habits of speckled trout in Ontario waters. American Fisheries Society Transactions 60: 64–72 (Brook Trout).

1932A. Studies of Speckled trout (*Salvelinus fontinalis*) in Ontario. University Toronto Studies, Biological Series, No. 36: Ontario Fisheries Research Laboratory Publication No. 44: 69–110.

1932B. Studies of trout-producing lakes and ponds. Idem. No. 45: 113–167.

1934. An ecological classification of certain Ontario streams. Idem. No. 49: 1–114.

1937. The food and food supply of sockeye salmon (*Oncorhynchus nerka* Walbaum) in Cultus Lake, British Columbia. Biology Board Canada Journal 3: 450–468.

1938. A comparison of several growth rates of young sockeye salmon and young squawfish in Cultus Lake. Pacific Biology Station Fish Experimental Station Progress Report 36: 3–5. Prince Rupert, British Columbia.

1941. The consumption of young sockeye salmon by predaceous fish. Fishery Research Board Canada Journal 5(3): 293–313.

1943. The rate of growth of bluegill sunfish in lakes of northern Indiana. Indiana Lakes Streams Investigations 2: 161–214.

1956. The marking of fish. Ecology 37(4): 655–670.

RICKER, WILLIAM E. and KARL F. LAGLER. 1943 (1942). The growth of spiny-rayed fishes in Foots Pond. Indiana Lakes Streams Investigations 2: 85–97.

ROBINS, C. RICHARD and ROBERT R. MILLER. 1957. Classification, variation, and distribution of the sculpins, genus *Cottus*, inhabiting Pacific slope waters in California and southern Oregon, with a key to the species. California Fish and Game 43(3): 213–233.

ROBINS, C. RICHARD and E. C. RANEY. 1956. Studies of the Catostomid fishes of the genus *Moxostoma*, with descriptions of two new species. Cornell University Memoirs 343: 1–56.

ROCKWOOD, A. P. 1874. The native fish of Utah. American Fisheries Society Transactions 4: 24–25.

ROSE, ROBERT H. 1938. Pleistocene deposits in southeastern Nevada. Geological Society America Proceedings 1937: 250–251.

ROSTLUND, ERHARD. 1952. Freshwater fish and fishing in native North America. University California Geography Publications 9: x–313 (fish as a natural resource utilized by the Amerind).

ROYCE, WILLIAM F. 1943. The reproduction and studies of the life history of the lake trout *Cristivomer namaycush namaycush* (Walbaum). Unpublished doctoral dissertation, Cornell University Library, Ithaca, New York. Pages 1–136.

1951. Breeding habits of lake trout in New York. United States Fish Wildlife Service Fish Bulletin 59: 59–76.
RUPP, ROBERT S. 1955. Studies of the Eastern brook trout populations and fishery in Sunkhaze stream, Maine. Wildlife Management Journal 19(3): 336–345.
RUPP, ROBERT S. and MARVIN C. MEYER. 1954. Mortality among brook trout, *Salvelinus fontinalis*, resulting from attacks of freshwater leeches. Copeia 1954(4): 294–295.
RUSSELL, ISRAEL C. 1883. Sketch of the geological history of Lake Lahontan. United States Geological Survey Annual Report 3: 189–235.
1885. Geological history of Lake Lahontan, A Quaternary lake of northwestern Nevada. United States Geological Survey Monograph 11: xiv–288.
1896. Present and extinct lakes of Nevada *IN* The Physiography of the United States. National Geographic Society Monograph 1896: 101–132.
RUSSELL, RICHARD J. 1927A. Landslide lakes of the northwestern Great Basin. University California Geography Publications 2: 231–254.
1927B. The land forms of Surprise Valley, northwestern Great Basin. University California Geography Publication 2: 323–358.
1928. Basin range structure and stratigraphy of the Warner Range, northeastern California. University California Geological Science Publication 17: 387–496.
RUTTER, CLOUDSLEY. 1903. Notes on fishes from streams and lakes of northeastern California not tributary to the Sacramento basin. United States Fish Commission Bulletin 22: 145–148 (1902) (*Agosia robusta* = *Rhinichthys*).
1908. The fishes of the Sacramento-San Joaquin basin, with a study of their distribution and variation. United States Bureau Fisheries Bulletin 27: 105–152.
SCATTERGOOD, LESLIE W. 1949. Notes on the Kokanee (*Oncorhynchus nerka kennerlyi*). Copeia 1949(4): 297–298.
SCHOFFMAN, ROBERT J. 1938. Age and growth of bluegills and largemouth black bass in Reelfoot Lake. Tennessee Academy Science Journal 13: 81–103.
1957. Age and rate of growth of the carp in Reelfoot Lake, Tennessee, for 1941 and 1956. Tennessee Academy Science Journal 32(1): 3–8.
1960. Age and rate of growth of the White Crappie in Reelfoot Lake, Tennessee, for 1950 and 1959. Tennessee Academy Science Journal 35(1): 3–8.
SCHRENKEISEN, RAY. 1938. Field book of freshwater fishes of North America north of Mexico. G. P. Putnam's Sons, New York. xii–312 pages.
SCHUCK, H. A. 1943. Survival, population density, growth and movement of the wild brown trout in Crystal Creek. American Fisheries Society Transactions 73: 209–230.
SCHULTZ, LEONARD P. and Students. 1935. The breeding activities of the Little Redfish, a landlocked form of the Sockeye Salmon, *Oncorhynchus nerka* [sic.] Pan-Pacific Research Institution Section Mid-Pacific Magazine, January-March 1935: 67–77.
SCHULTZ, LEONARD P. 1935. Species of salmon and trout in the northwestern United States. Fifth Pacific Science Congress Proceedings 1933: 3777–3782.
1936. Keys to the fishes of Washington, Oregon and closely adjoining regions. University Washington Biology Publications 2(4): 105–228.
1938. The breeding habits of salmon and trout. Smithsonian Institution Annual Report for 1937: 365–376.
SCHULTZ, LEONARD P. and A. C. DeLACY. 1935. Fishes of the American Northwest. Pan-Pacific Research Institution Journal 10: 365–380, October-December.

SCHULTZ, LEONARD P. and M. B. SCHAEFER. 1936. Descriptions of new intergeneric hybrids between certain cyprinid fishes of northwestern United States. Biological Society Washington Proceedings 49: 1–10.

SCHULTZ, LEONARD P. and EDITH M. STERN. 1948. The ways of fishes. D. Van Nostrand Company, Inc., New York. Pages xii–264 (good general classification in back).

SCOTT, W. B. 1954. Freshwater fishes of eastern Canada. University Toronto Press in cooperation with Royal Ontario Museum Zoölogy Paleontology. pp. xiv plus 128.

 1956. Wendigo the hybrid trout. Royal Ontario Museum, Division Zoölogy Paleontology pp. 1–7.

SEALE, ALVIN. 1897. Note on *Deltistes*, a new genus of catostomid fishes. California Academy Science Proceedings (2d series) 6: 269 (1896).

SELF, J. TEAGUE. 1940. Notes on the sex cycle of *Gambusia affinis affinis*, and on its habits and relation to mosquito control. American Midland Naturalist 23(2): 393–398.

SHAPOVALOV, LEO. 1941. The freshwater fish fauna of California. Sixth Pacific Science Congress Proceedings 3: 441–446.

 1947. Distinctive characters of the species of anadromous trout and salmon found in California. California Fish and Game 33(3): 185–190.

SHAPOVALOV, LEO and WILLIAM A. DILL. 1950. A checklist of the freshwater and anadromous fishes of California. California Fish and Game 36(4): 382–391.

SHAPOVALOV, LEO, WILLIAM A. DILL and ALMO J. CORDONE. 1959. A revised checklist of the freshwater and anadromous fishes of California. California Fish and Game 45(3): 159–180.

SHARP, ROBERT F. 1938. Pleistocene glaciation in the Ruby-East Humboldt Range, northeastern Nevada. Geomorphology Journal 1: 296–323.

 1939A. The Miocene Humboldt formation in northeastern Nevada. Geology Journal 47: 133–160.

 1939B. Basin-Range structure of the Ruby-East Humboldt Range, northeastern Nevada. Geological Society America Bulletin 50: 881–920.

 1940. Geomorphology of the Ruby-East Humboldt Range, Nevada. Geological Society America Bulletin 51: 337–372.

SHEBLEY, W. H. 1919. Hatchery notes. California Fish and Game 5: 37–39.

 1921. Report of the Department of Fish Culture. California Fish and Game Commission 26th Biennial Report 1918–1920: 1–149.

SHEPARD, M. P. 1955. Resistance and tolerance of young Speckled trout (*Salvelinus fontinalis*) to oxygen lack, with special reference to low oxygen acclimation. Fisheries Research Board Canada Journal 12(3): 387–446.

SHETTER, DAVID S. 1936. Migration, growth rate, and population density of brook trout in the North Branch of the Au Sable River, Michigan. American Fisheries Society Transactions 66: 203–210.

SHETTER, DAVID S. and A. S. HAZARD. 1940. Results from planting of marked trout of legal size in streams and lakes of Michigan. American Fisheries Society Transactions 70: 446–447.

SHETTER, DAVID S. and JUSTIN W. LEONARD. 1942. A population study of a limited area in a Michigan trout stream, September, 1940. American Fisheries Society Transactions 72: 35–51.

SHIELDS, G. O. 1892 (Editor). American game fishes, their habits, habitat and peculiarities; how, when and where to angle for them. Rand McNally and Company, Chicago and New York. Pages 1–580.

SHIRA, AUSTIN F. 1917. Notes on the rearing, growth, and food of the channel catfish, *Ictalurus punctatus*. American Fisheries Society Transactions 46(2): 77–88.

SHOMON, J. J. 1955. Freshwater fishing and fishlife in Virginia. Commonwealth Virginia Commission Game and Inland Fisheries, Richmond. pp. 1–104.

SIGLER, WILLIAM F. 1951. The life history and management of the mountain whitefish, *Prosopium williamsoni* (Girard) in Logan River, Utah. Utah State College Agricultural Experiment Station Bulletin 347: 1–21.

1952. Age and growth of the brown trout, *Salmo trutta fario* Linnaeus, in Logan River, Utah. American Fisheries Society Transactions 81(1951): 171–178.

1953. The rainbow trout in relation to other fish in Fish Lake, Utah. Utah State Agricultural College Bulletin 358: 1–26.

SIMON, JAMES R. 1951. Wyoming fishes. Wyoming Game and Fish Department Bulletin 4: 1–129 (First edition 1946).

SIMON, JAMES R. and R. C. BROWN. 1943. Observations on spawning of the sculpin, *Cottus semiscaber*. Copeia 1943: 41–42.

SIMON, JAMES R. and F. SIMON. 1939. Checklist and keys of the fishes of Wyoming. University Wyoming Publications 6(4): 47–62.

SIMPSON, GEORGE G. 1933. A Nevada fauna of Pleistocene type and its probable association with man. American Museum Novitates 667: 1–10.

SIMPSON, J. H. 1876. Report of explorations across the Great Basin of the Territory of Utah for a direct wagon route from Camp Floyd to Genoa, in Carson Valley, in 1859. United States Army Engineers Department, pp. 1–166, 208–241.

SLASTENENKO, E. P. 1954. The relative growth of hybrid char (*Salvelinus fontinalis* X *Cristivomer namaycush*). Fisheries Research Board Canada Journal 11(5): 652–659.

1958A. The freshwater fishes of Canada. Kiev Printers, Toronto, Ontario, Canada. pp. 5–363.

1958B. The distribution of freshwater fishes in the Provinces and main water basins of Canada. Shevchenko Scientific Society Bulletin 1(6): 3–11.

SMILEY, CHARLES W. 1884. A statistical review of the production and distribution to public waters of young fish, by the United States Fish Commission, from its organization in 1871 to the close of 1880. United States Commission Fish Fisheries Report 1881: 825–915.

1887. Loch Leven trout introduced into the United States. United States Fish Commission Bulletin 7: 28–32.

SMITH, B. G. 1922. Notes on nesting habits of *Cottus*. Michigan Academy Science, Arts Letters, Paper 1: 221–225.

SMITH, F. 1925. Variation in the maximum depth at which fish can live during summer in a moderately deep lake with a thermocline. United States Bureau Fisheries Bulletin 41: 1–7.

SMITH, H. W. 1929. The excretion of ammonia and urea by the gills of fish. Biological Chemistry Journal 81: 727–742.

SMITH, HOBART M. 1952. Definition of species.—II. Turtox News 30(10): 180–182 (mention of Hubbs and Miller *Gila orcutti-Siphateles mohavensis* problem).

SMITH, HUGH M. 1896. A review of the history and results of the attempts to acclimatize fish and other water animals in the Pacific States. United States Fish Commission Bulletin 15: 379–472 (1895).

1910. The United States Bureau of Fisheries. Its establishment, functions, organization, resources, operatings and achievements. United States Bureau Fisheries Bulletin 28: 1365–1411.

SMITH, LLOYD L., JR., RAYMOND E. JOHNSON and LAURENCE HINER. 1946. Fish populations in some Minnesota trout streams. American Fisheries Society Transactions 76: 204–214.

SMITH, M. W. 1951A. Further observations upon the movements of speckled trout in a Prince Edward Island stream. Canadian Fish Culturist No. 10: 1–3.

1951B. The speckled trout fishery of Prince Edward Island. Canadian Fish Culturist. No. 11: 1–6.

SMITH, OSGOOD R. 1941. The spawning habits of cutthroat and eastern brook trout. Wildlife Management Journal 5(4) : 461–471.

1947. Returns from natural spawning of cutthroat trout and eastern brook trout. American Fisheries Society Transactions 74(1944) : 281–296.

SMITH, OSGOOD R. and PAUL R. NEEDHAM. 1942. Problems arising from the transplantation of trout in California. California Fish and Game 28 : 22–27.

SMITH, O. H. and JOHN VAN OOSTEN. 1940. Tagging experiments with lake trout, whitefish, and other species of fish from Lake Michigan. American Fisheries Society Transactions 69(1939) : 63–84.

SMITH, SIDNEY I. 1874. The crustacea of the fresh waters of the United States. B—The crustacean parasites of the freshwater fishes of the United States. Appendix F, natural history. Part 25 IN Report of the Commissioner for 1872–1873. United States Commission Fish Fisheries, pp. cii–808.

SNEED, K. E. 1951. A method for calculating the growth of channel catfish, *Ictalurus lacustris punctatus*. American Fisheries Society Transactions 80 : 174–183.

SNYDER, JOHN OTTERBEIN. 1905. Notes on the fishes of the streams flowing into San Francisco Bay, California. United States Commission Fish Fisheries Report 1904 : 327–338.

1908A. The fauna of the Russian River, California, and its relation to that of the Sacramento. Science (new series) 27 : 269–271.

1908B. Relationships of the fish fauna of the lakes of southeastern Oregon. United States Bureau Fisheries Bulletin 27 : 69–102.

1908C. The fishes of the coastal streams of Oregon and northern California. United States Bureau Fisheries Bulletin 27 : 153–189.

1912. A new species of trout from Lake Tahoe. United States Bureau Fisheries Bulletin 32 : 25–28 (*Salmo regalis*).

1913. The fishes of the streams tributary to Monterey Bay, California. United States Bureau Fisheries Bulletin 32 : 47–72.

1914. The fishes of the Lahontan drainage system of Nevada and their relation to the geology of the region. Washington Academy Science Journal 4 : 299–300.

1915. Notes on a collection of fishes made by Dr. Edgar A. Mearns from rivers tributary to the Gulf of California. United States National Museum Proceedings 49 : 573–586 (1916).

1916. The fishes of the streams tributary to Tomales Bay, California. United States Bureau Fisheries Bulletin 34 : 375–381.

1917A. The fishes of the Lahontan system of Nevada and northeastern California. Ibid. 35(1915–1916) : 33–86(1918)[2] (new species *Catostomus arenarius* [= *C. tahoensis*], *Richardsonius microdon* [= hybrid between *Richardsonius egregius* and *Siphateles bicolor obesus*] ; new genus and species *Leucidius pectinifer* [= *Siphateles bicolor obesus*] ; new species *Salmo aquilarum* [= hybrid between *Salmo clarki henshawi* and *Salmo gairdneri irideus*] ; *Salmo smaragdus*, etc.).

1917B. An account of some fishes from Owens River, California. United States National Museum Proceedings 54 : 201–205.

1918. The fishes of Mohave River, California. United States National Museum Proceedings 54 : 297–299.

1919. Three new whitefishes from Bear Lake, Idaho and Utah. United States Bureau Fisheries Bulletin 36 : 3–9.

1921. Notes on some western fluvial fishes described by Charles Girard in 1856. United States National Museum Proceedings 59 : 23–28.

1923. The return of marked king salmon grilse. California Fish and Game 8(2) : 102–107.

1924. A second report on the return of king salmon marked in 1919, in Klamath River. California Fish and Game 9(1) : 1–9.

---

[2]Volume 35, covering the years 1915–1916, was issued in 1918, but Snyder's paper was issued as a separate in 1917.

1926. The trout of the Sierra San Pedro Martir, Lower California. University California Publications Zoölogy 21(17) : 419–426.
1927. Notes on certain catostomids of the Bonneville system, including the type of *Pantosteus virescens* Cope. United States National Museum Proceedings 64(2508) : 1–6.
1933. A new California trout. California Fish and Game 20: 105–112.
1940. The trouts of California. California Fish and Game 26(2) : 96–138.
SNYDER, RICHARD C. 1949. Vertebral counts in four species of suckers (Catostomidae). Copeia 1949(1) : 62–65.
SPEIRS, J. MURRAY. 1951. History of the original descriptions of Great Lakes fishes. Research Council Ontario technical session, London, Ontario, February 24, 1951: 1–38 (mimeographed).
1952. Nomenclature of the channel catfish and the burbot of North America. Copeia 1952 : 99–103.
SPENCER, W. P. 1939. Diurnal activity rhythm in freshwater fishes. Ohio Sciences Journal 39(3) : 119–132.
SPOOR, W. A. 1946. A quantitative study of the relationship between the activity and oxygen consumption of the goldfish, and its application to the measurements of respiratory metabolism in fishes. Biological Bulletin 91(3) : 312–325.
SPRUGEL, G. 1951. An extreme case of thermal stratification and its effect on fish distribution. Iowa Academy Science Proceedings 58: 563–566.
SPURR, JOSIAH E. 1903. Descriptive geology of Nevada south of the Fortieth Parallel and adjacent portions of California. United States Geological Survey Bulletin 208: 1–229.
SRINIVASACHAR, H. R. 1955. Osteology of catfishes. Current Science 24: 164.
STANSBURY, HOWARD. 1852. Exploration and survey of the Valley of the Great Salt Lake, Utah, including a reconnaissance of a new route through the Rocky Mountains. United States Senate Special Session, March, 1851 (Executive Document No. 3)·: 1–303, 423–487. Philadelphia.
STARKS, EDWIN C. 1907. *Chasmistes oregonus* Starks, new species, pp. 141–142 *IN* D. S. Jordan, 1907, The fossil fishes of California with supplementary notes on other species of extinct fishes. University California Geology Publications 5(7) : 95–144.
STEARNS, ROBERT E. C. 1901. The fossil freshwater shells of the Colorado Desert, their distribution, environment and variation. United States National Museum Proceedings 24: 271–299 (1902).
STENTON, J. E. 1952. Additional information on eastern brook trout x lake trout hybrids. Canadian Fish Culturist 13: 1–7.
STIRTON, REUBEN A. 1932. Correlation of the Fish Lake Valley and Adar Mountain beds in the Esmeralda Formation of Nevada. Science (new series) 76: 60–61.
STOKELY, PAUL S. 1952. The vertebral axis of two species of centrarchid fishes. Copeia 1952(4) : 255–261.
STONE, LIVINGSTON. 1874A. The introduction of eastern fish into the waters of the Pacific Slope, together with an account of operations of the United States salmon breeding establishment on the McCloud River, California. Forest and Stream 2: 100–102.
1874B. Report of operations during 1872 at the United States Salmon-Hatching Establishment on the McCloud River, and on the California Salmonidae generally; with a list of specimens collected. United States Commission Fish Fisheries. Part II. Report of the Commissioner for 1872 and 1873: cii + 808.
STRAHORN, A. T. and CORNELIUS VAN DUYNE. 1912. Soil survey of the Fallon area, Nevada. United States Department Agriculture, Field Operations, Bureau Soils 11: 1477–1516 (1909).
STROUD, R. H. 1948. Growth of the basses and black crappie in Norris Reservoir, Tennessee. Tennessee Academy Science Journal 23(1) : 31–99.

SUCKLEY, GEORGE. 1860. Report upon the fishes collected on the survey, No. 5. Chapter 1. Report upon the Salmonidae, pp. 307–349. Chapter 2. Report upon the fishes exclusive of the Salmonidae, pp. 350–368, in Route near the forty-seventh and forty-ninth parallels, explored by I. I. Stevens, Governor of Washington Territory, in 1853–1855, IN Reports of explorations and surveys to ascertain the most practicable and economical route for a railroad from the Mississippi River to the Pacific Ocean, made under the direction of the Secretary of War, in 1853–1855, according to act of Congress, of March 3, 1853, May 31, 1854, and August 5, 1854. Volume 12, Book 2. 36th Congress, 1st Session, Senate Executive Document. Washington, D. C. Thomas H. Ford, Printer.

1862. Notices of certain new species of North America Salmonidae, chiefly in the collection of the N. W. Boundary Commission, in charge of Archibald Campbell, Esq., Commissioner of the United States, by Dr. C. B. R. Kennerly, naturalist to Commission. Lyceum Natural History New York Annals 7: 306–313 (1861) (new genus *Oncorhynchus*; new species *Salmo kennerlyi* [= *Oncorhynchus*]).

1874. On the North American species of salmon and trout. Appendix B. Part 3: 91–160, IN Report of the Commissioner for 1872–1873, United States Commission Fish Fisheries, pp. cii–808.

SULLIVAN, CHARLOTTE and KENNETH C. FISHER. 1953. Seasonal fluctuations in the selected temperature of speckled trout, *Salvelinus fontinalis* (Mitchill). Fisheries Research Board Canada Journal 10 (4): 187–195.

SUMNER, FRANCIS B. and URLESS N. LANHAM. 1942. Studies of the respiratory metabolism of warm and cool spring fishes. Biological Bulletin 82(2): 313–327.

SUMNER, FRANCIS B. and M. C. SARGENT. 1940. Some observations on the physiology of warm spring fishes. Ecology 21(1): 45–54.

SUMNER, FRANCIS H. 1940. The decline of the Pyramid Lake fishery. American Fisheries Society Transactions 69: 216–224.

SUMNER, ROBERT C., V. KAY JOHNSON and DONALD J. KING. 1958A. Ryepatch Reservoir. Lahontan Project. Fisheries Management Report, Nevada Fish and Game Commission, iv–29 (mimeographed).

1958B. Stillwater Marsh. Lahontan Project. Fisheries Management Report, Nevada Fish and Game Commission, vi–92 (mimeographed).

1958C. Indian Lakes. Lahontan Project. Fisheries Management Report, Nevada Fish and Game Commission, vi–107 (mimeographed).

1958D. Lahontan Reservoir. Lahontan Project. Fisheries Management Report, Nevada Fish and Game Commission, vi–110 (mimeographed).

1958E. Washoe Lake. Lahontan Project. Fisheries Management Report, Nevada Fish and Game Commission, iv–28 (mimeographed).

SURBER, THADDEUS. 1920. Fish and fish-like vertebrates of Minnesota. Minnesota State Game and Fish Commission Biennial Report 1920: 1–92.

1933. Rearing lake trout to maturity. American Fisheries Society Transactions 63: 64–68.

SVÄRDSON, GUNNAR. 1945. Chromosome studies on Salmonidae. Swedish State Institute Fresh Water Fishery Research Report. Drottningholm.

TAFT, ALAN C. 1928. Age and growth of *Ptychocheilus grandis*, a western minnow. Master's thesis, Stanford University Library. pp. 1–27.

TAFT, ALAN C. and GARTH I. MURPHY. 1950. The life history of the Sacramento squawfish (*Ptychocheilus grandis*). California Fish and Game 36(2): 147–164.

TAIT, J. S. 1956. Nitrogen and argon in salmonoid swimbladders. Canadian Journal Zoölogy 34: 58–62.

TANNER, VASCO M. 1932. A description of *Notolepidomyzon utahensis*, a new catostomid from Utah. Copeia 1932(3): 135–136.

1936. A study of the fishes of Utah. Utah Academy Sciences, Arts Letters 13: 155–184.

1942. A review of the genus *Notolepidomyzon* with a description of a new species (Pisces-Catostomidae). Great Basin Naturalist 3(3) : 27–32 (*N. intermedius* = *Pantosteus*).

1950. A new species of *Gila* from Nevada (Cyprinidae). Great Basin Naturalist 10(1–4) : 31–36 (*Gila jordani*).

TANNER, VASCO M. and SHELDON P. HAYES. 1933. The genus *Salmo* in Utah. Utah Academy Science Proceedings 10 : 163–164.

TAYLOR, WILLIAM RALPH. 1954. Records of fishes in the John N. Lowe collection from the upper peninsula of Michigan. University Michigan Museum Zoölogy Miscellaneous Publication 87 : 1–50.

1955. A revision of the genus *Noturus* Rafinesque with a contribution to the classification of the North American catfishes. Unpublished Ph.D. thesis. University of Michigan Library, pp. 1–583.

THOMPSON, DAVID H. and LEVERETT A. ADAMS. 1936. A race of wild carp lacking pelvic fins. Copeia 1936(4) : 210–215.

THOMPSON, RICHARD B. 1959. Food of the squawfish *Ptychocheilus oregonensis* (Richardson) of the lower Columbia River. United States Fish Wildlife Service Fishery Bulletin 158: iv, 43–58 (from Vol. 60).

THOMPSON, WILL F. 1920. Investigation of the Salton Sea. California Fish and Game 6(2) : 83–84.

THOMPSON, WILL F. and HAROLD C. BRYANT. 1920. The mullet fisheries of Salton Sea. California Fish and Game 6(2) : 60–63.

THORNTHWAITE, C. W. 1948. An approach toward a rational classification of climate. Geographical Review 38 : 55–94.

THORPE, L. M., H. J. RAYNER, and D. A. WEBSTER. 1944. Population depletion in brook, brown and rainbow trout stocked in the Blackledge River, Connecticut, in 1942. American Fisheries Society Transactions 74 : 166–187.

THROCKMORTON, S. R. 1873. On the introduction of exotic food fishes into the waters of California. California Academy Science Proceedings 5 : 86–88.

TOWNSEND, LAWRENCE D. 1944. Variation in the number of pyloric caeca and other numerical characters in Chinook salmon and in trout. Copeia 1944(1) : 52–54.

TRAUTMAN, MILTON B. 1957. The fishes of Ohio. Ohio State University Press (in collaboration with Ohio Division Wildlife and Ohio State University Development Fund), pp. xvii + 683.

TREAT, ARCHIBALD. 1904. Trolling in deep water with the rod on Lake Tahoe. Western Field 3(6) : 864–868.

TRELEASE, THOMAS J. 1948. Report of field survey and investigations of the fisheries resources of Ruby Lake, Nevada. Nevada Fish and Game Commission Mimeographed Report, pp. 1–10.

1949. What lies ahead for Pyramid Lake? Nevada Hunting and Fishing 1(2) : 8–10.

1952. The death of a lake. Field and Stream 56(10) : 30–31, 109, 110–111.

TURNER, CHARLES L. 1946. A case of hermaphroditism in the cutthroat trout. Chicago Academy Sciences Natural History Miscellanea 1: 1–2.

TURNER, H. W. 1900. The Esmeralda Formation, a freshwater lake deposit. pp. 197–208. United States Geological Survey 21st Annual Report, Part II : 1–522.

TURNER, L. M. 1886. Contributions to the natural history of Alaska. No. 2. Arctic series of publications issued in connection with the Signal Service, United States Army, Washington, D. C. Government printing office, pp. 1–226 (Fishes: Part IV : 87–113).

UNITED STATES, BUREAU OF RECLAMATION.
1941A. Lake Mead density currents investigations. Vol. I : 1–327 (1937–1940).

1941B. Idem. Vol. 2: 328–453 (1937–1940).

1946. The Colorado River. A comprehensive report on the development of the water resources of the Colorado River basin for irrigation, power production, and other beneficial uses in Arizona, California, Colorado, Nevada, New Mexico, Utah and Wyoming. Pages vii–295.

1947. Idem. Vol. 3 : 454–904 (1940–1946).

UNITED STATES, Reports for Fish and Game for the years :

1895. Report on the propagation and distribution of food-fishes, by S. G. Worth. United States Commission Fish Fisheries Report 1893 : 78–138.

1919. The distribution of fish and fish eggs during the fiscal year 1918; by Henry O'Malley. United States Commission Fisheries Report 1918 : 1–82.

1920. Ibid. 1919, by G. C. Leach. Ibid. 1919: 1–76.

1927. Propagation and distribution of food fishes, fiscal year 1926, by G. C. Leach. Ibid. 1926 : 323–384.

1931. Ibid. 1930, by G. C. Leach. Ibid. 1930 : 1123–1191.

1932. Ibid. 1931, by G. C. Leach, Ibid. 1931 : 627–690.

1940. Ibid. 1939, by G. C. Leach, M. C. James and E. J. Douglass. Ibid. 1939 : 555–598.

1941. Ibid. 1940, by same authors. Ibid. 1940 : 555–603.

1942. Ibid. 1941, by same authors. United States Fish Wildlife Service Statistical Digest 3 : 1–25.

1943. Ibid. calendar year 1941, by same authors. Ibid. 6 : 1–26.

UNITED STATES, Water Supply Papers (Geological Survey).

(This vast compilation of water records contains much that is pertinent to the Great Basin area, and may be consulted in its entirety in such libraries as that of the University of Nevada, Reno.)

UNITED STATES FISH AND WILDLIFE SERVICE. 1955. Fishery Publication Index, 1920–1954. United States Government Printing Office, Washington, pp. x–254.

UPHAM, WARREN. 1922. Stages of the Ice Age. Geological Society America Bulletin 33 : 491–514.

UYENO, TERUYA. 1960. Osteology and phylogeny of the American cyprinid fishes allied to the genus *Gila*. University Michigan Ph.D. thesis.

VAN CLEAVE, HARLEY J. and JUSTUS F. MUELLER. 1932A. Parasites of Oneida Lake fishes. Part 1. Descriptions of new genera and new species. Roosevelt Wild Life Annals 3(1).

1932B. Parasites of Oneida Lake fishes. Part 2. Descriptions of new species and some general taxonomic considerations, especially concerning the trematode family Heterophyidae. Roosevelt Wild Life Annals 3(2).

VAN OOSTEN, JOHN. 1923. A study of the scales of whitefishes of known ages. Zoölogica 2(17) : 380–412.

1944. Lake trout. United States Fish Wildlife Service Fishery Leaflet 15 : 1–8 (first issuance 1943).

1950. Progress report on the study of Great Lakes trout. The Fisherman 18(5) : 5, 8–10, (6) : 5, 8.

1956. Biology of young lake trout (*Salvelinus namaycush*) in Lake Michigan. United States Fish Wildlife Service Research Report 42 : 1–88.

VAN OOSTEN, JOHN and HILARY J. DEASON. 1938. The food of the lake trout (*Cristivomer namaycush*) and of the lawyer (*Lota maculosa*) of Lake Michigan. American Fisheries Society Transactions 67(1937) : 155–177.

VESTAL, ELDEN H. 1942. Rough fish control in Gull Lake, Mono County, California. California Fish and Game 28 : 34–61.

1950. Chemical treatment of Upper Twin Lake, Robinson Creek, Mono County, California. California Division Fish and Game Bureau Fish Conservation, Typewritten Report, pp. 1–30.

VLADYKOV, VADIM D. 1954. Taxonomic characters of the eastern North American chars (*Salvelinus* and *Cristivomer*). Fisheries Research Board Canada Journal 11(6) : 904–932.

1956. Fecundity of wild speckled trout (*Salvelinus fontinalis*) in Quebec lakes. Fisheries Research Board Canada Journal 13(6) : 799–841.

VLADYKOV, VADIM D. and V. LEGENDRE. 1940. The determination of the number of eggs in ovaries of brook trout (*Salvelinus fontinalis*). Copeia 1940(4) : 218–220.
VON PIRKO, FRANZ. 1910. Naturalization of American fishes in Austrian waters. United States Bureau Fisheries Bulletin 28 : 977–982.
WALBAUM, JOHANN JULIUS. 1792. Petri Artedi renovati, i.e., bibliotheca et philosophia ichthyolegica. Ichthyologiae pars III. Petri Artedi Sueci genera piscium in quibus systema totum ichthyologiae proponitur cum classibus, ordinibus, generum characteribus, specierum differentiis, observationibus plurimis. Redactis speciebus 242 ad genera 52. Grypeswaldiae, Ant. Ferdin. Rose : 1–723.
WALES, JOSEPH H. 1930. Biometrical studies of some races of cyprinodont fishes from Death Valley region, with description of *Cyprinodon diabolis*, n. sp. Copeia 1930(3) : 61–70.
— 1946. Castle Lake trout investigation. First phase; interrelationships of four species. California Fish and Game 32(3) : 109–143.
— 1950. Swimming speed of the Western sucker, *Catostomus occidentalis* Ayres. California Fish and Game 36(4) : 433–434.
— 1957. Trout of California. California Department Fish and Game, pp. 1–56.
WALFORD, LIONEL A. 1931. Handbook of common commercial and game fishes of California. California Division Fish and Game Fish Bulletin 28 : 1–183.
WALLIS, ORTHELLO L. 1951. The status of the fish fauna of the Lake Mead National Recreational Area, Arizona-Nevada. American Fisheries Society Transactions 80 : 84–92 (1950).
WARD, HENRY B. 1904. A biological reconnaissance of some elevated lakes in the Sierras and the Rockies, with reports on the Copepoda by D. Dwight Marsh, and on the Cladocera by E. A. Birge. American Microscopical Society Transactions 25 : 127–152.
— 1932. The origin of the land-locked habit in salmon. United States National Academy Science Proceedings 18 : 569–580.
WARING, GERALD A. 1918. Ground water in Reese River Basin and adjacent parts of Humboldt River Basin, Nevada. United States Geological Survey Water Supply Paper 425–D : 95–129.
— 1920. Ground water in Pahrump, Mesquite and Ivanpah Valleys, Nevada and California. United States Geological Survey Water Supply Paper 450 : 51–81.
WEINREB, EVA LURIE. 1955. Histochemical demonstration of sites of acid phosphatase activity in mitochondria. Experimental Cell Research 8 : 159–162 (*Salmo gairdneri irideus*).
WEINREB, EVA LURIE and NELLIE M. BILSTAD. 1955. Histology of the digestive tract and adjacent structures of the rainbow trout, *Salmo gairdneri irideus*. Copeia 1955(2) : 194–204.
WEISEL, GEORGE F. 1943. A histological study of the testes of the sockeye salmon (*Oncorhynchus nerka*). Morphology Journal 73 : 207–229.
— 1954. A rediscovered cyprinid hybrid from western Montana, *Mylocheilus caurinus* X *Richardsonius balteatus balteatus*. Copeia. 1954(4) : 278–282.
— 1957. Fish guide for intermountain Montana. Montana State University Press, Missoula, Montana, pp. 1–88.
— 1960. The osteocranium of the catostomid fish, *Catostomus macrocheilus*. A study in adaptation and natural relationship. Morphology Journal 106(1) : 109–129.
WEISEL, GEORGE F. and H. WILLIAM NEWMAN. 1951. Breeding habits, development and early life history of *Richardsonius balteatus*, a northwestern minnow. Copeia 1951, (3) : 187–194. (= *Chenoda cooperi* Girard 1856).
WELCH, PAUL S. 1952. Limnology. McGraw-Hill Book Company, Inc. New York. Pages xi + 538, second edition.

WELLS, M. M. 1913. The resistance of fishes to different concentrations and combinations of oxygen and carbon dioxide. Biological Bulletin 25: 323–427.

WELLS, N. A. 1932. The importance of the time element in the determination of the respiratory metabolism of fishes. National Academy Sciences Proceedings 18: 580–585.

1935A. The influence of temperature upon the respiratory metabolism of the Pacific killifish, *Fundulus parvipinnis*. Physiological Zoölogy 8: 196–227.

1935B. Variations in the respiratory metabolism of the Pacific killifish, *Fundulus parvipinnis*, due to size, season and continued constant temperature. Ibid: 318–336.

WEMPLE, EDNA M. 1906. New cestraciont teeth from the West-American Triassic. University California Geology Publication 5(4): 71–73.

WEYMOUTH, FRANK W. 1923. The life history and growth of the Pismo clam (*Tivela stultorum* Mawe). California Fish and Game Fish Bulletin 7: 5–120.

WHEELER, GEORGE M. (assisted by D. W. Lockwood). 1875. Preliminary report upon a reconnaissance through southern and southeastern Nevada made in 1869. United States Geographical Survey West of 100th Meridian 1: 1–72.

WHEELER, HARRY E. 1939. *Helicoprion* in the anthracolithic (Late Paleozoic) of Nevada and California, and its stratigraphic significance. Paleontology Journal 13(1): 103–114.

WHEELER, SESSIONS S. and THOMAS J. TRELEASE. 1949. Nevada fisheries. Nevada Hunting and Fishing 1(4): 10–11, 21–23.

WHEELER, SESSIONS S., et al. 1949. Conservation in Nevada. Nevada State Department Instruction, Carson City, 131 pages.

1954. Nevada Conservation Adventure. Nevada State Department Public Instruction and Nevada State Fish and Game Commission, Carson City and Reno, 142 pages.

1959. Nevada Conservation Adventure. Nevada State Department Education, Nevada State Fish and Game Commission and Nevada State Department Conservation Natural Resources, Carson City and Reno, 125 pages.

WHITE, H. C. 1930. Some observations on the eastern brook trout (*Salvelinus fontinalis*) of Prince Edward Island. American Fisheries Society Transactions 60: 101–108.

1934. The spawning period of brook trout, *S. fontinalis*. American Fisheries Society Transactions 64: 356–357.

1940. Life history of sea-running brook trout (*Salvelinus fontinalis*) of Moser River, N. S. Fisheries Research Board Canada Journal 5(2): 176–186.

1941. Migrating behavior of sea-running *Salvelinus fontinalis*. Fisheries Research Board Canada Journal 5: 258–264.

1942. Sea life of the brook trout (*Salvelinus fontinalis*). Fisheries Research Board Canada Journal 5: 471–473.

WIEBE, A. H. 1933. The effect of high concentrations of dissolved oxygen on several species of pond fishes. Ohio Journal Science 33(2): 110–126.

WIEBE, A. H., A. M. McGAVOCK, A. C. FULLER and H. C. MARKUS. 1934. The ability of freshwater fishes to extract oxygen at different hydrogen-ion concentrations. Physiological Zoölogy 7: 435–448.

WIGGINS, W. G. B. 1950. The introduction and ecology of the brown trout (*Salmo trutta* Linnaeus) with special reference to North America. Unpublished Master's thesis, University of Toronto, Canada. pp.iii–109.

WILDER, D. G. 1952. A comparative study of anadromous and freshwater populations of brook trout (*Salvelinus fontinalis* (Mitchill)). Fisheries Research Board Canada Journal 9: 169–203.

WILDING, J. L. 1939. The oxygen threshold for three species of fish. Ecology 20: 253–263.

WILSON, WILLIAM D. 1957. Parasites of fishes from Leavenworth County State Lake, Kansas. Kansas Academy Science Transactions 60(4): 393–399.

WINN, HOWARD ELLIOTT and ROBERT R. MILLER. 1954. Native postlarval fishes of the lower Colorado River Basin, with a key to their identification. California Fish and Game 40(3): 273–285.

WITT, ARTHUR B., JR. 1952. Age and growth of the white crappie, (*Pomoxis annularis* Rafinesque), in Missouri. Ph.D. thesis, University of Missouri.

WOLF, H. and J. H. WALES. 1953. Color perception in trout. Copeia 1953 (4): 234–236.

WOOD, EDWARD M. 1953. A century of American fish culture, 1853–1953. Progressive Fish-Culturist 15(4): 147–162.

WORTH, S. G. See United States Reports for Fish and Game for the year 1895.

WRIGHT, JAMES E. 1955. Chromosome numbers in trout. Progressive Fish-Culturist 17(4): 172–176.

1956. Chromosome numbers in lake trout and splake. Unpublished report, Pennsylvania State University.

WRIGHT, SEWALL. 1940A. Breeding structure of populations in relation to speciation. American Naturalist 74: 232–248.

1940B. The statistical consequences of heredity in relation to speciation *IN* The New Systematics, Oxford University Press, pp. 161–183.

1942. Statistical genetics and Evolution. American Mathematical Society Bulletin 48: 223–246.

YARROW, H. C. 1874. On the speckled trout of Utah Lake. *Salmo virginalis*, Girard. Appendix B. Part 12: 363–368, *IN* Report of the Commissioner for 1872–1873. United States Commission Fish Fisheries, pp. cii + 808.

YARROW, H. C. and H. W. HENSHAW. 1878. List of marine fishes collected on the coast of California near Santa Barbara in 1875, with notes. Appendix K-1: 1623-1627 *IN* Annual Report of the Chief of Engineers to the Secretary of War for the year 1878, Part 3. 45th Congress, 3d Session, House Document No. 1, Part 2, Volume 2 (also issued, with the same title, as "Appendix K-1: 201–205, *IN* Appendix NN of the Annual Report of the Chief of Engineers for 1878 (Annual Report upon the geographical surveys of the territory of the United States west of the 100th Meridian, in the states and Territories of California, Colorado, Kansas, Nebraska, Nevada, Oregon, Texas, Arizona, Idaho, Montana, New Mexico, Utah, Washington, and Wyoming, by George M. Wheeler)).

YOUNG, G. J. 1914. Potash salts and other salines in Great Basin region. United States Department Agriculture Bulletin 61: 1–96.

YOUNGS, F. O. and E. J. CARPENTER. 1928. Soil survey of the Moapa Valley area, Nevada. United States Department Agriculture, Field Operations, Bureau Chemistry Soils 25: 749–774 (1923).

ZARBOCK, WILLIAM M. 1951. Age and growth and food habits of the sculpin, *Cottus bairdii semiscaber* (Cope) in the Logan River, Utah. Utah State Agriculture College Cooperative Wildlife Research Unit 16(2): 42–43 mimeographed abstract.

1952. Life history of the Utah sculpin, *Cottus bairdi semiscaber* (Cope), in Logan River, Utah. American Fisheries Society Transactions 81: 249–259.

POSTSCRIPT

For years the writer has collected early photographs of Pyramid, Winnemucca and Walker Lakes, particularly those showing water levels. If anyone reading this has such photographs, or knows of some, it would be greatly appreciated if they got in touch with the author at the University of Nevada.

PLUVIAL LAKES AND NOW DISCONNECTED PLUVIAL RIVERS OF THE GREAT BASIN AND OF THE DISRUPTED PARTS OF THE PACIFIC DRAINAGES, THE PAST CONNECTIONS OF THESE WATERS, AND THE FISH FAUNAL RELATIONSHIPS OF THE RECENT REMNANTS OF THE PLUVIAL WATERS

Pluvial Lake or River	Valley or Basin	Pluvial Connection	Most Plausible "Earlier Pluvial" Connection	Relationships of Recent Fish Fauna of Remnant Waters
Bonneville System (including Snake River above falls)				
1.—Lake Bonneville	Great Salt Lake, Sevier Lake, Great American Desert, etc.	Columbia R., via Snake R. (D)	Same as Pluvial (D)?; L. Lahontan and Colorado R. (C or A)	Uniform throughout; largely distinctive; in part with Columbia, Klamath, Sacramento and Colorado systems
1a.—Bear Lake	Bear Lake	L. Bonneville (D)	L. Bonneville (D)	Bonneville system (conclusive)
1b.—Little Salt Lake	Parowan Valley	L. Bonneville (D)	L. Bonneville (D)	Bonneville system (conclusive)
1c.—Pine Lake	Sage Brush Valley	L. Bonneville (D) (possibly)	L. Bonneville (D)	Basin now fishless
Lahontan System				
2.—Lake Lahontan	Pyramid Lake, Walker Lake, Humboldt River, etc.	Probably none (D); several surrounding basins (C or A)?	First toward SW.?; Klamath Lakes (D)?; surrounding basins ?(C or A)	Uniform throughout; in part distinctive; in part with Columbia, Klamath, Sacramento, Bonneville and minor systems
Lakes connected with central Lahontan basin				
3.—Lake Newark and Fish Creek	Newark and Little Smoky valleys	L. Lahontan, via S. Fork Humboldt R. (D)	L. Lahontan, via S. Fork Humboldt R. (D)	Lahontan system (evidence weak)
4.—Lake Diamond	Diamond, Kobeh, Monitor and Antelope valleys	L. Lahontan, via S. Fork Humboldt R. (D)	L. Lahontan, via S. Fork Humboldt R. (D)	Lahontan system (evidence moderately conclusive)
5.—Lake Gilbert	Grass Valley	L. Lahontan, via Crescent Valley (D)	L. Lahontan, via Crescent Valley (D)	Lahontan system (presumptive)
6.—Lake probably "earlier pluvial"	Crescent Valley	L. Lahontan, via Humboldt R. (D)	L. Lahontan, via Humboldt R. (D)	Lahontan system (evidence rather conclusive)
7.—Lake probably "earlier pluvial"	Carlco Valley	L. Lahontan, via Crescent Valley (D)	L. Lahontan, via Crescent Valley (D)	Basin now fishless
8.—Buffalo Lake	Buffalo Valley	L. Lahontan, via Reese R. (D); not certain	L. Lahontan, via Reese R. (D)	Basin now fishless
9.—Lake lacking or insignificant	Antelope Valley	L. Lahontan, via Reese R. (D)	L. Lahontan, via Reese R. (D)	Basin now fishless
10.—Lake Washoe	Washoe Lakes basin	L. Lahontan (D)	L. Lahontan (D)	Lahontan system (conclusive)
11A.—Lake Tahoe	Lake Tahoe basin	L. Lahontan (D)	L. Lahontan (D)	Lahontan system (conclusive)
11B.—Lake Truckee	In course, Truckee River	L. Lahontan (D)	Non-existent	Lahontan system (conclusive)
12.—Eagle Lake	Eagle Lake basin	L. Lahontan (D, probably, or S)	L. Lahontan (D)	Lahontan system (conclusive)
13.—Horse Lake	Horse Lake basin	L. Lahontan (D)	L. Lahontan (D)	Lahontan system (conclusive)
14.—High Rock Lake	High Rock Lake basin	L. Lahontan (D)	L. Lahontan (D)	Lahontan system (presumptive)
15.—Lake Lemmon	Lemmon Valley	L. Lahontan (D)	L. Lahontan (D)	Basin now fishless

FISHES AND FISHERIES OF NEVADA

Lakes in enclosed basins surrounded by Lahontan watershed

#	Lake	Location	Pluvial connection?	Pluvial source	Status
16.	Lake Wellington	Basin north of Wellington	None, or possibly L. Lahontan (D)	L. Lahontan, via Walker R. (D)	Basin now fishless
17A.	Lake Laughton	White Lake Valley	Probably none; possibly L. Lahontan (D)	L. Lahontan (D)	Basin now fishless
17B.	Lake Fred	Basin east of Fred Mt.	None	L. Lahontan (D) ?	Basin now fishless
18A.	Lake Kumiva	Kumiva (Kumavi) Valley	None	L. Lahontan (D) ?	Basin now fishless
18B.	Granite Springs Lake	Granite Springs Valley	None	L. Lahontan (D) ?	Basin now fishless

Minor Lakes Just South of Lake Lahontan System

#	Lake	Location	Pluvial connection?	Pluvial source	Status
19.	Fish Lake	Fish Lake Valley	Lake Columbus (D)	Lake Lahontan? (indirectly)	Lahontan system (dubious)
20.	Lake Dixie	Humboldt Salt Marsh, Dixie Valley	Probably none	Lake Lahontan	Lahontan system (moderately conclusive)
21.	Lake Labou	Fairview Valley	Lake Dixie (D)?	Lake Dixie (D)	Basin now fishless
22.	Lake Tolyabe	Big Smoky Valley, northern part	Lake Lahontan (C), via Lake Gilbert (problematical); Lake Tonopah (C) ?	Same as Pluvial, or with Lake Tonopah by discharge	Lahontan system (evidence weak)

23.—Area of Sterile Basins

#	Lake	Location	Pluvial connection?	Pluvial source	Status
A-G.	Basins adjacent to Lake Lahontan				
A.	Lake Edwards	Edwards Creek basin	Probably none	Dubious	Basin now fishless
B.	Lake Smith	Smith Valley	Probably none	Dubious	Basin now fishless
C.	Lake Gabb	Gabbs Valley	None	Probably none	Basin now fishless
D.	Lake Acme	Soda Springs Valley (west)	Possibly L. Lahontan (D)	Lake Lahontan (D) ?	Basin now fishless
D'.	Lake Luning	Soda Springs Valley (east)	L. Rhodes (D)	Lake Lahontan (D) ?	Basin now fishless
D".	Lake Rhodes	Rhodes Salt Marsh	None	Lake Lahontan (D) ?	Basin now fishless
E.	Lake Garfield	Garfield Flat	None	Probably none	Basin now fishless
F.	Lake Teel	Teels Marsh	None	Probably none	Basin now fishless
G.	Lake Huntoon	Huntoon Valley	None	Probably none	Basin now fishless
H-K.	Basins structurally connected with basins of Lake Tolyabe and Fish Lake				
H.	Lake Monte Cristo	Monte Cristo basin, north of Monte Cristo Mountains	Probably none	Possibly L. Tonopah (D)	Basin now fishless
I.	Lake Columbus	Columbus Salt Marsh	Possibly Fish Lake (T)	Fish Lake; perhaps also Lake Lahontan via Lake Rhodes	Basin now fishless
J.	Lake Tonopah	Big Smoky Valley, southern part	Lake Tolyabe (A) ?; Lake Clayton (A) ?	Lake Tolyabe (T) ?	Basin now fishless
K.	Lake Clayton	Silver Springs Marsh, Clayton Valley	Lake Tonopah (A) ?	Lake Tonopah (A) ?	Basin now fishless
L-T.	Basins structurally connected with Death Valley system				
L.	Lake ephemeral?	Ralston Valley	Possibly Stonewall Flat (D)	Stonewall Flat (D)	Basin now fishless

768   FISHES AND FISHERIES OF NEVADA

Pluvial Lake or River	Valley or Basin	Pluvial Connection	Most Plausible "Earlier Pluvial" Connection	Relationships of Recent Fish Fauna of Remnant Waters
M.—Lake ephemeral?	Salisbury Wash and Cactus Flat	Possibly Ralston Valley	Ralston Valley (D); Gold Flat (T)	Basin now fishless
N.—Lake ephemeral?	Gold Flat	Probably none	Cactus Flat (D)	Basin now fishless
O.—Lake ephemeral?	Alkali Spring Valley	Probably none	Ralston Valley (D)	Basin now fishless
P.—Lake ephemeral?	Stonewall Flat, NE. part	Possibly Ralston Valley (T); possibly SW. part, same flat	Ralston Valley (T); SW. part of same flat	Basin now fishless
P.—Lake ephemeral?	Stonewall Flat, SW. part	Possibly basins P and Q (T)	Basins P and Q (T); Sarcobatus Flat (D)	Basin now fishless
Q.—Lake ephemeral?	Basin west of Stonewall Flat	Probably none	Stonewall Flat (T)	Basin now fishless
R.—Lake ephemeral?	Sarcobatus Flat	Probably none	Stonewall Flat (T); Death Valley (D)	Basin now fishless
S.—Lake ephemeral?	Yucca Flat	Probably none	Frenchman Flat (D)	Basin now fishless
T.—Lake ephemeral?	Frenchman Flat	Probably none	Amargosa River (D) ?	Basin now fishless
U-V.—Basins adjacent to Railroad Lake system	Willow Creek basin	Probably none	Unknown	Basin now fishless
U.—Lake ephemeral?	Kawich Valley	Probably none	Railroad Lake (D) ?	Basin now fishless
V.—Lake ephemeral?				
W-Z.—Basins structurally connected with Pluvial White River				
W.—Lake ephemeral?	Penoyer or Sand Spring Valley	Probably none	Desert Valley (D)	Basin now fishless
X.—Lake ephemeral?	Papoose Dry Lake, Emigrant Valley	Perhaps Groom L. (D)	Groom Lake (D)	Basin now fishless
X'.—Lake ephemeral?	Groom Dry Lake, Emigrant Valley	Probably none	Desert Valley (D)	Basin now fishless
Y.—Lake ephemeral?	Desert Valley	Probably none	White River (D)	Basin now fishless
Z.—Lake ephemeral?	Valley just north of Dry Lake Valley	Dry L. Valley (D)?	White R., via Dry L. Valley (D)	Basin now fishless
Z'.—Lake ephemeral?	Dry Lake Valley	Probably none	White R., via California Wash (D)	Basin now fishless

Lakes Between Lahontan, Bonneville and Colorado Systems

24.—Lake Franklin	Ruby Valley and north part of Butte Valley	Almost certainly none (D); L. Waring (A)?	Clover L. (D)	Steptoe fauna (pp. 51-52)
25.—Lake Gale	Butte Valley, south part	Butte Bay of L. Franklin	Butte Bay of L. Franklin	Steptoe fauna
26.—Clover Lake	Clover and Independence valleys	Probably none	L. Franklin (T); L. Lahontan, via Humboldt R. (D)?	Steptoe fauna
27.—Lake Steptoe	Steptoe Valley, south part	L. Waring (D)	L. Waring (D or con- joined)	Steptoe fauna
28.—Lake Waring	Steptoe Valley, north part[1]	L. Franklin (A)?	Uncertain	Steptoe fauna
29.—Antelope Lake	Tippett basin of Antelope Valley	Probably none	Possibly L. Waring (D) or L. Bonneville (D)	Basin now fishless

[1] Including main, northern part of Antelope Valley.

FISHES AND FISHERIES OF NEVADA 769

30.—Spring Lake	Spring Valley	Probably none	Possibly Carpenter R. (D) or L. Bonneville (D)	Steptoe fauna; also Bonneville or Colorado
31.—Lake probably small and ephemeral	Long Valley	Probably none	L. Franklin?	Basin now fishless
32.—Lake Jake	Jakes Valley	Probably none	Possibly White R.	Basin now fishless
Lakes North of Lahontan System				
33.—Lake Madeline	Madeline Plains and Grasshopper Valley	Sacramento R., via Pit R. (C); L. Lahontan (S)	L. Lahontan (D or S)	Pit R. of Sacramento system (not very conclusive)
34.—Lake Alvord	Alvord Desert and connected basins	None	Possibly with Columbia R. (D)	Fauna distinctive (minnow) or uncertain (trout)
34a.—Lake Mann	Mann Lake	None, fide Russell	L. Alvord (D)	Basin now fishless
34b.—Juniper Lake	Juniper Lake	None, fide Russell	L. Alvord, via L. Mann Soldier Meadows arm of L. Lahontan (D)	Basin now fishless
35.—Summit Lake	Summit Lake basin	None	Columbia R., via Malheur R. (D)	Lahontan system (presumptive) or Alvord basin
35a.—Lake Catlow	Catlow Valley	None	L. Catlow (D)	Uncertain
36.—Guano Lake	Guano Valley	L. Catlow (D); with some doubt	L. Catlow (D)	L. Catlow (moderately conclusive)
37.—Lake Meinzer	Long Valley	None	Probably Warner L., via bay in Coleman Valley (D)	Warner L. (Inconclusive)
28.—Lake Surprise	Surprise Valley and Duck Flat	None	L. Warner (D or C) or L. Lahontan (D)	L. Warner and/or Lahontan system
39.—Lake Warner	Warner Valley	None (doubt raised by Van Winkle)	Columbia R., via Malheur R. (D)	Uncertain; probably various
40.—Cowhead Lake (not certainly Pluvial)	Cowhead Lake	L. Warner (D)	L. Warner (D)	L. Warner (presumptive) or one of surrounding basins
40a.—Alkali Lake	Alkali Lake and adjacent desert	Probably none	L. Chewaucan?	L. Chewaucan (presumptive) or one of other surrounding basins
40b.—Lake Chewaucan	Summer Lake, Chewaucan Marsh, and Abert Lake	Probably none	One or more of surrounding basins	Uncertain; probably closest with Fort Rock Lake and Alkali Lake faunas, but could be with Klamath Lakes, Goose Lake or Warner Lake fauna
41.—Klamath River System				
41a.—Klamath Lakes (several)	Klamath River system, upper part	Formerly interior drainage; later with ocean, via Klamath R. (D); Fort Rock L. (C)?	Probably L. Lahontan first; then probably with Sacramento and Columbia systems	Lahontan system; Columbia R.; Pit R. division of Sacramento system; surrounding lake systems
Caldera Lakes				
41b.—Crater Lake	Crater Lake	Lake existent?; certainly no connections	Lake not existent	Lake now fishless
41c.—Medicine Lake	Medicine Lake	None	Lake existent?	Lake now fishless
Disrupted Parts of Sacramento River System				
42.—Goose Lake	Goose Lake basin	Sacramento R., via Pit R. (D)	Very early with L. Lahontan??; later with Klamath L.?; finally with Sacramento R.	Upper Sacramento R. (conclusive); Klamath Lakes (probable); Lahontan system (possible)

Pluvial Lake or River	Valley or Basin	Pluvial Connection	Most Plausible "Earlier Pluvial" Connection	Relationships of Recent Fish Fauna of Remnant Waters
42a.—Clear Lake	Clear Lake	Sacramento R. (D); Russian R. (D)	Lake not existent?; streams tributary to Russian R.?	Sacramento system (conclusive)
42b.—Lake Tulare	Tulare and Buena Vista lakes	Pacific Ocean, via San Joaquin R. (D); Monterey Bay drainage (A or C)??	Same as Pluvial, if existent	Sacramento system (conclusive); Monterey Bay drainage?
42c.—**Minor Coastal Stream Systems**	Various streams	Sacramento and Columbia rivers (A and C)	Uncertain	Sacramento and Columbia river faunas, modified
42d.—Lake Carrizo (not certainly Pluvial)	Carrizo Plain	None	Salinas R. (of Monterey Bay drainage)	Basin now fishless
Disrupted Parts of Columbia River System				
42e.—Fort Rock Lake	Silver Lake, Fossil Lake, desert about Fort Rock, etc.	Columbia R., via Long Prairie and Deschutes R. (D); Klamath R. system (C)?	Same as Pluvial?	Uncertain; probably Columbia R. and Klamath R. faunas; also other lake basins?
42f.—Lake Malheur	Malheur or Harney basin	Columbia R., via Malheur R. (D)	Part of Columbia R. system	Columbia system (conclusive)
42g.—Cow Lake (Postpluvial?)	Cow Lakes basin	Columbia R., via Owyhee R. (D)	Part of Columbia R. system	Columbia system (conclusive)
42h.—Crab Creek	Crab Creek basin	Columbia R. (D)	Part of Columbia R. system	Columbia system (conclusive)
42i.—Lake Moses	Moses Lake	Columbia R., via lower Crab Creek (D)	Part of Columbia R. system	Columbia system (presumptive)
42j.—Many other lakes	Other lakes on Channelled Scablands	Columbia R. (D)	Part of Columbia R. system	Basins unexplored by ichthyologists
Isolated streams of Snake River Lava Plateau				
42k.—Mud Lake-Lost River group	Five streams, in part tributary to Mud L. (p. 76)	None?	Part of Snake R. system	Columbia system (presumptive)
Wood R. group	Wood River and tributaries	Snake R.	Part of Snake R. system	Upper Snake system (conclusive)
42l.—Lava Lake (Postpluvial?)	Lava Lake	Wood R., via Fish Creek?	Uncertain	Basin now fishless
42m.—Fish Creek	Fish Creek	Part of Wood R. system	Same as Pluvial?	Upper Snake system (conclusive)
42n.—Wood River	Wood River	Snake R. (D)	Snake R. (D)	Upper Snake system (conclusive)
Death Valley System				
Mono Lake basin				
43.—Lake Mono	Mono Lake basin	Perhaps none, or with Owens R. (D)	First into Lahontan system, via Walker R. (D); then into Death Valley system, via Owens R. (D)	Basin now fishless

FISHES AND FISHERIES OF NEVADA

44.—Lake Aurora	Aurora Valley	Perhaps none	Probably first in outlet from L. Mono to Walker R.; later into L. Mono (D)	Basin now fishless
Owens River system				
45.—Long Valley Lake	Long Valley, in headwaters of Owens River	Owens R. (D)	Owens R. (D), if lake existent	Owens R. system (conclusive)
46.—Adobe Lake	Adobe Valley	Owens R., via North Fork (D)	Owens R., via North Fork (D)	Basin now fishless
47.—Lake Owens	Owens Lake basin	Death Valley, via lakes Searles and Panamint (D)	Death Valley, via lakes Searles and Panamint (D)	Lahontan and Colorado systems
Owens River-Death Valley connectives				
48.—Lake Searles, higher stage	Indian Wells-Searles basin	L. Panamint (D, in Tahoe time)	L. Panamint (D)	Basin now fishless
A.—Lake China	Indian Wells-Salt Wells Valley	L. Searles (D, in Tioga time)	Part of higher L. Searles	Basin now fishless
B.—Lake Searles, lower stage	Searles basin	None	Part of higher L. Searles	Basin now fishless
49.—Lake Panamint	Panamint Valley	Death Valley (D)	Death Valley (D)	Basin now fishless
50.—Lake Manly, Death Valley	Death Valley	None (D); Owens, Amargosa and Mohave systems (T)	Same as Pluvial; probably also Colorado R. (D)	Colorado system
Amargosa River system				
51.—Lake Tecopa	Middle part of Amargosa River drainage	Death Valley (D)	Death Valley (D); Las Vegas R., via Indian Springs L. (C)	Death Valley; Colorado system
52.—Lake Pahrump	Pahrump Valley	Amargosa R., via Ash Meadows (D)?	Amargosa R., via Ash Meadows (D)	Ash Meadows of Amargosa R. drainage (conclusive)
Mohave River system				
53.—Lake Mohave	Playas of Soda and Silver dry lakes	Death Valley (D)	Death Valley (D); earlier, with Colorado R. (D)?	Mohave R.; remotely with Owens R.
54.—Little Lake Mohave	Cronise Valley	Pluvial Mohave R. (distributary)	Pluvial Mohave R. (distributary)?	Basin now fishless except when flooded by Mohave River
55.—Lake Manix	Middle part of Mohave River drainage	Lake Mohave, via Mohave R. (D)	Lake Mohave, via Mohave R. (D)	Mohave R.; remotely with Owens R.
56.—Small lakes in outlet of Lake Mohave	Outlet of Lake Mohave	Death Valley (D)	Death Valley (D)	Basins now fishless
Enclosed Fishless Basins Adjacent to Death Valley System				
57.—Lakes west of Mohave River				
A.—Lake Thompson	Antelope Valley	Lake Kane (D)	Lake Kane (D)?	Basin presumably fishless
B.—Lake Kane	Kane Dry Lake	Probably isolated	Possibly into Mohave R.	Basin now fishless
C.—Lake Harper	Harper Dry Lake	Probably isolated	Mohave R.	Basin now fishless
D.—Many other lakes, probably ephemeral	Dry lake basins west of Death Valley drainage	Probably none	Undetermined	Basins now all fishless
58.—Lakes between Death Valley system and Fish Lake Valley				
A.—Deep Springs Lake, probably ephemeral	Deep Springs Valley	Presumably none	Possibly into Eureka Valley	Basin now fishless

Pluvial Lake or River	Valley or Basin	Pluvial Connection	Most Plausible "Earlier Pluvial" Connection	Relationships of Recent Fish Fauna of Remnant Waters
B.—Eureka Lake	Eureka Valley	Presumably none	Undetermined	Basin now fishless
C.—Saline Lake	Saline Valley	Presumably none	Undetermined	Basin now fishless
D.—Butte Lake	Butte Valley	Saline Valley?	Undetermined	Basin now fishless
59.—Lakes between Death Valley system and Colorado River				
A.—Lake Mesquite	Mesquite Dry Lake	Probably none	Pahrump Valley (D)?	Basin now fishless
B.—Lake Ivanpah	Ivanpah Dry Lake	Probably none	Probably none	Basin now fishless
C.—Lake probably ephemeral	Jean Dry Lake	Possibly Lake Ivanpah (D)	Lake Ivanpah?	Basin now fishless
D.—Lake probably ephemeral	Dry lake just east of Jean Dry Lake	Possibly Jean Dry Lake (D)	Lake Ivanpah?	Basin now fishless
E.—Lake probably ephemeral	Mesquite Dry Lake or Opal Mountain basin	Presumably none	Possibly Colorado R. (D)	Basin now fishless
a.—Lake Amboy (basin labeled F on map 1)	Bristol–Cadiz basin		Colorado R. (D)?	Basin now fishless
b.—Lake Ward	Danby basin	Presumably none	Colorado R. (D)?	Basin now fishless
c.—Several other lakes, probably ephemeral	Dry lake basins, SE. Mohave Desert	Undetermined	Undetermined	Basins now fishless
d.—Lake Lucerne	Lucerne Dry Lake	Probably none	Uncertain	Basin now fishless
60.—**Railroad Lake System** (pp. 90–94)	Railroad Valley system	None (D); possibly L. Lahontan (A or C)	Colorado R. (D)?	Lahontan and Colorado systems
Disrupted Parts of the Colorado River System				
White River and probably connected lakes				
61.—White River	White River Valley, etc.	Colorado R. (D)	Colorado R. (D)	Colorado system (conclusive)
62.—Coal Lake	Coal Valley	Possibly White R. (D)	White R. (D)	Basin now fishless
63.—Lake Bristol	Bristol Valley	Probably White R., via L. Delamar (D)	White R. via L. Delamar	Basin now fishless
64.—Lake Delamar	Delamar Valley	Probably White R. (D)	White R. (D)	Basin now fishless
Carpenter River and Lake Carpenter				
65.—Carpenter River	Meadow Valley Wash, etc.	Colorado R. (D)	Colorado R. (D)	Colorado system (conclusive)
66.—Lake Carpenter	Duck Valley	Carpenter R. (D)	Carpenter R. (D)	Basin now fishless
Las Vegas River and probably connected lakes				
67.—Las Vegas River	Las Vegas Valley	Colorado R. (D)	Colorado R. (D); Amargosa R., via Indian Springs L. (C)	Colorado system (presumptive)
68A–B.—Unnamed lakes, probably ephemeral	Mcrmon Gulch	Probably East L. (D)	East L. (D)	Basin now fishless
68C.—East Lake	East Dry Lake Valley	Las Vegas R. (D)	Las Vegas R. (D)	Basin now fishless
69.—Indian Springs Lake	Indian Springs Valley	Probably none	Las Vegas R. (D); Amargosa R. (C)	Basin now fishless
69a.—Lake Pinto and Pinto River	Pinto basin	Colorado R. (D)?	Colorado R. (D)	Basin now fishless
Lakes LeConte, Clark and Pattie				
69b.—Lake LeConte (Lake Cahuilla)	Salton Sink	Colorado R. (distributary)	Colorado R. (distributary)	Colorado system (conclusive)
69b'.—Lake Clark	Clark Valley	None	L. LeConte	Basin now fishless
69c.—Lake Pattie	Pattie basin (Laguna Salada or Laguna Maquata)	Colorado R. (distributary)	Colorado R. (distributary)	Colorado system (conclusive)

Disrupted rivers and lakes on east side of Colorado River				
69d.—Sonoyta River	Sonoyta River			Colorado system (rather conclusive)
70.—Red Lake (existence in Pluvial time not certain)	Hualpal basin	Unknown	Unknown	Basin now fishless
70a.—Lake possibly Prepluvial	San Simon Valley	Colorado R., via Gila R. (D)	Colorado R., via Gila R. (D)	Colorado system
70b.—Lake Mormon	Mormon Lake	Unknown	Unknown	Fish fauna introduced
70c.—Red Desert lakes	Red Desert basin	Colorado R. (D) ?	Colorado R. (D) ?	Basin now fishless
70d.—Lake Animas	Animas Valley	Possibly Colorado R., via Gila R. (D)	Colorado R., via Gila R. (D)	Basin now fishless
Disrupted Parts of the Yaqui River System				
70e.—Lake Cochise	Cochise Valley	Yaqui R., via Whitewater R. (D); also possibly Colorado R., via Gila R. (C)	Yaqui R. and/or Colorado R. (D)	Insufficient fish data (see text, p. 115)
70f.—Lake Cloverdale	San Luis Valley, N. M.	Presumably Yaqui R. (D)	Yaqui R. and/or Colorado R., via Gila R. (D)?	Basin now fishless
Disrupted Parts of the Rio Grande System in the United States				
70g.—Rio Mimbres	Rio Mimbres	Laguna de Guzman (D)	Laguna de Guzman (D)	Rio Grande (conclusive)
70h.—Lake Playas	Playas Lakes	L. Hachita (D)?	L. Hachita (D)?	Basin now fishless
70i.—Lake Hachita	Hachita Valley	Laguna de Guzman (D)?	Laguna de Guzman (D)	Basin now fishless
70j.—Lake San Augustin	San Augustin Plains	Rio Grande (D)?	Rio Grande (D)	Basin now fishless
70k.—Lake Estancia and Encino	Estancia and Encino valleys	Rio Grande (D)?	Rio Grande (D)	Basin now fishless
70l.—Lake Otero	Tularosa Basin	Rio Grande (D)?	Rio Grande (D)	Rio Grande or Red R.
70m.—Salt Lakes	Salt Lakes Basin	Pecos R. (D)?	Pecos R. (D)	Basin now fishless
70n.—Coronado's Lakes	San Luis Valley, Colo.	Rio Grande (D)?	Rio Grande	Rio Grande (presumptive)

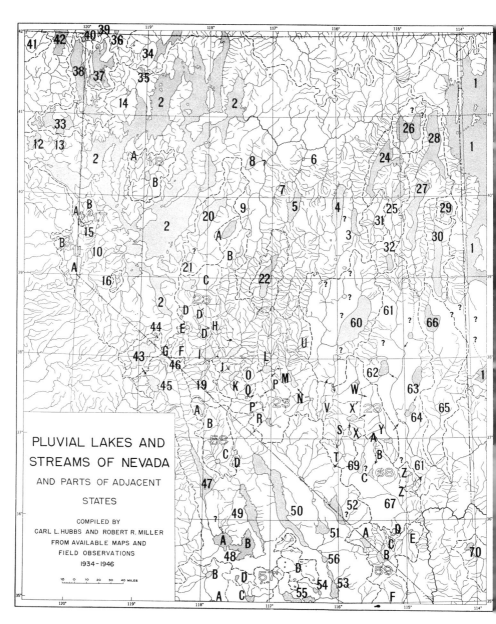

Legend begins on page 766.

INDEX

Abbott, Charles C., 17, 359
acidity, 611, 617, 619–620
Acrocheilus alutaceum, 17, 384–387
Acrodus alexandrae, 42
　oreodontus, 42–43
Agassiz, Louis, 16, 17, 386
Agosia, 432, 440
Alcorn, J. Ray, 29, 174, 379, 487
algae, 113, 135, 136, 153, 155, 180, 676–678, 679
alkalinity, 611, 617, 619, 620–622
　methyl orange, 600
　phenolphthalein, 599–600
Allan, Robert C. (Cal.), 143, 279, 297, 305, 357, 450, 592
Ambrysus mormon, 137–138, 674
Ameiuridae, 479
Amnicola micrococcus, 682
Amyzon mentalis, 18, 48–50
Antevs, Ernst, 79–81
Aphredoderidae, 51–53
Aphredoderus sayanus, 51
Applegates, Jesse and Lindsey, 15
Archoplites interruptus, 17, 18, 21, 545, 546, 547–553, 651
Bailey, Vernon, 20, 434, 514
Baird, Spencer Fullerton, 17, 20, 586
bass, see blackbass
Beckwith, Lieut. E. G., 17, 95–97
Behle, William, 84
Belding, L., 591
bibliography, 713
Bidwell-Bartleson party, 14
Bidwell, John, 97
blackbass, 25, 553
Blackbass, Largemouth, 24, 554–560, 649–650
　Smallmouth, 560–565, 649
Blackfish, Sacramento, 457–460, 592
Blake, James, 84
bluegill, see sunfish
Bonneville, Capt. Benjamin Louis Eulalie de, 13, 85
Bonneville system, Nevada, 116–118
bonytail, see Gila Chub
Bower's Mansion, 456
Bowman, J. Soulé, 15, 17, 407
Bridger, Jim, 12
Broecker and Orr, 83
Brown, Charles LeRoy, 136, 678
Brues, Charles T., 26, 519
budget, heat, 130–131
Buffalofish, Black, 50
bullfrog (see *Rana catesbeiana*), 434 441, 475, 505, 525
Bullhead, Black, 494–497, 648
　Northern, 497
　Southern, 497
　Brown, 18, 489–494, 648–649
Call, Richard Ellsworth, 19

Carassius auratus, 454–456
Carbine, William F., 27, 569–570
carbon dioxide, free, 599
Cárdenas, Garcia Lopez de, 112
carp and minnow family, 373
Carp, Asiatic, 18, 19, 20, 21, 448–453, 525
　mirror and leather varieties, 449
Carrington, Campbell, 430
Carson, Christopher (Kit), 13, 14, 121
Cary, W. M., 20, 452–453
catfish family, 479, 647–649
Catfish, Channel, 481–485, 647–648
　White, 485–489, 648
Catostomidae, 48–50, 56, 333, 647
Catostomus ardens, 344–349
　arenarius, 24, 32
　columbianus, 20, 342–344
　fecundus, 347
　latipinnis, 17, 350–352
　macrocheilus, 17, 340–342
　tahoensis, 18, 32, 84, 352–357
Cave, Lovelock, 11, 294
Cenozoic Era, 47–64, 76
Centrarchidae, 545, 649–652
Chabagno, Manuel, 57
charrs, 256
Chasmistes, 64
　cujus, 19, 24, 64, 81, 82, 363–372, 647
　liorus, 347
checklist of Nevada fishes, 193–198
Chiselmouth, 17, 384–387
Chub, Gila, 387–398
　Tui, 33, 64, 87, 89, 410–412, 697–700
　Lahontan, 15, 17, 412–421
Clark, William, 14
Clemmys marmorata, 20
Clogston, Bruce, 285
Clupeidae, 239
Clyman, Jim, 15
Colorado River system,
　lakes, 181–191
　rivers, 110–114
Contracaecum, 570
　multipapillatum, 305
Cooper, J. G., 18, 354
Cope, Edward Drinker, 18, 19, 48, 84, 124, 336, 412, 414, 427, 430
　and Yarrow, 346, 348
Coregonidae, 325, 647
Coregonus, 325
Corlett, Ray, 156, 263, 591
Cosmacanthus humboldtensis, 43–44
Cottidae, 584
Cottus bairdi, 585–586
　semiscaber, 586–588
　beldingi, 20, 25, 64, 588–592
Crappie, Black, 576–579, 652
　White, 580–583, 651–652
Crawfish, Pacific, 23, 151, 153, 159, 263

Creek, Baker, 116
  Big Springs, 116
  Bronco, 99
  Brown, 99
  Deadman, 116
  Deep Cañon, 116
  Dog, 99
  Donner, 98
  Evans, 94, 99
  Franktown, 99
  Galena, 99
  Goose, 116
  Gray, 99
  Hendrys, 116
  Hunter, 99
  Juniper, 99
  Kelly, 94
  Lake, 105, 116
  Lamoille, 95
  Lehman, 116
  Maggie, 94
  Mahogany, 173
  Martin, 94
  Martis, 99
  Ophir, 99
  Pine, 95, 175
  Prosser, 98
  Rock, 94
  Sagehen, 99
  Salmon Falls, 105
  Secret, 95
  Silver, 116
  Smith, 116
  Snake, 116
  Snow, 173
  Susie, 94
  Thomas, 99
  Thousand Springs, 116
  White, 99
Crenichthys, 87
  baileyi, 21, 31, 109, 512–516, 681–686
  nevadae, 27, 517–520
Creosote bush, 67
Cretaceous, 11, 44–47, 75
Creutzfeldt, F., 17, 401, 407
Cui-ui, see lakesucker
Cyprinidae, 51, 53–56, 373
Cyprinodon breviradius, 62
  diabolis, 26, 507–511
  nevadensis, 20, 31, 500–503
    mionectes, 504–505
    pectoralis, 505–506
Cyprinodontidae, 61–63, 499
Cyprinus carpio, 18, 19, 20, 21, 448–453, 525
Dace, Moapa, 30, 109, 436–441
  Speckle, 17, 90, 424–428
    Amargosa, 433–434
    Lahontan, 21, 89, 430–432
    Snake River, 428–430
    White River, 434–436
Dam, Boulder (see Hoover)
  Davis, 186, 187
  Derby, 25, 98, 158, 292

Dam—*Continued*
  Hoover, 181, 183
  (see various reservoirs)
David, Lore, 29
Davidson, Pirie, 25, 43
Desert, Carson, 93
  Forty-Mile, 15, 93
Desertfish, Soldier Meadows, 30, 441–447
Devil's Hole, 508–510
Dill, William A., 484, 485
Dimick and Merryfield, 342
Dingell-Johnson projects, 804–806
Disk, Secchi, 98, 132, 153, 155, 184
Dolomieu, M., 565
Donner party, 15
Dorosoma, 239
  petenense atchafalayae, 186, 240–243
Douglas, Philip A., 360–362
Drouet, Francis, 676, 678
Dymond, John R., 256, 450
Eastman, Charles R., 24, 61
Eddy and Surber, 423
Edestidae, 37
Eels, freshwater, 20
Eigenmanns, Carl and Rosa Smith, 20
Empetrichthys latos, 31, 523–527
  concavus, 528–530
  latos, 527
  pahrump, 527–528
  merriami, 21, 520–523
Eocottus, 584
epilimnion, 129
Eremichthys acros, 30, 441–447
Escalante, Silvestre Velez de, 12
Eureka Consolidated mine, 456
eutrophic, 135
Evermann, Barton Warren, 21, 26
Forbes and Richardson, 453, 484, 485, 496–497, 545, 570
Forel-Whipple lake classification, 183
Fowler, Henry W., 24, 338
Frantz, Ted, 117, 180, 271, 284, 297, 298, 397, 401, 402, 414, 456, 464, 519–520, 541, 549, 569, 587, 603–610
Fraser, Jack, 32
Frémont, John Charles, 13, 14, 15, 93, 95, 99, 121, 122, 145, 294–295
Fundulus curryi, 61–62
  davidae, 62
  eulepis, 62
  nevadensis, 25, 61–63
  parvipinnis, 62, 63
Gabrielson, Ira N., 32
Gairdner, Meredith, 14, 276, 304, 404
Gambusia affinis, 76, 531–536
Gammarus fasciatus, 188
Garcés, Francisco, 11, 112
Gasterosteidae, 56–61
Gasterosteus, 82
  aculeatus, 60
  doryssus, 23, 56–61, 63
  williamsoni leptosomus, 58

INDEX

Gibbons, W. P., 17
Gila, 440
  atraria, 17, 395–398
  cypha, 112, 389
  jordani, 32, 393
  robusta, 17, 388–390
    elegans, 391–393
    jordani, 393–395
Gila, Colorado, 17, 388–390
  Swiftwater, 391–393
  White River, 32, 393–395
  Utah, 17, 395–398
Gilbert, Charles H., 19, 21, 434, 513
  and Evermann, 249
Gill, Theodore, 18, 354
  and Jordan, 354
Girard, Charles, 15, 16, 17, 401, 402, 414, 427
glacial, 77
glossary, 701–712
goldfish, 454–456
Great Basin, 67–71
Greenwood, Caleb (Old Greenwood), 14, 15, 97
Griffin, Glen, 534
Grover, C. W., 25
Groves, Frank W., 32
Günther, Albert, 456
guppy, 531
Gyracanthidae, 43–44
Hall, E. Raymond, 84
Hanna and Grant, 677
Hansen, Donald F., 582–583
Hardman and Venstrom, 97–98
Harms, Clarence E., 484–485
Harlan-Young group, 15
Harry, Robert R., 418, 697–700
Hastings, Lansford Warren, 15
Hay, Oliver P., 23, 58
Headley, F. B., 64
Helicoprion, 11, 37
  bessonowi, 40
  ferrieri, 39, 40
  nevadensis, 28, 37–41
Henshaw, H. W., 18, 286, 287
Hessel, Rudolph, 451
Heterandria formosa, 531
Hiko pool, 107
Hobart Mills Lumber Company, 152
Hubbs, Carl L., 27, 298, 322, 513, 563
  and Bailey, 556, 557, 563, 564, 565
  Hubbs and Johnson, 337, 342, 351
  and Miller, 30, 32, 51, 56, 59, 78, 82, 85, 86, 87, 89, 90, 107, 174, 175, 176, 316, 360, 418, 446
  and Schultz, 344
Humboldt, Alexander von, 93, 97
Hutchinson, G. Evelyn, 27, 125, 131, 133, 145, 166, 176, 178, 675
Hutchinson, Samuel J., 380
Hybodontidae, 41–43
Hybodus nevadensis, 41–42
hybridization, 635–638
hydrogen-ion concentration, 627

hypolimnion, 129–130
Ice age, 77
Ictaluridae, 479, 647–649
Ictalurus catus, 485–489, 648
  melas, 494–497, 648
    catulus, 497
    melas, 497
  nebulosus, 18, 489–494, 648–649
  punctatus, 481–485, 647–648
Ictiobus niger, 50
index, 775
International Commission on Zoological Nomenclature, 386, 428
Jenkins, T. Travis, 305
Johnson, V. Kay, 125, 159, 163, 167, 297, 305, 371, 421
Jones, J. Claude, 24, 25, 81–83, 125, 135, 136
Jonez, Al, 125, 181, 187, 188
  and Sumner, 182, 305, 383, 390, 484, 534, 570, 574
Jordan, David Starr, 18, 19, 21, 23, 25, 346, 347, 349, 354, 364–365, 395, 584
  and Evermann, 287, 347, 381, 513
  Evermann and Clark, 407
  and Gilbert, 19, 346, 347, 348–349
  and Henshaw, 18, 286–287, 329, 330, 420
Juday, Chancey, 145, 146
Kemmerer, Bovard and Boorman, 25, 145, 146, 147
killifish family, 61, 499
Killifish, Nevada, 25, 61–63
Kimsey, J. Bruce, 31, 32, 33, 175, 263, 401, 417, 697
  Hagy and McCammon, 242
King, Clarence, 48
King, Donald, 163
Kispert, J. H., 534
kokanee, see salmon, kokanee
Kopec, John A., 31, 514–515, 681–686
Krumholz, L. A., 535
La Hontan, Baron Louis Armand, 78
Lahontan system, 93–104, 121–178
Lake, Adams-McGill (see Reservoir, Adams-McGill)
  Alvord, 90
  Baker, 271
  Boca (see reservoirs)
  Bonneville, 85
  Bridgeport (see reservoirs)
  Carson, 176
  Clear, 459, 489, 549–551
  Clover, 86
  Donner, 150–152
  Eagle, 174–176
  Fish, 89
  Franklin, 86
  Heenan, 284
  Humboldt, 93
  Independence, 152–154
  Indian, 161
  Johnson, 189

Lake—Continued
Klamath, 84
Lahontan, 19, 78–85, 93 (see reservoirs)
Liberty, 190
Manly, 88
Marlette, 146, 267
Mary's, 121
Mead, 28, 181–186, 558
Meinzer, 90
Mohave, 186–191
Moosehead, 133
Pahrump, 89
Prosser (see reservoirs)
Pyramid, 14, 19, 28, 97, 121–178
Railroad, 87
Rye Patch (see reservoirs)
Soda, 133, 176–178
Spring, 86
Steptoe, 86
Stillwater (see marshes)
Summit, 90, 172–174, 445
Sunnyside (see reservoirs)
Tahoe, 133–134, 144–150
Toiyabe, 89
Topaz, 156–157
Twin, 169–172
Walker, 15, 100, 101, 121, 141–143, 450
Warner, 84
Washoe, 165–168, 551
Webber, 154–156, 321
Wildhorse (see reservoirs)
Winnemucca, 121, 140
Lakesucker, 64, 362
Cui-ui, 19, 24, 64, 81, 82, 138, 363–372, 647
Lanham, Urless N., 29, 516
La Rivers, Ira, 31, 32, 59, 85, 110, 112, 113, 125, 132, 136, 143, 146, 147, 152, 166, 173, 175–176, 178, 263, 293, 315, 355, 356, 365, 417, 418, 428, 440, 456, 503, 505, 510, 511, 534, 551, 591
and Trelease, 32, 198, 420, 428, 459
Larrea divaricata, 67
Larson, E. Richard, 38, 39
Laythe, Leo L., 380, 381
Leach, Glen S., 541
Lebistes reticulatus, 531
Le Conte, John, 145, 146
Leonard, Zenas, 13, 67
Lepidomeda albivallis, 462–465
altivelis, 472–475
mollispinis, 465
mollispinis, 465–468
pratensis, 468–472
vittata, 18
Lepomis cyanellus, 571–575
macrochirus, 566–571, 650–651
Leptolepidae, 44–47
Leptolepis nevadensis, 29, 44–47
Lernaea, 157, 305, 592

Leucidius pectinifer, 24, 176, 416–417
Leuciscus lineatus, 55
turneri, 21, 54–56
Lewis, Meriwether, 296
lime-forming organisms, 135–138
Lipomyzon, 365
Lowder, Lyle Junior, 346, 347, 349
Lucas, Frederick A., 21, 54–55
McCammon, George W., 488, 489
McGarry, Fort (or Camp), 173, 174
McKay, Thomas, 13
Madsen, M. J., 27, 125
Malloch, P. D., 304–305
Margaritifera margaritifera, 151
Marsh, Ruby, 432, 558, 559
Stillwater, 160–163, 558, 559
Meadows, Ash, 536
Soldier, 174, 445
Merriam, C. Hart, 20, 435, 522
Merriam, John C., 58, 59
Mesozoic, 41–47
Micropterus dolomieui, 560–565, 649
salmoides, 554–560, 649–650
Miller, Richard G., 263, 591, 592
Miller, Robert R., 27, 31, 61, 175, 185, 258, 300, 310, 316, 348, 349, 393, 398, 407, 419, 423, 424, 459, 464–465, 468, 478, 502, 503, 508, 509, 511, 522, 525, 530, 590–591 (see Hubbs and Miller also)
and Alcorn, 29, 144, 198, 256, 258, 283–284, 297, 298, 397, 456, 483–484, 487, 497, 502, 534, 549, 565, 568–569, 575
and Hubbs, 461–475
and Miller, 31, 249, 342, 344, 427–428
Millin, Richard B., 173
Mills, George T., 20, 21, 23, 453
Mills, Vernon, 29, 487, 497, 534
Minnow, lake, 24
Miocene, 11, 47–63, 74
Moapa coriacea, 30, 109, 436–441
Moffett, James W., 29, 183, 184, 188
Mollienesia latipinna, 531, 536
Mollusca (see *Amnicola*, *Margaritifera* and *Tryonia*), 132, 137
Molly, Black, 531, 536
moonfish, 531
mosquitofish (see *Gambusia affinis*), 441
Mountainsucker(s), 18, 334
Bonneville, 336
Bluehead, Utah, 336
Lahontan, 334–337
White River, 337–339
Murphy, Garth I., 459, 549–551
Mussel, freshwater, 151
Nebria eschscholtzi, 85
Needham and Gard, 154, 638
Neuhold, John M., 398
Newlands Project, 157
Nidever, George, 13

INDEX

Nilsson, Nils, 667
Nomogram, Rawson, 623
North Pass, 15
Notemigonus crysoleucas, 421–424
Notolepidomyzon, 338–339
Notropis lutrensis, 185
Nyquist, David, 446
Ogden, Peter Skene, 12, 105
oligotrophic, 147
Olson, Richard H., 57
Oncorhynchus nerka, 251–256
   tshawytscha, 12, 247–251
Orthodon microlepidotus, 457–460, 592
Osino coal beds, 48
Otocoris alpestris lamprochroma, 84
overturn, fall, 127
   spring, 128
oxygen, dissolved (DO) (see Winkler, basic), 626–627
Pacifastacus leniusculus, 151, 153, 159, 263
paleontology, 37–64
Paleozoic, 37–41
Pantosteus, 18
   delphinus utahensis, 336, 351
   intermedius, 337–339
   lahontan, 334–337
   platyrhynchus, 336
Parafundulus nevadensis, 62
Parker, H. G., 18, 19, 20, 452
Pasco, I. D., 19, 452
peaks, 70–71
Pearse and Achtenberg, 541–543
Peer, Harold, 29
Perca flavescens, 538–543, 649
perch family, 537, 649
Perch, Nevada Pirate, 51–53
   Sacramento, 17, 18, 21, 545, 546, 547-553, 651
   Yellow, 538–543, 649
Percidae, 537, 649
Permian, 11, 37
pH (see hydrogen-ion concentration)
physiography, 65–90
Pike, Walleye, 160
Pister, E. P., 168, 169, 171, 172
Pithecomyzon, 24, 365
Placoderm, Humboldt, 43–44
Plagopterus argentissimus, 18, 457–478
Pleistocene, 11, 63–64
Pliocene, 11, 53–63, 74, 246
pluvial, 77
Poeciliidae, 531
poisoning programs, 157, 168
Pomoxis annularis, 580–583, 651–652
   nigromaculatus, 576–579, 652
Poolfish, Ash Meadows, 21, 520–523
   Pahrump, 31, 523–527
      Manse Ranch, 527
      Pahrump Ranch, 572–528
      Raycraft Ranch, 528–530
Poppe, R. A., 450
Powell, William A., Jr., 487

Procambarus clarkii, 505
Prosopium williamsoni, 17, 325–331, 647
Ptychocheilus, 17
   lucius, 17, 185, 381–384
   oregonensis, 14, 376–381
Pupfish, Amargosa, 20, 31, 500–503
   Big Spring, 504–505
   Lovell Spring, 505–506
   Devil, 26, 507–511
Purkett, Charles A., 452, 453, 575
Quaternary, 63–64
Radbruch, Dorothy H., 136
Rana catesbeiana (see bullfrog), 441, 475
   johnsoni, 55
Rawson and Elsey, 329
redfish, 290
Redshiner, Columbia, 14, 403–408, 687–695
   Bonneville, 408–410
   Esmeralda, 21, 54–56
   Lahontan, 17, 399–403
Regan, C. Tate, 531
Reservoir, Adams-McGill, 180–181, 514
   Boca, 172
   Bridgeport, 168–169
   Lahontan, 100, 157–160
   Prosser, 172
   Pruess, 116, 118
   Rye Patch, 163–165
   Sunnyside (see Adams-McGill)
   Wild Horse, 179
Rhinichthys osculus, 17, 424–428
   carringtoni, 428–430
   nevadensis, 433–435
   robustus, 21, 89, 430–432
   velifer, 434–436
Richardson, C. H., 23
Richardson, Sir John, 13, 14, 17
Richardsonius balteatus, 14, 15, 403–408, 687–695
   hydrophlox, 409–410
   egregius, 17, 399–403
   microdon, 24, 402
   turneri, 21, 54–56
River, Amargosa, 88, 114–116, 191
   Balm, 14
   Barren, 12, 13, 93
   Bruneau, 105
   Carpenter, 88
   Carson, 14, 99–100
   Colorado, 110–114
   Humboldt, 93–97
      Little, 94
      North Fork, 94
      South Fork, 95
   Jarbidge, 105
   Logan, 329, 588
   Mary's, 12, 13, 14, 93, 94
   Moapa, 111
   Ogden's, 13, 93
   Owyhee, 104, 105, 179, 342

River—*Continued*
  Paul's, 12, 93
  Reese, 95
  Salmon (see Salmon Falls Creek), 342
  Salmon Trout, 14, 97
  Swampy, 12
  Truckee, 14, 97–99, 125
    deflection of, 81
    Little, 99
    new channel, 139
  Unknown, 12, 93
  Virgin, 111
  Walker, 14, 100–101
    deflection of, 81
  White, 87, 88, 107–110
  Willamette, 342
roundtail (see Gila Chub)
Royce, William F., 259–261
Russell, Israel C., 19, 79, 124, 176
Rutter, Cloudsley, 21, 354
Sadler's, 432
salmon and trout family, 245, 639
Salmonidae, 245, 639
Salmo aguabonita, 317–318
  aquilarum (see *Salmo gairdneri aquilarum*), 24
  clarki, 14, 64, 90, 275–280, 645–646
    henshawi, 18, 174, 281–295
    lewisi, 295–297
    pleuriticus, 185, 299–300
    utah, 297–298
  gairdneri, 14, 300–306, 647
    aquilarum (see *Salmo aquilarum*)
    gairdneri, 23
    irideus, 17, 306–308
    kamloops, 308–311
    regalis, 311–313
    smaragdus, 313–317
  mykiss henshawi, 23
  trutta, 318–323, 643–645
Salmon, King, 247–251
  Kokanee Red, 254, 256
  Pacific, 246
Salvelinus fontinalis, 264–271, 639–643
  malma, 12, 32, 272–274, 647
  namaycush, 257–264
Sargent, M. C., 28, 515
Sauer, Leo, 165
Savage, James C., 29
scale readings, 631–634
Schoffman, R. J., 452
Schultz, Leonard P., 414, 427
Scott, James B., 39
sculpin family, 584
Sculpin, Baird, 585–586
  Bonneville, 586–588
  Belding, 20, 25, 64, 138, 588–592
seiches, 127
Semotilus atromaculatus, 55
shad family, 239
Shad, Mississippi Threadfin, 186, 240–243

Shark, Alexander Acrodus, 42
  Humboldt Acrodus, 42–43
  Humboldt Hybodus, 41–42
  Nevada Edestid, 28, 37–41
Sharp, Mrs. James, 456
Shiner, Golden, 421–424
  Plains Red, 185
  Tahoe, 24
shocking, electric, procedure, 603–604
shrimp, freshwater (see *Gammarus*)
Sierra Pacific Power Company, 152
Sigler, William F., 329, 331
Simon, James R., 348
Simpson, Capt. J. H., 17, 18
Sink, Carson, 93, 100
  Humboldt, 93
Siphateles bicolor, 33, 64, 87, 89, 410–412, 697–700
  obesus, 15, 17, 412–421
Smith, Jedediah S., 12
Snake system, lakes, 179
  rivers, 104–107
Snapp, Pete, 174
Snyder, John Otterbein, 22, 23, 24, 124–125, 145, 175, 176, 250–251, 264, 287, 289–291, 312–314, 316, 317, 321, 329, 331, 339, 347, 354, 355, 366–370, 371–372, 398, 401, 402–403, 407, 414–417, 420, 432, 591
Sodaville, 518
Speirs, J. Murray, 484
Spinedace, 18
  Big Spring, 468–472
  Colorado River, 465
  Pahranagat, 472–475
  Virgin River, 465–468
  White River, 462–465
spinefins, 460
Springfish, Railroad Valley, 27, 517–520
  White River, 21, 31, 109, 512–516, 681–686
Springs, Crystal, 515
  Fairbanks, 505
  Rogers, 536
Squawfish, Colorado, 17, 185, 381–384
  Northern, 14, 376–381
stagnation, summer, 128–129
  winter, 127–128
Steamboat, 99
Stenelmis calida, 509
Stickleback, Nevada, 23, 56–61
  Three-spined, 60
Stizostedion vitreum, 160
Stroud, R. H., 558–559
Sublette, Milton, 13
  Solomon, 15
sucker family, 333, 647
Sucker, Biglip, 17, 340–342
  Bridgelip, 20, 342–344
  Cui-ui (see Lakesucker)
  Flannelmouth, 17, 350–352

Sucker—*Continued*
  Nevada Amyzon, 18, 48–50
  Razorback, 17, 357–362
  Sandbar, 24, 32, 354
  Tahoe, 18, 32, 352–357
  Utah, 344–349
  (see also mountainsuckers)
Sumner, Francis B., 28, 29, 514, 515–516
  and Lanham, 29, 516
  and Sargent, 28, 506, 513, 514, 515–516
Sumner, Francis H. 28, 291
Sumner, Robert C., 159, 163, 167, 181, 183, 184, 185, 451, 453, 456, 484, 485, 536, 552–553, 558, 560, 570, 571, 574–575, 578, 579, 592
sunfish family, 545, 649–652
Sunfish, Bluegill, 566–571, 650–651
  Green, 571–575
Sutcliffe, 126
swordtail, 531, 536
Taft and Murphy, 380
Tanner, Vasco M., 29, 32, 346, 347, 349, 513
Terrapin, Pacific, 20
Tertiary, 47–63
thermal stratification, 127
thermocline, 129, 153, 170
Thompson, Richard B., 380–381
Thoreau, Henry David, 493
Tigoma humboldti, 15, 17
tommy, 290
Topminnow, American, 531
topminnow family, 531
Train, Percy, 26, 57
Trelease, Thomas Jarvis, Jr., 31, 32, 33, 125, 132, 136, 137, 152, 173, 200, 293, 355, 370, 371, 417, 494, 514, 518, 551, 558
Triassic, 11, 21, 41–44
Trichophanes hians, 51–53
Trout, Blackspotted (see Cutthroat Trout)
  Brook, 190, 264–271, 318, 639–643
  Brown, 318–323, 643–645
  Brownbow, 638
  Cutbow, 638
  Cutthroat, 14, 64, 90, 123, 138, 275–280, 645–646
    Colorado, 184, 299–300
    Lahontan, 18, 25, 31, 138, 174, 281–295
    Utah, 297–298
    Yellowstone, 295–297
  Dolly Varden, 12, 32, 272–274, 647
  Eagle Lake, 24, 175–176

Trout—*Continued*
  Emerald (see Pyramid Rainbow Trout), 24
  Golden, 317–318
  Lake, 257–264
  Loch Leven (see Brown Trout)
  Mackinaw (see Lake Trout)
  Rainbow, 14, 23, 138, 300–306, 647
    Kamloops, 308–311
    Pyramid, 24, 313–317
    Southcoast, 306–308
    Tahoe, 24, 311–313
  Royal Silver (see Tahoe Rainbow Trout), 24
  Sam Brown, 638
  Splake, 636
  Steelhead, 23
  Tiger, 637, 638
  Wendigo, 636
Tryonia clathrata, 682
tufa, 136, 137
Valley, Diamond, 456
  Jersey, 26, 57
Van Dyke, Edwin C., 85
Vestal, Eldon H., 169, 170, 171
Walbaum, Johann Julius, 12
Wales, Joseph H., 26, 509, 511
Walker, Joseph Reddeford, 13, 14, 15 99
Wallis, Orthello L., 32
Webber, Dr. David G., 154
Weisel and Newman, 407–408, 687–695
Wemple, Edna M., 21, 41, 42
Wheeler, Harry E., 28, 37, 38
Wheeler, Sessions S., 30, 32
whitefish family, 325, 647
Whitefish, Mountain, 17, 138, 325–331, 647
White River system, lakes, 180–181
  rivers, 107–110
Williamson, Lieut. R. S., 326, 327
Winkler, basic, dissolved oxygen procedure, 599, 614–616
  Rideal-Steward modification, 599, 616–617
Wirth, Willis W., 673
Wolfskill, William, 12
Wood, Norman, 156
Work, John, 12
worm, anchor, 125
Worth, S. G., 483
Woundfin, 18, 475–478
Xiphophorus helleri, 531, 536
  maculatus, 531, 536
Xyrauchen texanus, 17, 357–362
Yarrow, H. C., 18
Zarbock, William M., 587–588
Zimmer, Paul D., 380

CORRIGENDA

Page 14, line 4, read *oregonensis* for *oregonense*.
Page 41, lines 32 and 33, read cañon for canyon.
Page 45, line 6, read cañon for canyon.
Page 57, line 7 from bottom, read Olson for Olsen.
Page 58, line 4 from bottom, read *leptosomus* for *deptosomus*.
Page 157, line 12, read *Lernaea* for *Lernia*.
Page 305, line 2 from bottom, read *Contracaecum* for *Contracoecum*.
Page 401, line 6, insert 17 into blank page space.
Page 521, in figure legend, read Grace Eager for Silvio Santina.
Page 524, in figure legend, read Grace Eager for Silvio Santina.
Page 531, Regan, C. Tate, 1913. A revision of the Cyprinodont Fishes of the subfamily Poeciliinae. Zoological Society London Proceedings 1913(4) ; 977–1018.